COCHLEAR
IMPLANTS

AIP Series in
Modern Acoustics and Signal Processing

Editor-in-Chief

ROBERT T. BEYER
Department of Physics, Brown University, Providence, Rhode Island

Editorial Board

Books In The Series

(continued after index)

COCHLEAR IMPLANTS

Fundamentals and Applications

Graeme Clark

The University of Melbourne and The Bionic Ear Institute,
East Melbourne, Victoria, Australia

With 244 Illustrations

Graeme Clark
Department of Otolaryngology and
    The Bionic Ear Institute
The University of Melbourne
348-388 Albert Street
East Melbourne, Victoria 3002
Australia
g.clark@unimelb.edu.au

Series Editor:
Robert T. Beyer
Department of Physics
Brown University
Providence, RI 02912
USA

Library of Congress Cataloging-in-Publication Data
Clark, Graeme.
    Cochlear implants : fundamentals and applications / Graeme Clark.
        p.   cm. — (Modern acoustics and signal processing)
    Includes bibliographical references and index.
    ISBN 0-387-95583-6 (alk. paper)
        1. Cochlear implants.   2. Deaf—Rehabilitation.   I. Title.   II. AIP series in modern
    acoustics and signal processing.
RF305.C536 2003
617.8′9—dc21   2002030584

ISBN 0-387-95583-6   Printed on acid-free paper.

Printed in the United States of America.   (MV)

9 8 7 6 5 4 3 2

springeronline.com

This book is dedicated to my wife, Margaret, for her selfless support and wise counsel during the last 35 years while I was undertaking cochlear implant research at the University of Sydney, the University of Melbourne, and The Bionic Ear Institute. I would like also to express my appreciation to our children Sonya, Cecily, Roslyn, Merran, and Jonathan; their spouses Ian, Peter, and Marissa; and our grandchildren Elise, Monty, Daniel, Noah, and Rebekah for their encouragement and enriching our lives.

I have been very impressed by the emergence of the bionic ear as a practical proposition, but even more by the promise for the future that it seems to embody. It makes use of the arrangement in the cochlea for pitch recognition to bring electronic technology into direct functional relationship with the nervous system and the human consciousness. Maybe that unique relationship has no other parallel in the nervous system, and thus that direct link between electronics and physiology will find no other application to medicine. Nevertheless, I feel it may represent a new benchmark in the understanding of neural and mental function in terms of their physical components.

—Professor Emeritus Sir Macfarlane Burnett, A.K., O.M., K.B.E., M.D., Ph.D., Lond., F.A.A., F.R.S., Nobel Laureate (Physiology or Medicine)—The First Patron of the Bionic Ear Institute, 1985

Series Preface

Soun is nought but air y-broke
—Geoffrey Chaucer
end of the 14th century

Traditionally, acoustics has formed one of the fundamental branches of physics. In the twentieth century, the field has broadened considerably and become increasingly interdisciplinary. At the present time, specialists in modern acoustics can be encountered not only in physics departments, but also in electrical and mechanical engineering departments, as well as in mathematics, oceanography, and even psychology departments. They work in areas spanning from musical instruments to architecture to problems related to speech perception. Today, six hundred years after Chaucer made his brilliant remark, we recognize that sound and acoustics is a discipline extremely broad in scope, literally covering waves and vibrations in all media at all frequencies and at all intensities.

This series of scientific literature, entitled Modern Acoustics and Signal Processing (MASP), covers all areas of today's acoustics as an interdisciplinary field. It offers scientific monographs, graduate-level textbooks, and reference materials in such areas as architectural acoustics, structural sound and vibration, musical acoustics, noise, bioacoustics, physiological and psychological acoustics, speech, ocean acoustics, underwater sound, and acoustical signal processing.

Acoustics is primarily a matter of communication. Whether it be speech or music, listening spaces or hearing, signaling in sonar or in ultrasonography, we seek to maximize our ability to convey information and, at the same time, to minimize the effects of noise. Signaling has itself given birth to the field of signal processing, the analysis of all received acoustic information or, indeed, all information in any electronic form. With the extreme importance of acoustics for both modern science and industry in mind, AIP press, now an imprint of Springer-Verlag, initiated this series as a new and promising publishing venture. We hope that this venture will be beneficial to the entire international acoustical community, as represented by the Acoustical Society of America, a founding member of the American Institute of Physics, and other related societies and professional interest groups.

It is our hope that scientists and graduate students will find the books in this series useful in their research, teaching, and studies. As James Russell Lowell once wrote, "In creating, the only hard thing's to begin." This is such a beginning.

Robert T. Beyer
Series Editor-in-Chief

Preface

The cochlear implant is a device that bypasses a nonfunctional inner ear and stimulates the hearing nerves with patterns of electrical currents so that speech and other sounds can be experienced by profoundly deaf people. It is the culmination of investigations that started in the 19th century, and as such it is the first major advance in helping profoundly deaf children to communicate since the sign language of the deaf was developed at the Paris Deaf School 200 years ago. It is also the first direct interface to the central nervous system to restore sensory function for use on a regular clinical basis.

I became interested in helping deaf people hear when I was 10 years old, because my father had a severe hearing loss and I knew how difficult it was for him to cope as a pharmacist and as a family man. In 1966 I left my practice as an ear, nose, and throat surgeon in Melbourne to do research and to learn how it might be possible to help people with a profound hearing loss. These were the patients I had to turn away from my clinic, saying that a hearing aid would be of little help but that one day medical research might provide an alternative. For me this meant first undertaking basic studies to learn about the differences between acoustic and electrical stimulation of the auditory neural pathways.

When it became clear from these and other basic studies that the best chance of providing speech understanding was through multiple electrode stimulation, many scientific challenges were to lie ahead. As previous attempts to produce speech understanding with electrical stimulation had been unsuccessful, and as reproducing the coding of sound was not seen as feasible, the research faced rigorous scientific criticism. The first criticism came from auditory neuroscience, where research had shown the complexity of the inner ear and central brain pathways. Not surprisingly it was believed that inserting a relatively small number of electrodes into the inner ear to stimulate groups of nerve fibers would fail to produce sufficient information for speech understanding. The second criticism came from the biological and clinical disciplines. Here the concern was that implantation would damage the very nerves it was intended to stimulate. In addition, it was thought the electrode could be a pathway for middle ear infection to induce dangerous infection in the inner ear. These biological and clinical criticisms were also well founded. The delicacy of the inner ear had been appreciated in ear surgery, and the risk of infection was ever present in young children. The above two major criticisms required answers before clinical studies could be done on patients.

It was also essential to determine from a small group of volunteers how the complex signals of speech could be presented as patterns of electrical stimulation that could be understood. This seemed at the time an almost insurmountable challenge. Research that followed established that speech processing could in fact be achieved safely for profoundly deaf adults, who had hearing before going deaf. After the benefits were shown for adults, it was appropriate to initiate research to see if children born deaf or deafened early in life could obtain sufficient speech understanding to enable them to manage successfully in a hearing world. Would deaf children be able to develop the right central neural connections, as they had received no auditory stimulation during the plastic phase of brain development? Indeed children who were born deaf were shown to develop speech at a level comparable to that in adults who had prior exposure to sound. Furthermore, it was discovered that if they were operated on at a young age, they could develop good speech sounds as well as language.

Providing hearing and speech understanding for children born deaf then led to an intense ethical debate. The signing deaf community had developed an effective communication system and support network to help one another. Community members were upheld by a strong belief in their self-worth, which is so necessary to manage in a world of sound where people with good hearing did not fully appreciate the great difficulty they had. For a time the implant was seen as an ideological threat to their beliefs and as undermining this well-knit group, and for a number of years the efficacy of the procedure was questioned. It required many controlled studies and the opinion of educators who had experience with the achievements of children with hearing aids before the benefits of the implant for children were fully appreciated.

The cochlear implant has been the result of research in many disciplines, including surgical anatomy, surgical pathology, biology, biophysics, neurophysiology, psychophysics, speech science, engineering, surgery, audiology, rehabilitation, and education. Few medical advances have required the integration of so many disciplines. The scientific questions in these disciplines had to be addressed in a logical, systematic, and sequential manner, and are discussed in this book.

As a result of this research, cochlear implantation has grown from a small number of isolated experimental studies done by a few, to a diverse discipline investigated by many. Its scientific credibility has been recognized through its inclusion in international physiological, acoustical, surgical, otolaryngological, audiological, education, speech science and technology, and engineering society meetings. In addition, there are many international meetings devoted solely to the topic of cochlear implants. The growth in knowledge in the last 30 years has been rapid. This can be seen in the number of papers that include cochlear implants in the title, abstract, or subject heading: in the 1960s, one; in the 1970s, 72; in the

1980s, 679; in the 1990s, 1,935. There have been many other relevant publications. Not only have there been a very large number of scientific papers, but there also have been monographs and book chapters.

Initially the field drew on basic sciences for its development, and then gradually established its own body of scientific and clinical knowledge. This has continued to the point that now electrical simulation of the auditory system can justly claim to be making scientific contributions to the disciplines that helped establish it, in particular neurophysiology, biology, psychophysics, speech science, and the clinical disciplines of surgery, audiology, and rehabilitation.

One aim of this book is to show how the numerous disciplines have contributed and how they have interrelated. This book presents the fundamentals of the research as well as the clinical outcomes so that the reader will have a more complete understanding of the discipline. It is intended for a general reader, and those with a more specialized background can refer to the references. In presenting the fundamentals, research at the University of Melbourne/Bionic Ear Institute and elsewhere is cited. Clinical studies cannot be divorced from the basic research. The two must guide each other and the main aim should be to help people. This requires excellent basic research, but it should be focused and not an end in itself. In this book this interaction is presented at all opportunities.

Finally, the basic and clinical research would not have reached the wider community without the biomedical and engineering expertise of industry. The work has been much more demanding than developing a pacemaker, as more complex electronics have had to be encapsulated in a smaller implanted package. Furthermore, the interface with the auditory nervous system is a very intricate bioengineering achievement. For this reason this book not only presents the basic and clinical research, but also discusses how these have supported the industrial achievement.

Graeme Clark

Acknowledgments

Writing this comprehensive work has been a major task, but having been involved with the contributing disciplines since the 1960s has been of great help especially in understanding their evolution and interrelation. Research germane to this book, like the field itself, has been interdisciplinary, and has depended on a coordinated team approach. A team was also required to bring the elements of this book together.

I am greatly indebted to the tremendous support received from Sue Davine, David Lawrence, Helen Reid, and John Huigen in the preparation of this book. It has been a major undertaking, and Sue has completed many hours of typing and compiling the text. David has produced diagrams and figures of a very high standard, researched topics, and provided invaluable help. Helen has diligently searched for references and found them, and John has helped to coordinate this combined effort. I would also like to thank in particular Andrew Vandali, Anthony Burkitt, David Grayden, Ian Rutherfurd, Jim Patrick, Joanna Parker, Mark Harrison, Peter Busby, Peter Seligman, Richard Dowell, Thomas Stainsby, and Chris van den Honert for kindly reading sections of the text and for their helpful comments. The imperfections are mine alone. Thanks also to Russell Brooks for compiling the index.

The cochlear implant research in Australia has been a team effort, and it has been a privilege to have worked with young and talented research students and to have seen them develop into mature scientists. It has also been a valued experience to have been closely involved with the staff members of Cochlear Limited, a number of whom were research colleagues. Without the close relationship between the basic and focused research in Melbourne and the industrial research and development in Sydney, the Nucleus device would not have become available to tens of thousands of severely and profoundly deaf people in more than 120 countries.

I would also like to pay tribute to the many scientists and clinicians from other centers who have also worked with dedication to achieve hearing and speech for deaf people. There has been great collegiality internationally in this relatively small field that has crossed political and other divides to make the world a better place for people with a hearing disability.

Finally, the work would not have been possible without the belief and support of many benefactors, governments, trusts, and foundations. Sometimes simply having their encouragement at difficult times was enough.

Graeme Clark

Contents

Abbreviations

A	apical	BP	bipolar
AAA	auditory association area	C	capacitance
ABF	adaptive beam forming	CA	compressed analog
ABR	auditory brainstem response	CAP	compound action potential
		CASALA	Computer Aided Speech and Language Assessment
AC	alternating current		
AC	auditory cortex	CC	common cavity
ACE	advanced combination encoder	CD	characteristic delay
		CELF	clinical evaluation of language fundamentals
ADC	analog-to-digital converter		
ADRO	adaptive dynamic range optimization	CF	crista fenestra
		CG	common ground
AGC	automatic gain control	CI	cochlear implants
AM	amplitude modulation	CID	Central Institute for the Deaf
AN	auditory nerve		
ANF	auditory nerve fiber	CIS	continuous interleaved sampler
ANSI	American National Standards Institute		
		CM	cochlear microphonics
AP	action potential	CMOS	Complementary metal oxide semiconductor
APP	abnormal positive potential		
		CMV	cytomegalovirus
ARC	Australian Research Council	CN	cochlear nucleus
		CNC	consonant-nucleus-consonant
ASTM	American Society for Testing Materials		
		CNS	central nervous system
AVCN	anteroventral cochlear nucleus	CRC	Cooperative Research Center
B	basal turn	CSF	cerebrospinal fluid
BDNF	brain-derived neurotrophic factor	CSIRO	Commonwealth Scientific Industrial Research Organization
BKB	Bench-Kowal-Bamford		
BM	basilar membrane	CT	chorda tympani

CT (scan)	computed tomography
CUNY	City University of New York
DAC	digital-to-analog converter
dB	decibels
DC	direct current
DCN	dorsal cochlear nucleus
DD	data decoder
DDE	data decoder/encoder
DL	difference limen
DLS	development language scale
DNA	deoxyribonucleic acid
DNLL	dorsal nuclei lateral lemniscus
DRSP	differential rate speech processing strategy
DSP	digital signal processor
EABR	evoked auditory brainstem response
ECAP	electrically evoked compound action potential
EcoG	electrocochleography
ED	electrode decoder
EE	excitatory contralateral and excitatory ipsilateral
EEPROM	electrically erasable programmable read-only memory
EI	excitatory contralateral and inhibitory ipsilateral
ENT	ear, nose, and throat
EOWPVT	Expressive One-Word Picture Vocabulary Test
EPROM	erasable programmable read-only memory
EPSC	excitatory postsynaptic current
EPSP	excitatory postsynaptic potential
ES	electrode sheath
F	farad
F1	first formant frequency
F2	second formant frequency
FDA	Food and Drug Administration
FEP	fluoroethylene propylene
FET	field effect transistor
FFT	fast Fourier transform
FGF	fibroblast growth factor
FM	frequency modulation
FN	facial nerve
GASP	Glendonald Auditory Screening Procedure
H	helicotrema
HA	hearing aid
HCRC	Human Communication Research Center
HINT	Hearing in Noise Test
Hk	hook region
HL	hearing loss
HMM	hidden Markov model
HRP	horseradish peroxidase
Hz	hertz
I	current
IC	inferior colliculus
ICC	central nucleus of the inferior colliculus
IDE	Investigational Device Exemption
IE	inhibitory contralateral and excitatory ipsilateral
IHC	inner hair cell
IID	interaural intensity difference
ILD	interaural loudness difference
IMPEBAP	implant evoked brainstem auditory potential
IMSPACP	Imitated Speech Pattern Test
INLL	intermediate nuclei lateral lemniscus
INSERM	Institut National de la Santé et de la Recherche Médicale
IP	interleaved pulse
IPSP	inhibitory postsynaptic potential

IPSyn	index of productive syntax	MOS	metal oxide semiconductor
ITD	interaural time difference	MOSFET	metal oxide semiconductor field effect transistor
JFET	junction field effect transistor		
JLD	just discriminable level difference	MP	monopolar
		MPP	multiple pulse per period
JND	just noticeable difference	MRI	magnetic resonance imaging
KEMAR	Knowles Electronic Manikin for Acoustic Research	mRNA	messanger ribonucleic acid
kHz	kilohertz	MSO	medial superior olive
LARSP	language assessment, remediation, and screening procedure	MSP	miniature speech processor
		MSTP	Monosyllables, Spondees, Trochees, and Polysyllables Test
LDL	loudness discomfort level		
LIF	leukemia inhibitory factor	MTP	monosyllable, trochee, polysyllable
LiP	listening progress profile		
LL	lateral lemniscus	MUSL	Melbourne University Sentence Lists
LQ	language quotient		
LSO	lateral superior olive	NH&MRC	National Health and Medical Research Council of Australia
LVAS	large vestibular aqueduct syndrome		
M	middle turn	nHL	normal hearing level
MAA	minimum audible angle	NID	National Institute of Deafness
MAC	Minimal Auditory Capabilities (test)		
MAIS	Meaningful Auditory Integration Scale	NIH	National Institutes of Health
MAP	map for the threshold and maximum comfortable levels in the speech processor	NINCDS	National Institute of Neurological and Communicative Disorders and Stroke
MC	maximum comfortable level	NINDB	National Institute of Neurological Diseases and Blindness
MDL	minimum acceptable discomfort level	NINDS	National Institute of Neurological Diseases and Stroke
MGB	medial geniculate body		
MHz	megahertz	NRT	neural response telemetry
MLD	mesencephalicus lateralis dorsalis	NST	Nonsense Syllable Test
		NU	Northwestern University
MLU	mean length of utterance	OC	organ of Corti
MNTB	medial nucleus of the trapezoid body	OCG	output current generator
		OHC	outer hair cell

OHUI	Ontario Health Utilities Index	RM	Reissner's membrane
OVE II	Orator Verbis Electris (formant synthesizer)	RMS	root mean square
		RNID	Royal National Institute of the Deaf
OW	oval window	ROC	receiver-operating curve
P	promontory	ROM	read-only memory
PAA	polyacrylic acid	RTI	Research Triangle Institute
PAT	Parametric Artificial Talker (formant synthesizer)	RW	round window
		SA	stimulus artifact
PB	phonetically balanced	SAS	simultaneous analog system
PBK	phonetically balanced (kindergarten) monosyllables	SC	superior colliculi
PC	processus cochleariformis	SC	supporting cell
PDGF	platelet-derived growth factor	SEM	scanning electron microscope
PET	positron emission tomography	SERT	Sound Effects Recognition Test
PLD	programmable logic device	SG	spiral ganglion cells
		SI	Synchronization Index
PLE	phonetic level evaluation	SII	Speech Intelligibility Index
PLS	Preschool Language Scale		
PMA	premarket approval	SIT	Speech Intelligibility Test
PMMA	Primary Measures of Music Audiation	SL	spiral ligament
		SM	scala media
PP	ponticulus pyramidalis	SMSP	Spectral Maxima Sound Processor
PPLE	phonetic and phonologic level evaluations	SNR	signal-to-noise ratio
PPVT	Peabody Picture Vocabulary Test	SOC	superior olivary complex
		SP	summating potential
PTA	pure tone average	SPEAK	speech processing strategy: SMSP with 20 filters
PTFE	polytetrafluoroethylene (Teflon)		
PVA	polyvinyl alcohol	SPL	sound pressure level
PVCN	posteroventral cochlear nucleus	SpL	spiral lamina
		SPP	single pulse per period
QALY	quality-adjusted life-years	SPS	simultaneous pulsatile stimulation
R	resistance		
RAM	random access memory	SQUID	superconducting quantum interference device
RC	resistor capacitance		
RDLS	Reynell Developmental Language Scales	SRT	speech reception threshold
RF	Radiofrequency	SSEP	steady state evoked potential

ST	scala tympani	UCSF	University of California at San Francisco
StV	stria vascularis	V	volt
SUKL	Štátny Ústav pre Kontrolu Liečiv (State Institute for Drug Control, Slovakia)	V	vestibule
		VA	vestibular aqueduct
SUM	summation	VC	consonant-vowel
SV	scala vestibuli	VCN	ventral cochlear nucleus
T	threshold	VCV	vowel-consonant-vowel
TA	tactile aid	VNLL	ventral nuclei lateral lemniscus
TB	trapezoid body		
TESM	transient emphasis spectral maxima	VNTB	ventral nucleus of the trapezoid body
TM	tectorial membrane	VOT	voice onset time
TORCH	toxoplasmosis, rubella, cytomegalovirus, and herpes simplex	VRA	visual reinforcement audiometry
		XIC	commissure of the inferior colliculus
UCH	University College Hospital	Z	impedance

Introduction

Definition

The multiple-channel cochlear implant (bionic ear) is a device that restores useful hearing in severely to profoundly deaf people when the organ of hearing situated in the inner ear has not developed or is destroyed by disease or injury. It bypasses the inner ear and provides information to the hearing centers through direct stimulation of the hearing nerve.

Normal Hearing

Hearing occurs when sound is transmitted down the ear canal, through the middle ear, to the inner ear. The inner ear is a very small, coiled, snail-like structure embedded in bone that houses the sense organ of hearing (organ of Corti). The organ of hearing rests on a membrane (basilar membrane) lying across the coil. This membrane vibrates selectively to different sound frequencies, so that it acts as a sound filter. High frequencies produce maximal vibrations at the beginning of the coil near an opening from the middle ear called the round window. Low frequencies produce maximal vibrations at the other end of the coil.

The sense organ of hearing in the inner ear consists of cells with hairs that protrude into a gelatinous membrane. When these hairs move back and forth in response to sound, their vibrations are converted into electrical currents. This process results from chemical and physical changes in these hair cells. These electrical currents stimulate the hearing nerves and produce patterns of excitation. These patterns or stimulus codes are transmitted to the higher brain centers where they are interpreted as sound. The patterns of electrical responses are processed as pitch and loudness, as well as meaningful signals such as speech.

Deafness

A person who has a progressive sensorineural deafness loses the hair cells in the inner ear. As a result the hearing becomes faint and distorted and the sound has to be amplified for enough cells to respond. When most of the hair cells are absent, no amount of amplification with a hearing aid will help the person hear speech, as there is no hearing organ to excite the remaining hearing nerves leading to the brain centers. At best the person will hear muffled sounds. These people are profoundly deaf and were the first who stood to benefit from the bionic ear.

Overall Concept of the Bionic Ear

Research that commenced at the University of Sydney in 1967 and continued at the University of Melbourne in 1970 led to a multiple-electrode cochlear implant, which was developed industrially by Cochlear Proprietary Limited in 1982. As illustrated in Figure 1, it consists of a directional microphone (*a*) that converts sound into electrical voltages that are sent to a small speech processor worn behind the ear (*b*) or a larger, more versatile one attached to a belt (*c*). The speech processor filters this waveform into frequency bands. The outputs of the filters are referred to a map of the patient's electric current thresholds and comfortable listening levels for the individual electrodes. A code is produced for the stimulus parameters (electrode site and current level) to represent the speech signal at each instant in time. This code, together with power, is transmitted by radio waves via a circular aerial (*d*) through the intact skin to the receiver-stimulator (*e*) implanted

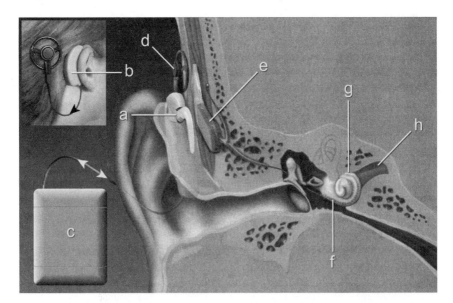

FIGURE 1. The cochlear implant. The components are as follows: a, microphone; b, behind-the-ear speech processor; c, body-worn speech processor; d, transmitting aerial; e, receiver-stimulator; f, electrode bundle; g, inner ear (cochlea); h, auditory or cochlear nerve. (Reprinted with permission from Clark, G.M. 2000b. Sounds from silence. St. Leonards, NSW, Allen & Unwin.)

in the mastoid bone. The receiver-stimulator decodes the signal and produces a pattern of electrical stimulus currents in a bundle of electrodes (*f*) inserted around the first turn of the inner ear (*g*) to stimulate the auditory nerve fibers (*h*). A pattern of hearing nerve activity in response to sound is produced, and provides a meaningful representation of speech and environmental sounds. The electrode bundle, lies close to, but not attached to, the spiral ganglion cells in the inner ear and their peripheral hearing nerve fibers.

Training in the Use of the Bionic Ear

After recovery from the cochlear implant operation, the patient attends training sessions in how to understand the sensations created by electrical simulation. The first task is to establish thresholds and maximum comfortable levels for electrical stimulation on each electrode pair. The thresholds and maximum comfortable levels are programmed into the map of the patient's speech processor. Auditory training exercises involve listening to speech and repeating what is heard. The speech material may be sentences, words, or vowels and consonants. The exercises allow the audiologist to assess the performance of the patient and at the same time provide training. The task must not be too difficult or the patient may be discouraged. The patient is also counseled on how to use the device, for example what to expect if the batteries become flat. Later, training is given in the use of the telephone. Auditory training for children concentrates on improving not only their ability to perceive and understand speech and environmental sounds, but also their speech production, receptive and expressive language, and communication. The speech material used for the training is age appropriate. The training is integrated into the child's educational program at either a preschool or school level. The children need to be taught by auditory-oral or auditory-verbal methods to take advantage of the new auditory information they are receiving. In certain situations the use of total communication where signed English is combined with an auditory stimulus will be required. Sign language for the deaf may also be used in certain children after individual assessment regarding their communication needs.

Fundamental Objections and Questions

In the 1960s and 1970s many believed that successful electrical stimulation of the hearing nerve to help people understand speech was not possible in the foreseeable future. A fundamental objection, which was reasonable, was that the inner ear hair cells and their nerve connections were too complex and numerous to reproduce the temporal and spatial pattern of responses in the hearing nerve by electrical stimulation with just a small number of electrodes. There are some 20,000 inner and outer hair cells required for normal hearing.

A second objection was that a bionic ear would destroy the very hearing nerves in the inner ear it was intended to stimulate. For example, a Teflon strip with sharp edges can cut through the inner ear basilar membrane and lead to near-total loss of the inner ear nerve cells in the vicinity of the injury. It was also believed that the electrode could be a pathway for middle ear infection to initiate infection of the inner ear, which could in turn spread to the meningeal lining of the brain.

A third objection was that speech was too complex to be presented to the nervous system by electrical stimulation for speech understanding.

A fourth objection was that there would not be enough residual hearing nerves in the inner ear after they died back due to deafness to transmit essential speech information. There can be an 80% loss of the hearing nerve ganglion cells and their fibers after the destruction of inner ear hair cells in deafness.

A fifth objection was that children born deaf would not develop appropriate nerve-to-brain cell connections, through lack of exposure to sound during the early critical phase of development, for electrical stimulation to give adequate hearing. The number of nerve connections on brain cells can be significantly reduced when compared to that in people with normal hearing.

There were other important questions: (1) Would the electrical stimulus currents damage the hearing nerves? (2) Were the candidate materials for the implantable electrodes and receiver/stimulator toxic to tissue? (3) Would middle ear infection spread along the electrode bundle to produce infection in the inner ear with possible life-threatening infection around the linings of the brain (meningitis)? (4) Could electrodes be inserted into the inner ear far enough so that the hearing nerves responsible for the place coding of speech frequencies would be stimulated? (5) What type of patients should be selected? (6) How should the operation be performed? (7) Would the perception of pitch on a multiple-electrode or place-coding basis be possible? (8) Would the perception of pitch on a time-coding basis be possible? (9) What electrical currents would produce loudness? (10) Would patients have memory for sounds and speech after prolonged deafness? (11) Could speech be processed so that patients could understand conversations? (12) Would speech and music sound natural? (13) If a speech-processing scheme was achieved for English, would it be effective in other languages? (14) How important a factor was the child's age at implantation with regard to learning to understand speech?

Answers to the Fundamental Objections

The first fundamental objection, that the inner ear hair cells and their nerve connections were too complex and numerous to reproduce the temporal and spatial pattern of responses in the hearing nerve by electrical stimulation with just a small number of electrodes, was studied by determining how well electrical stimulation could reproduce the coding of sound. The temporal coding of frequency was examined in the experimental animal by determining how well groups of brain cells could respond at increasing rates of stimulation. The voltages from brain cells and brainstem field potentials at increasing rates of stimulation showed the electrical activity in the auditory brainstem was markedly suppressed by stimulus rates at 100 pulses/second. Behavioral studies in the experimental animal showed that rates of stimulation in excess of 200 to 600 pulses/second could not be discriminated.

The experimental animal findings thus indicated that the reproduction of the temporal coding of frequency by electrical stimulation with a single-electrode cochlear implant could reproduce speech frequencies only from 200 to 600 cycles/second, which is much less than the 4000 cycles/second needed for speech intelligibility. Therefore, the best chance of helping deaf people understand speech was to use multiple-electrode stimulation to provided more information for speech understanding.

To achieve the place coding of frequency through multiple-electrode stimulation required determining where to place the electrodes in the inner ear so that the current would most easily pass through separate groups of hearing nerve fibers connected to the different frequency regions of the brain. Research showed that the compartment below the sense organ of hearing (scala tympani) and close to the ganglion cells at the center of the inner ear spiral was the correct location. Research also demonstrated that electrical currents could be partly localized to groups of nerve fibers within the inner ear without it short-circuiting away through fluid by pushing electrical current out one electrode and pulling it back from another (bipolar stimulation).

The animal experiments referred to above demonstrated that both temporal and place frequency coding or pitch perception could be only partially reproduced by electrical stimulation. In other words, a cochlear implant is like a bottleneck between the world of sound and the central hearing pathways of the brain.

The second fundamental objection was that if an electrode was implanted in the inner ear, which was particularly important for multiple-electrode stimulation, it would damage the very nerves it was intended to stimulate. It was found, however, in the experimental animal, that if no excessive force was used with its insertion, the hearing nerves were preserved. The risk of injury was reduced to a minimum if the electrode bundle had the right mechanical properties. It needed to be smooth, tapered, flexible at the tip, and stiffer toward the proximal end.

Infection could be restricted from entering the inner ear if the electrode entry point was sealed with a graft of fascia, and care was taken to prevent infection of the middle ear during the healing phase of the tissue over the first few weeks postoperatively.

The density of the electrical charge passing through electrodes with electrical stimulation was also known to damage nerve fibers. The safe limits for use with a cochlear implant had to be tested, too. It was found to be safe if the current had a positive and negative phase to reduce the buildup of direct current (DC), and the charge density was below approximately 32 microcoulombs per square centimeter per phase.

The third objection, that speech was too complex to be presented to the nervous system by electrical stimulation for speech understanding, would have to be addressed by multiple-electrode stimulation to transmit as much information as possible through the bottleneck. This required studies on patients to determine how effective multiple-electrode stimulation would be, as speech perception is an especially human skill and could not be evaluated on the experimental animal. Studies on patients required developing a fully implantable receiver-stimulator to receive information transmitted through the intact skin, rather than a plug and socket, which was more likely to break and become infected.

A prototype receiver-stimulator to use on patients was produced by the University of Melbourne from 1974 to 1978 using hybrid technology that connected a number of silicon chips together on a silicon substrate or wafer. The wafers were placed in a watertight or hermetically sealed container. The prototype receiver-stimulator was implanted in the first profoundly deaf adult patient on August 1, 1978, with the banded electrode array passing around the inner ear to lie near, but not in direct contact with, the nerves relaying speech frequency to the brain.

Perceptual studies were then undertaken on the first and subsequent patients to determine if the findings on the temporal and place coding of frequency in the experimental animal were applicable to humans. The patient studies confirmed that rate of stimulation was not effective in transmitting frequency or pitch information over the range required for speech understanding. Pitch ratios were plotted against repetition rate, and it was shown that when the pitch of a stimulus was compared with a reference rate of 100 pulses/s, the pitch ratios increase linearly up to 300 pulses/s, and then reached a plateau; 300 pulses/s is much less than the 4000 pulses/s needed for speech understanding.

The studies on the place coding of frequency showed that with localized electrical stimulation the patients could perceive only timbre, not true pitch. In the high-frequency areas of the inner ear the sensation was sharp, and on the lower frequency side it was dull. However, the patients could rank the timbre according to the site of stimulation.

The perceptual studies on the patients confirmed the findings on experimental animals that electrical stimulation with the cochlear implant was a bottleneck for information from the outside world to the central auditory pathways. The first research to transmit information through the bottleneck selected speech frequencies using fixed filters with similar properties to the tuning of the inner ear. When the outputs were used to stimulate the hearing nerves simultaneously, the result was poor. Simultaneous stimulation produced overlap of currents resulting in unpredictable variations in loudness. However, a speech-processing strategy was discovered that gave the patients the ability to understand connected or running speech when presented with speech reading or even using electrical stimulation alone. The clue to this speech-processing strategy came when the first patient reported vowel sounds when each electrode was stimulated on a place-coding basis. The vowels corresponded to those perceived by normal-hearing people when similar areas of the inner ear were excited by single-formant frequencies. Formants are concentrations of energy at particular frequencies or vocal tract resonances. They are important for intelligibility, especially the second formant.

This research led to the University of Melbourne's inaugural speech-processing strategy, which extracted the second formant frequency using a bank of filters. The voltages from the filters stimulated electrodes at appropriate frequency regions around the inner ear. The stimuli were perceived as timbre. The sound pressure for the formant frequency was coded as current level and perceived as loudness. The fundamental or voicing was perceived as pitch. This speech-processing strategy was tested at a number of centers in the United States and Europe, and in 1985 was the first multiple-electrode cochlear implant to be approved by the U.S. Food and Drug Administration (FDA). This inaugural speech processing strategy enabled patients who had hearing before going deaf to understand running speech when combined with lipreading and some speech using electrical stimulation alone.

The research at the University of Melbourne then focused on which further speech elements to extract and present on a place-coding basis. It was found that picking the energy in the first as well as the second formant peak, and presenting this nonsimultaneously on a place-coding basis, gave improved results. Then it was discovered that selecting energy in the high-frequency bands in the third formant region as well as the first and second formants (Multipeak) gave further improvement. The most recent strategies [spectral maxima speech processor (SPEAK) and advanced combination encoder (ACE)], implemented in the Nucleus 24 system, select the six to eight frequency bands with the greatest energy from a 16 to 20 band pass filter bank and present the information as a place code. As the strategy selects the six to eight maximal outputs from the band pass filters, the sites of stimulation within the cochlea may lie close together, leading to an overlap in the electrical current with unpredictable variations in loudness. This has been minimized by using a constant rate of stimulation on all electrodes. In this case, the rate of stimulation is not used to convey voicing, but voicing is conveyed through the amplitude variations in the signal. The present scores indicate that the average person can now communicate effectively over the telephone. Furthermore, the scores are now better than the average scores obtained by severely to profoundly deaf persons with some residual hearing using a hearing aid. Another strategy, the continuous interleaved sampler (CIS) developed at Research Triangle, North Carolina, produces similarly good results.

The fourth objection, that there would not be enough residual hearing nerves in the inner ear after they died back due to deafness for understanding speech, was partly resolved by finding the good results referred to above. However, the residual hearing nerve population could have been responsible for the significant variability in the scores. The relationship between the hearing nerve population and speech perception was studied by ranking speech perception scores versus the cause of deafness, and the hearing nerve population versus cause of deafness. The rankings for both speech perception and hearing nerve population versus cause of deafness were different. This suggested that speech perception is not strongly related to the population of nerves or ganglion cells. So die back after deafness is not a significant factor in performance with the present cochlear implant systems.

The final major objection was that children might not be able to develop the right nerve-to-brain cell connections for speech understanding through electrical stimulation when their brains are at a malleable stage. It was therefore of critical importance to learn whether speech perception performance in children who had hearing before going deaf was comparable to that for children who were born deaf and thus had no prior exposure to sound. It was important to determine in particular whether exposure to sound during a critical period when the brain connections are plastic would be a necessity for adequate perception or whether appropriate connections could develop in the absence of exposure to sound. The speech perception abilities of two groups of children (those born without exposure to sound and those becoming deaf after exposure to sound) were compared. Their best perception skills ranged from mere detection of sound to recognition of words in sentences from an open set. The recognition of closed and open sets of words by children born without hearing (prelinguistically deaf) and deaf after hearing (postlinguistically deaf) improved dramatically after operation. Although open-set recognition was better for the postlinguistic subjects, the performance of the children born deaf was sufficiently good to feel confident that prior exposure to sound was not necessary for good speech perception.

Although the results showed children born deaf could develop the right nerve-to–brain cell connections for speech understanding with a cochlear implant, it was still not clear how young they should be to get the best results. For this reason the sentence scores for a group of children at the clinic of the Royal Victorian Eye and Ear Hospital were plotted versus age at implantation. There was considerable variability in responses, but a curve fitting showed that performance improved the younger the age at operation. It also indicated that the scores might be better if the operation was performed when the child was under the age of 2 years.

There are special safety issues to be considered when implanting children under the age of 2: the effects of head growth; middle ear infection, which is especially common at this age; and electrical stimulation on a maturing nervous system. This research was undertaken through a 5-year contract to the U.S. National Institutes of Health and showed no cause for concern in operating on this group of children, provided that care was taken to seal the electrode entry point with a fascial graft, and that the operation did not take place in the presence of an incipient middle ear infection.

The research at the University of Melbourne and the Bionic Ear Institute has been germinal to the development of the multiple-electrode bionic ear manufactured by Cochlear Limited. The research was also fundamental in helping Cochlear Limited to achieve the largest share of the world market in the 1980s and through the 1990s.

1
A History

Pre-science

The challenges deaf people had in communicating were recognized by the Israelites in the Mosaic law where it was said, "Do not curse the deaf" (Leviticus 19:14). Then the "Judaic-Christian tradition, within which, along with the Arabic tradition and the insights of the Western Enlightenment, modern science grew up, has seen total health (like peace and a benign environment) as a future hope" (McCaughey 1995). There was to be a time when, according to the prophet Isaiah (35:5), "The eyes of the blind shall be opened and ears of the deaf unstopped." According to the physician and apostle Luke, Jesus said, "And the deaf hear" (Luke 7:22). Nevertheless, prior to the 18th century, deafness was a serious sensory disability. Apart from the ear trumpet there was little to help with communication. Children were particularly disadvantaged. They led sheltered, restricted lives in institutions, and were referred to as being deaf and dumb.

Eighteenth Century

The 18th century saw considerable social change as well as the flowering of science. This was the climate in which the first efforts to help deaf children were made by l'Abbé de l'Epée at the Paris Deaf School (Fig. 1.1) in approximately 1794, as well as Heineke in Germany. L'Abbé de l'Epée developed a language based on a system of signs, while Heineke saw learning to speech read as the better way to help. The need to help deaf people communicate could have been the reason count Alessandro Volta, an Italian physics professor, soon after developing the battery, carried out on himself in the late 1790s the first experiment on electrical stimulation of the auditory nerve. His results were read on June 26, 1800, before the Royal Society meeting in London presided over by the Rt. Hon. Sir Joseph Banks, who was the botanist on Captain Cook's voyage of discovery to Australia in 1770. The report is recorded in the *Philosophical Transactions of the Royal Society of London* for the year 1800, part I, p. 427:

FIGURE 1.1. A painting of l'Abbé de l'Epée teaching pupils at the Paris Deaf School. (Reproduced from the Paris Deaf School library, with permission.)

It only remains for me to say a word about hearing. I had tried without success to excite this sense with two single metallic plates, although they were the most active among all the movers of electricity, namely one of silver or gold and the other of zinc, but I finally managed to affect it with my new apparatus, made up of 30 or 40 pairs of these metals. I introduced right into both ears two probes or rods of metal with rounded ends; I linked them up immediately to the two extremities of the apparatus. At the moment when the circuit was completed in this way, I received a jolt in the head; and a few moments later (the circuit operating continuously without any interruption), I began to feel a sound, or rather a noise, in my ears which I cannot define clearly; it was a kind of jerky crackling or bubbling, as though some paste or tenacious matter was boiling. This noise continued without stopping and without increasing all the time the circuit was complete, etc. The disagreeable sensation of the jolt in the brain, which I feared might be dangerous, was such that I did not repeat this experiment several times. [Translated by E.C. Forsyth]

Because of the unpleasant sensation experienced, only sporadic attempts were carried out over the next 50 years to investigate the phenomenon.

Nineteenth Century

As sound is an alternating disturbance in an elastic medium, it was appreciated that stimulating the auditory system with a direct current (DC) could not reproduce a satisfactory hearing sensation. Consequently, Duchenne of Boulogne in 1855 stimulated the ear with an alternating current, which he produced by inserting a vibrator into a circuit containing a condenser and induction coil. The result, however, was still not satisfactory, as he experienced only a sound that resembled "the beating of a fly's wings between a pane of glass and a curtain."

A more extensive investigation was published by Brenner in 1868. He studied the effects of altering the polarity, rate and intensity of the stimulus, and placement of the electrodes on the hearing sensation produced (cited by Simmons 1966). One electrode was applied to saline in the ear canal, and the other at a remote location on the body. The results showed that hearing was better with an electrical stimulus that produced a negative polarity in the ear. The stimulus induced sensations that were metallic, but also resembled buzzing, hissing, whistling, ringing, etc., at various pitches. Although the sensations varied, it was the same for the one subject. The side effects of pain, vertigo, and facial nerve stimulation were reduced to a minimum with the correct placement of the electrodes. Brenner attributed the hearing sensations to electrical stimulation of the auditory nerve.

Twentieth Century

1900 to 1930s: Early Hearing Aids

In the first two decades of the 20th century efforts were directed at amplifying sound with a hearing aid rather than using electrical stimulation. Amplifying sound appeared to offer great promise to deaf people. The first carbon hearing aid was a table model called the Akoulallion, and it appeared in 1899 (Berger 1984; Dillon 2001). The first wearable models called the Akouphone and the Acousticon became available in about 1902. These carbon hearing aids consisted of a carbon microphone, battery, and magnetic receiver connected in series. The sound vibrations caused carbon particles to move closer and further apart, changing the electrical resistance. This change in resistance affected the current passing through a coil in the receiver, causing a fluctuating magnetic field. This field made the permanent magnet attached to a diaphragm move in and out in synchronization with the sound. The sound level from the receiver was 20 to 30 dB greater than at the microphone. This limited gain was increased by adding a second microphone and receiver. Carbon hearing aids continued to be used through to the 1940s.

The discovery of the vacuum tube electronic amplifier allowed more precision and greater amplification of sound. The electronic amplifier was invented in 1907 and applied to hearing aids in 1920 (Lybarger 1944). This amplifier permitted a small voltage coming from the microphone to control fluctuations in a large current, and this made it possible for a 70-dB gain and 130-dB sound pressure level (SPL) output. The electronic amplifier greatly increased the range of hearing levels that could be assisted. Vacuum tube hearing aids became practical during the 1930s, but until 1944 the batteries were so large that they had to be held separately, and in some cases they were worn strapped to the leg.

Hearing aids became much more effective with the introduction of the transistor invented by Bardeen, Brattain, and Shockley in 1947. By 1953 hearing aids used transistors. This allowed a reduction in battery size so that the hearing aid could be mounted on the head. Following the introduction of transistors, integrated circuit technology further reduced the aid in size and allowed more advanced processing to be used. Advances in microphone design included the piezoelectric microphone used in combination with a new type of transistor, the field effect transistor (FET).

1930s to 1940s: Initial Indirect Electrical Stimulation in the Human

The electronic amplifier also made it possible to control electrical stimulation of the auditory pathways, and was the technical advance needed to initiate studies on electrical stimulation of the cochlea and auditory nerve in the 1930s and 1940s. Renewed interest in reproducing hearing artificially through electrical stimulation occurred as only moderately and severely deaf people could benefit from the hearing aids being introduced. The thermionic (radio) valve enabled the auditory system to be stimulated electrically with much greater precision. This work was encouraged by the demonstration by Wever and Bray (1930) that the electrical response recorded from the vicinity of the auditory nerve of a cat was similar in frequency and amplitude to the sounds to which the ear had been exposed.

At this time, Russian investigators examined the effects of an alternating electrical stimulus on hearing. A report by Gersuni and Volokhov (1936) indicated that hearing for different pitches corresponded to the rate of stimulation. There was considerable debate about whether hearing different pitches was due to direct electrical stimulation of the auditory nerve or transduction of the electrical current into vibrations that then excited the hair cells in the cochlea. Evidence for indirect excitation of hair cells from a site external to the cochlea such as the tympanic membrane and middle ear ossicles came from Stevens and Jones (1939). They showed that changing the DC level of the stimulus altered the proportion of the harmonics formed, as would be expected for a mechanical system involving the ossicles. Hearing, however, occurred following the surgical removal of the tympanic membrane and ossicles. This suggested that these structures were not the only ones involved in the process. It was thought that the cochlea per se was the primary site for direct stimulation, although the mechanisms involved were not clear. Hearing induced this way was called "electrophonic."

Research by Jones et al (1940) led them to conclude that there were three mechanisms that produced hearing with electrical stimulation. First, the middle ear acted as a transducer that converted alterations in the strength of an electrical field into the mechanical vibrations that produced sound. It was thought that the tympanic membrane was attracted to and from the medial wall of the middle ear, and that these vibrations were responsible for the hearing sensation. Second, as sound could also be heard when the middle ear structures were absent, and the transducer was linear, this suggested that electrical energy could be converted into sound by a direct effect on the basilar membrane. In this case the membrane would vibrate maximally at a point determined by the frequency, and these vibrations would stimulate the hair cells. Third, a crude hearing sensation was produced in people with minimal or absent hearing, and this was probably due to direct stimulation of the auditory nerve.

Other body tissues were also found to act as transducers. Flottorp (1953) showed this applied to the skin over any part of the body with hearing produced by air or bone conduction. This meant that the middle ear was not the only transducer of sound external to the cochlea as postulated by Jones et al (1940), but the skin lining the external auditory meatus and the tissues overlying the mastoid could act equally well.

Further proof that hearing due to electrical stimulation could be produced by air or bone conduction, was advanced by Sommer and von Gierke (1964). In their experiments the whole head, or parts of its surface, were exposed to an alternating electrostatic field in the audiofrequency range, and threshold data obtained. These thresholds were compared with the sound pressures required to stimulate bone conduction, and the results were found to be comparable. When the electrode was inserted into the external auditory meatus, however, the thresholds were comparable to air rather than bone conduction.

It was thought hearing percepts might be induced in deaf patients using a radiofrequency signal, modulated to follow speech or other sounds. There have been a number of reports of people hearing sound in the vicinity of radio transmitters (cited by Frey 1961), but the phenomenon had not been fully investigated. Work by Frey (1961) was undertaken to determine whether the hearing produced in this way was due to sound generated in the environment, or to direct stimulation of the auditory apparatus, and if so, how it produced its effects. It was found that the radiofrequency signal could be detected only when the source was not shielded, and that a person with an air conduction loss of 50-dB hearing loss (HL), but with good bone conduction, had a normal threshold for the stimulus. Furthermore, a person with perceptive deafness was found to have an elevated threshold. This evidence pointed to the fact that the auditory nerve was not stimulated directly.

This research into the mechanisms involved in hearing with electrical stimulation indicated that audition was produced by transducing electrical energy into sound energy external to the inner ear, with it being transmitted to a functioning cochlea through bone or air conduction. Alternatively, electrophonic hearing occurred when vibrations in the basilar membrane were induced directly, with sound perception also mediated by intact hair cells. A rough hearing sensation was also produced by global excitation of the auditory nerve. Therefore, the data suggested that some hearing in total perceptive deafness might be possible, but little thought was given to how this might be achieved.

1950s to 1960s: Initial Direct Electrical Stimulation in the Human

One of the first recorded attempts to stimulate the auditory nerve directly was by Lundberg in 1950 (cited by Gisselsson 1950), who did so with a sinusoidal current during a neurosurgical operation. The patient, however, could hear only noise. A more detailed study was performed by Djourno and Eyriès (1957). As discussed by Eisen (submitted), Djourno approached the study from a background in stimulating nerves in general with inductive coils. Eyriès was a surgeon with an interest in facial nerve repair. He was asked to restore facial nerve function in a patient who was also totally deaf from extensive temporal bone surgery for bilateral cholesteatoma. The anatomist Delmas suggested stimulating the hearing nerve at the same time as the facial nerve graft. At surgery on February 25, 1957, the cochlear nerve stump was found shredded. A telecoil was placed on the nerve and stimulated with bursts of 100-Hz signals administered 15 to 20 times per minute. It was noted that intensity discrimination was good; on the other hand, the patient was able to detect differences in pitch only in increments of 100 pulses/s up to a rate of 1000 pulses/s (cited by Simmons 1966). Postoperative rehabilitation was undertaken, and the patient could distinguish certain words such as "papa," "maman," and "allo," but did not develop speech recognition. The device failed due to a breakage in the ground electrode in the temporalis muscle, and the same happened after a reimplantation. It is postulated by Eisen (submitted) that the hearing sensations were due to excitation of the cochlear nucleus, as the spiral ganglion cells would have degenerated as well as their axons. Djourno later carried out a second implant with otolaryngologist Maspétiol in a patient who this time went deaf from streptomycin but fared no better (Zollner and Keidel 1963). In this case an electrode was placed extracochlearly near the promontory. No further implants were performed, as it appeared that electrical stimulation of the auditory nerve could not reproduce the high frequencies, or the discrimination of small changes in rate necessary for speech comprehension.

In January 1961, House and Urban (1973) implanted single electrodes insulated with silicon rubber in the scala tympani of three profoundly deaf patients for a period of approximately 3 weeks. The first patient came because he had read a report of the work of Djourno, Eyriès, and Vallancien. Unfortunately, the electrodes had to be removed from the three patients because of local irritation. However, they experienced hearing sensations, although these were not analyzed. The work was in conjunction with neurosurgeon John Doyle and electronic engineer

James Doyle. An investigation was carried out by Simmons et al (1964), who stimulated the auditory nerve and inferior colliculus in a patient undergoing a craniectomy for a cerebellar ependymoma. The stimuli were varied in frequency from 1 to 1000 pulses/s, pulse duration from 0.1 to 1.0 ms, and amplitude from 0.01 to 2.0 V. When the auditory nerve was stimulated, the patient was able to differentiate different frequencies from 20 to 3500 pulses/s, and at 850 pulses/s had a difference limen (DL) of 5 pulses/s. Electrode placement was found to be critical, and hearing could not be experienced unless the two electrodes were aligned parallel to the nerve fibers. Stimulation of the inferior colliculus, however, produced no hearing sensation. The patient had cochlear function, and thus his ability to detect a frequency of 3500 pulse/s with auditory nerve stimulation was probably due to the fact that hearing was electrophonic in origin.

Michelson (1971) developed an interest in electrical stimulation of the hearing nerve when a hearing person in his operating room heard a high-pitch tone from the oscillation of a high-gain amplifier used to monitor the cochlear microphonic response. Michelson modified the microphone signal to emphasize the energy in some regions of the frequency spectrum, and provided an overall compression of its energy range. This was later upgraded to eight bipolar electrodes driven by a similar signal, but with the option to choose the most effective electrode or electrodes. This signal stimulated a single channel with bipolar electrodes placed in the scala tympani. All four profoundly deaf subjects could understand certain environmental sounds and modulate their voices better, but they were unable to recognize unaided speech.

All the above studies examined the effects of electrical stimulation on single electrodes. They were in fact assessing the possibility of reproducing the coding of frequency on the basis of the volley theory of Wever and Bray (1930), also demonstrated in the later physiological studies of Tasaki (1954), Katsuki and Kanno (1962), Rupert et al (1963), Kiang et al (1965), and Rose et al (1967). The studies did not investigate systematically the relation between the electrical stimuli and the hearing percepts, and no speech processing strategies were assessed.

Information about the frequency of a sound could also be conveyed to the brain on the basis of the place theory. The place coding of the spectrum of sound frequencies was first postulated by von Helmholtz (1863). It was demonstrated physiologically by Rose et al (1959) and Kiang et al (1965). With the place theory as initially postulated, it was not so much whether the nerve cells fired in time with the sine wave, but rather the spatial arrangement of the nerve cells and their best frequencies of response. The hearing receptors in the inner ear and the cells and fibers in the central nervous systems are arranged in an orderly anatomical way so that they will respond to only a limited range of frequencies along a scale from low to high. This meant that to reproduce a 5000-Hz tone artificially by electrical stimulation, those auditory nerve fibers that normally convey this frequency should be stimulated. This could be done surgically by placing multiple electrodes close to the auditory nerve fibers where they fan out to supply the organ of Corti.

An attempt to stimulate the hearing nerve with multiple electrodes was undertaken by Doyle et al (1963, 1964). They carried out experiments on 14 volunteers, placing electrodes in the middle ear to stimulate the auditory nerve. As now known, this would not have provided localized stimulation of separate groups of nerve fibers. They also placed four electrodes in the cochlea of a congenitally and profoundly deaf woman. The electrodes had speech signals superimposed upon them. The electrodes were not especially implanted to take advantage of the spatial distribution of auditory nerve fibers responding to different frequencies. The woman was able only to differentiate between male and female voices and to follow the rhythm of music and speech. The scientific reports generated media interest, but many otologists considered the publicity premature (House 1987).

A more detailed study was reported by Simmons et al (1965) and Simmons (1966), who described the results of implanting six electrodes into the modiolus (central axis of cochlea where the spiral ganglion cells of the auditory nerve lie) of a patient with complete perceptive deafness. The electrodes were placed in the modiolus with the aim of stimulating nerve fibers representing different frequencies selectively. The patient was then tested to determine the effect of alterations in the frequency and intensity of the signal. Monopolar and bipolar stimulation were compared, and the waveform of the stimulus was also varied. The results indicated that the patient could detect a change in pitch up to a frequency of 300 pulses/s. Single stimuli produced a pitch sensation, and this varied according to the position of the stimulating electrode. The relation of the electrodes to the position of the fibers in the auditory nerve was difficult to determine. This was evidence in support of the place theory of hearing. Simmons et al (1965) and Simmons (1966) then used a number of modified speech signals to drive a six-electrode implant in the cochlear nerve. Attempts were made to separate the speech spectrum into frequency bands, and process each band to produce pulse stimuli that would make more efficient use of the characteristic pitches of the various electrodes. They used an amplitude-compressed speech signal to provide two simultaneous channels of information (high and low frequency), or single channels of information (using frequency modulated or pulsatile signals). A single-channel implant using an 1800-Hz low-pass filtered version of the microphone signal to trigger and amplitude modulate rectangular current pulses produced a voice-like percept. Speech communication was not feasible with the initial implantee and the strategies used. Simmons did not attempt further implantations on people for approximately 10 years. He rated the chances of electrical stimulation of the auditory nerve providing uniquely useful communication at about 5% (Simmons 1966). He believed that tactile learning and intensively training residual hearing, especially in children, would be more likely to provide useful hearing.

The publication by Simmons (1966) of initial findings was the impetus for the author to initiate in 1967 basic electrophysiological studies to examine the neural responses to electrical stimulation in experimental animals to see how well they reproduced those for sound (Clark 1969b). These and other studies are described in the next subsection. In 1969 William House implanted a device with multiple-electrodes in three patients, but thereafter concentrated on a single electrode implant. The reason was not reported, but was presumably due to failure to see perceptual advantages because of widespread current spread.

In the case reports described above, experience with direct electrical stimulation of the auditory nerve and its terminal fibers in the cochlea had not clearly shown that the surgical treatment of perceptive deafness with speech understanding was possible. A number of problems would have to be solved, however, before satisfactory speech intelligibility could be achieved.

1960s: Fundamental Research in the Experimental Animal

In the 1960s there was considerable skepticism in scientific circles about the practical possibility of a cochlear implant. Merle Lawrence (1964) stated, "Direct stimulation of the auditory nerve fibers with resultant perception of speech is not feasible." He came to this conclusion on a number of grounds. First, previous studies had shown that there was only a 10-dB dynamic range for electrical stimulation, and that would be too small to be practicable when considering the 120-dB range of the normal ear. Second, the innervation patterns of dendrites in the cochlea were complex, and it was assumed that the discrimination of frequency and intensity depended on these patterns of innervation. It therefore would be difficult for electrical stimulation to reproduce the coding of sound. Third, there was evidence that if hair cells were destroyed, the peripheral processes or dendrites and cell bodies in Rosenthal's canal would degenerate. This assumption arose from studies that had shown that cochlear pathology could lead to the marked loss of auditory neurons (Schuknecht 1953). A significant number of spiral ganglion cells was thought necessary for adequate speech understanding (Schuknecht and Woellner 1955; Jerger 1960). As a result, concern was expressed that a majority of profoundly deaf people would not have sufficient residual ganglion cells to stimulate electrically and produce speech comprehension. This was based on data from Kerr and Schuknecht (1968) that only 25% of the cochleae they examined histologically had two thirds or more of the normal spiral ganglion cell population remaining. Their hypothesis was that a two-thirds proportion was required for speech comprehension. Furthermore, it was assumed that inserting an electrode into the cochlea would eventually lead to the loss of any remaining neurons. This view received support from clinical observations that the cochlea was very sensitive to trauma during a stapedectomy. As a result otologists treated the cochlea with great respect and avoided inserting anything into it if at all possible. In addition E.P. Fowler, Sr. had said, "Direct stimulation of the cochlear nerve will from time to time be discovered. There is no indication that it will ever succeed in enabling a patient to readily hear speech" (Fowler 1968, quoted by Goodhill 1979).

The above criticisms were largely physiological and biological, and were based on the supposition that the complex network of nerves in the cochlea was important for the coding of sound. The coding could not be reproduced with a relatively small number of electrodes or even 20. There were also important unresolved issues on the safety of inserting electrodes into the inner ear.

Fundamental research on hearing with electrical stimulation in the 1960s was first necessary in the experimental animal to help establish proof in principle for a single- or multiple-channel approach. The research in the 1960s, 1970s, and 1980s is discussed here by country of origin, as the funding and emphasis during these three decades was a national one. Since the industrial development in the 1980s, the emphasis has changed and has become more international, and includes commercially relevant work.

Australia

The inconclusive results obtained from the initial single- and multiple-electrode studies on patients indicated that fundamental research was needed to investigate a number of important questions. It was considered that a greater understanding of many of the questions would result from using animal experiments (Clark 1969b). Furthermore, data were required to answer the reservations against electrical stimulation of the auditory nerve by scientists (e.g., Lawrence 1964) before it would be clinically acceptable.

A study commenced in 1967 at the University of Sydney to determine especially the effect of rate of stimulation on the responses of cells in the auditory brainstem, rather than the auditory nerve (Clark 1969a–c). The processing of information in the brainstem was likely to be more closely correlated with the anticipated behavioral responses, as inhibitory mechanisms as well as the refractory periods are present at this level. Furthermore, it had been shown by Butler and Neff (1950), Butler et al (1957), and Goldberg and Neff (1961) in the cat that, after removal of the primary cortex and neural degeneration extending down to the medial geniculate body, the ability to discriminate changes in frequency was retained or relearned. This suggested that the processing of frequency information should be studied at a lower level, and that the brainstem nuclei were most suitable.

The above study by Clark (1969a,b, 1970a) showed sustained firing in time with the stimuli did not occur for most neurons above 200 pulses/s. A small number responded up to 500 pulses/s. To overcome the problem of sampling responses from a limited number of individual neurons, the changes in the amplitudes of field potentials were also determined for different rates of stimulation, as field potentials are the result of unit activity in a number of neurons. The results discussed in Chapter 5 show a marked reduction in amplitudes at various sites within the superior olive for stimulus rates of 300 pulses/s and above (Clark 1969a,b, 1970a). The field potential studies confirmed the cell data that there were limitations in maintaining unit responses to electrical stimulation above about 200 to 300 pulses/s. The research also compared the unit responses in the brainstem for acoustic and electrical stimulation (Clark 1969a). The study showed that the main differences between the responses to electrical square wave and acoustical stimulation were that neurons fired more synchronously or deterministically to electrical stimulation, and more asynchronously or stochastically to acoustical stimulation (Clark 1969a–c). This finding was consistent with that of Moxon (1971) for the auditory nerve fibers.

Following the above experimental animal research and literature review there were a number of conclusions (Clark 1969a, b):

"... the site and method of implantation are important as the neural pathways can be damaged and this would prevent the electrical signals being transmitted to the higher centers. Destruction of the cochlea can lead to trans-neuronal degeneration in the cochlear and superior olivary nuclei up to a year after the production of the lesions (Powell and Erulkar 1962).... a greater understanding of the encoding of sound is desirable. As emphasized by Lawrence (1964), the terminal auditory nerve fibers are connected to the hair cells in a complex manner, which could make it difficult for electrical stimulation to simulate sound.

The relative importance of the volley and place theories in frequency coding is also relevant to the problem. If the volley theory is of great importance in coding frequency, would it be possible for different nerve fibers, conducting the same frequency information, to be stimulated in such a way that they fired in phase at stimulus rates greater than 1000 pulse/s? If this were possible, it would then have to be decided whether this could be done by stimulating the auditory nerve as a whole, or whether local stimulation of different groups of nerve fibers in the cochlea would be sufficient. On the other hand, if the place theory is of great importance in coding frequency, would it matter whether the electrical stimulus caused excitation of nerve fibers at the same rate as an auditory stimulus, or could the nerve fibers passing to a particular portion of the basilar membrane be stimulated without their need to fire in phase with the stimulus?

If the answers to these questions indicate that stimulation of the auditory nerve fibers near their terminations in the cochlea is important, then it will be necessary to know more about the internal resistances and lines of current flow in the cochlea, and whether the electrical responses normally recorded are a reflection of the transduction of sound into nerve discharges, or directly responsible for stimulating the nerve endings.

The final criterion of success will be whether the patient can hear and understand speech. If pure tone reproduction is not perfect, meaningful speech may still be perceived if speech can be analyzed into its important components, and these used for electrical stimulation. More work is required, however, to decide which signals are of greatest importance in speech perception." [Clark 1969b]

As the research summarized above showed the limitations of coding speech frequencies from 300 to 3000 Hz on a rate basis, it was essential to confirm the findings on the behaviorally conditioned animal, to determine whether the speech frequencies should be represented by place of stimulation. To reproduce the place coding of frequency, it would be necessary to know more about the internal resistances and lines of current flow in the cochlea so the stimuli could be limited to separate groups of nerve fibers. These questions were the basis for further research at the University of Melbourne's Department of Otolaryngology.

United States

The cochlea was considered by some to be too delicate to tolerate surgical manipulation or the long-term placement of electrodes (Legouix 1957, cited by Simmons 1967). However, the possibility of implanting electrodes into the cochlea without significant loss of nerve fibers was investigated in the experimental animal by Simmons (1967). This research showed that the round window membrane could be incised, perilymph aspirated, and a metallic electrode inserted for weeks to months without widespread cochlear degeneration. Damage to cochlear tissues could be caused by the physical trauma of electrode insertions, but cellular degeneration was very localized unless infection occurred (Simmons 1967). In addition, Moxon (1971) studied the effect of electrical stimulation on the response rate of auditory nerve fibers in the cat for basic physiological reasons to determine the cause of their firing rate limitation with acoustic excitation, and this is discussed in Chapter 5. The auditory nerve fibers (not subject to inhibition as are units in the brainstem) responded initially at a rate of 900 pulses/s for a short time, and then fell to a plateau of 500 pulses/s. The firing was also deterministic.

1970s: Fundamental Research in the Experimental Animal and Human

The fundamental physiological and biological research in the 1970s was linked with engineering research for the development of prototypes to evaluate speech-processing strategies in humans. Special attention was required to develop the interface with the central nervous system through electrode arrays.

The 1970s saw the development of not only single-channel implants as an investigational tool, but also some multiple-channel implants. With multiple-channel processing the speech signal was filtered, and the output applied with different algorithms to separate electrodes. The psychophysics and speech research were important for assessing how best to help deaf people with cochlear implants.

Australia

In the 1970s funding from government agencies for cochlear implant research in Australia proved difficult, as the peer reviewers considered the research unlikely to be successful (Clark 2000a). Funding came for 3 years from nerve deafness telethons, as well as trusts and foundations.

Physiological and Biological Research

The research at the University of Melbourne focused at first on behavioral experiments in the alert animal to confirm the acute physiological findings (Clark 1969a–c, 1970a,b). The research showed the limitations of using rate of electrical stimulation to code speech and other frequencies. This research commenced in 1971, and was essential for making sure that a multiple- rather than a single-electrode implant should be developed. The results were published from 1972 to 1975 (Clark, Kranz et al 1972, 1973a,b; Clark, Nathar et al 1972, Williams et al 1974), and the findings were consistent with those from the cell studies and showed that the discrimination of rate was poorer than for sounds of the same frequency. Rate was unusable above about 600 pulses/s for a speech-processing strategy. The finding also helped establish the importance of the volley theory for coding low-frequency sounds. The discrimination of frequency modulation was also limited, and this stressed the difficulty of conveying the frequency transitions important for consonant recognition with a single-channel system. Another significant discovery was learning that rate discrimination was similar when stimulating the upper and lower basal turns (lower and higher frequency regions) of the cochlea. This would prove to be the means of presenting voicing frequencies at each electrode along the array while representing speech frequency spectral information on a place-coding basis in the case of the University of Melbourne's first three speech-processing strategies.

Research to develop an appropriate multiple-electrode interface with the auditory nervous system for psychophysical and speech research in patients also commenced in 1971. The aims were to determine (1) where to place the electrodes in the inner ear and the mode of stimulation to localize the current to separate groups of nerve fibers, (2) how to place the electrodes in the inner ear so that they would lie close to the auditory nerve fibers for the place coding of speech frequencies, (3) the design of the array so that it would be atraumatic and provide safe charge densities for neural stimulation, and (4) whether the electrode insertion would damage the inner ear and auditory neurons.

The results of computer modeling of electrical resistances in the cochlea and physiological studies showed that the most effective site for localized stimulation of the auditory nerve was the scala tympani with bipolar or common ground current flow (Black and Clark 1978, 1980). It was discovered that an array of uniform stiffness could be inserted into the apical and middle turns of the cochlea and could advance more easily to the round window than could an array of uniform stiffness pushed upward from the round window. Insertion depths of only 10 to 15 mm could be obtained this way. On the other hand, the electrodes and carrier could be inserted around the scala tympani from below upward if it was round (free-fitting) and had graded stiffness (Clark 1975; Clark, Hallworth et al 1975).

Biological safety studies were undertaken to determine the effects of implanting electrodes at different sites within the cochlea. Placing electrodes through holes drilled in the overlying bone was compared with inserting a bundle either upward or downward along a cochlear turn. It was discovered that the least trauma occurred when they were inserted upward along the scala tympani in the basal turn. They could be inserted without loss of auditory neurons provided the basilar membrane or spiral lamina were not torn or fractured, respectively, or infection introduced (Clark 1973a,b, 1977; Clark, Kranz et al 1975).

Engineering

The development of a receiver-stimulator for a 10-channel prosthesis commenced at the University of Melbourne in 1973. It was supported by the more basic research described above. The decision was made to develop a transcutaneous link (i.e., through intact skin) and use implanted electronics in a receiver-stimulator unit, rather than a plug and socket (percutaneous stimulation), which became infected in experimental animals. The electrode array referred to above would also have to allow electrical current to stimulate separate groups of cochlear nerve fibers in the speech frequency range. The history of its development is outlined in more detail elsewhere (Clark 1993, 1997, 2000a,b; Clark, Tong et al 1990); also see Chapter 8.

Research in the Departments of Otolaryngology and Electrical Engineering at the University of Melbourne to design the electronics for an implantable receiver-stimulator commenced in 1973 and was completed in 1976. The hermetically sealed package was finished in 1977. There were a number of questions to be answered with the engineering development:

1. How should information be transmitted through the skin to the implant? It was not clear whether this should be to the implanted electronics by infrared light shining through the ear drum, ultrasonic vibrations to the scalp, or radio waves through the tissues. The initial studies established that an electromagnetic link rather than ultrasound or infrared was the most efficient method of data transfer. The modulation of a high-frequency carrier wave gave the most error-free data transfer. Transmission was designed to occur via two circular matched aerials, one on the outside and the other on the implant. Hybrid circuitry was chosen to allow flexibility in design.
2. Should the implant have its own batteries or receive power from outside? Power to operate the implant and provide the electricity to stimulate the hearing nerves could either be obtained from the energy in the radio signals sent through the skin, or be provided by implanted batteries. Batteries were used in pacemakers, but they had to be replaced through a small skin incision every 5 years. Batteries would also increase the size of the implant. Therefore, radio signals were the better alternative.
3. Where should the implant be placed in the body? It was logical to site the microphone close to the ear as the head is turned in the direction of the sound receiving attention. The electronic package would need to be close to the microphone to avoid connecting wires breaking due to repeated flexing, and be small enough to place in the mastoid bone behind the ear.
4. What would be the design of the electronics to allow a wide range of stimulus currents to be produced? The receiver-stimulator electronics would have to provide enough flexibility to choose the right electrical stimuli to represent the speech signals, as a percutaneous plug was not considered suitable due to the risk of infection.
5. How should the electronics be packaged? This was also a key issue. The body is a very corrosive environment, and salts and tissue could find their way through very fine cracks and cause electronic failure. This was especially likely where the wires leave the package to join the electrode bundle passing to the inner ear. The sealing problem had bedeviled the heart pacemaker industry

until staff from the Australian firms Telectronics and AWA discovered that a ceramic mixture could be baked to fuse with a titanium case and platinum wires. With the University of Melbourne's prototype device, a Kovar steel container was used with glass to metal seals for the exiting wires. This developed cracks with fluid entry paths, and emphasized the need for the first Nucleus clinical trial device to use pacemaker-sealing technology.

6. Did the implant package need a connector in case it failed and another one had to be reattached to the electrodes in the inner ear? As it was not certain how often the electronics would fail, the replacement of the implanted receiver-stimulator had to be considered. It was also not clear whether it would be detrimental to remove the attached electrode from the inner ear. To be sure, a connector was designed so that only the package needed to be changed. With the University of Melbourne's prototype a connector using Elastomer conducting pads was produced. This was later found to be inadequate as the contacts would not stay compressed together. In the Nucleus clinical trial device, Silastic compression pads were used. However, in the later Nucleus CI-22 "mini" implant they were dispensed with, as it had by then been shown that the smooth free-fitting array could easily be removed and another reinserted (Clark, Pyman et al 1987).

With the University of Melbourne's prototype, an array of 20 electrodes was planned with 10 active electrodes and 10 to act as an interleaved common ground. A thin film array of platinum sputtered onto a strip of Teflon was first produced (Clark, Hallworth et al 1975; Clark and Hallworth 1976) to meet these requirements, but found to be unusable because bending stresses led to multiple fractures in the metal. A more conventional fabrication technique was developed.

As discussed above, it was found to be difficult to insert an electrode array of uniform stiffness upward in the scala tympani to lie opposite the cochlear nerves conveying speech frequencies. However, it was discovered in 1976 that with an array of graded stiffness an insertion from below upward into the scala tympani could be achieved and the insertion was deep enough for electrodes to lie opposite the speech frequency region. This minimized the frictional force on the outer wall limiting advancement, but was rigid enough at the base to apply force to the tip. It also had the advantage over discrete ball electrodes that it could be rotated without loss of effective contact with the cochlear nerves.

Psychophysics and Speech Research

On August 1, 1978, the University of Melbourne's multiple-channel receiver-stimulator (Fig. 1.2) and electrode array were implanted in the first postlinguistically deaf volunteer. The pitch perceived with rate of stimulation could be discriminated only up to approximately 300 pulses/s. This was the same limitation seen with the behavioral studies in the experimental animal. On the other hand, place pitch varied from sharp at the basal high-frequency end of the cochlea to dull at the lower frequency end and could best be described as timbre (Tong, Black et al 1979; Tong, Clark et al 1980a, 1982; Tong, Millar et al 1980). As

10 mm

FIGURE 1.2. The University of Melbourne's multiple-channel receiver-stimulator implanted in the first patient on August 1, 1978. The Kovar container for the electronics and the overlying receiver coils for the data and power are to the left. The connector is to the right. It was removed on July 26, 1983, and replaced with a Nucleus clinical trial device. (Clark G. M. et al, 1987. The University of Melbourne–Nucleus multi-electrode cochlear implant. Advances in Oto-Rhino-Laryngology Volume 38. Basel, Karger. Reprinted with permission.)

timbre varied with site of electrode stimulation, it showed that the electrical currents were being localized to separate groups of auditory nerve fibers for place coding.

An initial neurophysiological speech-processing strategy was tested and found to be unsuccessful due to simultaneous stimulation producing unpredictable variations in loudness. This stressed the importance of using nonsimultaneous stimulation. An effective speech-processing strategy was discovered in 1978 when the patient described vowel sounds that varied according to the site of stimulation. The site of stimulation corresponded with the frequency of a single formant vowel. Formants are concentrations of frequency energy that are important for intelligibility. The initial strategy coded the second formant as place of stimulation, the intensity of the formant as current level, and voicing as rate of stimulation. It resulted in a marked improvement in understanding open sets of speech when combined with speech reading (300% to 400% improvement), and also some open-set speech using electrical stimulation alone (Tong, Black et al 1979; Tong, Clark et al 1980a; Tong, Millar et al 1980; Clark and Tong 1981; Clark, Tong et al 1981a,b). This was demonstrated through the use of standardized audiological procedures.

Two further volunteers received the University of Melbourne's multiple-channel implant in 1979. Both showed the same psychophysical pitch percepts as the first volunteer. The second person also demonstrated a good memory for speech sounds after 13 years without exposure to sound. This indicated that other people with a long period of deafness could benefit. Finally, in 1979 a wearable speech processor was developed (Figure 1.3). It had the dimensions of 15 × 15 × 6.5 cm and a weight of 1.25 kg (Tong, Clark et al 1980b).

Austria

The biological, mechanical, and communications engineering research in Austria was undertaken at the Technical University of Vienna and funded by the Austrian Science Research Fund (Hochmair and Hochmair-Desoyer 1983).

Biological and Engineering Research

A main focus of the work was the development of a four-channel flexible intra-cochlear array as well as a pair of electrodes for extracochlear placement. The two alternatives were explored (Hochmair and Hochmair-Desoyer 1983), as it was stated that "no results have been found that could scientifically demonstrate

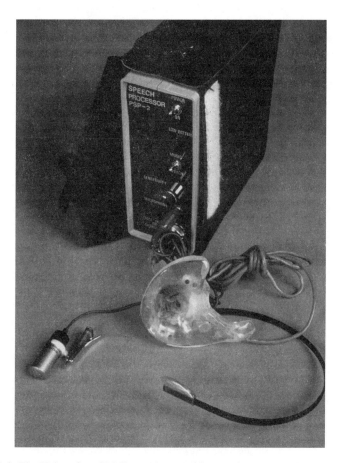

FIGURE 1.3. The University of Melbourne's wearable speech processor developed in 1979 with the headband and mold to keep the transmitting coil in place.

an advantage of multichannel stimulation." The intracochlear array had Teflon-coated platinum/iridium wires and was molded with silicone elastomer. The electrode wires terminated as balls with a diameter of 0.35 mm. They protruded from the surface with a usable area of 0.06 mm². The balls were on either side of the array to provide four bipolar pairs of stimuli. The wires were knurled to withstand stress. The electrical fields for stimulation between different pairs of electrodes were examined in tank studies, and a narrower field was found for bipolar rather than common ground stimulation. Initial experience with patients showed that the electrical stimuli produced only auditory sensations, and that over time there was no loss of threshold, indicating that the nerves had not degenerated producing a "physiological failure" as some had predicted.

Psychophysical and Speech Research

The Vienna group used a four-channel flexible intracochlear array as well as a pair of electrodes for extracochlear placement for evaluating speech-processing strategies in their patients as discussed above. Psychophysical research evaluated amplitude discrimination, gap detection, and pitch perception and discrimination versus rate of stimulation. Although a speech-processing system was developed with four stimulus channels, only one electrode could be stimulated at a time with bipolar pulses. A number of single-channel strategies were explored. "Speech-likeness" and some speech sound discrimination were possible, and the group reported that two patients had significant open-set scores for words and sentences for electrical stimulation alone (Hochmair et al 1979; Hochmair-Desoyer et al 1980, 1981).

France

Research in France was funded by the Institut National de la Santé et de la Recherche Médicale (INSERM), the Ministère de la Recherche Scientifique de France, and insurance companies.

Speech Research

From 1973 five to seven electrodes were implanted directly into the cochlea via a middle cranial fossa approach in the first of 42 deaf patients (Pialoux et al 1979). Insulating material was inserted into the scala tympani between each fenestra, with the intention of localizing the current flow for neural stimulation. The electrodes were connected percutaneously to the stimulating equipment via Teflon plugs. The plugs, however, had to be removed due to skin infection. Several electrodes elicited different pitches depending on their location (Chouard et al 1987). A processor was developed in which the speech signal was divided into frequency bands according to an empirically derived map gained from unilaterally deaf implantees through comparison with the opposite hearing ear. Energy in each frequency band was encoded as a repetition rate of fixed amplitude pulses.

Because of infection around the percutaneous plug, an implantable receiver was designed in conjunction with the company Bertin St. The device Chorimac originally had eight channels and later 12 channels. Further patients were implanted from October 1976 with Chorimac-8. As an alternative to the middle cranial fossa approach for multiple-electrode implantation (Chouard and Mac-Leod 1976; Chouard 1980), it introduced eight electrodes into the scala tympani through individual holes (fenestrae) drilled in the lateral wall of the cochlea in the middle ear. The patients were unselected for etiology or age when deafened. From 1976 to 1984 109 patients were implanted. The 54 prelinguistically deaf patients in the group could not achieve open-set speech understanding. Results on postlinguistically deaf patients showed that some phonemic discrimination could occur for electrical stimulation alone, but the patients required training. With closed sets of 17 words, a proportion could recognize them correctly. There was no report of identifying open sets of words with electrical stimulation alone.

United Kingdom

In the latter half of the 1970s a single-channel approach was developed in London (Douek et al 1977; Fourcin et al 1979). Extracochlear electrical stimulation of the auditory nerve was carried out by Douek et al (1977) and Fourcin et al (1979) in patients with open mastoid cavities. An extracochlear approach was used, as it was not clear whether damage from intracochlear electrodes would lead to loss of most of the auditory nerve fibers it was hoped to stimulate. With the speech-processing strategy, the glottal pulsing was picked up by a microphone beside the larynx, and the signals were transmitted to a receiver by infrared. The unit stimulated the cochlea in time with the glottal pulses. It was thus minimally invasive, with a single electrode placed on the promontory of the cochlea. Speech processing extracted a signal from speech that was a measure of the movement of the speaker's vocal cords, rather than from a microphone. Information from the vocal cords is normally completely denied to profoundly deaf people even if they are expert at processing visual cues. The performance of this device was measured in terms of its ability to transmit voicing information and the ability of the implantee to integrate this information with visual cues. Results showed considerable help as a lipreading aid, and there was improved vocalization.

United States

In the United States the National Institutes of Health (NIH) recognized the need for prostheses to restore neural function. Richard Maslan stated in the National Institute of Neurological Diseases and Blindness (NINDB) profile 1 in 1967: "New techniques of neurological control hold exciting possibilities for the development of an artificial eye, a hearing device for the deaf, and means of movement control for the physically handicapped." The NINDB set up a laboratory of neural control in 1967 headed by Dr. Karl Frank. It was renamed the National Institute of Neurological Diseases and Stroke (NINDS) on October 24, 1968, and the Sensory Prosthesis Program was formed in 1970, with Dr. Terry Hambrecht and Dr. Karl Frank as codirectors. Initially it focused on the feasibility of a sensory prosthesis for the blind. In 1971 the first Neural Prosthesis Workshop was formed as a meeting ground of ideas. At the third workshop in 1973 the challenges for a hearing prosthesis were discussed, including electrode materials, insertion trauma, and the acoustic elements of importance in understanding speech.

The NINDS and later in 1975 the National Institute of Neurological and Communicative Disorders and Stroke (NINCDS) awarded a number of grants and contracts for the development of a hearing prosthesis. The grants commenced in 1973 and continued throughout the 1970s at four centers. The centers and chief investigators were Stanford University (F. Blair Simmons/Robert L. White), University of California at San Francisco (UCSF) (Michael M. Merzenich), Oregon Health Sciences University (Richard A. Walloch), and the University of Washington (Joseph M. Millar).

Physiological, Biological and Engineering Research

University of Stanford. In 1973 research commenced at Stanford and focused on the hardware for a multiple-channel prosthesis, including thin film electrodes, packaging technology, and circuitry. The aim was to provide a more advanced electrode array (a bayonet thin-film array on sapphire) to interface current to the auditory nerve rather than the wires used by Simmons et al (1965) and Simmons (1966). The aims of the study were as follows: (1) ensure the preservation of the insulation over the array when the insulator was only a few microns thick, (2) prevent the layers from lifting off, (3) build sufficient metal mass in the electrode pads in case there was corrosion with long-term stimulation, and (4) achieve satisfactory lead attachment (White et al. 1983).

University of California at San Francisco. At the University of California at San Francisco (UCSF) in 1973 the research was physiologically based and aimed in particular at determining how to localize the electrical current to separate groups of nerve fibers, and the safety of long-term implantation. In the research by Merzenich et al (1973), recordings were made from units in the inferior colliculus of the cat, which receive input from both ears. This is an effective model as the spatial attenuation of current along the cochlea would be determined through knowing the site of the response from the unit's acoustic input from the other ear. The results showed that with bipolar stimulation from electrodes on a mold of the scala tympani, the current could be localized to separate groups of nerve fibers (Merzenich and Reid 1974). The current distribution for bipolar stimulation, however, varied depending on the orientation of the electrode pair and the presence or absence of peripheral processes (dendrites) of spiral ganglion cells (Merzenich et al 1979).

The studies on insertion trauma demonstrated that the peripheral processes of the auditory nerve would degenerate with a tear of the basilar membrane or a fracture of the spiral lamina, and there could be loss of the spiral ganglion cells (Schindler et al 1977; Leake-Jones and Rebscher 1983). The loss was localized to the region and did not affect the neurons at a more remote location in the cochlea.

University of Washington and Oregon Health Sciences University. Research at the University of Washington and Oregon Health Sciences University commenced in 1976 to study in particular the behavioral responses of the primate (University of Washington) and cat (Oregon Health Sciences University) to electrical stimulus parameters. The initial research was carried out to correlate psychophysical data in the monkey with histopathological findings, as this could indicate the possible limitations of implants on patients. The findings showed that the degree of pathology was related to the threshold currents and inversely related to the dynamic range (Pfingst et al 1979, 1981). Walloch et al (1980) found that the pulse width of electrical stimuli could affect perceived loudness in cats. Square-wave biphasic stimuli with pulse widths of 0.5 ms were perceived as louder than stimuli with 0.2-ms pulse widths and the same interstimulus interval (2 ms), although there was no difference between 0.5-ms and 1.0-ms pulse widths.

Huntington Research Laboratories. Research at the Huntington Research Laboratories was funded by the NIH to study the biophysics of the brain/electrode interface with special application to a visual cortical prosthesis. The research examined the electrical stimulus parameter range required to limit metal corrosion and neural damage. The findings (Brummer and Turner 1975, 1977) were relevant to cochlear prostheses and are discussed in Chapter 4.

Psychophysics and Speech Research

House Ear Institute. The House Ear Institute in Los Angeles developed a prosthesis using a single monopolar electrode in the scala tympani. It was used with just one time-varying signal to stimulate the electrode, and all the information transmitted was encoded in the amplitude of the waveform and its detailed shape over time. The signal was presented to a monopolar electrode that could have produced a large spread of current throughout the cochlea (House et al 1981). This was implanted in the first adult in October 1972, and by late 1978 it had been implanted in 33 patients. A band-pass filtered version of the signal controlled a nonlinear modulator that allowed speech signals in the range of 45 to 65 dB SPL to vary the amplitude of a 16-kHz carrier signal that provided the stimulating current to the electrode. The maximum level and the modulation depth of the signal were under the control of the user, and the 1973 design was described by Danley and Fretz (1982). Initially benefits were anecdotal, and no open-set speech understanding was achieved. However, there was some evidence that the patients' psychological state had improved. The NIH funded a study on the Los Angeles single-electrode patients to objectively assess performance, particularly examining their speech perception benefits (Bilger et al 1977). The patients achieved some improvements in discriminating elements of speech and assistance in modulating their own voices, and were able to hear a variety of sounds in their natural environment (Hannaway 1996).

Stanford University. Psychophysical and speech research at Stanford University commenced in 1977 on two volunteers who had four wires implanted directly into the auditory nerve in the modiolus and connected to a percutaneous plug behind the ear. In the same year the group began experiments to determine an optimal speech-processing strategy, but one patient had an explantation in 1979 because of infection. Following early negative results with strategies, psycho-physical studies confirmed the limitations of coding frequency as the rate of stimulation (Atlas et al 1983). Multiple-channel strategies for stimulating the cochlear nerve in the modiolus were pursued in the 1980s.

University of California at San Francisco. In 1971 Michelson reported the results of electrical stimulation in four profoundly deaf patients who had a single-channel bipolar electrode placed in the scala tympani through the round window. The four patients had severe hearing losses. The thresholds in one were 50 dB HL at 250 Hz and 500 Hz and 60 dB HL at 1000 Hz in both ears. The residual hearing in the operated ears was reported to be lost. In acute studies on two patients, both were able to hear signals up to 12,000 Hz, and to discriminate between tones only 50 Hz apart in the low to middle frequency range. The results could have been due to "electrophonic" hearing in the opposite ear and/or changes in loudness with frequency, as the stimuli were not reported as balanced. With the microphone voltage presented to the electrode, speech was not intelligible. After longer term implantation in two patients, a limited amount of closed-set speech understanding was obtained for electrical stimulation alone.

Following this early work on single-channel (though implemented via multiple electrodes) stimulation, the group at UCSF undertook electrophysiological re-search, funded by the NINCDS, to determine how best to localize the current to separate groups of nerve fibers, as discussed above. A molded multiple-electrode array was used in a patient study, and stimulated with band-pass filtered versions of the speech signal (Merzenich et al 1987). Biphasic pulsatile rather than analog stimuli were employed. A strategy was used where the energy of the speech signal was compressed, and then one band-pass filter output was used for low-frequency vocal resonances, two for midfrequency resonances, and one for higher frequencies.

University of Utah. The University of Utah in Salt Lake City (Eddington et al 1978a; Eddington 1980, 1983) undertook research with an implant with six mono-polar electrodes, with 4-mm separation between electrodes inserted into the scala tympani. These monopolar electrodes could have resulted in significant current spread. This system first had a Teflon percutaneous plug, but the incidence of infection was high, and warranted a change to a percutaneous pedestal of pyrol-ized carbon to allow better integration with the skin edges and bone.

The outputs of four band-pass channels, representing the speech spectrum, were used to control the level of analog signals stimulating four electrodes. The early research examined in particular pitch scaling and the parameters responsible for the growth of loudness (Eddington et al 1978b). Vowel and consonant identifi-cation in deaf subjects was then studied. The data suggested that the four band-pass fixed filter strategy might provide first formant frequency and intensity cues to help with speech understanding.

1980s: Fundamental Research, Industrial Development, and Clinical Trials

The open-set word and sentence results with multiple-channel speech processing, referred to above, produced heightened interest in cochlear implantation. But they also stimulated debate about the merits of multiple-channel versus single-channel stimulation. Furthermore, people still questioned whether the results justified its clinical use. This is reflected in a statement by Edgerton (1985): "Probably no area of health care has been more controversial during the past decade than the development and implementation of the cochlear implant." As open-set testing protocols varied across centers, results were hard to interpret. It was essential to standardize the test procedures and undertake controlled studies. The open-set speech perception results also stimulated research to see just how well speech could be analyzed and processed. Funding became more readily available for research to understand the relation between the neural and perceptual response to electrical stimulation, and the coding and psychophysics of sound. It saw wider acknowledgment of the importance of cochlear implantation scientifically as evidenced by the inclusion of electrical stimulation of the auditory nerve in the 10th International Congress of Acoustics in Sydney in 1980, and it was a major component of a satellite symposium (Mechanisms of Hearing) to the 23rd Congress of the International Union of Physiological Sciences in 1983 at Monash University, Melbourne. Research expanded from the original centers to others in a number of countries. In some cases linkages were formed with companies, and clinical trials commenced on larger groups of patients for health regulatory authorities.

The following relationships between research centers and industry developed: Telectronics (Cochlear Proprietary Limited, now Cochlear Limited) with the University of Melbourne, 3M with House Ear Institute and Technical University of Vienna, Symbion and Richards with the University of Utah, Storz with UCSF, Bertin St. with the University of Paris, Philips with the University of Antwerp, and Hortmann with the University of Düren. The strength and nature of the partnerships varied, and the evaluation of the results for the most commonly used devices is discussed below. Furthermore, it was important to establish the efficacy of cochlear implants on adults before surgery on children, for ethical and practical reasons. Even older children could not be expected to make decisions on a procedure when it was still experimental. Problems with the device needed to be resolved on adults, as children had more years ahead to cope with any adverse effects. With children it was especially important to know whether a multiple-channel implant was better than a single-channel system. If the speech results with a single-channel implant were only the same as a hearing aid or tactile vocoder, then would the risks of invading the inner ear outweigh the benefits of being aware of sound? (A vocoder separates two sources of information in the speech signal for efficient manipulation and transmission.) On the other hand, if the multiple-channel device was better, would it be the treatment of choice? In addition, a major question with children was, Could those who were born deaf (that is, without the sound exposure required to establish the neural connectivity for processing the speech information) be able to benefit? If so, as with a single-channel system, would the results be any better than a powerful and correctly fitted hearing aid or a tactile vocoder? In summary, by tackling and answering the above questions the 1980s was the decade when cochlear implant research flourished and made great progress.

Melbourne/Sydney, Australia

The open-set speech understanding achieved on the first postlinguistically deaf adults with the University of Melbourne's formant extraction speech processing strategy in 1978–79 led to funding from the Australian government through a public interest grant from 1979 to 1981 for research to develop it industrially. The National Health and Medical Research Council of Australia (NH&MRC) awarded project grants from 1978 to 1985 for basic research to continue the investigation on how best to use electrical stimulation of the auditory nerve for the treatment of a severe to profound deafness. With further advances in 1984 an NH&MRC program grant was awarded for 9 years. The University of Melbourne received from the NIH in 1984 a 3-year grant, "An Advanced Multi-channel Cochlear Implant for Deafness," to also study the relationship between the perception of complex electrical stimulation and advances in speech processing. In 1987 a 3-year NIH grant, "Improved Cochlear Implant: Psychophysics and Engineering," was awarded to extend the research. An NIH contract, "Speech Processors for Auditory Prostheses," extending over 7 years, was awarded in 1985. This aimed in particular at comparing formant extraction and fixed filter strategies and how best to optimize these. In 1987 the University of Melbourne received a 5-year NIH contract, "Studies on Pediatric Auditory Prosthesis Implants," to investigate all key biological safety issues before operations were carried out on children under 2 years of age. In 1988 an Australian Research Council's (ARC) Center of Excellence was awarded for a period of 9 years through which the Human Communication Research Center (HCRC) was established at the University of Melbourne's Department of Otolaryngology and Bionic Ear Institute. The HCRC not only studied the psychophysics of complex patterns of electrical stimulation and the relationship to improved speech processing strategies, but also commenced research on the psychophysics and speech processing for bilateral cochlear implants and bimodal stimulation (a hearing aid in one ear and implant in the other). The center undertook the initial research to develop a perimodiolar array that would lie closer to the neural elements in the modiolus. It was considered that an improved interface to the auditory nervous system would be needed for further advances. The center investigated perceptual plasticity of the nervous system in young children and its relationship to speech perception. It studied the development of speech perception and production as well as receptive and expressive language in implanted children and the factors responsible for best results. Studies were undertaken to determine safe electrical stimulus levels for brainstem implants. Research to establish an objective method for the early and objective diagnosis of a hearing loss in the newborn and infants using steady-state evoked potentials to amplitude modulated sounds was extended. A multiple-channel electrotactile aid as an alternative to a cochlear implant for children was also developed, and psychophysical and speech processing studies pursued.

The research of the HCRC was further developed industrially in the 1990s through a Cooperative Research Center for Cochlear Implant Speech and Hearing Research and, following this, the Cooperative Research Center for Cochlear Implant and Hearing Aid Innovation. In both cooperative centers the core partners have been the Bionic Ear Institute, the University of Melbourne, Cochlear Limited, and Australian Hearing Services. The centers have established collaborative research links internationally. The HCRC, however, became the main center for basic communication research at the University of Melbourne/Bionic Ear Institute. The projects discussed above were key investigations in the 1980s and 1990s, as was the neuroscience underlying the protection of nerve degeneration and neural regeneration.

Psychophysics and Speech Research

The psychophysics and speech research at the University of Melbourne/Bionic Ear Institute in the 1980s initially focused on detailed psychophysical studies on the first patient. The University of Melbourne's prototype receiver-stimulator had failed in the other two subjects. The research aimed to determine how complex electrical stimuli of relevance to speech processing were perceived. It was discovered that the rate of change in place of stimulation could convey the frequency transitions of consonants more effectively than changing the rate of stimulation (Tong, Clark et al 1982). Another important finding was that voicing could be transmitted independently of place of stimulation by varying the rate of stimulation on different electrodes and across electrodes (Tong, Blamey et al 1983). So rate of stimulation was perceived as voicing, voicing was perceived independently of place of stimulation, and voicing was perceived by varying the rate of stimulation across different nerve populations. Furthermore, the perceptual dimensions for rate and place pitch were found to be separate (Tong, Dowell et al 1983). Thus temporal and place information provided two components to the pitch of an electrical stimulus. It was also discovered that varying the amplitudes of two electrodes stimulated nearly simultaneously could shift the pitch to an intermediary one as well as alter the vowel perceived.

In the absence of research patients until after 1982 when the first Nucleus clinical trial device became available, acoustic models were developed for electrical stimulation so that different strategies could be simulated on normal-hearing volunteers. The acoustic simulation used pseudo-random white noise with the output fed through seven band pass filters corresponding to different electrode sites (Blamey et al 1984a,b). It was found to be a good model in predicting the pitch results with electrical stimulation, and by adding the first formant frequency (F1) as well as the second formant (F2), there was significant improvement in the perception of speech features and connected speech. The model results gave the impetus for later developing a cochlear implant speech processing strategy that coded F1 and F2 as place of stimulation, and used voicing (F0) as rate of stimulation (F0/F1/F2). This strategy was shown to produce better speech perception than the initial F0/F2 strategy and especially in noise (Dowell et al. 1986a, 1987; Franz et al 1987).

Biological Safety

In addition to the psychophysical and speech studies, the University of Melbourne's research was directed at determining the safety of long-term stimulation on the experimental animal. The parameters considered to be electrochemically safe were chosen, and their long-term effects on the survival of the auditory nerve and ganglion cell as well as the tissue effects on the cochlea as a whole were studied. Studies showed that if biphasic stimuli were used with a charge asymmetry between 0.01% and 0.1% to keep damaging DC current to a minimum 0.1 μA, and a charge density of less than 32 μC cm^{-2} geometric/phase, there was little risk of loss of the spiral ganglion cells after up to 2000 hours of continuous stimulation (Shepherd et al 1982, 1983a–c).

Industrial Development and Clinical Trial of F0/F2 Device for Postlinguistically Deaf Adults

In 1980 the Australian pacemaker firm Telectronics was selected by the Australian government under a public interest grant scheme to develop the University of Melbourne's prototype industrially. Therefore, some of the university's research was directed toward industry's more immediate needs. Nucleus Limited, the holding company for Telectronics and then its subsidiary, Cochlear Proprietary Limited, in 1980 commenced the industrial development of the University of Melbourne's second formant multiple-electrode prosthesis. Cochlear Proprietary Limited's multiple-channel cochlear implant was ready for clinical trial in 1982. It had 22 electrodes for common ground stimulation for the place coding of frequency.

The Nucleus implant and speech processor was first assessed on six postlinguistically deaf adults in Melbourne in 1982, and then an international trial commenced in 1983 for the U.S. Food and Drug Administration (FDA) at the University of Iowa; Baylor Medical College, Houston; the Mason Clinic, Seattle; New York University; Louisiana State University, New Orleans; the Good Samaritan Hospital, Toronto; the Medizinische Hochschule, Hannover; the University of Sydney; and the University of Melbourne. The trial at these nine centers included 40 subjects. By mid-1985, 85 patients had been implanted at 18 different centers. A study of 40 multiple-channel implant users by Dowell et al (1986b) showed a mean score of 55% for vowel identification and 42% for consonant identification. At 3 months postimplantation the patients had obtained a mean Central Institute of the Deaf (CID) sentence score of 87% (range 45–100%) for speech reading plus electrical stimulation, compared to a score of 52% (range 15–85%) for speech reading alone (Dowell et al 1986b). It was also evident that

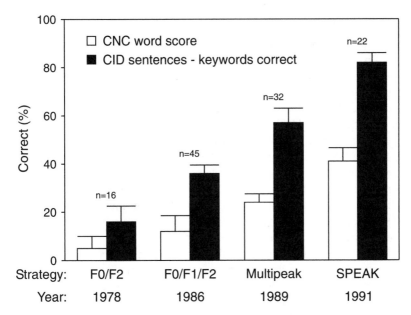

FIGURE 1.4. The open-set Central Institute for the Deaf (CID) sentence and consonant-nucleus-consonant (CNC) word scores for the Nucleus F0/F2, F0/F1/F2, Multipeak, and SPEAK (Spectral maxima speech processor) strategies. (Clark 2000a) Reprinted with permission from Whurr Publishers, Ltd.

the ability to understand speech with auditory input alone improved significantly between 3 and 12 months postoperatively. In a subgroup of 23 patients the mean CID sentence scores for electrical stimulation alone rose from 16% at 3 months postimplantation to 40% at 12 months (Dowell et al 1986b). The F0/F2 strategy WSP-II wearable speech processor was approved by the FDA in October 1985 as safe and effective for adults who had hearing before going deaf (postlinguistically deaf). The Premarket Approval (PMA) said Nucleus could claim that it provided help in understanding speech with lipreading, and some speech using electrical stimulation alone. At the time it was also important to determine if the F0/F2 strategy would be of benefit for tonal languages and this was first demonstrated for Mandarin by Xu et al (1987).

Further Cochlear Implant Speech-Processing Research

Further research in the HCRC at the University of Melbourne/Bionic Ear Institute through the 1980s demonstrated how the selection of additional speech information and its appropriate presentation to the nervous system led to better speech perception (Fig. 1.4). The combination of the first (F1) as well as second (F2) formants coded as place of stimulation together with voicing as rate of stimulation produced a quantum leap in scores. This F0/F1/F2 strategy and WSP-III processor was approved for use with postlinguistically deaf adults by the FDA in 1986.

Industrial Development and Clinical Trial of the F0/F1/F2 Device for Children

The research to help children born deaf or deafened early in life commenced with studies on prelinguistically deaf adults. The two who received the Nucleus (multiple-channel) implant had been educated almost entirely by signed English and sign language of the deaf. They were implanted in 1983 at ages 24 and 25 (Clark, Blamey et al 1987; Clark, Busby et al 1987). They performed well for current level identification and had satisfactory duration difference limens, but their abilities for pulse rate and electrode position identification were poor (Tong, Busby et al 1986, 1988; Clark, Blamey et al 1987; Clark, Bushby et al 1987). This was also reflected in their poor speech perception scores using the F0/F2 speech processing strategy (Busby et al 1986; Clark, Blamey et al 1987; Clark, Busby et al 1987; Tong, Busby et al 1988). It was concluded that the multiple-channel strategy used successfully for postlinguistically deaf people was probably not suitable for prelinguistically deaf people over approximately 20 years of age, especially if they had used sign language of the deaf beforehand. It may have also reduced their motivation to learn an auditory/oral-based system. The findings suggested that untreated deafness from an early age could lead to perceptual processing difficulties that could make speech processing using place and temporal cues unsatisfactory (Clark 1995).

Prior to implanting children, it was also realized that their training and assessment would need to be different from that for postlinguistically deaf adults. It was considered important to assess and train not only speech perception, but also speech production, as well as expressive and receptive language and communication skills (Nienhuys et al 1987). To this end, a protocol was developed that has been used subsequently for the management of all children in Melbourne, and was the basis for the trial set up for the FDA.

In view of the poor speech perception of the prelinguistically deaf adults it was necessary to operate on younger people as they would have better neural plasticity, and they preferably should have had an auditory/oral education. On January 8, 1985, a 14-year-old boy who had been taught with cued speech was implanted. His abilities in identifying electrode place and pulse rate were better than those of the two adult prelinguistically deaf subjects, but not as good as those generally obtained for postlinguistically deaf adults (Clark, Blamey et al 1987). Furthermore, his speech perception was better than that of the prelinguistically deaf adults, and he obtained some help in understanding running speech when the implant was used in combination with speech reading. Some months later (September 17, 1985) a 22-year-old prelinguistically deaf woman was implanted. She had received an auditory/oral education, and although born with a severe hearing loss, went profoundly or totally deaf over the first 18 years of her life. Interestingly, the speech perception results on this person were more similar to those obtained with postlinguistically deaf people (Clark, Blamey et al 1987). As the results on the child deafened early in life (the 14-year-old boy) were better than for the first two implanted as adults, it supported the desirability of operating on younger children.

Operating on younger children required the development of a new receiver-stimulator by Cochlear Proprietary Limited in association with University of Melbourne. It was smaller ("mini") and had a rare earth magnet at the center of the receiving coil (Dormer et al 1980), so the transmitting coil, also with a magnet, could be easily aligned by children. The connector was dispensed with as the biological research at University of Melbourne had shown the smooth free-fitting banded array could be easily removed and replaced with another if this was required. This made the package thinner and thus more suitable for placing in the skulls of children. When the device was ready and a clinical protocol established, the University of Melbourne's surgical staff implanted on August 20, 1985, the first child to have the Nucleus "mini" 22-electrode receiver-stimulator for the F0/F1/F2 WSP-III speech-processing system. The child was a 10-year-old who had a profound-total hearing loss at 3.5 years of age, and was educated by Total Communication. After establishing that the device performed to specifications, a series of psychophysical and speech perception studies were undertaken, which demonstrated an advantage for the perception of speech when the device was combined with speech reading, but little open-set speech recognition for electrical stimulation alone.

Thus it was decided to evaluate the prosthesis on younger patients, and a 5-year-old received a cochlear implant on April 15, 1986. This boy, who went deaf at 3 years of age from meningitis and was trained with cued speech, made excellent progress with the implant and after some months was able to get significant open-set speech identification scores from electrical stimulation alone.

Further children then received the implant at the University of Melbourne, and with support from the NH&MRC program grant the research focused on assessing the speech and language skills of the implantees and how best to rehabilitate them. The research also assessed the value of multiple-channel speech processing on children at different ages and as young as 2 years. The initial results suggested that both children born deaf and those deafened early in life could achieve similar benefits to adults who had been postlinguistically deaf (Dawson et al 1989), and a significant proportion could get open-set speech recognition for electrical stimulation alone (Dawson et al 1992). The prelinguistically deaf group, however, needed a longer period of training and appeared to do better if implanted at a young age.

The University of Melbourne's protocol for evaluating the children was modified for use in the United States with the FDA trials that commenced in 1987 with the F0/F1/F2 WSP-III and later the multipeak miniature speech processor (MSP) systems. The results for 80 children were presented to the FDA, and it was approved as safe and effective for children on June 27, 1990. When results on 142 children were analyzed (Staller et al 1991; Clark, Busby et al 1992), they confirmed the above findings from the University of Melbourne/Bionic Ear Institute (Dawson et al 1992) that pre- and postlinguistically deaf children could receive significant improvements when the device was combined with speech reading, and approximately 50% could get significant open-set speech scores for electrical stimulation alone. There were, however, some differences in the results for those born deaf and deafened early in life compared with those receiving an implant after developing language. The early-deafened children had slightly poorer open-set speech perception for electrical stimulation alone than those deafened after developing language. The F0/F1/F2 strategy was approved by the FDA on June 27, 1990, as safe and effective for children from 2 to 18 years of age. With more advanced speech-processing strategies developed in the 1990s and early operations, there is no difference in the results for children born deaf or postlinguistically deaf adults.

Further Psychophysics and Speech-Processing Research in Postlinguistically Deaf Adults

Further speech-processing research was undertaken at the University of Melbourne/Bionic Ear Institute in the mid- to late 1980s to improve the speech processing strategy. This focused on consonant recognition as it was still not ideal with the F0/F1/F2 processor, and in particular high-frequency spectral information was also coded as site of stimulation to help in the perception of frication. This led to the Cochlear Proprietary Limited's Nucleus Multipeak-MSP system. The additional presentation of two or three high-frequency filter band-pass filter outputs as well as F0/F1/F2 became the Multipeak strategy implemented on the processor MSP. It was found to give additional benefits (Fig. 1.4) over the F0/F1/F2 strategy (Dowell et al 1990), and was approved by the FDA in 1989 for postlinguistically deaf adults, and together with the F0/F1/F2 strategy for children in 1990.

In the mid- to late 1980s there was debate over the relative merits of the University of Melbourne/Bionic Ear Institute's formant and spectral extraction speech processing strategies versus fixed filter schemes. Studies in the HCRC at the University of Melbourne/Bionic Ear Institute were undertaken to compare the effects of selecting three, four, and six frequency peaks on speech perception. It was found that the extraction of six frequency peaks gave no better results than the selection of four peaks from the outputs of fixed filters. Thus a strategy that presented six spectral maxima from a bank of 16 filters was explored. A constant stimulus rate of 250 pulses/s was used so that voicing was coded as variations in amplitude. This strategy, called the spectral maxima sound-processing (SMSP) scheme, was tested in 1990 on an initial patient (McKay et al 1991), and found to give significant benefit (Fig. 1.4). It was then evaluated in two pilot studies on six patients and found to provide more speech understanding than the F0/F1/F2 and multipeak strategies (McKay et al 1992). This SMSP strategy was implemented by Cochlear Limited as SPEAK (McDermott et al 1992) in a processor referred to as Spectra-22. SPEAK-Spectra-22 (Seligman and McDermott 1995) differed from SMSP in being able to select six or more spectral maxima from 20 rather than 16 filters. A constant stimulus rate that varied adaptively from 180 to 300 pulses/s was used (Clark, Dowell et al 1996).

A multicenter comparison of the SPEAK-Spectra-22 and Multipeak-MSP systems was undertaken on 63 postlinguistically and profoundly deaf adults at eight centers in Australia, North America, and the United Kingdom (Skinner et al

1994). The mean scores for vowels, consonants, consonant-nucleus-consonant (CNC) words, and words in the City University of New York (CUNY) and Speech Intelligibility Test (SIT) sentences in quiet were all significantly better for SPEAK than for Multipeak-MSP (Fig. 1.4). SPEAK performed particularly well in noise. SPEAK-Spectra-22 was approved by the FDA for postlinguistically deaf adults on March 30, 1994.

Further Psychophysics Research in Early Deafened Children

The emphasis of the University of Melbourne/Bionic Ear Institute on psychophysical research in adults was extended to children. It was seen as very important to learn what were the percepts underlying speech perception, how these percepts developed, and to what extent plasticity was evident in their developing central nervous system. The work first commenced on early-deafened adults, and it was found that they had poor speech understanding, which appeared to be due to their inability to make satisfactory electrode place or rate discriminations. They had to rely a great deal on electrical current (loudness) discrimination. Then the research was extended to a younger person (a 14-year-old boy), which showed that he had improved place and rate discrimination. It was hypothesized this was the underlying reason for the better speech understanding. These psychophysical studies were extended in the 1990s to children implanted at young ages to determine their relationship to speech perception and issues of plasticity.

Biological Safety Studies for Infants and Very Young Children

Not only were the speech and language benefits with cochlear implants clearly demonstrated in children, but also there were improved results for children operated down to the age of 2 years. As with the restoration of hearing in a severely deaf child with an aid, would implants in children under 2 lead to even better results? However, there were additional safety concerns when operating on children of this age. These included the possibility that the high incidence of middle ear infection could lead to labyrinthitis and maybe meningitis, that head growth would lead to extraction of the electrodes from the inner ear or that drilling bone would affect head growth, and that electrical stimulation could have an adverse affect on the immature nervous system. These concerns were addressed through NIH contract 1-NS-7-2342, "Studies on Pediatric Auditory Prosthesis Implants," awarded in 1987 to the University of Melbourne for 5 years. The studies by the University of Melbourne aimed at investigating safety issues of importance for operations on children under the age of 2. Studies in cats using the pneumococcal, streptococcus, and staphylococcal bacteria, common invading organisms, showed that there was no significant difference in the incidence of infection in the inner ear compared to unimplanted controls, provided that the electrode entry point was sealed with a graft of fascia (see Chapter 3). Histological and x-ray studies in the monkey did not reveal any effect on head growth. Anatomical measurements in human temporal bones from approximately 2 months to adulthood demonstrated that there was no relative growth between the floor of the opening from the middle ear to mastoid and the inner ear, making this an excellent site for fixing the electrode to prevent head growth extracting the electrode out of the inner ear. X-ray studies in the monkey showed that a redundant loop would also allow lengthening. These investigations are discussed in more detail in Chapters 2 and 8. Finally, investigations on hearing and deafened kittens demonstrated that electrical stimulation had no adverse effects on the young cochlea or maturing auditory nervous system (see Chapter 4). When these studies were concluded at the beginning of the 1990s and showed no adverse effects, it was only then considered appropriate to operate on children under 2 years of age.

Vienna, Austria

It was not until the early 1980s that Hochmair and Hochmair-Desoyer (1983) reported their best results for a single-channel scheme that incorporated a spectrally modified signal that drove one of four electrodes in the scala tympani of the cochlea. This scheme not only limited loudness from becoming excessive, but also did so for frequency bands. The sensitivity range was optimized for each patient so that the loudness of sounds of different frequencies was preserved. This required compressing the speech intensity into a manageable range and then equalizing the intensities over the speech range 100 to 4000 Hz. The stimulus was mapped onto an equal loudness contour at a comfortable level. This single-channel strategy was compared with the House Ear Institute 3M single-channel, the University of Utah Ineraid multiple-channel, and the University of Melbourne Nucleus multiple-channel devices by Gantz et al (1987). Open-set speech recognition was not found for either the 3M, Vienna (see below), or House devices, but it did occur with the two multiple-channel devices from the University of Utah (Ineraid) and the University of Melbourne (Nucleus). This Vienna single-channel implant was developed industrially by 3M but with initial trials no open-set word scores were obtained with the U.S. studies. Approval for a PMA was not received from the FDA. Subsequently in the 1990s Med-El became a separate company and developed a multiple-electrode implant with an electrode array that had ball electrodes on a free-fitting carrier. A speech processor was developed to implement the CIS strategy (Wilson et al 1992, 1993). The research in the 1990s is discussed in more detail in Chapters 7 and 8.

Antwerp, Belgium

In Belgium the development of a multiple-electrode implant system (Laura) commenced in the early 1980s at the University of Louvaine (Peeters et al 1987). The system was designed with a radiofrequency (RF) power link between the external speech processor and the internal receiver stimulator. A link with infrared light for data was used to avoid interference. The transmission coil for power was in an ear mold, and the coil was a platinum wire placed around the external ear canal. The receiving power coil was welded to a feed through in the titanium box containing the electronics. The electronics were designed to permit simultaneous stimulation, but each channel required a space-consuming capacitor to prevent

DC. The first generation integrated circuit was fabricated in 1980, and a third generation in 1984 included telemetry.

The development of a titanium container for the receiver-stimulator electronics commenced in 1984. The bottom of the container had a ceramic substrate for the electrode feed-through, and the hybrid circuit was printed directly onto this substrate. However, there were technical difficulties due to leakage of electrical currents from the hybrid circuit to the titanium capsule. The titanium package also required a fitted glass window for the infrared transmission.

In addition to the titanium package a ceramic one was also developed in 1985, and the two were subjected to vacuum tests, temperature shocks, gravity tests, and submersion in physiological fluid. The ceramic packages showed leakage problems at some of their feed-throughs, while the titanium packages maintained their integrity. For this reason the titanium package development was continued for use in patients. It was covered with epoxy resin after welding the electrode power coil to the feed-through.

In 1981 efforts were directed toward improving electrodes based on thin-film technology originally developed at Stanford University. After 2 years the research was terminated because of problems in fixing platinum to the substrate, and problems with inserting the electrode into the cochlea. It was decided to develop an electrode array for bipolar radial stimulation similar to the one by UCSF. It was a straight electrode with a little bending at the tip to prevent axial torsion of the electrode during insertion. Such torsion would reduce the performance of the bipolar electrode. It was a multielectrode array with 16 platinum wires. The contacts were recessed into the silicon substrate to make the surface smooth.

Further design changes were made in 1986 to the receiver-stimulator and speech processor. The system used an electromagnetic transmission code of amplitude-modulated pulses to determine synchronization, channel selection, and the status of the current sources. It provided three modes of stimulation. The speech processor had a voice detector and provided fixed filtering of the speech signal. It had a programmable map for recording thresholds and uncomfortable levels for each of the electrodes. The internal microprocessor allowed enough flexibility to experiment with different coding systems for optimal speech discrimination by the subject. The processor detected a voiced signal and the four most important spectral frequencies. The outputs of those maxima stimulated electrodes that encoded the respective pitch at a rate corresponding with the voicing frequency.

Ottawa, Canada

The Canadian project in 1987 had focused on developing an advanced speech processor system and receiver-stimulator. The speech processor design had digital signal processors for extracting the speech frequency information using a fast Fourier transform and linear predictive coding. The receiver-stimulator had a titanium capsule rather than one made of ceramic, which was known to absorb water and develop leakages. It was thus necessary to have a receiving coil in a ferrite disk on the package as had been developed for the University of Melbourne's prototype device. The stimulator had eight monopolar or four bipolar pairs of electrodes and some flexibility in the choice of stimulus parameters. The stimulator circuit was implemented on a single custom-made silicon chip, with a microprocessor memory and capacitors to prevent DC stimulation. An electrode array was designed using thin-film technology and rolled on itself to produce an array similar to the University of Melbourne's banded array. This device was not used to any extent commercially.

Prague, Czechoslovakia

Three postlinguistically deaf adults were implanted at the ear, nose, and throat department at Charles University prior to May 1987. The device was developed at the laboratory of Electronic Sensory Substitutions of the Czechoslovak Academy of Sciences. It was a single-channel system with analog stimulation from an electrode placed in the round window. It was evaluated first in laboratory animals and met the standards of the SUKL, the Czechoslovak equivalent of the U.S. FDA. Studies were undertaken to determine thresholds, dynamic ranges, and frequency and amplitude difference limens. The patients were capable of distinguishing male and female voices and certain environmental sounds. It was used as an aid to speech reading, although it did not provide speech recognition, but overall was a help to the subjects (Valvoda et al 1987).

France

Bordeaux

An extracochlear single-channel implant was evaluated on 10 patients at the Department of Otolaryngology center in Bordeaux, and reported on by Negrevergne et al (1987). The device (Prelco) was designed by the Audiology Laboratory and marketed by the RACIA company in Bordeaux. In the clinical trial it was concluded that "the auditory sensations are too poor to give a complete vocal perception." It provided rhythm and intonation. Patients appreciated sounds even though not recognizable.

Grenoble

Research commenced in Grenoble in 1983 when a patient received a 12 electrode implant (Genin and Charachon 1984). A study examined the thresholds for sine wave stimuli ranging from 100 to 5000 Hz.

Lyon

The Chorimac implant and speech processing strategy (see below) was evaluated by Morgon et al (1984) in Lyon on two postlinguistically deaf adults. There was a trend for improved word recognition, but it was not statistically significant. Vowel recognition improved with training, but the recognition of the voiced/

unvoiced decision did not change i.e. whether the sounds were voiced /b/ or unvoiced /p/.

Research in Lyon (Berger-Vachon and Mouhssine 1985) also focused on developing a computer model for speech processing where the spectral composition could be varied and tested on subjects, with the view to adapting the information to suit the patient's place pitch perception variation.

Paris

The speech-processing strategy originating at the University of Paris first used eight fixed filters and was developed industrially by Bertin St. in the late 1970s as Chorimac-8. It was improved in 1982 as a 12-channel device (Chorimac-12). The work was undertaken by P. MacLeod, C.-H. Chouard, and J.P. Weber. It was the first speech-processing strategy to use a constant rate of stimulation on each electrode and was described by Chouard et al (1985). The authors aimed to transmit all possible speech signals to patients without selecting information or features. A bank of 12 filters, a stimulus rate of 300 pulses/s, and pulse duration to code intensity was used. The filter bands were one third of an octave. The technique of directly implanting 12 electrodes through 12 separate openings into the cochlea continued. A considerable variability in results was reported for this prosthesis from a very wide range of implantees, many of whom suffered multiple disabilities (Chouard et al 1983). There were 48 people evaluated (27 prelinguistically deaf; 21 postlinguistically deaf). The results of Fugain et al (1984) showed vowels were well recognized. With consonants, voicing was well differentiated (90%). Unfortunately, the Miller and Nicely (1955) classification of speech features was not used to make comparisons with other strategies possible. However, different fricatives could be distinguished, but place pitch information was poorly transmitted. Standardized open sets of words were not used, and so it was not clear to what extent open-set speech could be recognized. In 27 patients the percentage improvement in word recognition for a closed set of 20 words for speech reading with the implant compared to the implant alone varied from 0% to 40% (Chouard et al 1983).

Toulouse

The Nucleus system was used for the first time in France in Toulouse in 1987 at first on postlinguistically deaf adults and later on children. The results were grouped according to speech discrimination, and the factors that led to a good prognosis were analyzed. It was found that the length of auditory deprivation had a negative effect. Promontory test results were of value, and the length of the electrode insertion, the number of electrodes in use, and the dynamic range correlated positively with performance (Deguine et al 1993).

Germany

Aachen

The Nucleus device was implanted first in Aachen in 1986 (Doring et al. 1987). A rehabilitation program was developed for individual patients first to test that the training material met the specific requirements of the German language, and second to facilitate the adjustment of a sophisticated speech processor such as the Nucleus one. In particular, the study examined the discrimination experienced by each subject, and analyzed the speech processor output to see if the subjects were receiving the information required. If so, training was carried out to improve the discrimination. As a result of the specifically directed training, the recognition of consonants improved from 34% to 60% and vowels 45% to 72%.

Cologne-Düren

Extracochlear multiple-electrode stimulation was considered as an alternative to intracochlear stimulation on the grounds that it was less traumatic. The center at Cologne-Düren used extracochlear multiple electrode arrays to avoid invading the cochlea, thus minimizing the loss of spiral ganglion cells. Multiple electrodes were usually attached by drilling small holes in the outer wall of the cochlea down to the endosteal lining (Banfai et al 1984a,b, 1985). Initially a total of eight electrodes were placed in the bone over the three cochlear turns in holes drilled down to but not into the cochlea (Banfai et al 1984a,b). Each electrode carried speech information from third-octave band-pass filters. In 1984, 46 patients had the operation (Banfai et al 1986; Banfai 1987). Although studies showed that pitch discrimination could be obtained by extracochlear stimulation at different sites over the cochlea (Banfai et al 1986), this was not conclusive. In a larger group of patients the author reported that "11% of the first 129 implantees showed some signs of developing open-speech perception" (Banfai 1987). The group also reported in 1987 developing an extracochlear multiple-channel implant (Implex) that stimulated via up to 16 electrodes mounted on the medial wall of the middle ear. The localization of current using this form of stimulation was not well demonstrated experimentally. There was a risk with this approach that if the cochlea was entered, significant trauma would result, and the experiments by Clark, Shepherd et al (1983) also supported the view that the impedance of the bone was too high for effective stimulation.

Hannover

The research at the Medizinische Hochschule Hannover commenced in 1983, and by 1987 the group had operated on 75 patients with the Nucleus multiple-channel system with the F0/F2 and F0/F1/F2 strategies. The studies focused on the type of evaluation required in German for comparison with results in English. Patients were classified into three groups according to the test results. It was found that only those who had more than a 50% score for consonants could achieve open-set speech understanding. A small group of prelinguistically deaf adults and three with extracochlear implants for inner ears filled with new bone did not perform as well as those who were postlinguistically deaf and had the intracochlear array (Battmer et al 1987). Refinements were also carried out to the surgical approach that included the modification of the incision, which commenced in the ear canal and arched backward (Lehnhardt et al 1987).

Tokyo, Japan

The Japanese cochlear implant program commenced in January 1986 with the Nucleus multiple-channel system implanted in four postlinguistically deaf patients at the Tokyo Medical College and one at Toranomon Hospital. The main focus was to determine the benefits of the F0/F2 and F0/F1/F2 strategies for the Japanese language. The Japanese language uses only five vowels, and discrimination scores in the patients ranged from 60% to 93%. Consonant confusion as well as monosyllables, word and sentence recognition were carried out. The number of bunsetsus (phrasal units such as noun phrases or prepositional phrases in English), rather than words, were calculated, as they are the units of minimal meaning in Japanese. The number of bunsetsus for the F0/F1/F2 strategy was higher than for the F0/F2 strategy. The rising and falling of pitch was tested as tonal changes carry meaning. The scores ranged from 43% to 73%. The patients were also able to recognize a question because a sentence usually finishes with a /ka/ sound.

Stockholm, Sweden

In the Swedish Cochlear Implant program, 10 postlinguistically deaf patients had received the 3M/Vienna single-channel extracochlear implant by 1988. Testing and training of these patients was carried out at the Royal Institute of Technology. This single-channel device, discussed above in the section on Austria, provided help as a speech reading aid in Swedish. The average results for speech tracking with speech reading alone was 17 words/minute and with the implant 40 words/ minute. One patient had some open-set speech understanding. Overall the well-being of the patients improved (Bredberg et al 1987).

Research on these patients evaluated the relationship between electrical stimulation of the promontory preoperatively and postoperatively. The best predictor of results was a temporal difference limen (DL), with better results when the DL was less than 100 ms (Lindstrom 1987). A study also compared the relationship between speech perception and the results of psychoacoustic tests. The most significant finding was that frequency difference limens at 250 Hz over time reduced to 7% in one group (type I), but only to 20% to 30% in a second group (type II). The type I patients had improved ability to identify acoustic differences between speech sounds. The type I patients also had some residual hearing at the high-frequency area, and it was thought this could have contributed to their successful results. This was considered an argument in favor of extracochlear simulation rather than intracochlear when loss of hearing in this region occurs (Riseberg and Aglefors 1987).

Zurich, Switzerland

Research at the University Hospital in Zurich aimed to determine how to provide optimal single-channel stimulation (Dillier and Spillmann 1984). The authors stated, "In spite of considerable technical progress, electrode implantation is still an experimental surgical procedure." The single-channel pulsatile stimulator had parameters that could be varied: filter cutoffs, automatic gain control (AGC) level, zero crossing level, pulse duration, logarithmic compression amplitude, envelope modulation, and limiting output amplitude. In a psychophysical study, frequency and amplitude resolution were comparable for electrodes implanted at the round window, and electrodes in the scala tympani. The dynamic range for the round window electrode was the same as for intracochlear stimulation for both pulsatile and sinusoidal waveforms. Studies were also done to determine formant DLs for synthesized vowels. It was found that vowel identification improved by mapping F2 into the range of best frequency DLs. Words or sentences were better discriminated with fundamental or zero-crossing detection. Further studies to optimize the speech processing strategy were reported by Spillman and Dillier (1985), who evaluated a speech processor that encoded the microphone signal in a sequence of variable width and amplitude pulses. They also varied the type of energy compression and spectral modification of the microphone signal according to the psychophysical responses of the implantee.

Utrecht, The Netherlands

Studies were undertaken in Utrecht (Smoorenburg and van Olphen 1987) to determine the relationship between preoperative electrical stimulation findings and postoperative results for the House/3M cochlear implant. There had been uncertainty in the literature about the relative merits of the preoperative threshold and dynamic range as a predictor of results. It was found that the dynamic range measured with the House/3M device differed from the results obtained with preoperative electrical stimulation. The thresholds and uncomfortable stimulus levels were essentially independent of frequency and the dynamic range corresponded closely to the range found preoperatively for high-stimulus frequencies. It was therefore concluded that the House/3M device did not provide the patient with true low-frequency stimulation, but with bursts of high-frequency energy. The patient did not benefit from a large dynamic range found for low-frequency stimuli.

United Kingdom

Department of Phonetics and Linguistics, University College London, and Guys Hospital London

In the 1980s research on single-channel stimulation was the main emphasis of the group at the Department of Phonetics and Linguistics, University College London, and Guys Hospital London, in studying a speech pattern approach to electrical stimulation. Speech patterning meant selecting a pattern of temporal and amplitude information to represent speech for electrical stimulation on a single neural channel. It was considered that the explicit extraction of the speech pattern cues would be more effective than presenting the whole speech signal. This could also be done to match the residual sensory abilities of the patient. The studies were extended to a group of five subjects. The electrical stimulus patterns were varied, and with the optimal scheme it was confirmed as an effective speech reading aid, and provided voicing. As a speech reading aid the speech tracking went from 9.4 words/minute for speech reading alone to 18.2 words/minute when combined with single-channel electrical stimulation. It was found to give good speech reading assistance for Cantonese, a language rich in tonal changes. The strategy was compared with the single-channel system from the University of Vienna, but no open-set speech was seen for the latter (Walliker et al 1987).

The EPI group also continued its research to develop effective connection between the single-channel scheme and the auditory nerves. This was carried out with an electrical stimulator (Microstim). Work concentrated on three types of stimulating electrode: (1) A ball of platinum wire was placed in the round window niche of the cochlea. (2) A stainless steel ball was gently passed onto the promontory in an exposed mastoid and held in place with a spring. The spring was held in position by gold-plated plastic ear mold. (3) A percutaneous connector made from platinum and Bioglass, which was attached to the bone of the cochlear promontory. The Bioglass disk had a diameter of 5 mm and a concave platinum contact pad fused to the surface with a lead wire passing through it. The research continued with the development of prototype speech processors using a microcomputer with an electrically erasable programmable read-only memory (EPROM). This speech pattern research and electrical engineering was carried out in the Department of Phonetics and Linguistics, University College London; Department of Clinical Physics and the ENT Department of Guys Hospital; and the Department of Experimental Psychology, University of Cambridge, and funded by the British Medical Research Council.

University College Hospital London, Department of Physics and Bioengineering, and Royal National Institute for the Deaf

Another cochlear implant program in the U.K. was based at the University College Hospital (UCH) and carried out in collaboration with the Department of Physics and Bioengineering and the Phonetics Department at the Royal National Institute of the Deaf (RNID) (Fraser 1987). By 1987 the group had undertaken 31 implants in 24 patients. This included one UCSF multichannel device, 5 extracochlear, devices, and 25 UCH/RNID extracochlear implants. It was considered important to develop a device that could be produced at minimal cost. Furthermore, a round window system was advocated, as some patients suitable for implants had residual hearing and it was considered essential to preserve this. It was not known whether any long-term damage to surviving nerve tissue would result from implantation of intracochlear electrodes. These risks were avoided by the extracochlear system.

At the time the difficulty of diagnosing deafness in young children presented a special difficulty, and this was also one of the reasons that the extracochlear device was advocated. It was argued, too, that the significant improvements achieved with multiple channel implants in the laboratory had not been satisfactorily established in the everyday world against the background of environmental sound. The electrode was implanted near the round window, and the receiver-stimulator fixed through a small postauricular incision (Graham 1987).

The sound-processing system adopted was the same as the one developed by Hochmair and Hochmair-Desoyer (1983). The implanted circuit could also provide pitch extraction similar to the one used by Fourcin et al (1979). An RF carrier transmitted information. An implanted AM receiver circuit was encapsulated in a suitable polymer as developed by Donaldson (1987) for bladder controllers. The results were evaluated at the Royal Ear Hospital in London by H. Cooper and L. Carpenter. This study was carried out on 22 patients. The first four had the Vienna/Hochmair single-channel extracochlear device, the other 18 the UCH/RNID single-channel extracochlear system. It was found that a large proportion of the patients received voicing information when this was not possible with speech reading. In 10 of the 11 cases there was also greater information transfer for manner of articulation, and in approximately half the cases there was a small amount of additional information on place of articulation. In testing for vowels, the results were generally poor, and no better than expected when using duration as a cue for perception. The testing of voicing intonation was also undertaken, which showed that statements could be distinguished by a significant number of people. With perception of discourse tracking, there were significant improvements when it was used as a speech reading aid. These results indicated that for most patients the implant was useful for speech reading in quiet conditions, but background noise often drastically reduced its performance. In a questionnaire the patients' expectations of hearing environmental sound were met more often than their expectations for help in speech reading. However, more than half considered hearing any sound and relief from isolation as their primary reason for an implant, and to this extent the implant had been successful. All patients reported an improvement in their quality of life.

United States

In the mid-1980s, with the demonstration that a multiple-channel cochlear implant could provide open-set speech understanding, there was an increasing emphasis by the NIH on funding research into the restoration of hearing. The NINCDS was the dominant body in the U.S. funding cochlear implant research. The grant scheme supported more fundamental research, but not necessarily of immediate relevance to cochlear implants. The contract scheme was aimed at directing research to specific outcomes through facilitating collaborative research in a number of disciplines: otology, electronic engineering, physiology, psychophysics, audiology, and biology (Hannaway 1996).

Contracts from the 1970s continued to Stanford University (F. Blair Simmons/ Robert L. White), University of California at San Francisco (UCSF) (Michael M. Merzenich), and the University of Washington (Joseph M. Millar). In the 1980s the first additional recipients for contracts were the University of Melbourne (Graeme M. Clark) and the University of Rochester (Charles W. Parkins). The contract to the University of Rochester was for a printed circuit multichannel cochlear implant and stimulus coding of scala tympani cochlear prostheses. The University of Melbourne contracts were for speech processors for auditory prostheses and studies on pediatric auditory prosthesis implants. The outcomes for the contracts to the University of Melbourne are discussed in the section on research in Australia. The grant to the University of Melbourne was made as it was recognized that there was a need for more psychophysical evaluation of patients, and the development of advanced speech processors.

Research

Stanford University. The research to develop a thin-film array on a rigid implant for insertion into the modiolus that was funded by the NIH continued. The psychophysical and speech research continued into the 1980s on one patient with the four wires implanted into the modiolus in 1977. The most promising strategy involved the extraction of acoustic resonance frequencies. The values of the two lowest resonance frequencies were used to control the rate of pulsatile stimulation of two of the electrodes (Atlas et al 1983). This was based on psychophysical results indicating the ability of implantees to discriminate between different pulse rates on the basis of the pitch that they elicited. A third electrode was stimulated at a high rate whenever the ratio of high-to-low frequency energy exceeded a threshold. This approach produced results significantly above chance for discrimination between speech and noise signals that had the same energy pattern, and for identification of vowels and initial and final consonants from a limited set of alternatives. No significant results were obtained using a free-flowing speech input. This strategy was still not found to be as successful as some of their single-channel strategies. The NIH contract was also for the development of software and hardware for speech processors using the modiolar implant.

University of California at San Francisco. The biocompatibility of an improved electrode array for human insertion was studied through long-term implantation in the cat (Leake-Jones and Rebscher 1983). The changes to the original design were to make it (1) from a silicone rubber (MDX 4-4210) without stannous octoate as the catalyst, as it had been shown to be toxic (e.g., Clark (1987); (2) cylindrical in shape, so that it only filled the middle of the scala and did not fit snugly to the contours, as this had been shown to be traumatic (Sutton et al 1980); (3) with electrode mushroom-shaped contacts rather than balls, as the latter could damage the cochlea on removal, as was subsequently seen for the University of Utah array; and (4) from platinum/iridium (Pt/Ir) (90:10) wires rather than platinum/rhodium (Pt/Rh) (90:10) and with the wires also coated with Parylene-C. The study confirmed that the design changes had no adverse effects on the cochlea, and further showed that that chronic implantation in the scala tympani could occur without loss of neural elements.

The Coleman Laboratories received a 3-year NIH contract, "Studies on Pediatric Auditory Prosthesis Implants DC-7-2391," to investigate biological issues of importance for cochlear implant operations on children under the age of 2.

Studies on neonatally deafened cats were undertaken to determine the effects of localized and prolonged electrical stimulation of discrete areas of the cochlea on the extent of the spatial frequency input to the brain cells (Snyder et al 1990, 1991). The aim was to determine whether plasticity would lead to neural connections being altered by electrical stimulation. The animals were stimulated for periods up to 3 months, 5 days a week, and the results compared to controls. The distribution of the responses was mapped with extracellular recordings. It was found that the average area in the inferior colliculus activated by chronically stimulated electrode pairs at 6 dB above threshold was twice that of the unstimulated deafened animals. Results of the studies suggested that (1) acoustic input from the auditory periphery is not necessary for the development of some of the basic features of central auditory organization, in particular cochlear topic organization at the level of the midbrain; and (2) electrical stimulation of a very limited duration and intensity could expand central representation of that sector, and change the cochlear topic organization to the auditory central nervous system.

Basic biophysics of electrical stimulation of the auditory system was undertaken by Loeb et al (1983). This required measuring the neural excitability or chronaxie of the auditory nerve by recording response in the cochlear nucleus. The mean chronaxie of 478 μs was higher than the more usual 100 μs.

Research Triangle, North Carolina. Studies at the Research Triangle Institute (RTI) under an NIH contract from 1983 to 1985 were undertaken to design and evaluate speech processors for auditory prostheses to see if improvements could be made over the multiple-channel strategies already developed. The aims were as follows: (1) design a hardware interface between a computer and an implanted electrode; (2) develop and apply computer-based integrated field models of electrical stimulation of the auditory nerve; (3) identify promising approaches to the design of speech processors; (4) build a computer-based simulator that was capable of rapid and practical emulation of different speech processing strategies; (5) develop software support for basic psychophysical studies; (6) conduct tests with two patients at the UCSF to confirm the proper operation of the equipment and obtain basic psychophysical measures on the patients, and to compare different multichannel speech processing strategies; (7) build a portable real-time speech processor for use with single-channel auditory prostheses; (8) develop and apply models of the spatial and temporal patterns of neural discharges produced by intracochlear electrical stimulation of the neural pathways in the experimental animal; and (9) establish collaboration among UCSF, Duke Medical Center, Washington University Medical Center, and RTI.

Research from 1985 to 1989 helped establish an interleaved pulse (IP) strategy that overcame the difficulty of channel interaction that could account for the variable results seen with the fixed filter compressed analog (CA) strategy scheme with monopolar stimulation developed in Utah (Ineraid) (Eddington 1980, 1983). The rationale for interleaved pulses was similar to that for the Nucleus F0/F1/F2 and Multipeak strategies, as it had been found with the University of Melbourne's fixed filter physiological model (Laird 1979) that simultaneous stimulation produced unpredictable variations in loudness and that nonsimultaneous stimulation was indicated to avoid this problem. A comparison was made between the IP and CA strategies on two patients. The IP strategy produced much better results than the CA strategy. It was believed that this was due to the reduction in channel interaction. There was also a strong correspondence between the number of stimulus channels and performance. When the study was expanded to a larger group of patients, some performed better with the CA and some with the IP strategy. It was thought this was due possibly to variations in their neural populations.

A further NIH contract aimed to compare a number of speech-processing strategies. This was to be carried out with a percutaneous plug and socket. The strategies selected were the four channel UCSF/Storz (fixed filter with pulses), the single-channel extracochlear 3M/Vienna prosthesis, the Nucleus (University of Melbourne) F0/F1/F2 strategy, and the University of Utah/Ineraid fixed-filter compressed analog. Although a similar comparison was made by the University of Iowa, changes had been introduced, in particular the Nucleus F0/F2 strategy had been improved to the F0/F1/F2 one.

As there were limitations for the recognition of voicing with the IP strategy, it was considered that this would be better if the sampling rate were increased. The continuous interleaved sampler (CIS) strategy was developed from this idea. A high rate was applied to a fixed filter scheme that used IPs to avoid channel interaction. It was optimized as a processor with eight band-pass filters, and a constant stimulus rate between 833 and 1111 pulses/s per channel for bipolar or monopolar stimulus modes (Wilson et al 1992, 1993). The waveform envelopes from the band-pass filters modulated a high-pulse rate train (Wilson et al 1992, 1993). It used biphasic pulses rather than analog stimulation as occurred with the (CA) scheme. The outputs of six or more filters were sampled and used to stimulate the same number of electrodes on a place-coding basis. Lawson et al (1996) and Wilson (1997) found that increasing the number of electrodes to seven improved speech perception.

The CIS strategy was incorporated into the Advanced Bionics Clarion series devices with an electrode array that was derived from the one developed at UCSF. The CIS was also the main strategy offered with the Med El Combi-40 series, but with a multiple-electrode array arising out of previous studies by the Technical University in Vienna.

University of Utah. The strategy used in the 1980s was similar to the earlier version and presented the outputs of four fixed filters by simultaneous monopolar analog stimulation to six electrodes in the cochlea. Compression of the amplitude variations in speech, to bring them within the dynamic range of electrical stimulation, was achieved with a variable gain amplifier operating in compression mode. The studies were undertaken on the two initial subjects who had been operated on a few years earlier. It was reported that subjects obtained faultless discrimination of 4 closed-set consonant-vowel syllables and 3 vowels using electrical stimulation alone. "Although understanding free running speech by the profoundly deaf does not seem imminent, the results presented indicate that the multichannel system tested shows more promise of approaching this goal than the single-channel scheme" (Eddington 1980). The analysis of results showed that the first formant frequency was the most important cue being used by the patients in speech recognition (Eddington 1983).

University of Washington. Research at the University of Washington evaluated different electrode types, and their risk of trauma. The research at UCSF had shown that a molded array in the experimental cat model would localize current better with bipolar stimulation to the peripheral nerve elements than a free-fitting one. On the other hand, as it was a tighter fit it ran the risk of causing more trauma. This was confirmed on the macaque monkey, showing that a free-fitting array could be preferable. This center also looked at the effects of deafness and electrical stimulation on the viability of the central nervous pathways more centrally.

Miller et al (1983, 1985) also studied the effects of high stimulus rate with sine waves at 1000 Hz over a period of 3 hours. The high current levels (above 400 μA RMS) produced detrimental effects on neural function.

The results of research at the University of Washington (Pfingst and Rush 1985) showed difference limens for sinusoidal electrical stimuli in monkeys to range from 7% at 100 Hz (17 dB sensation level) to about 30% at 200, 300, and 600 Hz. These results supported those obtained previously on cats (Clark, Kranz et al 1972, 1973a,b; Clark, Nathar et al 1972; Williams et al 1976).

Huntington Research Laboratories. The biophysical research being undertaken at the Huntington Research Laboratories continued. This research was helping to establish fundamental knowledge about the interface between electrode and nervous tissue and the safe levels of current stimulation without gassing.

Industrial Development and Clinical Trial for Postlinguistically Deaf Adults

House Ear Institute 3M Single-Channel Implant. The single-electrode prosthesis from the House Ear Institute in Los Angeles was developed commercially by 3M (Fretz and Fravel 1985). It was evaluated for the FDA. The patients could not understand speech at conversational levels, but could identify environmental sounds with a fair degree of accuracy (Tyler et al 1984, 1987). They could discriminate some vowels based on fundamental frequency and first formant cues (Gantz et al 1989). A PMA was obtained in 1984 for its use in postlinguistically deaf adults to hear environmental sounds, but not as a device that provided speech understanding.

University of Utah Symbion Multiple-Channel Implant. The research on two subjects at the University of Utah used a strategy that filtered speech with four fixed filters and presented the outputs by simultaneous monopolar analog stimulation to four of six electrodes in the cochlea with analog current (Eddington 1980, 1983). The Symbion Ineraid device was tested for the FDA, but did not receive a PMA. This was primarily due to concerns about the percutaneous plug. A clinical study with the Symbion/Ineraid four-fixed-filter strategy examining the median score for open-sets of CID sentences with electrical stimulation alone found a mean word-in-sentence score of 45% (range 0–100%) (Dorman et al 1989).

University of California at San Francisco Storz Multiple-Channel Implant. Following the initial research at UCSF, Storz Medical Instruments collaborated with the research group to develop a prototype device (MiniMed) for evaluation. The group at RTI worked in collaboration with UCSF to develop speech processors for the UCSF/Storz cochlear implant. The collaboration with Storz commenced in 1984. The speech-processing strategy used one filter channel for low-frequency vocal resonances, two for midfrequency resonances, and one for higher frequencies. Bipolar electrodes located within the scala tympani on a molded array were stimulated by band-pass filtered versions of the speech signal (Merzenich et al 1987). Biphasic pulsatile rather than analog stimuli were employed. With visual enhancement (speech reading), the unaided preoperative scores out of 100 items of open speech were 4, 25, 43, and 15. At 6 to 8 weeks postoperatively, the four patients scored 17, 5, 63, and 23, respectively. Tests using spondees, CID sentences, speech in noise, and monosyllabic words were performed postoperatively. The CID sentence scores at 6 to 8 weeks postoperative using taped test material were 34%, 24%, 2%, and 25%.

Industrial Development and Clinical Trial for Children

3M-House Single-Channel Implant. In 1980 the FDA gave permission for the House Ear Institute to commence trials with its single-channel implant on children. Kirk (2000) stated, "The use of cochlear implants in children was quite controversial at the time. Many scientists believed that the limited auditory information provided by a single-channel cochlear implant was not sufficient for children to achieve speech perception and speech production." Most of the children were deafened from disease after birth; their mean age of deafening was 1.7 years, and age at implantation was 8.4 years (Berliner and Eisenberg 1985). The children could recognize environmental sounds, detect speech at conversational levels, and discriminate different speech patterns such as the number of syllables (Thielemeir et al 1985). A few of the early implanted children obtained some open-set speech understanding (Berliner et al 1989). The FDA PMA was not completed for this device for children.

1990s: Continuing Fundamental Research and Industrial Development

Cochlear implant research, industrial development, and clinical studies made considerable advances in the 1980s, as illustrated by the results in Fig. 1.4. At the beginning of the 1980s, it was not clear whether there were advantages in using multiple-channel versus single-channel stimulation for postlinguistically deaf adults, let alone for children. Advances in the processing of speech, however, demonstrated that multiple-channel stimulation would allow sufficient information to be transmitted for open-set speech understanding for both adults and children. The average open-set CNC word scores for the Nucleus F0/F1/F2 and Multipeak strategies for electrical stimulation alone in postlinguistically deaf people indicated that they could understand significant amounts of running speech without the need to speech read.

As the results continued to improve during the 1980s, this increased the incentive to explore the limits. Just how good could speech perception become? Had the results started to reach an asymptote, or would the linear rise continue further? There was also a great need to further improve the speech perception and production and language of implanted children. As speech understanding was improving, there was a need too for achieving high-fidelity sound and musical appreciation. This is turn would help in speech perception. One of the biggest challenges, however, was to achieve good speech understanding in the presence of background noise. Furthermore, in the 1980s Cochlear Proprietary Limited had increased its share of the international market to 80%, having had regulatory approval in the U.S. and other countries for speech-processing strategies in adults and children.

At the beginning of the 1990s two strategies (SPEAK and CIS) had evolved and appeared to have produced a stepwise improvement in speech perception. In the 1990s there was also a rationalization of the industrial development. This was in part due to the purchase by Cochlear Limited of 3M's interests in the single-channel House and Vienna devices, and the Symbion Ineraid device. The 1990s saw the emergence of three companies other than Cochlear Limited: Advanced Bionics, Med-El, and MXM.

There was also an expansion of the fundamental but relevant research as well as applied studies. The basic studies became more international, and the applied studies were oriented to the industrial needs of each company. The research in the 1990s is discussed in the chapters of this book, as it is not seen as part of the historical development of the device.

Fundamental research in Australia in the 1990s was to continue to be funded by the NIH and NH&MRC, and in other countries by their national funding agencies. However, it became more focused, through the close involvement of industry. The Australian government established cooperative research centers (CRCs) for developing research of importance to industry. The CRC for Cochlear Implant Speech and Hearing Research and subsequently the CRC for Cochlear Implant and Hearing Aid Innovation facilitated cochlear implant research of importance to industry through the core parties to the CRC, namely the University of Melbourne, Bionic Ear Institute, Cochlear Limited, and Australian Hearing Services. As cochlear implants reached clinical maturity at the end of the 1980s and beginning of the 1990s, a new phase in their evolution had been reached.

References

Atlas, L. E., M. K. Herndon, F. B. Simmons, L. J. Dent and R. L. White. 1983. Results of stimulus and speech-coding schemes applied to multichannel electrodes. Cochlear

Prostheses, an international symposium. Annals of the New York Academy of Sciences 405: 377–386.

Banfai, P., ed. Cochlear implant: current situation. Proceedings of the International Cochlear Implant Symposium, Sept. 7–12, 1987, Düren, West Germany.

Banfai, P. 1987. Cochlear implant: current situation. International Cochlear Implant Symposium, Duren.

Banfai, P., G. Hortmann and S. Kubik. 1985. Extracochlear eight-channel electrode system. Journal of Laryngology and Otology 99(6): 549–553.

Banfai, P., A. Karczag, S. Kubik, P. Luers and W. Sarth. 1986. Extracochlear sixteen channel electrode system. Otolaryngologic Clinics of North America 19: 371–408.

Banfai, P., A. Karczag and P. Luers. 1984a. Cochlear implants. Clinical results: the rehabilitation. Acta Oto-Laryngologica (suppl 411): 183–194.

Banfai, P., S. Kubik and G. Hortmann. 1984b. Our extra-scalar operating method of cochlear implantation. Experience with 46 cases. Acta Oto-Laryngologica (suppl 411): 9–12.

Battmer, R. D., E. Lehnhardt and R. Lasig. 1987. Speech perception for the German language. In: Banfai, P., ed. Cochlear implant: current situation. 467–471.

Berger, K. 1984. The hearing aid—its operation and development. 3rd ed. Livonia, MI, National Hearing Aid Society.

Berger-Vachon, C. and R. Mouhssine. 1985. Analysis of spectral models developed for the adaptation of a cochlear prosthesis. International AMSE Conference "Modelling and Simulation" 2: 205–219.

Berliner, K. I. and L. S. Eisenberg. 1985. Methods and issues in the cochlear implantation of children: an overview. Ear and Hearing 6(3 suppl): 6S–13S.

Berliner, K. I., L. L. Tonokawa, L. M. Dye and W. F. House. 1989. Open-set speech recognition in children with a single-channel cochlear implant. Ear and Hearing 10: 237–242.

Bilger, R. C., F. O. Black and N. T. Hopkinson. 1977. Evaluation of subjects presently fitted with implanted auditory prostheses. Annals of Otology, Rhinology and Laryngology 86(suppl 38): 1–176.

Black, R. C. and G. M. Clark. 1978. Electrical network properties and distribution of potentials in the cat cochlea. Proceedings of the Australian Physiological and Pharmacological Society 9: 71P.

Black, R. C. and G. M. Clark. 1980. Differential electrical excitation of the auditory nerve. Journal of the Acoustical Society of America 67(3): 868–874.

Blamey, P. J., R. C. Dowell, Y. C. Tong, A. M. Brown, S. M. Luscombe and G. M. Clark. 1984a. Speech processing studies using an acoustic model of a multiple-channel cochlear implant. Journal of the Acoustical Society of America 76: 104–110.

Blamey, P. J., R. C. Dowell, Y. C. Tong and G. M. Clark. 1984b. An acoustic model of a multiple-channel cochlear implant. Journal of the Acoustical Society of America 76: 97–103.

Bredberg, G., B. Ossean-Corp and B. Lindstrom. 1987. Results from 10 patients with extracochlear 3M/Vienna single electrode implant. In: Banfai, P., ed. Cochlear implant: current situation. 175–177.

Brummer, S. B. and M. J. Turner. 1975. Electrical stimulation of the nervous system: the principle of safe charge injection with noble metal electrodes. Bioelectrochemistry and Bioenergetics 2: 13–25.

Brummer, S. B. and M. J. Turner. 1977. Electrochemical considerations for safe electrical stimulation of the nervous system with platinum electrodes. IEEE Transactions on Biomedical Engineering 24: 59–63.

Busby, P. A., Y. C. Tong and G. M. Clark. 1986. Speech perception studies in the first year of usage of a multiple-electrode cochlear implant by prelingually deaf patients. Journal of the Acoustical Society of America 80(suppl 1): 30.

Butler, R. A., I. T. Diamond and W. D. Neff. 1957. Role of auditory cortex in discrimination of changes in frequency. Journal of Neurophysiology 20: 108–120.

Butler, R. A. and W. D. Neff. 1950. Role of the auditory cortex in the discrimination of changes in frequency. American Journal of Psychology 5: 474.

Chouard, C. H. 1980. The surgical rehabilitation of total deafness with the multichannel cochlear implant. indications and results. Audiology 19: 137–145.

Chouard, C. H., C. Fugain, B. Meyer and F. Chabolle. 1985. The Chorimac-12. A multichannel cochlear implant for total deafness. Description and clinical results. Acta Oto-Rhino-Laryngologica Belgica 39(4): 735–748.

Chouard, C. H., C. Fugain, B. Meyer and F. Chabolle. 1987. Indications for multi- or single-channel cochlear implant for rehabilitation of total deafness. Pacing and Clinical Electrophysiology 10: 237–239.

Chouard, C. H., C. Fugain, B. Meyer and H. Lacombe. 1983. Long-term results of the multichannel cochlear implant. Annals of the New York Academy of Sciences 405: 387–411.

Chouard, C. H. and P. MacLeod. 1976. Implantation of multiple intracochlear electrodes for rehabilitation of total deafness: preliminary report. Laryngoscope 86: 1743–1751.

Clark, G. M. 1969a. Hearing due to electrical stimulation of the auditory system. Medical Journal of Australia 1: 1346–1348.

Clark, G. M. 1969b. Middle ear and neural mechanisms in hearing and the management of deafness. PhD dissertation. University of Sydney.

Clark, G. M. 1969c. Responses of cells in the superior olivary complex of the cat to electrical stimulation of the auditory nerve. Experimental Neurology 24: 124–136.

Clark, G. M. 1970a. A neurophysiological assessment of the surgical treatment of perceptive deafness. International Audiology 9: 103–109.

Clark, G. M. 1970b. The surgical treatment of perceptive deafness. An experimental study. Australian and New Zealand Journal of Surgery 39: 319.

Clark, G. M. 1973a. Experimental studies on the surgical treatment of perceptive deafness. Journal of the Oto-Laryngological Society of Australia 3: 571–573.

Clark, G. M. 1973b. A hearing prosthesis for severe perceptive deafness—experimental studies. Journal of Laryngology and Otology 87: 929–945.

Clark, G. M. 1975. A surgical approach for a cochlear implant. An anatomical study. Journal of Laryngology and Otology 89: 9–15.

Clark, G. M. 1977. An evaluation of per-scalar cochlear electrode implantation techniques. An histopathogical study in cats. Journal of Laryngology and Otology 91: 185–199.

Clark, G. M. 1987. The University of Melbourne—Nucleus multi-electrode cochlear implant. Advances in Oto-Rhino-Laryngology Volume 38. Basel, Karger.

Clark, G. M. 1993. The University of Melbourne/Nucleus multiple-channel cochlear implant. Acoustics Australia 21: 91–97.

Clark, G. M. 1995. Cochlear implants: historical perspectives. In: Plant, G. and K.-E. Spens, eds. Profound deafness and communication. London, Whurr: 165–218.

Clark, G. M. 1997. Historical perspectives. In: Clark, G. M., R. S. C. Cowan and R. C. Dowell, eds. Cochlear implantation for infants and children: advances. San Diego, Singular Publishing Group: 9–27.

Clark, G. M. 2000a. The cochlear implant: a search for answers. Cochlear Implants International 1: 1–17.

Clark, G. M. 2000b. Sounds from silence. St. Leonards, NSW, Allen and Unwin.

Clark, G. M., P. J. Blamey, P. A. Busby, et al. 1987. A multiple-electrode intracochlear implant for children. Archives of Otolaryngology 113: 825–828.

Clark, G. M., P. A. Busby, R. C. Dowell, et al. 1992. The development of the Melbourne/ Cochlear multiple-channel cochlear implant for profoundly deaf children. Australian Journal of Oto-Laryngology 1: 3–8.

Clark, G. M., P. A. Busby, S. A. Roberts, et al. 1987. Preliminary results for the Cochlear Corporation multi-electrode intracochlear implants on six prelingually deaf patients. American Journal of Otology 8: 234–239.

Clark, G. M., R. C. Dowell, R. S. C. Cowan, B. C. Pyman and R. L. Webb. 1996. Multicenter evaluations of speech perception in adults and children with the Nucleus (Cochlear) 22-channel cochlear implant. In: Portmann, M., ed. Transplants and implants in otology III. Amsterdam, Kugler: 353–363.

Clark, G. M. and R. J. Hallworth. 1976. A multiple-electrode array for a cochlear implant. Journal of Laryngology and Otology 90: 623–627.

Clark, G. M., R. J. Hallworth and K. Zdanius. 1975. A cochlear implant electrode. Journal of Laryngology and Otology 89: 787–792.

Clark, G. M., H. G. Kranz and H. Minas. 1973a. Behavioral thresholds in the cat to frequency modulated sound and electrical stimulation of the auditory nerve. Experimental Neurology 41: 190–200.

Clark, G. M., H. G. Kranz and H. J. Minas. 1973b. Response thresholds to frequency modulated sound and electrical stimulation of the auditory nerve in cats. Proceedings of the Australian Physiological and Pharmacological Society 4: 134.

Clark, G. M., H. G. Kranz, H. J. Minas and J. M. Nathar. 1975. Histopathological findings in cochlear implants in cats. Journal of Laryngology and Otology 89: 495–504.

Clark, G. M., H. G. Kranz and J. M. Nathar. 1972. Behavioural responses in the cat to electrical stimulation of the cochlea and auditory neural pathways. Australian Journal of Experimental Biology and Medical Research 3: 202.

Clark, G. M., J. M. Nathar, H. G. Kranz and J. S. Maritz. 1972. A behavioral study on electrical stimulation of the cochlea and central auditory pathways of the cat. Experimental Neurology 36: 350–361.

Clark, G. M., B. C. Pyman, R. L. Webb, B. K.-H. G. Franz, T. J. Redhead and R. K. Shepherd. 1987. Surgery for safe the insertion and reinsertion of the banded electrode array. Annals of Otology, Rhinology and Laryngology 96(suppl 128): 10–12.

Clark, G. M., R. K. Shepherd, J. F. Patrick, R. C. Black and Y. C. Tong. 1983. Design and fabrication of the banded electrode array. Annals of the New York Academy of Sciences 405: 191–201.

Clark, G. M. and Y. C. Tong. 1981. Multiple-electrode cochlear implant for profound or total hearing loss: a review. Medical Journal of Australia 1: 428–429.

Clark, G. M., Y. C. Tong and L. F. Martin. 1981a. A multiple-channel cochlear implant. An evaluation using open-set CID sentences. Laryngoscope 91: 628–634.

Clark, G. M., Y. C. Tong, L. F. Martin and P. A. Busby. 1981b. A multiple-channel cochlear implant. An evaluation using an open-set word test. Acta Oto-Laryngologica 91: 173–175.

Clark, G. M., Y. C. Tong and J. F. Patrick. 1990. Cochlear prostheses. Edinburgh, Churchill Livingstone.

Danley, M. J. and R. J. Fretz. 1982. Design and functioning of the single-electrode cochlear implant. Annals of Otology, Rhinology, and Laryngology-supplement 91(2 pt 3): 21–26.

Dawson, P., P. J. Blamey, G. M. Clark, et al. 1989. Results in children using the 22 electrode cochlear implant. Journal of the Acoustical Society of America 86(suppl 1): 81.

Dawson, P. W., P. J. Blamey, L. C. Rowland, et al. 1992. Cochlear implants in children, adolescents and prelinguistically deafened adult: speech perception. Journal of Speech and Hearing Research 35: 401–417.

Deguine, O., B. Fraysse, A. Uziel, et al. 1993. Predictive factors in cochlear implant surgery. Advances in Oto-Rhino-Laryngology 48: 142–145.

Dillier, N. and T. Spillmann. 1984. Results and perspectives with extracochlear round window electrodes. Acta Oto-Laryngologica (suppl 411): 221–229.

Dillon, H. 2001. Hearing aids. Sydney, Australia, Boomerang Press.

Djourno, A. and C. Eyriès. 1957. Prosthese auditive par excitation electrique a distance du nerf sensoriel a l'aide d'un bobinage includ a demeure. Presse Medicale 35: 14–17.

Donaldson, P. E. 1987. Inductive RF link for an auditory prosthesis. Medical and Biological Engineering and Computing 25(3): 350–354.

Doring, W. H., S. Klajman, W. Huber and E. Becker. 1987. An interactive concept of speech processor fitting. In: Banfai, P., ed. Cochlear implant: current situation. 520–522.

Dorman, M. F., K. Dankowski and G. McCandless. 1989. Consonant recognition as a function of the number of channels of stimulation by patients who use the Symbion cochlear implant. Ear and Hearing 10: 288–291.

Dormer, K. J., G. Richard, P. E. Hough and J. V. D. Hough. 1980. The cochlear implant (auditory prosthesis) utilizing rare earth magnets. American Journal of Otology 2(1): 22–27.

Douek, E., A. J. Fourcin, B. C. J. Moore and G. P. Clarke. 1977. A new approach to the cochlear implant. Proceedings of the Royal Society of Medicine 70: 379–383.

Dowell, R. C., P. J. Blamey, P. M. Seligman, A. M. Brown and G. M. Clark. 1986a. Speech recognition performance with a two-formant coding strategy for a multi-channel cochlear prosthesis. Australian Journal of Audiology (suppl 2): 11.

Dowell, R. C., D. J. Mecklenburg and G. M. Clark. 1986b. Speech recognition for 40 patients receiving multichannel cochlear implants. Archives of Otolaryngology 112: 1054–1059.

Dowell, R. C., P. M. Seligman, P. J. Blamey and G. M. Clark. 1987. Speech perception using a two-formant 22-electrode cochlear prosthesis in quiet and in noise. Acta Oto-Laryngologica 104(5–6): 439–446.

Dowell, R. C., L. A. Whitford, P. M. Seligman, B. K.-H. Franz and G. M. Clark. 1990. Preliminary results with a miniature speech processor for the 22-electrode/cochlear hearing prosthesis. In: Sacristan, T., ed. Otorhinolaryngology, head and neck surgery. Amsterdam, Kugler and Ghedini: 1167–1173.

Doyle, J. B., H. D. Doyle, F. M. Turnbull, J. Abbey and L. House. 1963. Electrical stimulation in eighth nerve deafness. Bulletin of the Los Angeles Neurological Society 28: 148–150.

Doyle, J. H., J. B. Doyle and F. M. Turnbull. 1964. Electrical stimulation of eighth cranial nerve. Archives of Otolaryngology 80: 388–391.

Eddington, D. K. 1980. Speech discrimination in deaf subjects with cochlear implants. Journal of the Acoustical Society of America 68: 885–91.

Eddington, D. K. 1983. Speech recognition in deaf subjects with multichannel intracochlear electrodes. Annals of the New York Academy of Science 405: 241–258.

Eddington, D. K., W. H. Dobelle and D. E. Brackmann. 1978a. Auditory prostheses research with multiple channel intracochlear stimulation in man. Annals of Otology 87: 1–39.

Eddington, D. K., W. H. Dobelle, D. Brackmann, M. G. Mladejovsky and J. Parkin. 1978b. Place and periodicity pitch elicited by stimulation of multiple scala tympani electrodes in deaf volunteers. Transactions—American Society for Artificial Internal Organs 24: 1–5.

Edgerton, B. J. 1985. Implications of optimized single-channel cochlear implants. The Hearing Journal 38: 17–20.

Eisen, M. D. Submitted. History of the cochlear implant I: Djourno, Eyriès, and the first cochlear (?) implant. Otology and Neurotology.

Flottorp, G. 1953. Effect of different types of electrodes in electrophonic hearing. Journal of the Acoustical Society of America 25: 236–245.

Fourcin, A. J., S. M. Rosen, B. C. Moore, et al. 1979. External electrical stimulation of the cochlea: clinical, psychophysical, speech-perceptual and histological findings. British Journal of Audiology 13(3): 85–107.

Franz, B. K.-H. G., R. C. Dowell, G. M. Clark, P. M. Seligman and J. F. Patrick. 1987. Recent developments with the Nucleus 22-electrode cochlear implant: a new two formant speech coding strategy and its performance in background noise. American Journal of Otology 8: 516–518.

Fraser, J. G. 1987. UCH/RNID cochlear implant programme—an overview: patient selection and surgical technique. In: Banfai, P., ed. Cochlear implant: current situation. 273–279.

Fretz, R. J. and R. P. Fravel. 1985. Design and function: a physical and electrical description of the 3M House cochlear implant system. Ear and Hearing 6: 14S–19S.

Frey, A. H. 1961. Auditory system response to radio frequency energy—technical note. Aerospace Medicine 32: 1140–1142.

Fugain, C., B. Meyer, F. Chabolle and C. H. Chouard. 1984. Clinical results of the French multichannel cochlear implant. Acta Oto-Laryngologica-supplement 411: 237–246.

Gantz, B. J., B. F. McCabe, R. S. Tyler and J. P. Preece. 1987. Evaluation of four cochlear implant designs. Annals of Otology, Rhinology and Laryngology 96: 145–147.

Gantz, B. J., N. Tye-Murray and R. S. Tyler. 1989. Word recognition performance with single-channel and multichannel cochlear implants. American Journal of Otology 10: 91–94.

Genin, J. and R. Charachon. 1984. Electrical characteristics of a set of electrodes. Acta Oto-Laryngologica-supplement 411: 124–130.

Gersuni, G. V. and A. A. Volokhov. 1936. On the electrical excitability of the auditory organ: on the effect of alternating currents on the normal auditory apparatus. Journal of Experimental Psychology 19: 370–382.

Gisselsson, L. 1950. Experimental investigation into the problem of humoral transmission in the cochlea. Acta Oto-Laryngologica (suppl 82): 16.

Goldberg, J. M. and W. D. Neff. 1961. Frequency discrimination after bilateral section of the brachium of the inferior colliculus. Journal of Comparative Neurology 116: 265–290.

Goodhill, V. 1979. Progress in otology. Annals of Otology, Rhinology and Laryngology 88: 658–663.

Graham, J. M. 1987. Selection of patients for cochlear implantation: electrophysiological testing and promontory stimulation. Tinnitus; site of lesion; implants for children. In: Banfai, P., ed. Cochlear implant: current situation. 311–318.

Hannaway, C. 1996. The contributions of the National Institutes of Health to the development of cochlear prostheses. Personal communication, April 1996.

Hochmair, E. S. and I. J. Hochmair-Desoyer. 1983. Percepts elicited by different speech coding strategies. Annals of the New York Academy of Sciences 405: 268–279.

Hochmair, E. S., I. J. Hochmair-Desoyer and K. Burian. 1979. Investigations towards an artificial cochlea. International Journal of Artificial Organs 2(5): 255–261.

Hochmair-Desoyer, I. J., E. S. Hochmair and K. Burian. 1981. Four years of experience with cochlear prostheses. Medical Progress Technology 8: 107–119.

Hochmair-Desoyer, I. J., E. S. Hochmair, R. E. Fischer and K. Burian. 1980. Cochlear prostheses in use: recent speech comprehension results. Archives of Otorhinolaryngology 229: 81–98.

House, L. R. 1987. Cochlear implant: the beginning. Laryngoscope 97: 996–997.

House, W. F., K. I. Berliner and L. S. Eisenberg. 1981. The cochlear implant: 1980 update. Acta Oto-Laryngologica 91: 457–462.

House, W. F. and J. Urban. 1973. Long-term results of electrode implantation and electronic stimulation of the cochlea in man. In: Fields, W. S. and L. A. Leavitt, eds. Neural organization and its relevance to prosthetics. New York, Intercontinental Medical Book: 273–280.

Jerger, J. F. 1960. Audiological manifestations of lesions in the auditory nervous system. Laryngoscope 70: 417.

Johnsson, L.-G., W. F. House and F. H. Linthicum. 1982. Otopathological findings in a patient with bilateral cochlear implants. Annals of Otology Rhinology and Laryngology 91: 74–89.

Jones, R. C., S. S. Stevens and M. H. Lurie. 1940. Three mechanisms of hearing by electrical stimulation. Journal of the Acoustical Society of America 12: 281–290.

Katsuki, Y. and Y. Kanno. 1962. Neural mechanism of the peripheral and central auditory system in monkeys. Journal of the Acoustical Society of America 34: 1396–1410.

Kerr, A. and H. F. Schuknecht. 1968. The spiral ganglion in profound deafness. Acta Oto-Laryngologica 65: 568–598.

Kiang, N. Y.-S., R. F. Pfeiffer and W. B. Warr. 1965. Stimulus coding in the cochlear nucleus. Annals of Otology Rhinology and Laryngology 74: 2–23.

Kirk, K. I. 2000. Challenges in the clinical investigation of cochlear implant outcomes. In: Niparko, J. K., K. I. Kirk, N. K. Mellon, et al, eds. Cochlear implants: principles and practices. Philadelphia, Lippincott Williams & Wilkins: 225–259.

Laird, R. K. 1979. The bioengineering development of a sound encoder for an implantable hearing prosthesis for the profoundly deaf. Master of Engineering Science thesis, University of Melbourne.

Lawrence, M. 1964. Direct stimulation of auditory nerve fibers. Archives of Otolaryngology 80: 367–368.

Lawson, D. T., B. S. Wilson, M. Zerbi and C. C. Finley. 1996. Speech processors for auditory prostheses. Third quarterly progress report. NIH contract 1-DC-5-2103.

Leake-Jones, P. A. and S. J. Rebscher. 1983. Cochlear pathology with chronically implanted scala tympani electrodes. Annals of the New York Academy of Sciences 405: 203–223.

Legouix, J. P. 1957. Technique d'enrigistrement des potentials cochlearieres sur l'animal éveillé. Comptes Rendus des Seances de la Societe de Biologie et de Ses Filiales 151: 218–222.

Lehnhardt, E., R. Laszig, R. L. Webb, B. K.-H. G. Franz and G. M. Clark. 1987. Surgery for multielectrode cochlear implants. In: Banfai, P., ed. Cochlear implant: current situation. 477–480.

Lindstrom, B. 1987. Electric stimulation: pre and postoperatively. In: Banfai, P., ed. Cochlear implant: current situation. 179–181.

Loeb, G. E., M. W. White and W. M. Jenkins. 1983. Biophysical considerations in electrical stimulation of the auditory nervous system. Annals of the New York Academy of Sciences 405: 123–136.

Lybarger, S. F. 1944. U.S. patent application SN 543,278.

McCaughey, J. D. 1995. Cochlear implants-some considerations of a more or less ethical character. Annals of Otology, Rhinology and Laryngology 104: 16–17.

McDermott, H. J., C. M. McKay and A. Vandali. 1992. A new portable sound processor for the University of Melbourne/Nucleus Limited multi-electrode cochlear implant. Journal of the Acoustical Society of America 91: 3367–3371.

McKay, C. M., H. J. McDermott and G. M. Clark. 1991. Preliminary results with a six spectral maxima speech processor for the University of Melbourne/Nucleus multiple electrode cochlear implant. Journal of the Oto-Laryngological Society of Australia 6: 354–359.

McKay, C. M., H. J. McDermott, A. Vandali and G. M. Clark. 1992. A comparison of speech perception of cochlear implantees using the Spectral Maxima Sound Processor (SMSP) and the MSP (Multipeak) processor. Acta Oto-Laryngologica 112: 752–761.

Merzenich, M. M., D. K. Kessler, S. J. Rebscher and R. A. Schindler. 1987. Progress in development and application of the University of California at San Francisco/Storz multichannel cochlear implant. Annals of Otology, Rhinology and Laryngology 96(suppl 128): 122–125.

Merzenich, M. M., R. P. Michelson and C. R. Pettit. 1973. Neural encoding of sound sensation evoked by electrical stimulation of the acoustic nerve. Annals of Otology 82: 486–503.

Merzenich, M. M. and M. D. Reid. 1974. Representation of the cochlea within the inferior colliculus. Brain Research 77: 397–415.

Merzenich, M. M., M. White, M. C. Vivion, P. A. Leake-Jones and S. Walsh. 1979. Some considerations of multichannel electrical stimulation of the auditory nerve in the profoundly deaf: interfacing electrode arrays with the auditory nerve array. Acta Oto-Laryngologica 87: 196–203.

Michelson, R. P. 1971. Electrical stimulation of the human cochlea—a preliminary report. Archives of Otolaryngology 93: 317–323.

Miller, G. A. and P. E. Nicely. 1955. An analysis of perceptual confusions among some English consonants. Journal of the Acoustical Society of America 27(3): 338–352.

Miller, J. M., L. G. Duckert, D. Sutton, B. E. Pfingst, M. A. Malone and F. A. Spelman. 1985. Animal models: relevance to implant use in humans. In: Schindler, R. and M. Merzenich, eds. Cochlear implants. New York, Raven Press 35–54.

Miller, J. M., M. A. Malone and L. G. Duckert. 1983. Functional and histological effects of electrical stimulation of the cochlea. In: Webster, W., and L. Aitkin, eds. Mechanisms of hearing. Clayton, Victoria, Canada, Monash University Press.

Morgon, A., C. Berger-Vachon, J. M. Chanal, G. Kalfoun and C. Dubreuil. 1984. Cochlear implant: experience of the Lyon team. Acta Oto-Laryngologica-supplement 411: 195–203.

Moxon, E. C. 1971. Neural and mechanical responses to electrical stimulation of the cat's inner ear. PhD dissertation. Cambridge, MA, MIT.

Negrevergne, M., R. Dauman, P. Lagourgue and M. Bourdon. 1987. Extra-cochlear implant Prelco. In: Banfai, P., ed. Cochlear implant: current situations.

Nienhuys, T. G. W., G. N. Musgrave, P. A. Busby, et al. 1987. Educational assessment and management of children with multichannel cochlear implants. Annals of Otology, Rhinology and Laryngology 96(suppl 128): 80–82.

Peeters, I. S., J. Marquet, E. Offeciers, et al. 1987. The Laura cochlear prosthesis development description. In: Banfai, P., ed. Cochlear implant: current situations.

Pfingst, B. E., J. A. Donaldson, J. M. Miller and F. A. Spelman. 1979. Psychophysical evaluation of cochlear prostheses in a monkey model. Annals of Otology, Rhinology and Laryngology 88(5 pt 1): 613–625.

Pfingst, B. E. and N. L. Rush. 1985. Discrimination of simultaneous frequency and level changes in electrical stimuli. Annals of Otology, Rhinology and Laryngology 96(suppl 128): 34–37.

Pfingst, B. E., D. Sutton and J. M. Miller. 1981. Relation of psychophysical data to histopathology in monkeys with cochlear implants. Acta Oto-Laryngologica 92: 1–13.

Pialoux, P., C. H. Chouard and B. Meyer. 1979. Indications and results of the multichannel cochlear implant. Acta Oto-Laryngologica 87: 185–189.

Powell, T. P. and S. D. Erulkar. 1962. Transneuronal cell degeneration in the auditory relay nuclei of the cat. Journal of Anatomy 96: 249–268.

Riseberg, A. and E. Aglefors. 1987. Relation between speech perception ability and results on psychoacoustic tests for single channel cochlear implants. In: Banfai, P., ed. Cochlear implant: current situation. 187–193.

Rose, J. E., J. F. Brugge, D. J. Anderson and J. E. Hind. 1967. Phase-locked response to low-frequency tones in single auditory nerve fibers of the squirrel monkey. Journal of Neurophysiology 30: 769–93.

Rose, J. E., R. Galambos and J. R. Hughes. 1959. Microelectrode studies of the cochlear nuclei of the cat. Bulletin of John Hopkins Hospital 104: 211–251.

Rupert, A., G. Moushegian and R. Galambos. 1963. Unit responses to sound from auditory nerve of the cat. Journal of Neurophysiology 26: 449–465.

Schindler, R. A., M. M. Merzenich, M. W. White and B. Bjorkroth. 1977. Multi electrode intracochlear implants—nerve survival and stimulation patterns. Archives of Otolaryngology 103: 691–699.

Schuknecht, H. 1953. Techniques for study of cochlear function and pathology in experimental animals. Acta Oto-Laryngologica 58: 377.

Schuknecht, H. F. and R. C. Woellner. 1955. An experimental and clinical study of deafness from lesions of the cochlear nerve. Journal of Laryngology 69: 75.

Seligman, P. M. and H. J. McDermott. 1995. Architecture of the SPECTRA 22 speech processor. Annals of Otology, Rhinology and Laryngology 104(suppl 166): 139–141.

Shepherd, R. K., G. M. Clark and R. C. Black. 1983a. Chronic electrical stimulation of the auditory nerve in cats. Physiological and histopathological results. Acta Oto-Laryngologica-supplement 399: 19–31.

Shepherd, R. K., G. M. Clark and R. C. Black. 1983b. Physiological and histopathological effects of chronic intracochlear electrical stimulation. In: Webster, W. R. and L. M. Aitkin, eds. Mechanisms of hearing. Clayton, Melbourne, Australia, Monash University Press: 200–205.

Shepherd, R. K., G. M. Clark, R. C. Black and J. F. Patrick. 1982. Chronic electrical stimulation of the auditory nerve in cats. Proceedings of the Australian Physiological and Pharmacological Society 13: 211P.

Shepherd, R. K., G. M. Clark, R. C. Black and J. F. Patrick. 1983c. The histopathological effects of chronic electrical stimulation of the cat cochlea. Journal of Laryngology and Otology 97: 333–341.

Simmons, F. B. 1966. Electrical stimulation of the auditory nerve in man. Archives of Otolaryngology 84: 2–54.

Simmons, F. B. 1967. Permanent intracochlear electrodes in cats. Tissue tolerance and cochlear microphonics. Laryngoscope 77: 171–186.

Simmons, F. B., J. M. Epley, R. C. Lummins, et al. 1965. Auditory nerve: electrical stimulation in man. Science 148(366): 104–106.

Simmons, F. B., C. J. Monegeon, W. R. Lewis and D. A. Huntington. 1964. Electrical stimulation of acoustical nerve and inferior colliculus. Archives of Otolaryngology 79: 559–567.

Skinner, M. W., G. M. Clark, L. A. Whitford, et al. 1994. Evaluation of a new spectral peak coding strategy for the Nucleus 22 channels cochlear implant system. American Journal of Otology 15: 15–27.

Smoorenburg, G. F. and A. F. van Olphen. 1987. Pre-operative electrostimulation of the auditory nerve and post-operative results with the House/3M implant. In: Banfai, P., ed. Cochlear implant: current situation. 227–229.

Snyder, R. L., S. J. Rebscher, K. Cao, P. A. Leake and K. Kelly. 1990. Chronic intracochlear electrical stimulation in the neonatally deafened cat. I. Expansion of central representation. Hearing Research 50: 7–34.

Snyder, R. L., S. J. Rebscher, P. Leake, K. Kelly and K. Cao. 1991. Chronic intracochlear electrical stimulation in the neonatally deafened cat. II. Temporal properties of neurons in the inferior colliculus. Hearing Research 56: 246–264.

Sommer, H. C. and H. E. von Gierke. 1964. Hearing sensations in electric fields. Aeroscopic Medicine 35: 834–839.

Spillman, T. and N. Dillier. 1985. Cochlear implants in the deaf: indications, methods, results. Schweizerische Rundschau fur Medizin Praxis 74: 211–219.

Staller, S. J., R. C. Dowell, A. L. Beiter and J. A. Brimacombe. 1991. Perceptual abilities of children with the Nucleus 22-channel cochlear implant. Ear and Hearing 12(suppl 4): 34S–47S.

Stevens, S. S. and R. C. Jones. 1939. The mechanism of hearing by electrical stimulation. Journal of the Acoustical Society of America 10: 261–269.

Sutton, D., J. M. Miller and B. E. Pfingst. 1980. Comparison of cochlear histopathology following two implant designs for use in scala tympani. Annals of Otology Rhinology and Laryngology 89: 11–14.

Tasaki, I. 1954. Nerve impulses in individual auditory nerve fibres of the guinea pig. Journal of Neurophysiology 17: 97–122.

Thielemeir, M. A., L. L. Tonokawa, B. Petersen and L. S. Eisenberg. 1985. Audiological results in children with a cochlear implant. Ear and Hearing 6(3 suppl): 27S–35S.

Tong, Y. C., R. C. Black, G. M. Clark, et al. 1979. A preliminary report on a multiple-channel cochlear implant operation. Journal of Laryngology and Otology 93: 679–695.

Tong, Y. C., P. J. Blamey, R. C. Dowell and G. M. Clark. 1983. Psychophysical studies evaluating the feasibility of a speech processing strategy for a multiple-channel cochlear implant. Journal of the Acoustical Society of America 74: 73–80.

Tong, Y. C., P. A. Busby and G. M. Clark. 1986. Psychophysical studies on prelingual patients using a multiple-electrode cochlear implant. Journal of the Acoustical Society of America 80(suppl 1): S30.

Tong, Y. C., P. A. Busby and G. M. Clark. 1988. Perceptual studies on cochlear implant patients with early onset of profound hearing impairment prior to normal development of auditory, speech, and language skills. Journal of the Acoustical Society of America 84: 951–962.

Tong, Y. C., G. M. Clark, P. J. Blamey, P. A. Busby and R. C. Dowell. 1982. Psychophysical studies for two multiple-channel cochlear implant patients. Journal of the Acoustical Society of America 71: 153–160.

Tong, Y. C., G. M. Clark, P. M. Seligman and J. F. Patrick. 1980. Speech processing for a multiple-electrode cochlear implant hearing prosthesis. Journal of the Acoustical Society of America 68: 1897–1899.

Tong, Y. C., R. C. Dowell, P. J. Blamey and G. M. Clark. 1983. Two-component hearing sensations produced by two-electrode stimulation in the cochlea of a deaf patient. Science 219: 993–994.

Tong, Y. C., J. B. Millar, G. M. Clark, L. F. Martin, P. A. Busby and J. F. Patrick. 1980. Psychophysical and speech perception studies on two multiple-channel cochlear implant patients. Journal of Laryngology and Otology 94: 1241–1256.

Tyler, R. S., M. W. Lowder and S. R. Otto. 1984. Initial Iowa results with the multichannel cochlear implant from Melbourne. Journal of Speech and Hearing Research 27: 596–604.

Tyler, R. S., N. Tye-Murray, J. P. Preece, B. J. Gantz and B. F. McCabe. 1987. Vowel and consonant confusions among cochlear implant patients: do different implants make a difference? Annals of Otology, Rhinology and Laryngology 96(suppl 128): 141–144.

Valvoda, M., J. Betka and J. Hruby. 1987. The first experience with cochlear implantations in Czechoslovakia. In: Banfai, P., ed. Cochlear implant: current situations. International cochlear implant symposium. West Germany, Duren: 235–237.

von Helmholtz, H. L. F. 1863. Die Lehre von den tonempfindungen als physiologische grundlage fur die theorie der musik. Braunschweig, F Vieweg and Sohn.

Walliker, J. R., H. Carson, E. E. Douek, A. J. Fourcin and S. Rosen. 1987. Prosthesis development in the external pattern input group. In: Banfai, P., ed. Cochlear implant: current situation. 265–272.

Walloch, R. A., J. A. Fenwick and R. Boberg. 1980. A reaction-time analysis of electrocochlear stimulation in cats. Laryngoscope 90: 861–6. See Battmer et al 1987.

Wever, E. G. and C. W. Bray. 1930. Auditory nerve impulses. Science 71: 215.

White, R. L., L. A. Roberts, N. E. Cotter and O. H. Kwon. 1983. Thin-film electrode fabrication techniques. Annals of the New York Academy of Sciences 405: 183–90.

Williams, A. J., G. M. Clark and G. V. Stanley. 1974. Behavioural responses in the cat to simple patterns of electrical stimulation of the terminal auditory nerve fibres. Proceedings of the Australian Physiological and Pharmacological Society 5(2): 252.

Williams, A. J., G. M. Clark and G. V. Stanley. 1976. Pitch discrimination in the cat through electrical stimulation of the terminal auditory nerve fibers. Physiological Psychology 4: 23–27.

Wilson, B. S. 1997. The future of cochlear implants. British Journal of Audiology 31: 205–225.

Wilson, B. S., D. T. Lawson, M. Zerbi and C. C. Finley. 1992. Speech processors for auditory prostheses. Twelfth quarterly progress report, April 1992. NIH contract N01-DC-9-2401. Research Triangle Institute.

Wilson, B. S., D. T. Lawson, M. Zerbi and C. C. Finley. 1993. Speech processors for auditory prostheses. Fifth quarterly progress report, Oct 1993. NIH contract N01-DC-2-2401. Research Triangle Institute.

Xu, S., R. C. Dowell and G. M. Clark. 1987. Results for Chinese and English in a multichannel cochlear implant patient. Annals of Otology, Rhinology and Laryngology 96(suppl 128): 126–127.

Zollner, F. and W. D. Keidel. 1963. Gerorvermittlung durch elektrische erregung des nervus acousticus. Archiv Ohr Nas Kehlkopfheilk 181: 216–223.

2
Surgical Anatomy

Overview

The external ear or pinna (auricle) collects sound, which passes along the external auditory canal to the eardrum (tympanic membrane) (Fig. 2.1). The middle ear (tympanum) is a cavity containing three small, articulated bones (ossicles), and is closed externally (laterally) by the tympanic membrane. The cavity is connected by the eustachian tube to the pharynx at the back of the nose to allow the pressure between the middle ear and the outside to be equalized. The tympanic membrane vibrates in response to sound, and the vibrations are transmitted through the os-

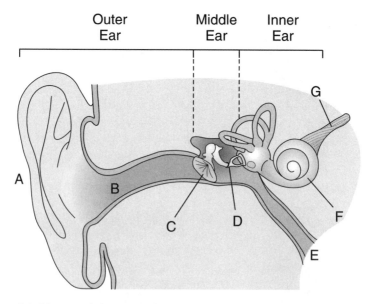

FIGURE 2.1. Diagram of the outer, middle, and inner ears and auditory nerve passing to the central auditory system. A, pinna; B, external auditory canal; C, eardrum; D, ossicles; E, eustachian tube; F, inner ear; G, auditory (cochlear) nerve.

sicles to the inner ear. The ossicles amplify the sound waves, and this overcomes the greater resistance of fluid in the inner ear to the transmission of sound. Thus the middle ear matches the impedances of sound in air and water for the effective transfer of sound energy. The inner ear contains the sense organ of hearing (organ of Corti) that converts sound waves into electrical signals. This electrical activity initiates patterns of action potentials in the auditory nerve and higher brain centers, thus encoding the sound. The encoded signal is finally decoded into the sensations of hearing. Deafness occurs if the structures or pathways are damaged or diseased. Deafness is referred to as conductive if the sound cannot be conducted to the inner ear. Deafness is sensorineural if the cochlea or auditory nerve is damaged or fails to develop properly.

Temporal Bone

The temporal bone houses the external, middle, and inner ears, and lies on the lateral side of the skull. It articulates with the frontal bone in front, the parietal bone above, the occipital bone behind, and the sphenoid bone internally.

Components

The temporal bone consists of four morphologically distinct parts (Fig. 2.2). These are the squamous, petromastoid, and tympanic parts as well as the styloid

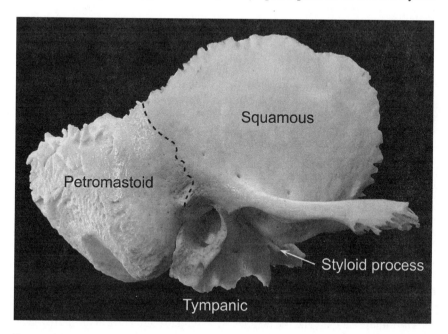

FIGURE 2.2. The human temporal bone with the squamous, petromastoid, and tympanic parts, and styloid process demarcated.

process. The squamous part, formed in the fetus from membrane, is the anterior and upper part, and it helps to enclose the cranial cavity. It articulates with other cranial bones, the parietal, and the greater wing of the sphenoid and the occipital bones. The petromastoid portion is formed in cartilage, and encloses the osseous and membranous labyrinth. The membranous labyrinth lying within the osseous labyrinth consists of the hearing and balance organs. The petromastoid part is divided into two sections: (1) The mastoid forms the posterior part, and has a protuberance behind the pinna (the mastoid process). The mastoid and process contain the greater number of the air cells, which have extended from the middle ear cavity through the aditus to the antrum. (2) The petrous part of the bone is wedged between the sphenoid and occipital bones at the base of the skull. It is directed medially and forward, and has a base, an apex, three surfaces, and three margins. The labyrinth lies within it. The anterior surface helps form the floor of the middle cranial fossa. The posterior surface forms the anterior part of the posterior cranial fossa. Near the center of the posterior surface is an orifice, the internal auditory meatus. This meatus transmits the cochlear and vestibular nerves from the labyrinth to the brain as well as the facial nerve. The inferior surface is irregular due to the attachments of muscles at the base of the skull. The tympanic section of the temporal bone develops to form part of the wall of the tympanic cavity and the external auditory meatus where it supports the ear drum. The styloid process is a slender structure protruding downward for the attachment of muscles of the pharynx.

Embryology

The four components of the temporal bone are ossified independently. In the squamous part this occurs in membrane from a single center at about the eighth intrauterine week. The petromastoid part, housing the labyrinth, is ossified from several centers that appear in the cartilaginous capsule around the ear from the 20th week. The tympanic part is ossified in membrane at about the 12th week. The styloid process is developed from the cranial end of the cartilage of the second branchial arch.

The bony labyrinth becomes adult in size by the 16th to 20th week of development (Bast 1942). Histological and radiological studies (Eby and Nadol 1986) showed no change in dimensions from birth to adulthood. The cartilage around the labyrinth is ossified to create three layers—endosteal, endochondral, and periosteal, from inside out. At birth ossification commences in the marrow of the endochondral layer, and by 3 years of age the bone overlying the cochlea is compact. Thus when creating an opening into the cochlea (cochleostomy) in young children the overlying bone will be more vascular than in the adult. After the cochleostomy is created, both in children and adults, and the electrode inserted, the surrounding space is filled with new bone.

The middle ear cleft and the ossicles are essentially adult size at birth. However, the tympanic bone, forming around the external auditory canal ossifies after birth moving laterally, taking the tympanic membrane with it. This changes the plane of the eardrum from primarily horizontal in the newborn to its adult position by 5 years of age (Birrell 1978).

Mastoid Air Cell System and Variations

The mastoid cells are first seen at 34 weeks of embryological development. Extension of the cells into the mastoid occurs as soon as it begins to form, and this is well before birth (Bast and Anson 1949). The air cell system commences as out-pouches from the middle ear cleft and antrum (the posterior extension of the cleft). However, it is not until the middle ear is aerated at birth that the pneumatization of the mastoid accelerates, and it continues throughout infancy and early childhood. Air-lined sacs extend into the connective tissue between spicules of bone. They pass in particular into the mastoid portion of the temporal bone and toward the tip. Some out-pouchings extend toward the apex of the petrous part of the temporal bone, and others insinuate themselves around structures such as the inner ear and carotid artery. The mastoid air cells are lined with a single layer of epithelial cells. The degree of pneumatization of the mastoid bone varies greatly.

With cochlear implant surgery it is necessary to remove the cells from part of the mastoid bone to provide space for the receiver-stimulator and lead wire assembly. Air cells must also be removed to expose the posterior (back) wall of the middle ear for access to the round window and basal turn of the cochlea for the insertion of the electrode array through a posterior tympanotomy.

Blood Supply and Innervation

The arterial blood supply for the temporal bone arises primarily from branches of the external carotid artery, in particular the superficial temporal, postauricular, and occipital arteries. These lie behind the pinna and are also the blood supply to the scalp. They must be considered when placing the incision in order to avoid cutting off the blood supply to the flap. This applies in particular if there has been an incision in the postauricular skin crease, and a later C-shaped incision was made behind it to insert the implant.

The internal carotid artery passes through the temporal bone en route to supply the brain. It lies just anterior to the cochlea, and care must be taken to avoid injuring it when drilling forward along a sclerosed basal turn. The carotid artery lies approximately 12 mm anterior to the round window membrane. The venous drainage from the brain passes through the mastoid bone where it has an S-shaped bend, the lateral sinus. In some bones it lies further forward than the average, and this can make it difficult to see the posterior wall of the middle ear for an opening (posterior tympanotomy) to inspect the basal turn of the cochlea. In this case it is necessary to remove a section of the bony posterior bony canal wall to provide access.

The sensory innervation of the middle ear is from a plexus of nerves on the medial wall of the middle ear (the tympanic plexus). It arises primarily from the ninth cranial nerve, the glossopharyngeal nerve. The tympanic branch of this cranial nerve (tympanic or Jacobson's nerve) enters the middle ear cavity through an opening in the lower (inferior) wall of the cavity. Consequently, if electrical current flows in an extracochlear direction either from an electrode placed outside the round window or through a low-impedance pathway created in the bone over the cochlea due to otosclerosis, pain will be experienced. The pain may be referred to the ear or throat, as these are regions supplied by the glossopharyngeal nerve.

The facial nerve, which innervates the muscles of the face, enters the temporal bone through the internal auditory meatus lying just above the cochlear nerve. From the medial wall of the middle ear it takes a sharp right-angled bend to pass backward above the oval window toward the lateral semicircular canal, where there is another knee-like bend or genu, and the nerve then travels in its third or vertical part down behind the middle ear to exit the skull through the stylomastoid foramen.

Infant and Young Child

The postnatal changes in the human temporal bone are of importance in undertaking cochlear implant surgery, in placing the receiver-stimulator package in the neonate, and in designing the lead wire assembly from the package to the electrode array.

Postnatal Growth Changes

After birth there is a variable increase in size in different regions of the bone, and molding takes place. Changes also occur due to the penetration of areas with the air cells. The mastoid part of the bone is at first flat, and the stylomastoid foramen and styloid process are immediately behind the tympanic rim. This makes the facial nerve lie superficial, and it is vulnerable when operating on an infant or young child. With the development of mastoid air cells the lateral part of the mastoid goes downward and forward, and the styloid process and the stylomastoid foramen come to lie on the undersurface of the bone. It is not until the latter part of the second year that the mastoid process forms a definite elevation. The first stage in the surgical access to the tympanic cavity through the mastoid in the child requires only a thin scale of bone to be removed from the suprameatal triangle to allow exposure of the antrum and mastoid air cells.

The main cause of growth in the temporal bone is its pneumatization. This affects the dimensions of the mastoid in particular. Pneumatization is completed from 6 to 19 months of age (Schillinger 1939; Rubensohn 1965). This is the first growth spurt that is followed by a second one at puberty. Eby and Nadol (1986) found these two growth periods applied to the length and width of the mastoid but not so much to its depth.

For implantation it was of importance to determine the growth of particular regions of the temporal bone. This was studied on 103 computed tomography (CT) scans of children of various ages (O'Donoghue et al 1986). The axes measured reflected both overall increase in skull size and components of the temporal bone. The measures showed the greatest increase in length over the first 2 years of life. A measure from the surface of the mastoid to the promontory was of special relevance with a total increase in length of 17 mm of which 8 mm (47%) occurred in the first 2 years of life.

A detailed analysis of human temporal bones was made by Dahm et al (1993) by dissecting 60 specimens from people ranging in age from 0.16 to 84 years. Measurements were made between a number of different anatomical landmarks to provide details of differential growth in three dimensions, as this was the most accurate method of determining the effect of growth changes on implant surgery. There were 35 separate measures made. Key findings were for growth between the sino-dural angle at the most lateral aspect of the mastoid bone and the round window, as well as the distance between the round window and the fossa incudis (floor of the mastoid antrum) (Fig. 2.3). The sino-dural angle best represented the site for the implant, and the distance from the round window increased on average 12 mm from birth to adulthood with a standard deviation of 5 mm. Therefore, a pediatric cochlear implant should allow up to 25 mm of lead wire lengthening. In addition, as there was no increase in the distance between the round window and the fossa incudis with age (Fig. 2.3), this indicated that fixation of the lead

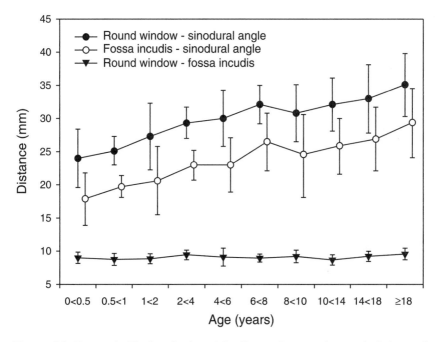

FIGURE 2.3. Top graph: The lengthening of the distance between the round window and the sino-dural angle versus age. Center graph: The distance between the fossa incudis and the sino-dural angle versus age. Bottom graph: The distance between the round window and the fossa incudis versus age (Dahm et al 1993. The postnatal growth of the temporal bone and its implications for cochlear implantation in children. Acta Oto-Laryngologica (suppl 505). Reprinted with permission.).

wire to the fossa (floor of the mastoid antrum) would be desirable as any growth changes would be transmitted to this point rather than pull the electrode from the round window. Furthermore, the lengthening of the bone from the fossa incudis to the sino-dural angle was the same as from the round window (Fig. 2.3).

Sutures

As the brain increases in size after birth due to the myelinization of the neural tracts, the skull has to expand to accommodate it. This takes place at the junctions between the skull bones (sutures). The skull expansion is the result of ossification and molding of each skull bone. This is by deposition and absorption of bone on the surface, and as the bones are displaced outward the suture also initiates bone growth to compensate (Enlow 1986). In some children there is early closure of the sutures leading to microcephaly. The suture lines need to be excised, otherwise mental retardation and deafness will ensue. In others there may be a localized early closure of the sutures, and if untreated there will be an asymmetrical deformity. Normal fusion of the bones with ossification of the sutures occurs at different ages. The petromastoid section fuses with the squamous part during the first year. This means that if an infection were to occur in the middle ear at this time there is a potential pathway for it to extend to the cranial cavity.

With cochlear implant surgery in infants and young children it will usually be necessary to drill down to the dura lining the brain to make a bed for the receiver-stimulator, as the bone is very thin at this age. The bed is usually placed in the mastoid and anterior segment of the occipital and parietal bones at the junction of the sutures between the mastoid, parietal, and occipital bones (asterion). As this requires drilling through the sutures it was thought this might stimulate early closure of the sutures and a skull deformity. For this reason a radiological study of head growth in the macaque monkey (Xu et al 1993), and the histological examination of the sutures 3 to 4 years after implantation was carried out, but no closure was seen (Burton et al 1992a,b, 1994).

External Ear

Pinna

The pinna or auricle collects the sound and funnels it down the external auditory meatus to the middle ear. It has a sculptured shape due to the underlying cartilage. These undulations introduce spectral changes in the sound depending on its position in relation to the head, and thus facilitate sound localization. The features of the pinna include the helix (rim), the concha (a large cavity leading into the external auditory canal), and the tragus (protruding cartilage protecting the external auditory meatus). The cleft between the tragus and the helix is the incisura terminalis, which is the site for an endaural incision to expose the middle ear. This endaural incision can be extended backward to expose the mastoid bone for the placement of the cochlear implant as undertaken by Lehnhardt et al (1987).

There is a crease (postauricular sulcus) behind the pinna where it is attached to the scalp. A good exposure of the mastoid, occipital, and parietal bones for cochlear implantation can be obtained through an incision in this sulcus with an upward extension, as described in Chapter 10.

External Auditory Meatus

Adult

The external auditory meatus extends from the concha to the tympanic membrane. In the adult it is approximately 2.5 cm long. The outer third is cartilaginous and the medial two thirds osseous. It forms an S-shaped curve. The meatus is ellipsoid, with the long axis directed downward and backward. It is important to note that the tympanic membrane is oblique to the main axis of the meatus, with the anterior wall and the floor of the meatus being slightly longer. To examine the eardrum it is important to pull the pinna upward and backward and use an ovoid speculum for viewing. The skin lining the meatus is closely attached to the periosteum of the canal, and if, when carrying out a cochlear implant operation, the opening from the mastoid into the middle ear is too far forward, the ear canal can be easily entered.

Infant and Young Child

In a newborn the meatus is almost entirely cartilaginous; the bony portion is represented by the tympanic bone that at this stage is a narrow ring articulating superiorly with the squamous part of the temporal bone. In the infant the meatus is directed medially and downward, with the eardrum being almost horizontal and thus facing inferiorly.

Middle Ear

The middle ear is an air-filled space that communicates with the back of the nose (nasopharynx) through the auditory or eustachian tube. The middle ear is box-like with a concave medial and lateral wall. It is divided into the tympanic cavity proper, which is opposite the eardrum, with the epitympanic recess above the drum, and the hypotympanum below the drum. The width of the epitympanum is 6 mm, and in the middle ear proper it is 2 mm at the center and 4 mm in the hypotympanum.

Ossicles

The middle ear, or tympanum, contains three ossicles: the malleus (or hammer), the incus (or anvil), and stapes (or stirrup) (Fig. 2.4). These are the smallest bones

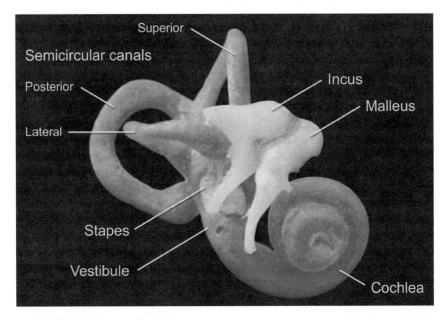

FIGURE 2.4. The middle ear ossicles (malleus, incus, and stapes) articulated and connected to a mold of the inner ear (osseous labyrinth). The cochlea is the anterior part of the labyrinth, the vestibule the center part, and the semicircular canals the posterior part.

in the body. The malleus has a head, neck, and three processes—the manubrium or handle, and the anterior and lateral processes. The manubrium passes downward and backward and is attached along its length to the eardrum. It thus makes the mechanical link between the eardrum and the ossicles. The head lies in the middle ear above the eardrum in the epitympanic space, where it articulates with the incus through a synovial joint. The incus has a body and a short and long process. The short process lies in the floor of the opening to the mastoid antrum, and the long process passes downward and backward to terminate as a rounded projection, the lenticular process, which articulates with the head of the stapes through another synovial joint. The head of the stapes is connected to the footplate via its anterior and posterior crus. The stapes footplate fits into an opening in the bony cochlea called the oval window or fenestra vestibulae.

Muscles

In the middle ear there are also two muscles: the stapedius and the tensa tympani. These are the smallest muscles in the body. They contract in response to a loud sound and stiffen the ossicular chain so that the inner ear is less likely to be damaged. The tensor tympani inserts into the neck of the manubrium of the malleus, and pulls the eardrum medially. The stapedius inserts into the posterior aspect of the neck of the stapes and tilts it backward.

Relationships

Superior

The roof of the tympanic cavity is a thin plate of bone, the tegmen tympani, which separates the middle ear from the middle cranial fossa. The lateral edge unites with the squamous temporal bone to form the petrosquamous suture, which is normally fused in adult life. In a child or adult the suture can be a route by which middle ear infection can spread to the brain.

Inferior

The floor of the tympanum has a thin plate of bone posteriorly overlying the superior bulb of the main vein draining the brain (internal jugular vein), and anteriorly the internal carotid artery. Between the carotid artery and the jugular bulb there is a small canal for the tympanic branch of the ninth cranial nerve (glossopharyngeal). As discussed above (see Blood Supply and Innervation), stimulation of this nerve electrically with a cochlear implant can lead to pain referred to the ear or throat.

Anterior

The anterior wall of the tympanum is taken up essentially by the openings of two canals. The upper canal is for the tensor-tympani muscle, and the lower one for the auditory or eustachian tube, which leads to the nasopharynx at the back of the nose.

Posterior

The posterior wall of the middle ear has the following structures, from above downward: (1) the opening leading from the epitympanum to the tympanic antrum and the mastoid air cells; (2) the fossa for the short process of the incus; (3) the pyramid, a conical bony projection that is perforated by the tendon of the stapedius muscle; and (4) the posterior canaliculus for the nerve of taste (chorda tympani). The chorda tympani leaves the vertical section of the facial nerve in the mastoid bone and passes through the posterior canaliculus across the middle ear. The chorda tympani and the vertical section of the facial nerve are the anterior and posterior boundaries, respectively, of the opening (posterior tympanotomy) for access to the middle ear from behind, so that the cochlear implant electrode can be inserted into the cochlea (see Chapter 10).

The middle ear cavity communicates posteriorly through the antrum with an extensive system of air cells that extend into the mastoid, squamous, and petrous parts of the temporal bone. Infections of the middle ear can produce mastoid infections with various complications. For example, if the infection spreads upward through the floor of the middle cranial fossa, the patient may develop meningitis or a temporal lobe abscess.

In some bones an air cell lies just below the basal turn of the cochlea. Its

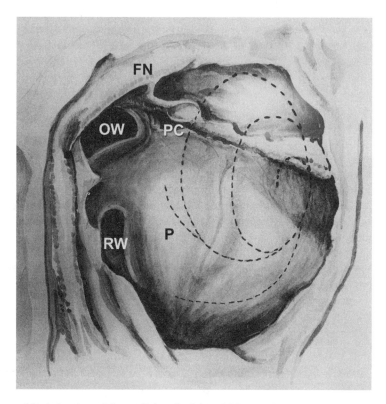

FIGURE 2.5. A drawing of the medial wall of the middle ear showing the underlying turns of the cochlea with the dashed line. P, promontory; OW, oval window; RW, round window; FN, facial nerve; PC, processus cochleariformis (Clark 1975. A Surgical Approach for a Cochlear Implant: An Anatomical Study. *Laryngology and Otology.* **89.** Reprinted with permission.).

opening can look like the round window niche (cavity leading to the round window). Implant electrodes have been placed extracochlearly into this cul-de-sac by mistake. However, the interior of the scala tympani has a different appearance, and the receding spiral of the scala is distinctive.

Lateral

The lateral wall of the tympanum consists primarily of the tympanic membrane that has a diameter of approximately 10 mm. A normal eardrum has a pearly gray appearance and reflects a triangle of light at a point in front and below the umbo or lower attachment of the handle of the malleus. The lower and central part is stiff (pars tensa), and the upper part flaccid (pars flaccida). The pars tensa is the larger and more visible section. It is thickened peripherally as a fibrocartilaginous ring that fits into a groove or sulcus in the tympanic bone of the ear canal. This sulcus is a landmark in the upper-anterior corner of the posterior tympanotomy opening into the middle ear.

Medial

The medial wall of the tympanum (Fig. 2.5) presents a round eminence called the promontory that is the bulge of the basal turn of the cochlea. It lies in front of the oval and round windows. Each turn of the cochlea was related to features on the medial wall of the middle ear by Clark (1975) to assist in creating an entry through the overlying bone. The study was undertaken on 10 human temporal bones. The medial wall has also been the site for drilling the overlying bone to site extracochlear electrodes (Banfai et al 1984, 1985). The relationships of the cochlear turns to the overlying inner wall of the middle ear are also of particular relevance to surgery for labyrinthitis ossificans. In this condition the cochlea is partly or completely replaced with bone as a chronic stage in the healing of infection after meningitis. The electrodes may need to be inserted through holes drilled over the middle and apical as well as the basal turns (Gantz et al 1989; Cohen and Waltzman 1993; Bredberg and Lindstrom 1995; Chouard et al 1995; Bredberg et al 2000). It should be noted that the upper portion of the basal turn lies under the horizontal portion of the facial nerve (Clark 1975) (Fig. 2.5). The upper portion of the middle turn, however, is more accessible. The processus cochleariformis, the bony fulcrum for the tensor tympani muscle as it swings laterally to insert into the malleus, is an important landmark to these turns. Furthermore, the helicotrema is along a line drawn through the center of the oval window and approximately at right angles to a line between the centers of the oval and round windows (Fig. 2.6). It should also be noted that as defined by Clark (1975) the start of the apical, middle, and basal turn is where they are crossed by the line between the center of the round window and the helicotrema. The distance from the oval window to the helicotrema is 4.6 mm, to the middle

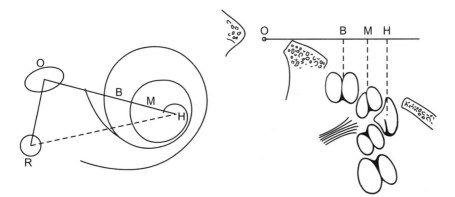

FIGURE 2.6. Left: A diagram of the cochlea and the oval and round windows showing how the position of the helicotrema (at right angles to a line between the oval and round windows) can be determined, and the start of the middle and apical turns demarcated (interrupted line). O, oval window; R, round window; B, basal turn; M, middle turn; H, helicotrema. Right: A diagram of a cross section of the human cochlea showing how the basal, middle and apical turns overlap each other (Clark 1975. A Surgical Approach for a Cochlear Implant: An Anatomical Study. *Laryngology and Otology.* **89**. Reprinted with permission.).

TABLE 2.1. Reference points for the apical, middle, and basal turns of the cochlea.

	Average (mm)	Range (mm)
Distance oval window to helicotrema	4.6	4.5–4.8
Distance oval window to distal portion middle turn	4.1	4.0–4.3
Distance oval window to distal portion basal turn	2.5	2.5–2.5
Depth of bone overlying proximal portion basal turn	1.3	0.8–1.7
Depth of bone overlying distal portion basal turn	1.9	1.3–2.0
Depth of bone overlying distal portion middle turn	1.7	1.3–2.1

turn 4.1 mm, and to the basal turn 2.5 mm (Clark 1975). In drilling down to the apical, middle, and basal turns, it is helpful to know the average depth before the turns will be encountered. The study (Clark 1975) showed the depth to the distal parts of the basal and middle turns to be 1.9 and 1.7 mm, respectively. A summary of the measurements, including the depth of the bone overlying each turn, is shown in Table 2.1.

The oval window for the footplate of the stapes lies above and behind the promontory, and the round window, a release valve for the sound pressure at the oval window, lies below and behind the promontory (Fig. 2.5). The round window, in particular, was the site for an intracochlear insertion of the electrode bundle around the basal turn of the cochlea (Clark, Pyman et al 1984). Subsequently, an opening drilled anteroinferior to the window has become the preferred site for entry and electrode insertion into the inner ear. The horizontal canal of the facial nerve passes just above the oval window. The vertical and horizontal sections lie within the field of cochlear implant surgery, and the anatomy has to be well understood and the nerve identified to prevent injury.

Posterior Tympanotomy

The surgical approach to the middle ear is to expose the round window from behind through a triangular space between the facial nerve, chorda tympani nerve, and the floor of the fossa for the short process of the incus (Fig. 2.7). It is referred to as a posterior tympanotomy. In cochlear implant surgery the facial nerve must be identified whenever possible before making the opening. The posterior tympanotomy is a V-shaped opening that has the vertical section of the facial canal along the posterior margin. The chorda tympani lies in the anterior margin, and the floor of the antrum in the roof. The facial nerve not only provides motor fibers for the face, but also is the conduit for sensory nerve fibers (chorda tympani) supplying taste to the front two thirds of the tongue. The fibers in the chorda tympani leave the vertical section of the facial nerve and pass superiorly and anteriorly, forming the anterior boundary of the posterior tympanotomy. The bundle of fibers then enters the middle ear posteriorly through a small canal to pass across the middle ear cavity just medial to the neck of the malleus before exiting through an anterior canal. These sensory fibers may be injured during the posterior tympanotomy, and lead to some disturbance of taste. If the nerve is abnormally

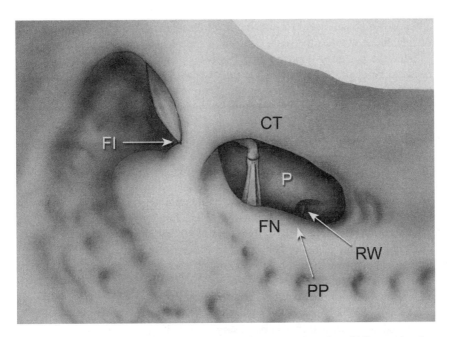

FIGURE 2.7. The posterior tympanotomy or posterior approach to the middle ear, showing the triangular-shaped opening with the vertical section of the facial nerve (FN) lying posterior, the ponticulus pyramidalis (PP), a bony ledge anteromedial to the facial nerve, the chorda tympani (CT) anterior, and the floor of the fossa for the short process of the incus (FI) superior. The round window (RW) and the basal turn of the cochlea under the promontory (P) can be seen through the tympanotomy. (Reprinted with permission from Cochlear Corporation, Englewood, CO.)

placed, and this compromises the exposure of the round window, it is in order for it to be sectioned. The posterior tympanotomy is discussed in more detail in Chapter 10.

The surgical dimensions of the facial recess were compared in 10 adult and 10 children bones to determine if there would be any limitations for the access to the round window and promontory for cochlear implant electrode insertion in young children. There were no statistical differences in the dimensions with age and hence with temporal bone growth (Bielamowicz et al 1988; Dahm et al 1992).

Round Window and Niche

The round window (cochlear fenestra), first described by the anatomist Antonio Scarpa in 1772 (Franz et al 1987), has importance for the insertion of the electrode array, and its anatomy is therefore relevant. Initially electrodes were inserted into the scala tympani through the window (Clark, Pyman et al 1984, 1987). Now it is an essential landmark for a cochleostomy that is placed approximately 1 mm anterior and just inferior to this window. The round window lies in a niche obscured to a variable degree by a bony overhang, which extends over the membrane anteriorly, and superiorly for a distance of up to 1 mm. The window sealed by its membrane overlies the scala tympani of the basal turn of the cochlea.

Once the posterior tympanotomy has been created the round window niche needs to be identified. It lies below the oval window and inferior and posterior to the promontory on the medial wall of the middle ear. The distance between the center point of the anterior rim of the oval window and the anterior and inferior sector of the round window rim is 4.1 to 4.5 mm, with a standard deviation of 0.34 mm (Dahm et al 1993). The round window niche varies in depth up to approximately 1 mm. In a 3-month fetus the round window is filled with tissue (mesenchyme). If this tissue is not absorbed during development, the outer region of the niche may be covered by a false membrane (Nomura 1984). This must not be mistaken for the true round window membrane when inserting an electrode. Failure to do so can lead to the electrode impacting on the true membrane, and damage to the electrode (Clark, Pyman et al 1979). The direction and shape of the round window can vary considerably. Proctor (1989) has found it may vary from a rectangular entrance that lies vertically to one that is square, ovoid, trapezoidal, round, semilunar, and even triangular. It usually has an anteroinferior and posterosuperior overhang.

The niche has a bony posterosuperior overhang that can obscure the view of the round window, and if it is necessary to drill it away, damage to the spiral lamina could occur. The true round window membrane is conical in shape with the apex lying superiorly where it is attached to the osseous spiral lamina. The round window membrane lies mostly in a horizontal plane, coming close to the spiral lamina posteriorly (Stewart and Belal 1981; Franz et al 1987), although the anterior portion is more vertical. The diameter of the membrane is about 1.5 mm across the base of the cone (Franz et al 1987). The mean horizontal diameter in 23 bones was 1.35 mm and the vertical diameter 1.79 mm (Clifford 1984). An anterior shelf of 0.72 mm obscured the view in some bones. The anterior and inferior margin of the round window overlies a crest (crista fenestra) (Fig. 2.8), which was found to project a mean distance of 0.2 mm (Clifford 1984). It needs

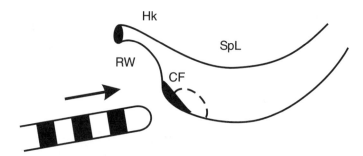

FIGURE 2.8. A drawing of the first part of the basal turn showing the bone anterior and inferior to the round window (RW) including part of the crista fenestra (CF) being drilled for good access through a cochleostomy. SpL, spiral lamina; Hk, hook region of the basal turn. (Franz et al 1987. Surgical Anatomy of the Round Window with Special Reference to Cochlear Implantation. *Laryngology and Otology,* **101(2).** Reprinted with permission.)

to be drilled away for a good exposure to allow the electrode to pass tangentially along the basal turn of the cochlea or for an adequate insertion through a cochle-ostomy especially for the precurved arrays such as the Nucleus Contour. Notice, too, the hook region of the basal turn of the cochlea that is a cul-de-sac lying posterior to the round window.

A hypotympanic air cell as discussed above (see Mastoid Air Cell System and Variations) may open immediately inferior to the round window niche. It can be readily mistaken for the niche if the round window is obscured, and particularly if the niche is obliterated. It is therefore important to visualize the round window membrane. If this is not possible, the stapes should be used as a guide, with the anterior and inferior margin of the round window being 4.1 to 4.4 mm below the anterior rim of the stapes footplate.

The structure of the round window membrane has been studied in the guinea pig by Kawabata and Paparella (1971), Richardson et al (1971), and Schachern et al (1984), and in the cat by Franz et al (1984). It consists of three layers. The external epithelial layer has mainly cuboidal cells that rest on a basement mem-brane. In places the cells are cylindrical, and ciliated cells with microvilli are occasionally seen at the periphery (Franz et al 1984). The cells are joined by tight junctions, which prevent the ingress of macomolecules such as albumin. However, with infection these tight junctions widen and toxins can enter the inner ear. The subepithelial layer consists on the outside of loose fibrous tissue, with capillaries and nerve fibers in this zone. There is an inner zone that is denser and made up of collagen fibers and fibrocytes, both oriented in the plane of the membrane. The subepithelial layer gradually changes into the endothelial layer and borders the perilymphatic space. The endothelium consists of flat cells, and due to a relatively large nucleus the cell bodies project into the perilymphatic space (Franz et al 1984).

Inner Ear

The inner ear or membranous labyrinth is divided into an anterior section for hearing (cochlea and saccule) and a posterior one for balance (utricle and semi-circular canals). The membranous labyrinth is encased in hard endochondral bone. The external case, the osseous labyrinth, consists of the cochlea, vestibule, and semicircular canals.

Osseous

The organ of hearing (the anterior part of the membranous labyrinth) lies within the osseous cochlea, and has 2½ to 2¾ turns (Fig. 2.4). The bony vestibule contains the saccule and utricle for sensing low-frequency vibrations and linear acceleration, respectively, and lies between the cochlea and the bony semicircular canals. The sense organ in the utricle responds to linear acceleration, and in the semicircular canals to angular acceleration.

In the cochlea the 2½ to 2¾ turns are wound around the central column (the modiolus). The turns overlap each other, as illustrated in Fig. 2.6, and are separated by thin bony partitions. The turns within are partially divided by an osseous spiral lamina attached centrally to the modiolus (Fig. 2.9). The partition is completed by the attachment of the basilar membrane to the spiral lamina and outer wall. The oval window (vestibular fenestra) is an opening into the vestibule between the saccule and utricle, and it receives the stapes bone for transmitting sound vibrations to the fluid in the cochlea. The round window (cochlear fenestra), opening into the scala tympani, acts as a release valve for the sound pressure transmitted through the oval window.

The osseous cochlea as a whole measures 5 mm from base to apex and its breadth across the base is about 9 mm. The apex of the cochlea is directed laterally and angled toward the upper and front part of the medial wall of the middle ear

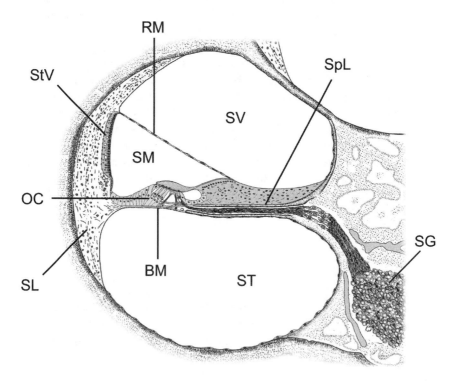

FIGURE 2.9. A cross section of the human cochlea showing the scala tympani (ST), the scala media (SM), and the scala vestibuli (SV), separated by the basilar membrane (BM) and the Reissner's membrane (RM). The basilar membrane is attached externally to the spiral ligament (SL) and internally to the spiral lamina (SpL). The stria vascularis (StV) generates the endolymphatic potential in the scala media. This powers the production of cochlear microphonics in the organ of Corti (OC), and the resulting action potentials pass along the peripheral processes in the spiral lamina to the spiral ganglion (SG) cells in the spiral canal in the modiolus.

cavity. Its base is at the bottom of the internal auditory meatus through which the cochlear nerves pass to the organ of hearing. The nerves for the first turn of the cochlea pass through a spiral of small holes, and those for the apical turn through a single central foramen. These foramina can be a route for infection to the cochlea from meningitis, and also in the reverse direction when the infection originates in the middle and inner ears. The modiolus (the central pillar of the cochlea) is broad at its base and receives the cochlear fibers. Within the modiolus there is an enlarged space, the spiral canal, that contains the spiral ganglion cells of the cochlear nerves. The peripheral processes of these ganglion cells pass out to the organ of Corti on the basilar membrane through small holes the (habenula perforata) in the spiral lamina.

The osseous vestibule (Fig. 2.4) contains receptors in the utricle for linear acceleration, and these assist in maintaining balance. The fibers of the utricular nerve, a branch of the superior division of the vestibular nerve, pass through tiny openings in the wall of the bony labyrinth known as the cribrose area as shown in Figure 2.10. This is another pathway for infection in the inner ear to pass from the vestibule to the cerebrospinal fluid (CSF) bathing the utricular and vestibular nerves. This was a well-defined route for acute inflammation due to *Streptococcus pneumoniae* and *Haemophilus influenzae* to spread to the meninges in patients who had previously had the stapes replaced with a prosthesis for conductive deafness. This was clearly shown in the histopathological findings in the temporal

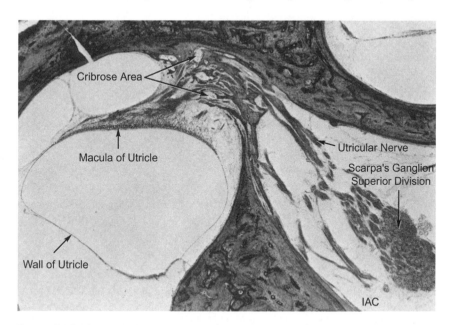

FIGURE 2.10. The vestibule and internal auditory meatus showing the fibers of the utricular nerve passing through the cribrose area to the superior vestibular ganglion. IAC, Internal auditory canal. (Schuknecht, Anatomy of the Temporal Bone with Surgical Implications. © Lea & Febiger, 1986. Reprinted with permission.)

bones of some patients who had fatal meningitis (Rutledge et al 1963; Wolff 1964; Matz et al 1968; Palva et al 1972; Benitez 1977).

There are two channels linking the inner ear to the subarachnoid space around the brain, the cochlear aqueduct (cochlear canaliculus) and the vestibular aqueduct (aqueduct of the vestibule), which could also be pathways for the spread of infection to the meninges. The most likely path is the cochlear aqueduct. This channel varies in size and passes from the scala tympani in the basal turn of the cochlea near the round window (Fig. 2.11) to the upper border of the jugular fossa, the groove for the internal jugular vein. It transmits the perilymphatic duct, a tubular prolongation of the dura mater, and contains a vein from the cochlea that joins the internal jugular vein. It is lined with epithelium and contains connective tissue with gaps that allow the transmission of fluid from the subarachnoid space to the perilymph in the scala tympani. It has been shown to be one of the pathways for the transmission of infection from the cochlea to the CSF leading to meningitis. It was examined by Palva (1970) in six infants as it was considered to be a pathway for middle ear infection leading to infection in the inner ear and then extending to the meninges. The study showed the width of the aqueduct was relatively large, being at least 150 μm at the narrowest point. It had an average length of 3.5 mm compared to the adult length of 6.2 mm. For this reason it could be a significant path especially with cochlear implants in young children. This

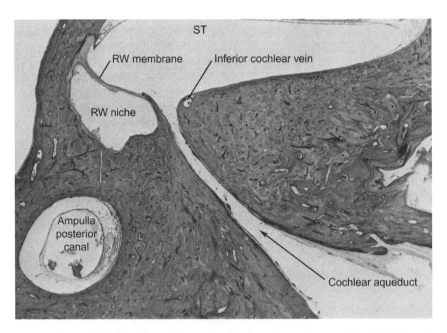

FIGURE 2.11. The cochlear aqueduct passing between the scala tympani and the internal auditory canal. Note the proximity of the inner ear opening near the round window (RW), which could account for it transmitting infection from the middle ear to the meninges. (Schuknecht, Anatomy of the Temporal Bone with Surgical Implications, © Lea & Febiger, 1986. Reprinted with permission.)

was seen in the case of an adult who had fatal meningitis 1 year after the stapes had been replaced with a strut entering into the inner ear (Rutledge et al 1963). The other canal is the vestibular aqueduct that passes from the medial wall of the vestibule to open as a slit just behind the internal auditory meatus in the posterior cranial fossa. It transmits the endolymphatic sac and duct as well as a small artery and vein.

Membranous

The membranous labyrinth is an enclosed system filled with endolymph containing the sense organs of hearing and balance and lying within the osseous labyrinth. In the cochlea the sensory organ of hearing (organ of Corti) rests on the fibrous basilar membrane, and is in a compartment enclosed by a thin membrane (Reissner's membrane) (Fig. 2.9). The cochlea has three compartments, and two of these, an upper (anterior) and lower (posterior) one, the scala vestibuli and scala tympani, respectively, are filled with perilymph and communicate with each other through the helicotrema at the apex of the cochlea. This system is the perilymphatic labyrinth (Schuknecht and Gulya 1986). The scala vestibuli is in continuity with the vestibule behind, and the scala tympani is a tube closed by the round window. If infection in the middle ear is transmitted around the cochlear electrode or between the electrode array and a second component through to the scala tympani, it may pass to the spiral canal and thence the meninges via small openings in the medial wall of the modiolus referred to as canaliculae perforantes. On the other hand, if the inserted electrode creates a tear through to the scala vestibuli, there is a potential pathway for infection to pass to the vestibule. It has been shown in three cases of fatal meningitis after the replacement of the stapes with a prosthesis that the resulting inner ear infection can be readily transmitted though the cribrose area and around the utricular nerve (Fig. 2.10).

The scala media, which contains the organ of Corti, lies between the scala vestibuli and the tympani. The scala media is part of the membranous cochlea, which is connected via a duct to the saccule in the anterior part of the vestibule. The saccule contains receptors for low-frequency vibration or linear acceleration. It communicates with the posterior-placed utricle that also has receptors for linear acceleration. The communication is through a Y-shaped tube. The apex of this tube becomes the endolymphatic duct ending in a blind pouch or sac under the dura mater or outer lining of the brain in the posterior cranial fossa. At the opening of the vestibular aqueduct the endolymphatic duct is sheathed by a short extension of the perilymphatic labyrinth. This is not so likely to be a path to the CSF as with the cochlear aqueduct. The semicircular canals are connected to the utricle and contain ducts that allow fluid to move in response to angular acceleration. This movement causes deflection in sensory receptors (cristae) in expanded areas of the three semicircular canals referred to as ampulae.

Knowledge of the anatomy of the basal turn of the cochlea is very important, as this turn is the site for the insertion of the multiple-electrode array. The scala tympani is narrowed just inside the round window membrane by an anteroinferior ridge, the crista fenestra referred to above (Franz et al 1987). This may need to be drilled away to facilitate the insertion of the electrode through the round window or to complete the cochleostomy. Beyond the crista the scala is fairly straight in an anteromedial direction for about 6 mm from the round window before starting to curve superiorly. There is a significant change of direction at 9 to 10 mm. This means that the spiral lamina forms the superior wall of the scala in the first 6 mm, but at 9 mm it forms more the posterior wall. The scala tympani rotates around the modiolus to form its spiral shape. This was studied when Silastic casts were made and sectioned (Fig. 2.12). The radius of curvature is about 4 mm (from the round window to a point 18 mm along the cochlea), and then 2.5 mm (from 18 to 25 mm). The cross section of the scala tympani changes in shape and is an approximate square with 0.8 mm sides near the round window, and a triangle with 0.4 mm sides 25 mm along the cochlea.

It is important to account for the rotation and change in shape when developing electrode arrays that are molded to the dimensions of the cochlea or free-fitting ones that have electrodes lying close to the spiral ganglion cells. The dimensions of human cochleae have been compared with those of experimental animals such as the cat and monkey (Igarashi et al 1968, 1976; Hatsushika et al 1990) for interpreting animal experiments used in the design of electrode arrays and the

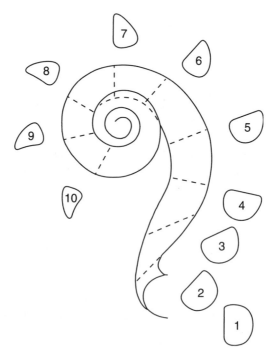

FIGURE 2.12. Diagram illustrating rotation of the cross section of the scala tympani. The crista fenestrae restricts access to the medial portion of the basal turn. (Reprinted from Shepherd et al 1990, with permission from Elsevier.).

prevention of trauma. The results showed that in the cat the height and width of the scala tympani became markedly reduced about 6 mm from the round window (Hatsushika et al 1990) (Fig. 2.13). In the human the height and width remained fairly constant with distance. This was also the case for the monkey (Igarashi et al 1968), indicating that it was a better model than the cat for deep electrode insertion studies. The data also indicated that in the human there was a need for only a small taper in the array. As the width was greater than the height, an array could be designed to move from a peripheral position to lie close to the modiolus (perimodiolar). On the other hand, the cross-sectional area became progressively smaller with distance and this was accompanied by a change in shape. The implication of these anatomical features on the design of advanced arrays is discussed in Chapter 8.

Histology of the Cochlea

Organ of Corti

The organ of Corti has inner and outer hair cells that respond to sound vibrations and convert them into patterns of neural discharge for coding by the brain. The hair cells have supporting cells beside and underneath them. The inner and outer hair cells are separated by two pillar cells, and these come together at their apices to form the tunnel of Corti. There are three rows of outer hair cells and one row of inner hair cells (Fig. 2.14). The outer cells contract and put energy into the vibration of the basilar membrane, while the inner cells convert mechanical vibrations into patterns of electrical activity in the cochlear nerve. The hair cells have cilia (hairs) protruding from the surface. Those from the outer cells are embedded in the gelatinous tectorial membrane. The shearing force between each hair cell and the tectorial membrane generates electrical activity that contributes to the processing of sound, as discussed in Chapter 5. The shearing force depends on the basilar membrane vibration, and the maximum vibration varies according to the frequency. The frequency maxima also vary with distance around the cochlear turns. A relationship between the hearing threshold for frequency and the site of the hair cell loss was found by Schuknecht (1953). A log-linear relation between the frequency maxima and the position along the basilar membrane was recorded by von Békésy (1960) using a stroboscope. This relationship was confirmed by Greenwood (1961), Bredberg (1968), and others. It is also discussed in Chapter 6.

Innervation

The peripheral processes (dendrites) of the spiral ganglion cells (afferent cochlear neurons) synapse on the base of the inner and outer hair cells, and take patterns of electrical activity to the ganglion cells and the higher central auditory pathways. The neurons from the outer hair cells cross through the tunnel of Corti (between the inner and outer pillar cells), and enter openings in the spiral lamina (habenula perforata) to reach the ganglion cells. In the tunnel of Corti they are bathed by

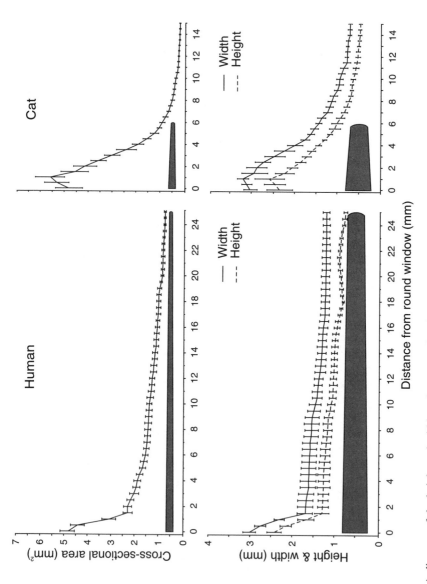

FIGURE 2.13. A diagram of the height and width and cross-sectional area of the cat and human cochleae (Hatsushika et al, Dimensions of the scala tympani in the human and cat with reference to cochlear implants. *Annals of Otology, Rhinology and Laryngology,* **99(11)**, 1990. Reprinted with permission.).

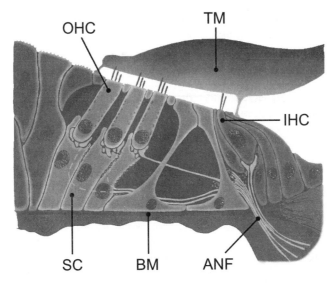

FIGURE 2.14. A cross section of the organ of Corti. BM, basilar membrane; IHC, inner hair cell; OHC, outer hair cell; TM, tectorial membrane; ANF, auditory nerve fiber; SC, supporting cell. (Reprinted from *Communicative Disorders: Hearing and Loss,* edited by J. L. Northern. Little, Brown & Co., Boston, 1976. With permission from Lippincott Williams & Wilkins.)

fluid with the same electrolyte composition as perilymph, making it comparable to the extracellular environment of other nerves. The neurons from the inner hair cells carry the information for the coding of sound, and together with those from the outer hair cells become thinner as they pass through the habenula and then receive a myelin sheath in the spiral canal of the modiolus before joining the spiral ganglion cells.

The cat innervation has been studied by Spoendlin (1975), who found that about 5% to 10% of the auditory nerve fibers synapse with the outer hair cells, and the fibers spiral about 0.6 mm toward the base. Each fiber thus diverges to supply a number of hair cells. On the other hand, 90% to 95% of the nerve fibers synapse on the inner hair cells. They converge so that a number of fibers connect with each cell. More details of the function of the cochlear innervation are outlined in Chapter 5.

The spiral ganglion lies in the modiolus beneath the internal wall of the scala tympani. Its position in relation to the spiral lamina varies with animal species. In the cat the ganglion is placed high in the scala tympani, but in the human it lies nearer the floor (Fig. 2.9). These variations need to be taken into consideration, for example, when applying cat data to the human on the spread of electrical current for different modes of stimulation. The peripheral processes pass to the nearby ganglion cells, and then their central processes (axons) become the cochlear (auditory) nerve. The axons from the lower frequencies in the more apical regions on their way to the base of the cochlea pass internal to the spiral ganglion cells, receiving peripheral processes from higher frequencies. Thus current from an electrode in the scala tympani can spread beyond the spiral ganglion cells opposite the electrode to stimulate fibers responsive to lower frequencies. This was found to be the case in psychophysical studies on implant patients (Blamey et al 1995), where the pitch percept was matched to those obtained for sound when there was some residual hearing in the opposite, nonimplanted cochlea. The pitch for electrical stimulation of the basal turn electrodes was lower than expected from the frequency-to-place distribution along the basilar membrane discussed above. Furthermore, although the basal membrane makes 2½ to 2¾ turns, the spiral ganglion only completes 1½ turns. Thus the peripheral processes from the most apical region pass to ganglion cells lying more basalward. As the majority of patients requiring cochlear implantation have lost their peripheral processes, it is not necessary to pass an electrode array beyond this region to excite the low-frequency cells.

The axons from the spiral ganglion cells enter the cochlear nerve where they spiral in the same direction as the cochlea. Lorente de Nó (1933) discovered by tracing stained fibers in the mouse that the number of turns was the same as that of the ganglion. Sando (1965) found in the cat, by making localized lesions in the cochlea and serially sectioning the nerve, that the apical fibers from the spiral ganglion made 1¾ turns around the axis of the trunk. There was progressively less twisting for units located basalward, so that the extreme basal turn fibers entered the auditory nerve trunk laterally and inferiorly, and maintained this inferior position to the cochlear nuclei. Another study in the cat by Arnesen et al (1978) showed results similar to those of Sando, but reported that the fibers from the apical turn were lying more centrally. The spiraling of cochlear nerve fibers is relevant surgically to the development of an intraneural array (Fig. 2.15).

Embryology

The inner ear commences to develop at approximately 3 to 4 weeks of intrauterine life. It arises from a thickened plate of surface ectoderm (the otic placode) that appears on each side of the developing hindbrain. This placode then sinks below the surface and the edges come together to form the otocyst. It becomes bilobed as the pars superior forms the vestibular system and the pars inferior forms the cochlea and saccule. Next the pars superior becomes molded into the three semicircular canals, and then the utricle and saccule form. At approximately 8 weeks the pars inferior starts to coil to form the cochlea. At 9 weeks the vestibular sense organs develop and at 10 weeks the organ of Corti.

As the auditory and vestibular labyrinths develop at an early stage of intrauterine life, if genetic control is defective or any noxious agent is present during this time, the deformity will be severe. Occasionally there is complete absence of the labyrinth (Michel's syndrome). If development is arrested at a later stage, the cochlea may only develop one turn or a common cavity (the Mondini syndrome). Other dysplasias are likely as illustrated by the cochlea described by Schuknecht and Gulya (1986), in which there was a deficiency of the modiolus and a marked confluence between the scala vestibuli and the CSF in the internal auditory canal (Fig. 2.16). This could cause a "perilymph gusher" at surgery, and

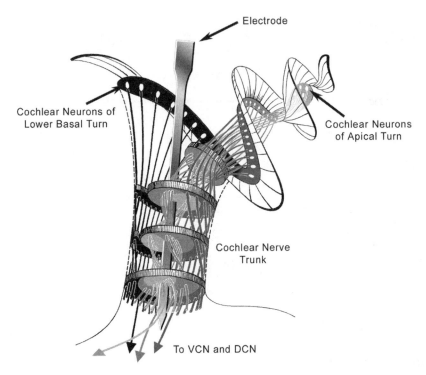

FIGURE 2.15. A diagram of the spiraling of the auditory nerve fibers from the cat cochlea, indicating how an intraneural electrode array might be used to provide place pitch. VCN ventral cochlear nucleus, DCN, dorsal cochlear nucleus. (From the data of Sando 1965.)

could be the conduit for infection from the middle and inner ears to the meninges. The Mondini dysplasia has been successfully implanted as the cochlear nerve is present. In addition, the osseous cochlea may develop completely, but the organ of Corti is absent (Scheibe's syndrome). This condition would be suitable for cochlear implantation.

Central Auditory System

Overview

The coding for frequency, intensity, and sound localization at a cellular level in the central auditory system are discussed in Chapter 5. However, the coding is ultimately carried out by the whole system of interacting central nervous centers and pathways. Specific neural patterns are selected and coded from the information arriving at the low centers. These patterns are refined and the coding becomes more complex from the brainstem to the auditory cortex. The structural network in the central neural pathways, required for this coding, is discussed below. The structure of the central auditory nervous system underlies the coding

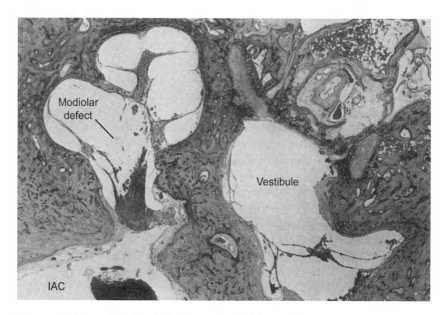

FIGURE 2.16. A cochlea in which there was a deficiency of the modiolus and a marked confluence between the scala vestibuli and the internal auditory canal (Schuknecht, Anatomy of the Temporal Bone with Surgical Implications. © Lea & Febiger, 1986. Reprinted with permission.).

of simple and complex sounds. In the human in particular the complex sounds can be categorized as speech signals.

There are seven main relay stations in the central auditory pathways for processing the information and interaction with other modalities (Fig. 2.17). The relay stations, from below up, are the cochlear nucleus (CN) in the medulla oblongata and the superior olivary complex (SOC) in the pons (both parts of the rhombencephalon); the nucleus of the lateral lemniscus (LL), the inferior colliculus (IC), and the superior colliculi (SC) in the midbrain (mesencephalon); the medial geniculate body (MGB) in the thalamus (part of the diencephalon or interbrain); and finally the information is processed in the auditory cortex (AC) (telencephalon).

The CN has a ventral (VCN) and dorsal (DCN) nucleus. The VCN is divided into the anteroventral (AVCN) and posteroventral nuclei (PVCN). The VCN is interconnected with the DCN through feedback circuitry. The SOC is a collection of nuclei, the main ones being the medial nucleus of the trapezoid body (MNTB), the medial superior olive (MSO), and the lateral superior olive (LSO). The lateral lemniscus has ventral (VNLL), intermediate (INLL), and dorsal (DNLL) nuclei. The IC has the central (ICC) and external nuclei (Cajal 1911). The MGB has two main divisions, the dorsolateral (principal nucleus) and the medial division (magnocellular). The AC consists of the primary cortex, Heschl's gyrus in the human, and it lies in the anterior temporal gyrus in the suprasylvian sulcus. The primary cortex is connected to the auditory association areas for the processing of speech and other tasks as well as to the AC on the other side.

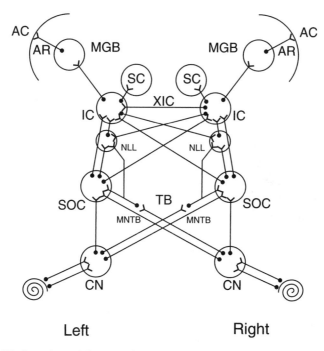

FIGURE 2.17. Overview of the central pathways. CN, cochlear nucleus; MNTB, medial nucleus of the trapezoid body; TB, trapezoid body; SOC, superior olivary complex; NLL, nucleus of the lateral lemniscus; IC, inferior colliculus; XIC, the commissure of the inferior colliculus; SC, superior colliculus; MGB, medial geniculate body; AR, auditory radiation; AC, auditory cortex.

The temporal and spatial patterns of excitation in the auditory nerve (AN) are received at the VCN situated in the brainstem. The information from both CNs passes either up the same side (ipsilateral) or the opposite side (contralateral) to reach the AC through the above centers. A greater proportion of fibers cross to the contralateral side. The CN is connected to the nuclei of the SOC on the opposite side through the trapezoid body (TB). In the SOC and other higher centers (LL, IC, MGB, AC) where there is bilateral innervation the sound processing for encoding directional hearing and hearing in noise can occur. From the SOC fibers pass along the tracts of the LL and some synapses occur in the nuclei of the LL where further crossing over takes place. The LL brings the afferent input to the IC. From the IC the fibers pass to both the superior colliculi SC and the MGB in the thalamus, and the IC on the other side. In the SC auditory information on the position of sound in space is integrated with the visual field. The MGB in particular receives input from the visual and tactile senses, as well as from the AC. Its fibers then pass directly to the primary AC. The primary cortex has connections with association areas that are for interrelating information from audition and vision in speech perception and production, as well as the higher levels of understanding. At the AC information is transferred from one cerebral hemisphere to the other via the corpus callosum, a large band of fibers connecting both halves of the cerebrum. This allows the transfer of information from the nondominant to the dominant AC for speech understanding. Speech is processed in the dominant hemisphere, and music in the nondominant one.

On the sensory side the auditory neural pathways are organized tonotopically from the CN right through to the AC. That means that they are spatially oriented so that the frequencies of best response of the units are arranged so that a frequency scale is preserved. Thus the filtering of frequencies by the cochlea is preserved right through the system enabling the place coding of frequency to take place. There is plasticity in the organization of these connections occurring at birth through exposure to sound. Furthermore, at the higher levels, in particular, there are interconnections with the other sensory modalities and systems. There is also auditory input to the cerebellum where the auditory tactile and other sensory modalities interact. At all levels of the auditory system there are inputs from both ears. This allows both the localization of sound as well as hearing in noise. Thus we know not only what the sound is but also where the sound is coming from.

In addition to the afferent or sensory system there is an efferent neural system that commences at the cortex and passes down to the CN and the cochlea. The efferent pathways provide feedback control of the information passing up the auditory system. From the SOC the ipsilateral and crossed olivary cochlear bundles pass to the CN and then become the bundle of Oort in the AN. They synapse on the base of the inner hair cells and are thought to improve the signal-to-noise ratio. This system is presumed to be especially important for attention and controlling the flow of information centrally.

For optimal perception of speech and environmental sounds, electrical stimulation of the central auditory nervous system with a cochlear implant requires a stimulus system and electrode array in the inner ear to reproduce the information normally presented to these pathways. For this reason the structure of each major center is discussed in some detail as their structure underlies the coding process.

Auditory Nerve

The AN fibers have uniform responses to sound and transmit information from different frequency regions of the cochlea. They fire in phase with the sound frequency. The auditory nerve transmits basic information to the brainstem on the site of basilar membrane excitation, as well as the temporal information for frequency. Intensity is transmitted through either an increase in rate of firing or a greater population of nerve fibers excited.

Some of the studies to reproduce the coding of sound with electrical stimulation were undertaken by recording from the AN, as the responses were uniform (van den Honert and Stypulkowski 1984; Javel et al 1987). On the other hand, as discussed in Chapter 5, initially recordings to electrical stimulation were made by Clark (1969a, c), Clark, Kranz et al (1972), and Clark, Nathar et al (1972) from the SOC, as this would better show the influence of central processing mechanisms such as inhibition on electrical stimulation.

The fibers in the AN spiral in the same direction as the turns of the cochlea. Of relevance to intraneural implantation of the AN in the human is that the nerve has an area of approximately 1.74 mm^2 (Clark, Shepherd et al 1988) that is large enough to insert an array along its length. The normal population of axons in the nerve varies from 35,000 to 40,000 (Rasmussen 1960). The afferent fiber diameters varied between 1 μm and 9 μm (Engstrom and Wersall 1958). On the other hand, in another study the mean diameters of mammalian auditory nerve fibers were approximately 2 to 4 μm (Spoendlin and Schrott 1989; Gleich and Wilson 1993). Even smaller diameters (0.1–0.7 μm) were measured at the nodes of Ranvier in the myelinated portion of the peripheral process proximal to the habenula perforata. This distribution of fiber diameters is not related to their site of origin in the cochlea. The diameters would affect their stimulus threshold and speed of conduction. The larger the diameter, the lower the threshold and the higher the rate of conduction. This variation would need to be taken into consideration in developing speech-processing strategies with an intraneural array.

Cochlear Nucleus

The CN is the first relay center in the central auditory system. As stated it is subdivided into a VCN and DCN. The VCN receives the fibers from the AN. These divide into anterior and posterior divisions, and they synapse with the neurons in the AVCN and PVCN so that tonotopic organization is preserved. Electrophysiological (Rose et al 1959; Evans and Nelson 1973; Goldberg and Brownell 1973) and mapping studies (Webster and Webster 1978; Ryan et al 1982) have shown that each frequency region of the cochlea is represented by layers of cells from high to low in a dorsal to ventral direction. This is in accord with the place theory of frequency coding. Fibers pass to the interstitial nucleus of Lorente de No. A small proportion of the fibers from the AN also pass to the DCN, but there is less evidence that they are tonotopically organized especially in the primate.

In the CN the processing of information is more complex than in the AN. This is carried out by different cell types distinguished by the shape of their bodies (somas), dendritic trees, connections, and membrane biophysical properties. These variations give them a variety of functions in the coding of sound. Brawer et al (1974) further subdivided the AVCN into subdivisions on the distribution of the cell types. Essentially, Golgi stains showed two distinct and large neurons in the AVCN—bushy cells and stellate cells. The classification of the cells, however, depends to a certain extent on the staining used, and the bushy and stellate cells correspond to the spherical and ovoid cells described by Cant and Morest (1979) from Nissl stains. The bushy cells commonly have a small number (one to three) of short stubby dendritic trunks that branch repeatedly (Cant and Morest 1979). Tufts of fine dendrites emerge from these trunks to give them the bushy appearance. In the immature animal the bushy cells usually only have one dendrite and some have none at all (Brawer et al 1974). This is evidence for dendritic sprouting during development. Compared to bushy cells the stellate cells have a larger number of long tapering and less branching dendrites.

The distribution of cell types varies within the subnuclei of the cochlear nucleus. The anterior part of the AVCN contains almost exclusively large bushy cells that correspond to Osen's (1969a, b) large spherical cells. In the posterior part of the anterior division, bushy and stellate cells intermingle in equal numbers. Electron microscopic studies have shown that Osen's globular cells correspond to bushy cells as well, hereafter referred to as globular bushy cells, and the multipolar cells correspond to the stellate cells.

The PVCN has stellate cells widely distributed through it. This nucleus also has distinctive "octopus" cells in the central part (Osen 1969a, b). They have four to six thick straight dendrites on one side of the cell, and these divide into secondary dendrites (Osen 1969a; Kane 1973). The octopus cells are also finely tuned to provide phase information, and could encode voicing.

The DCN has up to three layers of cells of variable types, the function of which is not clear. In the nonprimates the cells are tonotopically organized, but this is not the case in the human, where there is no layering of the cells. This raises doubts about the value of attempting to achieve place pitch discrimination with a brainstem implant that is normally placed over the dorsal nucleus. The most distinctive type is fusiform in shape with dendrites projecting into the peripheral and deep layers. The fusiform cells and their processes thus span the three layers of the DCN.

The large spherical or bushy cells in the VCN receive synapses called end bulbs of Held from AN fibers. These synapses resemble the calyces of Held, solitary endings surrounding a large proportion of each principal cell in the MNTB. The synapses on large spherical cells provide a secure transmission of timing information and thus have outputs that are very similar to the inputs from the AN fibers. The responses are described as primary-like. The primary-like response is one that commences with the onset of the tone burst, after which it continues for the duration of the stimulus with only a small decrement. The cells also fire in phase with the stimulus. These neurons provide low- to middle-frequency temporal information in particular to the MSO, and are thus thought to be important for the temporal coding of sound localization.

The other types of endings are smaller boutons. These also terminate on both the spherical and globular bushy cells, as well as the stellate cells. There is a predominance of type I boutons with spherical vesicles (considered to be excitatory) over type II boutons with flattened vesicles (considered to be inhibitory).

The end bulbs of Held, which contain large spherical vesicles, degenerate after cochlear ablation, but the other endings with smaller or flattened vesicles terminating on the bushy or stellate cells survive. This might explain why in deafness reproducing temporal information with electrical stimulation can be difficult especially the interaural timing differences that are important for sound localization and hearing speech in noise (van Hoesel et al 1990; van Hoesel and Clark 1995).

The stellate cells (see Chapter 5) may be responsible for the place coding of frequency. They fire in a regular periodic fashion while the tone is on ("chopper cells"), are finely tuned, and respond to excitation from a localized region of the cochlea. The spectral information and spectral tuning by these cells is transmitted via their axons through the VNLL to IC for further place encoding.

The CN neurons projecting centrally to the opposite or contralateral side send their axons across the midline in three main bands of fibers (striae) to the higher order nuclei. These bands are the trapezoid body (TB) as well as the overlying bands, the intermediate and dorsal acoustic stria. The connections of the CN have been studied using horseradish peroxidase and other methods of intravital labeling that allow dye to traverse the neurons. The projections from the three major CN divisions pass to the SOC, to the VNLL and the IC, and then to the higher centers. The AVCN projects via the ventral and dorsal components of the TB to the main nuclei of the SOC, whereas the PVCN projects through the caudal portion of the TB and the intermediate acoustic stria to the periolivary nuclei of the SOC. The afferent nerve fibers from the giant cells of the DCN project via the dorsal acoustic stria primarily to the IC.

The large spherical bushy cells in the AVCN project to the contralateral MSO, and transmit information on the low to middle frequencies. As the synapses on these cells are the end bulbs of Held, transmission of temporal information is secure and able to provide accurate timing information for the coding of sound localization by the MSO. The globular bushy cells project to the MNTB on the contralateral side and then to the LSO. They also project to the LSO on the ipsilateral side. The input from low to high frequencies from both ears provides differences in intensity to the LSO for coding sound localization.

The stellate cells have axons that essentially pass to the IC, but send collaterals to the SOC and VNLL. As the electrophysiological research of Paolini and Clark (1999) suggests that the stellate cells are important in the place coding of frequency (and hence the extraction of spectral information in speech), this information is passed to the IC for further integration.

The functions of the cell types in the cochlear nucleus in processing speech can be used in future studies to develop and evaluate speech-processing strategies for cochlear implants.

Superior Olivary Complex

The SOC is a collection of nuclei, consisting primarily of the MNTB, the MSO, and the LSO. There are marked differences between species in the relative sizes of these major nuclei that may relate to their functional requirements. In the mouse, dolphin, and some echo-locating bats, the MNTB/LSO is relatively large while the MSO is small (Irving and Harrison 1967; Harrison and Feldman 1970). On the other hand, in a group of primates including humans, the MSO is large and the MNTB/LSO is small. There have been a number of hypotheses to explain the differences, from the MSO being important for visual orientating reflexes, to a relation between the LSO and high-frequency hearing. It could be that the

MNTB/LSO is required for processing the high frequencies used in localization by the dolphin and bat.

The MSO is a flat plate of cells. Each cell has a single dendrite on the contralateral and ipsilateral sides. This allow the inputs from both ears to be integrated by the cell that then sends information rostrally to higher centers along the LL. The cells of the MSO are uniformly aligned and act as a dipole so that if an electrode is passed through this plate of cells there is a reversal of potential across the dipole. An electrical dipole consists of two charges of equal magnitude separated by some distance. The phenomenon is best explained by sources and sinks of current, and because the cells are uniformly aligned from cellular events and in particular excitatory postsynaptic potentials (EPSPs) (Clark 1969b).

The cells receive terminals with round (excitatory) and flat (inhibitory) vesicles (Clark 1969d, e). The anatomical arrangement of the MSO cells, their connections, and terminals allows the fine integration of excitation and inhibition from either side that is the basis for the coding of sound localization. The coding of interaural phase differences occurs at low to mid-frequencies, which are a substantial part of the frequency input to the MSO. The coding is undertaken in cells with an input from one ear providing excitation and the other ear inhibition. There is a relationship between the firing rate and the position of the sound in space (azimuth). The localization of an object in space is further coded at higher centers (IC, SC, and AC).

The principal cells of the MNTB have compact tufted dendrites. They are connected to the globular bushy cells in the AVCN by large-diameter fibers that make synaptic contact through a calyx of Held. It is a cup-shaped expansion of the axon that extends over the surface of the cell body in a number of thick, finger-like projections, and is probably the largest synaptic terminal in the mammalian brain. It is thought this synapse is important for the transmission of interaural timing and intensity differences. It is a secure synapse able to transmit temporal information accurately without degradation. Its response to electrical stimulation is of importance for the design of future cochlear implants. The principal cells project predominantly to the LSO and the VNLL. Some also pass to the MSO and periolivary nuclei. The MNTB is important in maintaining the timing for high-frequency information from the contralateral side.

The LSO is S shaped and has a central band of fusiform cells with emerging dendrites from each pole of the cell perpendicular to the curvature of the nucleus (Cajal 1909). As with the MSO these dendrites form a plexus on either side of the cells to receive the terminal axons from the ipsilateral CN on one side and the contralateral CN on the other side. The ipsilateral projections are direct from the globular bushy cells in the AVCN, and the contralateral ones from the AVCN via the MNTB. Cant (1984) has described three types of terminals with small, flat, and round vesicles. The LSO integrates information from the globular bushy cells of importance for the coding of interaural intensity differences. It receives an input from the higher as well as the lower frequencies. Because of the head shadow, intensity differences at the high frequencies are more important for sound localization. The response from cells was usually maximal for interaural intensity differences when the stimulus of the contralateral excitatory ear was more intense, and suppressed when the sound from the ipsilateral inhibitory ear was more intense (Goldberg et al 1963; Tsuchitani and Boudreau 1966, 1967, 1969).

Lateral Lemniscus

The LL is the tract of fibers from the VCN and the SOC to the IC; it also has ventral (VNLL), intermediate (INLL), and dorsal (DNLL) nuclei. The LL receives information from the VCN and the SOC and its output goes to the IC. The major input to the DNLL is from the ipsilateral MSO and ipsilateral and contralateral LSO. In marked contrast, the major source of afferents to the VNLL are the contralateral VCN. The pattern of afferent connections to the INLL is intermediate between those to the DNLL and the VNLL. The afferent innervation of the VNLL is largely from fibers that provide collaterals on their way to the IC. This may provide spectral tuning of information being transmitted to the IC for further place coding. However, in the ventral zone of the VNLL large-diameter axons terminate with calyceal endings that appear not to be derived from collaterals.

There are marked differences between the ascending projections from the DNLL and the VNLL. The DNLL has strong projections to the ipsilateral and contralateral IC and the contralateral DNLL. In contrast, the VNLL projects almost only to the ipsilateral IC and DNLL.

The nuclei of the lateral lemniscus are tonotopically organized (Aitkin et al 1970), and the cells encode bilateral interaural time and intensity differences or transmit information that originates in the SOC (Brugge et al 1970). The coding of bilateral interaural time and intensity differences applies in particular to the DNLL as its input is derived from the SOC (Aitkin et al 1970). Conversely, the majority of the VNLL units were monaural, consistent with the predominant input from the contralateral VCN units. It is thought the major innervation to the VNLL from collaterals from the VCN en route to the IC have an inhibitory effect and suppress noise and improve the signal-to-noise ratio.

In effect the LL, as with other nuclei, produces a partitioning of information. Binaural interaction at the SOC as stated is transmitted along the dorsal component (DNLL). In contrast, temporal and spectral frequency information passes from the CN to the IC with collaterals to the ventral division (VNLL).

Inferior Colliculus

The IC is the site for the termination of most of the fibers from the LL (Morest 1965), and the site of origin of the axons that ascend via the brachium of the IC to the MGB and also to the SC. All divisions project bilaterally to the MGB. The IC has a central (ICC) and external nuclei. The ICC is divided into a smaller dorsomedial division with large cells that receive efferent fibers from the AC, and a larger ventrolateral division receiving afferent sensory input from below. The ventrolateral division has a marked laminar arrangement of dendrites and axons (Morest 1964; Rockel and Jones 1973), required for tonotopic organization (Merzenich and Reid 1974). It receives tonotopically arranged afferent fibers from the LL for the place coding of frequency. Rose et al (1963) found in the ventrolateral division that the units' frequencies of best response were ordered from high to low from a dorsal, caudal, and lateral to a ventral, oral, and medial direction. The cells in the IC also phase-lock to the low to middle sound frequencies up to 5 kHz (Rose et al 1967), indicating its involvement in temporal processing of frequency. The units in the ICC respond to the fine time and intensity differences required for the coding of sound localization (Rose et al 1966).

The importance of the IC in coding frequency was demonstrated by Goldberg and Neff (1961) in behavioral experiments. The frequency discrimination was the same before and after sectioning the brachia of the inferior colliculi. Thus the animal was able to process frequency information without involvement of the higher centers. This study underpinned the importance of acute experiments on the brainstem to compare acoustic and electrical stimulation (Clark 1969c, 1970; Clark and Dunlop 1969).

The smaller dorsomedial division of the IC receives fibers not only from the ventrolateral division but also from the auditory cortex. This suggests cortical control of the input with facilitation or suppression as required for attention to a stimulus. In addition, the external nucleus has an extensive input from the somatosensory system (Aitkin et al 1978) for the interaction of audition with other senses. Descending projections through the pons supply the cerebellum.

The ICC is the predominant division that projects both to the SC and down to the brainstem. Because the IC is central to all of the streams of auditory activity, it could be thought of as a junction center.

Superior Colliculus

The neurons in the SC receive visual information in the peripheral division from the optic nerve for transmission to the lateral geniculate body and the visual cortex. The cells in the deeper division receive afferents from the auditory, visual, and somatosensory systems. The auditory neurons are sensitive to spatial localization (King and Palmer 1983). The multisensory input may thus play an important role in the movement of the eyes in relation to the direction of the sound.

Medial Geniculate Body

The MGB in the cat has two major divisions. These are a dorsolateral or principal division with medium-sized closely packed cells, and a medial or magnocellular division with larger, loosely packed cells (Rose and Woolsey 1949). The principal division is further subdivided into a dorsal and ventral subdivision. In the dorsal subdivision the neurons have radiating dendrites with larger dendritic fields than in the ventral subdivision. Recordings from the neurons in the dorsal subdivision show broadly tuned frequency responses (Aitkin and Prain 1974), and this is consistent with them having radiating dendrites with larger dendritic fields. In the ventral subdivision the cells have tufted dendrites arranged in laminae. The units in the ventral subdivision are sharply tuned (Aitkin and Webster 1971), as would be expected from the fact that they have tufted dendrites arranged in laminae. A number of studies have shown the principal division receives afferents from the midbrain and project to the AC (Morest 1965). They also exhibit binaural interaction, and this is supported by the labeling of similar frequency regions in both ipsilateral and contralateral ICs from retrograde labeling with horseradish peroxidase (HRP) injections in the MGB (Aitkin et al 1981).

Auditory Cortex

The AC in the human has a first acoustic area AI (primary cortex) that forms the anterior transverse temporal gyrus (Heschl's gyrus), and lies within the lateral sulcus. It also extends to a limited extent into the superior temporal gyrus. The second acoustic area AII lies below AI in the superior temporal gyrus. AII is flanked by the anterior and posterior ectosylvian areas (Ea and Ep). A similar organization holds for all primates, while in other mammals the AI and AII areas are located externally in the ectosylvian gyrus, the equivalent of the superior temporal gyrus.

There are two types of cells in the cortex: small stellate granular cells and both large and small pyramidal cells. They are arranged in six layers as with the sensory cortex as a whole. The AI and AII areas receive afferent fibers from the MGB through acoustic radiation, and also from the cortex on the opposite side via the corpus callosum. The major input to AI is from the ventral subdivision of the principal nucleus of the MGB (with the shortest latencies being in layers III to IV), and the anterior cortical field from the nucleus magnocellularis of the MGB. The AI area also receives afferent connections from AII and Ep areas as well as the hypothalamus and tegmentum. There is monosynaptic excitation from AII and Ep. Corticofugal fibers pass down to the divisions of the MGB, and in the ventral division they are highly tonotopic.

The AI region of the primary AC was shown to be tonotopically organized in the macaque monkey (Brugge and Merzenich 1973; Woolsey and Walzl 1982) as well as in other species. In the monkey low characteristic frequency units are rostrolateral and high-frequency units are caudomedial. In the human isotope studies of regional blood flow, positron emission tomography (Lauter et al 1985) and measurements of the magnetic field with the superconducting quantum interference device (SQUID) have shown tonotopic organization with the same layout as in the monkey.

Aitkin (1990) provides a comprehensive review of the auditory cortex.

References

Aitkin, L. M. 1990. The auditory cortex: structural and functional bases of auditory perception. London, Chapman and Hall.

Aitkin, L. M., D. J. Anderson and J. F. Brugge. 1970. Tonotopic organization and discharge characteristics of single neurons in nuclei of the lateral lemniscus of the cat. Journal of Neurophysiology 33: 421–440.

Aitkin, L. M., M. B. Calford, C. E. Kenyon and W. R. Webster. 1981. Some facets of the organization of the principal division of the cat medial geniculate body. In: Syka, J. and L. M. Aitkin, eds. Neural mechanisms of hearing. New York, Plenum Press: 163–181.

Aitkin, L. M., H. Dickhaus, W. Schult and M. Zimmermann. 1978. External nucleus of inferior colliculus: auditory and spinal somatosensory afferents and their interactions. Journal of Neurophysiology 41: 837–847.

Aitkin, L. M. and S. M. Prain. 1974. Medial geniculate body: unit responses in the awake cat. Journal of Neurophysiology 37: 512–521.

Aitkin, L. M. and W. R. Webster. 1971. Tonotopic organization in the medial geniculate body of the cat. Brain Research 26: 402–405.

Arnesen, A. R., K. K. Osen and E. Mugnaini. 1978. Temporal and spatial sequence of anterograde degeneration in the cochlear nerve fibers of the cat. A light microscopic study. Journal of Comparative Neurology 178: 679–696.

Banfai, P., G. Hortmann and S. Kubik. 1985. Extracochlear eight-channel electrode system. Journal of Laryngology and Otology 99(6): 549–553.

Banfai, P., S. Kubik and G. Hortmann. 1984. Our extra-scalar operating method of cochlear implantation. Experience with 46 cases. Acta Oto-Laryngologica-supplement 411: 9–12.

Bast, T. H. 1942. Development of the otic capsule. Annals of Otology, Rhinology and Laryngology 51:343–357.

Bast, T. H. and B. J. Anson. 1949. The temporal bone and the ear. Springfield, IL, Charles C. Thomas.

Benitez, J. T. 1977. Stapedectomy and fatal meningitis. ORL 39: 94–100.

Bielamowicz, S. A., N. J. Coker, J. H. A. and M. Igarashi. 1988. Surgical dimensions of the facial recess in adults and children. Archives of Otolaryngology–Head Neck Surgery 114:534–537.

Birrell, J. F. 1978. Paediatric otolaryngology. Chicago, A. John Wright and Sons.

Blamey, P. J., E. Parisi and G. M. Clark. 1995. Pitch matching of electric and acoustic stimuli. Annals of Otology, Rhinology and Laryngology 104: 220–222.

Brawer, J. R., D. K. Morest and E. Cohen Kane. 1974. The neuronal architecture of the cochlear nucleus of the cat. Journal of Comparative Neurology 55: 251–300.

Bredberg, G. 1968. Cellular pattern and nerve supply of the human organ of Corti. Acta Oto-Laryngologica-supplement 236: 1–138.

Bredberg, G. and B. Lindstrom. 1995. Insertion length of electrode array and its relation to speech communication performance and nonauditory side effects in multichannel-implanted patients. Annals of Otology, Rhinology and Laryngology 104(suppl 166): 256–258.

Bredberg, G., B. Lindstrom, H. Lopponen, M. A. Beltrame, W. Gstoettner and H. Skarzynski. 2000. A new approach for the treatment of ossified cochleas. In: Waltzman, S. and N. Cohen, eds. Cochlear implants. New York, Thieme Medical Publishers.

Brugge, J. F., D. J. Anderson and L. M. Aitkin. 1970. Responses of neurons in the dorsal nucleus of the lateral lemniscus of cat to binaural tonal stimulation. Journal of Neurophysiology 33: 441–458.

Brugge, J. F. and M. M. Merzenich. 1973. Responses of neurons in auditory cortex of the macaque monkey to monaural and binaural stimulation. Journal of Neurophysiology 36: 1138–1158.

Burton, M. J., L. T. Cohen, F. W. Rickards, K. I. McAnally and G. M. Clark. 1992a. Steady-state evoked potentials to amplitude modulated tones in the monkey. Acta Oto-Laryngologica 112: 745–751.

Burton, M. J., R. K. Shepherd, X. J., S. Xu, B. K.-H. G. Franz and G. M. Clark. 1994. Cochlear implantation in young children: histological studies on head growth, leadwire design and electrode fixation in the monkey model. Laryngoscope 104: 167–175.

Burton, M. J., R. K. Shepherd, J. Xu and G. M. Clark. 1992b. Cochlear implantation in young children: long term effects of implantation on the skull and underlying central nervous system tissues in a primate model. Otolaryngology Research Meeting, London: 1.

Cajal, S. R. 1909. Histology of the nervous system of man and vertebrates (translated by N. Swanson and L.W. Swanson). New York, Oxford University Press, 1995.

Cant, N. B. 1984. The fine structure of the lateral superior olivary nucleus of the cat. Journal of Comparative Neurology 227: 63–77.

Cant, N. B. and D. K. Morest. 1979. The bushy cells in the anteroventral cochlear nucleus of the cat. A study with the electron microscope. Neuroscience 4: 1925–1945.

Chouard, C. H., B. Meyer, C. Fugain and O. Koca. 1995. Clinical results for the Digisonic multichannel cochlear implant. Laryngoscope 105: 505–509.

Clark, G. M. 1969a. Hearing due to electrical stimulation of the auditory system. Medical Journal of Australia 1: 1346–1348.

Clark, G. M. 1969b. Middle ear and neural mechanisms in hearing and the management of deafness. PhD dissertation. University of Sydney.

Clark, G. M. 1969c. Responses of cells in the superior olivary complex of the cat to electrical stimulation of the auditory nerve. Experimental Neurology 24: 124–136.

Clark, G. M. 1969d. The ultrastructure of nerve endings in the medial superior olive of the cat. Brain Research 14: 293–305.

Clark, G. M. 1969e. Vesicle shape versus type of synapse in the nerve endings of the ct medial superior olive. Brain Research 15: 548–551.

Clark, G. M. 1975. A surgical approach for a cochlear implant. An anatomical study. Journal of Laryngology and Otology 89: 9–15.

Clark, G. M. and C. W. Dunlop. 1969. Response patterns in the superior olivary complex of the cat. Australian Journal of Experimental Biology and Medical Research 47: 5.

Clark, G. M., H. G. Kranz and J. M. Nathar. 1972. Behavioural responses in the cat to electrical stimulation of the cochlea and auditory neural pathways. Australian Journal of Experimental Biology and Medical Research 3: 202.

Clark, G. M., J. M. Nathar, H. G. Kranz and J. S. Maritz. 1972. A behavioral study on electrical stimulation of the cochlea and central auditory pathways of the cat. Experimental Neurology 36: 350–361.

Clark, G. M., B. C. Pyman and Q. R. Bailey. 1979. The surgery for multiple-electrode cochlear implantations. Journal of Laryngology and Otology 93: 215–223.

Clark, G. M., B. C. Pyman, R. L. Webb, Q. E. Bailey and R. K. Shepherd. 1984. Surgery for an improved multiple-channel cochlear implant. Annals of Otology, Rhinology and Laryngology 93(3 pt 1): 204–207.

Clark, G. M., B. C. Pyman, R. L. Webb, B. K.-H. G. Franz, T. J. Redhead and R. K. Shepherd. 1987. Surgery for safe the insertion and reinsertion of the banded electrode array. Annals of Otology, Rhinology and Laryngology 96(suppl 128): 10–12.

Clark, G. M., R. K. Shepherd, B. K.-H. G. Franz, et al. 1988. The histopathology of the human temporal bone and auditory central nervous system following cochlear implantation in a patient. Correlation with psychophysics and speech perception results. Acta Oto-Laryngologica-supplement 448: 1–65.

Clifford, A. and W. P. R. Gibson. 1987. Anatomy of the round window with respect to cochlear implant surgery. Annals of Otology, Rhinology and Laryngology 96 (suppl 128): 17–19.

Cohen, N. L. and S. B. Waltzman. 1993. Partial insertion of the nucleus multichannel cochlear implant: technique and results. American Journal of Otology 14(4): 357–361.

Dahm, M., H. L. Seldon, B. C. Pyman and G. M. Clark. 1992. 3D reconstruction of the temporal bone in cochlear implant surgery. In: Yanagihara, N. and J. Suziki, eds. Transplants and implants in otology II. Amsterdam, Kugler: 271–275.

Dahm, M., R. K. Shepherd and G. M. Clark. 1993. The postnatal growth of the temporal bone and its implications for cochlear implantation in children. Acta Oto-Laryngologica-Supplement 505: 1–39.

Eby, T. L. and J. B. Nadol. 1986. Postnatal growth of the human temporal bone—implications for cochlear implants in children. Annals of Otology Rhinology and Laryngology 95: 356–364.

Engstrom, H. and J. Wersall. 1958. The ultrastructural organisation of the organ of Corti. Experimental Cell Research (suppl 5).

Enlow, D. H. 1986. Normal craniofacial growth. In: Cohen, M. M. ed. Craniosynostosis: diagnosis, evaluation, and management. New York, Raven Press.

Evans, E. F. and P. G. Nelson. 1973. The responses of single neurones in the cochlear nucleus of the cat as a function of their locations and the anaesthetic state. Experimental Brain Research 17: 402–427.

Franz, B. K.-H. G., G. M. Clark and D. Bloom. 1984. Permeability of the implanted round window membrane in the cat—an investigation using horseradish peroxidase. Acta Oto-Laryngologica-supplement 410: 17–23.

Franz, B. K.-H. G., G. M. Clark and D. M. Bloom. 1987. Surgical anatomy of the round window with special reference to cochlear implantation. Journal of Laryngology and Otology 101(2): 97–102.

Gantz, B. J., B. F. McCabe and R. S. Tyler. 1988. Use of multichannel cochlear implants in obstructed and obliterated cochleas. Otolaryngology Head and Neck Surgery 98: 72–81.

Gleich, O. and S. Wilson. 1993. The diameters of guinea pig auditory nerve fibres—distribution and correlation with spontaneous rate. Hearing Research 71: 69–79.

Goldberg, J. M. and W. E. Brownell. 1973. Discharge characteristics of neurons in antero-ventral and dorsal cochlear nuclei of cat. Brain Research 64: 35–54.

Goldberg, J. M. and W. D. Neff. 1961. Frequency discrimination after bilateral section of the brachium of the inferior colliculus. Journal of Comparative Neurology 116: 265–290.

Goldberg, J. M., F. D. Smith and H. O. Adrian. 1963. Response of single units in the superior olivary cmplex of the cat to acoustic stimuli: laterality of afferent projections. Anatomy Record 145: 232.

Greenwood, D. D. 1961. Critical bandwidth and the frequency coordinates of the basilar membrane. Journal of the Acoustical Society of America 33: 1344–1356.

Harrison, J. M. and M. L. Feldman. 1970. Anatomical aspects of the cochlear nucleus and superior olivary complex. Contributions to Sensory Physiology 4: 95–142.

Hatsushika, S., R. K. Shepherd, Y. C. Tong and G. M. Clark. 1990. Dimensions of the scala tympani in the human and cat with reference to cochlear implants. Annals of Otology, Rhinology and Laryngology 99: 871–876.

Igarashi, M., R. C. Mahon and S. Konishi. 1968. Comparative measurements of cochlear apparatus. Journal of Speech and Hearing Research 11: 229–235.

Igarashi, M., M. Takahashi and B. R. Alford. 1976. Cross-sectional area of scala tympani in human and cat. Archives of Otolaryngology 102: 428–429.

Irving, R. and J. M. Harrison. 1967. The superior olivary complex and audition: a comparative study. Journal of Comparative Neurology 130(1): 77–86.

Javel, E., Y. C. Tong, R. K. Shepherd and G. M. Clark. 1987. Responses of cat auditory fibres to biphasic electrical current pulses. Annals of Otology, Rhinology and Laryngology 96: 26–30.

Kane, E. C. 1973. Octopus cells in the cochlear nucleus of the cat: heterotypic synapses upon homeotypic neurons. International Journal of Neuroscience 5: 251–279.

Kawabata, I. and M. M. Paparella. 1971. Fine structure of the round window membrane. Annals of Otology, Rhinology and Otolaryngology 80: 13–26.

King, A. J. and A. R. Palmer. 1983. Cells responsive to free-field auditory stimuli in guinea-pig superior colliculus: distribution and response properties. Journal of Physiology 342: 361–381.

Lauter, J. L., P. Herscovitch, C. Formby and M. E. Raichle. 1985. Tonotopic organization in human auditory cortex revealed by positron emission tomography. Hearing Research 20: 199–205.

Lehnhardt, E., R. Laszig, R. L. Webb, B. K.-H. G. Franz and G. M. Clark. 1987. Surgery for multielectrode cochlear implants. In: Banfai, P., ed. Cochlear implant: current situation, Proceedings of the International Cochlear Implant Symposium, September 7–12 1987. Duren, West Germany: 477–480.

Lorente de Nó, R. 1933. Anatomy of the eighth nerve. III. General plan of structure of the primary cochlear nuclei. Laryngoscope 43: 327–350.

Matz, G. J., H. B. Lockhart and J. R. Lindsay. 1968. Meningitis following stapedectomy. Laryngoscope 78: 56–63.

Merzenich, M. M. and M. D. Reid. 1974. Representation of the cochlea within the inferior colliculus. Brain Research 77: 397–415.

Morest, D. K. 1964. The neuronal architecture of the medial geniculate body of the cat. Journal of Anatomy, London 98: 611–630.

Morest, D. K. 1965. The lateral tegmental system of the midbrain and the medial geniculate body: study with Golgi and Nauta methods in cats. Journal of Anatomy 99: 611–634.

Nomura, Y. 1984. Otological significance of the round window. Advances in Oto-rhinolaryngology 33: 27–37.

O'Donoghue, G. M., R. K. Jackler, W. M. Jenkins and R. A. Schindler. 1986. Cochlear implantation in children: the problem of head growth. Otolaryngology Head and Neck Surgery 94: 78–81.

Osen, K. K. 1969a. Cytoarchitecture of the cochlear nuclei in the cat. Journal of Comparative Neurology 136(4): 453–484.

Osen, K. K. 1969b. The intrinsic organization of the cochlear nuclei in the cat. Acta Oto-Laryngologica 67: 352–359.

Palva, T. 1970. Cochlear aqueduct in infants. Acta Otolaryngology 70: 83–94.

Palva, T., A. Palva and J. Karja. 1972. Fatal meningitis in a case of otosclerosis operated upon bilaterally. Archives of Otolaryngology 96: 130–137.

Paolini, A. G. and G. M. Clark. 1999. Intracellular responses of onset chopper neurons in the ventral cochlear nucleus to tones: evidence for dual-component processing. Journal of Neurophysiology 81: 2347–2359.

Proctor, B. 1989. Surgical anatomy of the ear and temporal bone. New York, Thieme Medical.

Rasmussen, G. L. 1960. Efferent fibres of the cochlear nerve and cochlear nucleus. In: Rasmussen, G. L. and W. F. Windle, eds. Neural mechanisms of the auditory and vestibular systems. 105–115. Springfield, Illinois, Charles C. Thomas.

Richardson, T. L., E. Ishiyama and E. W. Keels. 1971. Submicroscopic structures of the round window membrane. Acta Otolaryngologica 71: 9–21.

Rockel, A. J. and E. G. Jones. 1973. The neuronal organization of the inferior colliculus of the adult cat, I. The central nucleus. Journal of Comparative Neurology 147: 11–60.

Rose, J. E., J. F. Brugge, D. J. Anderson and J. E. Hind. 1967. Phase-locked response to low-frequency tones in single auditory nerve fibers of the squirrel monkey. Journal of Neurophysiology 30: 769–93.

Rose, J. E., R. Galambos and J. R. Hughes. 1959. Microelectrode studies of the cochlear nuclei of the cat. Bulletin of John Hopkins Hospital 104: 211–251.

Rose, J. E., D. D. Greenwood, G. J. M. and J. E. Hind. 1963. Some discharge characteristics of single neurons in the inferior colliculus of the cat. I. Tonotopical organization, relation of spike-counts to tone intensity, and firing patterns of single elements. Journal of Neurophysiology 26: 294–320.

Rose, J. E., N. B. L. Gross, C. D. Geisler and J. E. Hind. 1966. Some neural mechanisms in the inferior colliculus of the cat which may be relevant to localization of a sound source. Journal of Neurophysiology 29: 288–314.

Rose, J. E. and C. N. Woolsey. 1949. The relation of thalamic connections, cellular structure and evocable electrical activity in the auditory region of the cat. Journal of Comparative Neurology 91: 441–446.

Rubensohn, G. 1965. Mastoid pneumatization in children at various ages. Acta Oto-Laryngologica 60: 11–14.

Rutledge, L. J., M. L. Lewis and F. Sanabria. 1963. Fatal meningitis related to stapes operation. Archives of Otolaryngology 78: 637–641.

Ryan, A. F., N. K. Woolf and F. R. Sharp. 1982. Tonotopic organization in the central auditory pathway of the Mongolian gerbil: a 2-deoxyglucose study. Journal of Comparative Neurology 207: 369–380.

Sando, I. 1965. The anatomical interrelationships of the cochlear nerve fibers. Acta Oto-Laryngologica 59: 417–436.

Schachern, P. A., M. M. Paparella, A. J. Duvall and Y. B. Choo. 1984. The human round window membrane. An electron microscopic study. Archives of Otolaryngology 110: 15–21.

Schillinger, R. 1939. Pneumatization of the mastoid. Radiology 33: 54–67.

Schuknecht, H. 1953. Techniques for study of cochlear function and pathology in experimental animals. Acta Oto-Laryngologica 58: 377.

Schuknecht, H. F. and J. Gulya. 1986. Anatomy of the temporal bone with surgical implications. Philadelphia, Lea and Febiger.

Shepherd, R. K., B. K.-H. G. Franz and G. M. Clark. 1990. The biocompatibility and safety of cochlear prostheses. In: Clark, G. M. Y. C. Tong and J. F. Patrick eds. Cochlear prostheses. London, Churchill Livingstone: 69–98.

Spoendlin, H. 1975. Neuroanatomical basis of cochlear coding mechanisms. Audiology 14: 383–407.

Spoendlin, H. and A. Schrott. 1989. Analysis of the human auditory nerve. Hearing Research 43: 25–38.

Stewart, T. J. and A. Belal. 1981. Surgical anatomy and pathology of the round window. Clinical Otolaryngology and Allied Sciences 6: 45–62.

Tsuchitani, C. and J. C. Boudreau. 1966. Single unit analysis of cat superior olive S segment with tonal stimuli. Journal of Neurphysiology 29: 684–697.

Tsuchitani, C. and J. C. Boudreau. 1967. Encoding of stimulus frequency and intensity by cat superior olive. Journal of the Acoustical Society of America 42(4): 794–805.

Tsuchitani, C. and J. C. Boudreau. 1969. Stimulus level of dichotically presented tones and cat superior olive S-segment cell discharge. Journal of the Acoustical Society of America 46: 979–88.

van den Honert, C. and P. Stypulkowski. 1984. Physiological properties of the electrically stimulated auditory nerve. II. Single fiber recordings. Hearing Research 14: 225–243.

van Hoesel, R. J. M. and G. M. Clark. 1995. Fusion and lateralization study with two binaural cochlear implant patients. Annals of Otology, Rhinology and Laryngology 104(suppl 166): 233–235.

van Hoesel, R. J. M., Y. C. Tong, R. D. Hollow, J. Huigen and G. M. Clark. 1990. Preliminary studies on a bilateral cochlear implant user. Journal of the Acoustical Society of America 88(Suppl 1): S193.

von Békésy, G. 1960. Experiments in hearing. New York, McGraw Hill.

Webster, D. B. and M. Webster. 1978. Cochlear nerve projections following organ of corti destruction. Otolaryngology 86: 342–353.

Wolff, D. 1964. Untoward sequelae eleven months following stapedectomy. Annals of Otology, Rhinology and Laryngology 73: 297–304.

Woolsey, C. N. and E. M. Walzl. 1982. Cortical auditory area of macaca mulatta and its relation to the second somatic sensory area (Sm II). In: Woolsey, C. N., ed. Cortical sensory organization. Clifton, NJ, Humana: 3: 231–256.

Xu, J., R. K. Shepherd, S. A. Xu, H. L. Seldon and G. M. Clark. 1993. Pediatric cochlear implantation: radiological observations of skull growth. Archives of Otolaryngology-Head and Neck Surgery 119: 525–534.

3
Surgical Pathology

Cochlear implants invade the inner ear and thus have the potential to produce complications. An electrode array implanted into the inner ear could result in degeneration of residual auditory nerve fibers, and predispose to middle ear infection extending to the cochlea and even the meningeal linings of the brain. Furthermore, with a profound hearing loss there could be too few residual hearing nerves in the cochlea to be stimulated for speech understanding. Implants in infants and young children introduced a new set of concerns, in particular the effect of head growth on the device and the effect of the device on head growth, as well as the increased risk of inner ear infection because of the high incidence of otitis media in young children. The effect of electrical stimulation per se could be damaging in the adult and child, and is addressed in Chapter 4.

The pathological effects of cochlear implantation should be kept to a minimum, as advanced speech-processing strategies would require an adequate population of neurons for good sound fidelity. Strategies might use residual hearing in association with direct electrical stimulation of the auditory nerve.

This chapter addresses specific pathological issues of importance for cochlear prostheses. The first is electrode insertion trauma and the histopathological consequences. The second is the biocompatibility of the device, including screening studies of candidate biomaterials and the assembled prosthesis. The third is infection of the middle ear and the risk of it spreading to the cochlea and meninges. This possibility is perhaps the most important issue for the safety of cochlear prostheses in children. The most common microorganisms associated with middle ear infections are summarized, and the development of suitable animal models discussed:

Inflammation

Inflammation is a response by the tissues of the body to an agent that is damaging, such as physical insults, chemicals, and toxins produced by bacteria. It is a defense mechanism to help nullify and contain the effects of the damaging agent.

Classification

Initially, there is an immediate response or acute phase, with the exudation of fluid and plasma proteins from the blood, as well as the emigration of white cells or polymorphonuclear leukocytes (in particular the neutrophils) through the smallest of the veins (venules). On the other hand, if healing does not happen and the agent persists, the inflammatory response will become subacute, and may pass into a chronic phase. In the subacute phase there are still signs of the initial inflammatory response, but this changes to one of containment and the development of fibrous tissue. In the chronic phase there is an accumulation of the white cells referred to as lymphocytes and macrophages, as well as the development of fibrosis and calcification and even new bone formation.

Etiology

Inflammation may be due first to the release of agents due to physical trauma. Second, chemicals can be released from the polymers used to make the implant or electrode. In this case the catalyst that polymerizes the filler may be toxic and exert its effects through leaching from the material. This was the case with stannous octoate used for some silicone rubbers (Clark 1987). Metals or ceramics used in the implant must also be checked to ensure that they have no deleterious effects. Third, bacteria may infect the tissue in the inner ear at the time of implantation, or extend into the inner ear from a middle ear infection. They exert their effect through the release of toxins. These toxins may be actively produced (exotoxins) or passively with the destruction of the bacterium (endotoxins).

Pathophysiology

Inflammation, as stated, is the body's response to a harmful agent. It destroys, dilutes, or walls off the agent. It also sets off a series of changes leading to healing and reconstitution of the tissue when possible. With repair, which begins during the early phase of the inflammation, the damaged tissue is replaced through regeneration from native cells, and/or the area is filled with fibrous or scar tissue.

With acute inflammation, agents (histamine, bradykinin, substance P) released as a result of the noxious agent, act on the wall of the venules, causing them to dilate and allow the passage of blood proteins through the wall. The released proteins include antibodies to counteract an infection. If the injurious agent is more severe, capillaries and small arteries may also become dilated.

Once released into the tissue, the white cells (neutrophils) migrate toward the site of injury down a chemical gradient. The white cells move when the chemically attracting agents act on cell membrane receptors, and this leads to the release of calcium ions into the cytoplasm that cause contractile movements in the cell (Snyderman and Uhuig 1992). Once arriving at the site of the inflammation, the white cells destroy bacteria through the process of phagocytosis (Fig. 3.1). This involves three distinct steps: (1) recognition and attachment of the particle,

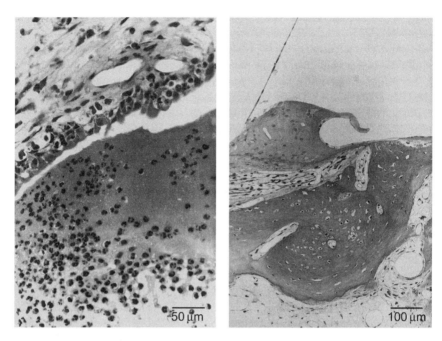

FIGURE 3.1. Left: Acute inflammation in the cochlea with the presence of poly-morphonuclear leukocytes (neutrophils). Right: Chronic inflammation with the development of fibrous tissue and new bone. New bone is particularly prone to occur beneath the spiral lamina.

(2) engulfment and inclusion in a vacuole in the cytoplasm, and (3) killing or degradation of the material.

With chronic inflammation there is infiltration with mononuclear white cells that include macrophages, lymphocytes, and plasma cells in response to persistent injury. There may be tissue destruction, and there is an attempt at healing through the proliferation of small blood vessels and the formation of fibrous tissue. The macrophage is the main cell initiating chronic inflammation. It is derived from monocytes that come from the bone marrow. The monocytes begin to migrate early in inflammation, that is, within the first 48 hours, and their spread into the tissue (extravasation) is governed by the same factors affecting the neutrophil white cells. Once in the tissue, the monocyte is transformed into a macrophage. The macrophages carry out phagocytosis and when activated produce fibrous tissue. Lymphocytes are also mobilized with continuing inflammation and together with plasma cells produce antibodies to infection.

Chronic inflammation may lead to the formation of not only fibrous tissue but also new bone (Fig. 3.1). The bone may arise from the actual bone of the organ (i.e., cochlea) when it is referred to as orthotopic, or it can come from the fibrous tissue. New bone formation is part of the healing phase in the tissue after physically and chemically induced inflammation. Bone formation commences in calcified tissue. Calcification results when calcium phosphate crystals are deposited in the tissue as an apatite. The process occurs in two phases: (1) initiation, and (2) propagation. They take place both intracellularly and extracellularly (Majno and Joris 1996). The initiation of intracellular calcification occurs in the mitochondria of the cells through the accumulation of calcium. The initiators of extracellular calcification are phospholipids bound to the membranes of vesicles. The calcium is concentrated in these vesicles through a series of biochemical steps. The propagation of the calcification depends on the concentration of calcium and phosphate ions as well as the presence of facilitating and inhibiting proteins in the extracellular space. These matrix proteins facilitate the interaction of the minerals and the cells to produce calcification in tissue that is then referred to as osteoid. There is phase delay of some days before bone formation commences in the osteoid tissue.

Bone formation commences with the activation of cells (osteoprogenitor cells) to produce osteoblasts and osteocytes. Their generation is regulated by cytokines and growth factors such as fibroblast growth factor (FGF), platelet-derived growth factor (PDGF), insulin-like growth factor, and transforming growth factor-β (TGF-β) (Mundy 1995). The osteoprogenitor cells are pluripotential mesenchymal stem cells located in the vicinity of all bony surfaces. When stimulated they undergo cell division and produce offspring that differentiate into osteoblasts. A specific transcription factor stimulates osteoblast specific gene expression. The osteoblasts synthesize, transport and arrange the many proteins of the matrix, and initiate further mineralization. They have cell surface receptors that bind many hormones to regulate activity, for example, parathyroid hormone, vitamin D, cytokines, and growth factors. Once the osteoblasts become surrounded by the matrix, they are known as osteocytes. The osteocytes communicate with each other via processes passing along an intricate network of tunnels in the matrix known as canaliculi. These processes transfer surface membrane potentials and substrates. Bone formation is an interplay between deposition and removal. The osteoclast is responsible for bone absorption, and is derived from hematopoietic progenitor cells that also give rise to monocytes and macrophages. Cytokines too are crucial for osteoclast differentiation and maturation, and these include interleukins, tumor necrosis factor, and granulocyte-macrophage colony-stimulating factor. These cells are multinucleated and the result of fusion of mononuclear nuclear precursors. They are intimately related to the bone surface, and form a localized seal around the surface of the bone. A hydrogen pump in the cell acidifies this localized area, and the regional bone is dissolved. The osteoclast also releases into the space a multitude of enzymes that help disassemble the matrix proteins into amino acids and liberate and activate growth factors. Thus as bone is broken down to its elemental units, substances are released that initiate its renewal. The proteins forming the matrix of bone are primarily type 1 collagens as well as other proteins from osteoblasts. The osteoblasts deposit the collagen for the matrix in a random weave know as woven bone or in an orderly layered manner referred to as lamellar. Woven bone is formed more quickly.

Insertion Trauma

Physical injury damages the cochlea through tears of the basilar membrane and fractures of the spiral lamina in particular, and this also induces an inflammatory response. The pathological responses of tissue to trauma are discussed in this chapter. The effects of the pathological changes induced by specific types of electrodes are described in the section on bioengineering in Chapter 8.

Tissue Responses in the Cochlea of the Experimental Animal

The effects of surgical trauma on the tissues in the cochlea can be studied in the experimental animal. The cat has been the model for most studies on the effects of injury to specific cochlear tissues. But the monkey cochlea is more similar to that of the human, and does not narrow markedly 6 to 7 mm from the round window, as occurs in the cat. Therefore, it has been more useful in studying the effects of different electrode designs.

Hair Cell Preservation

Hair cells were preserved in the cat and monkey cochleae implanted with scala tympani electrodes with and without electrical stimulation, unless there was infection or trauma to the basilar membrane and the spiral lamina in particular (Shepherd et al 1983b). In the studies by Shepherd et al (1983a,b), three quarters of the inner and outer hair cells were well preserved. Atrophy of hair cells in the organ of Corti was generally confined to the lower basal turn adjacent to the electrode array. Chronic electrical stimulation per se, did not contribute to the loss of hair cells. In addition, the preservation of spiral ganglion cells and the auditory nerve was associated with hair cell survival. This was consistent with the release of trophic factors from the hair cells for the health of the auditory neuron. In the cochleae, where there was widespread loss of hair cells, this was associated with severe inflammation. The preservation of hair cells in the cat was also seen by Xu et al (1997) in their study on the long-term effects of intracochlear electrical stimulation using high rate biphasic electrical pulses. An increase in the click-evoked auditory brainstem response (ABR) threshold was found following implantation in all cochleae, but recovery to near-normal levels occurred in approximately half of the stimulated cochleae. Frequency-specific stimuli indicated that the most extensive hearing loss occurred in the high-frequency or basal region of the cochlea (12,000 and 24,000 Hz) adjacent to the stimulating electrode. Thresholds at lower frequencies (2000, 4000, and 8000 Hz) appeared at near normal levels. There was no evidence of cochlear damage caused by high-rate electrical stimulation.

Physiological and psychophysical studies in the experimental animal have also demonstrated that functioning hair cells were present after cochlear implantation (Clark, Kranz et al 1973). The behavioral study by Clark, Kranz et al (1973) on cats showed that "electrophonic" hearing could be found in the chronically implanted animal at least up to 800 Hz. The findings were evaluated in some detail by Black et al (1983) and are discussed in Chapter 5.

Basilar Membrane Tears

In normal-hearing animals, damage to the basilar membrane can result in the loss of hair cells, which in turn leads to degeneration of the peripheral processes (dendrites) and spiral ganglion cells localized to the region of the trauma as discussed in the previous subsection. Transection of the peripheral processes with a tear can also result in a marked loss of auditory peripheral processes as well as the spiral ganglion cells (Simmons 1967). The loss is localized to the region, and does not affect the neurons at a more remote location in the cochlea. The tear disrupts the peripheral processes as they pass from synapses on both inner and outer hair cells to the spiral ganglion cells. The neuron degenerates both peripherally and centrally. Wallerian degeneration affects the peripheral part, and retrograde degeneration the more central part of the dendrite. The central loss can lead to degeneration of the spiral ganglion cells. Wallerian degeneration is due to the loss of trophic factors responsible for the health of the neuron peripherally and supplied from the nucleus. The degeneration centrally is due to the loss of trophic factors supplied by the hair cells as well as by the supporting and glial cells (Ylikoski et al 1993; Schecterson and Bothwell 1994). The loss of trophic factors can initiate a cell death cascade in the proteins in the neuron. A basilar membrane tear can create an opening (fistula) between the scala media and the scala tympani. The mixing of the perilymph and endolymphatic fluids as well as damage of the hair cells and the transection of the neuron leads to degeneration of the peripheral processes.

The loss of spiral ganglion cells following a tear of the basilar membrane is seen in Figure 3.2. In this case the Teflon strip, with sharp edges, cut the basilar membrane when it was inserted into the cochlea of the cat. There was almost total loss of spiral ganglion cells.

Spiral Lamina Fractures

With intracochlear insertions there can be micro- and macrofractures of the spiral lamina and damage to the endosteal lining of the scala tympani. A fracture of the spiral lamina causing transection of the peripheral processes of the spiral ganglion cells, as well as a tear of the basilar membrane, results in a loss of spiral ganglion cells. It was observed by Leake-Jones and Rebscher (1983) that a fracture could lead to a reduction in the density of the myelinated fibers close to the habenula perforata, but with normal numbers of spiral ganglion cells.

In addition, the fracture initiates bone growth (Fig. 3.3). The bone growth may be quite extensive and involve inflammatory tissue around the electrode. Bone growth itself does not lead to loss of hair cells or neurons (Shepherd et al 1983a). However, factors such as trauma and infection, which lead to bone growth, may themselves cause loss of the sensory and neural elements. Any toxic chemicals leaching from the electrode will also lead to hair cell death, and consequently the

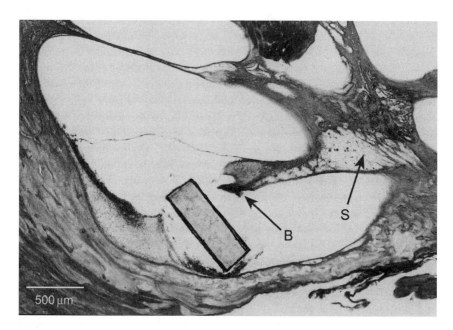

FIGURE 3.2. A photomicrograph of the cat cochlea after tearing of the basilar membrane with a Teflon strip showing the marked loss of spiral ganglion cells. B, basilar membrane; S, spiral ganglion. (Clark et al, 1987. The University of Melbourne–Nucleus multi-electrode cochlear implants. *Advances in Oto-Rhino-Laryngology* Volume 38. Basel, Karger. Reprinted with permission.)

loss of the trophic factors required for the health of the peripheral nerve processes. The loss of auditory neurons can have a significant effect on the function of the cochlear implant through either elevating the stimulus thresholds or restricting the ability to produce place pitch coding for effective speech-processing strategies.

Spiral Ligament Tears

The spiral ligament may be depressed or torn by an electrode. Unpublished findings in the University of Melbourne/Bionic Ear Institute, however, have shown that unless the basilar membrane is torn, the lesion produces no neuronal loss.

Auditory Neuron Survival

The two degenerative processes following injury of the peripheral processes (wallerian and retrograde degeneration) discussed above (see Basilar Membrane Tears) have different time courses. Wallerian degeneration occurs within a few days, and results in a complete breakdown in the distal (peripheral) portion of the fiber that has been separated from its cell body (perikaryon). Retrograde degeneration, affecting the proximal portion of the process, is observed over a longer period of time. Unlike other bipolar sensory neurons, up to 90% to 95% of spiral ganglion neurons in the cat exhibit retrograde degeneration that includes the cell

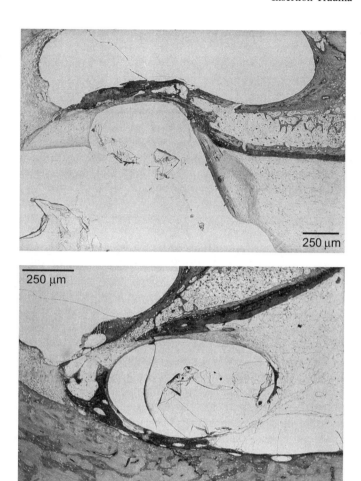

FIGURE 3.3. Top: A fracture of the spiral lamina of the cat with new bone formation near the round window from a circular Teflon felt disk around the array (Clark et al. Cochlear Prostheses: an international symposium. *Annuals of the New York Academy of Science,* **405**, 1983. Reprinted with permission.) Bottom: New bone surrounding the electrode carrier in the same animal.

body, therefore excluding the possibility of regeneration (Kerr and Schuknecht 1968; Spoendlin 1975). However, the extent and rate of neural loss depends on a number of factors, including the nature and extent of the cochlear trauma and the species under investigation. For example, the degenerative process following a surgical insult to the organ of Corti in guinea pigs is nearly complete within a 3-week period following the trauma (Webster and Webster 1978). In contrast, direct administration of an ototoxic drug in the same species requires a period of 60 days for the degenerative process to come to completion (Terayama et al 1979). There is also evidence of greater spiral ganglion cell survival in human cochleae than would be predicted from experimental animal data with similar pathologies (Ylikoski et al 1981).

The majority of animal studies investigating the histopathological effects of electrode insertion trauma used normal-hearing animal models, but it is uncertain whether an electrode array fracturing the osseous spiral lamina in a profoundly deafened cochlea would also produce a significant, further localized ganglion cell loss. Damage localized to the basilar membrane of a profoundly deafened cochlea would probably not contribute to further significant ganglion cell loss, as cochleae such as these would have few, if any, peripheral processes projecting to the basilar membrane. However, the increased fibrous tissue and new bone formation could reduce the effectiveness of the stimuli and prevent a more advanced array from being inserted. Thus the design of scala tympani electrode arrays and the surgical techniques used to insert them, should aim at minimizing this form of trauma.

New Bone Formation

As discussed above (see Pathophysiology), new bone formation is part of the healing phase in the tissue after physically and chemically induced inflammation, and commences in calcified tissue in the scala tympani. In my studies I have found a phase delay of some days before bone formation commences in osteoid tissue of the cat cochlea. This is also seen when operating on the cochlea of children after labyrinthitis due to meningitis. The calcified tissue is soft 4 to 6 weeks after recovery from the infection (William Gibson, personal communication). As seen in Fig. 3.3 the bone may be orthotopic or form within fibrous tissue. From an inspection of the implanted cat cochleae in the University of Melbourne's collection, the bone more usually arises from the spiral lamina and then extends into the fibrous tissue matrix. It may be associated with an obvious fracture or possible microfractures. In a number of cochleae there is no apparent reason for the new bone, but the area under the spiral lamina seems to have a propensity to form bone. The endothelial cells lining at this location swell and develop osteo-blastic activity. Woven bone is formed within the scala.

Unpublished studies at the University of Melbourne/Bionic Ear Institute have found that scraping the endosteum with the instillation of blood induces bone growth. Bone also formed around particles of bone that were introduced, and this indicates the need to avoid bone dust entering the cochlea when drilling an open-ing for the insertion of the electrode array. Bone or fibrous tissue increases the impedance of the tissue surrounding the electrode, thus resulting in the need for higher current levels. The bone formation can also alter the spread of current, and prevent more localized simulation of discrete groups of auditory nerve fibers.

Incision of the Round Window Membrane Versus Cochleostomy

One key issue was whether to introduce electrodes into the cochlea through an incision in the round window membrane or make a separate opening through the bone overlying the basal turn just anteroinferior to this window. The reason for considering the fenestration was that it provided a more direct approach to the basal turn to permit a deeper and possibly less traumatic insertion. It raised the question, however, whether there would be an adequate seal to protect the entry of infection. The two procedures were compared (Franz and Clark 1987), and it was found that with the fenestration, bony chips (sequestra) entered the cochlea; sometimes they remained inert but could be the focus for the development of calcification and bone formation. For that reason it is desirable to drill down to the endosteal lining of the inner ear or "blue-line" it and remove all the bone before entering into the scala vestibuli. The technique is referred to as "soft surgery" by Lehnhardt (1993), and is discussed in Chapter 10.

As discussed below (see Infection), the two procedures were compared regarding prevention of infection extending from the middle ear to the cochlea. They were both found to be equally effective, but the electrode tissue sheath with the cochleostomy was thicker and more uniform, and had a greater attachment to the surrounding tissue of the temporal bone (Fig. 3.4).

Tissue Responses in the Human

Similar tissue responses to those seen in the cat are evident in the human. As shown in Figure 3.5, a dense fibrous tissue capsule was found around the Nucleus array after it had been implanted for 27 months. This patient had previously had labyrinthitis at the age of 15 years. There was trabeculated bone in the scala tympani that was not dense enough to prevent the electrode from being inserted. New bone formed around the array, as shown in Figure 3.5. New bone occupied a large volume of the scala tympani close to the round window. However, the extent of this new bone decreased significantly with increasing distance from the round window. Bone formation may have been facilitated from the previous bone, and could be a sequela when implanting ears following labyrinthitis. No new bone was observed beyond the 12-mm region of the scala tympani. This was significant, as the most extensively used electrodes on the array were located apically, and therefore were not associated with new bone. This observation supported our previous findings from animal studies (Shepherd et al 1983a,b), and indicated that long-term electrical stimulation using carefully controlled, charge-balanced biphasic current pulses did not result in new bone. The new bone present in the basal region of this cochlea may have been the result of drilling associated with the implant surgery. However, the previous pathology (meningitis and labyrinthitis) certainly contributed to the findings as the contralateral unimplanted cochlea was almost completely filled with bone.

Extensive sensory hair cell loss was apparent throughout all turns of the cochlea. A small population of peripheral processes were present just apical to the tip of the electrode. Finally, the spiral ganglion cell population was approximately 10% of normal. Significantly, the largest number of surviving ganglion cells occurred adjacent to the electrode array. Extensive ganglion cell loss was reported in the apical turn of the cochlea, some distance from the array. This differential spiral ganglion survival pattern was probably a result of the previous pathology. Spiral ganglion cell loss was not associated with the intensity or extent of electrical stimulation as Rosenthal's canal, adjacent to the most extensively used

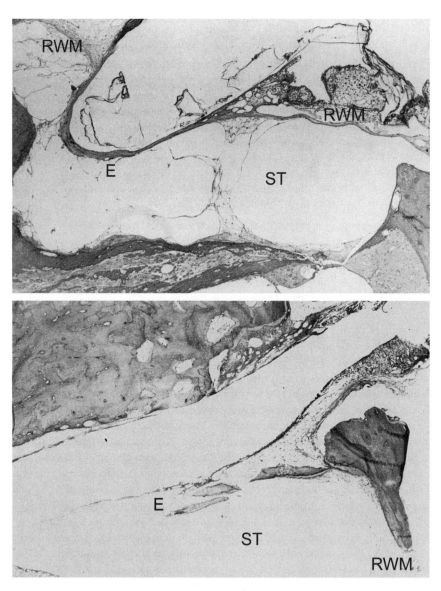

FIGURE 3.4. Top: A photomicrograph of an implant electrode (E) passing through the round window membrane (RWM) into the scala tympani (ST). Bottom: The electrode passing through a cochleostomy through the bone adjacent to the round window. (Franz et al 1987. Effect of experimentally induced otitis media on cochlear implants. *Annals of Otology, Rhinology and Laryngology,* **96(2)**. Reprinted by permission.).

electrodes on the array, (approximately 12 to 15 mm from the round window) exhibited some of the largest numbers of surviving ganglion cells within the cochlea. Electrodes in this region of the cochlea were stimulated at charge densities of up to 25.7 μC cm^{-2} geometric/phase. Finally, the mature fibrous tissue

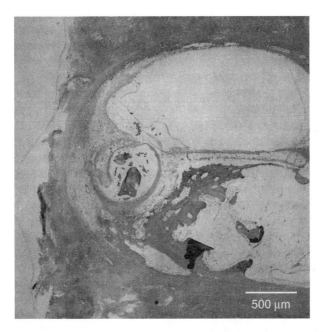

FIGURE 3.5. The human cochlea implanted for 27 months following labyrinthitis 40 years previously. There is old and new bone formation.

and new bone surrounding the point of entry of the electrode though the round window appeared to provide an effective seal between the cochlea and middle ear.

The temporal bones of 14 patients who had the cochlear implant were analyzed to determine the range of histopathological effects. In 10 of the 14 bones the insertion was in the scala tympani with some damage of different structures, most commonly indentation of the spiral ligament in the upper basal turn due to pressure from the electrode. The basilar membrane and in some cases the osseous spiral lamina showed localized damage with tearing or fracture. In three bones the cochleostomy and insertion of the array had occurred in the scala media and continued along this scala with rupture of Reissner's membrane in two. In the remaining bone there was extensive bone formation throughout the cochlea with no distinguishable scalae. This was presumed due to the fact this patient had previously had meningitis. In no specimen did the electrode array penetrate the basilar membrane or osseous spiral lamina into the scala vestibuli (Dahm et al 2000).

In another study on five implanted human temporal bones, the intracochlear pathological and anatomical factors were compared with psychophysical percepts (Kawano et al 1995, 1996, 1998). It was found that the threshold and comfortable levels were lower when the distance between the electrodes and the spiral canal containing the spiral ganglion cells (Rosenthal's canal) was reduced. Thresholds were elevated by the presence of intracochlear fibrous tissue and new bone, especially the former. This was in part due to the distance between the electrode and the modiolus increasing with greater amounts of fibrous tissue and new bone. The dynamic range was reduced especially in the presence of new bone, and this was probably the result of bone spreading the current pathways away from the neural elements with loudness summation as discussed in Chapter 6, or reduced ganglion cell numbers as discussed below. Maximum comfortable level and dynamic range but not the threshold were related to the spiral ganglion cell population. There was a wider dynamic range with a higher density of cells. Furthermore, spiral ganglion cell survival was decreased with the increasing levels of fibrous tissue and new bone formation. The experimental animal data would suggest that it was not the fibrous tissue or bone per se that led to the reduction, but rather a common cause such as trauma or operation.

Biocompatibility of Materials

The reaction of the body (host) to the implanted materials can be inflammation, a foreign body reaction, encapsulation, or hypersensitivity. The response may be the result of toxic chemicals from the device or just the physical shape and dimensions leading to foreign body reaction and encapsulation. The tissue responses should be studied through in vitro and in vivo tests. The in vivo tests are either acute or chronic and the chronic ones should include testing for carcinogenicity.

The toxicity of materials, used with the first Nucleus cochlear implant, were evaluated by the University of Melbourne from 1980 to 1982. Tissue responses were examined using an appropriate modification of the 1980 revision of the protocol in the United States Pharmacopeia. The research was undertaken according to the Good Laboratory Practice requirements of the U.S. Food and Drug Administration (FDA) for its premarket approval (PMA). Additional tests for cytotoxicity used cell culture and intracutaneous and systemic injections into laboratory animals. These were developed from recommendations of the American Society for Testing Materials (ASTM), subsequently outlined in the *Animal Book of ASTM Standards* (1986).

The PMA for the Nucleus device in 1985 provided the basis for later approvals of other devices on the grounds that their design, safety, and efficacy were comparable to those of their predecessor.

Methods of Investigation

The biocompatibility of the implant materials was determined first by the histological evaluation of the foreign body response. The presence of significant numbers of acute or chronic inflammatory cells following long-term implantation indicated the presence of an active tissue irritant. In contrast, biocompatible materials produce a fully resolved inflammation that is characterized by a thin, mature, fibrous tissue capsule. The extent and nature of the tissue response therefore reflected the degree of tissue biocompatibility of the implanted material (Homsy

1970). The evaluation of the tissue reaction to the biomaterials for the Nucleus receiver-stimulator and electrode array was performed by implanting the materials subcutaneously, intramuscularly, and within the scala tympani of experimental animals (Clark 1987). As stated, the procedures used to assess the toxicity of the biomaterials were a modified version of the U.S. Pharmacopeia (1980) recommendations. Differences included implantation in three mammalian species to ensure that there were no species differences; soft tissue implantation under vision into both subcutaneous tissue and paravertebral muscles via small skin incisions; the extension of the implant period beyond 7 days to ensure that the effects of surgery did not contribute to the tissue response; the use of at least four animals to test each material, rather than the two recommended, to better assess the degree of variation in tissue response between animals; the grading of the tissue reaction by microscopic examination rather than the macroscopic examination recommended; and finally, the use of stannous octoate, the catalyst for Silastic 382, as the reactive control, and medical grade Silastic tubing and Teflon as the nonreactive controls. All solid materials were meticulously cleaned and inserted aseptically for periods ranging from 14 to 120 days (Clark 1987). Each implant was removed together with its surrounding tissue capsule and prepared for light microscope histological examination. The tissue response to each implant and degree of inflammation was classified on the basis of its polymorphonuclear leukocyte, mononuclear leukocyte, and fibrous tissue reactions. The histopathologist recording the tissue reaction was given no information on the sections or the biomaterials used in the study.

Tissue Response

Inflammation following the implantation of a foreign body can proceed from an acute to a chronic response. As discussed this could be due to the physical effect of the material, the biotoxicity of the materials, or superimposed infection. The inflammation usually resolves with the encapsulation of the foreign body by a mature, fibrous tissue envelope. The tissue response to the foreign body is dynamic, with many stimuli acting simultaneously to maintain the inflammation. Mechanical irritation due to the mechanical properties of the implant can cause a prolonged foreign body reaction, particularly in sites subject to extensive movement such as muscle (Wood et al 1970). In addition, many implantable materials are not truly inert. Monomer or catalyst residues within the polymer can gradually leach into the surrounding tissue, resulting in inflammation or tissue necrosis. Other chemical irritants, such as metal ions, could also be released due to the dissolution of the implant within the corrosive environment of the body (Coleman et al 1974). The gradual release of these irritants would result in a prolonged inflammatory reaction. Finally, the presence of a foreign body predisposes to infection. Infection exacerbates the inflammation, and if unresolved results in widespread degeneration of residual auditory nerve fibers.

Suitable biomaterials for use in the initial Nucleus multiple-electrode cochlear prosthesis were screened to compare their inflammatory responses with those of control materials. This included implanting them in the site for which they were intended, that is, the cochlea. A second series of studies evaluated the biocompatibility of the assembled device. This was important, as the tissue response to the assembled prosthesis depends not only on the individual biomaterials but also on their interaction and on the manufacturing processes used in assembly. Excessive cold metal working of metallic electrodes during their manufacture, for example, resulted in localized corrosion, even though these metals were previously considered stable under in vivo conditions. Finally, the effects of using the prosthesis for electrical stimulation, which could be dissolution of the metal electrode, breakdown of insulating biomaterials, and adverse effects of electrical stimulation per se, were studied.

Subcutaneous and Intramuscular Implantation

The initial soft tissue screening studies evaluated the tissue toxicity of a number of candidate biomaterials. The extent of the tissue reaction depended on both the duration of implantation and the implant material. Lower polymorphonuclear leukocyte cell counts and more mature fibrous tissue responses were observed with longer periods of implantation. The study (Clark 1987) showed that polytetrafluoroethylene (PTFE) and polyurethane evoked the mildest tissue response. Silastic MDX-4-4210, Silastic tubing, Silastic adhesive type A, polyethylene, fluoroethylene propylene (FEP), and platinum also evoked mild tissue reactions. In contrast, Silastic 382 elastomer and PTFE sputter coated onto Silastic tubing exhibited greater than average mononuclear leukocyte and fibrous tissue responses and were consequently considered unsuitable for chronic implantation. The tissue toxicity associated with Silastic 382 was most likely due to the leaching of unpolymerized catalyst (stannous octoate), as the study showed that Silastic 382 catalyst evoked a very severe mononuclear leukocyte and fibrous tissue reaction. Other investigators have also reported an occasional adverse tissue response following implantation using Silastic 382 (Leake-Jones and Rebscher 1983). Loeb et al (1977) implanted 14 different dielectric polymers in the subdural space of cats for 8 to 30 weeks and considered Teflon, Epoxylite, Silastic, Kapton, and polyurethane safe. On the other hand Parylene and Mylar were doubtful. Rarely, an inflammatory reaction occurred for metals, but they are normally well tolerated. When present, this reaction was mostly due to hypersensitivity, evidenced by the accumulation of eosinophil leukocytes and plasma cells.

Electrode Sheath in the Cochlea

In addition to the above studies, 3- to 4-mm lengths of materials were implanted into the scala tympani of cochleae, and this showed that polyethylene (low density), polyurethane, and Silastic MDX-4-4210 evoked a minimal fibrous tissue response (Fig. 3.6). However, in two cochleae (one implanted with PTFE sputter coated onto Silastic tubing and one FEP implant) reduced spiral ganglion cell densities were observed in the basal turn. In the absence of any other obvious cause for this cell loss, it is possible that the reduced ganglion cell numbers were

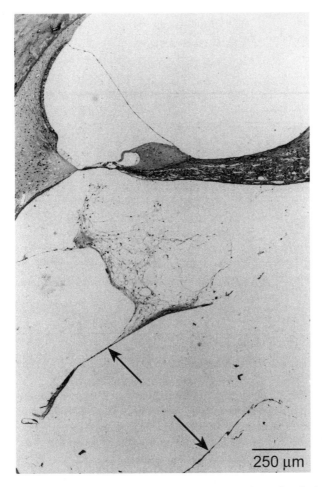

FIGURE 3.6. Photomicrograph of the basal turn of a cat cochlea following the implantation of a banded scala tympani electrode array of Silastic MDX-4-4210 for a period of 113 days. The location of the electrode array is indicated by the fine fibrous tissue capsule (arrows). A normal organ of Corti was apparent throughout the middle and apical turns of this cochlea. (From Shepherd et al 1984. Implanted material tolerance studies for a multiple-channel cochlear prosthesis. *Acta Oto-Laryngologica:* (suppl 144). Reprinted with permission.)

a result of leaching of toxic products from these materials. Otherwise, these scala tympani results were consistent with the results of the soft tissue study, and supported the choice of Silastic MDX-4-4210 as the electrode carrier material.

Other Tests of Cellular Responses

Independent evaluation of the materials was undertaken for Cochlear Proprietary Limited in 1982 by North American Science Associates. The materials were tested for cytotoxicity, mutagenicity, intracutaneous toxicity, hemolysis, inflammation, and sensitization. Further information on the evaluation of biomaterials for the FDA is discussed in Chapter 8.

Infection

Under certain circumstances the round window membrane can be an important avenue for the entry of microorganisms and toxins into the inner ear (Goycoolea et al 1980). An increased permeability of the round window membrane was observed in otitis media, and this could lead to an associated sensorineural hearing loss (Schachern et al 1981). The insertion of an electrode array into the scala tympani coupled with a poor seal at the round window or a fenestration through the endosteal bone overlying the cochlea (cochleostomy) could also provide a potential pathway for microorganisms and toxins to enter the inner ear, resulting in the degeneration of residual auditory neurons and spread to the meninges. This possibility is of particular concern when implanting children who are prone to recurrent middle ear infections.

Despite this fear, clinical experience (House et al 1985; Luxford and House 1985) and animal studies (Clark, Pyman et al 1984; Clark, Shepherd et al 1984; Franz et al 1984; Cranswick et al 1987; Dahm et al 1994, 1995) have shown that the implanted cochlea is capable of resisting the spread of infection similar to a nonimplanted one, provided a fascial autograft is placed all around the electrode at its entry point.

Otitis Media

Otitis media is inflammation of the middle ear. As the middle ear is connected to the postnasal space through the eustachian tube, it is most frequently infected from the nose and nasal sinuses. It is usually started by a viral infection. The initial infection of the nose and postnasal space may cause obstruction of the eustachian tube, and predispose the mucous membrane of the middle ear to secondary bacterial infection. Other predisposing factors for eustachian tube obstruction are adenoid enlargement, allergy, and sinus disease. Young children are more prone to develop otitis media, and in infants this is in part due to having a more patulous opening to the eustachian tube, as well as incompletely developed immune mechanisms. Otitis media can also be initiated by bacterial infection alone. The incidence of organisms producing otitis between 1983 and 1986 at the Pittsburgh Otitis Media Research Center was *Streptococcus pneumoniae* (29%), *Haemophilus influenzae* (23%), and *Moraxella catarrhalis* (13%). The incidence of *S. pneumoniae* has been reported to be as high as 55% (Luotonen et al 1981; Johnson et al 1991; Block 1997). The incidence of *S. pneumoniae* and *H. influenzae* infections declines with increasing age (Bluestone et al 1992). *Streptococcus pyogenes* (type A, *β*-hemolytic), and *Staphylococcus aureus* were each responsible for less than 10% of cases. In addition, *Pseudomonas aeroginosa* and *Proteus mirabilis* can be found in a small proportion.

In the acute stage of the infection the blood vessels of the middle ear and tympanic membrane are dilated. This leads to exudation of fluid into the middle ear cleft, and the infiltration of polymorphonuclear leukocytes or granulocytes (in particular neutrophils). The swelling of the tissue can impair the blood supply to the eardrum, and lead to necrosis and a perforation. The condition resolves normally within days with treatment. If not, it will become subacute or chronic with continued discharge. The mucous membrane becomes infiltrated with lymphocytes and monocytes; and then through the proliferation of fibrous tissue a condition called tympanosclerosis can occur.

If the acute infection is not treated early, the infection can spread to the mastoid air cells and cause mastoiditis. Rarely, labyrinthitis, meningitis, cerebral and cerebellar abscesses, facial nerve palsy, or lateral venous sinus thrombosis can ensue.

Labyrinthitis and Meningitis

Epidemiology

Bacterial meningitis is often a general illness, but can be secondary to an otitis media or an open head injury and surgery. As an illness in the community, its incidence varies according to age. As discussed in Chapter 10, the overall incidence in the United States has been reduced from 10 to 2.4 episodes per 100,000/ year over the last 10 years due to better prophylaxis. The incidences of organisms causing meningitis are *S. pneumoniae* (~50%), *Neisseria meningitidis* (~25%), group B streptococci (~10%), and *Listeria monocytogenesis* (~10%). *H. influenzae* was the most common cause, but has been reduced due to vaccination with *H. influenzae* type B (Hib). The incidence of organisms responsible also varies according to age. *N. meningitidis* accounts for up to 60% of cases in children and young adults, from the ages of 2 to 20 years, and occurs in episodes. *S. pneumoniae* is the most common cause in adults over 20 years. *L. monocytogenes* is more frequent in newborns, pregnant mothers, and those who are immunosuppressed.

N. meningitidis arises from colonization of the nasopharynx, and is a result of a bacteremia leading to invasion of meninges. The same applies to *S. pneumoniae*, but the bacteremia can lead to pneumonia, otitis media, or meningitis. In the epidemiological data from Victoria discussed in Chapter 10, there was a high incidence of bacteremia in children under 24 months of age, but only a small proportion led to a meningitis.

In addition, pneumococcal infections in the ear may cause meningitis, and this has been seen by Igarashi et al (1974) and Schuknecht (1974) in temporal bones of patients dying from meningitis, as discussed below (see Pathogenesis). However, the pathogens causing otitis media do not all lead to meningitis. This can be seen by comparing the spectrum of organisms producing these infections. The organisms causing otitis media are *S. pneumoniae* (29–50%), *H. influenzae*

(23%), *M. catarrhalis* (13%), and a small percentage are due to *Streptococcus β-hemolyticus* and *S. aureus.* These organisms are more adapted to invasion of the mucosa than the meninges, with the exception of *S. pneumoniae* and *H. influenzae.*

The range of organisms causing cochlear implant device-related meningitis has not been determined, although initial evaluation of patients with meningitis in the U.S. reported in Chapter 10 indicates that the *S. pneumoniae* is a frequent invader. Furthermore, as discussed in Chapter 10, it was also the most frequent cause of fatal meningitis as a complication of stapedectomy.

The organisms that lead to infection of the meninges following a neurosurgical procedure have a different spectrum from those in the community-related disease, and they may be gram-negative *Escherichia coli* or *P. mirabilis,* and also *S. aureus* as well as gram-negative staphylococci.

Pathogenesis

With community-based meningitis, particularly from *S. pneumoniae* and *N. meningitidis,* the organisms initially colonize the nasopharynx and become attached to the epithelial cells. They are then transported across these cells in vacuoles to the capillaries and venules through spaces they create between the tight junctions. Once in the bloodstream they can evade phagocytosis and complement-mediated bactericidal activity, because of their polysaccharide capsules. From the blood they reach the intraventricular choroid plexus. Invasion of the plexus cells allows the bacteria direct access to the cerebrospinal fluid (CSF). In addition, *S. pneumoniae* can adhere directly to cerebral capillary endothelial cells, and pass between these cells to reach the CSF. In the CSF it multiplies rapidly because of the absence of effective host immune defences. Furthermore, CSF contains few leukocytes, and only small amounts of complement proteins and immunoglobulins to control their multiplication.

Once in the CSF the bacteria produce tissue damage through an immune response to cell wall protein released following lysis. Inflammation is produced by cytokines and chemokines released by microglia, astrocytes, monocytes, and microvascular endothelial cells. As discussed by Roos and Tyler (2001) in experimental models of meningitis, cytokines including tumor necrosis factor and interleukin are present in the CSF within 1 to 2 hours of intracisternal inoculation of lipopolysaccharide molecules from gram-negative organisms. The elevated levels of CSF cytokines and chemokines increase the permeability of the blood–brain barrier, which leads to edema and leakage of serum proteins into the subarachnoid space. This in turn results in a diminished ability of arachnoid granulations in the dural sinuses to reabsorb fluid, and in turn hydrocephalus. In the early stages of meningitis there is an increased cerebral perfusion pressure, but later there is a loss of autoregulation of the blood flow, and together with edema, exudate, and vasculitis, there is ischemia and infarction of arteries and veins.

In contrast, studies indicate that meningitis, in particular when due to *S. pneu-*

moniae, may result from otitis media. Swartz and Dodge (1965) and Waring and Weinstein (1948) found that about 34% of patients with meningitis due to *S. pneumoniae* had an associated otitis media. In five pairs of temporal bones from patients who died from meningitis due to *S. pneumoniae* (Igarashi et al 1974), there was massive inflammatory cell invasion along the cochlear, vestibular, and facial nerves. There was some inflammation of scalae and the modiolus, with degeneration of the spiral ganglion cells. Furthermore, in this study many inflammatory cells were found in the scala tympani near the round window and along the cochlear aqueduct. In one set of bones there was hemorrhage and destruction of the stria vascularis due to hematogenous spread. In three of the five bones inflammation of the tympanic membrane and the middle ear was present, and in some a severe inflammatory reaction along the thin bony plate of the tegmen tympani, which separates the middle ear from the middle cranial fossa. This study showed that middle infections could lead to meningitis. It suggested that in some cases of meningitis there is spread of inflammation from the meninges to the cochlea, but in others it is from the middle ear to the meninges. If there is marked suppuration in the middle ear and the basal turn of the cochlea, this suggests it is from below up.

Schuknecht (1974) examined seven temporal bones from people who had died from pneumococcal meningitis, and although in some there was evidence that the infection spread directly from the middle ear to the meninges, for example, via the petromastoid canal of the mastoid bone, in others spread to the scala tympani could have occurred downward along the cochlear aqueduct or a pathway from the internal auditory canal. The evidence presented above shows that meningitis may result from the upward direct spread from the middle ear, or alternatively meningeal infection can extend down to the cochlea.

If infection gains access to the normal cochlea, it can pass along preformed anatomical pathways to the subarachnoid space, in particular through the cochlear aqueduct. As was shown in Chapter 2, this aqueduct enters the scala tympani close to the round window membrane. The infection may also spread from the vestibule along the utricular nerve to the CSF, as was discussed in Chapter 2. The infection may also spread through the tissue overlying the numerous channels in the bone covering the spiral canal (Rosenthal's canal), and from there to the CSF in the internal auditory meatus.

The importance of suppurative middle ear disease leading to labyrinthitis and then brain infection was emphasized by Friedmann (1974), who stated, "A cerebellar abscess frequently originates from purulent labyrinthitis." He also emphasized that the infection may occur due to spread from the endolymphatic sac, a fistula in the otic capsule, localized osteitis, or the petrous apex. Lateral sinus thrombophlebitis and empyema of the cochlear aqueduct are other possible sources. However, the dura is very resistant to the spread of infection, and a granulation tissue barrier can prevent a further extension of infection.

The spread of middle ear infection to the cochlea causing labyrinthitis can more easily extend to the meninges through a dehiscence between the scalae and the internal auditory meatus in a deformed cochlea. There are a variety of congenital abnormalities of the cochlea, including the Mondini dysplasia where there are pathways providing access for infection in the cochlea to pass easily to the subarachnoid space. With the Mondini dysplasia characterized by a common cavity (C. Suzuki et al 1998), a wide dehiscence between the scala tympani and the internal auditory canal was reported and led to the spread of labyrinthitis arising from otitis media leading to fatal meningitis in the unimplanted ear. Furthermore, Schuknecht and Gulya (1986) described a cochlea with a deficiency of the modiolus and a marked confluence between the scala vestibuli and the CSF in the internal auditory canal, as was illustrated in Chapter 2, which could be the conduit for infection.

When labyrinthitis occurred either secondary to spread of infection across from the middle ear or downward from the meninges, a proteinaceous precipitate was first seen in the scala tympani and vestibuli before involvement with inflammatory cells (Igarashi et al 1974). Next, an increase in volume occurred in the scala media (ectasia). Finally, when the Reissner's and basilar membranes were damaged, the infection affected the endolymphatic space. This same pattern of involvement of the cochlea was seen in the cats' temporal bones in the Department of Otolaryngology at the University of Melbourne by the author when the infection had been clearly shown to spread from the middle to the inner ear. As observed, after otitis media was induced with *S. pneumoniae, S. pyogenes,* and *S. aureus,* there was exudation of polymorphonuclear leukocytes within the scala tympani. This was associated with a marked dilatation of the capillaries and extravasation of red blood cells. The acute infection could extend to the scala vestibuli and even the scala media. The basilar membrane and the tight junctions of Reissner's membrane offered some protection against the spread to the scala media. If the infection involved the spiral canal in the modiolus, the capillaries became grossly dilated, and the spiral ganglion cells underwent autolysis. As discussed above (see Pathophysiology), the body's defenses attempt to reduce the inflammation with an outpouring of neutrophils, monocytes, lymphocytes, and plasma cells, and then a chronic stage may supervene.

The pathogenesis of device-related meningitis is different from the community-based disease, but more like that originating from middle ear infection. There are, as discussed for the implant patients in the U.S. with meningitis in Chapters 8 and 10, indications that infection can spread from the middle ear to the inner ear when there are difficulties in sealing the single element electrode array in cases of congenital anomalies such as the Mondini dysplasia.

Histopathological findings from patients who had stapedectomies and died when middle ear infections led to labyrinthitis and meningitis throw light on the pathogenesis of meningitis following cochlear implantation. With a stapedectomy the stapes is removed and replaced with a polyethylene, Teflon, wire, or steel strut. A graft of vein, fat, or Gelfoam is placed either in the oval window or around the strut. The histopathological findings in the temporal bones of five adult patients are summarized (Rutledge et al 1963; Wolff 1964; Matz et al 1968; Palva et al 1972; Benitez 1977). The first patient developed otitis media 2 years after the stapes was replaced with a polyethylene strut, and histopathological studies showed that the infection extended through a fistula beside the strut into the inner ear where there was severe inflammation of the cochlear turns. The infection appeared to spread to the meninges along the cochlear aqueduct (Rutledge et al 1963). The second patient had a polyethylene strut attached to the incus and Gelfoam placed in the oval window. Twelve months later she died from pneumococcal meningitis. The histopathology showed the strut disengaged from the incus, and the path for the spread of infection from the middle ear to the cochlea was not reported (Wolff 1964). The third patient had a polyethylene strut replace the stapes, and it rested on a depressed footplate. He developed otitis media 19 months later, and histological examination of the temporal bones showed the infection spread to the vestibule through the mucosa and the gap behind the stapes footplate. From there it extended through an anatomical pathway from the vestibule along the utricular nerve, as was shown in Chapter 2, to the subarachnoid space. There was also considerable inflammation of the cochlea (Matz et al 1968). A fourth patient had a polyethylene strut used to replace the stapes and it rested on a fascial graft placed into a large opening in the oval window. Four years later, following an upper respiratory tract infection, the patient developed pneumococcal meningitis. The middle ear contained frank pus, and the pus filled and surrounded the polyethylene tube. There was no fascial graft under the strut, but remnants were seen at the side. There was extensive inflammation in the cochlea. The granulocytes (these cells and debris are pus) were seen to extend from the scala tympani along the cochlear aqueduct to the subarachnoid space (Palva et al 1972). The fifth patient was the only one to have a Gelfoam wire prosthesis inserted, and this was placed on top of a footplate that had been dislodged at its posterior margin into the vestibule. Two weeks following surgery, he developed purulent otitis media that led to fatal meningitis from *H. influenzae*. Pathological findings showed a severe inflammation of the middle ear with pus. The infection had extended into the vestibule through a deficiency in the oval window created by the dislodged footplate of the stapes. Pus due to acute inflammation in the vestibule had extended along the branches of the vestibular nerve to the internal auditory meatus leading to the meningitis. The pus cells had then extended back to the modiolus of the cochlea, and invaded the scala tympani and vestibule. There was total loss of the spiral ganglion and hair cells. The cochlear aqueduct, however, contained only a few such cells (Benitez 1977).

The above studies on stapedectomies led to some important conclusions concerning cochlear implantation and the risk of developing meningitis. It is clear that otitis media can extend directly to the cochlea around or through a prosthesis in which there is inadequate sealing or, as in the case of the stapes prosthesis, one that extends only a short distance into the scala vestibuli. It is an advantage that single-element electrode arrays are likely to have a longer electrode sheath or path for infection to track along. The pathway from the inner ear for infection to gain access to the subarachnoid space is either around the branches of the vestibular nerve from the vestibule or along the cochlear aqueduct after extending to the cochlea. It is less common for it to extend directly from the cochlea, whereas with meningitis the infection is more likely to extend back through the internal auditory meatus to involve the scala vestibuli and tympani.

The above evidence only serves to stress the importance of ensuring a good seal around the electrode at its entry point and a sheath extending for some distance into the cochlea. If there is a gap, as would exist with a two-element array, there is a significant risk, as shown by these studies, of infection being conducted directly to the scala tympani at a point that is close to the cochlear aqueduct, providing a path to the CSF. It should also be stressed that if any electrode system were to lead to a path from the scala tympani to the scala vestibuli and thence to the vestibule, there is a direct route along the vestibular nerves to the internal auditory meatus likely to lead to meningitis. Trauma with fracture of the spiral lamina and perforation of the basilar membrane would allow infection to track to the scala vestibuli and the vestibule. Hence combinations of trauma and infection have serious consequences.

Experimental Animal Studies

Models of Otitis Media

There have been a number of approaches to the development of animal models for otitis media. In chinchillas and rats, transbullar and transtympanic inoculation procedures, or nasal inoculation followed by negative pressure applied to the middle ear, have been used successfully (Giebink et al 1980; Lewis et al 1980; Hodges et al 1984). The infection rate of these models is now almost 100%. Furthermore, these studies have shown that the most sensitive parameter for success is the concentration of bacteria in the inoculum.

Animal models of otitis media were adapted to study the effects of cochlear implants on the spread of infection from otitis media to the inner ear (Clark, Pyman et al 1984; Clark, Shepherd et al 1984; Franz et al 1984; Brennan and Clark 1985; Berkowitz et al 1987; Cranswick et al 1987; Purser et al 1991; Dahm et al 1994). These studies were first used to examine the effects of otitis media on the unimplanted normal cochlea (Franz et al 1984, 1987). In the study by Franz et al (1984), the otitis media occurred spontaneously, and in the study by Franz et al (1987), otitis media was induced after the inoculation with β-hemolytic, type A *S. pyogenes*. In the latter study there was subacute inflammation of the round window membrane 2 to 4 weeks after the inoculation, and cultures were negative. The membrane was three times the normal thickness due to the formation of capillaries and increased interstitial fluid. This was associated with infiltration of lymphocytes, monocytes, and plasma cells, and there were pockets of purulent material. The epithelium on the round window membrane was in part replaced with ciliated columnar cells.

With inflammation due to infection (otitis media) or trauma, there was an increase in the permeability of the round window membrane. In particular, opening of the tight junctions between the epithelial cells occurred. The resulting increased permeability was seen for macomolecules like albumin and staphylococcus exotoxin (Goycoolea et al 1980; Paparella et al 1980; Schachern et al 1981). The toxins alone, without the spread of bacteria to the inner ear, led to loss of hair cells and spiral ganglion cells in the basal turn of the cochlea. Horseradish peroxidase was suitable for investigating the increased permeability, as it passed easily through widely opened tight junctions (Terrahe and Westphal 1968; Tanaka and Motomura 1981). In the study by Franz et al 1987, horseradish peroxidase passed through the epithelial layer into subepithelial cells and intercellular spaces, indicating a loss of tight junctions between cells. However, the inflammation within the round window did not produce a reaction in the scala tympani, and the cochlea was normal.

Although the above studies by Franz et al (1984, 1987) did not show *S. pyogenes* passing though the round window membrane, this was not the case for *S. pneumoniae* in the studies by Berkowitz et al (1987) and Dahm et al (1994). Inflammation extended to the cochlea producing labyrinthitis in four of nine (44%) normal unimplanted kittens' ears when otitis media was induced with *S. pneumoniae* (Dahm et al 1994). In two there was serous labyrinthitis and in two suppurative labyrinthitis. When suppuration occurred, there was necrosis of the round window membrane as shown in Figure 3.7. A high incidence of labyrinthitis (56%) following *S. pneumoniae* otitis media was also found in the experimental animal by Meyerhoff et al (1980). The propensity for *S. pneumoniae* to produce labyrinthitis may be due not only to the increased permeability in the round window demonstrated with the horseradish peroxidase studies, but also to the nontoxic polysaccharide capsule that enables the organism to evade phagocytosis.

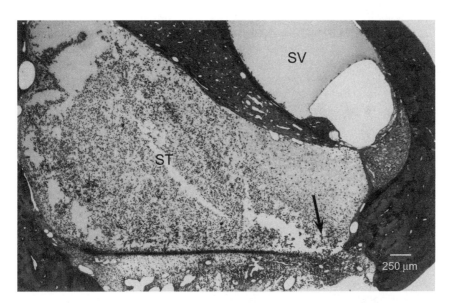

FIGURE 3.7. Labyrinthitis in one of the unimplanted controls inoculated with *Streptococcus pneumoniae*. The round window (incomplete) is indicated by the arrow. ST, scala tympani; SV, scala vestibuli. (Dahm et al 1994. Cochlear implants in children: labyrinthitis following pneumococcal otitis media in unimplanted and implanted cat cochleas. *Acta Oto-Laryngolocia:* (**suppl 114**). Reprinted with permission.)

Histopathology of the Cochlear Implant Entry Seal

A study was first undertaken on 12 implanted cat cochleae without inoculation of bacteria by (Franz et al 1984). The round window was incised and no graft placed around it. The electrodes were left in for periods that varied from 1 week to 5 months before the animal was sacrificed. In another study on 23 cochleae that were both uninoculated and inoculated, the electrode seal was also studied for an insertion through the round window without a graft (Cranswick 1984; Cranswick et al 1987). In both studies, after the electrode had been implanted for 1 to 2 weeks, the round window membrane was thickened, especially where it bordered the prosthesis, and it reached more than twice its former thickness. The epithelium had a single layer of squamous cells or two to three layers of cuboidal or columnar cells, and in many areas they were ciliated, especially near the electrode insertion site (Fig. 3.8). The subepithelial layer made the greatest contribution to the increased thickness of the membrane. This resulted from connective tissue, capillaries, and the accumulation of interstitial fluid. Acute polymorphonuclear cells were present in the first week, and in addition there were monocytes, lymphocytes, and plasma cells scattered throughout the tissue, as occurs in response to trauma (Clark 1987). The subepithelial layer formed a thick ring of connective tissue around the prosthesis, and here the membrane was slightly thicker than elsewhere. The inner, denser portion of the subepithelial layer remained intact throughout most of the membrane. It was not apparent whether this inner subepithelial layer was responsible for the connective tissue envelope that formed around the prosthesis. When a sheath commenced to form (type 1 seal), the collagen fibers appeared fragmented, and there were few vessels and inflam-

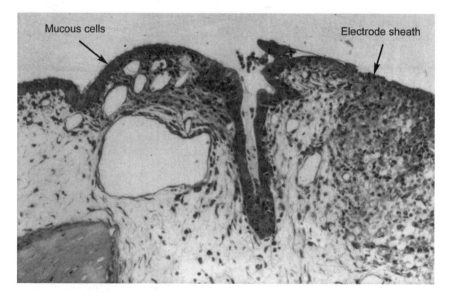

FIGURE 3.8. A photomicrograph of the electrode entry point to the scala tympani showing the mucous cells that have formed on the round window extending to the electrode sheath.

matory cells. This type 1 seal or electrode sheath was effective in preventing spontaneous middle ear infections in the cat extending to the inner ear several weeks after implantation, as illustrated in Figure 3.9. In other bones there were connective tissue strands (type 2 seal) extending from the mesothelial cells into the scala tympani, especially beneath the spiral lamina (Fig. 3.10).

The mucosa of the middle ear showed activity in the neighborhood of the implanted round window membrane. Mucosal buds called protuberances, consisting of connective tissue covered by columnar ciliated epithelium, appeared at the border of the bony window, and projected into the round window niche. Similar smaller protuberances were also formed by the round window membrane where it touched the prosthesis. They were filled with polymorphonuclear leukocytes and large collections of plasma cells and lymphocytes.

In the study by Cranswick (1984) and Cranswick et al (1987), the type 1 seal with a fibrous tissue envelope around the array (Figs. 3.9 and 3.11) was less frequently seen, and this was probably due to the fact that the insertion was through the round window, and no fascial graft was placed at the entry point. The tissue envelope around the electrode frequently had columnar mucus-secreting cells extending from the round window membrane (Fig. 3.8). Otherwise the track had a mesothelial lining of one or two layers of squamous cells. Around these

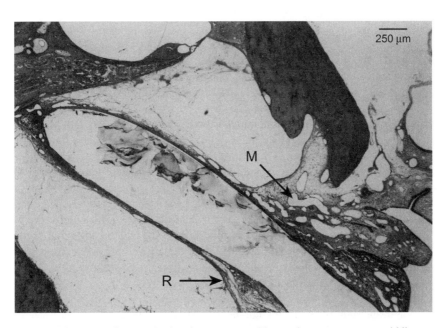

FIGURE 3.9. Photomicrograph showing a cat cochlea and a spontaneous middle ear infection several weeks after implantation. An envelope has formed around the prosthesis inside the scala tympani, the round window membrane is thickened (R), and the round window niche is filled with mucosal folds (M). The organ of Corti and neurons are intact (Reprinted from Cochlear Prostheses, Clark et al, © 1990, with permission from Elsevier Science.)

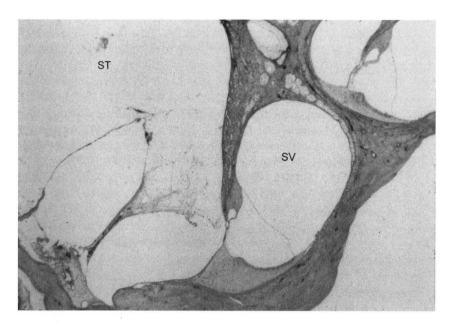

FIGURE 3.10. A photomicrograph of the electrode entry point where a sheath has not formed, but connective tissue extends from the round window membrane to the endothelial lining of the scala tympani (seal type 2).

lining cells there was a fibrous tissue capsule that varied in thickness composed of bundles of collagen parallel to the electrode. Localized collections of polymorphs, lymphocytes, and macrophages were seen, and some had penetrated the membrane to lie between the electrode envelope and the actual prosthesis. There were an increased number of inflammatory cells opposite the platinum bands.

In essence the type 1 seal externalized the short electrode in the cat, but would not probably extend to the end of the longer human array, as it is inserted a greater distance. Nevertheless, in the human temporal bones studied a sheath was frequently found extending a distance into the cochlea (Fig. 3.12). In the second or type 2 seal, as illustrated in Figure 3.10, the metaplastic epithelium or mucous cells passed along the electrode for a short distance into the cochlea before merging with connective tissue, which was connected to the endothelial lining of the scala tympani. For both type 1 and type 2 seals there was often connective tissue in the first part of the scala tympani, particularly beneath the spiral lamina. Sometimes the connective tissue in the basal turn was filled with many acute and chronic inflammatory cells. By 4 weeks the morphological changes of inflammation in the round window membrane had usually regressed (Franz et al 1984). It was, however, still thicker than normal. The membrane thickness was evenly distributed except close to the prosthesis where the connective tissue was still thicker. The epithelial layer had become thinner and almost regained its former appearance. In some the membrane was still covered by pseudostratified ciliated columnar cells with many interspersed goblet cells. In all cases the ciliated cells

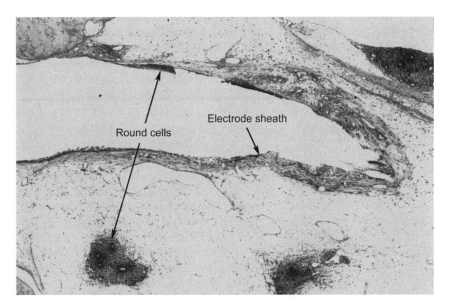

FIGURE 3.11. A photomicrograph of an electrode sheath of fibrous tissue (electrode seal type 1) with accumulations of round cells between the sheath and the array and within the loose connective tissue around the sheath. This demonstrates that with a single-element array the capillaries can bring the bodies defense mechanisms to the site of an infection. (Cranswick NE. 1984. Studies in the cochlear round window. B Med Sci thesis. The University of Melbourne. Reprinted with permission.)

extended down into the electrode track. The subepithelial tissue remained swollen for a further 2 to 3 weeks. The number of capillaries was reduced, and the white cells had all become mononuclear leukocytes. The distribution of horseradish peroxidase was becoming similar to that for the normal round window. The tracer substance was seen mainly in the epithelial layer. It crossed the basement membrane and appeared in the subepithelial cells, but was not evident in the intercellular space. The tracer passed easily into the gap between the prosthesis and its membranous sheath. It appeared in the connective tissue envelope and was still seen in the scala tympani. The type 1 seal at 4 weeks did not have a consistent lining of ciliated mucosal cells, and sometimes the prosthesis came in direct contact with the fibrous tissue. There were still localized collections of round cells between the sheath, and the prosthesis as illustrated in Figure 3.11.

After 1 month's implantation the membrane appeared only a little thicker than normal due to the presence of distended blood and lymph vessels. There were a few scattered inflammatory cells. The membrane had regained its former barrier to horseradish peroxidase, but the gap between the membrane and the prosthesis had not closed. The inflammatory changes in the type 1 and 2 seals had largely regressed. These studies thus demonstrated an increased permeability of the round window membrane that could apply to bacterial toxins and other macromolecules for the first 2 to 4 weeks after implantation (Franz et al 1984). Middle ear infec-

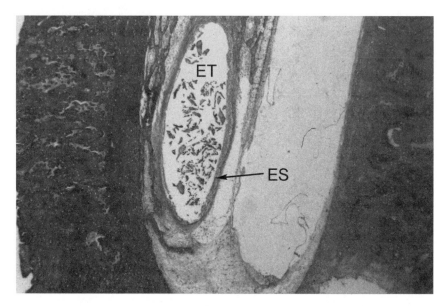

FIGURE 3.12. A photomicrograph of the scala tympani of the cochlea in a patient who had a cochlear implant 42 years after meningitis and used the implant regularly for 27 months before dying from unrelated causes. ET—electrode track; ES—electrode sheath.

tions during this time could lead to inner ear infection. If they occurred after this period, infection was less likely (Fig. 3.7). However, the gap between the prosthesis and the round window membrane persisted after 4 weeks, as demonstrated by the ease with which horseradish peroxidase could pass along this space. Nevertheless, a fibrous tissue envelope, especially the type 1 seal, and the mucous cells formed in the vicinity of the electrode entry point had the potential to be a barrier to infections in the middle ear extending to the cochlea.

Pathogenesis of Inner Ear Infection Postimplantation

Animal models of otitis media were adapted to study the effects of cochlear implants on infection from otitis media spreading to the cochlea, with the possibility of it leading to meningitis (Clark, Pyman et al 1984; Clark, Shepherd et al 1984; Franz et al 1984; Brennan and Clark 1985; Berkowitz et al 1987; Cranswick et al 1987; Purser et al 1991; Dahm et al 1994). Animals were implanted long term with single scala tympani electrode arrays, and the bulla inoculated at a specific period following the implant surgery. The period between implantation and inoculation was varied to study the vulnerability of the round window seal during its healing process. The effects of several types of microorganism were successfully investigated using the procedures in the cat. The organisms were *S. pneumoniae, S. pyogenes, S. aureus,* and *P. aeroginosa.* The studies indicated the most likely period for the spread of infection from the middle ear into the cochlea would be during the first few weeks after implantation. However, later disturbance of the seal by traction on the electrode during inoculation could be a factor. It was found that if there were a strong fibrous tissue response in the first part of the basal turn and the development of a fibrous sheath (type 1) around the electrode, this would impede the spread of infection. This created a physical and biological barrier. The biological barrier permitted the blood vessels to bring white cells and antibodies to the site. The development of mucus-secreting cells around the electrode entry point and along the electrode sheath could also have a bacteriostatic effect.

The importance of a cochlear implant entry seal was first seen in the study by Franz et al (1984), in which three cats developed acute spontaneous infection in the bulla at approximately 4, 8, and 12 weeks postimplantation. A seal had formed, and there was no spread to the cochlea, with loss of spiral ganglion cells. In these animals an incision was made through the round window membrane, but no graft inserted. There were similar responses of the round window membrane in all three cases, and they were similar in kind to the reaction of the tissue to implantation per se as discussed above. The round window membrane formed mucus-secreting cells, and there was edema and polymorphonuclear cell infiltrates in the middle layer and increased fibrous tissue production in the deeper portions of the membrane. In the round window niche, mucosal folds were a prominent feature. There was also an increased fibrous tissue response around the electrode. These responses to the acute infection effectively protected inner ear structures, as evidenced by an intact organ of Corti, and preserved neurons. In these infected cochleae horseradish peroxidase passed into the intercellular spaces of the subepithelial tissue and the electrode sheath. Its concentration, however, diminished toward the perilymphatic space, and there was less evidence of the substance in the scala tympani, where its concentration was similar to that in unimplanted cochleae 1 week postimplantation. The above findings indicated the protective effects of the fibrous tissue sheath as well as the mucus-secreting cells lining the membrane and sheath. The protection provided by the sheath is also illustrated in Fig. 3.11, which shows the presence of lymphocytes and monocytes between the electrode and the sheath and within the tissue of the sheath.

In the study by Cranswick (1984) and Cranswick et al (1987) on 15 ears with induced *S. pyogenes* otitis media, there were only two with acute inflammation throughout the cochlea. This occurred first in one where there was a type 2 seal, and *E. coli* was found at inoculation, indicating it had been introduced at the time of the initial surgery. Second, there was extensive inflammation throughout a cochlea when there was a fracture of the spiral lamina close to the round window. These two examples emphasize the risk of introducing infection at the initial operation before a seal has formed, and the hazard of labyrinthitis developing with an electrode design or procedure that leads to a fracture of the spiral lamina or tear of the basilar membrane. In this case the trauma provides a greater path for the infection to extend into the cochlea, the spiral canal, the scala vestibuli, and the vestibule, and from these sites to the CSF. But as explained above (see Pathogenesis), if it extends to the utricle via the scala tympani, there is a rapid and easy pathway along the utricular nerves to the CSF. In addition to the two ears with acute inflammation throughout the turns of the cochlea referred to above, there were two ears with mild inflammation in the distal part of the basal turn for a type 1 seal. All the other ears were infection free. Furthermore, in the eight unimplanted ears there were three in which acute inflammation from a spontaneous middle ear infection extended throughout all cochlear turns. The most acute inflammation was seen in two ears in which the inflammation occurred 2 weeks postimplantation, that is, before an impermeable or well-healed seal had occurred. The above research suggested that there is a need to improve the seal by creating a type 1 sheath and an electrode entry that facilitates this sheath.

Sealing Procedures and the Spread of Middle Ear Infection

The first study to examine methods of sealing the electrode entry point to prevent the ingress of infection was undertaken by Clark, Shepherd et al (1984). This research compared the use of a muscle autograft around the electrode and a Teflon felt disk glued to the array for round window insertions in 14 ears of uninoculated cats. This aimed to determine their effectiveness in the presence of naturally occurring otitis media, as well as in six ears following otitis media due to the inoculation of *S. aureus,* an organism that causes otitis media in children, but with an incidence of less than 10%. In the uninoculated group the animals were implanted for periods varying from 59 to 95 days before sacrifice. In the inoculated animals, otitis media was induced 68 to 70 days following implantation, and the animals were sacrificed after a further 14 days. The Teflon felt used had a tight weave that allowed only moderate penetration of fibrous tissue (Clark, Pyman et al 1980). The inflammatory response was rated according to the severity of the acute or chronic inflammation and its extent, as outlined by Clark, Shepherd et al (1984). In the uninoculated controls there were five ears with no localized inflammation in the first part of the basal turn, seven with it localized to the first part of the basal turn, and in two there was generalized chronic inflammation extending throughout the cochlea, one with a muscle autograft and another a Teflon felt disk. The inflammation in these two ears was due to a naturally occurring infection of the middle ear. In the other groups (inoculated with *S. aureus*) there was a more localized inflammatory response due to the insertion of the electrode per se as seen with other control ears. However, in the six inoculated ears there was no infection. Although the numbers were small, the results suggested that the seal might be adequate for the *S. aureus,* but not necessarily for other organisms and naturally occurring otitis media. This was consistent with the fact that *S. aureus* produces a coagulase enzyme that walls off the infection, preventing access from the body's defenses, and thus it may not be as invasive as other organisms, and thus more readily stopped by the seal. It was also noted that a muscle autograft remained avascular and necrotic and was a nidus for infection. This could have accounted for the one infection in the uninoculated group.

A second study (Clark, Shepherd et al 1984) was undertaken to compare Dacron velour and fascia as seals for the electrode entry point. A prior investigation by Clark, Pyman et al (1980) had shown, through implanting material in rat subcutaneous tissue, that Teflon felt had only a moderate ingrowth of fibrous tissue, while with Dacron velour there was a marked ingrowth of fibrous tissue and a pronounced foreign body reaction. It was considered that more penetration by fibrous tissue might greatly increase the path length for bacteria to reach the cochlea. Fascia was selected, as the avascular muscle grafts had provided a home for infection. The electrodes had four platinum bands, and the Dacron sleeve was glued to the Silastic tube with medical adhesive silicone type A. The creation of the electrode entry closely modeled the human insertion technique by retracting the round window membrane and drilling the bone on the lateral aspect of the round window niche. The Dacron sleeve was positioned within the round window niche rather than outside it. The facial autograft was also placed around the Dacron sleeve and superficial to it. An electrode without Dacron was inserted into the opposite ear, and also surrounded with a fascial autograft. There were three cats (six ears) in the uninoculated control group, and they were implanted for periods of 20 to 28 days. The other three cats (six ears) had otitis media induced with β-hemolytic type A *S. pyogenes* at 29 to 30 days postimplantation. *S. pyogenes* was selected, as it has different biological characteristics from *S. aureus,* with the production of lytic enzymes that dissolve tissue making it more invasive, and it also produced less than 10% of cases of otitis media in children. After inoculation the cats were sacrificed 14 days later.

The results in the uninoculated group showed extensive inflammation in two of three ears in the case of the Dacron sleeves, and the inflammation extended through the basal, middle, and apical turns. The fascia control electrode had slipped out, making comparison difficult. It was, however, significant that in the uninoculated ear there was extensive and acute inflammation associated with the Dacron sleeve placed within the electrode entry point. In the cats inoculated with *S. pyogenes,* there was evidence of acute or resolved inflammation for each cat inoculated with Dacron and fascia as illustrated in Figure 3.13, but a much less severe and a more localized inflammatory response when fascia was used alone. The data showed that acute and chronic inflammation in the first part of the basilar turn of the cochlea was more common with a Dacron seal than with a facial autograft. The same applied to the second half of the basal turn. Thus any foreign bodies that have a dead space for protein to collect and do not have a blood supply to bring antibodies to the region will be a home for infection. If they are placed between the middle and inner ears, infection will be transmitted across the boundary. Furthermore, it has been shown (see below) that tissue cages can make an organism more virulent (Zimmerli et al 1982). It was also noted in the above study (Clark, Shepherd et al 1984) that infection was more likely to track on the side of the thinner round window membrane, where a less extensive sheath had developed, than where the bone had been drilled. This suggested not only the importance of the electrode sheath, but also possibly siting the opening through the bone overlying the cochlea.

Further research was necessary to develop the most reliable method of ensuring that severe otitis media would occur in order to test the electrode seal. After inoculating the middle ear in the study by Clark, Shepherd et al (1984), *S. aureus*

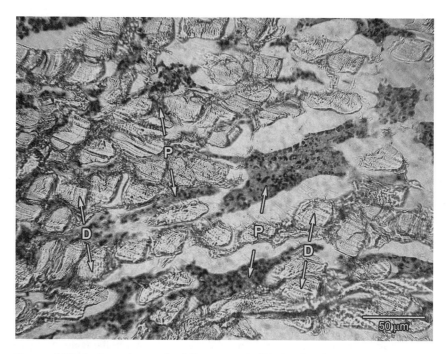

FIGURE 3.13. A photomicrograph of Dacron mesh (D) showing the pus (P) between the fibers. The Dacron mesh was implanted in the round window and an otitis media induced with *Streptococcus pyogenes*.

was cultured in three of five of the inoculated ears, but *S. pyogenes* was not found after 14 days. In most of the inoculated and uninoculated cochleae there was inflammation both outside and within the round window so that there was no clear-cut histological evidence of infection in the absence of positive cultures. In an experiment on four cats (Brennan and Clark 1985) it was shown that *S. aureus* otitis media was induced in each animal inoculated via the eustachian tube, and then the tube was occluded with a Teflon paste to prevent the inoculum draining out the tube. Furthermore, in the study it was noted on the histological examination of the tissue around the electrode that the sheath had a smooth mesothelial lining that could facilitate the passage of bacteria. It was observed, however, that this lining did not occur opposite a platinum band where there was greater adhesion to tissue. This could be an added advantage when using these bands for additional stiffening.

In light of the above findings, research was undertaken to investigate in more detail the tissue reaction to the prosthesis and the difference between tissue apposition to platinum or Silastic in limiting the spread of infection (Cranswick 1984; Cranswick et al 1987). The study examined the seal in the presence of induced *S. pyogenes* otitis media. This was carried out reliably by inoculating 10^8 organisms/mL, and packing the bulla with Gelfoam to retain the organisms. Direct inoculation had an advantage over eustachian catheterization in that the round window could be inspected prior to the inoculation. The study was carried out on 12 cats. Eight animals had *S. pyogenes* middle ear infections induced at 1, 2, 4, and 8 weeks postimplantation, and they were sacrificed 1 week later. In each cat a plain Silastic electrode was inserted in one ear and a Silastic electrode with bands in the other. In addition there were four uninoculated cats with the two types of electrodes in either ear. They were sacrificed at comparable times to the inoculated ones. The data showed, as discussed in the previous subsection, that there was generalized cochlear infection in two cochleae for the plain Silastic and three for the Silastic with platinum bands in both the uninoculated spontaneously infected and inoculated cats. In two of the acutely infected cochleae, gram-positive organisms were seen in the round window region. In a majority of inoculated ears the bacteria were not seen on Gram stain. As with the study by Clark, Pyman et al (1980), there was evidence of inflammation in the round window and basal turn in the uninoculated and inoculated ears, and no demonstrable correlation with culture of Gram stain results. The data nevertheless showed that there was no advantage in using a platinum band to assist with the electrode seal.

The above data showed there were deficiencies in the electrode seal with the electrode inserted through the round window without an accompanying fibrous tissue autograft. Furthermore, as a cochleostomy or opening drilled through the outer wall of the cochlea near the round window gave more direct access for the insertion of the straight but flexible Nucleus banded array than one through the round window membrane, a study was undertaken to determine the effectiveness of the two procedures in limiting the ingress of middle ear infection (Franz et al 1987). The research was undertaken on four cats, and the approaches compared on opposite ears. Again the entry was not grafted and banded electrodes were used. The middle ears were inoculated with *S. pyogenes* from 4 to 23 weeks after implantation. The cats were sacrificed at 3, 6, 7, and 9 weeks after inoculation. The *S. pyogenes* inoculation led to acute inflammation in the middle ears of the control unimplanted and the implanted ears. The histological effects of the infection on the round window membranes and particularly the electrode seal are discussed in the subsection above. With the round window insertion, both the round window membrane and electrode sheath were inflamed, but the organ of Corti and auditory nerves were preserved. A fractured spiral lamina occurred in one case in which there was a localized loss of neurons, but this had no effect on the spread of infection as it had not entered the cochlea. With the cochleostomy on four cats there was chronic inflammation of the mucosa around the electrode entry as well as in the neighboring round window membrane. The sheaths of the electrodes were thickened with organized connective tissue. There was, however, no spread of infection to the cochlea in these ears either. With the cochleostomy, because of the cat anatomy, there was a higher incidence of spiral lamina fractures with an associated neural loss. Horseradish peroxidase entered the basal turn of the cochlea between the prosthesis and the bony wall and from there via the envelope into the scala tympani. Hardly any trace was found in the organ of Corti. The study thus showed comparable results for a round window insertion or a cochleostomy, although the fibrous tissue sheath appeared better developed around the cochleostomy and there was a greater potential for using sealing procedures for this method of insertion (Fig. 3.4).

A pilot study was next undertaken with *S. pneumoniae* and *P. aeruginosa* (Berkowitz et al 1984, 1987). The electrodes were inserted through the round window without any graft to assist in sealing. *S. pneumoniae* was used as it is the most frequent organism causing otitis media in children. Its virulence as stated above is caused by the presence of a capsular polysaccharide that is nontoxic, but acts to neutralize antibodies before they can bind to the pneumococcus, and thus promote infection. *P. aeruginosa* produces exotoxins that can result in a virulent infection of the inner ear (labyrinthitis). Twelve weeks after implantation in two animals, *S. pneumoniae* otitis media was induced and the animals sacrificed 10 days later. The infection caused pathological changes in both the implanted and the control (nonimplanted) cochleae. In the control cochleae, there was a mild dilatation (hydrops) of the cochlear duct, collapse of the supporting cells, atrophy of the stria vascularis, mild protein precipitates in the perilymphatic space, and mild ganglion cell losses throughout the cochlea (Berkowitz et al 1987). This indicated that *S. pneumoniae* was more likely to invade the inner ear through a normal round window than *S. pyogenes,* as also demonstrated in the studies of Franz et al (1987). There was more severe inflammation in the implanted cochleae and widespread loss of neural elements. This study suggested that trauma could potentiate an infection with *S. pneumoniae,* and that a good seal was needed to prevent labyrinthitis. With the *Pseudomonas* infection more severe changes were found in the cochlea (Berkowitz et al 1984). The cochlear duct was collapsed, and Reissner's membrane was fused with the atrophic stria vascularis. The organ of Corti with its hair cells and supporting elements was absent not only in the basal turn but throughout the cochlea. In addition, mild ganglion cell degeneration was present and mainly in the basal turn. The cochlear aqueduct appeared closed with thick fibrous tissue and at its entrance new bone formation could be seen. The responses of the round window membrane also differed in these two inoculation studies. One week after inoculation with the pneumococcal inflammation, the membrane was thickened and showed the presence of secretory cells. There was a cellular infiltrate of polymorphs, lymphocytes, and fibroblasts. In contrast, the *Pseudomonas* infection was characterized by a nonproliferative inflammation.

In view of the propensity for *S. pneumoniae* to invade the inner ear from the middle ear, and the fact that in young children it is the commonest cause of otitis media, further research was undertaken to help ensure that labyrinthitis and meningitis would not be a significant risk for young children (Purser et al 1991; Dahm et al 1994, 1995). The research examined the most effective methods of sealing the electrode entry point. A study on five cats evaluated the effectiveness of a titanium collar as a seal for a cochleostomy. The gap between the array and collar was sealed with biocompatible glue (Silastic type A). Although there was no integration with bone, fibrous tissue adhered to the collar (Purser et al 1991) as illustrated in Chapter 8. The seal was effective in the five implanted cochleae, and there was no inflammation in any of these cochleae after inoculation with *S.*

pneumoniae 12 weeks postimplantation. On the control, unsealed side infection was present in one of four cochleae.

The above data demonstrated that a very good seal at the entry point or a type 1 seal or fibrous tissue sheath is important in preventing middle ear infection extending to the cochlea, especially against *S. pneumoniae*. The studies indicated it was not enough to allow a seal to form of its own accord. For this reason a study was undertaken in 21 kittens to compare different sealing techniques in the presence of pneumococcal otitis media. The ears were divided into those implanted with a fascial seal, implanted with Gelfoam seal, implanted with no seal, and unimplanted controls (Dahm et al 1994, 1995). An otitis media was initiated 8 weeks after implantation, and the animals sacrificed 1 week later. At the time of sacrifice the bullae were swabbed to determine whether there was ongoing infection. The temporal bones were sectioned to examine the degree of inflammation in the middle ear and its extension to the cochlea. Labyrinthitis was present in 44% of the unimplanted controls (Fig. 3.7), 50% of the implanted ungrafted cochleae (Fig. 3.14), and 6% of the implanted grafted (fascia and Gelfoam) cochleae (Fig. 3.15). There was no statistically significant difference between the unimplanted control and the implanted cochleae ($p < .05$). There was, however, a difference between the implanted ungrafted and the implanted grafted cochleae, but not between the use of fascia and Gelfoam. The data therefore indicated that cochlear implantation did not increase the risk of labyrinthitis following pneu-

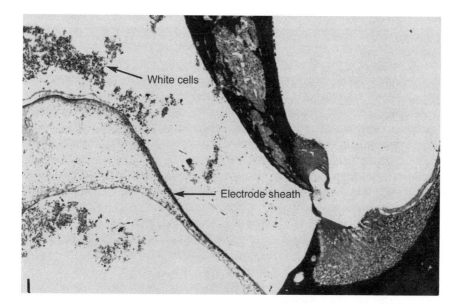

FIGURE 3.14. A photomicrograph of the basal turn of the cat cochlea where the electrode entry through the round window membrane was ungrafted, and an otitis media was induced with *Streptococcus pneumoniae*. The neutrophils reacting to the bacterial infection passing outside the thin electrode sheath are shown.

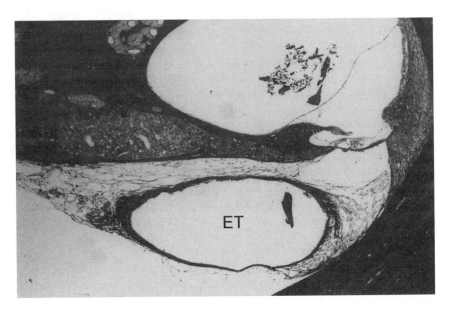

FIGURE 3.15. A photomicrograph of the basal turn of the cat cochlea where the electrode entry through the round window membrane was grafted with fascia, and an otitis media was induced with *Streptococcus pneumoniae*. This has formed a fibrous tissue sheath or type 1 seal. ET—electrode track.

mococcal otitis media, but there was a reduced incidence of infection when the entry point was grafted.

Therefore, for safety it is essential to place a graft around the electrode where it enters the cochlea. Although there was no statistically significant difference between fascia and Gelfoam, it is recommended that fascia and not Gelfoam be used. Gelfoam was used in the animal models to produce otitis media as described previously. If bacteria are introduced at surgery with Gelfoam around the electrode entry point, it could act as a home (nidus) for infection (Clark and Shepherd 1997). These experimental results apply to the Nucleus free-fitting array only. It must be stressed that a two-element array with members close to each other should not pass from the middle to the inner ear. A space between them is a conduit for infection, a home to allow pathogens to multiply, as well as a site to increase the pathogenicity of the organisms and reduce the ingress of antibodies and antibiotics. This is especially important considering the above studies showing the invasiveness of *S. pneumoniae*.

Host Factors and Foreign Bodies

Implanted foreign bodies, as discussed above (see Biocompatibility of Materials), are not totally inert, and should be evaluated for tissue toxicity. Foreign bodies markedly increase the pathogenic potential of organisms of low virulence, for example, *Staphylococcus epidermidis* (Lowy and Hammer 1983). Many studies have shown that a bacterial inoculum that is normally "subinfective" will lead to a severe infection in the presence of implanted material such as devitalized and crushed muscle and gelatin (Vaudaux et al 1994). Finally on the basis of the above evidence it is apparent that any dead space between the two members of a dual element array would not only be a pathway for infection to enter the inner ear and a home for the pathogens to multiply, but also would allow them to become more virulent.

The effect of a dead space either within a foreign body or between two bodies has been investigated by Zimmerli et al (1982) using Teflon perforated cylinders (tissue cages). With this and other implants producing a dead space (Bergan 1981; Marchant et al 1986), an inflammatory exudate accumulated within the cages within 2 to 4 weeks. If these tissue cages were infected with an organism at levels much below those normally causing infection, there would be a virulent inflammation with the ingress of polymorphonucleocytes and the formation of pus. This demonstrated that a dead space could make organisms more virulent. Furthermore, the tissue cage model also showed that parenteral antibiotics were ineffective against the organisms in the cage if administered more than 12 hours after the inoculation. This inefficacy of antibiotic therapy is commonly observed in the clinical context of foreign body infections (Vaudaux et al 1994).

Furthermore, it was shown that the phagocytic activity of neutrophils in the cages was markedly deficient and lower than observed with neutrophils from acute and chronic peritoneal exudates or blood. This suggested the neutrophils could be damaged through contact with the surface of foreign bodies, and this would reduce their antibacterial activity (Zimmerli et al 1982). Or alternatively it was associated with a reduced level of opsonins and complement in the tissue cages (Zimmerli et al 1982). In a later phase of infection, complement-mediated opsonic activity was reduced, and this too limited the ability of body to handle infection. Thus any dead space created within and across the inner ear is not only likely to be a path or home for infection, but also will increase the virulence of the organism and reduce the body's ability to deal with the infection either through phagocytic action or complement-mediated responses. It has also been shown with dead space that the access for antibiotics is significantly reduced. In addition, the studies with the infected tissue cages showed there was no associated bacteremia or spread by the bloodstream.

The penetration of antibiotics to infected locations almost always depends on passive diffusion. The rate is proportional to the concentrations of a drug in the plasma or extracellular fluid. Drugs that are extensively bound to protein may not penetrate to the same extent as those with lesser links. Drugs that are highly protein bound may have reduced activity because there is a smaller fraction to react with its target. For example, the drugs cefotaxime and ceftriaxone, both third-generation cephalosporins and the treatment of choice for *H. influenzae* and *S. pneumoniae* infections, have different degrees of binding. Ceftriaxone is used for adults and Cefotaxime in children. Ceftriaxone, however, is 90% to 95% protein bound, and that greatly reduces its efficacy. On the other hand, cefotaxime is only 36%. Vancomycin should be added to the therapy if the minimum inhibitory concentration (MIC) for these antibiotics is greater than 0.12 mg/L. Thus if there is a dead space as seen with a two-element array, the penetration of the antibiotics could be considerably reduced. In addition, in preventing infection spreading to the meninges many antibiotics that are polar and at a physiological pH do not cross the blood–brain barrier at all well. Some such as penicillin G are even actively transported out of the CSF by active transport mechanisms in the choroid plexus. The concentrations of penicillin and cephalosporins in the CSF are usually only 0.5% to 5% of the steady-state level in the plasma (Quagliarello et al 1986). The integrity of the blood–brain barrier, however, is diminished during bacterial infection.

With infections from *S. pneumoniae* and other pathogens, there is also the added problem of their developing a biofilm, a slime on the surface of the foreign material, and this will allow them to resist antibiotics and antibodies. Bacteria that adhere to implant materials by encasing themselves in a hydrated matrix of polysaccharide protein form a slimy layer known as a biofilm (Stewart and Costerton 2001). Bacteria in the biofilm are resistant to antibiotics. For example, a β-lactimase negative strain of *Klebsiella pneumoniae* had a MIC of 2 μg/mL of ampicillin in aqueous suspension, but when grown as a biofilm the organism was scarcely affected by 4 hours' treatment with 5000 μg/mL of ampicillin, a dose that would eradicate free-floating bacteria. The antibiotic resistance that normally occurs due to efflux pumps, modifying enzymes, and target mutations does not seem to apply to this mechanism of drug resistance with biofilms.

Furthermore, because active and inactive microbes are closely situated and because surviving bacteria can use dead ones as nutrients, the new cells remaining after antibiotic therapy can restore the biofilm to its original state in a matter of hours.

Single-Component Array and the Natural Defenses Against Infection

A single component array that is surrounded with a fibrous tissue sheath can, as described above, effectively work with the body's three defense mechanisms to prevent the ingress of infection from an otitis media to the cochlea and thence the meninges. The above studies demonstrated that the sheath around the single component array enabled three lines of defense to be used against the spread of infection. The first line of defense is the surface activity of mucus-secreting cells, and their extension around the electrode. The second line of defense is the mobilization of phagocytes in and around the sheath. The third line of defense is the mobilization of type B lymphocytes, and type T lymphocytes to the sheath and between the sheath and the electrode.

With the first line of defense against the spread of infection from otitis media, the surface cells around the electrode entry changed into mucus-secreting cells and extended around and along the electrode array. They produced mucus that is bacteriostatic, and the hairs of the mucous cells beat to and fro to sweep the bacteria away. Their growth around the electrode is illustrated in Figure 3.8.

The second line of defense operates when the bacteria release toxins into the sheath. The blood vessels dilate and bring the phagocytes to the site so they can digest the bacteria. This is illustrated in Figure 3.16.

The third line of defense is the production of type B and T lymphocytes, in response to the bacterial surface antigen. The B lymphocytes produce antibodies and the T lymphocytes are killer cells that pierce cells. Note that in Figure 3.11 the lymphocytes not only lie in the connective tissue around the sheath, but also enter between the sheath and the array.

Clinical Protocol

The results obtained from animal studies indicate that there is a risk of otitis media extending into the inner ear after implantation during the first few weeks, a period of increased vulnerability due to the increased permeability of the tissues and the need for the seal to form. To minimize the risk of infection spreading into the inner ear during or after implantation, it is recommended that surgery should be carried out under strict aseptic conditions, preferably using a laminar flow of filtered sterile air. Systemic antibiotics should be administered at the beginning and conclusion of the operation to eradicate organisms introduced during the procedure that could invade the inner ear during the period of increased vulnerability when the electrode seal is being established. As a further safeguard the operative wound should be irrigated with an antibiotic solution of ampicillin and

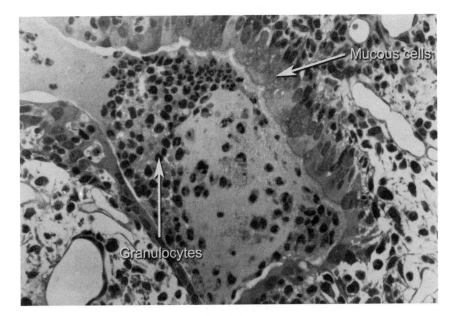

FIGURE 3.16. Phagocytosis of bacteria—the second line of defense. Photomicrograph shows granulocytes and debris in apposition to mucous lining cells.

cloxacillin. Although not the first-line antibiotics for the treatment of *S. pneumoniae* infections, they have a broad spectrum of activity. In the U.S., of the children who had meningitis, one child with a ventriculoperitoneal shunt developed the infection within a day or two of having the implant, and two with normal cochleae developed it within 24 hours. It is likely that the causal *S. pneumoniae* could have been introduced into the perilymph and thence the CSF. For this reason irrigation seems warranted. This is especially desirable as unpublished studies by Black and Clark showed that antibiotic concentrations were very low in the perilymph of the cat unless the cochlea was infected; furthermore the blood–brain barrier does not allow antibiotics to easily enter the CSF in the uninflammed condition. In children with the Mondini syndrome, special care should be taken as there can be a wide dehiscence between the scala tympani or the scala vestibuli and internal auditory canal. The data in the experimental animal presented in the sections above also demonstrate the necessity of a fascial autograft, which should be placed around the electrode in the cochleostomy. I have experimental unpublished data to suggest that if there are gaps between strips of fascia, they could be a passage for pathogens to enter the cochlea. If there is a "perilymph gusher" at surgery, then the fascia will need to be compressed quite firmly. The fascial autograft can be taken from the temporalis fascia. It is not desirable to use crushed muscle, as it can become necrotic and a home for infection. Bone paté provides spicules of bone that are not absorbed and may also be a nidus for infection, as may Gelfoam. Furthermore, as stated above, there are serious risks associated with the use of a two-element electrode array.

After the tissue around a cochleostomy or the implanted round window has healed, the response to infection appears similar to that of a nonimplanted cochlea. However, certain microorganisms could have a detrimental impact as seen with *S. pneumoniae* or *P. aeruginosa*. Improving the seal at the entry point still requires further research with other biocompatible materials and techniques.

Deafness and the Central Auditory Pathways

Spiral Ganglion

With the loss of hair cells there is a rapid and extensive reduction of the unmyelinated peripheral processes in the organ of Corti (Terayama et al 1977), and a more gradual degeneration of the myelinated portion of the peripheral processes within the spiral lamina as well as the spiral ganglion cells (Webster and Webster 1981; Spoendlin 1984; Leake and Hradek 1988; Shepherd and Javel 1997). Some surviving cells and processes may be demyelinated. These changes as discussed above are due to the loss of trophic factors from the hair cells, and vary according to the type of lesion and animal species. In the human there is better preservation of the spiral ganglion cells over longer periods of time than in other animals, for example, the guinea pig. Otte et al (1978) found 45% of cochleae from profoundly deaf people had at least a third or more of the number of ganglion cells found in the normal population. In 93 cochleae from profoundly deaf people Nadol et al (1989) found the main spiral ganglion population was half the normal. The loss was greater in older subjects, for longer durations of hearing loss, and in the basal turn. Etiology had the greatest impact and the depletion was most extensive in people with viral labyrinthitis, congenital or genetic deafness, or bacterial meningitis. The least extensive loss occurred after aminoglycosides and sudden idiopathic deafness (Nadol et al 1989; Nadol 1997). The physiological effects of these pathological changes and their impact on electrical stimulation with a cochlear implant are discussed in Chapter 5.

Cochlear Nucleus

Pathological changes in the central auditory pathways, as well as in the spiral ganglion, can follow loss of cochlear function. As distinct from spiral ganglion cell loss, occurring at any stage of life, transneuronal degeneration of higher order neurons only develops with the loss of cochlear function at a critical period early in life. Ablation of the cochlea in the experimental animal during a narrow time window near the onset of hearing is the only period when significant cell death is demonstrated in the anteroventral cochlear nucleus (AVCN) (Tierney et al 1997; Mostafapour et al 2000). With cochlear destruction in 6-day-old mice, the cochlear nucleus (CN) population was reduced to 34% of normal (Trune 1982). The changes were not due solely to ablation of the cochlea, but also to the loss of activity in the auditory nerve. Born and Rubel (1988) found transneuronal cell death and reduction in soma size also occurred when a sodium channel blocker was applied (Pasic and Rubel 1989). These changes could be prevented by electrical stimulation of the auditory nerve, but not by direct excitation of the neurons in the CN (Hyson and Rubel 1989; Zirpel and Rubel 1996). They were the result of presynaptic release of the transmitter shown to be glutamate (Zirpel and Rubel 1996). The effects were associated with reduced protein synthesis (Sie and Rubel 1992), and increased intracellular Ca^{2+} (Zirpel et al 1995). Mostafapour et al (2000) found evidence that suggested neuronal death was due to the inactivation of an antiapoptotic (anti–cell death) gene *bcl-2*. Early loss of hearing also led to a significant decrease in the expression of messenger RNA (mRNA)-encoded receptors to glutamate (Marianowski et al 2000). In addition, there was an increase in the expression of receptors to γ-aminobutyric acid, a major inhibitor (Marianowski et al 2000), as well as a long-term deficiency in glycinergic synaptic inhibition. In mammals the changes were most marked in the CN, but higher order effects could be observed. The significance of these events is not clear, but they presumably affect both place and temporal frequency codes, as discussed in Chapters 5 and 6. It is also unclear when and whether these changes occur in humans. They do, however, suggest the importance of early electrical stimulation of the auditory nerve.

If animals are deafened after the onset of hearing, there is no transneuronal degeneration, but a shrinkage in the soma size associated with downregulation of its metabolism, and a reduction in the neuropil (a complex mesh of terminal axons, dendrites, and neuroglial processes). The reduction in soma size was first demonstrated by Powell and Erulkar (1962), who destroyed the cochlea in mature cats, and reported neuronal shrinkage in the CN and superior olivary complex (SOC). In another study, a reduction in soma size by a third occurred within 1 week of the hearing loss (Pasic and Rubel 1989). The deafening had a marked effect on the metabolic activity (Wong-Riley et al 1978; Durham et al 1993). There was also a loss of the neuropil or axon terminals innervating the ventral cochlear nucleus (VCN) (Powell and Erulkar 1962; Trune 1982). This may have been due to a loss of spiral ganglion cells, and a reduction in the number of their axons converging on the AVCN cells. It resulted, too, in an increase in the packing density in the AVCN.

Cochlear ablation in adult experimental animals also led to a loss of synapses in the AVCN. This too could be related to the loss of auditory neurons converging on the AVCN cells. This was followed by the generation of synapses over a long period from the remaining afferent input (Benson et al 1997). The loss of hearing also affected the terminal boutons. For example, the end bulbs terminating on the globular bushy cells were smaller as were the end bulbs of Held terminating on spherical bushy cells (Ryugo et al 1997; Redd et al 2000). This effect could have been the result of a downregulation in the metabolism of the remaining spiral ganglion cells. The above changes were accompanied by a temporary reduction in the expression of mRNA receptors to glutamate (Sato et al 2000), the main excitatory neurotransmitter in the auditory pathway. There was also a deficiency in glycinergic synaptic inhibition (Willott et al 1997).

As the sensorineural hearing loss led to a loss of the terminal axons and synapses on the cells in the AVCN as well as soma size, this would limit the processing of temporal and place information as discussed in Chapter 5. As these effects were secondary to the loss of spiral ganglion cells it makes it essential to stimulate these cells electrically as soon as possible after deafening to preserve the input to the AVCN. The connections could thus be preserved for improved strategies that may be developed later to provide fine temporal spatial patterns of excitation for the temporal coding of frequency.

Chouard et al (1983) found the soma size of octopus cells in the VCN of the guinea pig was preserved with electrical stimulation. In a study by Xu et al (1990), kittens were deafened 37 to 40 days after birth with ototoxic drugs. The animals were stimulated 80 to 90 days after birth on one side. The mean soma areas in the AVCN were significantly greater than the unstimulated control side. There was a weaker trend for the cells to be larger in the posteroventral cochlear nucleus on the stimulated side.

Pons and Midbrain

A bilateral sensorineural hearing loss at the onset of hearing resulted in a significant reduction in synaptic density in the central nucleus of the inferior colliculus

(ICC) (Hardie et al 1998). In view of the loss of neurons in the AVCN discussed above, this would lead to a loss of input and synaptic connections at the IC. A unilateral loss, however, did not lead to a loss in density. This was associated with an increase in the proportion of neurons projecting from the ipsilateral side (Nordeen et al 1983; Moore and Kitzes 1985). This suggests that the relative level of neural activity in the pathway from each VCN determines the success of each side in forming or retaining synapses in the auditory midbrain (Moore 1990).

A sensorineural loss after the onset of hearing also affected the higher brain centers in the pons and midbrain. There was a reduction in the soma area of neurons in the trapezoid body (Pasic et al 1994), the SOC and nucleus of the lateral lemniscus (Powell and Erulkar 1962), and ICC (Nishiyama et al 2000).

Human Brainstem

There are few studies on the human central auditory pathways following a profound hearing loss. A reduction in the soma area was found in the CN by Clark, Shepherd et al. (1988) Seldon and Clark (1991), and Moore et al (1997), but also in the medial superior olive (MSO) and IC (Moore et al 1997). There was also a reduction in the volume of the CN, especially the VCN, in the studies by Clark, Shepherd et al (1988) and Seldon and Clark (1991). These findings, also discussed in Chapter 4, were essentially consistent with those from experimental animals.

The brainstem and temporal bone of the first University of Melbourne/Bionic Ear Institute patient to have a bilateral cochlear implant were also studied (Yukawa et al 2001a,b). The sections were compared with those from a second person who had a cochlear implant on one side. The bilateral patient died at the age of 59 years. He went profoundly deaf in the left ear at 31 years due to a head injury, and became profoundly deaf in the right ear at 36 years. At 46 years he had a right cochlear implant and a left cochlear implant at 51 years. Thus the right ear was implanted for 13 years and the left for 8 years. He had only fair speech discrimination with the right implant and satisfactory results with the left. Binaural psychophysical studies showed there was a marked reduction in the interaural temporal discrimination difference limens for electrical stimulation. It was well below that for normal hearing, as discussed in Chapter 6. The unilateral subject died at the age of 62 years. She suffered a hearing loss due to mumps and then had a 27-year history of a slow progressive loss and had a profound hearing loss for 5 years prior to implantation. She had the implant for 1.5 years in the left ear. The brainstems were sectioned and the MSO analyzed, as it is considered an important nucleus for coding interaural time differences (see Chapter 5). The trigeminal nucleus was also examined as a control for tissue fixation and processing artifacts. The cell density and volume were determined for each nucleus. Cell numbers and volume were determined by a technique in which a criterion was established to ensure that the cells were not counted twice.

The results are shown in the Table 3.1 for cell density and volume, and statistical significance was determined with the Mann-Whitney *U* test. There was a

TABLE 3.1. The cell density and volume measures for the right and left medial superior olive from one patient with a bilateral and another with a unilateral cochlear implant.

	Side	Cell density ($\times 10^{-5}/\mu m^3$)	Mean
Bilateral	Right	1.23	1.17*
	Left	1.1	
Unilateral	Right	2.2	1.96
	Left	1.71	

*$p < .05$, Mann-Whitney U test.

	Side	Cell volume (μm^3)	Mean
Bilateral	Right	2241	2005**
	Left	1693*	
Unilateral	Right	2688	2577
	Left	2511*	

*$p < .05$, **$p < .0001$, Mann-Whitney U test.

significant difference between the bilateral and the unilateral subjects. For the combined right and left sides there was a lower cell density and volume for the bilateral compared with the unilateral subject. This suggests that the MSO was affected by the hearing loss occurring well after the onset of hearing, and this is the reason the patient did not have satisfactory interaural temporal difference limens. It is also of interest that for both patients the cell volume was lower on the left side. This is consistent with the fact that the first patient received more help from the left implant, and the second unilateral patient had a left implant. In both cases there would have been more contralateral stimulation to the right, thus helping to preserve its function. This is consistent with the experimental animal studies showing that electrical stimulation maintains cell viability and size (Miller and Altschuler 1995).

Prenatal (Congenital) and Postnatal Hearing Loss

Deafness may occur before or during birth (prenatal and perinatal, respectively) when it is referred to as congenital. It can also occur after birth (postnatal). Congenital deafness may arise from genetic causes, chromosomal abnormalities, or diseases affecting the mother during pregnancy. In about two thirds of children with prelinguistic severe or profound sensorineural deafness without syndromes (before language develops), the cause is thought to be genetic (Morton 1991). Postnatal deafness is mostly from disease or injury, but may also be the result of delayed genetic effects.

Genetic and Chromosomal

Body cells contain 46 chromosomes, and the genes are located at different points along the chromosomes. In the male the body cell divides into two germ cells;

the sperms each contain 23 chromosomes. The same occurs in the female for the ova. When the two germ cells containing 23 chromosomes unite, they form a new cell with 46 chromosomes. Two chromosomes determine the sex of the individual. In the male, one of the two sex chromosomes is small (Y chromosome) and inherited from the father, and the other, the X chromosome, is inherited from the mother. The female has two X chromosomes, one being inherited from the father and one from the mother. The other 22 chromosomes are referred to as autosomes.

If a parent passes on a dominant gene causing deafness, it only requires one chromosome of the pair to have the deafness gene for the child to be affected. If it is a recessive gene, the child needs to have one on each chromosome pair. A sex-linked inheritance may occur in the male when the X gene is affected, and thus without protective effects from the Y or male chromosome. Genetic deafness may be classified thus as dominant or recessive. Most genetic deafness presenting congenitally is transmitted as a recessive, and about half those with recessive deafness have no accompanying abnormalities.

Congenital, Genetic Deafness

Nonsyndromic

As stated, genetic deafness frequently occurs alone without other abnormalities (nonsyndromic). In about 80% of children with nonsyndromic deafness, the inheritance is autosomal recessive (Dahl et al 2001). Using DNA markers, genetic linkage studies have shown over 20 genes for nonsyndromic deafness (Van Camp and Smith 2002). A mutation of the connexin 26 gene has been found to account for up to 50% of cases of nonsyndromic deafness in children of European descent (Maw et al 1995; Denoyelle et al 1997). In addition 50% to 90% of chromosomes on which a connexin 26 mutation has been determined have the same specific mutation (deletion of a guanine nucleotide at position 35, i.e., 35delG) (Denoyelle et al 1997). A similar incidence to the European data was found for a group of Australian deaf children (Dahl et al 2001). Furthermore, over 40 connexin 26 mutations have been reported (Denoyelle et al 1999). On the other hand, the incidence of connexin mutations is very low in Asian-American and African-American populations (Morell et al 1998).

Connexin 26 belongs to a family of proteins that mediate the exchange of molecules between adjacent cells. The number refers to the size of the protein in thousands of daltons. Connexin is highly expressed in the cells lining the cochlear duct and the stria vascularis. It is thought that it is important for the recycling of (K^+) ions from sensory hair cells into the endolymph in the process of transduction of sound to electrical signals. The slope of the hearing loss (over 2000 to 8000 Hz) was greater than in children without connexin 26 mutant alleles (Wilcox et al 2000). It is not known to what extent cochlear implants benefit children with connexin 26 and other genetic disorders.

Nonsyndromic deafness has variable anatomical and histological features. First, there may be total lack of development of the inner ear, and the x-ray will show complete absence. It can be difficult to distinguish this condition from bony labyrinthitis. This condition is called the Michel syndrome, and it is inherited as autosomal dominant. It will not be possible to implant children with this condition, but fortunately it only accounts for a small proportion of genetic deafness. Second, only 1½ turns of the cochlea may develop, rather than the normal 2½ turns. This condition is often associated with underdevelopment of the vestibular structures, and is called the Mondini syndrome. Endolymphatic hydrops (distention of the endolymphatic system) is often present, and there may be some residual hearing. It is inherited as autosomal dominant and is characterized pathologically by an absence of the septum (interscalar) between the apical and the middle turns, thus creating a common cavity. In a child with the Mondini syndrome who died from infection in the nonimplanted ear as discussed above, the temporal bones showed a wide dehiscence between the scala tympani and the internal auditory canal that could have accounted for the CSF leak at surgery (C. Suzuki et al 1998). The histology also showed there was a wide vestibular aqueduct, and expansive communication between the cochlea and vestibule that could lead to a misplaced electrode. There was a hypoplastic modiolus, and the spiral ganglion cell population was 10,826. In the unimplanted ear there was inflammatory necrosis of the round window membrane, and many polymorphonuclear leukocytes in the adjacent scala tympani, indicating the route for the spread to the cochlea. The extension to the meninges probably occurred through the abnormally patent modiolus. This 6-year-old child was developing speech and language, and this indicates the importance of providing an implant. However, because of meningitis it stresses the need to ensure there is an adequate seal around the electrode entry into the inner ear, and the aggressive treatment of any middle ear infection.

In other cases the modiolus may be better developed, and this is apparent on the computed tomography (CT) scan. A perimodiolar electrode array could be used, as the spiral ganglion cells lie centrally. On the other hand, the modiolus may be deficient, and the cochlear nerve fibers lie peripherally. When this happens it is preferable to use the straight but flexible Nucleus array. This array produces less trauma, and lies closer to the nerve fibers. Schmidt (1985) examined eight bones and found a significantly reduced population in those where there had been a severe hearing loss.

The Mondini dysplasia may be associated with a wide cochlear aqueduct (perilymph gusher) (Nadol 1984). This is seen on the CT scan, and indicates that a large outflow of CSF (perilymph gusher) may occur when an opening is made into the cochlea for the insertion of the electrode array. So in summary, satisfactory to good results have been reported for cochlear implants with the Monidini dysplasia (Silverstein et al 1988; Turrini et al 1997; M. Suzuki et al 1998).

A related condition is the large vestibular aqueduct syndrome (LVAS) first described by Valvassori and Clemis (1978) on radiological findings. An autosomal-recessive or X-linked inheritance was suggested by Griffith et al (1996). A profound hearing loss was reported in 39% of patients (Jackler and De La Cruz 1989).

Children with the Mondini dysplasia have a higher risk of meningitis whether they have an implant or not. Phelps et al (1994) report an incidence of four of 20 children with congenital dysplasia (unimplanted) developed meningitis. In an analysis of the 19 Nucleus patients who developed meningitis out of 16,500 implantees in North America (see Chapter 10), at least 9 had a deformity of the cochlea. It is unclear if any were device-related, but it again serves to emphasize the importance of sealing the round window entry point together with extreme care in the antibacterial management.

Finally, if the development of the osseous cochlea is complete, but the sensory elements have failed to develop, they may be represented only by mounds of undifferentiated cells. This is referred to as the Scheibe syndrome. It is the commonest of all inherited congenital deafness, and is autosomal recessive.

Syndromic

In a number of children deafness is associated with other abnormalities, and hearing loss may be the first symptom. With Waardenburg's syndrome, the features other than deafness are a lateral displacement of the inner canthus of the eye, heterochromia of the iris, and a white forelock. It is inherited as autosomal dominant. Pathologically there is atrophy of the organ of Corti and stria vascularis, and a reduction in the number of ganglion cells. In albinism, where there is loss of pigmentation resulting in fair skin and poor vision, the deafness is bilateral and severe. It is inherited as an autosomal-dominant or -recessive or sex-linked trait. With onchodystrophy there is sensorineural deafness and nail dystrophy. Pendred's syndrome may account for 10% of recessive deafness. In this syndrome there is abnormal iodine metabolism. It is often associated with a Mondini deformity of the cochlea. In Jervell's syndrome there is a bilateral severe hearing loss and cardiac abnormality (prolonged Q-T interval) that can lead to sudden death (Stokes-Adams attacks). It is inherited as autosomal recessive. Usher's disease is a congenital condition in which there is combined sensorineural hearing loss and retinitis pigmentosa. It is inherited as sex linked or autosomal dominant, and there is a recessive form. So it is in fact a collection of conditions. There are a number of other syndromes that have associated deafness, and more details can be obtained from standard texts.

Deafness may also occur due to chromosome abnormalities. Normally the 22 pairs of autosomal chromosomes are grouped according to similar morphologies from A to G. Trisomy 13 to 15 (D) have an extra chromosome located in the group D, and trisomy 18 (E) in group E. These conditions are often associated with other ear or body defects, and the children die early.

Delayed

Delayed sensorineural deafness coming on sometime after the baby is born can also be genetic, and deafness is commonly the only abnormality. It is inherited as an autosomal-dominant condition, and there is a progressive sensorineural hearing loss. A delayed sensorineural loss may also be associated with other abnormalities and there are a number of these conditions (see Chapter 9).

Acquired

Prenatal and Perinatal

There are a number of nongenetic causes of congenital deafness. These are infective agents; trauma, in particular drugs; and metabolic conditions. The most common infective agents are toxoplasmosis, rubella, cytomegalovirus (CMV), and herpes simplex, together referred to as TORCH. O'Sullivan et al (1997) showed that the most common viral causes of a hearing loss in the Melbourne Cochlear Implant Clinic were CMV and rubella. Rubella and other viruses cross the placental barrier to infect the fetus, and this impairs the development of the cochlea and other organs. With rubella the hearing loss is more severe if the infection affects the mother in the first 3 months of pregnancy (first trimester), but it may occur following infections in the second or third trimesters. It is necessary to make the diagnosis by detecting the virus in the pharynx, urine, or CSF, and the presence of a specific immunoglobulin M (IgM) antibody in the chord blood or body serum. There are also persistent elevated levels of rubella IgG in the serum. In a review of 300 children with congenital rubella, 50% of the mothers had no clinical evidence of the disease, so there was a high incidence of subliminal infections. Maternal rubella infection during pregnancy must be confirmed by viral isolation or serological tests. Prospective studies based on laboratory diagnosis show the incidence of deafness to be from 50% to 70%. Rubella can also be associated with cardiac defects or mental retardation. The hearing loss is predominantly bilateral, but may be asymmetrical. In a small proportion the deafness became more severe with time. This was probably due to persistent infection, indicated by the continued shedding of the rubella virus after birth. The central auditory pathways may also be affected, and this could account for the lack of language development. This could also apply to results with cochlear implantation. It has been shown, too, that a child is more likely to develop deafness from rubella when there is a genetic predisposition shown by a positive family history.

CMV results in deafness that is often severe to profound, and like rubella can progress. It too affects the central nervous system, with impaired vision, cerebral palsy, epilepsy, and intellectual disability. CMV infections are highly prevalent and can be detected in 0.5% to 2.4% of all live births (Pass, Stagno et al. 1980). Rasmussen (1990) estimated that 10% of infected newborns are at risk from hearing loss, impaired vision, or neuromuscular abnormalities. Although 90% of patients are without symptoms, there may be swollen lymph nodes and enlargement of the liver and spleen. In children presenting with a severe-to-profound hearing loss, it is considered important to undertake serological tests on both the mother and child, as well as viral cultures from the saliva and urine up to the age of approximately 4 years (S. Locarnini, personal communication). This helps in deciding whether the child has had a CMV infection, and whether it was of congenital origin when the effects are more severe. The children with CMV at the University of Melbourne's Cochlear Implant Clinic have not had as good results as other children, and this may be due to the involvement of central auditory pathways.

Herpes simplex encephalitis is a viral infection usually from genital herpes. It may present neonatally as a localized mucocutaneous or disseminated infection. When it is disseminated there is a 30% risk of meningoencephalitis that is likely to occur in the second or third week of life. Herpes simplex infections leading to sensorineural hearing loss may also involve the central auditory pathways.

The pre- and perinatal viral infections infect the fetus and neonate in the same way as a postnatal invasion. Lindsay (1973) has shown that the spread of the virus by the bloodstream to the endolymph produces a different pathological picture from the one where the spread is from the meninges or lining of the brain via the cochlear aqueduct to the perilymph in the scala tympani of the basal turn. With an endolymphatic involvement there is often a normal spiral ganglion or cochlear nerve population. With spread to the perilymph there is more often degeneration of spiral ganglion cells and nerve fibers and variable changes in the cochlear duct including hydrops. Malformations such as a rudimentary organ of Corti and underdeveloped stria vascularis and tectorial membrane are rare. In most cases the lesions are due to small hemorrhages that are probably the result of increased coagulability produced by the viruses. The vestibular system is only affected in a small number of cases.

The only specific bacterial cause of deafness is syphilis. This organism cannot cause malformations of the cochlea, as the treponema is not able to pass through the placental blood barrier before the fifth month. Its effects are either through inflammation of the meninges and nerve or due to labyrinthitis. The latter is more common and the hearing loss increases in a steplike fashion. Loss of the spiral ganglion cells is more likely to occur with this condition.

Parasitic protozoa are also agents that can lead to a severe sensorineural hearing loss in the fetus. They are single-celled motile organisms. In particular *Toxoplasma gondii* infections in the mother (toxoplasmosis) can pass through the placenta after 6 weeks. It is acquired either by contact with oocyte-shedding kittens or by eating cyst-ridden undercooked meat. It is a common condition, and some 87% of the population over 30 years of age have serologically positive tests. The infection of the mother is generally not apparent. The diagnosis is made from serological tests, the Savin's lysis, and complement fixation tests. The significance of the serological test depends on the age of the child. A positive result at 3 to 12 months would indicate congenital toxoplasmosis. The deafness is associated with chorioretinitis and the characteristic deterioration of the fundus of the eye, hydrocephalus or microcephaly, with calcification of the brain seen on x-ray. In the cochlea there is calcification of the stria vascularis and spiral ligament, and inflammation of the whole vestibule. Physical trauma in the form of misapplied forceps during delivery may lead to loss of hearing through fractures of the skull base. Chemical agents during pregnancy, such as ototoxic antibiotics, can lead to profound hearing loss. Poor blood supply to the fetus (anoxia), through a hemorrhage behind the placenta (antepartem hemorrhage) or the placental cord around the neck during delivery, is a factor. Other anoxic and metabolic causes are hypertension, toxemia, diabetes, renal disease, and Rh blood incompatibility. It is especially important to assess the condition of the baby after birth and to calculate an Apgar score, which is based on the color, reflex responses, respiratory effort, heart rate, and muscle tone.

Child and Adulthood

The viral causes of a severe postnatal hearing loss are mumps, measles, influenza, and chicken pox. These produce a viral labyrinthitis that affects the endolymphatic duct. Pathological changes are more pronounced in the basal cochlea and include degeneration of the organ of Corti, atrophy of the stria vascularis, displacement and distortion of tectorial membrane, and distortion and degeneration of the saccule. The utricle and semicircular canals are seldom involved.

The most common bacterial cause of a hearing loss in the newborn (neonate) and in later childhood is labyrinthitis following meningitis. A study by Goodhill (1950) on 904 deaf children showed 10% had meningitis as the cause of their hearing loss. When deafness occurs it is mostly a very severe or a total loss, and is usually due to infection of the inner ear (labyrinthitis). The incidence in meningitis normally varies from 5% to 30%, depending on the causal organism. Antibiotics have now reduced the incidence.

With meningitis the infection is transmitted to the perilymph either through the internal auditory canal or via the cochlear aqueduct. When through the internal auditory canal the spread is via the perineural and perivascular spaces. In the cochlea the pathological changes are the formation of serofibrinous exudate, infiltration with pus cells (polymorphonuclear leukocytes), and then the formation of granulation tissue followed by healing characterized by fibrosis and ossification. Ossification is usually more marked near the round window, where the perilymphatic spread to the basal turn occurs.

Personal studies in the cat show that osteoid tissue commences within 2 weeks. Therefore, in the human once the infection is controlled and the hearing loss established, surgery should be considered to ensure that the electrode array can be inserted an adequate distance. From the study of Blamey et al (1992) it has been shown that up to 21 channels of stimulation are important (see Chapter 7). Furthermore, as discussed above, if electrical stimulation is commenced early, spiral ganglion cells will be preserved. Sometimes there is only fibrous tissue rather than bone in the scala tympani, and for this reason magnetic resonance imaging (MRI) should be carried out before operating on a patient with a history of meningitis.

A head injury can produce fractures at the base of the skull. Sensorineural hearing loss is more likely with a transverse than a longitudinal fracture of the temporal bone. However, the fracture lines are not easily categorized as transverse and longitudinal. With a transverse fracture of the temporal bone the cochlear nerve may have been sectioned in which case the results will be unsatisfactory, and this may be seen with a CT scan and the status of the cochlear nerve observed with MRI. Ototoxic drugs such as neomycin, kanamycin, polymyxin, and chloramphenicol, as well as loop diuretics, cause a hearing loss both in children and in adults. The antibiotics usually have their effects on the outer hair cells. With antibiotics ototoxicity may occur suddenly after a few injections and can continue after the withdrawal of the drug. It may continue for many months after treatment. The effect of ototoxic drugs on spiral ganglion cell numbers varies with species, and there is a marked loss in the guinea pig within weeks (Webster and Webster 1978). However, in the human, as discussed above, there is greater spiral ganglion cell survival than would be predicted from experimental animal data (Ylikoski et al 1981). The results of implantation in these patients can be expected to be good.

References

Benitez, J. T. 1977. Stapedectomy and fatal meningitis. ORL 39: 94–100.

Benson, C. G., J. S. Gross, S. K. Suneja and S. J. Potashner. 1997. Synaptophysin immunoreactivity in the cochlear nucleus after unilateral cochlear or ossicular removal. Synapse 25: 243–257.

Bergan, T. 1981. Pharmacokinetics of tissue penetration of antibiotics. Reviews of Infectious Diseases 3: 45–66.

Berkowitz, R. G., B. K.-H. G. Franz, R. K. Shepherd, G. M. Clark and D. Bloom. 1984. Cochlear implant and otitis media. A pilot study to assess the feasibility of pseudomonas aeruginosa and streptococcus pneumoniae infection in the cat. Journal of the Oto-Laryngological Society of Australia 5: 297–299.

Berkowitz, R. G., B. K.-H. G. Franz, R. K. Shepherd, G. M. Clark and D. Bloom. 1987. Pneumococcal middle ear infection and cochlear implantation. Annals of Otology, Rhinology and Laryngology 96(suppl 128): 55–56.

Black, R. C., G. M. Clark, R. K. Shepherd, S. J. O'Leary and C. W. Walters. 1983. Intracochlear electrical stimulation of normal and deaf cats investigated using brainstem response audiometry. Acta Oto-Laryngologica-supplement 399: 5–17.

Blamey, P. J., B. C. Pyman, M. Gordon, et al. 1992. Factors predicting postoperative sentence scores in postlinguistically deaf adult cochlear implant patients. Annals of Otology, Rhinology and Laryngology 101: 342–348.

Block, S.L. 1997. Causative pathogens, antibiotic resistance and therapeutic considerations in acute otitis media. Pediatric Infectious Disease Journal 16: 449–456.

Bluestone, C.D., J.S. Stephenson, L.M. Martin. 1992. Ten-year review of otitis media pathogens. Pediatric Infectious Disease Journal 11(suppl): S7–S11.

Born, D. E. and E. W. Rubel. 1988. Afferent influences on brain stem auditory nuclei of the chicken: presynaptic action potentials regulate protein synthesis in nucleus magnocellularis neurons. Journal of Neuroscience 8: 901–919.

Brennan, W. J. and G. M. Clark. 1985. An animal model of acute otitis media and the histopathological assessment of a cochlear implant in the cat. Journal of Laryngology and Otology 99: 851–856.

Chouard, C. H., B. Meyer, P. Josset and J. F. Buche. 1983. The effect of the acoustic nerve chronic electric stimulation upon the guinea pig cochlear nucleus development. Acta Otolaryngologica 95: 639–645.

Clark, G. M. 1987. The University of Melbourne–Nucleus multi-electrode cochlear implant. Advances in Oto-Rhino-Laryngology. Vol. 38. Basel, Karger.

Clark, G. M., H. G. Kranz and H. Minas. 1973. Behavioral thresholds in the cat to frequency modulated sound and electrical stimulation of the auditory nerve. Experimental Neurology 41: 190–200.

Clark, G. M., B. C. Pyman and R. E. Pavillard. 1980. A protocol for the prevention of infection in cochlear implant surgery. Journal of Laryngology and Otology 94(12): 1377–1386.

Clark, G. M., B. C. Pyman, R. L. Webb, Q. E. Bailey and R. K. Shepherd. 1984. Surgery for an improved multiple-channel cochlear implant. Annals of Otology, Rhinology and Laryngology 93(3 pt 1): 204–207.

Clark, G. M. and R. K. Shepherd. 1997. Biological safety. In: Clark, G. M., R. S. C. Cowan and R. C. Dowell, eds. Cochlear implantation for infants and children: advances. San Diego, Singular: 29–46.

Clark, G. M., R. K. Shepherd, B. K.-H. G. Franz and D. Bloom. 1984. Intracochlear electrode implantation. Round window membrane sealing procedures and permeability studies. Acta Oto-Laryngologica-supplement 410: 5–15.

Clark, G. M., R. K. Shepherd, B. K.-H. G. Franz, et al. 1988. The histopathology of the human temporal bone and auditory central nervous system following cochlear implantation in a patient. Correlation with psychophysics and speech perception results. Acta Oto-Laryngologica-supplement 448: 1–65.

Clark, G. M., R. K. Shepherd, J. F. Patrick, R. C. Black and Y. C. Tong. 1983. Design and fabrication of the banded electrode array. Annals of the New York Academy of Sciences 405: 191–201.

Coleman, D. L., R. N. King and J.D. Andrade. 1974. The foreign body reaction: a chronic inflammatory response. Journal of Biomedical Materials Research 8: 199–211.

Cranswick, N. E. 1984. Studies in the cochlear round window. B. Med. Sci thesis, University of Melbourne.

Cranswick, N. E., B. K.-H. G. Franz, G. M. Clark and R. K. Shepherd. 1987. Middle ear infection postimplantation: response of the round window membrane to Streptococcus pyogenes. Annals of Otology, Rhinology and Laryngology 96(suppl 128): 53–54.

Dahl, H. H. M., K. Saunders, T. M. Kelly, et al. 2001. Prevalence and nature of connexin 26 mutations in children with non-syndromic deafness. Medical Journal of Australia 175: 191–194.

Dahm, M., G. M. Clark, B. K.-H. Franz, R. K. Shepherd, M. J. Burton and R. Robins-Browne. 1994. Cochlear implantation in children: labyrinthitis following pneumococcal otitis media in unimplanted and implanted cat cochleas. Acta Oto-Laryngologica 114: 620–625.

Dahm, M., R. L. Webb, G. M. Clark, et al. 1995. Cochlear implants in children: the value of cochleostomy seals in the prevention of labyrinthitis following pneumococcal otits media. Australian Journal of Oto-Laryngology 2: 90–92.

Dahm, M., J. Xu, M. Tykocinski, R. K. Shepherd and G. M. Clark. 2000. Intracochlear damage following insertion of the Nucleus 22 standard electrode array: a post mortem study of 14 implant patients. CI 2000, the 6th International Cochlear Implant Conference, Miami Beach, Florida: 149.

Denoyelle, F., S. Marlin, D. Weil, et al. 1999. Clinical features of the prevalent form of childhood deafness, DFNB1, due to a connexin-26 gene defect: implications for genetic counselling. Lancet 353: 1298–1303.

Denoyelle, F., D. Weil, M. A. Maw, et al. 1997. Prelingual deafness; high prevalence of a 30delG mutation in the connexin 26 gene. Human Molecular Genetics 6: 2173–2177.

Durham, D., F. M. Matschinsky and E. W. Rubel. 1993. Altered malate dehydrogenase activity in nucleus magnocellularis of the chicken following cochlear removal. Hearing Research 70: 151–159.

Franz, B. K.-H. G. and G. M. Clark. 1987. Refined surgical technique for insertion of banded electrode array. Annals of Otology, Rhinology and Laryngology 96(suppl 128): 15–16.

Franz, B. K.-H. G., G. M. Clark and D. Bloom. 1984. Permeability of the implanted round window membrane in the cat—an investigation using horseradish peroxidase. Acta Oto-Laryngologica-supplement 410: 17–23.

Franz, B. K.-H. G., G. M. Clark and D. Bloom. 1987. Effect of experimentally induced otitis media on cochlear implants. Annals of Otology, Rhinology and Laryngology 96: 174–177.

Friedmann, I. 1974. Pathology of the ear. Oxford, Blackwell Scientific.

Giebink, G. S., I. K. Berzins and P. G. Quie. 1980. Animal models for studying pneumococcal otitis media and pneumococcal vaccine efficacy. Annals of Otology Rhinology and Laryngology 89: 339–343.

Goodhill, V. 1950. The nerve-deaf child: significance of RH, maternal rubella and other etiologic factors. Annals of Otology and Rhinology 59: 1123–1147.

Goycoolea, M. V., M. M. Paparella, G. Goldberg and A. M. Carpenter. 1980. Permeability of the round window membrane in otitis media. Archives of Otolaryngology 106: 430–433.

Griffith, A. J., A. Arts, C. Downs, et al. 1996. Familial large vestibular aqueduct syndrome. Laryngoscope 106: 960–965.

Hardie, N. A., A. Martsi-McClintock, L. M. Aitkin and R. K. Shepherd. 1998. Neonatal sensorineural hearing loss affects synaptic density in the auditory midbrain. Neuro-Report 9: 2019–2022.

Hodges, K. B., J. E. Penny, D. Brown and C. Herley. 1984. Scanning electron microscopy of the cochlea in rats with Streptococcus pneumoniae otitis media. Archives of Otolaryngology 110: 429–436.

Homsy, C. A. 1970. Biocompatibility in selection of materials for implantation. Journal of Biomedical Materials Research 4: 341–356.

House, W. F., W. M. Luxford and B. Courtney. 1985. Otitis media in children following the cochlear implant. Ear and Hearing 6(3 suppl): 24S–26S.

Hyson, R. L. and E. W. Rubel. 1989. Transneuronal regulation of protein synthesis in the brain-stem auditory system of the chick requires synaptic activation. Journal of Neuroscience 9: 2835–2845.

Igarashi, M., R. Saito, B. R. Alford, M. V. Filippone and J. A. Smith 1974. Temporal bone findings in pneumococcal meningitis. Archives of Otolaryngology 99: 79–83.

Jackler, R. K. and A. De La Cruz. 1989. The large vestibular aqueduct syndrome. Laryngoscope 99: 1238–1243.

Johnson, C. E., S. A. Carlin, D. M. Super, et al. 1991. Cefixime compared with amoxicillin for treatment of acute otitis media. Journal of Pediatrics 119: 117–122.

Kawano, A., H. L. Seldon and G. M. Clark. 1996. Computer-aided three-dimensional reconstruction in human cochlear maps: measurement of the lengths of organ of corti, outer wall, inner wall, and Rosenthal's canal. Annals of Otology, Rhinology and Laryngology 105: 701–709.

Kawano, A. S., H. L. Seldon, G. M. Clark, R. Madsen and C. Raine. 1995. Intracochlear factors contributing to psychophysical percepts following cochlear implantation: a case study. Annals of Otology, Rhinology and Laryngology 104: 54–57.

Kawano, A., H. L. Seldon, G. M. Clark and R. Ramsden. 1998. Intracochlear factors contributing to psychophysical percepts following cochlear implantation. Acta Oto-Laryngologica 118: 313–326.

Kerr, A. and H. F. Schuknecht. 1968. The spiral ganglion in profound deafness. Acta Oto-Laryngologica 65: 568–598.

Leake, P. A. and G. T. Hradek. 1988. Cochlear pathology of long term neomycin induced deafness in cats. Hearing Research 33: 11–34.

Leake-Jones, P. A. and S. J. Rebscher. 1983. Cochlear pathology with chronically implanted scala tympani electrodes. Annals of the New York Academy of Sciences 405: 203–223.

Lehnhardt, E. 1993. Intracochlear placement of cochlear implant electrodes in soft surgery technique. HNO 41(7): 356–359.

Lewis, D. M., J. L. Schram, S. J. Meadema and D. J. Lim. 1980. Experimental otitis media in chinchillas. Annals of Otology Rhinology and Laryngology 68: 344–350.

Lindsay, J. R. 1973. Histopathology of deafness due to postnatal viral disease. Archives of Otolaryngology 98: 258–264.

Loeb, G. E., A. E. Walker, S. Uematsu and B. W. Konigsmark. 1977. Histological reaction to various conductive and dielectric films chronically implanted in the subdural space. Biomedical Materials Research 11: 195–210.

Lowy, F. D. and S. M. Hammer. 1983. Staphylococcus epidermitis infections. Annals of Internal Medicine 99: 834–839.

Luotonen, J., E. Herva, P. Karma, M. Timonen, M. Leinonen and P.H. Makela. 1981. The bacteriology of acute otitis media in children with special reference to Streptococcus pneumoniae as studied by bacteriological and antigen detection methods. Scandinavian Journal of Infectious Disease 13: 177–183.

Luxford, W. M. and W. F. House 1985. Cochlear implants in children: medical and surgical considerations. Ear and Hearing 6(3 suppl): 20S–23S.

Majno, G. and I. Joris 1996. Cells, tissues and disease: principles of general pathology. Cambridge, MA, Blackwell Science: 229–246.

Marchant, R. E., J. M. Anderson, E. Castillo and A. Hiltner. 1986. The effects of an enhanced inflammatory reaction on the surface properties of cast Biomer. Journal of Biomedical Materials Research 20: 153–168.

Marianowski, R., W.-H. Liao, T. Van Den Abbeele, et al. 2000. Expression of NMDA, AMPA and GABA A receptor subunit mRNAs in the rat auditory brainstem. I. Influence of early auditory deprivation. Hearing Research 150: 1–11.

Matz, G. J., H. B. Lockhart and J. R. Lindsay. 1968. Meningitis following stapedectomy. Laryngoscope 78: 56–63.

Maw, M. A., D. R. Allen-Powell, R. J. Goodey, et al. 1995. The contribution of the DFNB 1 locus to neurosensory deafness in a Caucasian population. American Journal of Human Genetics 57: 629–635.

Meyerhoff, W. L., D. A. Shea and G. S. Giebink. 1980. Experimental pneumococcal otitis media: a histopathologic study. Otolaryngology-Head and Neck Surgery 88: 606–612.

Miller, J. M. and R. A. Altschuler. 1995. Effectiveness of different electrical stimulation conditions in preservation of spiral ganglion cells following deafness. Annals of Otology, Rhinology and Laryngology 104(suppl 166): 57–60.

Moore, D. R. 1990. Auditory brainstem of the ferret: early cessation of developmental sensitivity of neurons in the cochlear nucleus to removal of the cochlea. Journal of Comparative Neurology 302: 810–823.

Moore, D. R. and L. M. Kitzes. 1985. Projections from the cochlear nucleus to the inferior colliculus in normal and neonatally cochlear-ablated gerbils. Journal of Comparative Neurology 240: 180–195.

Moore, J. K., J. K. Niparko, L. M. Perazzo, M. R. Miller and F. H. Linthicum. 1997. Effect of adult-onset deafness on the human central auditory system. Annals of Otology Rhinology and Laryngology 106: 385–390.

Morell, R. J., H. J. Kim, L. J. Hood, et al. 1998. Mutations in the connexin 26 gene (GJB2) among Ashkenazi Jews with nonsyndromic recessive deafness. New England Journal of Medicine 339: 1500–1505.

Morton, N. E. 1991. Genetic epidemiology of hearing impairment. Annals of the New York Academy of Science 630: 16–31.

Mostafapour, S. P., S. L. Cochran, N. M. Del Puerto and E. W. Rubel. 2000. Patterns of cell death in mouse anteroventral cochlear nucleus neurons after unilateral cochlea removal. Journal of Comparative Neurology 426: 561–571.

Mundy, G. R. 1995. Local control of bone formation by osteoblasts. Clinical Orthopaedics and Related Research 313: 19–26.

Nadol, J. B. 1984. Histological considerations in implant patients. Archives of Otolaryngology 110: 160–163.

Nadol, J. B. 1997. Patterns of neural degeneration in the human cochlea and auditory nerve: implications for cochlear implantation. Otolaryngology Head and Neck Surgery 117: 220–228.

Nadol, J. B., Y.-S. Young and R. B. Glynn. 1989. Survival of spiral ganglion cells in profound sensorineural hearing loss: implications for cochlear implantation. Annals of Otology, Rhinology and Laryngology 98: 411–416.

Nishiyama, N., N. A. Hardie and R. K. Shepherd. 2000. Neonatal sensorineural hearing loss affects neurone size in cat auditory midbrain. Hearing Research 140: 18–22.

Nordeen, K. W., H. P. Killackey and L. M. Kitzes. 1983. Ascending projections to the inferior colliculus following unilateral cochlear ablation in the neonatal gerbil, Meriones unguiculatus. Journal of Comparative Neurology 214: 144–153.

O'Sullivan, P., S. Ellul, R. C. Dowell, B. C. Pyman and G. M. Clark. 1997. The relationship between aetiology of hearing loss and outcome following cochlear implantation in a paediatric population. In: Clark, G. M., ed. Cochlear implants. XVI World Congress of Otorhinolaryngology Head and Neck Surgery. Bologna, Monduzzi. 169–172.

Otte, J., H. Schuknecht and A. Kerr. 1978. Ganglion cell populations in normal and pathological human cochleae. Implications for cochlear implantation. Laryngoscope 88: 1231–1246.

Palva, T., A. Palva and J. Karja. 1972. Fatal meningitis in a case of otosclerosis operated upon bilaterally. Archives of Otolaryngology 96: 130–137.

Paparella, M. M., M. V. Goycoolea and W. L. Meyerhoff. 1980. Inner ear pathology and otitis media: a review. Annals of Otology, Rhinology and Laryngology 89 (suppl 68): 249–253.

Pasic, T. R., D. R. Moore and E. W. Rubel. 1994. Effect of altered neuronal activity on cell size in the medial nucleus of the trapezoid body and ventral cochlear nucleus of the gerbil. Journal of Comparative Neurology 348: 111–120.

Pasic, T. R. and E. W. Rubel. 1989. Rapid changes in cochlear nucleus cell size following blockade of auditory nerve electrical activity in gerbils. Journal of Comparative Neurology 283: 474–480.

Pass, R. F., S. Stagno, G. J. Myers and C. A. Alford. 1980. Outcome of symptomatic congenital cytomegalovirus infection: results of long-term congenital follow-up. Pediatrics 66: 758–762.

Phelps, P.D., A. King, L. Michaels, F.R.C. Path. 1994. Cochlear dysplasia and meningitis. American Journal of Otology 15: 551–557.

Powell, T. P. and S. D. Erulkar. 1962. Transneuronal cell degeneration in the auditory relay nuclei of the cat. Journal of Anatomy 96: 249–268.

Purser, S., R. K. Shepherd and G. M. Clark. 1991. Evaluation of a sealing device for the intracochlear electrode entry point. Journal of the Oto-Laryngological Society of Australia 6: 472–480.

Quagliarello, V. J., W. J. Long and W. M. Scheld. 1986. Morphologic alterations of the blood-brain barrier with experimental meningitis in the rat. Temporal sequence and role of encapsulation. Journal of Clinical Investigation 77: 1084–1095.

Rasmussen, L. 1990. Immune responses to human cytomegalorvirus infection. In: McDougall J., ed. Cytomegalovirus. Berlin, Springer-Verlag: 221–254.

Redd, E. E., T. Pongstaporn and D. K. Ryugo. 2000. The effects of congenital deafness on auditory nerve synapses and globular bushy cells in cats. Hearing Research 147: 160–174.

Roos, K. L. and K. L. Tyler. 2001. Bacterial meningitis and other suppurative infections. In: Braunwald, E., A. S. Fuaci, D. L. Kasper, et al., eds. Harrison's Principles of Internal Medicine. 15th ed. New York, McGraw-Hill: 2462–2471.

Rutledge, L. J., M. L. Lewis and F. Sanabria. 1963. Fatal meningitis related to stapes operation. Archives of Otolaryngology 78: 637–641.

Ryugo, D. K., T. Pongstaporn, D. M. Huchton and J. K. Niparko. 1997. Ultrastructural analysis of primary endings in deaf white cats: morphological alterations in endbulbs of Held. Journal of Comparative Neurology 385: 230–244.

Sato, K., S. Shiraishi, H. Nakagawa, H. Kuriyama and R. A. Altschuler. 2000. Diversity and plasticity in amino acid receptor subunits in the rat auditory brainstem. Hearing Research 147: 137–144.

Schachern, P. A., M. M. Paparella, M. Goycoolea, B. Goldberg and P. Schlievert. 1981. The round window membrane following application of staphylococcal exotoxin: an electromicroscopic study. Laryngoscope 91: 2007–2017.

Schecterson, L. C. and M. Bothwell. 1994. Neurotrophin and neurotrophin receptor mRNA expression in developing inner ear. Hearing Research 73: 92–100.

Schmidt, J. M. 1985. Cochlear neuronal populations in developmental defects of the inner ear. Implications for cochlear implantation. Acta Oto-Laryngologica 99: 14–20.

Schuknecht, H. F. 1974. Pathology of the ear. Cambridge, MA, Harvard University Press.

Schuknecht, H. F. and J. Gulya. 1986. Anatomy of the temporal bone with surgical implications. Philadelphia, Lea and Febiger.

Seldon, H. L. and G. M. Clark. 1991. Human cochlear nucleus: comparison of Nissl-stained neurons from deaf and hearing patients. Brain Research 551: 185–194.

Shepherd, R. K., G. M. Clark and R. C. Black. 1983a. Chronic electrical stimulation of the auditory nerve in cats. Physiological and histopathological results. Acta Oto-Laryngologica-supplement 399: 19–31.

Shepherd, R. K., G. M. Clark, R. C. Black and J. F. Patrick. 1983b. The histopathological effects of chronic electrical stimulation of the cat cochlea. Journal of Laryngology and Otology 97: 333–341.

Shepherd, R. K., B. K.-H. G. Franz and G. M. Clark. 1990. The biocompatibility and safety of cochlear prostheses. In: Clark, G. M., Y. C. Tong and J. F. Patrick, eds. Cochlear prostheses. London, Churchill Livingstone: 69–98.

Shepherd, R. K. and E. Javel. 1997. Electrical stimulation of the auditory nerve: I. Correlation of physiological responses with cochlear status. Hearing Research 108: 112–144.

Shepherd, R. K., R. L. Webb, G. M. Clark, et al. 1984. Implanted material tolerance studies for a multiple-channel cochlear prosthesis. Acta Oto-Laryngologica-supplement 411: 71–81.

Sie, K. C. Y. and E. W. Rubel. 1992. Rapid changes in protein synthesis and cell size in the cochlear nucleus following eighth nerve activity blockade or cochlear ablation. Journal of Comparative Neurology 320: 501–508.

Silverstein, H., E. Smouha and N. Morgan. 1988. Multichannel cochlear implantation in a patient with bilateral Mondini deformities. American Journal of Otology 9(6): 451–455.

Simmons, F. B. 1967. Permanent intracochlear electrodes in cats. Tissue tolerance and cochlear microphonics. Laryngoscope 77: 171–186.

Snyderman, R. and R. J. Uhuig. 1992. Chemoattractant stimulus-response coupling. In: Gallin, J. I., ed. Inflammation: basic principles and clinical correlates. New York, Raven Press: 421–441.

Spoendlin, H. 1975. Neuroanatomical basis of cochlear coding mechanisms. Audiology 14: 383–407.

Spoendlin, H. 1984. Factors inducing retrograde degeneration of the cochlear nerve. Annals of Otology Rhinology and Laryngology 112: 76–82.

Stewart, P. S. and J. W. Costerton. 2001. Antibiotic resistance of bacteria in biofilms. Lancet 358: 135–138.

Suzuki, C., I. Sando, J. J. Fagan, D. B. Kamerer and A. S. Knisely. 1998. Histopathological features of a cochlear implant and otogenic meningitis in Mondini dysplasia. Archives of Otolaryngology–Head and Neck Surgery 124: 462–466.

Suzuki, M., K. C. Cheng, M. S. Krug and T. J. Yoo. 1998. Successful prevention of retrocochlear hearing loss in murine experimental allergic encephalomyelitis with T cell receptor Vbeta8-specific antibody. Annals of Otology, Rhinology and Laryngology 107: 917–927.

Swartz, M. N. and P. R. Dodge. 1965. Bacterial meningitis—a review of selected aspects. New England Journal of Medicine 272: 725–731, 779–787, 842–848, 898–902, 1003–1010.

Tanaka, K. and S. Motomura. 1981. Permeability of the labyrinthine windows in guinea pigs. Archives of Otolaryngologica 233: 67–73.

Terayama, Y., K. Kaneko and K. Tanaka. 1979. Ultrastructural changes of the nerve elements following disruption of the organ of corti. II. Nerve elements outside the organ of corti. Acta Oto-Laryngologica 88: 27–36.

Terayama, Y., Y. Kaneko, K. Kawamoto and N. Sakai. 1977. Ultrastructural changes of the nerve elements following disruption of the organ of Corti. I. Nerve elements in the organ of Corti. Acta Otolaryngologica 83: 291–302.

Terrahe, K. and U. Westphal. 1968. Elektronenmikroskopische resorptionsstudien an der gesunden und entzundlichen mittelohrschleimhaut. Archiv fur Klinische und Experimentelle Ohren, Nasen und Kehlkopfheilkunde 191: 458.

Tierney, T. S., F. A. Russel and D. R. Moore. 1997. Susceptibility of developing cochlear nucleus neurons to deafferentation-induced death abruptly ends just before the onset of hearing. Journal of Comparative Neurology 378: 295–306.

Trune, D. R. 1982. Influence of neonatal cochlear removal on the development of mouse cochlear nucleus: I. Number, size and density of its neurons. Journal of Comparative Neurology 209: 409–424.

Turrini, M., E. Orzan, M. Gabana, E. Genovese, E. Arslan and U. Fisch. 1997. Cochlear implantation in a bilateral Mondini dysplasia. Scandinavian Audiology Supplementum 46: 78–81.

Valvassori, G. E. and J. D. Clemis. 1978. The large vestibular aqueduct and associated anomalies of the inner ear. Laryngoscope 88: 723–728.

Van Camp, G. and R. J. Smith. 2002. Hereditary hearing loss home page. http://www.uia.ac.be/dnalab/hhh (June 5, 2002).

Vaudaux, P. E., D. P. Lew and F. A. Waldvogel. 1994. Host factors predisposing to and influencing therapy of forming body infections. In: Bisno, A. L. and F. A. Waldvogel, eds. Infections associated with indwelling medical devices. 2nd ed. Washington, DC, ASM Press.

Waring, G. W. and L. Weinstein. 1948. Treatment of pneumococcal meningitis. American Journal of Medicine 5: 402–418.

Webster, D. B. and M. Webster. 1978. Cochlear nerve projections following organ of corti destruction. Otolaryngology 86: 342–353.

Webster, M. and D. B. Webster. 1981. Spiral ganglion neuron loss following organ of Corti loss: a quantitative study. Brain Research 212: 17–30.

Wilcox, S. A., K. Saunders, A. H. Osborn, et al. 2000. High frequency hearing loss correlated with mutations in the GJB2 gene. Human Genetics 106: 399–405.

Willott, J., J. Milbrandt, L. Bross and D. Caspary. 1997. Glycine immunoreactivity and receptor binding in the cochlear nucleus of C57BL16J mouse. Hearing Research 141: 12–18.

Wolff, D. 1964. Untoward sequelae eleven months following stapedectomy. Annals of Otology, Rhinology and Laryngology 73: 297–304.

Wong-Riley, M. T. T., M. M. Merzenich and P. A. Leake-Jones. 1978. Changes in endogenous enzymatic reactivity to DAB induced by neuronal inactivity. Brain Research 141: 185–192.

Wood, N. K., E. J. Kaminski and R. J. Oglesby. 1970. The significance of implant shape in experimental testing of biological materials: disc vs rod. Journal of Biomedical Materials Research 4: 1–12.

Xu, J., R. K. Shepherd, R. E. Millard and G. M. Clark. 1997. Chronic electrical stimulation of the auditory nerve at high stimulus rates: a physiological and histopathological study. Hearing Research 105: 1–29.

Xu, S. A., R. K. Shepherd and G. M. Clark. 1990. Rapid and permanent hearing loss in cats following co-administration of kanamycin and ethacrynic acid. Proceedings of the Australian Physiological and Pharmacological Society 21: 44P.

Ylikoski, J., A. Belal and W. F. House. 1981. Morphology of human cochlear nerve after labyrinthectomy. Acta Oto-Laryngologica 91: 161–166.

Ylikoski, J., U. Pirvola, M. Moshnyakov, J. Palgi, U. Arumae and M. Saarma. 1993. Expression patterns of neurotrophin and their receptor mRNAs in the rat inner ear. Hearing Research 65: 69–78.

Yukawa, K., S. O'Leary, M. Clarke and G. M. Clark. 2001a. Histopathology of the brainstem of a binaural cochlear implant subject. Program and abstracts of the Third International Congress of Asia Pacific Symposium on Cochlear Implant and Related Sciences, Osaka, Japan: 48.

Yukawa, K., S. J. O'Leary, M. Clarke and G. M. Clark. 2001b. Histopathology of the binaural cochlear implant patient. Abstract for the Australian Society for Otolaryngology Head and Neck Surgery Annual Scientific Meeting.

Zimmerli, W., F. A. Waldvogel, P. Vaudaux and U. E. Nydegger. 1982. Pathogenesis of foreign body infection: description and characteristics of an animal model. Journal of Infectious Diseases 146: 487–497.

Zirpel, L., E. A. Lachica and W. R. Lippe. 1995. Deafferentation increases the intracellular calcium of cochlear nucleus neurons in the embryonic chick. Journal of Neurophysiology 74: 1355–1357.

Zirpel, L. and E. W. Rubel. 1996. Eighth nerve activity regulates intracellular calcium concentration of avian cochlear nucleus neurons via a metabotropic glutamate receptor. Journal of Neurophysiology 76: 4127–4139.

4
Neurobiology

Overview

Electrical stimulation as well as the toxicity of materials, trauma, and infection may cause spiral ganglion cell and auditory nerve loss. Electrical stimulation within physiological limits can also help preserve the auditory nerve and spiral ganglion cells and prevent degeneration. This chapter discusses the effects of acute and chronic intracochlear electrical stimulation as well as the effects of stimulating the cochlear nucleus. The possible neural damage mechanisms associated with electrical stimulation at high stimulus intensities and rates are appraised. Electrical stimulation may result in adverse effects through the direct effect of electrical charge on the biochemistry of the neuron. Electrochemical reactions associated with the electrical stimulus, such as the release of platinum ions into the biological environment, can also be the cause of the above responses (Agnew et al 1977). In addition, scanning electron microscopy of the electrode array, together with chronically recorded electrode impedance data, provide an important insight into the long-term performance of the stimulating electrodes and the tissue response. The clinical implications of neural and other tissue damage with electrical stimulation are reviewed. Finally, the results from several studies on the histopathology of human temporal bones following cochlear implantation are described.

Definition of Terms

Current and Charge

Electrical current is the passage of electrons in a conducting material. Current, I, is defined as the rate at which the number of electrons or charge, Q, measured in coulombs, passes a given point. Consequently $I = dQ/dt$, expressed in coulombs/s or amperes. Furthermore, if the current is steady it is referred to as DC (direct current), or if it alternates, AC (alternating current).

160

Voltage

The energy required to accelerate electrons from one point to another, is the potential difference between the two points. The potential difference, V, is measured in terms of work, W, per unit charge. Thus $V = W/Q$, expressed in volts, after Alessandro Volta, who made key discoveries in electricity and was the first person to attempt to produce hearing by electrical stimulation as discussed in Chapter 1.

Resistance

Resistance, R, occurs when electrons collide with atoms and lose energy. The resistance depends on the properties of the material referred to as resistivity, ρ. The resistance to electron flow also depends on the relation between cross-sectional area, A, and length, L. Thus $R = \rho L/A$. The unit of resistance is the ohm (Ω), after Georg Ohm, who discovered the relationship among current, voltage, and resistance. It was found that in order to maintain a large current in a conductor, more energy, that is, a greater potential difference, was required than for a smaller current. This constant relationship is $V = IR$, and it is referred to as Ohm's law.

Capacitance

If two metal plates are separated by an insulator, with one plate positively charged and the other negatively, the capacity C of the two plates to hold the charge is proportional to the voltage. Thus $Q = CV$ where the capacitance is a constant for the proportionality. The unit of capacitance is the farad, named in honor of Michael Faraday. With a sinusoidal current the relation among voltage, current, and capacitance and angular velocity (ω) or ($2\pi f$) is $V = \dfrac{1}{2\pi fC} \times I$. The current in the capacitor leads the voltage with a phase angle of $\pi/2$. The equation above is similar to Ohm's law with the proportionality factor $\dfrac{1}{2\pi fC}$ being the capacitive reactance; the larger the capacitance the smaller the reactance or impedance. The impedance is also inversely proportional to the frequency f, that is, the lower the frequency the greater the impedance. For this reason the capacitor is used to prevent DC (in theory of zero frequency) in the electronic circuit of a receiver-stimulator producing current at the electrode-tissue interface.

Impedance

Impedance is opposition to current flow and is made up of reactance and resistance. Reactance impedes current flow through storing energy as discussed, and is frequency dependent. On the other hand, with resistance the energy is dissipated. Reactance and resistance cannot simply be summed to achieve a total resistance; rather, due to their phase differences they must be treated like vectors. The vector addition of reactance and resistance results in a total value known as impedance.

Electrode/Tissue Interface

Polarization

When an electrode is placed in an electrolyte solution (e.g., perilymph) electrical charges become distributed at the electrolyte interface and in the neighboring solution, so that an electrical potential is developed and the electrode is polarized.

Charge Transfer

There are two mechanisms for the transfer of electrical charge across the electrode tissue interface. These mechanisms are referred to as capacitive and faradic. Capacitive charge injection occurs through the charging and discharging of the "double layer" of ions at the interface. This double layer is capacitor-like, and is due to ions in solution being attracted to excess charge on the surface of the metal electrode (Crow 1988). When a small voltage is applied, no actual transfer of electrons between the electrode and the solution takes place, and therefore capacitive injection is the ideal method of charge delivery. The maximum charge density for capacitive charge delivery is only about 20 μCcm^{-2} (Robblee and Rose 1990). With charge densities above this limit, charge is transferred by faradic means. With faradic stimulation charge is transferred through chemical reactions. If the charge density is not too high, the reactions are reversible, and are usually oxidation and reduction, or H-atom plating, as in the case of a platinum (Pt) electrode (Robblee and Rose 1990). At the positive anode, OH^{1-} ions are attracted to the platinum to form PtO leaving H^{1+} in solution and the release of electrons (e).

$$Pt + H_2O \leftrightarrow PtO + 2H^{1+} + 2e^-$$

At the negative cathode, positive hydrogen ions are attracted to produce H atom plating with the absorption of electrons and (OH^{1-}) ions in solution.

$$Pt + H_2O + e^- \leftrightarrow PtH + OH^{1-}$$

As the reactions are reversible and chemically neutral, they are considered safe. But if the charge density becomes too high, the reactions are irreversible with the evolution of oxygen or hydrogen.

$$2H_2O + 2e^- \rightarrow H_2 \uparrow + 2OH^{1-}$$
$$2H_2O \rightarrow O_2 \uparrow + 4H^{1+} + 4e^-$$

Irreversible reactions can lead to undesirable changes in the pH, dissolved metal, or the formation of metal protein complexes (Brummer and Turner 1977b). With long-term stimulation these toxic materials could build up in the perilymph (Clark, Tong et al 1977). The products are toxic to neural tissue, and therefore the charge density must be kept below the level at which irreversible reactions occur. The resulting electrolysis can be avoided if a biphasic stimulus pulse is used so that the double layer capacitors can be alternately charged and discharged.

Future electrode arrays with more electrodes will have their surface area reduced, and so there will be a need to ensure that the charge density remains within the safe limit. The method of capacitive transfer of charge should be enhanced to avoid irreversible faradic effects. A high voltage results in irreversible changes that can occur at localized regions of the electrode surface that should ideally be homogeneous. The passage of electrical current through tissue is electrolytic via positively and negatively charged ions, and not the flow of electrons. If the metal of the stimulating electrode does not pass readily into solution, a higher voltage is required to transfer charge to the neighboring ions. This voltage can be reduced, for example, if iridium is plated onto platinum electrodes (Blau et al 1997; Weiland and Anderson 2000). Such materials can allow smaller electrodes to be used without toxic metal ions passing into solution or generating toxic by-products.

Charge Density

To determine the charge density with electrical stimulation either by modifying the electrical double layer at the electrode-tissue interface (capacitance) or coupling via a surface layer oxidation-reduction process (faradic), it is necessary to know the real active area of the electrode. The real area, as distinct from the geometric area, is measured by the deposition of hydrogen ions on its surface. When a cathode pulse is applied, H_2 is evolved from H_2O. Just prior to the release of H_2 there is a monolayer of H atoms. The charge involved is the hydride formation charge Q_{HF}. One real cm² holds a charge of $210\,\mu C$. Usually 1.0 geometric cm² of a shiny Pt electrode is equivalent to 1.4 real cm² (Brummer and Turner 1977a).

Equivalent Circuits

When a potential difference is applied across the electrode–tissue interface, as discussed above, charge is transferred either by varying the polarization at the double layer (capacitance) or by producing a flow of charge across the interface if the potential difference is higher (faradic stimulation). The equivalent circuit for the interface is shown in Figure 4.1. The resistance R_e is due to the conducting track as well as the properties of the electrode. The resistance of the electrode depends on the resistivity of the metal and its geometric size. The electrical properties of the electrode–tissue interface can be modeled as a leaky capacitor (Weinman and Mahler 1964; Dymond 1976). That means there is a resistor in parallel

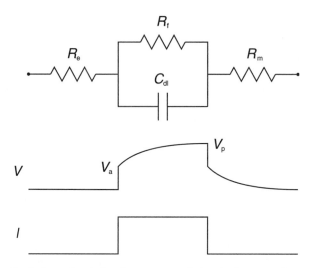

FIGURE 4.1. Equivalent circuit for the passage of electrical current from an electrode through the tissue interface to the return electrode. R_e is the resistance of the electrode track and terminal. The electrode tissue interface is modeled by a capacitor C_{dl} (the double-layer capacitance) in parallel with a resistor R_f (the faradic impedance). R_m is the resistance of the medium. V is the voltage developed across a pair of electrodes, and I is the current. The voltage due to the series resistance $R_e + R_m$ is also referred to as the access voltage is V_a, and the peak voltage after the addition of that due to capacitance is V_p. (Reprinted from Cochlear Prostheses, Clark et al, © 1990 Churchill Livingstone, with permission of Elsevier Science.)

with the capacitor. The capacitor C_{dl} is the double layer of atoms responsible for capacitive charge injection, and R_f is the resistance that represents the Faraday current path. This circuit only applies for a perfectly smooth electrode. Surface roughness causes the double layer of charge to deviate from an ideal capacitor, when it is referred to as distributed capacitance, and requires a modified equivalent circuit.

The voltage V, developed across a pair of electrodes in response to a current pulse (Vaadia et al 1995), can be determined from the equivalent circuit. The start of the current pulse initiates an abrupt rise in the voltage across the electrodes, V_a, and is equal to $I \times (R_e + R_m)$. The series resistance is made up of the electrode, R_e, and medium, R_m, containing electrolyte in solution, and fibrous tissue or bone. If the resistance of the electrode track is negligible, R_e is primarily due to the electrode surface area. This purely resistive component of the electrode impedance produces the instantaneous voltage developed across the electrodes. Following this rapid initial rise in V_a, the voltage across the electrodes gradually rises for the duration of the current pulse to a peak voltage V_p. This is the over-voltage and is due to the capacitor or double layer. The shape of this portion of the voltage depends on the reactive properties of the electrode–tissue interface. The resistive and reactive components of the electrode impedance therefore can be determined from the electrode voltage and current waveforms.

The impedance varies with frequency and can be measured with impedance spectroscopy. In this case a low-voltage sine wave is applied to the electrode so that the system is minimally affected. At high frequencies (e.g., 1000 to 100,000 Hz) the double layer acts as a short circuit, and so the series resistance ($R_e + R_m$) is the main component. The frequency at which the double layer is shorted varies between electrodes (material, area, roughness). At low frequencies, however, the electrode impedance is also affected by C_{dl} and R_f. If the potential difference is small, then C_{dl} is the main factor, since a small potential difference means that faradic reactions are minimal and R_f is very large.

For an electrode with a rough surface area, the degree of roughness also contributes to the capacitance. At high frequencies there is insufficient time for a charge to cover the rough areas, so the geometric surface area is the main component. At low frequencies the impedance depends on the geometric surface area and the surface roughness. So a rougher surface will cause a larger double-layer capacitance. To avoid faradic current, the potential on the electrodes should be kept as small as possible. To do this without sacrificing the delivery of charge, the impedance of the electrode should also be kept low. With small electrodes one method of keeping the impedance low is to produce surface roughening techniques. The surface roughness should be designed to lower the impedance at the stimulus rates to be used (Parker 2002).

Impedance

When an electrical current flows across the electrode–tissue interface, it results in an alteration in the charge distribution at the interface and hence the impedance to current flow. With electrical stimulation it is preferable that current be passed as a result of double-layer charging at the electrode electrolyte interface. Otherwise electrolysis will occur with the production of toxic substances. The impedance of the electrode acting as a capacitor is $Z_c = 1/j\omega C$, where Z_c is the impedance, C is the capacitance, ω the angular frequency, and j is the imaginary unit $\sqrt{-1}$. If the surface is roughened, then the relationship needs to be modified. As the impedance varies not only with frequency but also with the size of the current, a constant current stimulator is required to produce a known current.

Constant Current and Voltage Stimulators

There are two general classes of neural stimulators. The relative performance of these stimulators depends on changes in electrode impedance. The most common type of stimulators for neural prostheses are constant current devices, which are designed to output a constant current amplitude (and therefore a constant charge) irrespective of variations in electrode impedance. An increase in electrode impedance results in a concomitant increase in the voltage developed across the electrodes ($I = V/R$; Ohm's law). The current amplitude, and therefore the amount of charge per phase, remains constant. Constant voltage devices form the second class of stimulators used in neural prostheses, and as their name suggests they are designed to output a constant voltage that is independent of variations in electrode impedance. An increase in electrode impedance results in a decrease in the amplitude of the stimulus current (Ohm's law) and thus a decrease in the amount of charge delivered per phase of the current pulse. Constant voltage devices have two advantages over constant current stimulators in that they are simpler to design and they operate using lower power consumption. As stated above, the performance of both classes of stimulator depends on the variations in electrode impedance. Constant current stimulators are generally very effective in environments where there are small or moderate changes in electrode impedance. However, when stimulating with high-impedance electrodes, these devices can lose voltage compliance. Under these conditions the stimulator can produce asymmetrical current pulses and deliver net DC charge to the adjacent tissue. This can result in tissue damage. Alternatively, changes in the impedance of electrodes being driven by a constant voltage stimulator would significantly affect the amount of charge being injected into the tissue, therefore varying the extent of neural excitability and reducing the effectiveness of this class of stimulator.

Monitoring Impedance In Vivo and In Vitro

The impedance of the stimulating electrodes should be periodically monitored as they reflect changes at the electrode–tissue interface due to, for example, the degree of fibrous tissue and bone formation or an increase in the surface area of the electrodes as a result of Pt dissolution (Babb et al 1977; Harrison and Dawson 1977; Agnew et al 1983).

Representative examples of total electrode impedance, $Z = V_p/I$, (Fig. 4.1) and series (access) resistance for the Melbourne/Cochlear banded electrode array, are illustrated in Figure 4.2 (Shepherd et al 1990). These data were collected during long-term stimulation in the study reported by Shepherd et al (1983a). They illustrate variability among electrodes. For comparison, data from electrodes stimulated in inorganic saline are also included. Although there was a wide range of variability among electrodes, there were consistencies. First, in vitro access resistance and electrode impedance measurements, recorded just prior to and following the completion of the chronic stimulation period (illustrated in Fig. 4.2 as solid symbols), were consistently lower than the measurements made in vivo (open symbols). Second, the similarity of the in vitro access resistance and electrode impedance measurements recorded prior to and following in vivo stimulation suggested that the variations recorded in vivo reflected changes in the environment adjacent to the electrodes, rather than permanent changes to the electrode surface itself. This was supported by the observation that electrodes stimulated in inorganic saline exhibit stable access resistance and electrode impedance values. The fact that electrode impedance generally changed in concert with the access resistance supported the view that the variation in electrode impedance reflected changes in the tissue environment adjacent to the electrodes. Third, in vivo access resistance and electrode impedance values showed a gradual increase over the first 12 to 30 days following implantation for two of the electrodes, after which they remained relatively constant except, however, for some significant

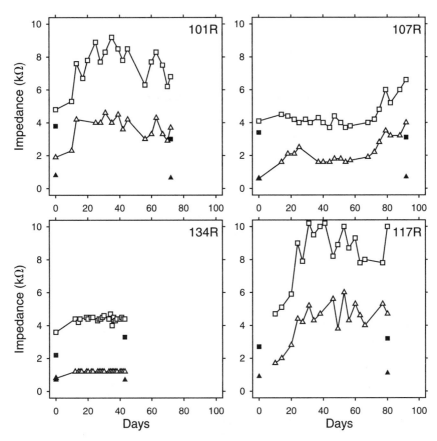

FIGURE 4.2. Electrode impedance (□) and access resistance (△) as a function of implantation time for four stimulated electrodes. Solid symbols indicate in vitro measurements in saline prior to implantation and following explantation. Dashed lines indicate recordings made prior to the start of the continuous electrical stimulation (Shepherd et al 1983a, 1990). (Reprinted from Cochlear Prostheses, Clark et al, © 1990 Churchill Livingstone, with permission of Elsevier Science.)

short-term fluctuations. This gradual increase was probably due to the change in the tissue surrounding the electrode. This was seen in the study by Clark, Shute et al (1995), in which impedance correlated with the grading of the tissue around the track, and was particularly high with densely packed round cell inflammatory cells. An increase in round cells occurs in the first few weeks of inflammation, and this time course is similar to the increase in impedance seen in the study by Shepherd et al (1983a, 1990). Finally, long-term electrical stimulation (illustrated in Fig. 4.2 by the solid lines connecting data points) did not appear to have any significant effect on the in vivo measured resistance and electrode impedance, as they remained reasonably constant during the stimulus period after the initial increase.

The marked short-term fluctuations of both access resistance and electrode impedance observed for some electrodes in this study were also reported by other investigators using both scala tympani electrodes (Byers et al 1981) and electrodes implanted in other sites (Babb et al 1977; Agnew et al 1981). Although the origin of these fluctuations is not clear, they were observed only among in vivo electrodes that have relatively high impedances.

These short-term changes in impedance highlight the importance of using constant current rather than constant voltage stimulators for neural prostheses, as constant voltage devices cannot control the amount of charge being injected in environments in which there are significant variations in electrode impedance.

Correlation with Tissue Response

The changes in impedance in the study by Shepherd et al (1990) were compared with the nature of the pathological changes around the array. This suggested that the increases were due to the growth of fibrous tissue. For example, electrodes from 134R showed minimal changes in impedance and access resistance over the implant period. The results suggested minimal tissue reaction adjacent to the electrode pair. Histological analysis of the cochlea and scanning electron microscope (SEM) analysis of the electrodes confirmed that this array did not evoke a fibrous tissue capsule. In contrast, electrode pair 117R exhibited the highest in vivo impedance and resistance values measured. Histological examination of this cochlea revealed a compact fibrous tissue capsule that completely encapsulated the electrode array. SEM analysis of the electrode array also showed an extensive covering of fibrous tissue over both electrodes. A comparison of electrode impedance and cochlear histology indicated that the most obvious association between fibrous tissue and electrode impedance was the density and continuity of the fibrous tissue capsule surrounding the electrode array. These results were consistent with those of Clark, Shute et al (1995) referred to above, who found in particular that the density of the round cell infiltrate was the key factor contributing to the impedance. This would also lead to later dense fibrous tissue.

It is possible that fibrous tissue may reduce the effective surface area of the electrodes, elevating charge densities and resulting in the production of an electrochemical reaction product, such as the formation of localized gas bubbles or a protein complex on the electrode surface. The formation and removal of this coating could account for the short-term impedance changes.

Corrosion-Stimulus Parameters

Corrosion is the loss of metal through electrochemical processes. A stimulus current producing a negative or positive charge at the electrode tissue interface induces electrolysis.

Mechanisms

Corrosion occurs through faradic stimulation, discussed above, when the current or charge density reaches a level where the changes at the electrode-tissue interface are no longer reversible and metal is lost. The toxic products produced and the degree of metal lost will depend on the type of metal and other factors such as current density and wave shape. Furthermore, the so-called noble metals like platinum are not immune from these processes, and careful consideration must be given to minimizing the production of toxic products and metal loss in any metal electrodes used with the cochlear implant.

When a platinum electrode is placed in a solution with the composition of perilymph, the toxic products produced and the degree of metal lost depend on whether the electrode is positive or negative with respect to ground. If it is positive, (OH^{1-}) will be attracted and oxygen will be released as a gas (Greatbatch et al 1969). This will lower the pH and make the solution more acidic. Chloride ions (Cl^{1-}) will also be attracted with chorine (Cl_2) and other chloride oxidation products $(ClO^{1-}, ClO3^{1-}$, etc.) being released. These would be toxic to nerve fibers if produced in quantities that could not be buffered or removed by the perilymph. Studies also show that platinum ions (Pt^{1+}) will pass into solution, and this will lead to metal loss (Brummer and Turner 1975). Trace analyses have demonstrated that Pt electrodes in saline can show dissolution with biphasic currents even when faradic reactions with the release of (Cl_2) and (O_2) and so on do not occur (McHardy et al 1980). Furthermore, when the electrode is negative with respect to ground, it will attract hydrogen ions (H^+) that will be absorbed and released as hydrogen (H_2). This will lead to an increase in the pH and the solution becoming more alkaline.

Stimulus Parameters

As it was shown that even noble metals such as Pt could corrode when an electrical voltage was applied, it became necessary to know the stimulus parameters that would minimize the loss, and what levels would be damaging to tissue. Studies by Brummer and Turner (1975) demonstrated that Pt dissolution was greater when the first part of the biphasic wave was positive (phase lead) with respect to ground rather than negative (phase lag). This would be expected electrochemically as discussed above, and also the (Pt^{1+}) would form a complex protein that would not reverse during the phase lag. This Pt-protein complex could also be toxic to tissue (Agnew et al 1977). It is harder to explain why there would be some loss of Pt when the electrode was negative. It was probably due to an irreversible reduction of (O_2) to (OH^{1-}) ions, raising the average potential of the electrode above the open circuit value (McHardy et al 1980). For the phase lead, charge density was the main factor controlling dissolution, and for phase lag it was pulse duration.

The loss of Pt was directly related to the charge density, and Brummer et al (1977) showed that the charge density should be kept below $300 \, \mu C \, cm^{-2}$ geometric to avoid gassing. To reduce the charge density, it is necessary to increase the surface area, and this was done by platinization that roughens the surface (platinum black). This increased the real but not the geometric surface area. The Pt dissolution for the same current was less for the platinized electrodes (50 ng/C).

However, it was demonstrated by Schwan (1963) that platinum black electrodes could lose part of the platinum during mechanical insertion, and deteriorate in biological fluids, probably due to the entry of protein molecules into the porous surface. This was one of the main reasons that smooth platinum rather than platinum black electrodes were used for the University of Melbourne cochlear implant (Clark, Tong et al 1977). The electrodes also needed to be smooth to avoid cochlear damage both on implantation and explantation.

In designing a cochlear implant to present stimulus parameters that minimized corrosion, it was also necessary to optimize the pulse duration. An in vitro study by Black and Clark (1977) and Black and Hannaker (1979) using spectrophotometry to detect small concentrations of Pt showed that the dissolved Pt was greatly reduced for pulse durations less than 500 μs (Fig. 4.3). It was found platinum dissolution occurred for biphasic current pulses at charge densities as low as 25 μC cm^{-2} geometric. The average dissolution of Pt with a current density of 2 mA mm^{-2} was 20 ng at a pulse repetition rate of 1000 Hz. Furthermore, to achieve a maximum current density no greater than 2 mA mm^{-2}, the University of Melbourne's prototype implant was designed with the electrodes having a surface

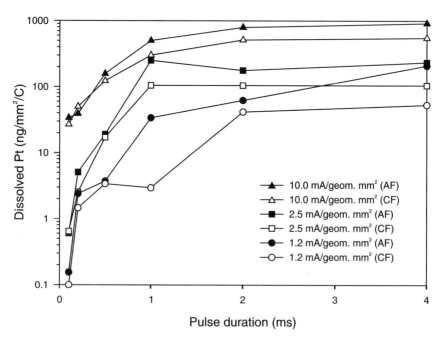

FIGURE 4.3. Platinum dissolution for biphasic pulse stimulation at current densities plotted as a function of pulse width per phase. Each point is the mean of two measurements. AF, anodic first; CF, cathodic first. (Reprinted from Black, R. C., Hannaker, P. 1979 Appl. Neurophysiol. 42: 366–374. With permission from S. Karger AG, Basel.)

area of 0.5 mm². In fact the University of Melbourne prototype and Nucleus clinical trial devices had banded electrodes with surface areas from 0.4 to 0.6 mm². The surface area of stimulating electrodes needed to be large enough to reduce charge density to a level that would avoid electrode corrosion and damage to neural tissue, but still provide localized stimulation of auditory nerve fibers.

As protein could affect metal corrosion, with an increase for some metals and a decrease for others, it was important to determine what would happen to Pt in vivo especially when a voltage was applied. The effect of protein on a charge-induced dissolution of Pt was examined in vitro by Robblee et al (1980). The study showed that the dissolution rate decreased with time and approximated zero. A critical level of 0.15 mg/mL of protein was required, but higher levels did not enhance the protection. The effects were probably due to adsorption of the protein on the surface.

Scanning Electron Microscope Evaluation of Electrodes

Further evidence for the protective effect of protein on corrosion was seen in the SEM evaluation of the banded Pt electrodes after a long-term (chronic) stimulation study in cats (Shepherd et al 1985). Although these electrodes had been stimulated for periods of up to 2000 hours at charge densities of up to 32 μC cm^{-2} geometric/phase, the surface features of the Pt electrodes could not be distinguished from those of unstimulated control electrodes. Furthermore, there was no evidence of degradation of the Silastic carrier. Had Pt corrosion occurred, there would have been tissue irritation (Dymond et al 1970; Bernstein et al 1977) and a more extensive fibrous tissue reaction than was evident in the study by Shepherd et al (1985).

These experimental results were corroborated following the SEM analysis of a banded electrode array that had been removed from a patient 27 months following implantation (Clark et al 1988). It was estimated that the device was used for approximately 10,000 hours, developing charge densities of up to 25.7 μC cm^{-2} geometric/phase. Examination of the surface features of the 22 active electrodes showed no evidence of corrosion. An example of the surface condition of one stimulated electrode from this array is shown in Figure 4.4. There was no difference between the surface features of electrodes on this array and those of unstimulated control electrodes. In addition, the Silastic MDX-4-4210 carrier appeared normal. The surface scratches in this earlier machined electrode could have been the sites for corrosion, as the charge density is high at these points. This is especially likely when the surface as a whole is resistant (Mears and Brown 1941). The fact that no corrosion was seen indicates the strong protective effect of protein.

The surface of the Pt banded electrodes from the Nucleus multiple-channel cochlear implant remained essentially unchanged during long-term electrical stimulation using charge balanced biphasic constant current pulses, and charge densities of up to approximately 30 μC cm^{-2} geometric/phase.

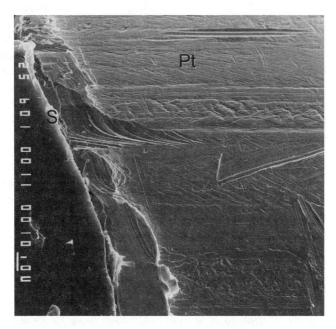

FIGURE 4.4. Scanning electron microscope (SEM) micrograph of a platinum (Pt) electrode from an array that had been removed from a patient following an implant period of 27 months. This electrode was one of the most extensively used on the array, and developed charge densities of up to 25.7 μC cm^{-2} geometric/phase. The Pt surface shows no evidence of corrosion S,—Silastic. Original magnification: 1200$\times$. (Reprinted from Clark, GM et al, 1988. The histopathology of the human temporal bone and auditory central nervous system following cochlear implantation in a patient, *Acta Oto-Laryngologica,* (suppl **448**), with permission.)

Electrical Parameters and Neural Stimulation

Electrical stimulation of the auditory nerve is due to the electrical charge. The charge required to excite a neural population is delivered via a series of reversible electrochemical reactions at the electrode–tissue interface, as discussed above. When the electrical charge passes through the neural membrane, a number of sites depolarize (see Chapter 5). These depend on the geometry of the electrode and the polarity of the charge delivered (McNeal 1976). The membrane becomes permeable to sodium ions, and when a threshold level of depolarization is reached the ions cascade across the membrane, initiating an action potential. The stimulus parameters selected may cause neural damage through the mechanisms that lead to corrosion and release of toxic products or the biochemical effects of over-stimulation.

Electrochemically Safe Stimulus Parameters

The electrochemically safe stimulus parameters are below the level discussed above that leads to corrosion and the release of toxic products from the electrode.

To this end nondamaging electrical stimulation can be achieved using short-duration charge-balanced biphasic current pulses (Lilly 1961; Mortimer et al 1970). With these stimuli the localized electrochemical reactions are reversed during the second phase of the biphasic current pulse, therefore ensuring that no net electrochemical products are formed. An electrochemically safe stimulus regime was demonstrated for Pt electrodes using bipolar current pulses (Brummer et al 1977). The stimuli consisted of short-duration $(100-200\,\mu s)$ biphasic current pulses, with a maximum charge density of $300\,\mu C\;cm^{-2}$ geometric/phase. Electrolysis of water occurred at charge densities above this limit (Brummer et al 1983), as well as the production of (Pt^{1+}), which would damage neural tissue (Pudenz 1942; Agnew et al 1977). The degree of Pt dissolution in saline is linearly related to the aggregate charge delivered (McHardy et al 1980). As discussed above (see Corrosion-Stimulus Parameters), protein significantly reduced the extent of this dissolution (Robblee et al 1980). Other materials can be used to increase the safe charge density, and at the same time reduce the electrode overvoltage.

Charge Density and Charge per Phase

While appropriate electrode materials and stimuli minimize electrochemical damage, electrical stimulation per se may have a deleterious effect on neural tissue. Charge density, charge per phase, and total charge delivered are important parameters (Yuen et al 1981; Agnew et al 1983), but their safe stimulus levels have not been clearly defined for the auditory nerve (Walsh and Leake-Jones 1982). Other related parameters are the location of the electrodes with respect to the neural population being stimulated, the electrical stimulus regime used, the electrode geometry, and the nature of the biological environment adjacent to the electrodes. A relationship between charge density and charge per phase in producing neural damage was determined by McCreery et al (1988, 1990, 1994) for electrodes in direct contact with the surface of the cortex. Their data suggested charge density and charge per phase covaried in producing neural damage. Thus a high charge density required a low charge per phase to avoid damage and vice versa. However, for electrodes in the posteroventral cochlear nucleus of the cat, McCreery et al found only damage that correlated with the charge per phase for biphasic pulses at 500 pulses/s. There was little correlation with geometric charge density, the amplitude of the cathodic phase, or pulse duration. The damage was severely edematous axons. It commenced at approximately 3 nC/phase compared to a stimulus threshold of 1 nC/phase.

Biochemical Effects

The biochemical changes in the neuron depend on stimulus rate and intensity. This was demonstrated by Sokoloff (1983) for the rates of glucose utilization. Energy is required for metabolic processes such as cellular homeostasis, protein synthesis, tissue repair, and axoplasmic transport. It is especially required for the maintenance and restoration of ionic gradients across the neural membrane, primarily following spike activity.

The effects of rate of stimulation on the neuron were studied by Ochs and Smith (1971). They reported temporary stimulus-induced reductions in the rate of fast axoplasmic transport with increased electrical stimulus rate. Increases in the extracellular (K^{1+}) concentration not returning rapidly to normal during electrical stimulation at high rates, and charge densities per phase (100 μC cm^{-2}/ phase or 1 μC/phase at 50 Hz) have been reported by Heinemann and Lux (1977), Nicholson et al (1978), Urbanics et al (1978), Stockle and Ten Bruggencate (1980), Agnew et al (1983), and McCreery and Agnew (1983). These increases in the extracellular (K^{1+}) concentration corresponded with a reduction in the excitability of the neural population, and were interpreted as an inability of local cellular metabolism to maintain homeostasis (McCreery and Agnew 1983). The inability of a neuron to maintain homeostasis not only may result in a reduction in neural excitability, but also could ultimately lead to permanent neural damage due, for example, to the accumulation of metabolic products (Shepherd et al 1990).

Neural Preservation

After a sensory hearing loss the spiral ganglion cells degenerate. This is within weeks to months for animals such as guinea pigs (Webster and Webster 1978) and years for humans (Otte et al 1978; Hinojosa and Marion 1983; Nadol et al 1989). This could be due to the loss of trophic factors from the hair cells and loss activity in the cochlear nerve. Restoration of neural activity with electrical stimulation from an implanted electrode array reduced spiral cell degeneration following inner hair cell loss (Lousteau 1987; Hartshorn et al 1991; Leake et al 1992) as well as central auditory changes (Miller et al 1992).

Significant recovery in the soma volume of cochlear nucleus neurons following chronic electrical stimulation was reported in long-term deafened cats and guinea pigs (Chouard et al 1983; Matsushima et al 1991; Lustig et al 1994).

It was also shown by Miller and Altschuler (1995) that survival could be enhanced even if stimulation was delayed until a substantial number of ganglion cells had begun degenerating. The degree or preservation was a function of the intensity level of the current. In another study with continuous pulsatile stimuli 6 dB above threshold presented a few days after deafening, it was found that preservation was much better for low (250 Hz) rather than high (1000 Hz) stimulus rates. Even low stimulus levels of 5 μC cm^{-2}/phase were very effective in preservation, provided they were presented with little delay after the loss of hearing.

In a study on the human brainstem of a bilateral and a unilateral patient, the medial superior olive (MSO) had increased volume on the side opposite the one most frequently stimulated in the bilateral patient and opposite the implant in the unilateral patient (Yukawa et al 2001a,b). This helped confirm the experimental animal findings that chronic stimulation improves brain cell function.

The restoration of function has been demonstrated at a biochemical level. Wong-Riley et al (1981) found the reduction in the metabolic enzyme cytochrome oxidase seen in deafness was partially restored with chronic electrical stimulation of the auditory nerve. A similar improvement with [^{14}C] 2-deoxyglucose labeling indicating a reversal in metabolic activity with electrical stimulation.

Electrical Stimulation of the Cochlear Nerve

The above variables, in particular, charge density, charge per phase, rate, DC, and electrode geometry, can induce a pathological response in the neurons to electrical stimulation. The responses also vary depending on the neural tissue excited. Charge densities below approximately 40 to 50 μC cm^{-2} geometric/phase were found safe for electrical stimulation of a variety of neural sites that did not include the auditory nerve (Agnew et al 1975; Pudenz et al 1975 a,b, 1977; Yuen et al 1981; Walsh and Leake-Jones 1982).

Electrical stimulation with a cochlear implant differs with respect to the stimulus regime, the size and materials of the electrodes, and the location of the electrodes with respect to the neural population stimulated. All these variations can influence the response of the auditory neurons and cochlear tissue to long-term electrical stimulation. Therefore, it was necessary that the neurobiological response be evaluated specifically for each cochlear prosthesis. In the following subsections a number of studies that have investigated safety with the University of Melbourne/Bionic Ear Institute/Nucleus scala tympani electrode arrays are reviewed.

Acute Studies on the Effects of Low Rates of Stimulation

The safe electrical parameters for stimulating the cochlear nerve had in the first instance to be less than those causing electrochemical reactions. The initial studies on the safety of electrical stimulation were carried out at low rates (e.g., 200 to 500 pulses/s). This was appropriate as human (Simmons 1966) and electrophysiological and animal behavioral research (Clark 1969, 1970; Clark and Dunlop 1969; Clark, Nathar et al 1972; Clark, Kranz et al 1973; Williams et al 1974) showed these rates were the upper limit for reproducing the temporal coding of frequency. The stimuli for the Nucleus device were primarily biphasic pulses as the parameters referred to above could be more precisely controlled than for analog pulses and charge balance achieved. With analog pulses the current would be integrated over each phase of the sine wave.

Acute studies were undertaken to provide initial guidance on the safe levels for low rates of electrical stimulation. Deleterious effects were considered to occur if there was a prolonged reduction in the electrically evoked auditory brainstem response (EABR) amplitude or a raised threshold for the duration of the experiment. These changes suggested loss of viability of the nerve fibers.

The University of Utah (Ineraid) implant, in contrast to the University of Melbourne/Nucleus device, provided analog stimulation from four fixed filters that provided higher stimulus rates. For this reason an acute study by Duckert and Miller (1982) and Duckert (1983) examined the effects of sinusoidal stimuli on guinea pigs at a rate of 1000 Hz. The threshold for damage was 400 μA root mean square (RMS) (70 μC cm^{-2}/phase) for 3 hours of continuous stimulation. As the voltage varies with a sine wave, it was quantified as the RMS, that is, the square root of the mean of the squares of the sine wave voltages. In a later study the damage threshold for a 100 Hz stimulus at 200 μA RMS was less than for a rate of 1000 Hz (Duckert and Miller 1984).

The Nucleus implant provided pulsatile stimuli because of the more precise control of current flow with a pulse, and at low rates because of the perceptual limits on rate coding. Studies by Shepherd (1983a,c) in the cat examined the effects of electrical stimulation with biphasic pulses (200 μs/phase) at rates of 200 to 500 pulses/s on auditory nerve responsiveness using 0.3-mm-wide bands around a Silastic tube with outside diameter of 0.64 mm. This was done to help determine the current or charge densities that could be safely used with the Nucleus banded electrodes and speech processors. This study showed that a charge density of approximately 33.2 μC cm^{-2} geometric/phase at 200 Hz could suppress activity in the auditory neurons (Fig. 4.5). This is of interest, as the long-term study reported below showed that continuous stimulation with charge densities of 32 μC cm^{-2}/phase at 500 pulses/s did not lead to loss of neurons.

Acute stimulation was also carried out to determine mechanisms responsible for the suppression of neuronal responses (Fig. 4.6). The animals were stimulated for 10 minutes, at which time the responses stabilized, and then they were made anoxic for a further 10-minute period. There was a greater depression of the EABR for the stimulated cochlea compared with the unstimulated control. These findings support the above conclusions (Nicholson et al 1978; Stockle and Ten Bruggencate 1980; Agnew et al 1983; McCreery and Agnew 1983) that stimuli with high amplitude and rate suppress neural activity through overstimulation of biochemical pathways. The relation between this suppression and neural degeneration is not clear.

Chronic Studies on the Effects of Low Rates of Stimulation

The acute studies referred to above showed the stimulus levels that led to a short-term depression in neural activity, and these levels might lead to neuronal loss. The studies, however, did not show this would be the case for long-term (chronic) stimulation in the implanted experimental animal or patient. Chronic studies were required, and it was necessary to monitor not only the EABR changes but also the histopathological effects of the stimulation.

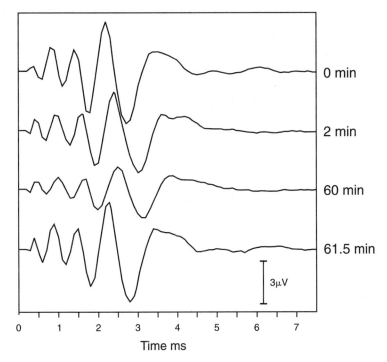

FIGURE 4.5. Evoked auditory brainstem responses (EABRs) recorded prior (0), during (2 and 60 minutes), and 1.5 minutes following completion of 60 minutes of continuous electrical stimulation with biphasic pulses at an intensity of 1 mA, 100 μs/phase, and rate of 200 pulses/s using a banded scala tympani electrode array. Note the reduction in amplitude and increase in latency of all EABR waves. Significantly greater changes in the EABR were observed following stimulation at higher stimulus intensities and rates (Shepherd et al 1983a,c). Reprinted from Cochlear Prostheses, Clark et al, © 1990 Churchill Livingstone, with permission of Elsevier Science.

Experimental Animal

Adult Animals

Research to define the safe operating range for electrical stimulation of the auditory nerve was carried out on the experimental animal. A study by Leake-Jones et al (1981) on cats using high current levels (4 to 8 mA) with charge-balanced biphasic pulses (100 μs/phase) at a rate of 20 to 25 pulses/s for up to 500 hours of continuous stimulation showed significant damage to cochlear structures. There was degeneration of the peripheral processes and spiral ganglion cells, loss of the organ of Corti, and new bone growth in the scala tympani of the basal turn. It was not clear in particular if the latter was due to trauma or toxicity of materials. In contrast, Leake-Jones et al (1981), Walsh et al (1981), and Walsh and Leake-Jones (1982) reported that charge densities of 20 of 40 μC cm^{-2} geometric/phase were within biologically safe limits.

To ensure that the middle to upper range of charge densities used with the

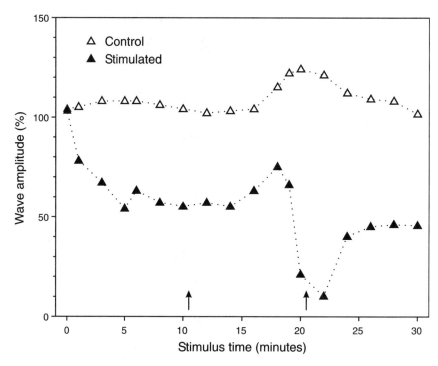

FIGURE 4.6. Effects of anoxia on EABR wave IV amplitudes and latencies during 30 minutes of electrical stimulation at an intensity of 1.5 mA, 200 μs/phase and a rate of 200 pulses/s. After a 10.5-minute stimulus period to allow stimulus-induced changes in the EABR to stabilize, the animal was gradually made anoxic over a period of 10 minutes (triangles). Significantly greater changes in both amplitude and latency of the EABR of the stimulated cochlea (▲) compared with the unstimulated control cochlea (△) were apparent (Shepherd et al 1990). Reprinted from Cochlear Prostheses, Clark et al, © 1990 Churchill Livingstone, with permission of Elsevier Science.

University of Melbourne/Nucleus multiple-channel cochlear prosthesis (18–32 μC cm^{-2} geometric/phase) were safe, a study was undertaken on 10 normal-hearing adult cats that were chronically stimulated at 500 pulses/s for continuous periods of up to 2000 hours (Shepherd et al 1983a–c). The stimulus level was midway between threshold and a level that evoked an aversive response. The study aimed to determine whether charge densities of 18 to 32 μC cm^{-2} geometric/phase would damage not only spiral ganglion cells but also hair cells in the organ of Corti. In addition, the effect of electrical stimulation on the inflammatory response and new bone formation was assessed, as well as the relation between these pathological changes and impedance measures. EABRs were periodically recorded during each animal's chronic stimulation program to monitor auditory nerve function, and allow the results to be compared with subsequent histopathological findings. Electrode impedance was monitored periodically by recording the electrode voltages and current waveforms. An increase in the impedance of an electrode pair could indicate a significant increase in the degree of fibrous tissue adjacent to the electrodes, and the time at which this occurred.

Following chronic stimulation, the cochleae were examined under the light microscope for pathological changes. Furthermore, the electrode arrays were viewed under the scanning electron microscope for evidence of platinum dissolution as discussed above (see Scanning Electron Microscope Evaluation of Electrodes) (Shepherd et al 1985). The stimulated electrodes were compared with unstimulated controls implanted in the opposite cochleae.

In the majority of the cochleae (15 out of 20) there was a mild tissue response (Fig. 4.7), and in the majority inner and outer hair cells were well preserved. In these cases, atrophy of hair cells and the organ of Corti was generally confined to the lower basal turn adjacent to the electrode array. It was significant that chronic electric stimulation per se did not contribute to the loss of hair cells, as there were a number of cochleae with hair cell survival adjacent to the stimulating electrodes (Shepherd et al 1990).

Furthermore, hair cell survival was associated with the preservation of spiral ganglion cells and auditory nerve fibers (Fig. 4.7). Adjacent to the implanted electrode array ganglion cell densities varied from 10% to normal. There was no statistical correlation between spiral ganglion cell density and the duration or intensity of electrical stimulation. Furthermore, spiral ganglion cells subjected to chronic electrical stimulation appeared histologically normal (Fig. 4.7). This was consistent with the EABR data that showed that chronically stimulated auditory nerve fibers remained physiologically viable for the duration of the stimulus regime.

On the other hand, five cochleae had moderate to severe inflammatory reactions including widespread loss of hair cells. The survival of the organ of Corti and hair cells depended on the severity of the inflammation. Extensive loss was observed throughout all turns in cochleae with moderate to severe inflammation from infection. This response was not associated with either the intensity or the duration of the electrical stimulus, but was due to the presence of infection. This finding highlights the importance of minimizing the risk of infection both during and following cochlear implant surgery. Significantly, EABRs recorded from these animals exhibited elevated thresholds. New bone growth was observed in nearly half the cochleae, but was not related to the degree of electrical stimulation since half the cochleae with new bone growth were in the unstimulated controls.

In these cochleae the spiral ganglion cell densities, measured in regions adjacent to the implanted electrode array, also varied from 10% to normal. This variation too was directly related to the degree of inflammation, and there was no significant correlation between spiral ganglion cell density and the duration or intensity of electrical stimulation. Furthermore, spiral ganglion cells that were stimulated chronically were normal in appearance when viewed microscopically. This was consistent with the EABR data that showed that chronically stimulated auditory nerve fibers remained functional for the duration of the stimulus regime.

This study indicated that long-term intracochlear electrical stimulation, using carefully controlled charge-balanced, biphasic current pulses at charge densities

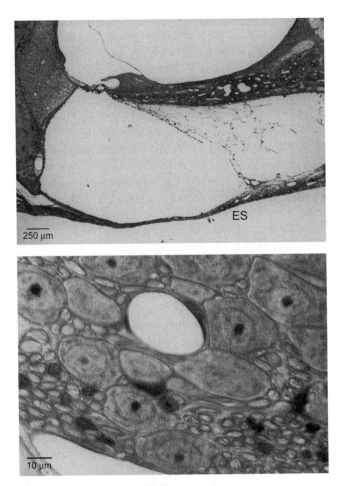

FIGURE 4.7. Typical histopathological response of a cat cochlea to long-term implantation and electrical stimulation. Top: Low-power micrograph illustrating the tissue sheath associated with the electrode array (ES, electrode sheath) (Shepherd et al 1983a). Bottom: High-power micrograph of the spiral ganglion cells adjacent to the stimulating electrodes. These ganglion cells appear histologically normal at the limit of the light microscope (Clark, Tong et al 1990). Reprinted from Cochlear Prostheses, Clark et al, © 1990 Churchill Livingstone, with permission of Elsevier Science.

from 17.7 to 32 μC cm^{-2} geometric/phase did not adversely affect the spiral ganglion population or the cochlea in general. The results also did not show a correlation between aggregate charge injection and spiral ganglion cell loss. These levels were significantly less than those leading to toxic electrochemical products. If neural damage were to occur below the electrochemical limit of $300\mu C$ cm^{-2} geometric/phase for Pt electrodes, it would be the result of electrical stimulation per se.

Young Animals

Prior to implantation in children under 2 years of age, research was undertaken at the University of Melbourne/Bionic Ear Institute through an NIH contract ("Studies on Pediatric Auditory Prosthesis Implants," NIH contract No. 1-NS-7-2342) to examine the effect of chronic electrical stimulation on the young experimental animal. The studies were undertaken with normal-hearing animals as well as those deafened to help ensure that the effects on the cochlea were no different from the adult. The initial study examined the effects on the cochlea and auditory neurons in the normal-hearing kitten (Ni et al 1992a,b). This showed results consistent with those from normal-hearing adult cats (Shepherd et al 1983a–c). Again, provided that the cochlea was free of insertion trauma or infection, stimuli with charge densities up to 52 μC cm^{-2} geometric/phase and charge balanced with DC levels less than 0.1 μA did not adversely affect the adjacent organ of Corti, peripheral processes, and spiral ganglion cells.

The effect of electrical stimulation on the developing auditory central nervous system was also examined in both the normal-hearing and deafened kitten to help determine whether the changes seen in other species as a result of unilateral deafening occurred. In the ferret it had been shown by Moore (1990) that there was a marked loss of neurons in the cochlear nucleus (CN) after ablation of the cochlea 5 days after birth. Would continuous stimulation of the cochlear nerve have a deleterious or beneficial effect? Furthermore, Nordeen et al (1983) showed there was an increase in the number of projections from the CN to the ipsilateral inferior colliculus (IC) when the cochlea on the opposite side was destroyed in the neonatal gerbil, and there was a critical period of up to 90 postnatal days for this. A similar phenomenon was demonstrated in the ferret (Moore and Kowalchuk 1988) where the critical period extended well beyond the onset of hearing.

The effects of electrical stimulation on the morphology of the CN were studied in the young, deafened animal to see if there was an interaction between these changes and electrical stimulation. This was considered the most appropriate model for deaf children. Furthermore, it had been shown by Powell and Erulkar (1962), Trune (1982), Anniko et al (1989), Hashisaki and Rubel (1989), and Moore (1990) that a sensorineural hearing loss resulted in a reduction in the size of neurons in the ipsilateral ventral cochlear nucleus (VCN). Would electrical stimulation either exacerbate or ameliorate the effects?

Four kittens were deafened with kanamycin and ethacrinic acid 37 to 40 days after birth. The administration of these drugs according to the protocol of Xu et al (1990) had been shown to produce a bilateral, complete loss of inner and outer hair cells, and a symmetrical moderate-to-severe loss of spiral ganglion cells (Shepherd et al 1989). The animals were stimulated 80 to 90 days after birth with charge densities that varied from 12 to 26.2 μC cm^{-2} geometric/phase for periods from 1100 to 1600 hours. In the anteroventral cochlear nucleus (AVCN) the mean soma areas of the cells were significantly greater than for the unstimulated control. The same applied to the spherical cells in the anterior part of the AVCN. There was also a significant but weaker trend for the cells to be larger in the stimulated posteroventral cochlear nucleus (PVCN). These data are consistent with the finding of Chouard et al (1983), who found the soma area in the octopus cell area of the VCN was maintained in guinea pigs following electrical stimulation. In contrast, Leake et al (1990) and Hultcrantz et al (1991) did not find any difference between the stimulated and control sides in kittens deafened soon after birth. The discrepancy between their results and those of Chouard et al (1983) and Matsushima et al (1991) could have been due to differences in sample size (10,000 cells were sampled in the study by Matsushima et al). In the dorsal cochlear nucleus (DCN) there was no difference between the soma areas of the stimulated and unstimulated sides. This is consistent with the data that spontaneous activity, of importance in maintaining neuronal size, is retained in the DCN following a profound hearing loss (Koerber et al 1966; Wong-Riley et al 1978; Durham et al 1989). This study showed that morphological changes (soma size) in the central auditory pathways due to a sensorineural deafness could be partially reversed by functional electrical stimulation of the auditory nerve (AN). These are presumed due to the induced neural activity, as other studies (e.g., Born and Rubel 1988) have suggested that even in the absence of acoustic excitation the spontaneous activity can maintain the cell volume. Any pharmacological agent that blocks this activity leads to a rapid reduction in the size of second-order neurons (Born and Rubel 1988; Pasic and Rubel 1989), and thus postsynaptic excitation is important.

With improved results, children suitable for implants were likely to have residual hearing. It was becoming appropriate to operate on the poorer ear even in the presence of some residual hearing. Thus a study was undertaken on the normal-hearing kitten to see whether electrical stimulation could affect the maturation of neurons activated by residual hair cells (Ni et al 1993). Both intracochlear and extracochlear stimulation were used. There was, however, no difference between the stimulated and control VCN. Therefore, chronic electrical stimulation had no positive or negative trophic effect on neurons connected to functioning hair cells.

Human (Patient)

The first human temporal bone examined after long-term stimulation with biphasic charge-balanced pulses from the Nucleus stimulator was reported on by Clark, Shepherd et al (1988). One of the most important findings from the study was that there were equal numbers of ganglion cells in the spiral canal (Rosenthal's canal) both underlying as well as beyond the area implanted and stimulated electrically. This finding indicated that the electrical stimuli from the Nucleus cochlear implant receiver-stimulator did not lead to loss of the spiral ganglion cells. The speech processor was used for over 2 years, during which time it operated daily. It was calculated that the patient received 10,000 hours of stimulation. The minimum and maximum charge densities were calculated from the current amplitude at the thresholds and maximum comfortable levels recorded on the speech processor MAPs. These showed the minimum charge density to be 6.1 μC cm^{-2} geometric/phase and the maximum charge density to be 25.7 μC cm^{-2} geometric/phase. These findings are consistent with the above animal experimental data.

Acute Studies on the Effects of High Rates of Stimulation

Overview

Following success with low-rate speech-processing strategies (Clark, Tong et al 1978; Tong et al 1979, 1980; Eddington 1980, 1983), research was directed toward the use of high rates of stimulation (Clark, Tong et al 1985). Speech-processing research suggested that electrical stimulation of the auditory nerve with a cochlear implant at high rates of 800 pulses/s and even up to 2000 pulses/s could lead to an improved detection of some temporal components in speech (Wilson et al 1992, 1993; Lawson et al 1996; Wilson 1997; Grayden and Clark 2000). Research therefore was needed to make sure that neuronal damage did not occur at these rates. Acute studies were initially required to help narrow down the parameter values that could be assessed in chronic studies prior to their use in patients, as was the practice with the low rate Nucleus strategies. EABRs were monitored in cats prior, during, and following the completion of periods of acute electrical stimulation at increasing rates, intensities, and durations (Shepherd and Clark 1986; Shepherd et al 1987, 1990). The EABR, as discussed above, was a measure of the excitability of the auditory nerve after stimulation. Studies were also carried out in the guinea pig (Miller et al 1983; Killian et al 1994; Tykocinski et al 1995b; Huang et al 1998; Huang and Shepherd 1999), and the rat (Vischer et al 1997; Haenggeli et al 1998). Stimulus-induced changes in EABR included a reduction in amplitude, a raised threshold, and an increase in latency of all waves in the EABR. The extent of these changes depended on stimulus rate and intensity. At the completion of the experiments the animals were sacrificed and the cochleae sectioned, stained, and examined for the spiral ganglion cell population and pathological changes in the cochlea.

The research also revealed that with the Nucleus device at high stimulus rates and intensities, there was a significant level of residual DC despite the use of charge-balanced, biphasic current pulses accompanied by electrode shorting. The initial Nucleus devices used circuitry that shorted any residual current between pulses to minimize any buildup of DC, as discussed in Chapter 8. With low rates of stimulation there was time for this to occur. Capacitors in the circuit were effective, but were large and not used in the initial Nucleus devices. They were subsequently used in the Nucleus-24, Clarion, and Combi 40 systems. Lilly (1961) and Mortimer et al (1970) showed that stimulation with DC could damage biological tissue. This was demonstrated for the cochlea by Shepherd et al (1991), who found a DC level of 2 μA caused extensive pathological changes in the cochlea. The research at the University of Melbourne was therefore expanded to investigate whether or not there was a relationship between the reduction in neural excitability following acute high-rate stimulation and the level of DC. Studies at high stimulus rates and intensities (Tykocinski et al 1999) were also undertaken to provide an insight into neural damage mechanisms associated with electrical stimulation, as well as help define maximum safe stimulation levels for the auditory nerve.

Any adverse effects of high rates of stimulation were not likely to be due to a greater degree of corrosion of the Pt electrode, as the study by Black and Hannaker (1979) showed that PT dissolution was less when the stimulus rate rose from 100 pulses/s to 1000 pulses/s. Thus any effects of high stimulus rates on neural tissue were not likely to be due to increased loss of Pt.

Stimulus Parameters

Research by Miller et al (1983, 1985) had demonstrated that high rates could suppress neural activity for continuous sinusoidal electrical stimulation at 1000 Hz over a period of 3 hours. Current levels above 400 μA RMS (70 μC cm^{-2}/phase) produced an increase in EABR thresholds, a reduction in amplitude, and an increase in latency.

In the research reported by Shepherd and Clark (1986) and Shepherd et al (1990) using acute stimulation in the cat, at medium to high stimulus rates (e.g., 400 or 800 pulses/s) and high stimulus intensities (approximately 53–70 μC cm^{-2}/phase), there were significant long-term reductions in the amplitude of all waves of the EABR. With higher stimulus rates of 1000 pulses/s the depression of neural activity was greater than at 400 pulses/s. The changes in the EABR were probably due to overstimulation of the neuron and downregulation of the biochemistry. In the guinea pig with a stimulus intensity of 0.34 μC/phase there was a severe decrement in the EABR at 400 and 1000 pulses/s. This was particularly severe at 1000 pulses/s (Tykocinski et al 1995a). If a 50% duty cycle was introduced into the stimulus paradigm, resulting in a reduction in the total number of stimuli being presented over a given time, the degree of stimulus-induced changes observed in the EABR were reduced (Tykocinski et al 1995a). Stimulus-induced changes were potentiated by anoxia, also suggesting they were due to overstimulation.

In another study, the effects of continuous stimulation and a 50% duty cycle were compared with an amplitude modulated (AM) pulse train at 0.34 μC/phase and rates of 400 and 1000 pulses/s. The AM train was based on the amplitude changes at the most intensively activated electrode for the SPEAK speech processing strategy with four-talker babble. This was done to mimic a severe real-life situation. There was marked suppression of the response as seen previously for continuous stimuli. But there was a more rapid recovery for the 50% duty cycle and AM stimuli. There was no significant difference between these two.

Current Versus Charge Density

It was not clear if the above effects were due to the current (charge/phase) or charge density. To further examine the stimulus parameters required for safe stimulation at high rates, research was undertaken by Tykocinski et al (1997,

1998) in which stimulus rates varied from 100 to 1000 pulses/s and stimulus intensities from 0.16 to 1.0 μC/phase.

To monitor the recovery in excitability of the auditory nerve following acute stimulation, EABR thresholds, wave I and III amplitudes, and their latencies were determined for periods of up to 12 hours following the acute stimulation. Higher stimulus rates and, to a lesser extent higher intensities, led to greater decrements in the poststimulus EABR amplitude and prolonged the recovery period. While continuous stimulation at 100 pulses/s induced no decrement in the EABR, stimulation at 200 and 400 pulses/s produced an increasingly significant poststimulus reduction of the EABR amplitude, which showed only partial recovery during the monitoring period. No EABR response could be evoked immediately following stimulation at 1000 pulses/s, using a probe intensity 16 to 19 dB below the stimulus intensity. However, partial EABR recovery was observed for wave III following stimulation at the lowest stimulus intensity (0.16 μC/phase). These stimulus-induced reductions in the EABR amplitude were also reflected in increased thresholds and latencies. The data also indicated that the effects were primarily due to the current per phase. Significantly, the introduction of a 50% duty cycle into the stimulus pulse train resulted in a more rapid and complete poststimulus recovery of the EABR compared to continuous stimulation (Tykocinski et al 1995a).

In a study by McCreery et al (1988) on the cerebral cortex charge density and charge per phase covaried in producing neural damage, that is, a charge density at safe levels would produce injury if the charge per phase was increased. Shannon (1992) analyzed the data and found there was a linear relation between electrode perimeter and maximum allowable charge for safe stimulation. Damage was probably due to the concentration of charge around the edges, creating localized increase in charge density. On the other hand, charge density decreased hyperbolically as both the perimeter and area increased. It is postulated that in the cochlea there is a near-field effect, and charge density would be the prime cause of damage. However, as the spiral ganglion cells are at a distance from the electrode, a far-field effect would operate where the important parameter for damage would be the amount of charge per phase in each pulses as demonstrated by the above studies by Tykocinski et al (1997, 1998).

DC Levels

Although the neural suppression at high stimulus rates in the above studies was most likely due to metabolic overload, DC buildup at the electrode/tissue interface could still have been responsible. It was found that for 1000 pulses/s at intensities within the normal clinical range, the DC level was 2.8 μA when shorting between pulses, but at intensities above those used clinically the DC level rose to 7 μA (Tykocinski et al 1995b). The unsafe levels of DC had not been clearly established, but the acute study on guinea pigs by Tykocinski et al (1995b) demonstrated that a 2 μA DC current level applied for 2 hours had no more effect on the EABR than that of an unstimulated control. Depression of the EABR and compound action potential (CAP) only occurred for DCs of 7 μA and 12 μA. This was in a short-term study, and therefore long-term effects might not have been brought to light. Long-term stimulation (200 hours) at a DC level of 2 μA in one cat was reported by Shepherd et al (1991) to lead to severe effects on the spiral ganglion cells and the organ of Corti, and extensive new bone growth. But the effects were consistent with that of a resolved chronic infection so the question was still an open one.

The DC level was likely to be a function of stimulus rate, intensity, charge density, and the charge recovery system. To distinguish between the effects of rate and DC, two acute studies on guinea pigs by Huang et al (1998) and Huang and Shepherd (1999) compared the effects of different rates of stimulation with either electrode shorting alone (study I) or electrode shorting with capacitive coupling to remove all DC (study II). In study I the DC current levels were measured in vivo in the guinea pig cochlea (scala tympani), subcutaneously, and the wound area outside the cochlea, and in vitro in saline and human albumin with the University of Melbourne DC monitoring technique. The experiment showed the DC current level in vivo was approximately three times higher than in vitro, and DC levels as expected were minimized by the use of electrode shorting when combined with coupling capacitors. It was not clear why the DC levels recorded in vivo were greater than identical measurements taken in vitro for a common stimulus waveform. In study I a group of six guinea pigs were bilaterally implanted and unilaterally stimulated by using charge-balanced bi-phasic current pulses at stimulus rates of 200, 400, and 1000 pulses/s and a stimulus intensity of 0.34 μC/phase (2.0 mA, 170 μs/phase). The intensity was well above that used clinically. To monitor the status of the auditory nerve, the EABR was recorded before and periodically following 2 hours of stimulation. There was no significant difference in the extent of reduction in EABR amplitudes for stimulus rates of 200 pulses/s and 400 pulses/s, but the difference was significant at 1000 pulses/s. The results thus indicated that stimulating at high rates and intensities and/or the presence of DC could lead to reduced neural activity.

In study II a group of six guinea pigs were bilaterally implanted and unilaterally stimulated using charge-balanced biphasic current pulses at the same stimulus rates and intensities as for study I, but with capacitor coupling and shorting techniques (Huang et al 1998; Huang and Shepherd 1999). The study showed a decrease in EABR amplitude and increase in threshold after the acute simulation period. The higher the stimulus rate, the more significant were the changes. While some recovery of the EABR was observed, higher stimulus rates resulted in more extensive changes associated with more gradual recovery. When the results were compared with those obtained in study I, where there was shorting between stimuli but no capacitive coupling, there was a statistically significant difference in the extent of the EABR recovery using stimulus rates of 400 pulses/s and 1000 pulses/s. No statistically significant difference in recovery was observed at lower stimulus rates (100 and 200 pulses/s). There were also more extensive stimulus-induced changes observed at high rates and intensities for both studies. The results indicated that high rates of stimulation using stimulus intensities at and beyond maximum clinical levels could result in temporary or permanent auditory nerve damage. The degree to which this was due to charge imbalance and damaging DC levels or high levels of neural activity resulting in metabolic downregulation was not clear, but the data suggested that metabolic overstimulation was an important factor.

The effect of rate, intensity, and DC levels were further evaluated in the guinea pig cochlea after 2-hour periods of stimulation using a Nucleus standard array and another roughened electrolytically to increase the real surface area by a factor of 70. There was less reduction in EABR amplitude and increase in thresholds with the etched electrode for stimuli 12 dB above threshold at rates of 400 and 1000 pulses/s and for stimuli 22 to 30 dB above threshold (Huang and Shepherd 2000). For the standard Nucleus array 12 dB above the EABR threshold the DC levels were 0.05 μA at 400 pulses/s and 0.06 μA at 1000 pulses/s, and for the roughened array 0.004 μA at 400 pulses/s and 0.016 μA at 1000 pulses/s. On the other hand for stimuli 22 to 30 dB above threshold the DC levels were 1.28 ìA at 400 pulses/s and 2.34 μA at 1000 pulses/s, and for the roughened array 0.016 μA at 400 pulses/s and 0.026 μA at 1000 pulses/s. The reduction in DC due to the roughening could explain the greater fall in the EABR with the standard electrode at high rates and for the more intense stimuli (22–30 dB), but not the reduction in EABR at the smaller levels (12 dB) as the DC was significantly lower than the safe levels used in the other studies above. These data thus indicate that charge density is an important parameter and should be kept low by increasing the surface area of perimodiolar electrodes to lie close to the peripheral processes and spiral ganglion cells.

Chronic Studies on the Effects of High Rates of Stimulation

The effects of high rates of stimulation on the cochlear nerve finally had to be determined in chronic studies on the experimental animal before application of high rate strategies to patients. Furthermore, although the acute studies described above were important in understanding the mechanisms of stimulus-induced neural damage, they were performed using stimulus regimes outside the normal clinical limits. A study was therefore designed to evaluate the physiological and histopathological effects of long-term intracochlear electrical stimulation using high-rate biphasic electrical pulses at stimulus intensities within clinical limits (Xu et al 1997). In this study 13 normal-hearing adult cats were bilaterally implanted with scala tympani electrode arrays and unilaterally stimulated for periods of up to 2100 hours using either two pairs of bipolar or three monopolar stimulating electrodes. Both modes of stimulation were evaluated, as either could be provided by cochlear implant receiver-stimulators. Stimuli consisting of short duration (25–50 μs/phase) charge-balanced biphasic current pulses were presented at 1000 pulses/s per channel for monopolar stimulation and 2000 pulses/s per channel for bipolar stimulation. The electrodes were shorted between current pulses to minimize any residual DC, and the pulse trains were presented using a

500-ms on/off duty cycle to simulate speech. Acoustically evoked auditory brain-stem responses (ABRs), in addition to EABRs, were recorded periodically during chronic stimulation to assess the effects of implantation and stimulation on residual hearing.

An increase in the click-evoked ABR threshold was found following the implant surgery in all cochleae, but recovery to near-normal levels occurred in approximately half of the stimulated cochleae 1 month postoperatively. The use of frequency-specific stimuli indicated that the most extensive hearing loss generally occurred in the high-frequency basal region of the cochlea (12,000 and 24,000 Hz) adjacent to the stimulating electrode. However, thresholds at lower frequencies (2000, 4000, and 8000 Hz), appeared at near-normal levels despite long-term electrode implantation and electrical stimulation. The longitudinal EABR results showed a statistically significant increase in threshold in nearly 40% of the chronically stimulated electrodes evaluated; however, the amplitude of the evoked response generally remained quite stable throughout the chronic stimulation period. Histopathological examination of the cochleae showed no statistically significant difference in ganglion cell densities between cochleae using monopolar and bipolar electrode configurations, and no evidence of cochlear damage caused by high-rate electrical stimulation when compared with control cochleae. Indeed, there was no statistically significant relationship between spiral ganglion cell density and electrical stimulation, or between the extent of loss of inner or outer hair cells and electrical stimulation. Spiral ganglion cell loss, however, was influenced by the degree of inflammation and electrode insertion trauma. These histopathological findings were consistent with the physiological data. Finally, the electrode impedance, measured daily during the chronic stimulation program, showed close correlation with the degree of tissue response adjacent to the electrode array. These results indicated that chronic intracochlear electrical stimulation, using charge-balanced biphasic current pulses at stimulus rates of up to 2000 pulses/s per channel and operating at stimulus intensities within clinical limits, did not adversely affect residual auditory neurons or the cochlea in general. This study provided an important basis for the safe application of improved speech-processing strategies based on high-rate electrical stimulation. Comparable results were also obtained in a study by Mitchell et al (1997). The above results should be contrasted with those from another chronic study on eight normal-hearing cats where a charge imbalance at 2000 pulses/s was maintained, leading to DC levels of 0.6 to 2.8 μA, but the DC levels for 500 pulses/s were 0.3 μA or less. For the cochleae stimulated at 2000 pulses/s there was a statistically significant reduction in spiral ganglion cell density and an elevation of EABR thresholds during the experiment compared to those stimulated at 500 pulses/s (Linahan et al 1998). The study indicated the importance of charge balance, and keeping the DC levels below 0.3μA and preferably at 0.1μA. This was particularly likely to occur with electrode-shorting techniques at rates of 1000 pulses/s or in excess.

Although the clinical benefits for rates higher than 2000 pulses/s had not been established, they were considered for research studies. High rates were also possible with the digital representation of simultaneous analog stimulation. Before embarking on any such clinical research study, a chronic experiment was next undertaken on four normal-hearing cats for rates of 5000 pulses/s using charge-balanced monopolar stimuli on an intracochlear banded electrode over a period of up to 2700 hours. After sacrifice, however, no significant difference was seen between the spiral ganglion cell numbers on the stimulated and control sides (Linahan et al 1999). A study was also done using a rate of 14,493 pulses/s. The cats were stimulated with charge-balanced pulses for up to 2000 hours. ABRs and compound action potentials (CAPs) were monitored during the experiment (Tykocinski et al 1998). The thresholds were elevated immediately after the implant, but partially recovered to 0 to 40 dB. A subsequent deterioration of thresholds occurred on the stimulated side, suggesting that chronic stimulation at very high rates may decrease residual hearing.

Electrical Stimulation of the Cochlear Nucleus

Acute Studies on the Effects of Low Rates of Stimulation

The efficacy and safety of the surface brainstem auditory prosthesis has been studied electrophysiologically and histologically in acute experiments on guinea pigs (Liu 1997a,b, 1998). Biphasic charge-balanced pulses were used. The result showed that in the stimulated cochlear nucleus the neuronal excitability did not change and no pathological change was observed. Only the auditory pathway was activated. No changes in respiration and body temperature were found during the stimulation.

The comparison of the electrophysiological and histological effects of surface and penetrating brainstem prostheses was performed in the cat and guinea pig. The results showed that in the guinea pig the penetrating electrodes had a functional advantage over surface electrodes, but there was a trade-off with its safety. The EABR threshold was lower and dynamic range wider with the penetrating prosthesis. However, the neuron density was decreased, but the soma area increased in the cochlear nucleus stimulated with a penetrating prosthesis. It is not clear if the decrease in neuronal density was due to trauma or to electric stimulation per se.

Shannon (1992) undertook a study on the cochlear nucleus to determine the effects of both charge per phase and charge density. The data indicated that safe stimulation was linearly related to electrode diameter, not electrode area. Thus for this tissue a high charge density with an electrode with a small surface area may not be as damaging as the same charge density on an electrode with a larger area. It was also suggested that damage limits be considered in terms of near-field, mid-field, and far-field stimulation.

Chronic Studies on the Effects of Low Rates of Stimulation

The efficacy and safety of the surface brainstem prosthesis has also been studied electrophysiologically and histologically in chronic experiments (Liu et al 1998).

The results showed that the neuronal excitability in the stimulated cochlear nucleus did not change during 3 months of continuous stimulation (16 hours per day). No pathological change was observed. Only the auditory pathway was activated. No changes in respiration and body temperature were found during the course of the stimulation.

References

Agnew, W. F., L. Bullara, T. G. H. Yuen, D. B. McCreery and D. B. Jacques. 1981. The effects of electrical stimulation on the central and peripheral nervous systems. NIH contract NO-1-NS-0-2319. Pasadena, CA, Huntington Institute of Applied Medical Research.

Agnew, W. F., T. G. H. Yuen and D. B. McCreery. 1983. Morphological changes after prolonged electrical stimulation of the cats cortex at defined charge densities. Experimental Neurology 79: 397–411.

Agnew, W. F., T. G. H. Yuen, R. H. Pudenz and L. A. Bullara. 1975. Electrical stimulation of the brain. IV. Ultrastructural studies. Surgical Neurology 4: 438–448.

Agnew, W. F., T. G. H. Yuen, R. H. Pudenz and L. A. Bullara. 1977. Neuropathological effects of intracerebral platinum salt injections. Journal of Neuropathology and Experimental Neurology 36: 533–546.

Anniko, M., B. Sjostrom and D. Webster. 1989. The effects of auditory deprivation on morphological maturation of the ventral cochlear nucleus. Archives of Otolaryngology 246: 43–47.

Babb, R. L., H. V. Soper, J. P. Lieb, W. J. Brown, C. A. Ottino and P. H. Crandall. 1977. Electrophysiological studies of long-term electrical stimulation of the cerebellum in monkeys. Journal of Neurosurgery 47: 353–365.

Bernstein, J. J., P. F. Johnson, L. L. Hench, G. Hunter and W. W. Dawson. 1977. Cortical histopathology following stimulation with metallic and carbon electrodes. Brain Behavior and Evolution 14: 126–157.

Black, R. C. and G. M. Clark. 1977. Electrical transmission line properties of the cat cochlea. Proceedings of the Australian Physiological and Pharmacological Society 8: 137.

Black, R. C. and P. Hannaker. 1979. Dissolution of smooth platinum electrodes in biological fluids. In: Gildenberg, L., ed. Applied neurophysiology. Basel, Karger: 366–374.

Blau, A., C. Ziegler, M. Heyer, et al. 1997. Characterization and optimization of microelectrode arrays for in vivo nerve signal recording and stimulation. Biosensors and Bioelectronics 12: 883–892.

Born, D. E. and E. W. Rubel. 1988. Afferent influences on brain stem auditory nuclei of the chicken: presynaptic action potentials regulate protein synthesis in nucleus magnocellularis neurons. Journal of Neuroscience 8: 901–919.

Brummer, S. B., J. McHardy and M. J. Turner. 1977. Electrical stimulation with Pt electrodes. Trace analysis for dissolved platinum and other dissolved electrochemical products. Brain Behavior and Evolution 14: 10–22.

Brummer, S. B., L. S. Robblee and F. T. Hambrecht. 1983. Criteria for selecting electrodes for electrical stimulation. Theoretical and practical considerations. Annals of the New York Academy of Sciences 405: 159–171.

Brummer, S. B. and M. J. Turner. 1975. Electrical stimulation of the nervous system: the principle of safe charge injection with noble metal electrodes. Bioelectrochemistry and Bioenergetics 2: 13–25.

Brummer, S. B. and M. J. Turner. 1977a. Electrical stimulation with Pt electrodes: I. A method for determination of 'real' electrode areas. IEEE Transactions on Biomedical Engineering 24: 436–439.

Brummer, S. B. and M. J. Turner. 1977b. Electrochemical considerations for safe electrical stimulation of the nervous system with platinum electrodes. IEEE Transactions on Biomedical Engineering 24: 59–63.

Byers, C. L., P. A. Leake-Jones, S. T. Rebscher and M. M. Merzenich. 1981. Development of multichannel electrodes for an auditory prosthesis. December quarterly report. NIH contract NS-0-2337.

Chouard, C. H., B. Meyer, P. Josset and J. F. Buche. 1983. The effect of the acoustic nerve chronic electric stimulation upon the guinea pig cochlear nucleus development. Acta Otolaryngologica 95: 639–645.

Clark, G. M. 1969. Responses of cells in the superior olivary complex of the cat to electrical stimulation of the auditory nerve. Experimental Neurology 24: 124–136.

Clark, G. M. 1970. Middle ear and neural mechanisms in hearing and the management of deafness. PhD dissertation, University of Sydney.

Clark, G. M. 1973. A hearing prosthesis for severe perceptive deafness—experimental studies. Journal of Laryngology and Otology 87: 929–945.

Clark, G. M. and C. W. Dunlop. 1969. Response patterns in the superior olivary complex of the cat. Australian Journal of Experimental Biology and Medical Research 47: 5.

Clark, G. M., H. G. Kranz and H. Minas. 1973. Behavioral thresholds in the cat to frequency modulated sound and electrical stimulation of the auditory nerve. Experimental Neurology 41: 190–200.

Clark, G. M., J. M. Nathar, H. G. Kranz and J. S. Maritz. 1972. A behavioral study on electrical stimulation of the cochlea and central auditory pathways of the cat. Experimental Neurology 36: 350–361.

Clark, G. M., R. K. Shepherd, B. K.-H. G. Franz, et al. 1988. The histopathology of the human temporal bone and auditory central nervous system following cochlear implantation in a patient. Correlation with psychophysics and speech perception results. Acta Oto-Laryngologica-supplement 448: 1–65.

Clark, G. M., S. A. Shute, R. K. Shepherd and T. D. Carter. 1995. Cochlear implantation: osteogenesis, electrode-tissue impedance, and residual hearing. Annals of Otology, Rhinology and Laryngology 104: 40–42.

Clark, G. M., Y. C. Tong, Q. R. Bailey, et al. 1978. A multiple-electrode cochlear implant. Journal of the Oto-Laryngological Society of Australia 4: 208–212.

Clark, G. M., Y. C. Tong, R. C. Black, I. C. Forster, J. F. Patrick and D. J. Dewhurst. 1977. A multiple electrode cochlear implant. Journal of Laryngology and Otology 91: 935–945.

Clark, G. M., Y. C. Tong and P. J. Blamey. 1985. Studies to develop sensory prostheses for deaf children and adults. National Health and Medical Research Council Program grant application.

Clark, G. M., Y. C. Tong and J. F. Patrick. 1990. Cochlear prostheses. Edinburgh, Churchill Livingstone.

Crow, D. R. 1988. Principles and applications of electrochemistry. Victoria, Blackie Academic and Professional.

Dewhurst, D. J. 1966. Physical instrumentation in medicine and biology. Oxford, Pergamon.

Duckert, L. G. 1983. Morphological changes in the normal and neomycin perfused guinea pig cochlea following chronic prosthetic implantation. Laryngoscope 93: 841–855.

Duckert, L. G. and J. F. Miller. 1984. Morphological changes following cochlear implantation in the animal model. Acta Oto-Laryngologica-supplement 411: 28–37.

Duckert, L. G. and J. M. Miller. 1982. Acute morphological changes in guinea pig cochlea following electrical stimulation—a preliminary scanning electron microscopy study. Annals of Otology Rhinology and Laryngology 91(1): 33–40.

Durham, D., E. W. Rubel and K. P. Steel. 1989. Cochlear ablation in deafness mutant mice: 2-deoxyglucose analysis suggests no spontaneous activity of cochlear origin. Hearing Research 43: 39–46.

Dymond, A. M. 1976. Characteristics of the metal tissue interface of stimulation electrodes. IEEE Biomedical Engineering 23: 274–280.

Dymond, A. M., L. E. Kaechele, J. M. Jurist and P. H. Crandall. 1970. Brain tissue reaction to some chronically implanted metals. Journal of Neurosurgery 33: 574–580.

Eddington, D. K. 1980. Speech discrimination in deaf subjects with cochlear implants. Journal of the Acoustical Society of America 68: 885–891.

Eddington, D. K. 1983. Speech recognition in deaf subjects with multichannel intracochlear electrodes. Annals of the New York Academy of Science 405: 241–258.

Grayden, D. B. and G. M. Clark. 2000. The effect of rate stimulation of the auditory nerve on phoneme recognition. In: Barlow, M., ed. Proceedings of the Eighth Australian International Conference on Speech Science and Technology. Canberra, Australian Speech Science and Technology Association: 356–361.

Greatbatch, W., B. Piersma, F. D. Shannon and S. W. Calhoon. 1969. Polarization phenomena relating to physiological electrodes. Annals of the New York Academy of Sciences 167: 722–744.

Haenggeli, A., J. S. Zhang, M. W. Vischer, M. Pelizzone and E. M. Rouiller. 1998. Electrically evoked compound action potential (ECAP) of the cochlear nerve in response to pulsatile electrical stimulation of the cochlea in the rat: effects of stimulation at high rates. Audiology 37: 353–371.

Harrison, J. M. and W. W. Dawson. 1977. The visual cortex during chronic stimulation. Brain Behavior and Evolution 14: 87–102.

Hartshorn, D. O., J. M. Miller and R. A. Altschuler. 1991. Protective effect of electrical stimulation on the deafened guinea pig cochlea. Otolaryngology Head and Neck Surgery 104: 311–319.

Hashisaki, G. T. and E. W. Rubel. 1989. Effects of unilateral cochlea removal on anteroventral cochlear nucleus neurons in developing gerbils. Journal of Comparative Neurology 283: 465–473.

Heinemann, U. and H. D. Lux. 1977. Ceiling of stimulus induced rises in extracellular potassium concentration in the cerebral cortex of the cat. Brain Research 120: 231–249.

Hinojosa, R. and M. Marion. 1983. Histopathology of profound sensorineural deafness. Annals of the New York Academy of Sciences 405: 459–484.

Huang, C. Q. and R. K. Shepherd. 1999. Reduction in excitability of the auditory nerve following electrical stimulation at high stimulus rates. IV. Effect of stimulus intensity. Hearing Research 132: 60–68.

Huang, C. Q. and R. K. Shepherd. 2000. Reduction in excitability of the auditory nerve following electrical stimulation at high stimulus rates: V. Effects of electrode surface area. Hearing Research 146: 57–71.

Huang, C. Q., R. K. Shepherd, P. M. Seligman and G. M. Clark. 1998. Reduction in excitability of the auditory nerve following electrical stimulation at high stimulus rates.

III. Capacitive versus non-capacitive coupling of the stimulating electrodes. Hearing Research 116: 55–64.

Hultcrantz, M., R. Snyder, S. Rebscher and P. Leake. 1991. Effects of neonatal deafening and chronic intracochlear electrical stimulation on the cochlear nucleus in cats. Hearing Research 54: 272–280.

Killian, M. J., S. F. Klis and G. F. Smoorenburg. 1994. Adaptation in the compound action potential response of the guinea pig VIIIth nerve to electric stimulation. Hearing Research 81: 66–82.

Koerber, K. C., R. R. Pfeiffer, W. B. Warr and N. Y. S. Kiang. 1966. Spontaneous spike discharges from single units in the cochlear nucleus after destruction of the cochlea. Experimental Neurology 16(2): 119–130.

Lawson, D. T., B. S. Wilson, M. Zerbi and C. C. Finley. 1996. Speech processors for auditory prostheses. Third quarterly progress report. NIH contract No1-DC-5-2103.

Leake, P. A., M. Hultcrantz, R. L. Snyder, G. T. Hradek and S. J. Rebscher. 1990. Effects of chronic electrical stimulation in neonatally deafened cats. A.R.O. Midwinter Research Meeting 13: 328.

Leake, P. A., R. L. Snyder, G. T. Hradek and S. J. Rebscher. 1992. Chronic intracochlear electrical stimulation in neonatally deafened cats: effects of intensity and stimulating electrode location. Hearing Research 64: 99–117.

Leake-Jones, P. A., S. J. Rebscher and C. L. Byers. 1981. Development of multichannel electrodes for an auditory prosthesis. NIH Contract No 1 NS-7-2367, 1–23.

Lilly, J. C. 1961. Injury and excitation by electrical currents. A. The balanced pulse pair waveform. In: Sheer, D. E., ed. Electrical stimulation of the brain. Austin, University of Texas Press. 60–64.

Linahan, N., R. K. Shepherd, J. Xu, S. Araki and G. M. Clark. 1998. Chronic electrical stimulation of the auditory nerve using non-charge stimuli. Proceedings of the Australian Neuroscience Society 9: 163.

Linahan, N., M. Tykocinski, R. K. Shepherd and G. M. Clark. 1999. Physiological and histopathological effects of chronic monopolar stimulation on the auditory nerve using very high stimulus rates. Proceedings of the Australian Neuroscience Society 10: 181.

Liu, X., G. McPhee, H. L. Seldon and G. M. Clark. 1997a. Histological and physiological effects of the central auditory prosthesis surface versus penetrating electrodes. Hearing Research 114: 264–274.

Liu, X., G. McPhee, H. L. Seldon and G. M. Clark. 1998. Histological and physiological effects of the central auditory prosthesis: surface versus penetrating electrodes. Proceedings of the Australian Neuroscience Society 9: 164.

Liu, X., H. L. Seldon, G. McPhee and G. M. Clark. 1997b. Acute study on the neuronal excitability of the cochlear nuclei of the guinea pig following electrical stimulation. Acta Oto-Laryngologica 117: 363–375.

Lousteau, R. J. 1987. Increased spiral ganglion cell survival in electrically stimulated, deafened guinea pig cochleae. Laryngoscope 97: 836–842.

Lustig, L. R., P. A. Leake, P. A. Snyder and S. J. Rebscher. 1994. Changes in the cat cochlear nucleus following neonatal deafening and chronic intracochlear electrical stimulation. Hearing Research 74: 29–37.

Matsushima, J., R. K. Shepherd, H. L. Seldon, S. Xu and G. M. Clark. 1991. Electrical stimulation of the auditory nerve in deaf kittens: effects on cochlear nucleus morphology. Hearing Research 56: 133–142.

McCreery, D. B. and W. F. Agnew. 1983. Changes in extracellular potassium and calcium concentration and neural activity during prolonged electrical stimulation of the cat cerebral cortex at defined charge densities. Experimental Neurology 79: 371–396.

McCreery, D. B., W. F. Agnew, T. G. Yuen and L. A. Bullara. 1988. Comparison of neural damage induced by electrical stimulation with faradaic and capacitor electrodes. Annals of Biomedical Engineering 16: 463–481.

McCreery, D. B., W. F. Agnew, T. G. H. Yuen and L. Bullara. 1990. Charge density and charge per phase as cofactors in neural injury by electrical stimulation. IEEE Transactions on Biomedical Engineering 537: 996–1001.

McCreery, D. B., T. G. Yuen, W. F. Agnew and L. A. Bullara. 1994. Stimulus parameters affecting tissue injury during microstimulation in the cochlear nucleus of the cat. Hearing Research 77(1–2): 105–115.

McHardy, J., L. S. Robblee, J. M. Marston and S. B. Brummer. 1980. Electrical stimulation with Pt electrodes: IV Factors influencing Pt dissolution in inorganic saline. Biomaterials 1: 129–134.

McNeal, D. R. 1976. Analysis of a model for excitation of myelinated nerve. IEEE Transactions on Biomedical Engineering 23: 329–337.

Mears, R. B. and R. H. Brown. 1941. Causes of corrosion currents. Industrial and Engineering Chemistry 33(8): 1001–1010.

Miller, J. M. and R. A. Altschuler. 1995. Effectiveness of different electrical stimulation conditions in preservation of spiral ganglion cells following deafness. Annals of Otology, Rhinology and Laryngology 104 (suppl 166): 57–60.

Miller, J. M., R. A. Atschuler, J. K. Niparko, D. O. Hartshorn, R. H. Helfert and J. K. Moore. 1992. Deafness-induced changes in the central nervous system: reversibility and prevention. In: Marshall, D., ed. Noise induced hearing loss. St. Louis, Mosby: 130–145.

Miller, J. M., L. G. Duckert, B. M. Clopton. 1983. Stimulation-induced changes in auditory system function and structure. In Webster, W. and L. Aitkin, eds. Mechanisms of hearing. Clayton, Victoria, Monash University Press.

Miller, J. M., L. G. Duckert, D. Sutton, B. E. Pfingst, M. A. Malone and F. A. Spelman. 1985. Animal models: relevance to implant use in humans. In: Schindler, R. and M. Merzenich, ed. Cochlear implants. New York, Raven Press. 35–55.

Mitchell, A., J. M. Miller, P. A. Finger, J. W. Heller, Y. Raphael and R. A. Altschuler. 1997. Effects of chronic high-rate electrical stimulation on the cochlea and eighth nerve in the deafened guinea pig. Hearing Research 105(1–2): 30–43.

Moore, D. R. 1990. Auditory brainstem of the ferret: early cessation of developmental sensitivity of neurons in the cochlear nucleus to removal of the cochlea. Journal of Comparative Neurology 302: 810–823.

Moore, D. R. and N. E. Kowalchuk. 1988. Auditory brainstem of the ferret: effects of unilateral cochlear lesions on cochlear nucleus volume and projections to the inferior colliculus. Journal of Comparative Neurology 272: 503–515.

Mortimer, J. T., C. N. Shealy and C. Wheeler. 1970. Experimental non destructive electrical stimulation of the brain and spinal cord. Journal of Neurosurgery 32: 553–559.

Nadol, J. B., Y.-S. Young and R. B. Glynn. 1989. Survival of spiral ganglion cells in profound sensorineural hearing loss: implications for cochlear implantation. Annals of Otology, Rhinology and Laryngology 98: 411–416.

Ni, D., H. L. Seldon, R. K. Shepherd and G. M. Clark. 1993. Effect of chronic electrical stimulation on cochlear nucleus neuron size in normal hearing kittens. Acta Oto-Laryngologica 113: 489–497.

Ni, D., R. K. Shepherd and G. M. Clark. 1992a. A physiological investigation of chronic electrical stimulation with scala tympani electrodes in kittens. Acta Academiae Medicinae Sinicae 14: 402–407.

Ni, D., R. K. Shepherd, H. L. Seldon, S. Xu and G. M. Clark. 1992b. Cochlear pathology following chronic electrical stimulation of the auditory nerve. I: Normal hearing kittens. Hearing Research 62: 63–81.

Nicholson, C., G. Ten Bruggencate, G. Stochle and R. Steinberg. 1978. Calcium and potassium changes in the extracellular microenvironment of cat cerebellar cortex. Journal of Neurophysiology 41: 1026–1038.

Nordeen, K. W., H. P. Killackey and L. M. Kitzes. 1983. Ascending projections to the inferior colliculus following unilateral cochlear ablation in the neonatal gerbil, Meriones unguiculatus. Journal of Comparative Neurology 214: 144–153.

Ochs, S. and C. Smith. 1971. Effect of temperature and rate of stimulation on fast axoplasmic transport in mammalian nerve fibres. Federation Proceedings 30: 2627.

Otte, J., H. Schuknecht and A. Kerr. 1978. Ganglion cell populations in normal and pathological human cochleae. Implications for cochlear implantation. Laryngoscope 88: 1231–1246.

Parker, J. 2002. Studies towards a micromachined cochlear implant electrode array. PhD thesis.

Pasic, T. R. and E. W. Rubel. 1989. Rapid changes in cochlear nucleus cell size following blockade of auditory nerve electrical activity in gerbils. Journal of Comparative Neurology 283: 474–480.

Powell, T. P. and S. D. Erulkar. 1962. Transneuronal cell degeneration in the auditory relay nuclei of the cat. Journal of Anatomy 96: 249–268.

Pudenz, R. H. 1942. The use of tantalum clips the hemostasis in neurosurgery. Surgery 12: 791–797.

Pudenz, R. H., W. F. Agnew, T. G. H. Yuen and L. A. Bullara. 1977. Electrical stimulation of the brain. Light and electron microscopy studies. In: Hambrecht, F. and J. Reswick, eds. Applications in neural prostheses. New York, Marcel Dekker.

Pudenz, R. H., L. A. Bullara, D. Dru and A. Talalla. 1975a. Electrical stimulation of the brain II. Effects on the blood barrier. Surgical Neurology 4: 265–270.

Pudenz, R. H., L. A. Bullara, S. Jacques and F. T. Hambrecht. 1975b. Electrical stimulation of the brain. III. The neural damage model. Surgical Neurology 4: 389–400.

Robblee, L. S., J. McHardy and J. M. Marston. 1980. Electrical stimulation with Pt electrodes. V. The effect of protein on Pt dissolution. Biomaterials 1: 135–139.

Robblee, L. S. and T. L. Rose. 1990. Electrochemical guidelines for selection of protocols and electrode materials for neural stimulation. In: Agnew, W. F. and D. B. McCreery, eds. Neural Prostheses: Fundamental Studies. Englewood Cliffs, NJ, Prentice Hall.

Schwan, H. P. 1963. Determination of biological impedances. In: Nastuk, W. L., ed. Physical techniques in biological research. New York, Academic Press: 323–406.

Shannon, R. V. 1992. A model of safe levels for electrical stimulation. IEEE Transactions on Biomedical Engineering 39: 424–426.

Shepherd, R. K. and G. M. Clark. 1986. Electrical stimulation of the auditory nerve: effects of high stimulus rates. Proceedings of the Australian Physiological and Pharmacological Society 17: 16P.

Shepherd, R. K., G. M. Clark and R. C. Black. 1983a. Chronic electrical stimulation of the auditory nerve in cats. Physiological and histopathological results. Acta Oto-Laryngologica-supplement 399: 19–31.

Shepherd, R. K., G. M. Clark and R. C. Black. 1983b. Physiological and histopathological effects of chronic intracochlear electrical stimulation. In: Webster, W. R. and L. M.

Aitkin, eds. Mechanisms of hearing. Clayton, Victoria, Monash University Press: 200–205.

Shepherd, R. K., G. M. Clark, R. C. Black and J. F. Patrick. 1983c. The histopathological effects of chronic electrical stimulation of the cat cochlea. Journal of Laryngology and Otology 97: 333–341.

Shepherd, R. K., B. K.-H. G. Franz and G. M. Clark. 1990. The biocompatibility and safety of cochlear prostheses. In: Clark, G. M., Y. C. Tong and J. F. Patrick, eds. Cochlear prostheses. London, Churchill Livingstone: 69–98.

Shepherd, R. K., J. Matsushima, R. E. Millard and G. M. Clark. 1991. Cochlear pathology following chronic electrical stimulation using non charge balanced stimuli. Acta Oto-Laryngologica 111: 848–860.

Shepherd, R. K., M. T. Murray, M. E. Houghton and G. M. Clark. 1985. Scanning electron microscopy of chronically stimulated platinum intracochlear electrodes. Biomaterials 6: 237–242.

Shepherd, R. K., S. A. Xu, B. K.-H. Franz, et al. 1987. Studies on pediatric auditory prosthesis implants. Progress reports. NIH contract No1-NS-7-2342.

Shepherd, R. K., S. A. Xu, B. K.-H. G. Franz, et al. 1989. Studies on pediatric auditory prosthesis implants. Seventh quarterly progress report. NIH contract No1-NS-7-2342. University of Melbourne.

Simmons, F. B. 1966. Electrical stimulation of the auditory nerve in man. Archives of Otolaryngology 84: 2–54.

Sokoloff, L. 1983. Measurement of local glucose utilization in the central nervous system and its relationship to local functional activity. In: Lajtha, A., ed. Handbook of neuro-chemistry. Vol. 3. (2nd Ed.) New York, Plenum Press: 225–257.

Stockle, H. and G. Ten Bruggencate. 1980. Fluctuation of extracellular potassium and calcium in the cerebellar cortex related to climbing fibre activity. Neuroscience 5: 893–901.

Tong, Y. C., R. C. Black, G. M. Clark, et al. 1979. A preliminary report on a multiple-channel cochlear implant operation. Journal of Laryngology and Otology 93: 679–695.

Tong, Y. C., G. M. Clark, P. M. Seligman and J. F. Patrick. 1980. Speech processing for a multiple-electrode cochlear implant hearing prosthesis. Journal of the Acoustical Society of America 68: 1897–1899.

Trune, D. R. 1982. Influence of neonatal cochlear removal on the development of mouse cochlear nucleus: I. Number, size and density of its neurons. Journal of Comparative Neurology 209: 409–424.

Tykocinski, M., N. Linahan, R. K. Shepherd and G. M. Clark. 1998. Electrical stimulation of the auditory nerve: chronic monopolar stimulation using very high stimulus rates. Proceedings of the Australian Neuroscience Society 9: 163.

Tykocinski, M., N. Linahan, R. K. Shepherd and G. M. Clark. 1999. Physiological and histopathological effects of chronic monopolar high rate stimulation of the auditory nerve. 1999 Conference on Implantable Auditory Prostheses, California.

Tykocinski, M., R. K. Shepherd and G. M. Clark. 1995a. Acute effects of high-rate stimulation on auditory nerve function in guinea pigs. Annals of Otology, Rhinology and Laryngology 104(suppl 166): 71–74.

Tykocinski, M., R. K. Shepherd and G. M. Clark. 1995b. Electrophysiologic effects following intracochlear direct current stimulation of the guinea pig cochlea. Annals of Otology, Rhinology and Laryngology 104(suppl 166): 68–71.

Tykocinski, M., R. K. Shepherd and G. M. Clark. 1997. Reduction in excitability of the auditory nerve following electrical stimulation at high stimulus rates. II. Comparison of fixed amplitude with amplitude modulated stimuli. Hearing Research 112: 147–157.

Urbanics, R., E. Leninger-Follert and W. Lubbers. 1978. Time course of changes of extracellular H + and K + activities during and after direct stimulation of the brain cortex. Pflugers Archiv–European Journal of Physiology 378: 47–53.

Vaadia, E., I. Haalman, M. Abeles, et al. 1995. Dynamics of neuronal interactions in monkey cortex in relation to behavioural events. Nature 373: 515–518.

Vischer, M. W., V. M. Bajo, J. S. Zhang, E. Calciati, C. A. Haenggeli and E. M. Rouiller. 1997. Single unit activity in the inferior colliculus of the rat elicited by electrical stimulation of the cochlea. Audiology 36: 202–27.

Walsh, S. M. and P. A. Leake-Jones. 1982. Chronic electrical stimulation of the auditory nerve in cat: physiological and histological results. Hearing Research 7: 281–304.

Walsh, S. M., P. A. Leake-Jones, L. S. Vurek and M. M. Merzenich. 1981. Chronic electrical stimulation with intracochlear electrodes: electrophysiological results. Annals of Otology, Rhinology, and Laryngology 90(suppl 82): 27–29.

Webster, D. B. and M. Webster. 1978. Cochlear nerve projections following organ of corti destruction. Otolaryngology 86: 342–353.

Weiland, J. D. and D. J. Anderson. 2000. Chronic neural stimulation with thin-film, iridium oxide electrodes. IEEE Transactions on Biomedical Engineering 47: 911–919.

Weinman, J. and J. Mahler. 1964. An analysis of electrode properties of metal electrodes. Medical Electronics and Biological Engineering 2: 299–310.

Williams, A. J., G. M. Clark and G. V. Stanley. 1974. Behavioural responses in the cat to simple patterns of electrical stimulation of the terminal auditory nerve fibres. Proceedings of the Australian Physiological and Pharmacological Society 5(2): 252.

Wilson, B. S. 1997. The future of cochlear implants. British Journal of Audiology 31: 205–225.

Wilson, B. S., D. T. Lawson, M. Zerbi and C. C. Finley. 1992. Speech processors for auditory prostheses. Twelfth quarterly progress report, April 1992. NIH contract N01-DC-9-2401. Research Triangle Institute.

Wilson, B. S., D. T. Lawson, M. Zerbi and C. C. Finley. 1993. Speech processors for auditory prostheses. Fifth quarterly progress report, October 1993. NIH contract N01-DC-2-2401. Research Triangle Institute.

Wong-Riley, M. T. T., M. M. Merzenich and P. A. Leake-Jones 1978. Changes in endogenous enzymatic reactivity to DAB induced by neuronal inactivity. Brain Research 141: 185–192.

Wong-Riley, M. T. T., S. M. Walsh, P. A. Leake-Jones and M. M. Merzenich. 1981. Maintenance of neuronal activity by electrical stimulation of unilaterally deafened cats demonstrable with cytochrome oxidase technique. Annals of Otology Rhinology and Laryngology 90(suppl 82): 30–32.

Xu, J., R. K. Shepherd, R. E. Millard and G. M. Clark. 1997. Chronic electrical stimulation of the auditory nerve at high stimulus rates: a physiological and histopathological study. Hearing Research 105: 1–29.

Xu, S. A., R. K. Shepherd and G. M. Clark. 1990. Rapid and permanent hearing loss in cats following co-administration of kanamycin and ethacrynic acid. Proceedings of the Australian Physiological and Pharmacological Society 21: 44P.

Yuen, T. G. H., W. F. Agnew, L. A. Bullara, S. Jacques and D. B. McCreery. 1981. Histological evaluation of neural damage from electrical stimulation: considerations for the selection of parameters for clinical application. Neurosurgery 9: 292–299.

Yukawa, K., S. O'Leary, M. Clarke and G. M. Clark. 2001a. Histopathology of the brainstem of a binaural cochlear implant subject. Program and abstracts of the Third International Congress of Asia Pacific Symposium on Cochlear Implant and Related Sciences. Osaka, Japan: 48.

Yukawa, K., S. J. O'Leary, M. Clarke and G. M. Clark. 2001b. Histopathology of the binaural cochlear implant patient. Abstract for the Australian Society for Otolaryngology Head and Neck Surgery Annual Scientific Meeting.

5
Electrophysiology

Cochlear implants should aim to reproduce the coding of sound in the auditory system as closely as possible, for best sound perception. The cochlear implant is in part the result of reverse engineering from systems neurophysiology and thus attempts to reproduce the coding of sound. The coding of sound, however, cannot be easily reproduced, as sound produces a complex temporal and spatial pattern of nerve discharges in approximately 20,000 auditory neurons, and only about 22 electrodes can be used with current technology, for example the Nucleus/Cochlear Limited (University of Melbourne/Bionic Ear Institute) systems (Clark 1996, 1997). Nevertheless, by partially reproducing the coding with multiple-electrode stimulation, as well as extracting appropriate information from the speech signal, good speech perception is still possible (Clark 1996, 1997).

General Neurophysiology

The cochlear implant bypasses the transduction of sound vibrations into electrical signals in the cochlea, and directly stimulates cochlear nerve fibers. These in turn excite the higher centers in the central auditory nervous system. Electrical stimulation aims to reproduce the pattern of excitation in the nervous pathways produced by sound, and to manipulate the mechanisms underlying the generation of neural activity to produce this pattern.

Action Potentials

Information is transmitted throughout the central nervous system by action potentials. These electrical events are initiated in dendrites and cell bodies and transmitted along axons, which make connections (synapses) on the dendrites and body of the next cell, as illustrated in Figure 5.1, and so on. The synapses are localized connections that are highly specialized. A more complete account of neural function and the structure of the brain can be found in texts such as Levitan and Kaczmarek (1997) and Shepherd (1998).

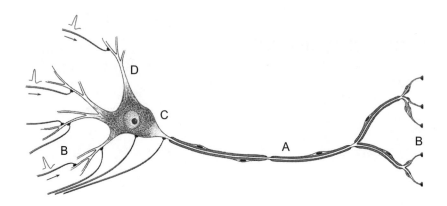

FIGURE 5.1. A diagram of a neuron and connections illustrating the dendrites (D) bringing information to the cell's body (C), and the axons (A) taking information to other neurons via the terminal button-shaped endings or boutons (B).

The action potentials are initiated when the resting potential of the neuron (the voltage across the neural membrane) is decreased (depolarized) to a threshold level as illustrated in Figure 5.2. There is a higher concentration of (K^{1+}) inside and (Na^{1+}) outside the nerve membrane. The resting potential is due to selective permeability of the nerve membrane to the potassium ion (K^{1+}) with a concentration gradient driving it outside the cell membrane. The ion flow across the nerve membrane can be represented as an electrical current that obeys Ohm's law (Mortimer 1990):

$$I_{ion} = g_{ion} (V_m - E_{ion})$$

where I_{ion} is the flow of ions, g_{ion} is the membrane conductance; V_m is the membrane potential; and E_{ion} is the equilibrium potential for a particular ion and given by the Nernst equation:

$$E_{ion} = (RT/zF)\ln(C_o/C_i)$$

where R is the gas constant, T is the absolute temperature, z is the valence of the ion, F is the Faraday constant, C_o is the concentration of the ion on the outside of the membrane, and C_i the concentration on the inside.

Under resting conditions the inward sodium current (I_{Na}) is equal and opposite to the outward potassium current (I_K):

$$I_{Na} = -I_K$$

This can be written as

$$g_{Na}(V_m - E_{Na}) = -g_K (V_m - E_K)$$

where E_{Na} is the equilibrium potential for (Na^{1+}) and E_K for (K^{1+}).
The equation can be solved for V_m:

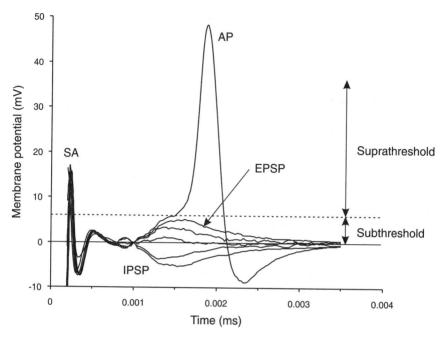

FIGURE 5.2. An intracellular recording of excitatory postsynaptic potentials (EPSPs) and the action potential (AP) generated in a globular bushy cell in the anteroventral cochlear nucleus of the rat. The stimulus artifact (SA) results in EPSPs and when threshold is reached an AP occurs. An inhibitory postsynaptic potential (IPSP) is also shown (Paolini and Clark 1998a).

$$V_{m} = \frac{E_{Na}\,(g_{Na}/g_{K}) + E_{K}}{(g_{Na}/g_{K}) + 1}$$

Thus the membrane potential is regulated by the ion conductance.

If the sodium conductance is very small during the resting state, the membrane potential approaches -60 mV (i.e., positive outside the membrane), and if the sodium conductance becomes large, the potential approaches the sodium potential of $+55$ mV (i.e., negative outside the membrane).

An excitatory transmitter acts on the postsynaptic membrane to produce an excitatory postsynaptic potential (EPSP), and an inhibitor transmitter an inhibitory postsynaptic potential (IPSP). With excitation when the potential reaches a threshold level the sodium gates in the nerve membrane open and the sodium ions (Na^{1+}), which have a higher concentration outside the cell, rush inward down a potential gradient to produce the action potential. The current due to the passage of (Na^{1+}) produces a further depolarization that in turn increases the conductance of (Na^{1+}). This positive feedback continues until equilibrium is reached with the concentration and voltage gradients of ions balancing each other. The creation of an action potential makes the voltage on the outer surface of the cell membrane negative relative to the inside of the cell. The EPSPs producing an action potential are generated when acetylcholine and amino acids such as glutamic and aspartic acid are released from the boutons on the presynaptic side of the synaptic space. Hyperpolarization occurs through the release of transmitters that open chloride or potassium channels, and produce IPSPs. In this case chloride ions (Cl^{1-}) move inward or potassium ions (K^{1+}) outward so the outside of the cell membrane becomes more positive relative to the inside. Important inhibitory amino acid transmitters are γ-aminobutyric acid and glycine.

The chemical transmitters for excitation and inhibition are released when an action potential reaches the presynaptic terminal. This induces an increase in the local (Ca^{2+}) concentration close to the open (Ca^{2+}) ion channels. A (Ca^{2+}) sensor binds the (Ca^{2+}) before the release of the transmitter into the synaptic cleft (Bollman et al 2000). The transmitter is released from synaptic vesicles, each being in a docking position close to the presynaptic membrane. The transmitter then initiates an EPSP or IPSP in the postsynaptic membrane.

The EPSPs and IPSPs take a certain period of time to reach a maximum, and then decay over a longer period, as illustrated in Figure 5.2. The postsynaptic potentials sum to produce a potential large enough to initiate an action potential. The summing can be spatial or occur over time as each potential takes a defined period to decay. The response of each neuron is thus the sum total of the excitatory and inhibitory inputs. When the postsynaptic potential reaches a threshold level, the sodium gates in the nerve membrane open wide and sodium ions rush in from the extracellular fluid to produce a rapid voltage change or spike, as illustrated in Figure 5.2. After a short period of time, which depends on the properties of the nerve membrane, the sodium gates close, and the potassium gates open to allow the resting nerve membrane potential to be reestablished. In the meantime, the voltage change or spike is propagated along the neuron by this active self-regenerating process.

If a stimulus arrives during the stage when the sodium gates are open, a second action potential will not occur. The period when a second action potential cannot be initiated is the absolute refractory period. This was shown to be approximately 0.5 ms for the auditory nerve (Moxon 1967). After the absolute refractory period there is a relative refractory period during which time a stronger stimulus is required to excite the neuron. A study by Roberts et al (2000) suggested this can have a mean value of approximately 0.2 ms. During this time the sodium gates have not completely returned to their normal state.

Strength-Duration Curves

A well-defined relation between the electrical current amplitude and the duration of the pulse was found for threshold responses (Hill 1936; Katz 1939). This relationship is a strength-duration curve, and is illustrated for the cat auditory nerve in Figure 5.3. Notice that as the pulse duration decreases, the threshold for an action potential increases in an exponential fashion. A cochlear implant stimulator would need to be able to produce high current amplitudes for the short-

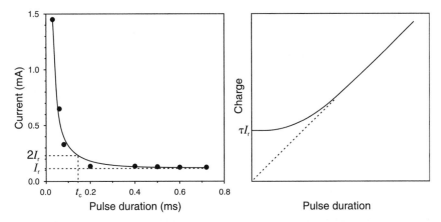

FIGURE 5.3. Left: Strength-duration curve from a cat auditory nerve fiber. I_r, rheobase current; t_c, chronaxie (Clark et al 1977. A multiple electrode hearing prosthesis for cochlear implantation in deaf patients. *Medical Progress through Technology* **5**: 127–140. Reprinted with permission of Springer-Verlag.) Right: Charge-duration curve (Mortimer 1990).

duration pulses required for high stimulus rates. The curve also shows that pulse amplitude and width can be traded in exciting a threshold response. If the current is to be minimized, the pulse duration must be increased. Note that as the pulse duration increases, the threshold current reaches an asymptote—the rheobase.

The mathematical relation between the threshold current (I_{th}) and the stimulus duration (t) can be calculated (Hill 1936; Katz 1939) and the values predicted from the following expression:

$$I_{th} = \frac{I_r}{(1 - e^{-t/\tau})}$$

where I_{th} is the threshold current, I_r the rheobase current, τ the excitation time constant, t the pulse duration and e is the base of natural logarithms. The excitation constant is related to the chronaxie (t_c), which is the pulse width for a current twice the rheobase current.

Depending on the dimensions of the electrode pads and the distance from the cochlear nerve, it may be necessary to minimize charge per phase. The relation between charge per phase and pulse width can be derived from the above equation by multiplying the current by the pulse width. Figure 5.3 shows the amount of charge that must be injected to initiate an action potential. Note that for very short pulse widths the charge approaches an asymptote that has a constant value.

Electrical Models of the Nerve Membrane

The first model of the electrical activity of the nerve membrane was developed by Hodgkin and Huxley (1952). Their studies were undertaken on the squid axon, and the voltage across the membrane could be set and maintained at a constant level through a feedback loop. The voltages and ionic gradients could thus be manipulated allowing the ionic conductance to be studied. With small voltage changes there was an early inward current due to the influx of (Na^{1+}), and with an increase in voltage a delayed current with an outward flow of (K^{1+}). They applied nonlinear differential equations to determine a fiber's response to a stimulus. For subsequent studies by others on the effects of electrical stimulation on a network of neurons, the equations of Hodgkin and Huxley (1952) required considerable computational facilities, and the effects of stochastic or random processes in the nerve membrane were difficult to include in the calculations. See Neural Models, below for further discussion.

Another model was developed by Rubinstein (1991, 1993) on the biophysics of the voltage sensitive sodium channels, leakage currents, and membrane capacitances to provide a stochastic representation of the responses of the neural membrane as applied to the nodes of Ranvier of the peripheral processes of the auditory nerve. It had a deterministic representation of the internode. It was also computationally very demanding for the study of responses in a neural network.

The responses in a neural network have been more simply studied with point process models (Perkel et al 1967; Miller 1971), which gave the statistical properties of responses over time, but do not calculate the biophysical properties of the membrane. Their application to electrical stimulation of the central auditory pathways is discussed below (see Neural Models).

Convergence and Divergence

Information in the central nervous system is transmitted as patterns of action potentials along nerve fibers, and received at neurons along convergent or divergent pathways. The convergent pathways serve to bring information from widespread areas to a more localized point for coding, while divergent pathways distribute information from a localized point for more widespread effects. When the information from convergent or divergent pathways arrives at the cells, it is further encoded as the result of temporal and spatial summation of EPSPs and IPSPs. These in turn lead to temporal and spatial patterns of action potentials that are propagated up the auditory pathways to other nuclei for processing. A schematic diagram of convergent innervation is shown in Figure 5.4. In reproducing the coding of sound with electrical stimulation, the degree of convergence of neurons from the cochlea to the cochlear nucleus needs to be considered to determine the interaction required for the coding of frequency and the spacing of the electrodes and stimulus current. The information converging on a cell as a temporal and spatial pattern of action potentials can be coded by detecting the coincidences in the time of their arrival (Carney 1994). As illustrated in Figure 5.4, if the incoming action potentials arrive within a certain time window, the resultant EPSPs can sum temporally to initiate a response from the cell. The time window is determined by the biophysical properties of the cell. The timing of the action potentials going out from the cell is determined by the time between the arriving stimuli.

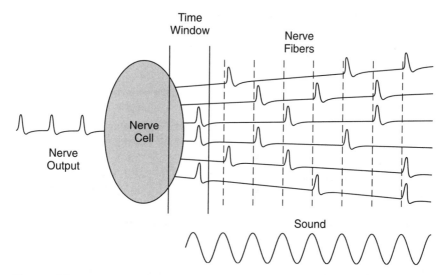

FIGURE 5.4. Temporospatial patterns of action potentials in an ensemble of neurons in response to a low to middle acoustic frequency. Top: Nerve action potentials in a population of neurons. Bottom: Pure tone acoustic stimulus. This demonstrates the phase locking of neurons to the sound wave, but note that the action potentials do not occur each cycle. The diagram also demonstrates convergent pathways on a cell. The convergent inputs initiate an action potential in the cell only if they arrive within a time window (coincidence detection).

Auditory Physiology

The auditory system is a functional unit comprising the external and middle ears, the cochlea, and the central brain pathways and centers. Auditory physiology and in particular neurophysiology, is one of the key disciplines underlying the application of electrical stimulation of the cochlear nerve [auditory nerve (AN)] for bringing hearing and speech understanding to deaf people. Studies were needed to show how well electrical stimulation could reproduce the coding of frequency and intensity. The limitations in reproducing the temporal coding of frequency (Clark 1969a–c, 1970a,b; Clark, Nathar et al 1972; Williams et al 1974; Merzenich, White et al 1979), led to an emphasis on the place coding of frequency through multiple-electrode stimulation of separate groups of auditory nerves in the cochlea. The possibility of electrically reproducing place pitch was supported by studies that showed how well current could be localized to separate groups of nerve fibers (Merzenich, White et al 1979; Black and Clark 1980).

Although both temporal and place coding were limited with electrical stimulation, effective speech-processing strategies for profoundly deaf people were achieved by also selecting the crucial speech information for transmission through what is essentially an electroneural bottleneck (Tong et al 1979, 1980; Eddington 1980; Merzenich and White 1980; Dowell et al 1987, 1990; McKay et al 1991;

McDermott and McKay 1992). In turn, electrical stimulation of the auditory pathways has helped in understanding the normal coding of sound (Clark 1996), as it has enabled both temporal and place of stimulation to be studied separately. This cannot be so readily achieved with sound.

Physics of Sound

Sound results from compression and rarefaction of molecules in an elastic medium, and this is propagated through the medium. It has the parameters of frequency, amplitude, and velocity. Sound vibrations are due to the movement of acoustic particles back and forth around their resting point, and this can be described by simple harmonic motion. This can be analyzed mathematically using a planar projection of a point undergoing circular motion as illustrated in Figure 5.5. The motion can be described using sine and cosine rules:

$$y = a \cos(2\pi f t)$$

where a is the amplitude, f the frequency, and t the time. $2\pi f$ is the angular velocity around the circle; thus the equation can be expressed as

$$y = a \cos(\omega t)$$

where ω is the angular velocity.

A complex wave is made up of a number of sound waves of variable intensity and frequency added together. The resulting wave is also determined by the phase relations of the composite sound frequencies. The intensity of a sound is measured in bels or decibels (i.e., a tenth of a bel). This is a convenient unit as it enables the wide range of sound pressures and intensities heard from threshold to discomfort to be reduced to a smaller more manageable scale. A bel is a log ratio to base 10:

$$I = \log_{10} I_1/I_0$$

where $I =$ the intensity in decibels, $I_1 =$ the measured intensity, and $I_0 =$ the reference intensity, which is 10^{-12} W/m^2.

Sound has not only energy but also pressure. The relation between energy and pressure is as follows:

$$E = P^2/\rho_0 c$$

where $E =$ the sound energy, $P =$ the sound pressure, $\rho_0 =$ the density of the medium; and $c =$ the speed of sound.
Therefore:

$$P = 10\log_{10} P_1^2/P_0^2$$
$$P = 20\log_{10} P_1/P_0$$

where $P =$ the pressure in decibels, $P_1 =$ the measured pressure, and $P_0 =$ the reference pressure, which is 2×10^{-5} Pa.

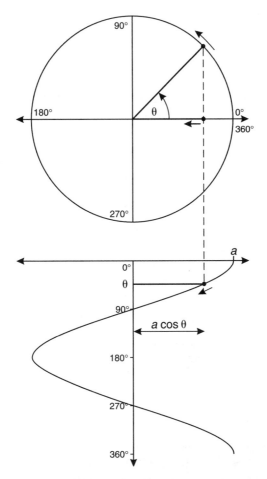

FIGURE 5.5. The cosine representation of the projection of a point moving around a circle describing the simple harmonic motion of an acoustic particle around its resting point.

External and Middle Ear Function

The external ear or pinna and external auditory canal collect sound and are its conduit to the middle ear. The pinna, through its scalloped shape, introduces a series of resonances in the sound depending on whether it arrives from the front or back. This occurs for the sound spectrum in the 5000- to 7000-Hz region. This helps in localizing the source in particular when there is only one hearing ear. The external auditory canal acts as a resonating tube, and in the human this is at approximately 3430 Hz. It is not a perfect pipe, and the resonance is broad from 2000 to 5000 Hz. The canal provides a gain of up to 20 dB in this frequency band.

The middle ear is a cavity closed externally by the eardrum (tympanic membrane). The sound induces vibrations in this membrane, which are transmitted via the middle ear ossicles (malleus, incus, and stapes) and then through the oval window to the fluid in the cochlea (see Fig. 2.1 in Chapter 2). The sound energy passing down the external auditory canal causes the eardrum or tympanic membrane to vibrate. This membrane is attached to the malleus, and the vibrations are transmitted via the incus and stapes through the oval window, a small opening in the bony cochlea, to the fluid in the inner ear. These vibrations stimulate the cochlea.

The middle ear matches the impedances of sound in air and water. If the middle ear did not exist, most of the sound energy would be reflected. The proportion of sound energy transmitted can be calculated from the following formula:

$$X = 4Z_b \times Z_a/(Z_b + Z_a)^2$$

where X is the proportion transmitted, Z_b is the impedance for sound in air (415 Pa·s/m), and Z_a is the impedance for water (1.5 × 10⁶ Pa·s/m). From this it can be calculated that 3% of sound energy is transmitted and 97% reflected. The middle ear improves the impedance match by increasing the pressure of sound at the oval window. The tympanic membrane and ossicles amplify the sound energy by approximately 20 times. This is achieved through (1) a hydraulic ratio between the eardrum and the footplate of the stapes, (2) a lever ratio through the head of the incus, and (3) a phase difference between the sound arriving at the oval and round windows.

Cochlea

Sound energy is transduced into electrical signals by the sense organ of hearing (organ of Corti) in the membranous cochlea, and filtered into its frequency components. The membranous cochlea lies within a spiral tube in the temporal bone, the bony or osseous cochlea (see Fig. 2.9 and 2.12 in Chapter 2).

Endolymph and Perilymph

The central tube (scala media) in the membranous cochlea, containing the organ of Corti, is filled with the fluid endolymph. The scala media is surrounded above and below by two tubes, the scala vestibuli and tympani, respectively, that contain perilymph fluid.

Perilymph, is a filtrate of blood plasma or a result of diffusion of cerebrospinal fluid (CSF) into the cochlea from the cochlear aqueduct. Accordingly perilymph has a high sodium ion concentration (approximately 140 mM). The cochlear aqueduct connects the subarachnoid space lying around the brain with the scala tympani in the basal turn of the cochlea near the round window. The percolation of CSF along the aqueduct has been demonstrated by Silverstein et al (1969) in the cat, with the passage of particulate matter from the CSF to the perilymph within 24 hours. The reverse flow could occur with the alteration of the CSF pressure, and facilitate the spread of bacteria to the meninges from the cochlea. Furthermore, as perilymph is formed as a transudate as well as from the CSF, it will continue to be produced if the cochlear aqueduct is blocked from a fibrous tissue sheath formed around the electrode near the entry of the aqueduct and the round window.

In contrast, endolymph has a high potassium ion concentration (154 mM) and a high positive potential (80 mV) (Smith et al 1954; Bosher and Warren 1968; Sterkers et al 1984). This requires a metabolic pump in the basal cells of the stria vascularis to maintain it. A radial flow of endolymph results, but there is also a longitudinal flow of 0.1 μL/minute driven by an osmotic force.

Traveling Wave

Sound frequencies induce traveling waves (Fig. 5.6) that pass along the basilar membrane of the membranous cochlea to produce maximal displacements at resonant sites specific to each frequency. The lowest frequencies produce peaks in the apical turn, and the highest frequencies produce peaks in the basal turn. This filtering of frequencies is due to a gradual decrease in stiffness in the basilar membrane from the basal to apical turns (von Békésy 1960). The acoustic compliance per unit length typically varies by a factor of nearly 10^5 over the membrane length. The traveling waves pass rapidly along the membrane at first, and slow down at the point of resonance where most of the energy is dissipated as lateral displacement. Calculations have shown the wave travels along the basilar membrane at approximately 8.7 m/s, but the peaks at the site of maximal displacement have a velocity of only 1.3 m/s. (L. Cohen, personal communication).

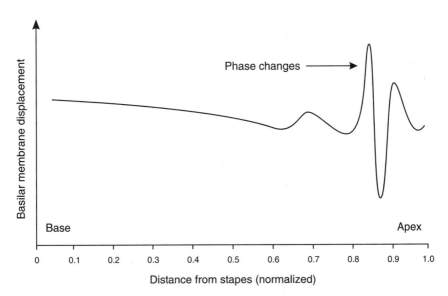

FIGURE 5.6. Traveling wave generated from a cochlear model (Neely and Kim 1986; Au et al 1995). Reprinted from Clark, Cochlear implants in the third millennium, Am J Otol 20: 4–8, 1999, with permission.

Von Békésy (1960) determined the place of maximum vibration of the human basilar membrane versus frequency through stroboscopic illumination, and plotted the distance between these maxima and the stapes footplate as illustrated in Fig. 5.7. This was compared with data from Stevens and Davis (1938) derived from hearing loss measurements, and they show good agreement. These data were consistent with those from experimental studies in the cat by Schuknecht and Neff (1952) and Schuknecht (1953). They were important not only in understanding the physiology of the cochlea, but also subsequently in knowing where to place the electrodes to reproduce the place coding of frequency (see the section on bioengineering in Chapter 8).

Organ of Corti

The organ of Corti transduces sound vibrations in the basilar membrane into electrical signals. Its microscopic structure is described in more detail in Chapter 2 (see Fig. 2.12). It has rows of inner and outer hair cells with supporting cells. They rest on a vibrating membrane (the basilar membrane).

Sound vibrations result in a shearing force on the hairs (stereocilia) of the cells in the sense organ of hearing (organ of Corti). The outer hair cells are embedded in the tectorial membrane and the force causes them to bend. The bending of the hairs opens and closes ion channels, resulting in the modulation of the receptor potential (Russell et al 1986). There is a large potential difference between the

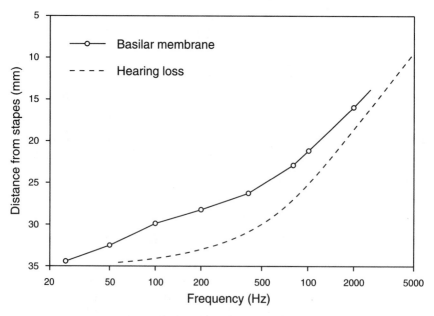

FIGURE 5.7. Solid line: the site of basilar membrane maxima for frequency versus distance from the stapes footplate (von Békésy 1960). Dashed line: localization of frequency versus distance from hearing loss studies (Stevens and Davis 1938). (Reprinted with permission.)

positive endolymph potential (80 mV) and the negative intracellular fluid (-60 to -80 mV). This potential drives a current mainly of potassium ions through the apical cell membrane. The response is rectified with current flowing with contraction and saturating in elongation (Evans et al 1989; Santos-Sacchi 1989). The outer hair cells are contractile, and this amplifies the vibration of the basilar membrane. Isolated outer hair cells contract when stimulated by transcellular current (Brownell et al 1985). The current in the inner hair cells initiates the action potentials in the auditory nerve fibers.

Cochlear Models

The function of the cochlea has been modeled mathematically to study in more detail how sound is filtered to produce a spectrum of frequencies. Cochlear models could provide directions for improved methods of extracting speech signals for coding sound through electrical stimulation. These models predict the mechanical events in the basilar membrane at each instant in time. An accurate model of cochlear function is also required as the input to a mathematical model of the neural coding in the AN and neural pathways in the auditory brainstem. The models are also leading to a better understanding of how the brain processes information.

The basilar membrane traveling wave can be modeled knowing (1) the impedance to sound at points along the membrane, (2) the difference in fluid pressure across the membrane, and (3) the frequency of the stimulus. Models make certain assumptions, for example, that the volume flow along the upper scala vestibuli is balanced by the return flow along the scala tympani, although there is an oscillating volume exchange across the basilar membrane due to its displacement. Models also assume the cross-sectional areas of the scalae do not vary with distance, and ignore frictional forces, as they are small. Nevertheless, these models quite accurately predict basilar membrane movement. They thus indicate which physical parameters are of most importance in generating the traveling wave. More refined models (Neely and Kim 1986) have included active elements for reproducing the high sensitivity and sharp tuning of the mammalian cochlea. They represent the motile action of the outer hair cells powered by electrochemical energy.

With future cochlear implant speech processing, the models could be used to determine not only the shape of the frequency filter but also the temporal and spatial pattern of electrical stimulation needed to reproduce the normal pattern at the point of resonance.

Auditory Neurophysiology

The physiology of the central auditory pathways depends on the organization and innervation of the cochlea as well as the brain centers. As distinct from vision, the auditory system is designed to primarily process temporal rather than spatial information. The difference in processing could account for the greater number of centers in the auditory system, and the marked increase in the population of neurons from the periphery to the cortex. The complex temporal patterns may require stage-by-stage coding.

There is considerable evidence that the coding of the sound frequencies is both through time/period (rate), and place cues (Wever and Bray 1930; Tasaki 1954; Katsuki and Kanno 1962; Rupert et al 1963; Kiang et al 1965a; Rose et al 1967; Sachs and Young 1979, 1980; Geisler 1981; Sinex and Geisler 1981). Reproducing the temporal and place coding of frequency with electrical stimulation underlies the development of cochlear implants, and improvements are necessary for further advances.

Temporal Frequency Coding

The temporal coding of frequency is the result of action potentials locked to the same phase of the sine wave and the detection of this information through coincidences in the arrival of the action potentials at neurons in the cochlear nucleus (CN) and higher nuclei from sites in the cochlea.

Phase Locking of Action Potentials and Excitatory Postsynaptic Potentials

The phase locking results in the intervals between the action potentials being a multiple of the period of the sound wave, and so representing the frequency. The period is the time taken for one complete cycle of the sine wave. As can be seen at the left of Figure 5.8, a unit from the region of the medial nucleus of the

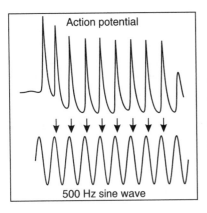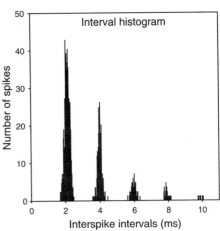

FIGURE 5.8. Action potentials and phase locking. Left: Unit responses in the medial nucleus of the trapezoid body (MNTB) in the brainstem of the cat to a sound frequency of 500 Hz (top, brainstem action potential; bottom, sine wave). Note the action potentials fire in phase with the sine wave (Clark 1969b). Right: Interspike interval histogram from a unit in the anteroventral cochlear nucleus (AVCN) of the cat to a sound of 500 Hz. Note the predominant interval between action potentials is the same as the period of the sine wave. The other intervals are multiples of the predominant one.

trapezoid body (MNTB) of the cat brainstem fired in phase with the sine wave (approximately at the same point in time) (Clark 1969b). This is illustrated for a frequency of 500 Hz. The intervals between successive action potentials in response to sound can be plotted as an interspike interval histogram, as illustrated at the right of Figure 5.8 and in Figure 5.9.

The phase locking to sounds was strong in the low- to midfrequency range and occurred to a limited extent up to a high frequency of 5000 Hz, where phase locking disappeared. The phase locking of the units in the central nucleus of the inferior colliculus (ICC) to the low to middle sound frequencies was studied by constructing a histogram of unit responses to each half of the sine wave (period histogram). There was a statistically significant difference in the number of responses in each half up to 5000 Hz (Rose et al 1967). However, the degree of phase locking was much greater at the lower frequencies. The data supported the time-period code for frequency, but did not answer the question of up to what frequency this was important. The relevance of the time intervals between the neural spikes for coding frequency was also demonstrated by Rose et al (1969)

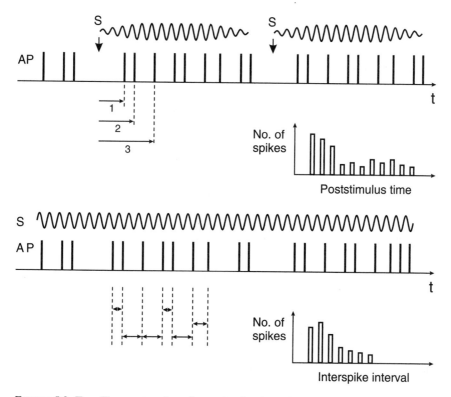

FIGURE 5.9. Top: The construction of per-stimulus (poststimulus) histogram; 1, 2, and 3 refer to the time from the stimulus to the first three action potentials. Bottom: The construction of interspike interval histogram. AP, action potential; S, a burst or continuous tone.

when they presented a two-tone complex and found the patterns of interspike intervals corresponded to the change in the waveform as the phase relationships between the frequencies were varied.

Studies were undertaken to see if electrical stimulation of the AN could reproduce these temporal firing patterns to encode the range of speech frequencies with a single-channel cochlear implant, but this was shown not to be the case in the experimental animal (Clark 1969b).

Primary-like units in the CN and higher nuclei have similar response characteristics to those of AN fibers. The first interval is often the predominant one, and is equal to the period of the sound wave for frequencies below approximately 600 Hz (Clark 1996; Clark, Irlicht et al 1996). For frequencies above 600 Hz and below 5000 Hz, the intervals are second and higher order multiples of the period of the sound wave. At these higher frequencies the first interspike interval is not of the same duration as the period of the sound wave. Then there is no one-to-one correspondence between the interspike intervals, and the period of the sound wave. For example, the first interspike interval for a frequency of 1000 Hz could be 2 ms, and that is the same as the first and predominant interval for a tone of 500 Hz.

For this reason it was postulated by Rose et al (1967), and subsequently by others, that the intervals in a population of neurons, and not just individual neurons, are important in the coding of frequency (Fig. 5.4). Although the interspike intervals on individual fibers are not necessarily the same as the period of the sound wave (they are however multiples of it), in a population of neurons there will always be some firing in phase with each sine wave.

Furthermore, Figure 5.8 on the right shows that the intervals are not precise multiples of the period of the stimulus; that is, there is some jitter around the mode for each of the peaks, referred to as stochastic firing. This jitter increases and the phase locking decreases, with an increase in frequency. Although Rose et al (1967) showed phase locking in AN fibers could occur up to 5000 Hz, a more detailed examination by Johnson (1980) and Clark, Carter et al (1995) demonstrated well-defined phase locking only to frequencies of 1500 to 3000 Hz. In addition, Joris et al (1994a,b) showed for frequencies below 2000 Hz that synchronization was greater in the CN than the AN, but the reverse applied for frequencies above 2000 Hz. This finding suggests higher processing for temporal information below 2000 Hz.

It is important for the temporal coding of frequency that information is not lost along the central pathways. Phase-locking, however, may be compromised up the auditory pathway due to the additive effect of jitter from one stage of processing to the next (Koppl 1997). It may become ineffective at high frequencies when an additive jitter approaches the period of the sound wave. The jitter is due to variations in both the axonal propagation times (Anderson 1973), and fluctuation in the number and the timing of the presynaptic inputs. This will result in variability in the time it takes for the postsynaptic membrane potential to reach threshold (Koppl 1997; Burkitt and Clark 1999a–c). The jitter can also be due to the stochastic nature of the mechanisms involved in synaptic transmission (Koppl 1997).

However, the system is designed with end bulbs and membrane specializations for fast synaptic transmission, as seen in "bushy" cells that can maintain temporal information. For example, Manis and Marx (1991) showed that bushy cells in the anteroventral cochlear nucleus (AVCN) had short membrane time constants. The cells also have been shown to have very nonlinear voltage-current functions (Oertel 1983), thought to be due to the ion channel opening and lowering the membrane resistance (Rhode and Greenberg 1994). This would keep the membrane voltage near the resting potential so the cell could respond to the second stimulus more rapidly.

Phase locking and temporal coding may be affected by inhibition as it suppresses, in particular, the response of units at their best frequency (Caspary et al 1994; Paolini and Clark 1998a). Inhibition effectively raises the threshold, and this prevents noise affecting temporal firing, thus sharpening temporal integration. Paolini et al (2001) found loss of inhibition was correlated with a loss of synchronization of the responses in the CN of the deafened rat.

To further understand the temporal coding of sound frequencies, studies have recorded both the EPSPs, and the resulting action potentials from electrodes inside CN cells (Paolini et al 1997). Most previous studies have placed electrodes outside brain cells, where they detect action potentials but not EPSPs. The findings have demonstrated that the EPSPs occur mostly on successive sine waves up to a frequency of 2500 Hz. This is illustrated for 1000 Hz in Figure 5.10. This one-to-one correspondence with the interval between EPSPs and the period of the

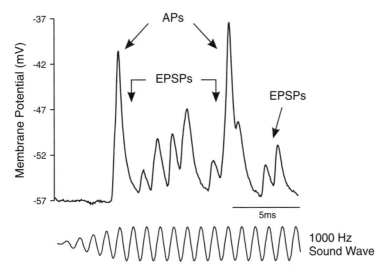

FIGURE 5.10. Temporal responses of excitatory postsynaptic potentials (EPSPs). Intracellular responses from globular bushy cells in the rat anteroventral cochlear nucleus. EPSPs or fast synaptic potentials and action potentials (APs) are phase locked to the stimulus, a pure tone of 1000 Hz at 90 dB sound pressure level (SPL). The EPSPs but not the APs usually occur on successive sine waves (Reprinted with permission from Paolini and Clark 1998b).

sound waves does not occur with action potentials as discussed above. Furthermore, the one-to-one correspondence holds over a wide intensity range. This is important because the perception of pitch is related to the coding of frequency, and pitch stays relatively constant as the loudness of the sound increases. The data provide more information on the importance of the temporal coding of frequencies up to approximately 2500 Hz.

The EPSPs not only initiate action potentials in the brain cells on which the nerve fibers converge, but also enable their time locking to the sound wave to be maintained and transmitted to other nerve cells in higher brain centers. The EPSPs also provide a link with the biomolecular status of the cell (Clark 1997).

As discussed below there were limitations in how well electrical stimulation could reproduce the temporal coding of frequency, and this required an emphasis on place coding and the need for multiple-electrode stimulation.

Coincidence Detection

It has been postulated that the coincident arrival of inputs at the nerve cell is important for the processing of a temporal code (Carney and Friedman 1998; Joris and Yin 1998). When the inputs to a neuron arrive during a defined time window, this allows temporal summation of the EPSPs and an action potential to occur. The time window defines, in particular, the maximum allowable phase delay between the inputs from the AN fibers to the CN cells for an action potential to be initiated (Fig. 5.4). In this way the firing of the neuron is related to the time window, and in turn the timing of acoustic events in the inner ear. Modeling studies have also shown that the convergence of inputs on a neuron, and variables such as synaptic efficacies, thresholds, and time constants can affect the degree of phase locking (Rothman et al 1993; Rothman and Young 1996).

Histological studies (Brawer and Morest 1975; Lorente de No 1981; Smith and Rhode 1987; Ryugo and Sento 1991) demonstrated the AN fibers converge on brain cells in the CN, and this provides a structural basis for the temporal coding of frequency information through the coincident arrival of potentials during a specific time window (Fig. 5.4). It was also necessary to know the number of excitatory inputs to produce an action potential during this time window, and the distribution of the converging fibers from sites on the basilar membrane. Studies have estimated that between two and six AN fibers converge onto a bushy cell (Brawer and Morest 1975; Lorente de No 1981; Ryugo and Sento 1991). The number of boutons and synapses, but not necessarily the number of AN fibers (each fiber may branch into a number of terminals), has been recorded from histological sections, and varies from 5 to 50 depending on the type of cell (Smith and Rhode 1987).

The length of the basilar membrane sending convergent fibers to each cell has been studied physiologically and with mathematical models by Fitzgerald et al (2000, 2001), Paolini et al (2001), and Burkitt (2000). It was first necessary to consider the time taken (phase delay) for a single crest of a steady-state sine wave to make its way along the basilar membrane to a specific site. The phase delay is calculated as follows:

$$D_{ph} \ (\omega) \ = \ \phi/\omega$$

where D_{ph} is the phase delay, ϕ is the corrected phase, and ω is the angular velocity $(2\pi f)$.

In contrast, group delay (D_{gr}) is due to the movement of the wave envelope, and corresponds to the first spike latency of the tone burst. It is the result of wave mechanics. Thus group delay corresponds to the local gradient of the phase versus frequency plot. The group delay is calculated as follows:

$$D_{gr} \ (\omega) \ = \ \frac{d\phi}{d\omega}$$

Ruggero (1980) describes the "signal front delay" (D_{sf}) as the closest measure to travel time:

$$D_{sf} \ = \ \lim_{\omega \to \infty} \frac{d\phi}{d\omega}$$

Ruggero (1980) emphasized that the delay is rarely constant; for instance, the basilar membrane's resonant properties introduce a frequency-dependent component to the phase of the responses. A study by Fitzgerald et al (2000) examined the average group delay from click latencies for units in the AN, AVCN, trapezoid body (TB), and MNTB, and found results similar to those of Ruggero (1980). A mathematical relationship was developed between the temporal code on which delay calculations are made and the response to spectrally complex stimuli such as clicks, and was able to predict the difference. This enabled a plot to be made of how synchronization varied (synchronization index) versus the frequency of stimulation and the summation distance of the neural input from the cochlea as shown below in Figure 5.32. The distance varied with frequency from 1.25 mm at 500 Hz to 0.75 mm at 1,000 Hz. These distances would be sufficient for future arrays to provide finer temporal and spatial information for the coding of frequency. This is discussed in more detail below (see Integrate-and-Fire Model).

The development of electrical stimulation patterns that can better utilize coincidence detection and provide stimuli to reproduce the phase delays of the basilar membrane traveling wave are likely to lead to better temporal coding of frequency with cochlear implants, as discussed below (Clark 1998a,c, 1999a–c).

Place Frequency Coding

The theory of place coding for the spectral analysis of sound frequencies was first annunciated by the physicist Hermann Helmholtz. It was demonstrated physiologically by Rose et al (1959) and Kiang et al (1965a), and has been well reviewed by Evans (1978, 1981), Aitkin (1985), and Irvine (1986). It has been especially important for the development of the multiple-channel cochlear implant, and is illustrated in Figure 5.11. Place coding depends on the place or site of neural excitation.

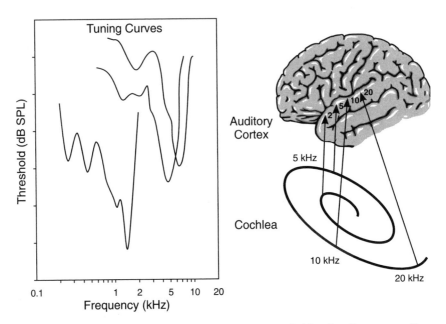

FIGURE 5.11. Place coding of frequency. Left: The thresholds of auditory nerve fibers versus frequency (tuning curves) demonstrating the filtering of frequencies by the cochlea. Right: Connections from the regions of best frequency response in the cochlea to neurons in the primary auditory cortex, demonstrating how a frequency scale is preserved (tonotopic organization). The intermediary centers are not shown for simplicity, but a tonotopic scale is preserved in each one (Reprinted from Clark GM, 1996, Electrical stimulation of the auditory nerve, *Clinical and Experiment Pharmacology and Physiology* **23**: 766–776, p 769, with permission of Blackwell Publishing.)

Cochlear Filtering

Sound is filtered by the basilar membrane of the cochlea, with the maximal displacement of the membrane for the high frequencies in the basal turn moving to the apical turn for the low frequencies. The filtering is due to the mechanical properties of the membrane that is stiffer at the basal end, as was discussed above (see Traveling Wave). This is the first filter. There is a metabolically active second filter (the outer hair cells) that acts to 50 to 60 dB sound pressure level (SPL) above threshold, to sharpen the filter.

Tonotopic Organization and Tuning Curves

Place coding is preserved throughout the auditory system via connections between the cochlea and each of the centers in the central auditory pathways so that an orderly frequency scale is preserved. The spatial or tonotopic organization of the brain is used with a cochlear implant electrode array to stimulate different fibers conveying place frequency information. The higher centers—CN, superior olivary complex (SOC), lateral lemniscus (LL), inferior colliculus (IC), superior colliculus (SC), medial geniculate body (MGB), and auditory cortex (AC)—have a cytoarchitecture, as discussed above and in Chapter 2, that preserves this frequency scale spatially, that is, it is tonotopically organized.

Frequency can be coded spatially by a number of mechanisms. (1) It may be simply the site in the brain where the fibers fire most rapidly (highest mean firing rate). This mechanism would have a restricted dynamic range for intensity, and be ineffective for the spatial representation of complex stimuli at high stimulus levels where there is considerable overlap in the excited populations of neurons. Alternative hypotheses, which overcome these objections, are coding through (2) the fibers that have the greatest change in mean discharge rate compared to their spontaneous rate (Kim et al 1990); (3) the fibers with the greatest difference in mean discharge rate across a local population (Whitfield 1979; Kim and Parham 1991); and (4) the degree of synchronization in a local population of fibers, which provides good representation of the frequency components of complex stimuli even at high intensities (Young and Sachs 1979; Jenison et al 1991).

Place coding as discussed above was first demonstrated physiologically by sampling cells in auditory nuclei and plotting their frequencies of best response (characteristic frequencies) spatially. The spatial distribution of a population of responding cells can also be seen by the administration of [^{14}C] 2-deoxyglucose, a radioactive tracer for regional cerebral metabolism (Serviere and Webster 1981). Restricted bands of activity are seen corresponding to each frequency and ordered sequentially (Fig. 5.12).

Lateral Suppression and Inhibitory Side Bands

Auditory neurons are often suppressed by a sound near their frequency of best response (Figs. 5.12 and 5.13). This lateral suppression sharpens the spatial excitation. With AN fibers the suppression is the result of cochlear mechanisms. The motion of the basilar membrane in the region of the neighboring frequency interacts negatively to reduce the drive to the inner hair cells (Rhode 1971; Patuzzi and Sellick 1984). In the CN and higher centers the suppression is more pronounced in some cells than in the AN fibers, and this is assumed to be due to inhibition (Greenwood and Maruyama 1965; Greenwood and Goldberg 1970; Evans and Nelson 1973). This is supported by the demonstration of inhibitory transmitter synapses and vesicles (Kane 1973), receptors (Cant 1981), and postsynaptic potentials (Oertel et al 1988) in the CN and higher brain centers.

In the ventral cochlear nucleus (VCN) a major source of inhibition is from the dorsal cochlear nucleus (DCN). This is frequency specific and mediated by glycine (Wu and Oertel 1986; Osen et al 1991; Saint-Marie et al 1991; Wickesberg et al 1994). This inhibition may "sharpen" frequency tuning by inhibiting the neural activity from units on either side. This enhances the presentation of spectral information on a place-coding basis.

The role of inhibition in shaping neural responses to sound was first studied with intracellular recordings by Starr and Britt (1970) and Britt and Starr (1976a,b). Intracellular experiments by Paolini and Clark (1999) have also demonstrated the presence of inhibition. When tones were presented above the high-

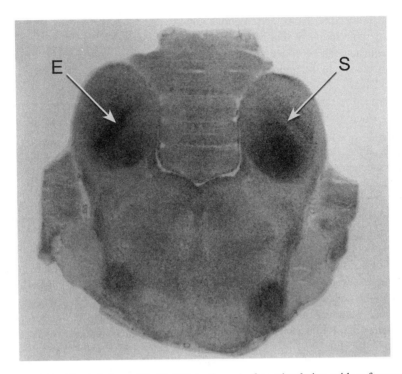

FIGURE 5.12. The inferior colliculi (IC) in the cat after stimulation with a frequency followed by [^{14}C] 2-deoxyglucose as a tracer for regional cerebral metabolism. This demonstrates place coding. Note the excitation band (E) in the left IC and a suppression band (S) in the right IC accompanied by spontaneous activity. (Reprinted with permission from Webster et al 1985).

frequency edge of the neuron's response area, strong inhibitory influences were seen. In combination with a tone presented at the best frequency, inhibition decreased the spike discharge rate and lowered the membrane potential toward resting levels.

Lateral inhibition is seen with stellate cells in the VCN (these cells have sustained chopping or onset chopping responses) that are excited from a localized region of the cochlea, and are finely tuned. The chopper units fire in a regular periodic fashion while the tone is on. The period is not related directly to the frequency, but can vary from cell to cell. Intracellular recordings from these cells demonstrate afterhyperpolarization, indicating inhibition as an explanation for the fine tuning. Lateral inhibition may also produce sharpening or better synchronization in the presence of noise. These stellate cells could thus play a role in the place coding of frequency and the spectral information in speech.

Suppressive bands have been shown to reduce the average firing rate of neurons in the CN in the presence of broadband noise (Rhode and Greenberg 1994). By suppressing the responses of neighboring cells, they enhance the presentation of spectral information on a place-coding basis. Temporal information of great im-

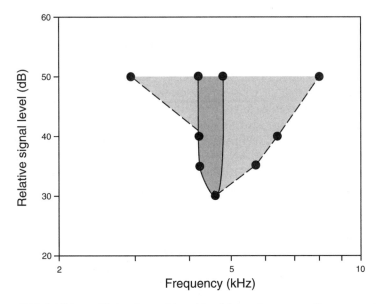

FIGURE 5.13. Inhibitory side bands on either side of the excitatory area for a neuron in the dorsal cochlear nucleus. (Evans and Nelson 1973, with permission of Springer-Verlag.)

portance in detecting speech sounds in the presence of noise is also enhanced by inhibition as explained above (see Temporal Frequency Coding). Furthermore, a small number of stellate cells have onset responses that are very broadly tuned (three octaves), and they are innervated from a wide area and terminate broadly in the DCN in particular. They are excited by a wide band of frequencies and consequently are thought to act as broadband inhibitors.

The technique of using [14C] 2-deoxyglucose as a tracer for regional cerebral metabolism (Serviere and Webster 1981) has also been used to demonstrate lateral suppression in the central auditory pathways. When a pure tone was presented in the presence of broadband noise (Brown et al 1992), there was reduced background activity around the frequency band (Fig. 5.12).

The inhibitory side bands of units in the central auditory pathways could be used to advantage in the development of future cochlear implant speech-processing strategies and electrode arrays to improve the transmission of temporal and place information. This would require stimulation of electrodes on either side of the one providing the spectral information. Inhibitory mechanisms, however, may be lost in deafness, and this may account for the poor place coding in children not exposed to sound at an early age.

Intensity Coding

The coding of sound intensity, reviewed by Irvine (1986) over the normal range of 100 to 120 dB, in part may be due to the mean rate of unit firing of cochlear nerve fibers or the population of neurons excited.

Rate of Firing

One difficulty for rate of firing as the sole coding mechanism is that central auditory neurons have only a 20- to 50-dB range from threshold to the point where the firing saturates (Kiang et al 1965b). Nevertheless, the normal dynamic range could be coded if there is a population of units with low and high thresholds. In fact, auditory nerves have been classified into two groups: those with high spontaneous rates having low thresholds, and those with low spontaneous rates having high thresholds (Liberman 1978). Further evidence that higher auditory neurons receive inputs from AN fibers with different thresholds comes from recordings from neurons in the DCN that respond with increasing firing rates over a dynamic range up to 100 dB above threshold (Evans and Palmer 1975).

Population of Responses

There is good evidence that the number of neurons excited is very important in the coding of intensity or its psychophysical correlate loudness. Dobie and Kimm (1980) and Charlet de Sauvage et al (1983) found the amplitude of the field potential increased linearly with the current level in monkeys and guinea pigs, respectively. Smith et al (1994) also discovered that there was good correlation between the electrically evoked brainstem response and the behavioral threshold. The same is also seen for the auditory brainstem responses (ABRs) in the human. The growth of the field potential recorded from the experimental animal in response to increasing intensities of electrical stimulation is illustrated in Figure 5.14. These and other data support the hypothesis that intensity is coded according to the population of neurons excited.

Complex Sound and Speech Coding

The coding of complex sounds is required for communication in many vertebrates. This is also necessary for the "bottom-up" processing for human speech. Processing occurs first in the VCN, where the homogeneous response patterns of the AN fibers are received, and then in the higher centers where the responses are more complex.The coding depends on the patterns of innervation (including inhibitory inputs), the degree of convergence, and intrinsic membrane properties. As a result the neuronal networks have different capacities to extract the temporal and spectral features of a complex stimulus, such as speech. Even at the level of the VCN there are diverse response patterns from different types of cells, which can be classified as primary-like, chopper, and onset.

Complex Sounds

Frequency and amplitude-modulated sounds are one group of complex signals of importance to the coding of speech. For this reason it is desirable they be replicated with electrical stimulation in speech-processing strategies for deafness. In the CN responses of most cells to modulated sound can be predicted from the response to steady tones (Evans 1975). Thus modulation of frequency from the

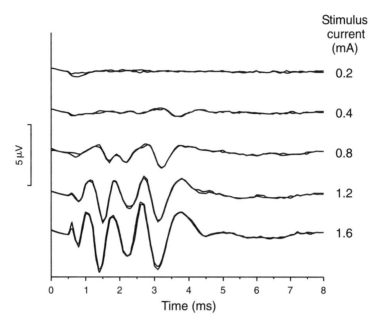

FIGURE 5.14. Population responses. A series of electrically evoked brainstem responses (EABRs) produced by a bipolar + 1 stimulating electrode close to the inner wall of the scala tympani of the cat and recorded differentially with subcutaneous scalp needle electrodes. The amplitude of the waves is plotted from increases in current level from 0.2 to 1.6 mA (Reprinted from Shepherd et al. 1993, with permission from Elsevier.)

high frequencies into the excitatory area of the unit produced a roughly sinusoidal distribution of discharge rates. A small number of units, however, in the DCN produced asymmetrical responses (Erulkar et al 1968; Moller 1971). This was due to the asymmetry of the inhibitory side bands. At higher modulation rates (50–300 Hz) the bandwidth of the unit's response area becomes narrower. Therefore, at the first auditory nucleus there are cells that demonstrate some selectivity in their response to modulated sounds. This becomes much more marked in the higher levels of the auditory system, particularly in the AC, where there are units that are very specific in their sensitivities to the direction and depth of modulation (Bogdanski and Galambos 1960; Suga 1963, 1965; Evans and Whitfield 1964). Phillips and Hall (1987) discovered units responsive not only to amplitude modulated (AM) tones but also to the rate of change and the base sound intensity level. The response could be explained from the inhibitory side bands for the unit. As the sound level increased, it excited the neighboring area with a suppression of the response. Whitfield and Evans (1965) also showed that certain cortical cells responded to frequency modulated (FM) tones and not to pure tones, and the response was directionally selective.

Furthermore, in the AC the tuning curves were found by Oonishi and Katsuki (1965) to vary in shape from flat through irregular and multipeaked to sharp. This would indicate multifrequency input and more complex processing. However, the tuning curves were found to be similar in shape and bandwidth to those at lower levels by Calford, Webster et al (1983).

Speech

Much of the research on the coding of speech as a whole has been carried out on the AN, and many studies have used synthetic speech. A lot less is known about how natural speech is processed and transformed by the central auditory nuclei, and how the critical temporal and spectral features that identify a speech sound segment (phoneme) are extracted.

The responses of AN fibers to the synthesized vowels /ɑ/,/ɛ/,/ɪ/,/u/ were examined by Sachs and Young (1979). This was done to see how the responses to the spectrally more complex vowels compared to the responses to two-tone stimuli. The study showed that there were formant frequency peaks for normalized discharge rate. At high intensities the peaks disappeared due to rate saturation and two-tone suppression. This raised the question of how place coding alone could convey speech information at high intensities, and suggested that temporal coding was also involved.

A study was undertaken by Delgutte (1984) and Delgutte and Kiang (1984) to help determine whether the formant pattern and fundamental (voicing) frequency could be represented in the fine time patterns of the discharges of the AN fibers. Results of the analysis of period histograms showed the intervals between action potentials were almost always harmonics of the vowel fundamental frequency. They were either the fundamental frequency or one of the formants or the fiber's characteristic frequency. The relative contribution of these frequencies depended on a fiber's characteristic frequency relative to the formant frequencies. It was found that (1) if the characteristic frequency was below the first formant, the largest response components were harmonics of the fundamental frequency closest to unit's characteristic frequency; (2) in the region around the characteristic frequency of the first formant, this formant and its harmonics were the largest components; (3) an intermediate region between the first and second formants had prominent components of both the fibers characteristic frequency and the fundamental frequency; (4) in a region centered around the second formant, the harmonics closest to the second formant were dominant; and (5) in a high-frequency area there were multiple intervals at both the formant and fundamental frequencies. These results suggested that in addition to place coding, the temporal coding of speech information is fundamentally important and is likely to be so in noise. It also indicated that for electrical stimulation and speech-processing strategies for implant patients, information on the fundamental frequency should be presented across the electrodes used for place coding of frequency.

Much less is known about the coding of the complex temporal and spectral features of consonants. There is a need to examine the ability of the VCN to extract consonant features from naturally spoken speech. Research by Clarey and Clark (2001) and Clarey et al (2001) has shown the chopper cell in the VCN codes the voice onset time (VOT) of syllables with great accuracy. The VOT, as discussed in Chapter 7, is the time between the release in the closure of the vocal tract to produce a plosive such as /b/ or /g/ and to the onset of the voicing. The intracellular recordings of Clarey et al show hyperpolarization during the period of the VOT, which could result from inhibitory side bands that sharpen the discharge peak at the onset of the burst and thus the salience of the VOT. The octopus cells in the posteroventral cochlear nucleus (PVCN) are also finely tuned to provide phase information. They are sensitive to phase and could be coding voicing.

The studies on the AC by Evans and Whitfield (1964) laid the foundation for studies on species vocalizations. A strong response to a natural call does not mean the unit has extracted this feature, as it may evoke strong responses regardless. One way to overcome this difficulty is to present the stimulus backward. In a study by Wang et al (1995), it was found that natural vocalizations of marmoset monkeys produced stronger responses in the primary AC than did equally complex sounds such as a time-reversed call. The population of cells also had a clearer representation of the spectral shape of the call.

Sound Localization

The direction of a sound is coded primarily through interaural differences in intensity or the time of arrival of the signal (phase). The spectral differences introduced by the pinna for sound from various locations are also important, especially if a person has hearing in only one ear. See Chapter 6 for more details. The coding takes place in cells that have binaural inputs, so that an interaction can occur as a result of interaural intensity or timing differences.

Cells that code the information have predominantly inhibitory (I) inputs from either the contralateral (IE cell) or ipsilateral ears (EI cell), and excitatory (E) inputs from the other ear. The convention is to refer to the input from the contralateral ear first. Coding may also occur through excitatory inputs from both the contralateral and ipsilateral ears (EE) cells (Goldberg and Brown 1969).

Interaural Intensity Differences

The binaural coding of inter-aural intensity differences (IIDs) by units in the SOC was demonstrated by Goldberg and Brown (1969). EI and IE units were relatively insensitive to the base binaural intensity, but sensitive to IIDs (Hall 1965). The sensitivity of EI cells to IIDs was seen in the lateral superior olive (LSO) (Tsuchitani and Boudreau 1967; 1969; Boudreau and Tsuchitani 1968, 1970; Caird and Klinke 1983). EI units in the ICC were found to respond over a range of IIDs (Hind et al 1963; Rose et al 1963; Geisler et al 1969; Semple and Aitkin 1979). A curve was fitted to normalized IID functions from 43 EI cells deep in the SC of the cat (Fig. 5.15) (Wise and Irvine 1985). This shows changes in response to increases in intensity from the contralateral side, thus coding for sound localization in the contralateral azimuth. The explanation is that as they were EI cells with a strong stimulus from the ipsilateral inhibitory ear, there would be no response when stimulating this side, but a graded response occurred to variations

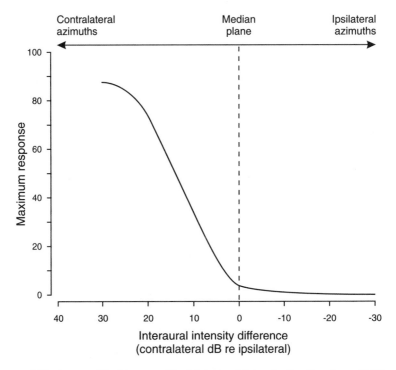

FIGURE 5.15. A curve fitted to normalized interaural intensity function from 43 EI cells deep in the SC of the cat. The maximal brain cell response is plotted for interaural intensity differences for the contralateral and ipsilateral side. When there is a strong ipsilateral input the cell does not fire. When the strength of the excitatory input from the contralateral side increases, then the cell fires with a graded response to each portion in the contralateral field. (Wise and Irvine 1985, reprinted with permission of the American Physiological Society.)

in the intensity from the excitatory contralateral input. In the SOC and IC there was a smaller proportion of IE cells to process information from the ipsilateral side of the body.

In addition to the EI and IE cells that were responsive to IIDs, Goldberg and Brown (1969) found that EE cells were generally not responsive to IIDs, but sensitive to changes in the overall intensity. They had sharper rate/intensity functions, and a wider dynamic range than normal stimuli.

The studies referred to above show there are units in both the SOC and IC that code information from either half of the field midway between each ear. Studies have also shown that units in the LL, SC , MGB, and AC also code IIDs.

Interaural Time Differences

The processing of interaural time differences (ITDs) involves disparity in the arrival of transients as well as the phase of the ongoing pure tones. There is evidence that these are both processed by two different mechanisms. The cells in the SOC sensitive to transients are those that are excited by one ear and inhibited by the other (i.e., EI or IE cells). As with intensity, the cell is excited maximally when the excitatory ear leads, and suppressed maximally when the inhibitory ear leads (Galambos et al 1959; Moushegian et al 1964a,b; Hall 1965). Some cells (EI/IE) were sensitive to both ITDs and IIDs, and with these one could be traded against the other; that is, shortening the time of arrival at one ear could be counterbalanced by a reduction in the intensity (time/intensity trading). Caird and Klinke (1983) found that some IE cells in the LSO had similar IID and ITD functions, the range for IID being 30 dB and the ITD range being 2 ms, which was greater than the 300 to 400 μs for sound localization. Kuwada and Yin (1983) report that most cells in the IC that were sensitive to interaural phase were insensitive to interaural transients, and this further supported the view that the coding of interaural transients and phase are through different mechanisms. But only a small proportion responded differentially as a function of the direction of the ITD variation, and together with the data from Yin and Kuwada (1983) this suggests that the coding of the direction of sound movement is at a higher level, and presumed in the SC and AC.

The coding of interaural phase difference was reported in the IC by Rose et al (1966), as was done previously for the medial superior olive (MSO) by Moushegian et al (1964a,b). Rose et al (1966) found that when sine waves were presented to each ear there was an optimal phase difference that gave a maximal response [the characteristic delay (CD)]. This was consistent with the coincidence detection model of Jeffress (1948). The model is illustrated in Figure 5.16. The model postulates there are units with different delay lines from each ear. When the delay line is such that a certain phase difference between each ear provides maximal excitation, phase difference is coded on a place basis. Furthermore, a study of MSO and ventral nucleus of the trapezoid body (VNTB) units in the dog by

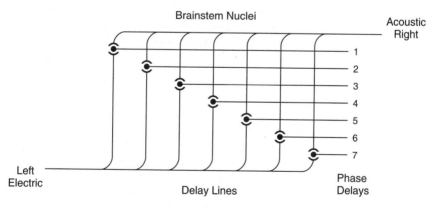

FIGURE 5.16. A delay line where phase differences between each ear are converted to place of excitation. This is the basis of the model of Jeffress (1948). This model is relevant to bilateral cochlear implants or bimodal speech processing with hearing in one ear and electrical stimulation of the other.

Goldberg and Brown (1969) showed the maximum discharge occurred for a binaural phase delay that corresponded to the difference between the monaural phase locking for each ear. These data were also consistent with the coincidence detector model for interaural phase differences proposed by Jeffress (1948). According to these studies and the coincidence detection model of Jeffress, the timing and site of origin of the inputs to each ear are important for binaural processing of temporal information with electrical stimulation.

Evidence for the role of the MSO in coding ITDs comes from a study on a patient who had bilateral cochlear implants and poor interaural temporal discrimination. The patient died 13 years after the first implant, and the section of the brainstem showed that the cell density and volume of the MSO were significantly less than the MSO of a person of the same age with a single cochlear implant (Yukawa et al 2001) (see Chapter 3).

In addition, Yin and Kuwada (1983) found low-frequency units in the IC with interaural periodic delay functions responding to stimuli with small differences in frequency to produce beating. Most of these cells were insensitive to onset or transient disparity. The majority of the cells were excitatory (EE), but there were a variety of other types. Furthermore, only a small proportion of the binaural phase-sensitive cells exhibited monaural phase locking (Kuwada et al 1984), and this suggested earlier processing in the SOC. The majority of cells responded when the contralateral ear was leading and thus coded the localization of sound in the contralateral half of space.

When low-pass noise was presented to low-frequency units in the IC, the phase delay varied the interaural response curves and had a periodicity that followed the cell's characteristic or best frequency (Geisler et al 1969). However, when uncorrelated noise with the same spectral composition was presented, there was no evidence sensitivity to the delays (Yin et al 1986). This indicates phase sensitivity depends on the fine structure of the signal, and cross-correlation of the signal could explain the coding. It also serves to emphasize the importance of reproducing the fine temporospatial patterns of response with electrical stimulation to localize and understand speech in the presence of background noise. This is especially relevant to the design of bilateral cochlear implants or bimodal speech processing.

Higher Level Processing

The physiological studies on the SOC and IC for the coding of sound localization referred to above showed basic mechanisms for the processing of IIDs and ITDs. They did not reveal a specific response for a particular location or demonstrate how a moving sound could be detected. Evidence was seen for this in the barn owl by Knudsen and Konishi (1978). Recordings were made from the mesencephalicus lateralis dorsalis (MLD), the avian homologue of the IC, and this showed cells that responded to sounds arising from restricted areas in space.

In mammals a map of auditory space was found in the deep and intermediate layers of the SC in the guinea pig (Palmer and King 1982; King and Palmer 1983) and the cat (Middlebrooks and Knudsen 1984). This map for auditory space resembled that for visual space in the superficial layers of the SC. In a study on monkeys a similar organization was found. For some cells the position of the auditory receptive field was affected by the eye position. Thus a discrepancy between the position of a sound and the visual axis was mapped onto the SC.

The responses of many single units in the primary cortex of monkeys and cats were influenced by the interaural differences in time (ITD) and intensity (IID) (Brugge et al 1969; Brugge and Merzenich 1973). The unit's responses may be facilitatory (EE) or suppressive (EI or IE).

Behavioral studies after bilateral ablation of the primary AC in the cat, monkey, and other experimental animals have helped establish the importance of the AC in sound localization. The physiological studies referred to above have demonstrated how sound is converted from bottom-up processing into appropriate information for final coding by the AC. Studies by Neff (1968) and Strominger (1969) demonstrated that after bilateral ablation of the primary AC the animal's ability to localize sound was grossly impaired or reduced to chance level. Sound localization and lateralization were affected, as were detection of temporal patterns and order, change in duration of sound, and change in the spectra of complex sounds (Neff et al 1975).

Coding and Perception

Pitch and Loudness

Frequency and intensity coding correlate predominantly with pitch and loudness perception, respectively, and these sensations underlie the perception of speech and environmental sounds, although the relation is not well understood. Nevertheless, an adequate representation of frequency and intensity coding using electrical stimulation is important for cochlear implant processing of speech and other sounds. The time/period (rate) code for frequency results in temporal pitch, and the place code in place (spectral) pitch.

With sound it is difficult to determine the relative importance of the time/period and place codes in the perception of pitch, as the underlying frequency codes operate together when sound stimulates the cochlea. With electrical stimulation of auditory nerves the two codes can be reproduced separately to study their relative importance.

Although frequency and intensity coding correlate predominantly with the perception of pitch and loudness, respectively, frequency coding may have an effect on loudness, and loudness coding on pitch. For example, increasing intensity increases the loudness, and there may be a small change in pitch. For frequencies below 2000 Hz there can be a maximum 5% decrease in pitch with an increase in intensity, and a 5% increase in pitch for frequencies above 4000 Hz (Moore 1997).

Sound Localization

The responses of units in the auditory pathways to IIDs and ITDs are consistent with the findings from psychophysical studies, and form the basis for analyzing the effects of electrical stimulation on the cochlear nerve for the restoration of hearing in deafness with bilateral cochlear implants or bimodal speech processing (an implant in one ear and a hearing aid in the other). For example, the perception of IIDs was better preserved than ITDs for electrical stimulation with bilateral cochlear implants (see Chapter 6). Understanding the coding of binaural excitation is increasingly important with the introduction of bilateral cochlear implants, and bimodal speech processing, especially to improve hearing signals in noise. The data indicate the importance of stimuli being presented from the same site in each ear. They also suggest that coding strategies need to emphasize the interaction of both IIDs and transient ITDs if the phase ITDs cannot be readily transmitted.

Neural Plasticity

Learning to use the perceptual information provided by the cochlear implant, which only partially reproduces the coding of sound, depends in part on the plasticity of the responses in the central auditory nervous system, especially in children born with little hearing. In the experimental animal there are two types of plasticity. The first is the development of neural connections within a critical period after birth—developmental plasticity. The second results from a change in the central representation of neurons in the mature animal after neural connectivity has been established—postdevelopmental plasticity.

Developmental Plasticity

Evidence of a critical period for changes in the central auditory system in response to surgical destruction of the cochlea was demonstrated in the ferret, where a marked loss of neurons in the CN occurred after its ablation 5 days after birth (Moore 1990b). However, ablation of the cochlea 24 days postpartem (i.e., a week before the onset of hearing) had little effect. This was discussed in more detail in Chapter 3.

   An example of developmental plasticity is the increase in the number of projections from the CN to the ipsilateral IC when the cochlea on the opposite side was destroyed in the neonatal gerbil (Nordeen et al 1983). In this case there was a critical period that extended to 40 to 90 postnatal days, but did not occur in the adult animal. A similar phenomenon was demonstrated in the ferret (Moore and Kowalchuk 1988), where the critical period for the neural modeling extended to postnatal days 40 to 90, that is, well beyond the onset of hearing. There was a substantial increase in the expression of the growth-associated protein GAP-43, an indicator of synaptic remodeling (Illing et al 1997). Evidence that the changes were not due to ablation of the cochlea per se was demonstrated by Born and Rubel (1988), who found that they still occurred when there was lack of neural activity in the auditory nerve due to a neural blocker. This was supported in another study in which ferrets were unilaterally deafened with a conductive loss without damage to the cochlea, and again the same changes were seen (Moore et al 1989).

The neural modeling changes described by Nordeen et al (1983) and Moore and Kowalchuk (1988) were accompanied by lower response thresholds, greater peak discharge rates, and shorter minimum response latencies (Kitzes 1984; Kitzes and Semple 1985). In addition, the marked developmental changes in the neural pathways were not seen after bilateral ablation of the cochleae (Moore 1990a), indicating that with unilateral loss there was an upregulation of connectivity on the active side. Furthermore, downregulation with loss of stimulation was seen by Hardie et al (1998), who found the density of synapses on central neurons was halved in animals with bilateral experimentally induced deafness, but not if the deafness was unilateral. Evidence was also found that hearing loss also involved changes in the type of transmitters at the synapses (Redd et al 2000). Furthermore the biological basis for these changes could be the effect of an anti-apoptotic gene *bcl-2* (Mostafapour et al 2000).

The fact that an increase in connections occurred with loss of hearing in one ear rather than in both indicates that the innervation of the cells in the IC was due to a competitive interaction between the afferent projections from each ear during development. This suggests that if a person had a congenital hearing loss in one ear and then became deaf, it would be preferable to insert a cochlear implant in the more recently deafened ear. However, an early patient in Melbourne had the congenitally deaf ear implanted, and her speech perception results with the University of Melbourne's F0/F2 strategy were above average. This indicates that there were sufficient connections for transmitting information through electrical stimulation, that higher processing above the level of the IC was of great importance, or that the results on the experimental animal do not apply to the human.

The experimental animal findings could also indicate that implanting one ear in a child during the developmental stage could later limit the ability to use information from two ears, should that be shown to be of benefit. This question is unresolved. The above results, however, do support the clinical findings that psychophysical and speech results are better if an implant is undertaken at an early age (Dowell et al 1986, 1995; Clark, Blamey et al 1987).

Evidence of plasticity is also seen in the cortex. When cats are visually deprived at birth, they are superior in auditory location tasks than sighted cats (Korte and Rauschecker 1993; Rauschecker and Korte 1993). This is discussed further in a review (Kral et al 2001b). The physiological basis for this effect was increased tuning in the anterior ectosylvian area (a higher order region of the cortex), and the auditory area expanded to areas normally receiving only visual stimuli. However, only a few units responded to auditory stimuli in the primary visual cortex (Yaka et al 1999). Furthermore, the higher auditory cortex (for example, AII in cats) has greater plasticity than the primary auditory cortex (Diamond and Weinberger 1984; Weinberger et al 1984), and in congenital auditory deprivation may be recruited for the processing of other sensory modalities. This is supported by the observation that deaf subjects perform better in visual tests than do hearing subjects (Neville and Lawson 1987a,b; Levänen et al 1998; Marschark 1998; Parasnis 1998). It has also been observed with implantation in prelinguistically deaf patients that if the amount of activation in higher order auditory centers is increased with visual stimuli, in particular sign language, as determined by positron emission tomography (PET), then speech perception will be poor (Lee et al 2001), and the older the person the poorer the speech results (Dowell et al 1985; Blamey et al 1992).

The mechanisms underlying the above changes are assumed to be long-term potentiation (Bliss and Lomo 1973) and long-term depression (Ito 1986). The different susceptibilities to long-term potentiation and depression are based on change in glutamate receptors. Inhibition seems to be related to the sensitive periods in development. In the auditory cortex the γ-aminobutyric acid receptor cell count increases at the end of the sensitive period (Gao et al 1999), and is responsible for its termination. Nerve growth factors and brain-derived neurotrophic factors are crucial for cortical development and influence the duration of sensitive periods in cats and rats (Galuske et al 1999; Pizzorusso et al 1999; Sermasi et al 1999). They participate in stimulus-dependent postnatal development. Their production depends on activity, and they affect synaptic plasticity and dendritic growth (Boulanger and Poo 1999; Caleo et al 1999). Further understanding of the plasticity of the central nervous system is revealed through experimental animal studies using electrical stimulation, and are also directly relevant to cochlear implantation (see Plasticity, below).

Postdevelopmental Plasticity

Postdevelopmental plasticity was demonstrated in the mature guinea pig when an area of the cochlea was destroyed, and the corresponding area of the brain, in particular the cortex, was shown to have increased representation from the neighboring frequency regions (Robertson et al 1989). This postdevelopmental plasticity was probably due to the loss of inhibition that normally suppresses the input from neighboring frequency areas. It was shown in the cat that there was reorganization of the topographical map in the primary AC contralateral to the lesioned side, but the cortical field was normal for ipsilateral excitation from the unlesioned cochlea (Rajan et al 1993). This reorganization could also have been due to an increase in dendrite length in spiny-free neurons. McMullen et al (1988) found the dendrite length increased by 27% in the contralateral cortex compared to littermate controls. In addition, it was found by Snyder et al (2000a) that changes occur at the level of the IC and soon after a lesion of the spiral ganglion. IC units previously tuned to the frequency corresponding to the site of the lesion became less sensitive to that frequency, but tuned to the frequencies at the edge of the lesion. Furthermore, behavioral training can modify the tonotopic organization of the primary AC in the primate. Recanzone et al (1993) report an increase in cortical representation for frequencies where there was improved discrimination. These data underpin the clinical findings that speech perception results improve in postlinguistically deaf adults up to at least 2 years postoperatively.

Electrophonic Hearing (Electrical Stimulation of the Cochlea)

Electrophonic hearing results from electrical stimulation of ears with residual inner ear function. Studies in the 19th and early 20th centuries, as discussed in Chapter 1), were undertaken to see whether electrical stimulation could induce hearing. Although these early studies showed that electrophonic responses occurred with a functioning cochlea, more research was needed to see specifically how they were generated. In particular, were they due to direct stimulation of inner hair cells by the spread of current along the cochlea, or to indirect stimulation through the propagation of a traveling wave arising near the electrodes? In the latter case, was a basilar membrane traveling wave produced by direct excitation of outer hair cells near the electrode array or by electromechanical transduction? This also had clinical significance in determining whether it would be possible to stimulate residual hearing in an implanted ear electrophonically. Alternatively, would electrophonic hearing interact constructively or destructively if a hearing aid were used in addition to the implant?

Mechanisms

To see if electrophonic hearing was due to direct electrical stimulation of the inner hair cells from the distant spread of current within the cochlea, the masking of acoustic probes with electrical stimuli presented at different rates and intensities was examined by McAnally et al (1993) with the experimental design shown in Figure 5.17. If direct stimulation of inner hair cells occurred, then the degree of masking would decrease with current spread along the cochlea and be less for lower probe frequencies.

The masking results (McAnally et al 1993) using sinusoidal monopolar electrical stimuli at the round window showed peaks of masking at the probe fre-

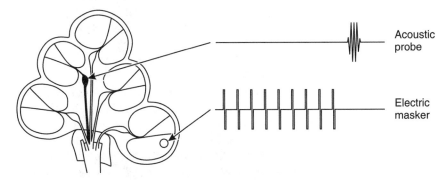

Acoustic probe

Electric masker

FIGURE 5.17. The masking of acoustic probe tones of different frequencies by the electrical masker consisting of a burst of electrical pulses at different rates (McAnally and Clark 1994, reprinted with permission of Taylor and Francis).

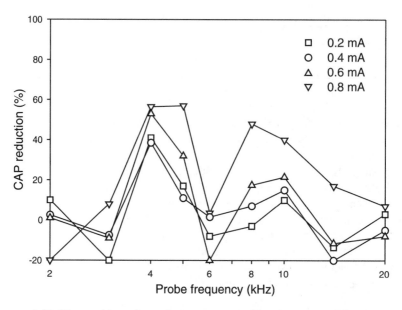

FIGURE 5.18. The masking of a probe tone at a masking frequency of 4000 Hz. The percentage reduction of the compound action potential (CAP) (field potential) from the auditory nerve and brainstem as a measure of the degree of masking of probe tones by the electrical stimulus at 4000 Hz is shown. Notice that the masker frequency produces maximum masking for a 4000-Hz tone and its first harmonic (McAnally et al 1993, reprinted with permission of Elsevier Science).

quencies (e.g., 4000 Hz) (Fig. 5.18). As the degree of masking did not diminish with a decrease in the probe frequency the results suggested that electrophonic hearing was due to the propagation of a wave along the basilar membrane, and not direct spread of current and stimulation of inner or outer hair cells.

Further evidence that electrophonic hearing was not due to direct stimulation of the outer and inner hair cells was provided by a similar study using intracochlear bipolar stimulation. As the current spread is more localized with bipolar than monopolar round window stimulation, masking would be expected to be maximal near the site of stimulation if there was a direct effect on the hair cells. However, again the masking of the probe was maximal when the probe frequency was the same as the rate of the masker, and not related to the current spread.

The next question was: could electrophonic hearing be due to local excitation of the outer hair cells and then indirect excitation of inner hair cells due to a traveling wave propagated to the corresponding frequency site? To help answer this question, the masking study was performed after damaging the hair cells in the region of the electrode by overstimulation with a 10,000-Hz tone (McAnally and Clark 1994). The results showed there was little change to the peak of masking after overstimulation. This suggested that electrophonic hearing was not due to direct stimulation of outer hair cells with the propagation of a traveling wave, but a traveling wave produced by another mechanism, for example, electromechanical transduction, whereby the electrical field induced a vibration in the basilar membrane.

Electrophonic Hearing and Cochlear Implantation

The above studies were undertaken not only to elucidate the mechanisms underlying electrophonic hearing but also to determine their relevance to cochlear implantation. In the first instance electrophonic hearing could affect the interpretation of results from direct electrical stimulation with a cochlear implant. The perception of sound in part could be due to the residual hearing. On the other hand, it might be possible to take advantage of electrophonic hearing to improve results by implanting ears with some hearing. Alternatively, if a hearing aid was used in combination with electrical stimulation, the two could interact constructively or destructively.

Biphasic Intracochlear Electrical Stimulation

The research on the mechanism of electrophonic hearing (McAnally and Clark 1994) used sinusoidal electrical stimuli chiefly because sound has the same waveform. It was assumed the mechanism producing electrophonic hearing was influenced by the wave shape, and therefore a sine wave would be more effective.

With cochlear implants, however, it is preferable to use biphasic pulsatile stimuli for more precise control of the stimulus parameters, and to ensure charge balance. The above experiments were therefore repeated with pulsatile stimuli (McAnally and Clark 1994). The results were very similar to those for sinusoidal stimuli. This suggested that it might be possible to use electrophonic hearing with a speech-processing strategy for a patient with an implant in an ear with some residual hearing. The strategy would stimulate not only cochlear nerve fibers directly but also residual hair cells. Alternatively, if an acoustic stimulus was used in addition to the implant, electrophonic hearing could interfere with the acoustic signal.

Preservation of Hair Cells with Cochlear Implantation

The use of residual hearing in an implanted ear through electrophonic hearing or independently from associated acoustic excitation is biologically feasible because of research, as was discussed in more detail in Chapter 2, where the preservation of hair cells has been demonstrated in the implanted cochlea of the experimental animal (Shepherd et al. 1983a,b, 1994; Ni et al 1992). In cats and monkeys, hair cells were preserved apical to scala tympani arrays with and without electrical stimulation unless there was infection or marked trauma. Physiological and psychophysical studies in the experimental animal also demonstrated that functioning hair cells were present after cochlear implantation (Clark, Kranz et al 1973; Black et al 1983a; O'Leary et al 1992; McAnally and Clark 1994). The behavioral study by Clark, Kranz et al (1973) on cats showed that electrophonic hearing could be induced in the chronically implanted animal at least up to 800 Hz. Research by Black et al (1983a), using brainstem response audiometry, indicated first that if implantation was carried out gently without loss of perilymph postoperative tone, pip thresholds were within 10 dB of preoperative values. It was found that click-evoked ABRs exhibited a normal latency-frequency function following implantation. This indicated that implantation per se did not disturb the frequency-place mechanics of the basilar membrane.

Pulse Shape and Current Spread to the Apical Turn of the Cochlea

Research was needed to determine the energy transfer to the low and middle frequency regions of the cochlea from an electrode in the basal turn, to assess the extent to which electrophonic hearing could be induced in this frequency range. This was crucial for those likely to require an implant. Although they usually have some hearing in the low and middle frequencies, the thresholds are higher than normal and sufficient energy from electrophonic hearing would be needed to excite these hair cells. A study was carried out by forward masking a probe tone with an electrical masker (McAnally et al 1997), as described in the studies above. The data indicated that the mechanical response would not be confounded by electrical stimulation of the cochlear nerve. Furthermore, it had been shown that the spectrum of a train of biphasic pulses is composed of harmonics of the pulse rate, and that the distribution of the energy among these harmonics depended on the duration of each pulse. Thus to maximally excite hair cells tuned to 250 Hz required a pulse duration of 4 ms. With this duration there would be less time to stimulate other channels nonsimultaneously if the F0/F1/F2 or multipeak strategies were used. It was postulated that stimulation with charge-balanced asymmetric pulses could lead to nonlinear electromechanical transduction and to significant energy in the low frequencies. The study compared the masking of both symmetrical and asymmetrical pulses, but there was no significant difference. The findings, therefore, suggest that in patients implanted with some residual hearing in the low frequencies, shaping the pulses does not make use of the residual hearing, and this requires the presentation of sound as well, provided the hair cells are not damaged by the implant. However, the data suggest that with the use of low constant rates of stimulation as occurs with the SPEAK strategy (see Chapter 7), electrophonic stimulation might be possible and might improve the quality of the percept.

Electrical Stimulation of the Cochlear Nerve

Neurophysiological studies on the coding of sound were the original basis for investigations on the effects of direct electrical stimulation of the auditory pathways. Electrical stimulation has also become an important tool for studying neural mechanisms independently of acoustic stimuli.

Temporal Coding

Reproducing the temporal coding of the frequency of sound with electrical stimulation is of fundamental importance in developing speech-processing strategies for cochlear implants, and studies in the experimental animal were essential before doing clinical studies on patients.

Initial Unit and Field Potential Studies

Experimental studies were first undertaken to determine to what extent the neurons in the brainstem centers could follow rate of electrical stimulation of AN fibers in the cochlea without being suppressed by inhibitory mechanisms (Clark 1969c, 1970a,c). This was necessary, as the coding of frequency and its discrimination in the behavioral experimental animal had been shown to occur up to the level of the ICs in the midbrain (Goldberg and Neff 1961). It was important to know if electrical stimulation with a cochlear implant could reproduce a rate or timing code for speech frequencies. If this were the case, a simpler single channel would have only been required rather than a more complex multiple-channel implant. With this aim in mind, studies were undertaken in the experimental animal (Clark 1969c, 1970a,c) to record unit responses in the nuclei of the SOC including the MNTB to different rates of electrical stimulation. Responses were recorded from the SOC in the brainstem rather than in the AN, as this permitted an evaluation of the effects of central processing mechanisms important in coding frequency, in particular inhibition. The data showed that electrical stimulation above 200 to 500 pulses/s did not reproduce the same sustained firing rate or firing patterns (per-stimulus-time histograms) as a tone of the same frequency, and that the firing was deterministic (Clark 1969c, 1970a,c). The inability of the brainstem neurons to respond at rates higher than 500 pulses/s was most likely due to the fact that electrical stimulation produced strong inhibition that suppressed neural activity.

With this research it was also thought that if the cochlear nerve fibers were excited so that the fibers were stimulated in an asynchronous or stochastic manner, it would be possible to reproduce a rate or time-period code for higher frequencies. For this reason, square waves and sine wave voltages were compared, as the waveform of an electrical stimulus could affect a neuron's responsiveness. A more asynchronous or stochastic manner of firing was considered possible, with the more gradual rise in voltage from a sine wave. However, in the research study described above (Clark 1969c, 1970a,c), no significant differences were observed between the cell responses to square wave and those to sine wave voltage stimuli.

It was also of interest that although electrical stimulation of the cochlear nerve above 200 to 500 pulses/s could not reproduce the continuous firing of a cell, some cochlear nerves responded only to the rise and decay of a tone burst discharged to an electrical stimulus at its onset or offset in a similar way over a wide range of stimulus rates. The cell responses in the above study were only a small sample of the total population in the SOC. In contrast, field potentials are the summed electrical activity of many action potentials from a population of cells. Consequently, the AN was stimulated electrically and the field potentials recorded from electrodes placed within the TB (Clark 1969c, 1970a,c). The nerve was stimulated in four cats with 0.1-ms voltage square waves at various rates. The field potentials were markedly suppressed at rates from 100 to 300 pulses/s (Fig. 5.19).

The field potential recordings from the brainstem thus showed that electrical stimulation that attempted to reproduce a rate or time-period code would probably not convey adequate frequency information for a cochlear implant to help deaf patients understand the range of speech frequencies required for speech understanding, which is up to 4000 Hz. Consequently, local stimulation of nerve endings along different parts of the basilar membrane in accord with the place theory would be needed (Clark 1969c, 1970a,c).

The conclusions from the field potential study by Clark (1969c, 1970a,c) on the limitation of rate of stimulation reproducing acoustic information were supported by a later field potential study by Glattke (1974). In this study repetitive stimulation by acoustic transients and electrical pulses was used. The evoked responses to clicks followed the stimulus rate up to 800 clicks/s, and in some cases up to 1500 clicks/s. An autocorrelation analysis of the results for electrical stimulation showed that above 250 pulses/s the one-to-one firing with each stimulus was lost.

At about the same time recordings were made from the brainstem in response to electrical stimulation (Clark 1969c, 1970a,c), research was being undertaken by Moxon (1967) to record auditory nerve fiber's responses to electrical stimulation of the cochlea to help understand the physiology of acoustic stimulation. The research was to determine the refractory period of AN fibers, as it had been shown by Sachs (1967) that 200 spikes/s was the maximum rate for an acoustic

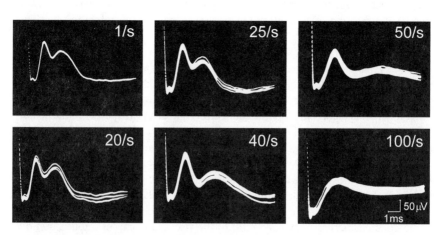

FIGURE 5.19. Field potentials from the superior olivary complex (SOC) in the auditory brainstem of the cat for different rates of stimulation of the cochlea and auditory nerve (Clark 1969c).

stimulus. Was this limitation due to the refractoriness of the nerve fibers or to mechanisms within the cochlea? The absolute refractory period of the auditory nerve fibers was found, by presenting pairs of electrical stimuli, to be as short as 0.5 ms (Moxon 1967). Although the refractory period put a limit on the maximum possible instantaneous discharge rate, the fibers could not fire at the maximum rate over time. For bursts of electrical stimuli the initial maximum rate of stimulation was 900 pulses/s, but this fell to 500 pulses/s over 2 minutes. Furthermore, the maximum rate of stimulation possible with electrical stimulation depended on the stimulus intensity, being greater with a higher current level. The study also showed that the dynamic range from threshold to a 100% response rate was much narrower with electrical compared to acoustic stimulation.

This research was later extended by Moxon (1971) to provide information on the site of origin of the AN responses to electrical stimulation of the cochlea. It helped establish the earlier findings of Jones et al (1940), discussed in Chapter 1, that there were two types of AN responses: (1) those due to direct electrical stimulation of the auditory nerve, and (2) those due to electromechanical stimulation of the cochlear hair cells (electrophonic hearing). The responses due to direct stimulation (α response) were at low thresholds, of short latency, and more deterministic (highly synchronized to the stimulus) than those for acoustic clicks. Also, the responses due to electrophonic hearing were stochastic with longer latencies, similar to those induced by acoustic stimuli.

The ability of electrical stimulation to reproduce the acoustic patterns in the auditory nerve and brainstem neurons was also studied using interspike interval histograms by Merzenich et al (1973). Recording the intervals between spikes (interspike interval histogram) is a more direct method of measuring the synchrony of responses and phase locking compared with the measurement of intervals after the onset of the stimulus (per-stimulus histogram). The responses of units in the IC to different stimulus rates were recorded, and it was reported that the sinusoidal electrical stimuli were encoded in the discharges of the neurons up to a stimulus rate of 400 to 600 pulses/s.

Rate Discrimination in the Experimental Animal

As the per-stimulus and interspike interval histograms from units and field potential data had been recorded directly from auditory pathways in the anesthetized cat, it could still not be concluded that they were applicable to the coding of frequency in the intact alert animal. For this reason rate discrimination for electrical stimulation of the auditory nerve in the behaviorally conditioned animal (Clark, Nathar et al 1972; Clark, Kranz et al 1973; Williams et al 1974) became essential in helping to establish whether a time/period (rate) code could be used to convey temporal information in speech. The behavioral findings were complementary to the data recorded from cells, not only in understanding the limitations of reproducing temporal coding, but also for the coding itself.

Rate discrimination was measured using difference limens (DLs; the just detectable differences in rate that could be perceived). These studies were crucial in deciding whether to develop a single- or multiple-electrode (channel) implant. If a rate code could be used to code speech frequencies from approximately 125 to 4000 Hz, a single-electrode implant would have been all that was needed, and cheaper.

The first study (Clark, Nathar et al 1972) showed that cats had only a very limited ability to discriminate changes in rate of stimulation above 200 pulses/s, even though loudness differences associated with rate of stimulation were not controlled for. At rates of 100 and 200 pulses/s the DLs for rate of stimulation were also considerably poorer than for sounds of the same frequency. The results showed that for electrical stimuli of 100, 200, and 400 pulses/s, the DLs varied from 50% and above. These DLs were greater than those obtained by Shower and Biddulph (1931) for acoustic stimulation in humans at 125, 250, and 500 Hz. The DLs were 3%, 1%, and 1%, respectively. The frequency DLs in cats for acoustic stimulation at various frequencies were between 1% and 5% (Butler et al 1957; Kranz 1971). At the conclusion of the experiments the temporal bones were sectioned and the cochleae examined to ensure that there were no residual hair cells that could have led to false results from electrophonic hearing.

A second behavioral study on cats (Clark, Kranz et al 1973) was carried out to help confirm the results of the first study. It was undertaken also to compare the responses of time-varying rate of electrical stimulation with those of comparable acoustic stimuli. This was done as a great deal of information in speech is conveyed by variations in frequency over time, and it was considered important to see how well a rate or time-period code could convey this information along a single stimulus channel. The study was undertaken by conditioning cats to respond to sound and electrical stimuli that were frequency modulated by sine or triangular waves. The measurements were made before and after any residual hair cells were destroyed using an ultra–high-frequency electron beam to exclude electrophonic hearing. This was confirmed by the failure to record cochlear microphonics from the animals and by the absence of hair cells after sacrificing the animals and examining the cochleae under light microscopy.

The results showed that the rates of stimulation that could be discriminated for electrophonic hearing were 1600 and 2400 pulses/s in the two cats in the study. For direct electrical stimulation of the cochlear nerve the upper limit was 600 pulses/s and 800 pulses/s. The findings thus not only confirmed the limitation of using rate of stimulation to code frequency, but also demonstrated that hair cell function could be preserved in an implanted ear for frequencies below 2400 Hz. The electrode was inserted through the round window into the scala tympani of the basal turn of the cochlea. As with the other studies reported above (see Electrophonic Hearing), this study supported the view that care must be taken when interpreting results for electrical stimulation of deaf implant patients with residual hair cells in the implanted ear. Electrophonic hearing could lead to better speech perception scores than when stimulating cochlear nerve fibers alone. But the study suggested that electrical stimulation in patients with some residual hearing could be used to advantage.

The results of these studies established that there would be serious limitations in using rate of electrical stimulation in coding the higher frequencies required for speech intelligibility. Thus place coding in addition to the limited rate coding possible with electrical stimulation would be needed. In fact, the abilities of the cats in discriminating rate of stimulation were similar to the psychophysical findings later obtained from the first cochlear implant patients (Tong and Clark 1985) (see Chapter 6). They had difficult discriminating rates above 200 to 600 pulses/s (Tong and Clark 1985).

The second part of the above behavioral study on cats measured more accurately their ability to detect a changing stimulus rate, as it was also relevant to the design of a cochlear implant that would help patients understand running speech. The study was undertaken by varying the slope or rate of change in the stimulus frequency for sound at 200 and 2000 Hz and electrical stimuli at 200 pulses/s and 2000 pulses/s. The carrier frequencies were modulated by triangular waves to produce graded changes in frequency over the duration of 500 ms. The results for sound at 200 Hz and electrical stimuli at 200 pulses/s were similar. The thresholds at a 50% response level were 97 Hz/s for sound and 85 pulses/s/s for electrical stimuli. The ability of the cats to detect changes in high rates of stimulation (2000 pulses/s) was poor compared to that for sound at the same frequency.

It is of interest to compare the result in the cat, where a threshold of 85 pulses/s/s for a change in a stimulus rate of 200 pulses/s was obtained, with that subsequently found on cochlear implant patients (Tong et al 1982). The assessment procedure used for the patients was different from that for the cat study, but a low estimate of the rate of change in stimulus rate that could be detected was 300 pulses/s/s (Clark and Tong 1990).

The ability of cats to detect only changes in rate of stimulation at low frequencies subsequently supported the use of rate of stimulation in coding voicing; where a change in 200 Hz could be expected over the duration of a sentence of 2 s (100 Hz/s). On the other hand, their inability to detect changes in stimulus rate at high frequencies indicated that rate was inappropriate for conveying the rapid frequency changes in consonants that can be as high as approximately 10,000 Hz/s.

A third study was undertaken to help confirm the findings on rate of stimulation in the above behavioral studies (Williams et al 1976). The effect of rate of stimulation was evaluated by requiring the cats to make comparative judgments of whether a stimulus was higher or lower than a reference rather than requiring "same" or "different" judgments as in the first two studies. With this experimental design it was considered the cats were more likely to respond to the psychophysical correlate of pitch. The intensity variations that occur with changes in stimulus rate were controlled for, constant current stimulation was used, and current levels were set halfway between those resulting in a threshold and an aversive response.

In this series of experiments signal detection theory was the basis for determining thresholds for different combinations of stimuli. The thresholds were obtained by scoring responses as hits, misses, false alarms, and correct rejections, and plotting a receiver-operating curve (ROC). The thresholds recorded by this method were independent of decision criterion levels, and intensity was randomized. In the study electrophonic hearing was controlled for by administering neomycin in dose levels needed to destroy hair cells. The destruction of hair cells was subsequently confirmed by serially sectioning the cochleae, and examining the sections under light microscopy.

The results of this third study showed that, when electrophonic hearing was excluded by the destruction of hair cells, the cats could at least discriminate stimulus rates that varied from 348 to 490 pulses/s (a DL of 41%). These stimulus rates could be discriminated with electrodes placed in either the apical or basal turns of the cochlea. The ability to discriminate changes in low stimulus rates at the basal as well as apical turns was subsequently seen in the initial implant patients (Tong et al 1982, 1983), who used a speech-processing strategy in which the low frequencies of voicing were used to stimulate at different sites along the basal turn. The site depended on the electrode chosen to represent the higher second formant frequency.

Results from the above three behavioral studies on experimental animals showed that the rate code, as distinct from the place code, could convey only temporal information for low rates of stimulation up to 600 pulses/s. The limitations in replicating the time/period code for frequencies above 600 Hz by rate of electrical stimulation indicated that there were deficiencies in how well electrical stimulation reproduced acoustic stimulation. Furthermore, place of stimulation could have been the main code for frequencies above about 600 Hz.

As the behavioral research indicated that higher frequencies (up to 4000 Hz), of importance in speech intelligibility could not be conveyed by variations in stimulus rate, it was considered that multiple-electrode stimulation would be needed to present these frequencies on a place-coding basis. There would be serious limitations in using electrical stimulation on a single electrode cochlear implant for speech understanding.

A behavioral study to determine frequency DLs for sinusoidal electrical stimuli was also undertaken by Pfingst and Rush (1985) in monkeys. The frequency DLs were measured at equal loudness points. These were defined as the points at which the discrimination of a frequency change was minimal when frequency and intensity level were varied simultaneously. The results showed DLs that ranged from 7% at 100 Hz (17 dB sensation level) to about 30% at 200, 300, and 600 Hz (7–9 dB sensation level).

Further Unit Studies

A further study to that of Moxon (1971) on the site of excitation in the AN by electrical stimulation of the cochlea was done by van den Honert and Stypulkowski (1984). They firstly found there were two peaks (N0 and N1) in the compound action potential (CAP) recorded from the AN. To explain these peaks and see if the one with the shorter latency (N0) was due to excitation of the dendrites, and N1 from the axons, an investigation was carried out by recording from individual nerve fibers. It was discovered that as the threshold stimulus intensity was doubled, not only did the firing jitter become less, but the mean latency was reduced from 685 μs (α response) to 352 μs (δ response). Evidence for the δ response arising from the axon came when recordings with similar latencies were obtained after stimulating the exposed stump of the AN. In the study there was no discontinuous shift in latency that would indicate a discrete delay across the soma from α to δ. Consequently, a more gradual excitation along the dendrite and across the soma was the likely explanation. This research also confirmed that electrical stimulation produced deterministic rather than stochastic responses.

The above studies were extended by Javel et al (1987). They showed that with low-intensity electrical stimulation there was less synchronous firing and a response B with a latency of 0.6 ms (Fig. 5.20). As the intensity of the pulse increased, the B response was gradually replaced by a shorter latency (0.3 ms), the A response. It was considered that the B response was due to stimulation of peripheral processes and the A response to spiral ganglion cells. There was a D

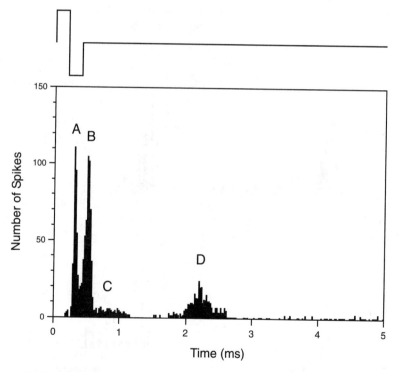

FIGURE 5.20. A composite period histogram showing the four latencies at which auditory nerve fibers discharged in response to a biphasic current pulse. They are labeled A to D: A, spiral ganglion cell; B, peripheral process; C, unexplained; D, hair cell mediated (Javel et al 1987, reprinted with permission).

response that had a long latency (1.5–2.5 ms) and was not deterministic but stochastic. These D responses were considered to be due to electrophonic hearing.

In addition, unit studies were undertaken to confirm and extend the initial research on the synchrony of the responses of ANs to electrical stimulation. Hartmann et al (1984a) confirmed that the phase locking of the responses from the AN to electrical stimulation was much stronger than for acoustic stimulation. They also demonstrated that with interstimulus periods equivalent to repetition rates above 200 pulses/s, there was a decrease in the probability of the nerve firing. When stimulus periods equivalent to rates of 200 to 500 pulses/s occurred, the intensity of the stimulus had to be increased to maintain the probability of firing. It was also shown that AN fibers had significant phase locking to electrical stimuli at higher frequencies (Stypulkowski and van den Honert 1984). It was found that phase locking of the AN responses could occur up to 800 pulses/s (Javel et al 1987; Clark 1998b,c) (Fig. 5.21). In another study (Hartmann et al 1984b) the unit responses of AN fibers in the cat to both biphasic current pulses and sinusoidal current waveforms were compared. The only difference between the response characteristics was that the peaks of the period histograms were smaller with electrical pulses than sinusoids. A study by Clopton and Glass (1984) on unit responses in the CN of the guinea pig to sinusoidal electrical stimuli indicated that if complex stimuli consisting of two and five sinusoids were used,

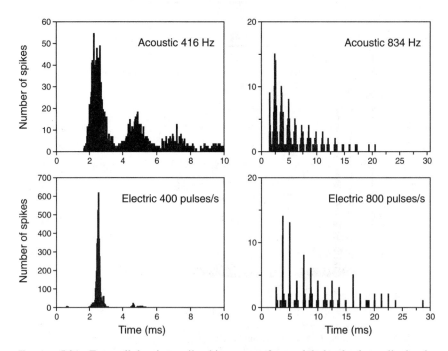

FIGURE 5.21. Extracellular interspike histograms from globular bushy cells in the anteroventral cochlear nucleus for acoustic stimulation at 416 and 834 Hz and electrical stimulation at 400 and 800 pulses/s (Clark 1998b,c. Reprinted with permission of Elsevier Science.)

the units responded primarily to the more intense peaks. This indicated that the amplitude envelope could also be used to effectively code variations in amplitude in speech sounds.

Furthermore, studies in the IC (Snyder et al 1991, 1995; Shepherd et al 1999) showed a decrease in temporal resolution with rate of electrical stimulation. Units in the IC responded in a temporally synchronized manner typically up to 120 pulses/s (Shepherd et al 1999). This is much lower than that observed in the AN, but more similar to that demonstrated in the TB and the SOC (Clark 1969a,c, 1970b,c; Clark and Dunlop 1969). This indicates that processes such as inhibition limiting temporal resolution were located centrally (Clark 1996; Snyder et al 2000b).

Comparison of Acoustic and Electrical Responses with Physiological and Behavioral Studies

Unit Responses in the Experimental Animal

To determine how best to code frequency on a rate or time period code, further studies were needed to compare the pattern of responses in interspike interval histograms for sound and electrical stimulation up to higher rates of stimulation than previously presented (Clark 1996, 1998b,c). As shown in the histograms in Figure 5.21 from units in the cat AVCN, the pattern of responses was less similar for the lower rate of 400 pulses/s and more similar for stimulus rates of 800 pulses/s. For an acoustic stimulus of 416 Hz, the peaks or modes of the interspike intervals were multiples of the period of the sound wave, and the intervals were distributed around the mode. There was jitter, and the action potentials did not time lock precisely with a particular phase on the sound wave. This pattern is referred to as stochastic firing. On the other hand, with electrical stimulation at 400 pulses/s, there was essentially only a single population of intervals, which was the same as the period of the stimulus. There was also very little jitter around the mode, and this is referred to as deterministic firing. The pattern of unit responses for electrical stimulation varied with intensity, and at low intensities was more similar to that for sound. Furthermore, as illustrated in Figure 5.21, at higher frequencies (800 Hz) the interspike interval histograms were more similar for acoustic and electrical stimulation. With electrical stimulation at 800 pulses/s there were more peaks and an increase in jitter than for lower rates of stimulation. The per-stimulus time and period histograms for primary-like units in the AVCN for electrical stimulation at different intensities were also compared (Javel et al 1987). They too were more similar at high frequencies and low intensities.

At increasingly higher rates of stimulation the time between pulses becomes shorter than the relative refractory period of the AN fibers and CN brain cells, and then less than the absolute refractory period. For example, Figure 5.22 shows the interspike interval histograms of the intracellular responses from the primary-like responding cells in the cat AVCN for rates of 200, 800, 1200, and 1800 pulses/s (Paolini and Clark 1997; Clark and Lawrence 2000). It is assumed that the absolute refractory period is 0.5 ms (Moxon 1967), and the relative refractory

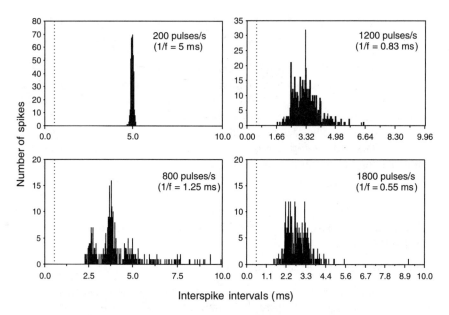

FIGURE 5.22. Intracellular interspike histograms from globular bushy cells in the anteroventral cochlear nucleus for increasingly higher rates of electrical stimulation (200, 800, 1200, and 1800 pulses/s). Broken line indicates the absolute refractory period (Paolini and Clark 1997).

period 0.2 ms (Roberts et al 2000). At 200 pulses/s, with a period of 5 ms, the pulses result in deterministic firing in the neuron at the same rate as the stimulus. At 1200 pulses/s (period 0.83 ms) it becomes harder to excite the nerve fiber in time with the stimulus. The neuron does not fire at this rate, but the predominant intervals between spikes are multiples of the period. Finally, at 1800 pulses/s where the period is 0.55 ms and close to the absolute refractory period, the firing pattern is a Poisson distribution, and thus the response of the unit is not related to the stimulus rate.

Behavioral and Psychophysical Responses

The physiological studies referred to above showed a better correspondence between the unit responses for acoustic and electrical stimulation at higher rather than low frequencies. In contrast, the behavioral findings from the experimental animal showed that rate discrimination with electrical stimulation was poorer for high rates. The same applied to implanted patients, as discussed in more detail in Chapter 6, whose ability to discriminate low rates of stimulation (approximately 250 pulses/s) was similar to that of a sound at the same frequency, but this did not occur for higher frequencies. Furthermore, pitch differences could only be discriminated up to approximately 500 pulses/s (Tong et al 1982), with a plateau in pitch estimation at this rate.

Why, then, if the single unit responses for acoustic and electrical stimulation at higher frequencies were apparently similar, was pitch perception poor, and if the physiological responses at lower frequencies were different, why were the pitch percepts more similar? It is hypothesized that the discrepancy is due to the fact that the timing relationships of individual fibers in transmitting phase information is critical in conveying frequency information (Clark 1995, 1996; Clark, Carter et al 1995). "With acoustic stimulation the interval histograms for each individual fiber may be the same, but the phase relations could be different, and the probability of interconnected neurons firing simultaneously is not the same. In using electrical stimulation to simulate the temporal coding of frequency, it may be important to not only produce an interval histogram similar to that for sound, but also produce a pattern of responses in an ensemble of neurons that is similar" (Clark, Carter et al 1995). This is illustrated in Figure 5.23. Thus not only should the overall population of fibers fire in phase with the sound wave and produce the same interval histogram as illustrated, but the phase information conveyed by the traveling wave should be coded as well.

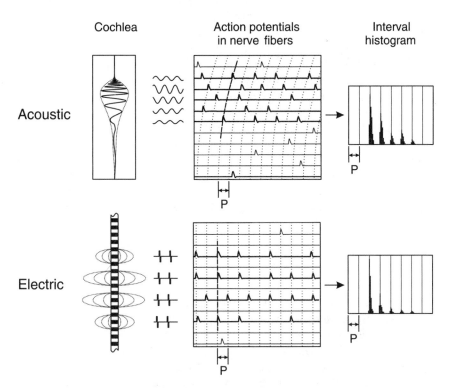

FIGURE 5.23. A diagram of the unit responses or action potentials in an ensemble of auditory neurons for electrical and acoustic stimulation showing the effects of the phase of the basilar membrane traveling wave. Top: The probability of firing in an ensemble of neurons to acoustic excitation due to phase delays along the basilar membrane. Bottom: The probability of firing due to electrical stimulation (Clark 2001. Reprinted with permission.)

Fine Temporal and Spatial Patterns of Response

A temporal and spatial pattern of responses to explain the discrepancy between
the physiological and psychophysical findings may well be due to the time delay
from the basilar membrane traveling wave, and in particular the rapid phase
changes in the membrane around its point of maximal vibration. The input from
the AN fibers to the CN would be coded through coincidence detection, requiring
coherence in the spike times in the afferent fibers. This hypothesis requires determining (1) the effect of the phase of the traveling sound wave along the basilar
membrane at its region of maximal vibration on an ensemble of neurons; (2) the
spread of neural excitation and convergence of neurons from this region of maximum vibration; (3) the statistical properties of inputs to the CN; (4) the threshold
of individual inputs; and (5) the integration of the input to the CN cells resulting
from this excitation, and its output. This has been studied physiologically in the
globular bushy cells in the CN (Paolini et al 2001) and with mathematical models
(Bruce, Irlicht et al 1998a; Kuhlmann et al 2002). The intracellular recording of
EPSPs as well as action potentials has enabled the effects of phase delays to be
assessed, together with the area of the basilar membrane, bringing this phase
information through convergent fibers. If the phase information is from a restricted area of the basilar membrane, the summing of the potentials will be well
synchronized. But if they are converging from a wider area, there will be a smear-
ing of the EPSPs as illustrated in Figure 5.24.

The physiological issues have been studied using in vivo intracellular recordings from the VCN of the rat. This has led to the analysis of the latency, amplitude,
number, time course, synchrony, and duration of postsynaptic events. Globular
bushy cell responses to both intracochlear electrical and acoustic stimuli have
been investigated. Intracellular recordings to sound have fast EPSPs corresponding to the period of the sound wave up to 2500 Hz (Figure 5.25). Data from
intracochlear electrical stimulation of AN fibers have demonstrated uniform conduction velocities for the inputs to the globular bushy cells (Fig. 5.26). The amplitudes of the EPSPs are stepped, indicating discrete synaptic inputs (Fig. 5.26,
left). A mathematical fit can be made to each of these inputs to determine the
time constant of excitation (Fig. 5.26, right).

These physiological data suggest that (1) the different phase relationships of
these inputs are not compensated for by conduction delays, and (2) the spread of
excitation is from a sufficiently narrow portion of the basilar membrane that will
enhance temporal summation. The data have been used to develop a mathematical
model to explain the events (see Integrate-and-Fire Model, below). The data also
indicate that to improve the fidelity of pitch perception, more electrodes need to
be placed close to the cochlear nerve fibers to reproduce the fine temporospatial
patterns of neuronal firing.

Addition of Noise

Noise or spontaneous activity in the AN is seen in the normal ear but not in the
deafened ear (Kiang et al 1979; Liberman and Dodds 1984). Studies by Ehren-

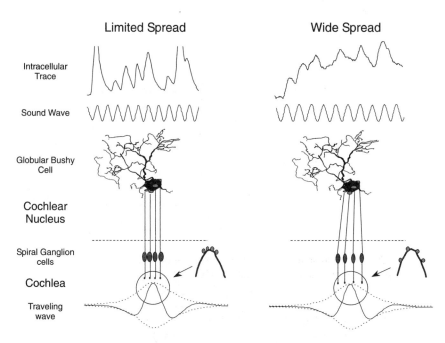

FIGURE 5.24. The effect of spread of excitation of the basilar membrane on the processing of phase information in the first stage of auditory brain processing (globular bushy cells in the CN). Left: Limited spread. Right: Wide spread (Reprinted with permission from Paolini et al 2000).

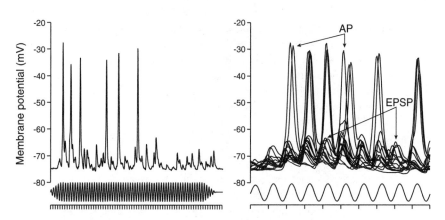

FIGURE 5.25. Left: Intracellular traces from the globular bushy cells in the cochlear nucleus show responses to a 1000-Hz acoustic stimulus. Right: Superimposed traces on an expanded time base are seen. AP, action potential; EPSP, excitatory postsynaptic potential. (Reprinted with permission from Paolini et al 2000.)

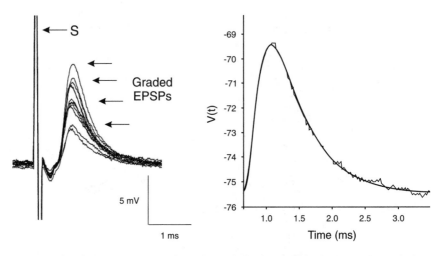

FIGURE 5.26. Intracellular traces from the globular bushy cells in the cochlear nucleus show responses to intracochlear electrical stimulation. Left: Superimposed traces indicate graded excitatory postsynaptic potentials (EPSPs) steps. S, stimulus artifact. Right: Single trace of an excitatory postsynaptic potential with overlaid mathematical fit used to calculate the membrane time constant. (Reprinted with permission from Paolini et al 2000).

berger et al (1999) and Zeng et al (2000) support the idea that stochastic activity in the AN may play a functional role in hearing. In addition, studies by Morse and Evans (1996, 1999a,b) and Morse and Roper (2000) suggest that introducing noise in cochlear implants may enhance the coding of speech. In their study, Morse and Evans (1996) electrically stimulated the amphibian sciatic nerve with analog signals from the output of a speech filter. The recorded CAP was both deterministic and dominated by the fundamental frequency and its harmonics. When noise was added, the output became stochastic and more like the response to sound, and the instantaneous stimulus amplitude determined the probability that the threshold would be reached. However, the synchronization of the fundamental frequency and harmonics was suppressed. With the optimal addition of noise, first formant temporal information became evident. As the noise level increased further, the firing became random.

Neural Models

Models of electrical stimulation of AN fibers are important in determining how cochlear implants may more effectively reproduce temporal coding of frequency. There are a number of issues that must be addressed when deciding on the most appropriate way to model the responses of AN fibers individually or as the input to a neural network in the brainstem. One of the most important is the stochastic nature of neuronal spikes. Significant variance has been measured in the response of nerve fibers to current pulses and pulse trains (Verveen 1960; O'Leary et al 1997). Derksen and Verveen (1966) showed that the variance in the response to electrical pulses can be attributed to fluctuations in the voltage across the neural membrane, and their amplitudes have a gaussian distribution. For these reasons the deterministic models are not suitable for studies to investigate coding strategies for sound using electrical stimulation. Furthermore, the Hodgkin-Huxley model (Hodgkin and Huxley 1952; Frankenhauser and Huxley 1964) referred to above (see Electrical Models of the Nerve Membrane), being deterministic, does not naturally incorporate a description of stochastic (random) processes for a network, since connectivity and the input currents are rarely known with certainty. Moreover, these models possess such a degree of physiological detail that in practice they are too computationally laborious to address questions about the cooperative behavior of large groups of neurons. Consequently, there has been considerable effort devoted to the study of somewhat simpler neural models, such as the integrate-and-fire neural model, that capture the essential features of the neuronal processing, including stochastic input, while remaining mathematically tractable. The degree of biophysical detail that a model requires depends on the particular phenomenon that is being studied or explained and the accuracy that is required, as will become apparent in the following sections.

A Probabilistic Model

To answer the question posed above on the discrepancy between the acoustic and electrical effects for low and high rates of stimulation in the physiological and psychophysical studies, a probabilistic model was first used. It was assumed that coincidence detection, as discussed by Carney (1994), was the coding mechanism. The probability of coincidences occurring in brain cells for acoustic and electrical stimulation at low and higher frequencies was modeled (Clark, Carter et al 1995). The model detected two or more inputs when they arrived during a specified time window (Fig. 5.4).

With the analysis of the probability of coincidences occurring for acoustic and electrical stimulation at low and higher frequencies the following were assumed: (1) The coincidence time window (T_m) was small enough so that the distances between pulses along a single fiber were greater than the coincidence time window (T_m). (2) The timing of the action potentials formed an inhomogeneous Poisson process with average rate R. (3) It was sufficient to measure the coincidences across two fibers, and each had identical Poisson rates. Making these assumptions it was calculated that the average rate of coincidences would be

$$2T_m \times R \times c$$

where T_m = the coincidence time window, R = the average Poisson rate, and c = the average spike rate for the duration over which neural responses occur for each sine wave or pulse interval.

When the coincidences per spike for electrical and acoustic stimulation at different frequencies were calculated, it was found there were more coincidences for electrical than acoustic stimulation at low frequencies. At higher frequencies there were equal numbers of coincidences for acoustic and electrical stimulation.

At low frequencies the proportionately greater number of coincidences seen with electrical stimulation should have increased the probability of the appropriate patterns of stimulation being produced, and hence explain the closer pitch match obtained between acoustic and electrical stimulation at these low frequencies. However, from the result for high frequencies, it can be concluded that if a co-incidence detector encodes frequency, it is not enough to have the same number of coincidences with electrical stimulation as for sound as the pitch percepts for electrical and acoustic stimulation at high frequencies are very dissimilar. This model did not examine stochastic firing, neural synchronization, and the coding of temporal and spatial patterns of intervals and phase information.

Point Processing Model

The importance of stochastic processes in the central nervous system generally has been recognized for some time and is well reviewed by Tuckwell (1988a,b). The AN's response to sound can be approximated as a series of stochastically distributed action potentials (Snyder and Miller 1991), and this has been modeled with a point process model. Point process models (Perkel et al 1967; Miller 1971) give the statistical properties of neural events over time. They do not calculate biophysical properties such as channel conductance. The rate of occurrence of action potentials over time (intensity) is a function of the properties of the neuron, the applied stimulus, and the neuron's firing history. The response history of the nerve is important, as there will be a lowered probability of firing if the stimulus occurs within the nerve's refractory period. The effect of the refractory period on the AN firing probability has been studied for sound by Gaumond et al (1982), Johnson and Swami (1983), Jones and Tubis (1985), and Bi (1989).

Point process models have been used to study the firing statistics of fibers to determine how electrical stimulation can reproduce the stochastic responses for sound (Irlicht et al 1995; Irlicht and Clark 1995a,b, 1996; Bruce et al 1997a–c, 1998a,b, 1999a–d, 2000). The multiplicative stimulus/refractory model was se-lected as it had provided a good approximation of neuron responses (Mark and Miller 1992). It was used to determine how well the statistics of the neural re-sponse to electrical stimuli could be made to approximate those for sound. For this reason it could be useful for designing stimuli for electrical stimulation with a cochlear implant.

The maximum likelihood algorithm (Mark and Miller 1992) was applied to optimize the stimulus/refractory function to reproduce the neural responses ob-tained experimentally for acoustic and electrical stimulation. Because the neural responses to acoustic and electrical stimuli were different (Javel et al 1987), elec-trical stimulus strategies were examined that would force the neuron to have the same temporal response as that for sound. This required modifying the stimulus function as the refractory function remained fixed. It was also used to predict a neuron's per-stimulus time histogram (Irlicht and Clark 1996). This modeling study showed that simply applying an electrical signal that is the same shape as an acoustic signal would not evoke the same per-stimulus time histogram as the one from acoustic stimulation.

As the physiological and psychophysical research had shown that the firing patterns in an ensemble of neurons (Rose et al 1967; Clark 1996), rather than in single neurons as investigated above, were most important in the temporal coding of frequency, the firing statistics have also been modeled in a population of neurons. For electrical stimulation a model of auditory function based on that of Colombo and Parkins (1987) was used. In this model the signal was separated into its frequency components by band-pass filters that approximated the filter characteristics of the basilar membrane. There was a phase lag to simulate the phase delay of the basilar membrane. This model was similar to that employed by Laird (1979) and Kiang et al (1979). In addition, the band-pass signal was half-wave rectified to simulate the hair cell directionality in response, and then low-pass filtered to allow for a 1-ms refractory period.

For hearing, the model first used the basilar membrane model of Au et al (1995), in turn derived from that of Neely and Kim (1986), and a computational model of the inner ear auditory nerve synapse (Meddis et al 1990). The values from points along the cochlea were normalized and fed into the Meddis et al model. The statistical properties of the nerves modeled were the summed period histograms of the responses of all the fibers (200 fibers/cm length of basilar membrane). Normally each fiber responds differently due to the speed of its conduction, which depends on the fiber diameter, and the phase information from the basilar membrane traveling wave. However, these variables were not taken into consideration with the summed period histograms. The standard biphasic bipolar pulse of 250 μs provided a poor simulation of the acoustic period histogram.

It was found by experimentation that by presenting multiple-pulses per period (Fig. 5.27) and varying the amplitude, there was better correlation, as shown in Fig. 5.28. The intensity of the pulses needed to increase for each period. Thus the central fibers were first simulated and then the more peripheral ones during the time that the central ones remained in a refractory state.

Biophysical Model

The failure of electrical stimulation to adequately reproduce frequency coding may be not only due to the inadequate representation of the stochastic firing in individual neurons, as explained above (see Comparison of Acoustic and Electrical Responses), but also related to the independence of information across fibers. Sound produces a progressive increase in synchronization within and across fibers as intensity is increased (Schoonhoven et al 1997). Replication of this across-fiber synchronization should at least allow a greater dynamic range and more orderly loudness growth.

A stochastic/deterministic model was developed by Rubinstein (1991, 1993, 1995) and was used to determine how to increase stochastic independence across fibers in response to electrical stimulation. The model is based on the biophysics of the voltage sensitive sodium channels, leakage currents, and membrane capacitances. It provided a stochastic representation of the responses of the neural membrane as applied to the nodes of Ranvier of the peripheral processes of the AN in the cochlea, and a deterministic representation of the internode. It is computationally very demanding. A study by Rubinstein et al (1999) demonstrated

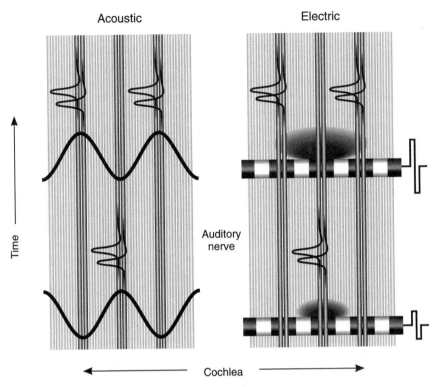

FIGURE 5.27. Diagrams of a temporospatial pattern of responses to acoustic and electrical stimulation in an ensemble of AN fibers. Left: The action potentials to two phases of a sound wave showing how the directional sensitivity of hair cells leads to excitation in different groups of nerve fibers depending on the phase. Right: The action potentials to a smaller amplitude pulse followed by a larger amplitude pulse within the refractory period for the first group of fibers. Thus there is some reproduction of the temporospatial firing of the pattern produced by acoustic stimulation.

in models as well as through electrically evoked compound action potentials (ECAPs), recorded from intracochlear electrodes in implant patients, that an interpulse interval between 0.9 and 1.0 ms (1000-Hz stimulus rate) substantially decreased the slope of the neural input/output function and ECAP growth function (Matsuoka et al 1997). This was due to an increase in the noise of the voltage-resistance sodium channel during the relative refractory period (Rubinstein et al 1997). This could make the prediction of events in a neural population difficult.

An alternative approach was to modulate pulses at high stimulus rates. At low rates the ECAPs showed an alternating pattern suggesting refractory effects due to the high degrees of synchronization (Wilson et al 1994, 1997a–c; Abbas et al. 1997; Wilson 1997). At rates above 2000 Hz the response amplitudes were constant, and this was consistent with an increased stochastic independence of the

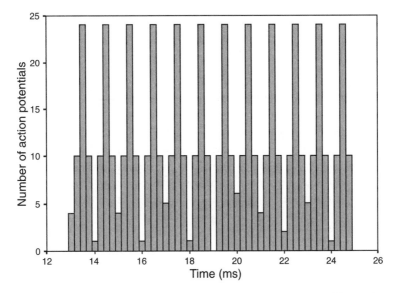

FIGURE 5.28. Summed period histogram for electrical stimuli with multiple pulses per period. (Reprinted with permission from Irlicht et al 1995.)

firing patterns of the neural population. Small ECAPs were produced, suggesting only a limited number of fibers were not in a refractory state at any one time.

Producing a statistical independence of responses in fibers, however, does not show how to produce the correct temporal and spatial response pattern of excitation in an ensemble of fibers assumed to be important for the coding of frequency.

The model of Rubinstein et al (1999) was used to re-create noise seen in AN fibers. The rationale for reproducing noise was discussed above. Per-stimulus and interval histograms were plotted for stimuli at 5000 Hz. The per-stimulus histogram showed a small degree of synchronization (0.26) as measured by the vector strength procedure (Goldberg and Brown 1969). The interval histogram showed responses consistent with a Poisson process following a dead time for the refractory period and a renewal process. This strongly resembled spontaneous activity in the intact AN. The spike times were determined by the relative refractory period. This was similar to the experimental findings of Dynes and Delgutte (1992) on single units. It was different from the findings of Paolini et al (2000), in which there was a loss of synchronization between 1200 and 1800 pulses/s. This was consistent with an absolute refractory period of 0.5 ms (Moxon 1967) and a relative refractory period of 0.2 ms (Roberts et al 2000). The conditional mean histogram (Johnson 1996) was constant, indicating that the firing probabilities were not affected by the intervals prior to the previous spike. The hazard function was also constant, indicating the same. As there was statistical similarity between these electrically evoked responses and spontaneous activity in the normal nerve, it was referred to as "pseudo-spontaneous" firing. Like normal spontaneous activity (Johnson and Kiang 1976), it was independent across neurons.

The effect of noise on temporal coding in cochlear implant patients was examined by recording the ECAP (Rubinstein et al 1999). Many variables in the response have been seen for different stimulus rates and amplitudes. The response at 1016 pulses/s, using a subtraction technique, showed some synchrony. As the rate was increased to 3049 pulses/s, the synchrony was lost. This is consistent with the intracellular responses from the globular bushy cells (Paolini et al 2000). When a pulse rate of 1016 pulses/s was superimposed on a high rate conditioner of 5081 pulses/s, the temporal synchrony to 1016 pulses/s was reestablished at increasing intensities. Whether this provides better synchrony than stimuli without added noise has to be determined. The advantage of a response due to stochastic resonance, however, may only lie close to threshold as determined by Hohn and Burkitt (2001a) using an integrate-and-fire model. There are biological safety concerns when stimulating at high rates (see Chapter 4).

Integrate-and-Fire Model

A difficulty with the point process model is that it determines only the average firing statistics in a population of neurons. It does not examine the interaction of events, or provide information on the neural processes involved in coding. The biophysical model could do this, but it requires such large computing power that it is not appropriate. For this reason an integrate-and-fire model has been used to study a number of questions. This model is based on the generation of an action potential (spike) when the sum of the incoming postsynaptic potentials reaches a threshold (Tuckwell 1988a,b). One of the earliest threshold models that incorporated stochastic inputs approximated the subthreshold potential of a spontaneously active neuron by a random walk process (Gerstein and Mandelbrot 1964). The model was extended by Stein (1965) to incorporate the exponential decay of the membrane potential using stochastic differential equations. With this model, for each arriving EPSP there is a step increase in potential, which then decays. The decay in the potential is referred to as a leaky integrator in analogy with electrical circuits. The integrate-and-fire model has been extended using the gaussian approximation (Burkitt and Clark 1999a; Burkitt 2001) in which the membrane voltage is described as a Taylor's series expansion in the amplitude of the incoming postsynaptic potentials. As the amplitude of the incoming excitatory potentials are small relative to the spiking threshold, the Taylor's series allows higher order terms to be neglected, making analytic computation possible. The model took into consideration the properties of the EPSPs and IPSPs, and is thus more realistic physiologically. It enables the calculation of the probability density function of the membrane potential reaching threshold, and the probability of the output spikes.

The integrate-and-fire model has been used to examine the relationship between the input and output of a nerve cell when the inputs have a firing rate that has a Poisson distribution (Burkitt and Clark 1998, 2000). More importantly, it has been used to study the synchronization of the responses in a neuron, in particular the relationship between the spike input and output (Burkitt and Clark 1999a,b; Burkitt 2001).

Synchronization increases the reliability of transmitting information along the nervous system. A neuron that receives information simultaneously is much more likely to generate a spike than one that receives fewer inputs or the same number of inputs distributed over a longer period of time. Synchronization allows the grouping of neurons that respond to the same stimulus features, and these groupings will be more resistant to amplitude fluctuations.

Synchronization was studied with (1) a perfect integrator model, in which the decay in the potential across the membrane was neglected; (2) the Stein model; and (3) the alpha model of Burkitt and Clark (1999a). With the Stein and alpha models the output jitter (σ_{out}) was substantially less than the input jitter (σ_{in}) over a large range of inputs and threshold ratios. This was consistent with the physiological data where synchronization was improved from the AVCN (Joris et al 1994) to the nucleus of the MNTB (Fitzgerald et al 2000). The modeling study by Burkitt and Clark (1999a) also showed that synchronization improved with the number of synaptic inputs, but was unaffected by the threshold ratios (i.e., the variation in the amplitudes on the EPSPs relative to the spiking threshold). This extended the findings of Marsalek et al (1997), who used a perfect integration model (in which the leakage of the membrane potential is neglected) that was therefore less physiological. The study by Burkitt and Clark (1999a) also showed that synchronization was reduced with an increase in the proportion of inhibitory synaptic potentials in the input.

Stochastic resonance may also be used to improve the temporal coding of frequency, using electrical stimulation essentially at low thresholds. This was studied with the point process and biophysical models, discussed above, with the addition of noise. Stochastic resonance is a phenomenon illustrated in Figure 5.29,

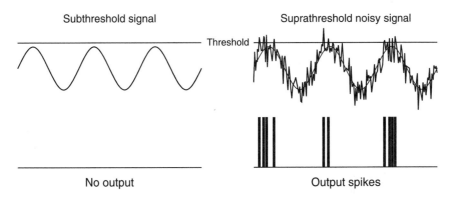

FIGURE 5.29. A diagram illustrating the principles underlying stochastic resonance and the effect of noise on a subthreshold stimulus. Left: A subthreshold stimulus without noise evokes no response. Right: The addition of noise evokes output spikes that provide information about the periodicity of the subthreshold stimulus (Reprinted with permission from Hohn and Burkitt 2001b).

describing how the detection of a weak signal can occur by the addition of noise. On the left of the figure there is a subthreshold stimulus. When noise is added, the threshold is exceeded, and output spikes are generated at times that on average provide a good representation of the periodicity of the subthreshold stimulus. When studied with the integrate-and-fire model, it was found that the amount of noise that is added to the signal is critical, and this is illustrated in Figure 5.30, where the coherence between the output and the input signal is plotted against the noise level (Hohn and Burkitt 2001b). As the noise level rises, so does the coherence, but further increases cause a decrement in response. So in summary, the addition of noise to the speech-processing strategy using subliminal random stimuli may help the perception of speech when it is close to threshold or just above, but not at suprathreshold levels, when the noise would interfere with perception.

A temporal and spatial pattern of action potentials in a group of nerve fibers is illustrated in Figure 5.4, which shows that the individual fibers in a group do not respond with an action potential each sine wave, but when a spike occurs it is at the same phase on the sine wave. However, the physiological, psychophysical, and mathematical modeling studies (Clark, Carter et al 1995) indicate that it is not enough for electrical stimulation to simply model this temporal and spatial pattern without taking into consideration the correct temporal and spatial patterns

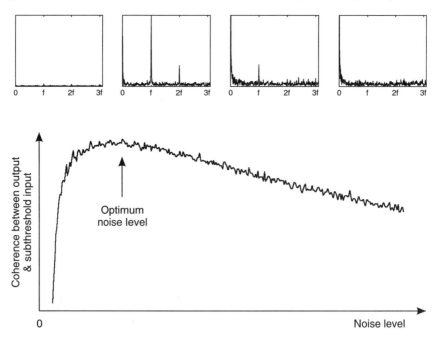

FIGURE 5.30. The coherence between the output and subthreshold input versus noise level, showing an optimal coherence at a nonzero level of noise, as well as a decrease in coherence as the noise level is further increased (Reprinted with permission from Hohn and Burkitt 2001b).

of excitation resulting from phase delays due to the traveling wave passing along the basilar membrane to the site of maximal vibration, or the changes in the vibratory peaks at the site of maximum resonance.

How phase delays along the membrane can affect the temporal and spatial pattern of responses in a group of fibers (Au et al 1995; Irlicht et al 1995) is illustrated in Figure 5.23, which shows the AN fiber firing probabilities over time, versus distance from the stapes along the basilar membrane for sound and electrical stimulation. The firing probabilities have been calculated from cochlear and hair cell auditory neuron models (Neely and Kim 1986; Au et al 1995). Notice that the probability of action potentials occurring on neighboring fibers is shifted in time according to the basilar membrane phase delay for sound and not for electrical stimulation. The phase delays occurring at the site of maximal vibration are illustrated in Figure 5.6. The replication of these latter delays in a small group of nerve fibers with electrical stimulation is possible, because as stated above (see Organ of Corti), the velocity of the traveling wave at the point of maximum vibration is approximately 1.3 m/s, and this is within the capacity of the neural pathways to process the information.

The integrate-and-fire model has been used to study the effect of basilar membrane phase on the relationship between the output synchrony of CN neurons and the site of stimulation in the cochlea. The model takes into account the stochastic nature of AN fiber firing statistics (Tuckwell 1988a,b). In this single neuron model the cell begins at resting membrane potential, and as inputs arrive, individual synaptic input currents create postsynaptic responses in the membrane potential by summing at a single point, considered to be the site of action potential initiation (the axon hillock). When the membrane potential, corresponding to the sum of the input currents, reaches a threshold, an action potential is initiated. This is demonstrated in Figure 5.2. The integrate-and-fire model is defined by several parameters: (1) input frequency, (2) input synchronization index, (3) mean input firing rate, (4) baseline input firing rate, (5) spatial spread of the inputs, (6) postsynaptic potential amplitude of each input, (7) the threshold of the model neuron, and (8) the absolute refractory period. By using (intracellular) electrophysiological data to estimate values for most of these parameters, it becomes possible to use the model to gain estimates of the spatial spread over the basilar membrane from which globular bushy cell input AN fibers originate.

The integrate-and-fire model has been shown to provide a good qualitative approximation of globular bushy cells of the CN (Kuhlmann et al 2002). Figure 5.31 shows examples of a periodic firing rate function that describes the individual inputs to the model (left) and a comparison between the output period histograms of the model and physiological data from a globular bushy cell (right). From Figure 5.31 (right) it is clear that globular bushy cells show phase locking, that is, they have the ability to generate action potentials at a certain phase of the cycle of a component frequency of a sound stimulus. The degree of phase locking to a particular input frequency can be measured by the synchronization index (SI), which takes on the values between 0, when a cell is not phase locked at all, and 1, when a cell is fully phase locked.

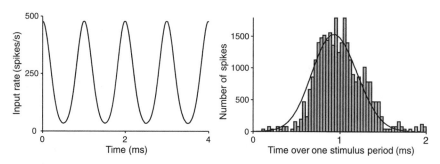

FIGURE 5.31. Left: The rate of inputs, corresponding to the rate of spikes in a single auditory nerve fiber. Right: The period histogram of a globular bushy cell compared to the output of the model. (Kuhlmann et al 2002. Summation of spatiotemporal input patterns in leaky integrate-and-fire neurons: application to neurons in the cochlear nucleus receiving converging auditory nerve fiber input. *Journal of Computational Neuroscience* 12: 55–73. Reprinted with permission of Kluwer Academic Publishers.)

To investigate the relationship between the output synchronization index of CN neurons (namely globular bushy cells) and the site of stimulation in the cochlea, phase differences between the periodic inputs of the model were incorporated, to mimic how the traveling wave consecutively activates primary afferent fibers originating over a spatial spread of the basilar membrane. This is shown in Figure 5.6.

The model has thus provided an approximate understanding of the relationship between the output synchrony of CN neurons and the site of stimulation in the cochlea. Analysis of the model found that output SI decreased with an increase in frequency and spatial spread. In addition, enhancement of the output SI relative to the input SI occurred for small spatial spreads of the basilar membrane over which input primary afferent fibers originate. Neural noise and refractory effects were also incorporated into the model. This is shown in Figure 5.32, where output SI is plotted against frequency and spatial spread (a), and critical distance for the enhancement of output SI relative input SI is plotted against frequency (b). This model has predicted well the behavior of globular bushy cells. However, other variables, such as inhibition, may need to be considered in order to improve the model. Research to implement this temporal and spatial pattern of action potentials with electrical stimulation needs to be done first to investigate the relationship of spike events between pairs of nerve fibers with different spatial separations and propagation delays. Speech-processing strategies will also need to retain phase information. This requires improved electrode arrays that lie close to the cochlear nerve fibers and ganglion cells, and have a greatly increased density of electrodes for stimulation as discussed in Chapters 8 and 14.

Neural Responses After Deafening

Auditory Nerve

Deafening produced a loss of peripheral processes and spiral ganglion cells, as well as demyelinization of some of the remaining neurons. The loss of hair cells

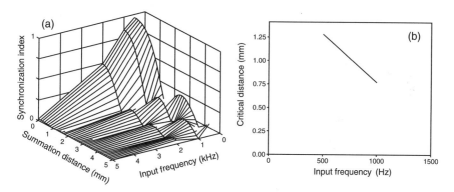

FIGURE 5.32. (a) Plot of output synchronization index against frequency of stimulation and summation distance. (b) The critical summation distance for which there is an enhancement of synchronization (Kuhlmann et al 2002 Summation of spatiotemporal input patterns in leaky integrate-and-fire neurons: application to neurons in the cochlear nucleus receiving converging auditory nerve fiber input. *Journal of Computational Neuroscience* **12**: 55–73. Reprinted with permission of Kluwer Academic Publishers.)

removed the electrophonic activity, and there was thus an absence of the higher threshold (D) responses of Javel et al (1987). The AN responses had higher thresholds and were more deterministic than for the normal hearing ear. There was less localized stimulation of neurons for place coding, as the peripheral processes are better arrayed for this. The loss of peripheral processes and ganglion cells led to elevated thresholds, longer latencies, and a reduction in the amplitude of the evoked potentials (Zhou et al 1995; Shepherd and Javel 1997b; Hardie and Shepherd 1999). Demyelinization reduced the efficiency of transmission and increased the risk of a conduction block (Smith and McDonald 1999). This is the probable explanation for the observation that some fibers in the deafened cochlea exhibited bursting alternating between periods of decreased activity and inactivity (Shepherd and Javel 1997a). Otherwise there were monotonic rate/intensity functions as seen in normal fibers. Long term, there was a reduction in the transmission of temporal information compared to the normal (Shepherd and Javel 1997a; Javel and Shepherd 2000), and this was also related to an increase in the absolute refractory period (Shepherd and Hardie 2001; Shepherd, Roberts, and Paolini, unpublished observations).

Cochlear Nucleus

In vivo intracellular recordings in the AVCN evoked by electrical stimulation of the cochlear nerve in long-term deafened animals showed that the arrival of synchronized EPSPs was compromised (Paolini, Roberts, Clark, and Paolini, unpublished observations). This was probably due to changes in synapses induced by the deafness and would have significant effects on the temporal processing in the higher pathways in view of the observations of Clark (1996), Irlicht and Clark (1996), Burkitt and Clark (1998), and Fitzgerald et al (2000) that were discussed in detail above (see Addition of Noise).

Midbrain

The limitations in the temporal transmission of information in the lower pathways is reflected in the midbrain. Shepherd et al (1999) reported increases in response latency and temporal jitter, in long-term versus short-term deafened animals. This is likely due to the reduced processing of information in the VCN due to the effects of demyelinization. Studies have been undertaken on the temporal responsiveness of neurons in bilaterally deafened animals (Snyder et al 1995; Shepherd et al 1999), but the data and design do not lead to definitive conclusions about the importance of using unilateral or bilateral implants.

Ganglion Cell Population and Frequency Coding

A behavioral study (Black et al 1981b, 1983a) was undertaken on the experimental animal to help determine whether the population and density of residual ganglion cells had an effect on the detection of rate of stimulation. Experimentally deafened cats with differing populations of residual spiral ganglion cells were implanted with cochlear electrodes and stimulated electrically. They were conditioned to respond to changes in electrical pulse rate, and electrical pulse rate DLs were determined. It was found that although there were some variations in DLs between animals, there appeared to be no correlation between DLs and residual ganglion cell populations over a range of 8% to 44% of the normal.

Greatly reduced ganglion cell numbers may also not affect DLs for rate of stimulation in humans. For example, in a patient, who had a cochlear implant and died due to heart disease, the ganglion cell population in the stimulated cochlea was on average 10% of the normal, and his rate difference limen of 10% was in the upper 13th percentile level for all patients (Clark, Shepherd et al 1988). Nevertheless, rate discrimination may not have been the major factor accounting for this patient's lower than average speech perception scores. His inability to discriminate place of stimulation may have been more important, especially as his speech processing strategy required the first and second formants of speech to be presented as place of stimulation.

Place Coding

Studies in the experimental animal have been important in determining whether electrical stimulation could convey frequency information not only on a temporal or rate basis but also as a place code. Initially, it was not certain whether current could be localized to separate groups of nerve fibers, as it could short-circuit through the fluid and electrolytes in the cochlea.

Mode of Stimulation

Studies were undertaken to see how effective bipolar, common ground, and monopolar stimulation (Fig. 5.33) would be in localizing electrical current to separate groups of nerve fibers for place coding. Bipolar stimulation occurs when a po-

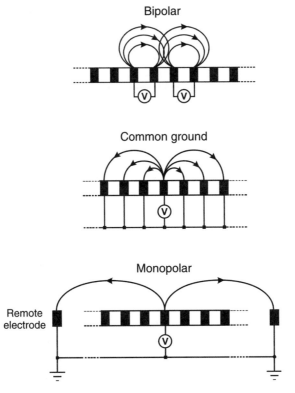

FIGURE 5.33. A diagram of the voltage fields for bipolar, pseudo-bipolar (common ground), and monopolar stimulation.

tential difference is created between neighboring electrodes to allow current to flow between the two. Common ground stimulation occurs between one active electrode and the others all connected electronically. The current therefore flows from the active electrode to all the others. Finally, monopolar stimulation occurs when a potential difference is created between one active electrode and a distant ground outside the cochlea.

Experimental Animal Studies and Mode of Stimulation

Studies were required to investigate current spread for electrodes at different sites within the cochlea and various electrode geometries to determine how best to localize the current to separate groups of cochlear nerve fibers. This applied in particular when scar and bone had formed after implantation, and in diseased cochleae. The electrodes also had to be designed for biological safety.

A study by Merzenich et al (1973), recording from units in the IC, allowed the spatial attenuation of current along the cochlea to be calculated. This provided evidence that current could be localized to separate groups of cochlear nerve fibers to make the coding of frequency on a place basis possible. The spread of current for bipolar and monopolar stimulation was determined by recording the thresholds for units in the IC. By also recording their tuning curves or frequencies of best response for acoustic stimulation of the opposite ear, it was possible to plot the threshold to electrical stimulation versus distance along the cochlea. The results showed that with bipolar stimulation using a carrier molded to the shape of the scala tympani, current would be localized to separate groups of nerve fibers (Merzenich and Reid 1974). The localization of current for bipolar stimulation with a molded array containing small electrodes placed beneath the basilar membrane was shown to be better than for the electrodes lying freely in the scala, and considerably better than for monopolar stimulation (Schindler et al 1977). The current distribution for bipolar stimulation, however, varied depending on the orientation of the electrode pair and the presence or absence of peripheral processes (dendrites) of spiral ganglion cells (Merzenich et al 1979).

The orientation of the electrodes was relevant when the peripheral processes of the spiral ganglion cells were present, as they pass across the basilar membrane where they lie close to the electrodes on the molded array. The data did not apply if the dendrites had retracted ("died back"), which usually occurs following a hearing loss with degeneration of the sensory hair cells. In this case only ganglion cells remain in the modiolus.

It should also be noted that the anatomy of the cat is different from the human. In the human the spiral ganglion cells abut the scala tympani lower down the inner wall (Clark 1987). Consequently, it is difficult to design an optimal array and mode of stimulation from cat data alone. The site of expected neural stimulation was important, and was relevant for the molded array as it quite markedly limited the spread of current (Merzenich et al 1973; Merzenich and Reid 1974). It had the electrodes beneath the basilar membrane to provide bipolar stimulation of the peripheral processes after they emerged from the habenula perforata in the osseous spiral lamina. Physiological studies reported above (Stypulkowski and van den Honert 1984; Javel et al 1987) showed that with an electrode in the scala tympani, low-stimulus intensities probably excited the peripheral processes, but higher intensities were required to stimulate the spiral ganglion cells. In contrast, pathological studies showed that only a small proportion of patients with a profound-total hearing loss have significant numbers of peripheral processes (Paparella and Sugiura 1967; Hinojosa et al 1987).

The above studies by Merzenich and Reid (1974) were all carried out acutely on cats with normal cochleae that did not have either the fibrous tissue sheath or new bone formation that develop some weeks after implantation and alter current flow. The surrounding tissue has a very significant effect on current flow, and must be taken into consideration for cochlear implant electrode arrays in patients. In fact when a molded array with small electrode pads and bipolar pulses was used in patients, it was not always possible to stimulate over the dynamic range, as the stimulator would run out of voltage compliance (Wilson et al 1988). Studies by Osberger and Fisher (1999) found 95% of patients were able to use pulsatile stimulation only in the monopolar mode. A modified bipolar stimulus mode was devised to cover the dynamic range for bipolar stimulation. Current was passed from a medial electrode to lateral electrode of the next pair.

However, it was later shown with psychophysical data (Busby et al 1994) that monopolar stimulation between an active and distant reference electrode in the cochlea could allow localized stimulation. It further emphasizes the problem of extrapolating from the experimental animal to the human, and considering the anatomy and pathology.

Resistance Models of Current Spread

The research at the University of Melbourne first aimed to develop a three-dimensional discrete resistance computer model of the cochlea to determine the current spread within the cochlea and AN for a variety of stimulus sites and electrode geometries before the validation of the findings on the experimental animal (Black and Clark 1978, 1980). The electrical resistances of cochlear structures in the model had been previously measured to help determine how voltages to sound stimuli are distributed (von Békésy 1951; Johnstone et al 1966) (Fig. 5.34). Kirchhoff's rules were important to reduce the network to simple series/parallel combinations. Six equations specified each model section, and one equation the stimulus loop current. The model results showed that the distribution of current within the cochlea varied greatly, and depended on the site and orientation of the stimulating electrodes. It also showed that the attenuation of current in the peripheral processes and cochlear nerve fibers was different from the voltage attenuation in the scalar fluids.

To assess the model in predicting the spread of current to the peripheral processes and spiral ganglion cells, physiological experiments measured current spread for bipolar and monopolar stimulation with scala tympani electrodes, and bipolar stimulation between the scala vestibuli and scala tympani. The thresholds of IC units to stimulation were plotted as a function of the characteristic frequency

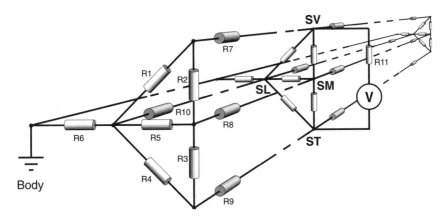

FIGURE 5.34. The three-dimensional discrete resistance model of the cochlea. SM, scala media; SV, scala vestibuli; ST, scala tympani; SL, spiral lamina. (Reprinted with permission from Black, R.C., Clark G.M., "Differential electrical excitation of the auditory nerve," *Journal of the Acoustical Society of America* **67**(3), 1980, pp. 863–874, Acoustical Society of America.)

of the units and their computed distances from the round window. The results confirmed that the current was much more localized for bipolar compared to monopolar stimulation in the scala tympani, falling off at 3 to 4 dB/mm and 1 dB/mm, respectively. Alternatively, when expressed as length constants, the results were 12 and 3 for monopolar and bipolar stimulation in the scala tympani, and 7.5 for stimulation between the scala tympani and scala vestibuli. The length constant (λ) was the inverse of the natural logarithm of the voltage 1 mm from the recording site (V_1) divided by the voltage at the site (V_0).

$$\lambda = \left(\ln \frac{V_1}{V_0} \right)^{-1}$$

The results thus showed that bipolar stimulation in the scala tympani provided better localization of current than bipolar stimulation between the scala vestibuli and scala tympani.

A resistance model, similar to that of Black and Clark (1980), was evaluated by Spelman et al (1980). They used a four-electrode array to inject current into the cochlea, and measured potential differences in current flow between various electrode pairs. Results agreed with von Békésy's conclusion that the principal grounding path out of the cochlea was through the vascular supply.

Banded Free-Fitting Electrode Array

The above model and experimental results for bipolar stimulation within the scala tympani were obtained with electrodes with small surface areas lying freely within the scala tympani. In the study by Black and Clark (1978, 1980), the spread of current was greater than that obtained by Merzenich and Reid (1974) for bipolar stimulation using small circular electrodes on a molded array, where the attenuation ranged between 3.3 and 8 dB/mm.

However, as discussed in Chapter 4, the surface area of stimulating electrodes needs to be large enough to reduce charge density to a level that would avoid electrode corrosion and damage to neural tissue. The dimensions to minimize corrosion were established from in vitro research (Black and Clark 1978; Black and Hannaker 1979). It was found platinum dissolution occurred for biphasic current pulses at charge densities as low as 25 μC cm^{-2} geometric. Therefore, the electrodes needed to be large, but still provide localized stimulation of cochlear nerve fibers. As investigations to be discussed in Chapter 10 had shown that the electrodes needed to be smooth to avoid cochlear damage, they were made of bands of polished platinum. This design was also found to make it easy for the electrodes to be explanted without avulsing the fibrous tissue sheath formed around the array and thus minimize damaging cochlear structures (see Chapter 3). The platinum dissolution seen in vitro was later not found to be as great in vivo (Robblee et al 1980). Protein had a protective effort on the surface of the metal. The width of the bands was optimized to 0.3 mm with an interelectrode distance of 0.45 mm. With an electrode diameter at the tip of 0.4 mm increasing to 0.6 mm, the surface area thus varied from 0.4 to 0.6 mm^2.

Further studies were then needed to determine the voltage attenuation along the scala tympani for this banded electrode, as well as the localization of the current within the cochlear nerve fibers. The voltage attenuation was first determined in saline tubes used to model cochleae of different dimensions, and then in the cat and human (Black et al 1981a; Lukies et al 1986, 1987). In the saline tube (cochlear model), the current distributions for pseudo-bipolar or common ground stimulation (Fig. 5.33) were determined. They were found to be similar to those measured in the human cochlea (Black et al 1981a). Another study to measure the voltage attenuation in a cochlear model (5 mm diameter saline tube) showed that for bipolar stimulation using the banded electrode the field was bell-shaped (Black et al 1983b). The attenuation was 6 to 10 dB/mm.

The spread of current to the peripheral processes and spiral ganglion cells for bipolar stimulation with the banded electrode was measured in the cat scala tympani (O'Leary et al 1985). This was undertaken by determining the extent to which the electrically evoked response from electrode pairs at different sites along the basal turn of the cochlea were masked by a stimulus from a fixed electrode pair. The degree of masking produced a plot of the distribution of the excitatory field. The results showed an attenuation of 2.2 to 2.9 dB/mm (3- to 4-mm length constant).

Current spread in the human cochlea after implantation and time for fibrous tissue and bone to form was studied later for the banded array using psychophysical masking procedures (Tong et al 1987) (see Chapter 6). The results showed the spread of neural excitation for comfortable loudness to be 3 to 9 dB/mm for bipolar stimulation.

Although this research demonstrated that current localization with the free-fitting banded electrode was satisfactory, an investigation was carried out to see if current localization could be improved with radial rather than longitudinal stimulation using a half-band instead of a full-band electrode (Clark, Shepherd et al 1983). Better current localization would allow an increase in the number of stimulus channels to be used to convey place pitch.

The half-band and full-band electrodes for radial and longitudinal stimulation in the cat cochlea are shown in Figure 5.35 (Clark, Shepherd et al 1983; Clark et al 1987). A_1 and A_2 are the two half-bands making up one full-band electrode, and B_1 and B_2 are the two half-bands making up an adjacent full-band electrode. The amplitude of the brainstem responses at different current intensities was recorded for each combination of bands. As shown in Figure 5.35, with longitudinal stimulation between the two half-band electrodes, A_1 to B_1, close to the spiral lamina, the threshold was lower than for the two half-band electrodes, A_2 to B_2, further from the spiral lamina. Furthermore, the results obtained during longitudinal stimulation between the two full-band electrodes (A_1 and A_2 to B_1 and B_2 joined together) lay between those for the two half-band electrodes. During radial stimulation (A_1 to A_2) the threshold was high and the growth of response with current was small. This latter finding implies that radial stimulation between half-band electrodes provided improved current localization. It was assumed the gra-

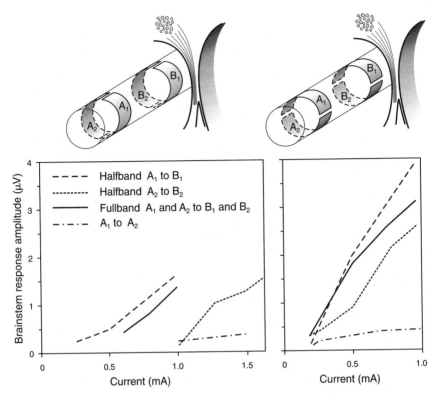

FIGURE 5.35. Brainstem response amplitude growth functions in the cat for full (A_1 and A_2 to B_1 and B_2), longitudinal half-band (A_1 to B_1 and A_2 to B_2) and radial half-band (A_1 to A_2) stimulation. (Clark, Shepherd et al 1983, Cochlear Prostheses: an international symposium. *Annals of the New York Academy of Science* **405**: 191–201. Reprinted with permission.)

dient was flat, as the localized radial stimulation did not recruit as many neurons with an increase in intensity.

After making the recordings in the positions shown on the left of Figure 5.35, the electrode array was then rotated through 90 degrees, and the thresholds and amplitude growth responses measured again. This resulted in a change in the thresholds and slope of the growth curves. The full-band growth curve had changed significantly and had a lower threshold, indicating that the electrode had moved closer to the nerve fibers. The results showed less difference for longitudinal stimulation using the two pairs of half-band electrodes A_1 to B_1 and A_2 to B_2. The threshold for radial stimulation A_1 to A_2 dropped significantly, as would be expected if the electrode had moved closer to the nerve fibers, and its growth response function was still very flat. The results show that for half-band electrodes current localization is improved with radial rather than longitudinal stimulation, as was seen by Merzenich and White (1980) for small circular electrodes in a cochlear mold. However, longitudinal stimulation between pairs of half-band electrodes did not provide better current localization than full-band stimulation.

The small advantage in current localization for half-band free-fitting electrodes lying around the periphery of the scala tympani is more than offset by the fact that during insertion they could be misaligned due to rotation of the electrode as it passes around the basal turn (see Chapters 2 and 8). In addition, rotation may also be carried out deliberately by the surgeon to facilitate its insertion, but disease and anatomical variations prevent accurately locating the electrode array to take advantage of either half-band longitudinal or half-band radial stimulation. Consequently, these electrophysiological findings together with biological safety advantages of a free-fitting electrode made the Melbourne/Cochlear banded electrode array an optimal solution for 20 channels of multiple-electrode intracochlear stimulation.

In addition, the administration of [^{14}C] 2-deoxyglucose while stimulating has also helped demonstrate the degree of current localization within the IC for bipolar stimulation with the Nucleus banded array (Brown et al 1992).

Finite Element Analysis Models

With the development of computer processing power, more detailed modeling of current spread became possible. As a result finite difference and finite element models were used to examine the current spread for different electrode geometries. With a finite difference model equations are integrated at fixed points. In contrast, a finite element model integrates equations over a specified surface or area to produce a shape function. Girzon and Eddington (1987) refined a three-dimensional finite difference model of the cochlea. By comparing the predictions of the model with direct measurements of current spread, they found good agreement between the model and the experimental findings, especially for the basal and central areas of the scala tympani. They concluded that the anatomical model of the cochlea could be used to predict the magnitude and direction of current density in any structure of interest. Finite element analysis was used by Wilson et al (1988) to develop an integrated field neuron model. This model coupled a three-dimensional finite element model of the implanted cochlea with a lumped element model of an auditory neuron. The rationale behind the model was to split the current and potential distributions from implant stimulation and neural excitation to directly relate various strategies to neural excitation patterns. Different electrode configurations and shapes were tested in the model. They were the pure longitudinal and pure radial point electrodes, offset radial configuration (the University of California at San Francisco), and the University of Melbourne banded longitudinal configuration. The model showed that the predictions of pure radial and longitudinal and offset radial configurations were reasonably well modeled. However, the predictions for the integrated field of the University of Melbourne's array were not in agreement with extensive clinical experiences for the Melbourne system.

A more representative model of electrical resistances and current flow within the cochlea was developed by Frijns et al (1995). This was a rotationally symmetrical volume conductor model using the boundary element method. The calculated potential distribution in the cochlea due to stimulating electrodes was combined with a multiple nonlinear model of AN fibers. The model predicted that the threshold, spatial selectivity, and dynamic range depended on the exact location of the electrode within the scala tympani. The results were in agreement with electrically evoked auditory brainstem response (EABR) data of Shepherd et al (1993). The model demonstrated that an active modeling of nerve fibers was essential to obtain correct predictions for biphasic pulses. Previous models that unrolled the cochlea could have produced errors. The model was applied to the guinea pig cochlea and compared to electrophysiological data from the cat (Frijns et al 1996; van den Honert and Stypulkowski 1987). The data indicated that asymmetric biphasic pulses approximately doubled the number of independent channels that could be used with longitudinal electrodes.

Plasticity

It was important to know whether people born deaf would have the basic neural pathways for place coding of frequency, and to what extent this could be modified through plasticity. Unit recordings by Snyder et al. (1990) suggested that chronic electrical stimulation might increase cortical representation through wider excitation due to the spread of current from the stimulus site. To better understand if such a response depended on maturation of neurons, research examined the uptake of $[^{14}C]$ 2-deoxyglucose after stimulating animals of different ages (Seldon et al 1996). There was, however, no difference in uptake with age or with whether the animal was previously stimulated or unstimulated. This suggested that basic neural connections were maintained regardless of stimulation, and that other factors such as the position of the stimulating electrode and current spread, may need to be considered. This conclusion is consistent with findings in altricial animals that gross connections in the auditory system are in place before the onset of hearing (Young and Rubel 1986; Friauf and Kandler 1990). So the evidence suggested that the basis for place coding and presumably temporal coding is present at birth, but the sensory input at an early stage leads to modification in fine neural connectivity. Further evidence that the immature cortex could be activated on a place-coding basis, was provided in congenitally deaf cats by Hartmann et al (1997) from electrically evoked field potentials and unit responses. This was also demonstrated in neonatally deafened animals (Raggio and Schreiner 1999). However, in short-term deafness tonotopic organization was present in parts of the primary cortex, but was largely absent in long-term deafness. This could have been due to the loss of inhibition. However, with the positron emission tomography (PET) data on deaf subjects (Truy et al 1995), with electrical stimulation of the cochlea the cortical blood flow increased in people with a hearing loss of up to 26 years' duration.

Plasticity has been demonstrated at a cellular level in the auditory cortex particularly in the middle cortical layers III/IV. This applies to cats (Friauf et al 1990; Friauf and Shatz 1991) as well as to humans (Huttenlocher and Dabholkar 1997). Layer III remains plastic well into adulthood (Snyder et al 1995; Kaczmarek et al 1997). Field potentials in each layer can be analyzed to measure synaptic currents (Mitzdorf 1985). Furthermore, infragranular layers project down to the thalamus and, as corticothalamocortical loops, may be involved in memory. Lack of synaptic activity in this layer may lead to defects in auditory memory.

Exposure of normal-hearing cats to meaningful stimuli produced long latency responses from the primary auditory cortex (>150 ms) (Eggermont 1992; Dinse et al 1997; Kral et al 1999). Long latency responses were also recorded from units in response to chronic electrical stimulation (Klinke et al 1999). This suggested that electrical stimuli could carry meaning similar to sound. The long latency responses are possibly mediated by corticothalamic loops (Contreras et al 1996; Steriade 1997, 1999), and thus involved in short-term memory. The higher order auditory areas involved in these responses project to cross-modal regions (Pandya and Yeterian 1985). Maturation of the long latency responses is thus of crucial importance for wider auditory cognitive function.

Electrical stimulation of the cochlea in experimental animals can lead to synaptic activation of cortical layers, referred to above, comparable to normal hearing cats (Kral et al 2001b). The greatest increases in synaptic currents were seen in the layers where cortical plasticity was the greatest (Kaczmarek et al 1997). This is consistent with the findings of Eggermont et al (1997) on deaf children, which showed the latencies of P1 and P2 under electrical stimulation approached those with normal-hearing children. With the increase in response amplitude, there was an extension of the area of the cortex activated (Klinke et al 1999) presumably due to the spread of current within the cochlea. The activated area was the one producing middle latency responses (10 to 50 ms). This expansion was associated with enlargement of the dendritic trees and axonal sprouting. These were the neuro-architectural changes associated with developmental plasticity (Kral et al 2001b). In addition, there is a rapid development of synapses that is genetically determined up to 1 year postnatally. Then there is a slow stabilization or elimination of synapses from 2 years to adolescence (Conel 1939–1967; Huttenlocher and Dabholkar 1997). This phase depends on auditory stimuli. The fastest synaptic pruning occurs between the fifth and twelfth year. Further evidence for the effects of electrical stimulation on plasticity were obtained by Leake et al (2000) who demonstrated that non-simultaneous stimulation of two adjacent sectors of the AN led to a sharpening of the central representation of these sectors, but if the stimuli were simultaneous then there was an expansion in the area of activity.

In a study by Kral et al. (2001a) the area of cortical activation increased up to the age of five months. This correlated with the time course for the maturation of the cat auditory cortex (Eggermont 1996). The time course however is longer in humans. Evoked potential studies by Ponton et al. (1999, 2000a,b) indicate the sensitive period in the development of human auditory potential ends around 12 years.

The above findings are consistent with the clinical data for electrical stimulation of the cochlear nerve in children from the University of Melbourne's Cochlear Implant Clinic at the Royal Victorian Eye and Ear Hospital. It has been shown that speech perception continues to improve for children implanted under 5 years of age, but reaches a plateau in children operated on at an older age, approximately 5 years post-implantation.

Intensity Coding

Stimulus Parameters (Strength Duration)

The electrical stimulus parameters responsible for coding intensity are studied through their effects in depolarizing the nerve membrane. The nerve membrane acts as a leaky capacitor. In particular, there is a relationship between the current strength and its duration in initiating a neural response, as discussed in Chapter 4. A minimum quantity of electrical charge (current intensity multiplied by time) must pass through the membrane with each pulse to change its potential. Intensity and duration can be traded against each other. This relationship is plotted as a strength-duration curve for the AN (Fig. 5.3) (Black and Clark 1977). The chronaxie, a measure of neural excitability and twice the rheobase current, was 180 μs for the AN. This was seen as an optimal pulse duration for the initial University of Melbourne's multiple-electrode receiver-stimulator, as the range in current amplitude for the threshold and dynamic range would not be excessive. A behavioral study in cats at the University of Melbourne (Johnson 1980) demonstrated thresholds were lowest for biphasic pulses with phase widths of 100 μs, suggesting the need to use an even lower pulse width.

Rate and Latency of Unit Responses

The discharge rate in AN fibers for variations in the intensity of an electrical stimulus was studied by Moxon (1971) and Kiang and Moxon (1972). They observed that the dynamic range of the fibers was approximately 4 dB from threshold to maximum firing. This is illustrated in Figure 5.36 by Javel et al (1987). This is a very narrow range when compared to that for AN fiber responses to sound where there is a 20 to 50 dB range from threshold to the point where the firing saturates (Kiang et al 1965b).

The dynamic range of single unit response in the cat has been shown to increase for low-frequency modulated high-frequency (5000 pulses/s) pulse trains. The range averaged 18 to 25 dB compared to the 2 dB without the high rates (Litvak 2002). Runge-Samuelson (2002) (quoted by Hong et al, submitted) recorded sinusoidal electrically evoked compound action potentials with a 5000-pulses/s conditioner, and the decrease in the slope of the input/output function also indicated an increase in the dynamic range. This is consistent with the model of Rubinstein discussed above (see Biophysical Model).

Population of Neural Responses

Firing rate, however, is not the only coding mechanism for intensity. The coding of intensity is also due to the number of neurons excited or a combination of the population of neurons excited and their discharge rate. Studies on field potentials (potentials that reflect the activity of a number of neurons) have shown that with electrical stimulation their amplitude grows rapidly with increases in intensity.

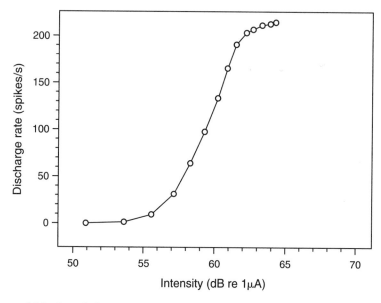

FIGURE 5.36. Growth function for unit responses. Rate-intensity function for fiber in response to 200 ms/phase pulses presented at 200 Hz in 100-ms bursts (Javel et al 1987. Responses of cat auditory nerve fibers to biphasic electrical current pulses. *Annals of Otology, Rhinology and Laryngology* **96** (1 part 2) Supplement 128: 26–28. Reprinted with permission.).

Recordings made from the IC of the cat in response to electrical pulses and acoustic clicks (Simmons and Glattke 1972; Glattke 1974) have shown that the relative growth of the response to electrical stimulation (threshold to maximum amplitude) occurs for a stimulus range of 10 to 20 dB. In contrast, the stimulus intensity range for acoustic clicks was 60 dB. Similarly, electrically evoked responses recorded from the scalp of cats also show this same steep slope from the amplitude/intensity function (Black et al 1983a). The findings from these and other field potential studies suggest that intensity coding may be due to rate of firing as well as the number of nerves excited. This is consistent with the data from acoustic studies (Dobie and Kimm 1980; Charlet de Sauvage et al 1983), where there was a direct relation between the amplitude of the field potential and the current level in monkeys and guinea pigs, respectively. There was also good correlation between the electrically evoked brainstem response and the behavioral threshold (Smith et al 1994).

Behavioral Studies

As unit and field potential responses to the intensity of an electrical stimulus do not reveal what is perceived, behavioral studies in the experimental animal are also of value. Behavioral studies on intensity discrimination in monkeys (Pfingst et al 1983) have shown that DLs decrease as a function of the initial stimulus intensity from values of 1.5 to 3 dB near threshold to as low as 0.5 dB near the upper limit of the dynamic range. They also showed that DLs for electrical stimuli were roughly half those for acoustic stimuli. This latter finding demonstrates that the reduced dynamic range for electrical versus acoustic stimulation can be partially compensated for by the smaller DLs with electrical stimuli. It was also of importance in designing a cochlear prosthesis to know whether current pulse width or amplitude was preferable for coding intensity. An animal behavioral study at the University of Melbourne (Johnson 1980) indicated that the two could be traded, and that total charge was the important parameter.

Ganglion Cell Population and Amplitude Coding

The behavioral study on rate of stimulation referred to above (Black et al 1981b; Black et al 1983a) was also undertaken on the experimental animal to help determine whether the population and density of residual ganglion cells had an effect on the detection of amplitude of stimulation. The animals were conditioned to respond to changes in electrical amplitude, and amplitude DLs were determined. Although there were some variations in DLs between animals, there appeared to be no correlation between DLs and residual ganglion cell populations over a range of 8% to 44% of normal.

References

Abbas, P. J., C. M. Miller, A. Matsuoka and J. T. Rubinstein. 1997. The neurophysiologic effects of simulated auditory prosthesis stimulation. Fourth quarterly progress report. NIH contract NO1-DC-6-2111. Neural Prosthesis Program, NIH, Bethesda, MD.

Aitkin, L. M. 1985. The auditory midbrain. Clifton, New Jersey, Human Press.

Anderson, D. J. 1973. Quantitative model for the effects of stimulus frequency upon synchronization of auditory nerve discharges. Journal of the Acoustical Society of America 54: 361–364.

Au, D., I. Bruce, L. Irlicht and G. M. Clark. 1995. Cross-fiber interspike interval probability distribution in acoustic stimulation: a computer modeling study. Annals of Otology, Rhinology and Laryngology 104: 346–349.

Bi, Q. 1989. A closed-form solution for removing the dead-time effects from the poststimulus time histograms. Journal of the Acoustical Society of America 85: 2504–2513.

Black, R. C. and G. M. Clark. 1977. Electrical transmission line properties of the cat cochlea. Proceedings of the Australian Physiological and Pharmacological Society 8: 137.

Black, R. C. and G. M. Clark. 1978. Electrical network properties and distribution of potentials in the cat cochlea. Proceedings of the Australian Physiological and Pharmacological Society 9: 71P.

Black, R. C. and G. M. Clark. 1980. Differential electrical excitation of the auditory nerve. Journal of the Acoustical Society of America 67(3): 868–874.

Black, R. C., G. M. Clark and J. F. Patrick. 1981a. Current distribution measurements within the human cochlea. IEEE Transactions on Biomedical Engineering 28: 721–724.

Black, R. C., G. M. Clark, R. K. Shepherd, S. J. O'Leary and C. W. Walters. 1983a. Intracochlear electrical stimulation of normal and deaf cats investigated using brainstem response audiometry. Acta Oto-Laryngologica-supplement 399: 5–17.

Black, R. C., G. M. Clark, Y. C. Tong and J. F. Patrick. 1983b. Current distributions in cochlea stimulation. Annals of the New York Academy of Sciences 405: 137–145.

Black, R. C. and P. Hannaker. 1979. Dissolution of smooth platinum electrodes in biological fluids. In: Gildenberg, L., ed. Applied neurophysiology. Basel, Karger: 366–374.

Black, R. C., A. C. Steel and G. M. Clark. 1981b. Frequency discrimination and the spiral ganglion cell population in cats. Proceedings of the Australian Physiological and Pharmacological Society 12: 173P.

Blamey, P. J., P. W. Dawson, S. J. Dettman, et al. 1992. Speech perception, production and language results in a group of children using the 22-electrode cochlear implant. Journal of the Oto-Laryngological Society of Australia 1: 105–109.

Bliss, T. V. and T. Lomo. 1973. Long-lasting potentiation of synaptic transmission in the dentate area of the anaesthetized rabbit following stimulation of the perforant path. Journal of Physiology 232: 331–356.

Bogdanski, D. F. and R. Galambos. 1960. In: Rasmussen, G. L. and W. F. Windle, ed. Neural mechanisms of auditory and vestibular systems. Springfield, IL, Charles C. Thomas: 143–148.

Bollman, J. H., B. Sakmann and J. G. G. Borst. 2000. Calcium sensitivity of glutamate release in a calyx-type terminal. Science 289: 953–957.

Born, D. E. and E. W. Rubel. 1988. Afferent influences on brain stem auditory nuclei of the chicken: presynaptic action potentials regulate protein synthesis in nucleus magnocellularis neurons. Journal of Neuroscience 8: 901–919.

Bosher, S. K. and R. L. Warren. 1968. Observations on the electrochemistry of the cochlear endolymph of the rat: a quantitative study of its electrical potential and ionic composition as determined by means of flame spectrophotometry. Proceedings of the Royal Society of London 171B: 227.

Boudreau, J. C. and C. Tsuchitani. 1968. Binaural interaction in the cat superior olive S segment. Journal of Neurophysiology 31(3): 442–454.

Boudreau, J. C. and C. Tsuchitani. 1970. Cat superior olive S-segment cell discharge to tonal stimulation. In: Neff, W. D., ed. Contributions to sensory physiology. New York, Academic 4: 143–213.

Boulanger, L. and M. Poo. 1999. Presynaptic depolarization facilitates neurotrophin-induced synaptic potentiation. Nature Neuroscience 2: 346–351.

Brawer, J. R. and D. K. Morest. 1975. Relations between auditory nerve endings and cell types in the cat's anteroventral cochlear nucleus seen with the Golgi method and Nomarski optics. Journal of Comparative Neurology 160: 491–506.

Britt, R. and A. Starr. 1976a. Synaptic events and discharge patterns of cochlear nucleus cells. II. frequency-modulated tones. Division of Neurosurgery, Stanford University: 179–194.

Britt, R. and A. Starr. 1976b. Synaptic events and discharge patterns of cochlear nucleus. I. Steady-frequency tone bursts. Division of Neurosurgery, Stanford University: 162–178.

Brown, M., R. Shepherd, W. Webster, R. Martin and G. Clark. 1992. Cochleotopic selectivity of a multi-channel scala tympani electrode array using the 2-deoxyglucose technique. Hearing Research 59: 224–240.

Brownell, W. E., C. R. Bader, D. Bertrand and Y. de Ribaupierre. 1985. Evoked mechanical responses of isolated cochlear outer hair cells. Science 227: 194–196.

Bruce, I. C., L. S. Irlicht and G. M. Clark. 1998a. A mathematical analysis of spatiotemporal summation of auditory nerve firings. Information Sciences: Applications 111: 303–334.

Bruce, I. C., L. S. Irlicht, S. Dynes, E. Javel and G. M. Clark. 1997a. A stochastic model of the electrically stimulated nerve designed for the analysis of large-scale population. In: Popelka, G. R., ed. Abstracts of the Twentieth Midwinter Research Meeting of the Association of Research in Otolaryngology. Des Moines, Association for Research in Otolaryngology: 57.

Bruce, I. C., L. S. Irlicht, M. W. White, S. J. O'Leary and G. M. Clark. 2000. Renewal-process approximation of a stochastic threshold model for electrical neural stimulation. Journal of Computational Neuroscience 9: 119–132.

Bruce, I. C., L. S. Irlicht, M. White, et al. 1997b. An improved model of electrical stimulation of the auditory nerve. In: Clark, G. M., ed. Cochlear implants. XVI World Congress of Otorhinolaryngology Head and Neck Surgery. Bologna, Monduzzi Editore: 125–130.

Bruce, I. C., L. S. Irlicht, M. W. White, et al. 1999a. A stochastic model of the electrically stimulated auditory nerve: pulse-train response. IEEE Transactions on Biomedical Engineering 46: 630–637.

Bruce, I. C., L. S. Irlicht, M. W. White, et al. 1997c. Electrical stimulation of the auditory nerve: prediction of psychophysical performance by a model including stochastic aspects of neural response. In: Popelka, G. R., ed. Abstracts of the Twentieth Midwinter Research Meeting of the Association of Research in Otolaryngology. Des Moines, Association for Research in Otolaryngology: 81.

Bruce, I. C., M. White, L. Irlicht, S. J. O'Leary and G. M. Clark. 1998b. Advances in computational modeling of cochlear implant physiology and perception. In: Greenberg, S., and M. Slaney, eds. Proceedings of the NATO Advanced Study Institute on Computational Hearing. July 1–July 12 1998, Il Ciocco (Tuscany), Italy: 31–36.

Bruce, I. C., M. W. White, L. S. Irlicht, S. J. O'Leary and G. M. Clark. 1999c. Advances in relating cochlear implant physiology and psychophysics. Program and Abstracts of 1999 Conference on Implantable Auditory Prostheses: 5.

Bruce, I. C., M. W. White, L. S. Irlicht, S. J. O'Leary and G. M. Clark. 1999d. The effects of stochastic neural activity in a model predicting intensity perception with cochlear implants: low-rate stimulation. IEEE Transactions on Biomedical Engineering 46: 1393–404.

Bruce, I. C., M. W. White, L. Irlicht, et al. 1999b. A stochastic model of the electrically stimulated auditory nerve: single-pulse response. IEEE Transactions on Biomedical Engineering 46: 617–629.

Brugge, J. F., N. A. Dubrovsky, L. M. Aitkin and D. J. Anderson. 1969. Sensitivity of single neurons in auditory cortex of cat to binaural tonal stimulation: effects of varying interaural time and intensity. Journal of Neurophysiology 32: 1005–1024.

Brugge, J. F. and M. M. Merzenich. 1973. Responses of neurons in auditory cortex of the macaque monkey to monaural and binaural stimulation. Journal of Neurophysiology 36: 1138–1158.

Burkitt, A. N. 2000. Interspike interval variability for balanced networks with reversal potentials for large numbers of inputs. Neurocomputing 32–33: 313–321.

Burkitt, A. N. 2001. Balanced neurons: analysis of leaky integrate-and-fire neurons with reversal potentials. Biological Cybernetics 85: 247–255.

Burkitt, A. N. and G. M. Clark. 1998. New method for analyzing the synchronization of synaptic input and spike output in neural systems. In: Downs, T., M. Frean and M. Gallagher, eds. Proceedings of the Ninth Australian Conference on Neural Networks (ACNN '98). Brisbane, University of Queensland: 94–98.

Burkitt, A. N. and G. M. Clark. 1999a. Analysis of integrate and fire neurons: synchronization of synaptic input and spike output. Neural Computation 11: 871–901.

Burkitt, A. N. and G. M. Clark. 1999b. The dependence of synchronization upon stimulus frequency for integrate and fire neurons. Proceedings of the Australian Neuroscience Society 10: 173.

Burkitt, A. N. and G. M. Clark. 1999c. Modeling the response of neurons to auditory stimuli: differences between acoustical and electrical stimulation. In: Lithgow, B. and I. Cosic, eds. Biomedical Research in the 3rd Millennium. Proceedings of the Inaugural Conference of the Victorian chapter of the IEEE Engineering in Medicine and Biology Society. Caulfield, Victoria, Monash University: 46–49.

Burkitt, A. N. and G. M. Clark. 2000. Calculation of interspike intervals for integrate and fire neurons with Poisson distribution of synaptic inputs. Neural Computation 12: 1789–1820.

Busby, P. A., L. A. Whitford, P. J. Blamey, L. M. Richardson and G. M. Clark. 1994. Pitch perception for different modes of stimulation using the Cochlear multiple-electrode prosthesis. Journal of the Acoustical Society of America 95: 2658–2669.

Butler, R. A., I. T. Diamond and W. D. Neff. 1957. Role of auditory cortex in discrimination of changes in frequency. Journal of Neurophysiology 20: 108–120.

Caird, D. and R. Klinke. 1983. Processing of binaural stimuli by cat superior olivary complex neurons. Experimental Brain Research 52: 385–399.

Caleo, M., C. Lodovichi and L. Maffei. 1999. Effects of nerve growth factor on visual cortical plasticity require afferent electrical activity. European Journal of Neuroscience 11: 2979–2984.

Calford, M. B., W. R. Webster and M. M. Semple. 1983. Measurement of frequency selectivity of single neurons in the central auditory pathway. Hearing Research 11: 395–401.

Cant, N. B. 1981. The fine structure of two types of stellate cells in the anterior division of the anteroventral cochlear nucleus of the cat. Neuroscience 6: 2643–2655.

Carney, L. H. 1994. Spatiotemporal encoding of sound level: models for normal encoding and recruitment of loudness. Hearing Research 76: 31–44.

Carney, L. H. and M. Friedman. 1998. Spatiotemporal tuning of low-frequency cells in the anteroventral cochlear nucleus. Journal of Neuroscience 18: 1096–1104.

Caspary, D. M., P. M. Backoff, P. G. Finlayson and P. S. Palombi. 1994. Inhibitory inputs modulate discharge rate within receptive fields of anteroventral cochlear nucleus neurons. Journal of Neurophysiology 72: 2124–2133.

Charlet de Sauvage, R., Y. Cazals and J.-P. Erre. 1983. Acoustically derived auditory nerve action potential evoked by electrical stimulation: an estimation of the waveform of single unit contribution. Journal of the Acoustical Society of America 73: 616–627.

Clarey, J. and G. M. Clark. 2001. Inhibition underlies the encoding of short voice onset times in the ventral cochlear nucleus. Proceedings of the Australian Neuroscience Society 12: 219.

Clarey, J., A. G. Paolini and G. M. Clark. 2001. Brainstem encoding of short voice onset times in natural speech. Proceedings of the Australian Neuroscience Society, Brisbane 12: 218.

Clark, G. M. 1969a. Hearing due to electrical stimulation of the auditory system. Medical Journal of Australia 1: 1346–1348.

Clark, G. M. 1969b. Middle ear and neural mechanisms in hearing and the management of deafness. PhD dissertation. University of Sydney.

Clark, G. M. 1969c. Responses of cells in the superior olivary complex of the cat to electrical stimulation of the auditory nerve. Experimental Neurology 24: 124–136.

Clark, G. M. 1970a. A neurophysiological assessment of the surgical treatment of perceptive deafness. International Audiology 9: 103–109.

Clark, G. M. 1970b. The surgical treatment of perceptive deafness. An experimental study. Australian and New Zealand Journal of Surgery 39: 319.

Clark, G. M. 1987. The University of Melbourne–Nucleus multi-electrode cochlear implant. Advances in Oto-Rhino-Laryngology. Vol. 38. Basel, Karger.

Clark, G. M. 1995. Cochlear implants: future research directions. Annals of Otology, Rhinology and Laryngology 104: 22–27.

Clark, G. M. 1996. Electrical stimulation of the auditory nerve: the coding of frequency, the perception of pitch and the development of cochlear implant speech processing strategies for profoundly deaf people. Clinical and Experimental Pharmacology and Physiology 23: 766–776.

Clark, G. M. 1997. Electrical stimulation of the auditory nerve with a cochlear implant and the temporal coding of sound frequencies: a brief review. Australian Journal of Oto-Laryngology 2: 543–546.

Clark, G. M. 1998a. Cochlear implants. In: Wright, A. and H. Ludman, eds. Diseases of the ear. London, Edward Arnold: 149–163.

Clark, G. M. 1998b. Cochlear implants in the second and third millennia. In: Mannell, R. H. and J. Robert-Ribes, eds. Proceedings of ICSLP '98 Fifth International Conference on Spoken Language Processing. Canberra, Australian Speech Science and Technology Association: 1–16.

Clark, G. M. 1998c. Research advances for cochlear implants. Auris Nasus Larynx 25: 73–87.

Clark, G. M. 1999a. The Bionic Ear in the second and third millennia. Proceedings of the Australian Neuroscience Society 10: 4.

Clark, G. M. 1999b. The Bionic Ear towards 2000 and beyond. Taralye Bulletin 17: 4–10.

Clark, G. M. 1999c. Cochlear implants in the third millennium. American Journal of Otology 20: 4–8.

Clark, G. M. 2001. Editorial. Cochlear implants: climbing new mountains. The Graham Fraser Memorial Lecture 2001. Cochlear Implants International 2(2): 75–97.

Clark, G. M., R. C. Black, D. J. Dewhurst, I. C. Forster, J. F. Patrick and Y. C. Tong. 1977. A multiple-electrode hearing prosthesis for cochlear implantation in deaf patients. Medical Progress through Technology 5: 127–140.

Clark, G. M., P. J. Blamey, P. A. Busby, et al. 1987. A multiple-electrode intracochlear implant for children. Archives of Otolaryngology 113: 825–828.

Clark, G. M., T. D. Carter, C. L. Maffi and R. K. Shepherd. 1995. Temporal coding of frequency: neuron firing probabilities for acoustic and electric stimulation of the auditory nerve. Annals of Otology, Rhinology and Laryngology 104: 109–111.

Clark, G. M. and C. W. Dunlop. 1969. A technique for a wide approach to the medulla and bullae of the cat. Journal of Auditory Research 9: 189–193.

Clark, G. M., L. Irlicht and T. D. Carter. 1996. A neural model for the time/period coding of frequency for acoustic and electric stimulation. Proceedings of the Australian Neuroscience Society 7: 115.

Clark, G. M., H. G. Kranz and H. Minas. 1973. Behavioral thresholds in the cat to frequency modulated sound and electrical stimulation of the auditory nerve. Experimental Neurology 41: 190–200.

Clark, G. M. and D. Lawrence. 2000. Technical features of the Nucleus, Med-El and Clarion cochlear implants. Australian Journal of Oto-Laryngology 3: 516–522.

Clark, G. M., J. M. Nathar, H. G. Kranz and J. S. Maritz. 1972. A behavioral study on electrical stimulation of the cochlea and central auditory pathways of the cat. Experimental Neurology 36: 350–361.

Clark, G. M., R. K. Shepherd, B. K.-H. G. Franz, et al. 1988. The histopathology of the human temporal bone and auditory central nervous system following cochlear implantation in a patient. Correlation with psychophysics and speech perception results. Acta Oto-Laryngologica-supplement 448: 1–65.

Clark, G. M., R. K. Shepherd, J. F. Patrick, R. C. Black and Y. C. Tong. 1983. Design and fabrication of the banded electrode array. Annals of the New York Academy of Sciences 405: 191–201.

Clark, G. M. and Y. C. Tong. 1990. Electrical stimulation, physiological and behavioural stimulation. In: Clark, G., Y. Tong and J. Patrick, eds. Cochlear prostheses. London, Chuchill Livingstone: 15–32.

Clopton, B. M. and I. Glass. 1984. Unit responses at cochlear nucleus to electrical stimulation through a cochlear prosthesis. Hearing Research 14: 1–11.

Colombo, J. and C. Parkins. 1987. A model of electrical excitation of the mammalian auditory-nerve neuron. Hearing Research 31: 287–312.

Conel, J. L. 1939–1967. The postnatal development of human cerebral cortex. Cambridge, Harvard University Press: I–VIII.

Contreras, D., A. Destexhe, T. J. Sejnowski and M. Steriade. 1996. Control of spatiotemporal coherence of a thalamic oscillation by corticothalamic feedback. Science 274: 771–774.

Delgutte, B. 1984. Speech coding in the auditory nerve II. Processing schemes for vowel-like sounds. Journal of the Acoustical Society of America 75: 879–886.

Delgutte, B. and N. Y. S. Kiang. 1984. Speech coding in the auditory nerve I. Vowel-like sounds. Journal of the Acoustical Society of America 75: 866–878.

Derksen, H. E. and A. A. Verveen. 1966. Fluctuations of resting neural membrane potential. Science 151: 1388–1389.

Diamond, D. M. and N. M. Weinberger. 1984. Physiological plasticity of single neurons in auditory cortex of the cat during acquisition of the pupillary conditioned response. II. Secondary field (AII). Behavioral Neuroscience 98: 189–210.

Dinse, H. R., K. Kruger, A. C. Akhavan, F. Spengler, G. Schoner and C. E. Schreiner. 1997. Low-frequency oscillations of visual, auditory and somatosensory cortical neurons evoked by sensory stimulation. International Journal of Psychophysiology 26: 205–227.

Dobie, R. A. and J. Kimm. 1980. Brainstem responses to electrical stimulation of the cochlear. Archives of Otolaryngology 106: 573–577.

Dowell, R. C., P. J. Blamey and G. M. Clark. 1995. Potential and limitations of cochlear implants in children. Annals of Otology, Rhinology and Laryngology 104(suppl 166): 324–327.

Dowell, R. C., G. M. Clark, S. J. Dettman and P. W. Dawson. 1986. Results for children and adolescents using the multichannel cochlear prosthesis. Australian Journal of Audiology (suppl 5): 13.

Dowell, R. C., L. F. A. Martin, P. J. Blamey and A. M. Brown. 1985. Assessment of implant patient speech discrimination. In: Schindler, R. and M. Merzenich, eds. Cochlear implants. New York, Raven Press: 465–468.

Dowell, R. C., P. M. Seligman, P. J. Blamey and G. M. Clark. 1987. Speech perception using a two-formant 22-electrode cochlear prosthesis in quiet and in noise. Acta Oto-Laryngologica 104(5–6): 439–446.

Dowell, R. C., L. A. Whitford, P. M. Seligman, B. K.-H. Franz and G. M. Clark. 1990. Preliminary results with a miniature speech processor for the 22-electrode/Cochlear hearing prosthesis. In: Sacristan, T., ed. Otorhinolaryngology, Head and Neck Surgery. Amsterdam, Kugler and Ghedini: 1167–1173.

Dynes, S. B. and B. Delgutte. 1992. Phase-locking of auditory-nerve discharges to sinusoidal electric stimulation of the cochlea. Hearing Research 58(1): 79–90.

Eddington, D. K. 1980. Speech discrimination in deaf subjects with cochlear implants. Journal of the Acoustical Society of America 68: 885–891.

Eggermont, J. J. 1992. Stimulus induced and spontaneous rhythmic firing of single units in cat primary auditory cortex. Hearing Research 61: 1–11.

Eggermont, J. J. 1996. Differential maturation rates for response parameters in cat primary auditory cortex. Auditory Neuroscience 2: 309–327.

Eggermont, J. J., C. W. Ponton, M. Don, M. D. Waring and B. Kwong. 1997. Maturational delays in cortical evoked potentials in cochlear implant users. Acta Otolaryngologica 117: 161–163.

Ehrenberger, K., D. Felix and K. Svozil. 1999. Stochastic resonance in cochlear signal transduction. Acta Oto-Laryngologica 119(2): 166–170.

Erulkar, S. D., R. A. Butler and G. L. Gerstein. 1968. Excitation and inhibition in cochlear nucleus. II. Frequency modulated tones. Journal of Neurophysiology 31: 537–548.

Evans, B. M., P. Dallos and R. Hallworth. 1989. Asymmetries in motile responses of outer hair cells in simulated in vivo conditions. In: J. P. Wilson, ed. Cochlear mechanisms. New York, Plenum: 205–206.

Evans, E. F. 1975. Cochlear nerve and cochlear nucleus. In: Keidel, W. D. and W. D. Neff, eds. Handbook of sensory physiology. Auditory system, part 2. New York, Springer-Verlag: 1–108.

Evans, E. F. 1978. Place and time coding of frequency in the peripheral auditory system: some physiological pros and cons. Audiology 17: 369–420.

Evans, E. F. 1981. The dynamic range problem: place and time coding at the level of cochlear nerve and nucleus. In: Syka, J. and L. Aitkin, eds. Neuronal mechanisms of hearing. New York, Plenum: 69–85.

Evans, E. F. and P. G. Nelson. 1973. The responses of single neurones in the cochlear nucleus of the cat as a function of their locations and the anaesthetic state. Experimental Brain Research 17: 402–427.

Evans, E. F. and A. R. Palmer. 1975. Responses of units in the cochlear nerve and nucleus of the cat to signals in the presence of bandstop noise. Journal of Physiology 252: 60–62.

Evans, E. F. and I. C. Whitfield. 1964. Classification of unit responses in the auditory cortex of the unanaesthetised and unrestrained cat. Journal of Physiology 171: 476–493.

Fitzgerald, J. V., A. N. Burkitt, G. M. Clark and A. G. Paolini. 2001. Delay analysis in the auditory brainstem of the rat: comparison with click latency. Hearing Research 159: 85–100.

Fitzgerald, J. V., A. G. Paolini, A. N. Burkitt and G. M. Clark. 2000. Delay analysis in an investigation of auditory temporal coding. Proceedings of the Australian Neuroscience Society 11: 216.

Frankenhauser, B. and A. Huxley. 1964. The action potential in the myelinated nerve fiber of Xenopus laevis as computed on the basis of voltage clamp data. Journal of Physiology 171: 302–315.

Friauf, E. and K. Kandler. 1990. Auditory projections to the inferior colliculus of the rat are present by birth. Neuroscience Letters 120: 58–61.

Friauf, E., S. K. McConnell and C. J. Shatz. 1990. Functional synaptic circuits in the subplate during fetal and early postnatal development of cat visual cortex. Journal of Neurophysiology 66: 2601–2613.

Friauf, E. and C. J. Shatz. 1991. Changing patterns of synaptic input to subplate and cortical plate during development of visual cortex. Journal of Neurophysiology 66: 2059–2071.

Frijns, J. H., S. L. de Snoo and R. Schoonhoven. 1995. Potential distributions and neural excitation patterns in a rotationally symmetric model of the electrically stimulated cochlea. Hearing Research 87(1–2): 170–186.

Frijns, J. H., S. L. de Snoo and J. H. ten Kate. 1996. Spatial selectivity in a rotationally symmetric model of the electrically stimulated cochlea. Hearing Research 95: 33–48.

Galambos, R., J. Schwartzkopff and A. Rupert. 1959. Microelectrode study of superior olivary nuclei. American Journal of Physiology 197(3): 527–536.

Galuske, R. A. W., D.-K. Kim and W. Singer. 1999. The role of neurotrophins in developmental cortical plasticity. Restorative Neurology and Neuroscience 15: 115–124.

Gao, W. J., D. E. Newman, A. B. Wormington and S. I. Pallas. 1999. Development of inhibitory circuitry in visual and auditory cortex of postnatal ferrets: immunocytochemical localization of GABAergic neurons. Journal of Comparative Neurology 409: 261–273.

Gaumond, R. P., C. E. Molnar and D. O. Kim. 1982. Stimulus and recovery dependence of cat cochlear nerve fiber spike discharge probability. Journal of Neurophysiology 48: 856–873.

Geisler, C. D. 1981. A model for discharge patterns of primary auditory-nerve fibers. Brain Research 212: 198–201.

Geisler, C. D., W. H. Rhode and D. W. Hazelton. 1969. Responses of inferior colliculus neurons in the cat to binaural acoustic stimuli having wide-band spectra. Journal of Neurophysiology 32: 960–974.

Gerstein, G. L. and B. Mandelbrot. 1964. Random walk models for the spike activity of a single neuron. Journal of Biophysics 4: 41–68.

Girzon, G. and D. K. Eddington. 1987. A three dimensional, electro-anatomical model of the implanted cochlea. Proceedings of the Ninth Annual Conference of the IEEE Engineering in Medicine and Biology Society. New York, IEEE 4: 1904–1905.

Glattke, T. 1974. Electrical stimulation of the auditory nerve in animals. In: Merzenich, M., R. Schindler and F. Sooy, eds. Proceedings of the First International Conference on Electrical Stimulation of the Acoustic Nerve as a Treatment for Profound Sensorineural Deafness in Man. San Francisco, Velo-Bind: 105–121.

Goldberg, J. M. and P. B. Brown. 1969. Response of binaural neurons of dog superior olivary complex to dichotic tonal stimuli: some physiological mechanisms of sound localization. Journal of Neurophysiology 32: 613–636.

Goldberg, J. M. and W. D. Neff. 1961. Frequency discrimination after bilateral section of the brachium of the inferior colliculus. Journal of Comparative Neurology 116: 265–290.

Greenwood, D. D. and J. M. Goldberg. 1970. Response of neurons in the cochlear nuclei to variations in noise bandwidth and to tone-noise combinations. Journal of the Acoustical Society of America 47: 1022–1040.

Greenwood, D. D. and N. Maruyama. 1965. Excitatory and inhibitory response areas of auditory nucleus. Journal of Neurophysiology 28: 863–892.

Hall, J. L. 1965. Binaural interaction in the accessory superior-olivary nucleus of the cat. Journal of the Acoustical Society of America 37(5): 814–823.

Hardie, N. A., A. Martsi-McClintock, L. M. Aitkin and R. K. Shepherd. 1998. Neonatal sensorineural hearing loss affects synaptic density in the auditory midbrain. Neuroreport 9: 2019–2022.

Hardie, N. A. and R. K. Shepherd. 1999. Sensorineural hearing loss during development: morphological and physiological response of the cochlea and auditory brainstem. Hearing Research 128: 147–165.

Hartmann, R., R. K. Shepherd, S. Heid and R. Klinke. 1997. Response of the primary auditory cortex to electrical stimulation of the auditory nerve in the congenitally deaf white cat. Hearing Research 112: 115–133.

Hartmann, R., G. Topp and R. Klinke. 1984a. Discharge patterns of cat primary auditory fibres with electrical stimulation of the cochlea. Hearing Research 13: 47–62.

Hartmann, R., G. Topp and R. Klinke. 1984b. Electrical stimulation of the cat cochlea-discharge pattern of single auditory fibres. Advances in Audiology 1: 18–29.

Hill, A. V. 1936. Excitation and accommodation in nerves. Proceedings of the Royal Society 119: 305–355.

Hind, J. E., J. M. Goldberg, D. D. Greenwood and J. E. Rose. 1963. Some discharge characteristics of single neurons in the inferior colliculus of the cat. II. Timing of the discharges and observations on binaural stimulation. Journal of Neurophysiology 26: 321–341.

Hinojosa, R., R. Blough and E. Mhoon. 1987. Profound sensorineural deafness: a histopathologic study. Annals of Otology, Rhinology and Laryngology 96(suppl 128): 43–46.

Hodgkin, A. and A. Huxley. 1952. A quantitative description of membrane current and its application to conduction and excitation in nerve. Journal of Physiology 117: 500–544.

Hohn, N. and A. N. Burkitt. 2001a. Modeling the neural response to speech: stochastic resonance and coding vowel-like stimuli. In: Lithgow, B. and I. E. Cosic, eds. Biomedical Research in 2001. Proceedings of the 2nd Conference of the Victorian Chapter of the IEEE Engineering in Medicine and Biology Society. Clayton, Monash University: 46–49.

Hohn, N. and A. N. Burkitt. 2001b. Shot noise in the leaky integrate-and-fire neuron. Physical Review E63: 031902.

Hong, R.S., J.T. Rubinstein, D. Wehner and D. Horn. Submitted. Dynamic range enhancement for cochlear implants. Otology and Neuro-Otology.

Huttenlocher, P. R. and A. S. Dabholkar. 1997. Regional differences in synaptogenesis in human cerebral cortex. Journal of Comparative Neurology 387: 167–178.

Illing, R. B., M. Horvath and R. Laszig. 1997. Plasticity of the auditory brainstem: effects of cochlear ablation on GAP-43 immunoreactivity in the rat. Journal of Comparative Neurology 382: 116–138.

Irlicht, L. S., D. Au and G. M. Clark. 1995. New temporal coding scheme for auditory nerve stimulation. Annals of Otology, Rhinology and Laryngology 104(suppl 166): 358–360.

Irlicht, L. S. and G. M. Clark. 1995a. Control strategies for nerves modeled by self-exciting point processes. Annals of Otology, Rhinology and Laryngology 104(suppl 166): 361–363.

Irlicht, L. S. and G. M. Clark. 1995b. Control strategies for neurons modeled by self-exciting point processes. Journal of the Acoustical Society of America 98: 2927.

Irlicht, L. S. and G. M. Clark. 1996. Control strategies for neurons modeled by self-exciting point processes. Journal of the Acoustical Society of America 100: 3237–3247.

Irvine, D. R. F. 1986. The auditory brainstem. A review of the structure and function of auditory brainstem processing mechanisms. Berlin, Springer-Verlag.

Ito, M. 1986. Long-term depression as a memory process in the cerebellum. Neuroscience Research 3: 531–539.

Javel, E. and R. K. Shepherd. 2000. Electrical stimulation of the auditory nerve. III. Response initiation sites and temporal fine structure. Hearing Research 140: 45–76.

Javel, E., Y. C. Tong, R. K. Shepherd and G. M. Clark. 1987. Responses of cat auditory fibres to biphasic electrical current pulses. Annals of Otology, Rhinology and Laryngology 96: 26–30.

Jeffress, L. A. 1948. A place theory of sound localization. Physiological Psychology 41: 35–39.

Jenison, R. L., S. Greenberg, K. R. Kluender and W. S. Rhode. 1991. A composite model of the auditory periphery for the processing of speech based on the filter response functions of single auditory-nerve fibers. Journal of the Acoustical Society of America 90: 773–786.

Johnson, D. and A. Swami. 1983. The transmission of signals by auditory-nerve fiber discharge patterns. Journal of the Acoustical Society of America 74: 493–501.

Johnson, D. H. 1980. The relationship between spike rate and synchrony in responses of auditory-nerve fibers to single tones. Journal of the Acoustical Society of America 68: 1115–1122.

Johnson, D. H. 1996. Point process models of single-neuron discharges. Journal of Computational Neuroscience 3: 275–299.

Johnson, D. H. and N. Y. S. Kiang. 1976. Analysis of discharges recorded simultaneously from pairs of auditory nerve fibers. Journal of Biophysiology 16: 719–734.

Johnstone, B. M., J. R. Johnstone and I. D. Pugsley. 1966. Membrane resistance in endolymphatic walls of the first turn of the guinea-pig cochlea. Journal of the Acoustical Society of America 40: 1398–1404.

Jones, K. and A. Tubis. 1985. On the extraction of the signal-excitation function from a non-Poisson cochlear neural spike train. Journal of the Acoustical Society of America 78: 90–94.

Jones, R. C., S. S. Stevens and M. H. Lurie. 1940. Three mechanisms of hearing by electrical stimulation. Journal of the Acoustical Society of America 12: 281–290.

Joris, P. X., L. H. Carney, P. H. Smith and T. C. T. Yin. 1994a. Enhancement of neural synchronization in the anteroventral cochlear nucleus I. Responses to tones at the characteristic frequency. Journal of Neurophysiology 71: 1022–1036.

Joris, P. X., P. H. Smith and T. C. T. Yin. 1994b. Enhancement of neural synchronization in the anteroventral cochlear nucleus. II. Responses in the tuning curve tail. Journal of Neurophysiology 71(3): 1037–1051.

Joris, P. X. and T. C. T. Yin. 1998. Envelope coding in the lateral superior olive. III. Comparison with afferent pathways. American Journal of Physiology 79: 253–269.

Kaczmarek, L., M. Kossut and J. Skangiel-Kramska. 1997. Glutamate receptors in cortical plasticity: molecular and cellular biology. Physiology Review 77: 217–255.

Kane, E. C. 1973. Octopus cells in the cochlear nucleus of the cat: heterotypic synapses upon homeotypic neurons. International Journal of Neuroscience 5: 251–279.

Katsuki, Y. and Y. Kanno. 1962. Neural mechanism of the peripheral and central auditory system in monkeys. Journal of the Acoustical Society of America 34: 1396–1410.

Katz, B. 1939. Electric excitation of nerve. London, Oxford University.

Kiang, N. Y.-S., D. K. Eddington and B. Delgutte. 1979. Fundamental considerations in designing auditory implants. Acta Oto-Laryngologica 87: 204–218.

Kiang, N. Y.-S. and E. C. Moxon. 1972. Physiological considerations in artificial stimulation of the inner ear. Annals of Otology 81: 714–729.

Kiang, N. Y.-S., R. F. Pfeiffer and W. B. Warr. 1965a. Stimulus coding in the auditory nerve and cochlear nucleus. Acta Otolaryngologica (Stockh) 59: 186–200.

Kiang, N. Y.-S., R. F. Pfeiffer and W. B. Warr. 1965b. Stimulus coding in the cochlear nucleus. Annals of Otology Rhinology and Laryngology 74: 2–23.

Kim, D. O., S. O. Chang and J. G. Sirianni. 1990. A population study of auditory-nerve fibers in unanesthetized decerebrate cats: response to pure tones. Journal of the Acoustical Society of America 87: 1648–1655.

Kim, D. O. and K. Parham. 1991. Auditory nerve spatial encoding of high-frequency pure tones: population response profiles derived from d' measure associated with nearby places along the cochlea. Hearing Research 52: 167–180.

King, A. J. and A. R. Palmer. 1983. Cells responsive to free-field auditory stimuli in guinea-pig superior colliculus: distribution and response properties. Journal of Physiology 342: 361–381.

Kitzes, L. M. 1984. Some physiological consequences of neonatal cochlear destruction in the inferior colliculus of the gerbil. Brain Research 306: 171–178.

Kitzes, L. M. and M. N. Semple. 1985. Single-unit responses in the inferior colliculus: effects of neonatal unilateral cochlear ablation. Journal of Neurophysiology 53: 1483–1500.

Klinke, R., A. Kral, S. Heid, J. Tillein and R. Hartmann. 1999. Recruitment of the auditory cortex in congenitally deaf cats by long-term cochlear electrostimulation. Science 285(5434): 1729–1733.

Knudsen, E. I. and M. Konishi. 1978. A neural map of auditory space in the owl. Science 200: 795–797.

Koppl, C. 1997. Phase locking to high frequencies in the auditory nerve and cochlear nucleus magnocellularis of the barn owl, Tyto alba. Journal of Neuroscience 17: 3312–3321.

Korte, M. and J. P. Rauschecker. 1993. Auditory spatial tuning of cortical neurons is sharpened in cats with early blindness. Journal of Neurophysiology 70: 1717–1721.

Kral, A., R. Hartmann, J. Tillein, S. Heid and R. Klinke. 2001a. Auditory developmental plasticity in cats: a sensitive period of 6 months. Association for Research in Otolaryngology 24: 128.

Kral, A., R. Hartmann, J. Tillein, S. Heid and R. Klinke. 2001b. Delayed maturation and sensitive periods in the auditory cortex. Audiology and Neuro-Otology 6: 346–362.

Kral, A., J. Tillein, R. Hartmann and R. Klinke. 1999. Monitoring of anaesthesia in neurophysiological experiments. Neuroreport 10: 781–787.

Kranz, H. G. 1971. The role of the place and volley principles in pitch perception. MA preliminary dissertation. Melbourne, University of Melbourne.

Kuhlmann, L., A. N. Burkitt, A. G. Paolini and G. M. Clark. 2002. Summation of spatio-temporal input patterns in leaky integrate-and-fire neurons: application to neurons in the cochlear nucleus receiving converging auditory nerve fiber input. Journal of Computational Neuroscience 12: 55–73.

Kuwada, S. and T. C. Yin. 1983. Binaural interaction in low-frequency neurons in inferior colliculus of the cat. I. Effects of long interaural delays, intensity, and repetition rate on interaural delay function. Journal of Neurophysiology 50: 981–999.

Kuwada, S., T. C. Yin, J. Syka, T. J. Buunen and R. E. Wickesberg. 1984. Binaural interaction in low-frequency neurons in inferior colliculus of the cat. IV. Comparison of monaural and binaural response properties. Journal of Neurophysiology 51: 1306–1325.

Laird, R. K. 1979. The bioengineering development of a sound encoder for an implantable hearing prosthesis for the profoundly deaf. Master of Engineering science thesis. University of Melbourne.

Leake, P. A., R. L. Snyder, S. J. Rebescher, C. M. Moore and M. Vollmer. 2000. Plasticity in central representations in the inferior colliculus induced by chronic single vs. two channel electrical stimulation by a cochlear implant after neonatal deafness. Hearing Research 147: 221–241.

Lee, D. S., J. S. Lee, S. H. Oh, et al. 2001. Cross-modal plasticity and cochlear implants. Nature 409: 149–150.

Levänen, S., V. Jousmäki and R. Hari. 1998. Vibration-induced auditory cortex activation in a congenitally deaf adult. Current Biology 8: 869–872.

Levitan, I. B. and L. K. Kaczmarek. 1997. The neuron: cell and molecular biology. New York, Oxford University Press: 543.

Liberman, M. C. 1978. Auditory-nerve response from cats raised in a low-noise chamber. Journal of the Acoustical Society of America 63: 442–455.

Liberman, M. C. and L. W. Dodds. 1984. Single-neuron labeling and chronic cochlear pathology. II. Stereocilia damage and alterations of spontaneous discharge rates. Hearing Research 16: 43–53.

Litvak, L. 2002. Towards a better speech processor for cochlear implants: auditory nerve responses to high rate electric pulse trains. PhD thesis. Massachesetts Institute of Technology.

Lorente de No, R. 1981. The primary acoustic nuclei. New York, Raven Press.

Lukies, P. M., Y. C. Tong and G. M. Clark. 1987. Current distributions produced by the banded electrode array: an experimental study conducted with a tank model. Annals of Otology, Rhinology and Laryngology 96(suppl 128): 24.

Lukies, P. M., Y. C. Tong, G. M. Clark and P. A. Busby. 1986. Modeling studies on current distributions produced by an intracochlear electrode array. Journal of the Acoustical Society of America 80(suppl 1): S30.

Manis, P. B. and S. O. Marx. 1991. Outward currents in isolated ventral cochlear nucleus neurons. Journal of Neuroscience 11: 2865–2880.

Mark, K. E. and M. I. Miller. 1992. Bayesian model selection and minimum description length estimation of auditory-nerve discharge rates. Journal of the Acoustical Society of America 91: 989–1002.

Marsalek, P., C. Koch and J. Maunsell. 1997. On the relationship between synaptic input and spike output jitter in individual neurons. Proceedings of the National Academy of Science USA 94: 735–740.

Marschark, M. 1998. Memory for language in deaf adults and children. Scandinavian Audiology Supplementum 49: 87–92.

Matsuoka, A. J., P. J. Abbas, J. T. Rubinstein and C. A. Miller. 1997. Temporal properties of the electrical evoked compound action potentials with pulse train stimulation. Conference on Implantable Auditory Prostheses, Asilomar, CA.

McAnally, K. I., M. Brown and G. M. Clark. 1997. Comparison of current waveforms for the electrical stimulation of residual low frequency hearing. Acta Oto-Laryngologica 117: 831–835.

McAnally, K. I. and G. M. Clark. 1994. Stimulation of residual hearing in the cat by pulsatile electrical stimulation of the cochlea. Acta Oto-Laryngologica 114(4): 366–372.

McAnally, K. I., G. M. Clark and J. Syka. 1993. Hair cell mediated responses of the auditory nerve to sinusoidal electrical stimulation of the cochlea in the cat. Hearing Research 67: 55–68.

McDermott, H. J. and C. M. McKay. 1992. Place pitch perception with multiple cochlear implants: the use of concurrent activity of nearby electrodes to produce additional pitch percepts. Australian Journal of Audiology (suppl 5): 18.

McKay, C. M., H. J. McDermott and G. M. Clark. 1991. Preliminary results with a six spectral maxima speech processor for the University of Melbourne/Nucleus multiple electrode cochlear implant. Journal of the Oto-Laryngological Society of Australia 6: 354–359.

McMullen, N. T., B. Goldberger, C. M. Suter and E. M. Glaser. 1988. Neonatal deafening alters nonpyramidal dendrite orientation in auditory cortex: a computer microscopic study. Journal of Comparative Neurology 267: 92–106.

Meddis, R., M. J. Hewitt and T. M. Shackleton. 1990. Implementation details of a computation model of the inner hair-cell/auditory-nerve synapse. Journal of the Acoustical Society of America 87: 1813–1816.

Merzenich, M. M., R. P. Michelson and C. R. Pettit. 1973. Neural encoding of sound sensation evoked by electrical stimulation of the acoustic nerve. Annals of Otology 82: 486–503.

Merzenich, M. M. and M. D. Reid. 1974. Representation of the cochlea within the inferior colliculus. Brain Research 77: 397–415.

Merzenich, M. M. and M. White. 1980. Coding considerations in design of cochlear prostheses. Annals of Otology 89: 84–87.

Merzenich, M. M., M. White, M. C. Vivion, P. A. Leake-Jones and S. Walsh. 1979. Some considerations of multichannel electrical stimulation of the auditory nerve in the profoundly deaf; interfacing electrode arrays with the auditory nerve array. Acta Oto-Laryngologica 87: 196–203.

Middlebrooks, J. C. and E. I. Knudsen. 1984. A neural code for auditory space in the cat's superior colliculus. Journal of Neuroscience 4: 2621–2634.

Miller, R. K. 1971. Nonlinear volterra integral equations. Mathematics Lecture Note Series. New York, Benjamin.

Mitzdorf, U. 1985. Current source density method and application in cat cerebral cortex: investigations of evoked potentials and EEG phenomena. Physiology Review 65: 37–100.

Moller, A. R. 1971. Unit responses in the rat cochlear nucleus to tones of rapidly varying frequency and amplitude. Acta Physiologica Scandinavica 81: 540–556.

Moore, B. C. J. 1997. An introduction to the psychology of hearing. San Diego, Academic Press.

Moore, D. R. 1990a. Auditory brainstem of the ferret: bilateral cochlear lesions in infancy do not affect the number of neurons in the cochlear nucleus to the inferior colliculus. Brain Research. Developmental Brain Research 54: 125–130.

Moore, D. R. 1990b. Auditory brainstem of the ferret: early cessation of developmental sensitivity of neurons in the cochlear nucleus to removal of the cochlea. Journal of Comparative Neurology 302: 810–823.

Moore, D. R., M. E. Hutchings, A. J. King and N. E. Kowalchuk. 1989. Auditory brain stem of the ferret: some effects of rearing with a unilateral ear plug on the cochlea, cochlear nucleus, and projections to the inferior colliculus. Journal of Neuroscience 9: 1213–1222.

Moore, D. R. and N. E. Kowalchuk. 1988. Auditory brainstem of the ferret: effects of unilateral cochlear lesions on cochlear nucleus volume and projections to the inferior colliculus. Journal of Comparative Neurology 272: 503–515.

Morse, R. P. and E. F. Evans. 1996. Enhancement of vowel coding for cochlear implants by addition of noise [see comments]. Nature Medicine 2(8): 928–932.

Morse, R. P. and E. F. Evans. 1999a. Additive noise can enhance temporal coding in a computational model of analogue cochlear implant stimulation. Hearing Research 133: 107–19.

Morse, R. P. and E. F. Evans. 1999b. Preferential and non-preferential transmission of formant information by an analogue cochlear implant using noise: the role of the nerve threshold. Hearing Research 133: 120–32.

Morse, R. P. and P. Roper. 2000. Enhanced coding in a cochlear-implant model using additive noise: aperiodic stochastic resonance with tuning. Physical Review E 61(2): 5683–5692.

Mortimer, J. T. 1990. Electrical excitation of nerve. In: Agnew, W. F. and D. B. McCreery, eds. Neural prosthesis: fundamental studies. Englewood Cliffs, NJ, Prentice Hall: 68–83.

Mostafapour, S. P., S. L. Cochran, N. M. Del Puerto and E. W. Rubel. 2000. Patterns of cell death in mouse anteroventral cochlear nucleus neurons after unilateral cochlea removal. Journal of Comparative Neurology 426: 561–571.

Moushegian, G., A. Rupert and M. A. Whitcomb. 1964a. Brain-stem neuronal response patterns to monaural and binaural. Journal of Neurophysiology 27: 1174–1191.

Moushegian, G., A. Rupert and M. A. Whitcomb. 1964b. Medial superior-olivary-unit response patterns to monaural and binaural clicks. Journal of the Acoustical Society of America 36: 196–202.

Moxon, E. C. 1967. Electric stimulation of the cat's cochlea: a study of discharge rates in single auditory nerve fibers. M.Sc. dissertation. Cambridge, MA, MIT.

Moxon, E. C. 1971. Neural and mechanical responses to electrical stimulation of the cat's inner ear. Ph.D. dissertation. Cambridge, MA, MIT.

Neely, S. T. and D. O. Kim. 1986. A model for active elements in cochlear biomechanics. Journal of the Acoustical Society of America 79: 1472–1480.

Neff, W. D. 1968. Localization and lateralization of sound in space. In: de Reuch, A. V. S., J. Knight and A. Churchill, eds. Ciba Foundation Symposium on Hearing Mechanisms in Vertebrates. London: 207–231.

Neff, W. D., I. T. Diamond and J. H. Casseday. 1975. Behavioral studies of auditory discrimination: central nervous system. In: Keidel, W. D. and W. D. Neff, eds. Handbook of sensory physiology. Volume 2. Auditory system. Berlin, Springer-Verlag: 307–400.

Neville, H. J. and D. Lawson. 1987a. Attention to central and peripheral visual space in a movement detection task. II. Congenitally deaf adults. Brain Research 405: 268–283.

Neville, H. J. and D. Lawson. 1987b. Attention to central and peripheral visual space in a movement detection task. III. Separate effects of auditory deprivation and acquisition of a visual language. Brain Research 405: 284–294.

Ni, D., R. K. Shepherd, H. L. Seldon, S. Xu and G. M. Clark. 1992. Cochlear pathology following chronic electrical stimulation of the auditory nerve. I: Normal hearing kittens. Hearing Research 62: 63–81.

Nordeen, K. W., H. P. Killackey and L. M. Kitzes. 1983. Ascending projections to the inferior colliculus following unilateral cochlear ablation in the neonatal gerbil, Meriones unguiculatus. Journal of Comparative Neurology 214: 144–153.

Oertel, D. 1983. Synaptic responses and electrical properties of cells in brain slices of the mouse anteroventral cochlear nucleus. Journal of Neuroscience 3(10): 2043–53.

Oertel, D., S. H. Wu and J. A. Hirsch. 1988. Electrical characteristics of cells and neuronal circuitry in the cochlear nuclei studies with intracellular recordings from brain slices. In: Edelman, G. M., ed. Auditory function: neurobiological bases of hearing. New York, Wiley: 313–336.

O'Leary, S. J., R. C. Black and G. M. Clark. 1985. Current distributions in the cat cochlea. A modeling and electrophysiological study. Hearing Research 18: 273–281.

O'Leary, S. J., L. S. Irlicht, I. C. Bruce, M. W. White and G. M. Clark. 1997. Prediction of variance in neural response to cochlear implant stimulation and its implications for perception. XVI World Congress of Otorhinolaryngology Head and Neck Surgery: 190–191.

O'Leary, S. J., Y. C. Tong and G. M. Clark. 1992. Excitation and inhibition in responses of cochlear nucleus single units to electrical stimulation of the auditory nerve. Proceedings of the Australian Physiological and Pharmacological Society 21: 45P.

Oonishi, S. and Y. Katsuki. 1965. Functional organization and integrative mechanisms in the auditory cortex of the cat. Japanese Journal of Physiology 15: 342–365.

Osberger, M. J. and L. Fisher. 1999. SAS-CIS preference study in postlingually deafened adults implanted with the CLARION cochlear implant. Annals of Otology, Rhinology, and Laryngology 108(suppl 177): 74–79.

Osen, K. K., D. E. Lopez, T. A. Slyngstad, O. P. Ottersen and J. Storm-Mathisen. 1991. GABA-like and glycine-like immunoreactivities of the cochlear root nucleus in rat. Journal of Neurocytology 20: 17–25.

Palmer, A. R. and A. J. King. 1982. The representation of auditory space in the mammalian superior colliculus. Nature 299: 248–249.

Pandya, D. N. and E. H. Yeterian. 1985. Architecture and connections of cortical association areas. In: Peter, A. and E. G. Jones, eds. Cerebral cortex. New York, Plenum Press 4: 3–62.

Paolini, A. G. and G. M. Clark. 1997. The effect of pulsatile intracochlear electrical stimulation on intracellularly recorded cochlear nucleus neurons. In: Clark, G. M., ed. Cochlear implants. XVI World Congress of Otorhinolaryngology Head and Neck Surgery. Bologna, Monduzzi Editore: 119–124.

Paolini, A. G. and G. M. Clark. 1998a. Intracellular responses of onset neurones in the ventral cochlear nucleus to acoustic stimulation. Proceedings of the Australian Neuroscience Society 9: 51.

Paolini, A. G. and G. M. Clark. 1998b. Intracellular responses of the rat anteroventral cochlear nucleus to intracochlear electrical stimulation. Brain Research Bulletin 46: 317–327.

Paolini, A. G. and G. M. Clark. 1999. Intracellular responses of onset chopper neurons in the ventral cochlear nucleus to tones: evidence for dual-component processing. Journal of Neurophysiology 81: 2347–2359.

Paolini, A. G., G. M. Clark and A. N. Burkitt. 1997. Intracellular responses of the rat cochlear nucleus to sound and its role in temporal coding. Neuroreport 8: 3415–3421.

Paolini, A. G., J. V. Fitzgerald, A. N. Burkitt and G. M. Clark. 2001. Temporal processing from the auditory nerve to the medial nucleus of the trapezoid body in the rat. Hearing Research 159: 101–116.

Paolini, A. G., J. V. Fitzgerald and G. M. Clark. 2000. Responses of bushy cells to tones:

implications for place and temporal sound coding. Proceedings of the Australian Neuroscience Society 11: 76.

Paparella, M. M. and S. Sugiura. 1967. The pathology of suppurative labyrinthitis. Annals of Otology Rhinology and Laryngology 76(3): 554–586.

Parasnis, I. 1998. Cognitive diversity in deaf people: implication for communication and education. Scandinavian Audiology Supplementum 49: 109–115.

Patuzzi, R. and P. M. Sellick. 1984. The modulation of the sensitivity of the mammalian cochlea by low frequency tones III. Basilar membrane motion. Hearing Research 13: 19–27.

Perkel, D., G. Gerstein and G. Moore. 1967. Neuronal spike trains and stochastic point processes. I. The single spike train. Journal of Biophysics 7: 391–418.

Pfingst, B. E., P. A. Burnett and D. Sutton. 1983. Intensity discrimination with cochlear implants. Journal of the Acoustical Society of America 73(4): 1283–92.

Pfingst, B. E. and N. L. Rush. 1985. Discrimination of simultaneous frequency and level changes in electrical stimuli. Annals of Otology, Rhinology and Laryngology 96(suppl 128): 34–37.

Phillips, D. P. and S. E. Hall. 1987. Responses of single neurons in cat auditory cortex to time-varying stimuli: linear amplitude modulations. Experimental Brain Research 67: 479–492.

Pizzorusso, T., N. Berardi, F. M. Rossi, et al. 1999. TrkA activation in the rat visual cortex by antirat trkA IgG prevents the effect of monocular deprivation. European Journal of Neuroscience 11: 204–212.

Ponton, C. W., J. J. Eggermont, M. Don, et al. 2000a. Maturation of the mismatch negativity: effects of profound deafness and cochlear implant use. Audiology and Neuro-Otology 5: 167–185.

Ponton, C. W., J. J. Eggermont, B. Kwong and M. Don. 2000b. Maturation of human central auditory system activity: evidence from multi-channel evoked potentials. Clinical Neurophysiology 111: 220–236.

Ponton, C. W., J. K. Moore and J. J. Eggermont. 1999. Prolonged deafness limits auditory system developmental plasticity: evidence from an evoked potentials study in children with cochlear implants. Scandinavian Audiology 28(suppl 51): 13–22.

Raggio, M. W. and C. E. Schreiner. 1999. Neural responses in cat primary auditory cortex to electrical stimulation. III. Activation patterns in short- and long-term deafness. Journal of Neurophysiology 82: 3506–3526.

Rajan, R., D. R. F. Irvine, L. Z. Wise and P. Heil. 1993. Effect of unilateral partial cochlear lesions in adult cats on the representation of lesioned and unlesioned cochleas in primary auditory cortex. Journal of Comparative Neurology 338: 17–49.

Rauschecker, J. P. and M. Korte. 1993. Auditory compensation for early blindness in cat cerebral cortex. Journal of Neuroscience 13: 4538–4548.

Recanzone, G. H., C. E. Schreiner and M. M. Merzenich. 1993. Plasticity in the frequency representation of primary auditory cortex following discrimination training in adult owl monkeys. Journal of Neuroscience 13: 87–103.

Redd, E. E., T. Pongstaporn and D. K. Ryugo. 2000. The effects of congenital deafness on auditory nerve synapses and globular bushy cells in cats. Hearing Research 147: 160–174.

Rhode, W. S. 1971. Observations of the vibration of the basilar membrane in squirrel monkeys using the Mossbauer technique. Journal of the Acoustical Society of America 49: 1218–1231.

Rhode, W. S. and S. Greenberg. 1994. Lateral suppression and inhibition in the cochlear nucleus of the cat. Journal of Neurophysiology 71: 493–514.

Robblee, L. S., J. McHardy and J. M. Marston. 1980. Electrical stimulation with Pt electrodes. V. The effect of protein on Pt dissolution. Biomaterials 1: 135–139.

Roberts, L. A., R. K. Shepherd, A. G. Paolini, G. M. Clark and A. N. Burkitt. 2000. Effects of a sensorineural hearing loss on the refractory properties of auditory nerve fibres. Proceedings of the Twentieth Annual Meeting of the Australian Neuroscience Society 11: 144.

Robertson, D., G. K. Yates and I. M. Winter. 1990. Primary afferent dynamic ranges and cochlear mechanics. In: Rowe, M. and L. M. Aitkin eds. Information processing in mammalian auditory and tactile systems. New York, Wiley-Liss: 61–71.

Rose, J. E., J. F. Brugge, D. J. Anderson and J. E. Hind. 1967. Phase-locked response to low-frequency tones in single auditory nerve fibers of the squirrel monkey. Journal of Neurophysiology 30: 769–93.

Rose, J. E., J. F. Brugge, D. J. Anderson and J. E. Hind. 1969. Time structure of discharges in single auditory nerve fibers of the squirrel monkey in response to complex periodic sounds. Journal of Neurophysiology 32: 386–401.

Rose, J. E., R. Galambos and J. R. Hughes. 1959. Microelectrode studies of the cochlear nuclei of the cat. Bulletin of John Hopkins Hospital 104: 211–251.

Rose, J. E., D. D. Greenwood, G. J.M. and J. E. Hind. 1963. Some discharge characteristics of single neurons in the inferior colliculus of the cat. I. Tonotopical organization, relation of spike-counts to tone intensity, and firing patterns of single elements. Journal of Neurophysiology 26: 294–320.

Rose, J. E., N. B. L. Gross, C. D. Geisler and J. E. Hind. 1966. Some neural mechanisms in the inferior colliculus of the cat which may be relevant to localization of a sound source. Journal of Neurophysiology 29: 288–314.

Rothman, J. S. and E. D. Young. 1996. Enhancement of neural synchronization in computational models of ventral cochlear nucleus bushy cells. Auditory Neuroscience 2: 47–62.

Rothman, J. S., E. D. Young and P. B. Manis. 1993. Convergence of auditory nerve fibers onto bushy cells in the ventral cochlear nucleus-implications of a computational model. Journal of Neurophysiology 70: 2562–2583.

Rubinstein, J. T. 1991. Analytical theory for extracellular electrical stimulation of nerve with focal electrodes. II. Passive myelineated axon. Biophysical Journal 60: 538–555.

Rubinstein, J. T. 1993. Axon termination conditions for electrical stimulation. IEEE Transactions on Biomedical Engineering 40: 654–663.

Rubinstein, J. T. 1995. Threshold fluctuations in an N sodium channel model of the node of Ranvier. Biophysical Journal 68: 779–785.

Rubinstein, J. T., A. J. Matsuoka, P. J. Abbas and C. A. Miller. 1997. The neurophysiology effects of simulated auditory prosthesis stimulation. Second quarterly progress report. NIH contract NO1-DC-6-2111. Neural Prosthesis Program, NIH, Bethesda, MD.

Rubinstein, J. T., B. S. Wilson, C. C. Finley and P. J. Abbas. 1999. Pseudospontaneous activity: stochastic independence of auditory nerve fibers with electrical stimulation. Hearing Research 127: 108–118.

Ruggero, M. A. 1980. Systematic errors in indirect estimates of basilar membrane travel times. Journal of the Acoustical Society of America 67: 707–710.

Runge-Samuelson, C. 2002. Response of the auditory nerve to sinusoidal electrical stimulation: effects of high-rate pulse trains. PhD thesis. University of Iowa.

Rupert, A., G. Moushegian and R. Galambos. 1963. Unit responses to sound from auditory nerve of the cat. Journal of Neurophysiology 26: 449–465.

Russell, I. J., A. R. Cody and G. P. Richardson. 1986. The responses of inner and outer hair cells in the basal turn of the guinea-pig cochlea grown in vitro. Hearing Research 22: 199–216.

Ryugo, D. K. and S. Sento. 1991. Synaptic connections of the auditory nerve in cats: relationship between endbulbs of Held and spherical bushy cells. Journal of Comparative Neurology 305: 35–48.

Sachs, M. B. 1967. Auditory nerve fiber responses to two-tone stimuli. PhD dissertation. Cambridge, MA, MIT.

Sachs, M. B. and E. D. Young. 1979. Encoding of steady-state vowels in the auditory nerve: representation in terms of discharge rate. Journal of the Acoustical Society of America 66: 470–479.

Sachs, M. B. and E. D. Young. 1980. Effects of nonlinearities on speech encoding in the auditory nerve. Journal of the Acoustical Society of America 68: 858–875.

Saint-Marie, R. L., C. G. Benson, E. M. Ostapoff and D. K. Morest. 1991. Glycine immunoreactive projections from the dorsal to the anteroventral cochlear nucleus. Hearing Research 51: 11–28.

Santos-Sacchi, J. 1989. Asymmetry in voltage-dependent movements of isolated outer hair cells from the organ of Corti. Journal of Neuroscience 9: 2954–2962.

Schindler, R. A., M. M. Merzenich, M. W. White and B. Bjorkroth. 1977. Multi electrode intracochlear implants-nerve survival and stimulation patterns. Archives of Otolaryngology 103: 691–699.

Schoonhoven, R., V. F. Prijs and J. H. M. Frijns. 1997. Transmitter release in inner hair cell synapses: a model analysis of spontaneous and driven rate properties of cochlear nerve fibers. Hearing Research 113: 247–260.

Schuknecht, H. 1953. Techniques for study of cochlear function and pathology in experimental animals. Acta Oto-Laryngologica 58: 377.

Schuknecht, H. and W. D. Neff. 1952. Hearing losses after apical lesions in the cochlea. Acta Oto-Laryngologica 42: 263–274.

Seldon, H. L., A. Kawano and G. M. Clark. 1996. Does age at cochlear implantation affect the distribution of 2-deoxyglucose label in cat inferior collicullus? Hearing Research 95: 108–119.

Semple, M. N. and L. M. Aitkin. 1979. Representation of sound frequency and laterality by units in central nucleus of cat inferior colliculus. Journal of Neurophysiology 42: 1626–1639.

Sermasi, E., D. Tropea and L. Domenici. 1999. A new form of synaptic plasticity is transiently expressed in the developing rat visual cortex: a modulatory role for visual experience and brain-derived neurotropic factor. Neuroscience 91: 163–173.

Serviere, J. and W. R. Webster. 1981. A combined electrophysiological and 2-deoxyglucose study of the frequency organization of the inferior colliculus of the cat. Neuroscience Letters 27(2): 113–118.

Shepherd, G. M., ed. 1998. The synaptic organization of the brain. 4th ed. Oxford, Oxford University Press.

Shepherd, R. K., J. H. Baxi and N. A. Hardie. 1999. Response of inferior colliculus to electrical stimulation of the auditory nerve in neonatally deafened cats. Journal of Neurophysiology 82: 1363–1380.

Shepherd, R. K., G. M. Clark and R. C. Black. 1983a. Chronic electrical stimulation of the auditory nerve in cats. Physiological and histopathological results. Acta Oto-Laryngologica-supplement 399: 19–31.

Shepherd, R. K., G. M. Clark, R. C. Black and J. F. Patrick. 1983b. The histopathological effects of chronic electrical stimulation of the cat cochlea. Journal of Laryngology and Otology 97: 333–341.

Shepherd, R. K. and N. A. Hardie. 2001. Deafness-induced changes in the auditory pathway: implications for cochlear implants. Audiology and Neuro-Otology 6: 305–318.

Shepherd, R. K., S. Hatsushika and G. M. Clark. 1993. Electrical stimulation of the auditory nerve: the effect of electrode position on neural excitation. Hearing Research 66: 108–120.

Shepherd, R. K. and E. Javel. 1997a. Electrical stimulation of the auditory nerve single fibre responses in normal and pathological cochleae. Proceedings of the Australian Neuroscience Society 8: 82.

Shepherd, R. K. and E. Javel. 1997b. Electrical stimulation of the auditory nerve: I. Correlation of physiological responses with cochlear status. Hearing Research 108: 112–144.

Shepherd, R. K., J. Xu, R. E. Millard and G. M. Clark. 1994. Chronic electrical stimulation of the auditory nerve at high stimulus rates: preliminary results. Australian Journal of Oto-Laryngology 1: 453.

Shower, E. G. and R. Biddulph. 1931. Differential pitch sensitivity of the ear. Journal of the Acoustical Society of America 2: 275–287.

Silverstein, H., D. G. Davies and W. L. Griffin. 1969. Cochlear aqueduct obstruction. Changes in perilymph biochemistry. Annals of Otology, Rhinology and Laryngology 78: 532–541.

Simmons, F. B. and T. J. Glattke. 1972. Comparison of electrical and acoustical stimulation of the cat ear. Annals of Otology, Rhinology and Laryngology 81: 731–738.

Sinex, D. G. and C. D. Geisler. 1981. Auditory-nerve fiber responses to frequency-modulated tones. Hearing Research: 127–148.

Smith, C. A., O. H. Lowry and M. L. Wu. 1954. The electrolytes of the labyrinthine fluids. Laryngoscope 64: 141.

Smith, D. W., C. C. Finley, C. van den Honert, V. B. Olszyk and K. E. Konrad. 1994. Behavioral and electrophysiological responses to electrical stimulation in the cat. I. Absolute thresholds. Hearing Research 81: 1–10.

Smith, K. J. and W. I. McDonald. 1999. The pathophysiology of multiple sclerosis: the mechanisms underlying the production of symptoms and the natural history of the disease. Philosophical Transactions of the Royal Society of London Series B 354: 1649–1673.

Smith, P. H. and W. S. Rhode. 1987. Characterization of HRP-labeled globular bushy cells in the cat anteroventral cochlear nucleus. Journal of Comparative Neurology 266: 360–375.

Snyder, D. L. and M. I. Miller. 1991. Random point processes in time and space. New York, Springer.

Snyder, R. L., S. Rebscher and R. Beitel. 1995. Temporal resolution of neurons in cat inferior colliculus to intracochlear electrical stimulation: effects of neonatal deafening and chronic stimulation. Journal of Neurophysiology 73: 449–467.

Snyder, R. L., S. J. Rebscher, K. Cao, P. A. Leake and K. Kelly. 1990. Chronic intracochlear electrical stimulation in the neonatally deafened cat. I. Expansion of central representation. Hearing Research 50: 7–34.

Snyder, R. L., S. J. Rebscher, P. Leake, K. Kelly and K. Cao. 1991. Chronic intracochlear electrical stimulation in the neonatally deafened cat.II. Temporal properties of neurons in the inferior colliculus. Hearing Research 56: 246–264.

Snyder, R. L., D. G. Sinex, J. D. McGee and E. W. Walsh. 2000a. Acute spiral ganglion lesions change the tuning and tonotopic organization of cat inferior colliculus neurons. Hearing Research 147: 200–220.

Snyder, R. L., M. Vollmer, C. M. Moore, S. J. Rebscher, P. A. Leake and R. E. Beitel. 2000b. Responses of inferior colliculus neurons to amplitude-modulated intracochlear electrical pulses in deaf cats. Journal of Neurophysiology 84(1): 166–183.

Spelman, F. A., B. E. Pfingst, J. M. Miller, W. Hassul, W. E. Powers and B. M. Clopton. 1980. Biophysical measurements in the implanted cochlea. Otolaryngology and Head and Neck Surgery 88(2): 183–7.

Starr, A. and R. Britt. 1970. Intracellular recordings from cat cochlear nucleus during tone stimulation. Journal of Neurophysiology 33(1): 137–147.

Stein, R. B. 1965. A theoretical analysis of neuronal variability. Journal of Biophysics 5: 173–194.

Steriade, M. 1997. Synchronized activities of coupled oscillators in the cerebral cortex and thalamus at different levels of vigilance. Cerebral Cortex 7: 583–604.

Steriade, M. 1999. Coherent oscillations and short-term plasticity in corticothalamic networks. Trends in Neuroscience 22: 337–345.

Sterkers, O., G. Saumon and P. Tran Ba Huy. 1984. Electrochemical heterogeneity of the cochlear endolymph: effect of acetazolamide. American Journal of Physiology 246: 47.

Stevens, S. S. and H. Davis. 1938. Hearing: its psychology and physiology. New York, John Wiley: 315–316.

Strominger, N. L. 1969. Localization of sound in space after unilateral and bilateral ablation of auditory cortex. Experimental Neurology 25: 521–533.

Stypulkowski, P. H. and C. van den Honert. 1984. Physiological properties of the electrically stimulated auditory nerve. I. Compound action potential recordings. Hearing Research 14: 205–223.

Suga, N. 1963. Single unit activity in cochlear nucleus and inferior colliculus of echolocating bats. Journal of Physiology 172: 449–474.

Suga, N. 1965. Analysis of frequency-modulated sounds by auditory neurones of echolocating bats. Journal of Physiology 179: 26–53.

Tasaki, I. 1954. Nerve impulses in individual auditory nerve fibres of the guinea pig. Journal of Neurophysiology 17: 97–122.

Tong, Y. C., R. C. Black, G. M. Clark, et al. 1979. A preliminary report on a multiple-channel cochlear implant operation. Journal of Laryngology and Otology 93: 679–695.

Tong, Y. C., P. J. Blamey, R. C. Dowell and G. M. Clark. 1983. Psychophysical studies evaluating the feasibility of a speech processing strategy for a multiple-channel cochlear implant. Journal of the Acoustical Society of America 74: 73–80.

Tong, Y. C. and G. M. Clark. 1985. Absolute identification of electric pulse rates and electrode positions by cochlear implant patients. Journal of the Acoustical Society of America 77: 1881–1888.

Tong, Y. C., G. M. Clark, P. J. Blamey, P. A. Busby and R. C. Dowell. 1982. Psychophysical studies for two multiple-channel cochlear implant patients. Journal of the Acoustical Society of America 71: 153–160.

Tong, Y. C., G. M. Clark and H. H. Lim. 1987. Estimation of the effective spread of neural excitation produced by a bipolar pair of scala tympani electrodes. Annals of Otology, Rhinology and Laryngology 96(suppl 128): 37–38.

Tong, Y. C., J. B. Millar, G. M. Clark, L. F. Martin, P. A. Busby and J. F. Patrick. 1980.

Psychophysical and speech perception studies on two multiple-channel cochlear implant patients. Journal of Laryngology and Otology 94: 1241–1256.

Truy, E., M. P. Deiber, L. Cinotti, F. Mauguiere, J. C. Froment and A. Morgon. 1995. Auditory cortex activity changes in long-term sensorineural deprivation during crude cochlear electrical stimulation: evaluation by positron emission tomography. Hearing Research 86(1–2): 34–42.

Tsuchitani, C. and J. C. Boudreau. 1967. Encoding of stimulus frequency and intensity by cat superior olive. Journal of the Acoustical Society of America 42(4): 794–805.

Tsuchitani, C. and J. C. Boudreau. 1969. Stimulus level of dichotically presented tones and cat superior olive S-segment cell discharge. Journal of the Acoustical Society of America 46: 979–88.

Tuckwell, H. C. 1988a. Introduction to theoretical neurobiology. Vol. 1. Linear cable theory and dendritic structure. Cambridge, Cambridge University Press.

Tuckwell, H. C. 1988b. Introduction to theoretical neurobiology. Vol. 2. Nonlinear and stochastic theories. Cambridge, Cambridge University Press.

van den Honert, C. and P. Stypulkowski. 1984. Physiological properties of the electrically stimulated auditory nerve. II. Single fiber recordings. Hearing Research 14: 225–243.

van den Honert, C. and P. H. Stypulkowski. 1987. Single fiber mapping of spatial excitation patterns in the electrically stimulated auditory nerve. Hearing Research 29: 195–206.

Verveen, A. A. 1960. On the fluctuation of threshold of the nerve fibre. In: Tower, D. B., ed. Structure and function of the cerebral cortex. The Netherlands, Elsevier: 282–288.

von Békésy, G. 1951. The coarse pattern of the electrical resistance in the cochlea of the guinea pig (electroanatomy of the cochlea). Journal of the Acoustical Society of America 23: 18–28.

von Békésy, G. 1960. Experiments in hearing. New York, McGraw Hill.

Wang, X., M. M. Merzenich, R. Beitel and C. E. Schreiner. 1995. Representation of a species-specific vocalization in the primary auditory cortex of the common marmoset: temporal and spectral characteristics. Journal of Neurophysiology 74: 2685–2706.

Webster, W. R., J. I. Serviere, et al. 1985. Uncrossed and crossed inhibition in the inferior colliculus. Journal of Neuroscience 5: 1820–1832.

Weinberger, N. M., W. Hopkins and D. M. Diamond. 1984. Physiological plasticity of single neurons in auditory cortex of the cat during acquisition of the pupillary conditioned response. I. Primary field (AI). Behavioral Neuroscience 98: 171–188.

Wever, E. G. and C. W. Bray. 1930. Auditory nerve impulses. Science 71: 215.

Whitfield, I. C. 1979. Periodicity, pulse interval and pitch. Audiology 18: 507–512.

Whitfield, I. C. and E. F. Evans. 1965. Responses of auditory cortical neurons to stimuli of changing frequency. Journal of Neurophysiology 28: 655–672.

Wickesberg, R. E., D. Whitlon and D. Oertel. 1994. In vitro modulation of somatic glycine-like immunoreactivity in presumed glycinergic neurons. Journal of Comparative Neurology 339: 311–327.

Williams, A. J., G. M. Clark and G. V. Stanley. 1974. Behavioural responses in the cat to simple patterns of electrical stimulation of the terminal auditory nerve fibres. Proceedings of the Australian Physiological and Pharmacological Society 5(2): 252.

Williams, A. J., G. M. Clark and G. V. Stanley. 1976. Pitch discrimination in the cat through electrical stimulation of the terminal auditory nerve fibers. Physiological Psychology 4: 23–27.

Wilson, B. S. 1997. The future of cochlear implants. British Journal of Audiology 31: 205–225.

Wilson, B. S., C. C. Finley and D. T. Lawson. 1988. Speech processors for cochlear prostheses. Proceedings IEEE 76: 1143–1153.

Wilson, B. S., C. C. Finley, D. T. Lawson and M. Zerbi. 1997a. Speech processors for auditory prostheses. Eleventh quarterly progress report. NIH contract NO1-DC-2-2401. Neural Prosthesis Program, NIH, Bethesda, MD.

Wilson, B. S., C. C. Finley, D. T. Lawson and M. Zerbi. 1997b. Temporal representations with cochlear implants. American Journal of Otology 18: S30–S34.

Wilson, B. S., C. C. Finley, D. T. Lawson, M. Zerbi and C. van den Honert. 1997c. Speech processors for auditory prostheses. Seventh quarterly progress report. NIH contract NO1-DC-5-2103. Neural Prosthesis Program, NIH, Bethesda, MD.

Wilson, B. S., C. C. Finley, M. Zerbi and D. T. Lawson. 1994. Speech processors for auditory prostheses. Seventh quarterly progress report. NIH contract NO1-DC-2-2401. Neural Prosthesis Program, NIH, Bethesda, MD.

Wise, L. Z. and D. R. Irvine. 1985. Topographic organization of interaural intensity difference sensitivity in deep layers of cat superior colliculus: implications for auditory spatial representation. Journal of Neurophysiology 54: 185–211.

Wu, S. H. and D. Oertel. 1986. Inhibitory circuitry in the ventral cochlear nucleus is probably mediated by glycine. Journal of Neuroscience 6: 2691–2706.

Yaka, R., U. Yiono and Z. Wollberg. 1999. Auditory activation of cortical visual areas in cats after early visual deprivation. European Journal of Neuroscience 11: 1301–1312.

Yin, T. C., J. C. Chan and D. R. Irvine. 1986. Effects of interaural time delays of noise stimuli on low-frequency cells in the cat's inferior colliculus. I. Responses to wideband noise. Journal of Neurophysiology 55: 280–300.

Yin, T. C. and S. Kuwada. 1983. Binaural interaction in low-frequency neurons in inferior colliculus of the cat. III. Effects of changing frequency. Journal of Neurophysiology 50: 1020–1042.

Young, E. D. and M. B. Sachs. 1979. Representation of steady-state vowels in the temporal aspects of the discharge patterns of populations of auditory-nerve fibers. Journal of the Acoustical Society of America 66: 1381–1403.

Young, S. R. and E. W. Rubel. 1986. Embryogenesis of arborization pattern and topography of individual axons in N. laminaris of the chicken brain stem. Journal of Comparative Neurology 254: 425–459.

Yukawa, K., S. O'Leary, M. Clarke and G. M. Clark. 2001. Histopathology of the brainstem of a binaural cochlear implant subject. Program and abstracts of the Third International Congress of Asia Pacific Symposium on Cochlear Implant and Related Sciences. Osaka, Japan: 48.

Zeng, F. G., Q.-J. Fu and R. Morse. 2000. Human hearing enhanced by noise. Brain Research 869: 251–255.

Zhou, R., P. J. Abbas and J. G. Assouline. 1995. Electrically evoked auditory brainstem response in peripherally myelin-deficient mice. Hearing Research 88: 98–106.

6
Psychophysics

Understanding how simple and complex electrical stimuli are perceived and their relationship with speech perception has been important in the development of cochlear implants, and was fundamental to the University of Melbourne's approach. Although the relationship between psychophysics and speech perception is not well defined, the perception of temporal and place pitch perception and loudness are key elements. Furthermore, the relationship between both pitch and loudness and speech features has become better understood, in part due to cochlear implant research. The research has also helped explain the effectiveness of the formant-based as well as fixed filter speech-processing strategies (Tong and Clark 1982; Tong et al 1982, 1983a,b), and has guided research leading to improvements in speech processing (Tong et al 1983b). It has also aided in explaining plasticity in children, and differences in their performance. This research will be needed in the development of bimodal and bilateral speech processing. The perceptual studies discussed below are important also in understanding the underlying electrophysiology.

Acoustic Stimulation

Frequency coding correlates predominantly with pitch perception. The time/period (rate) code for frequency results in temporal pitch, and the place code in place (spectral) pitch. As the perception of pitch is highly relevant to the perception of speech, early research at the University of Melbourne focused on the relative importance of the rate and place coding of frequency, and how well they could be reproduced by electrical stimulation.

With sound it is difficult to determine the relative importance of the time/period and place codes in the perception of pitch as the two codes operate together. The temporal responses of the neurons vary according to the site of excitation. On the other hand, with electrical stimulation of auditory nerves the two codes can be reproduced separately to study their relative importance.

Psychoacoustic data were also the basis for studying the percepts obtained with simple and complex electrical stimulation (Tong et al 1979; Tong et al 1982; Tong et al 1983). For more comprehensive descriptions of psychoacoustics the reader is referred to the texts of Stevens (1975), Plomp (1976), and Moore (1997).

Pitch and Timbre

Pitch is a basic percept underlying speech recognition, and is related to its intelligibility. Timbre is also important in speech perception and in musical appreciation.

Definition

Pitch is the subjective attribute of tones that correlates most closely with the physical dimension of frequency. It is also related to duration and intensity. In 1960 the American Standards Association defined pitch as "that attribute of auditory sensation in terms of which sounds may be ordered on a musical scale."

Timbre relates to the quality of the sound. It depends primarily on the spectra of the stimuli (i.e., their harmonic content) and to a lesser extent on their relative phases. Tones with strong lower harmonics sound more mellow. The American Standards Association defined timbre as "that attribute of auditory sensation in terms of which a listener can judge that two sounds similarly presented and having the same loudness and pitch are dissimilar."

Measurement

Pitch is measured in units referred to as mels (Stevens and Volkmann 1940). The pitch in mels is determined by scaling the pitch of a frequency relative to the pitch of the reference frequency of 1000 Hz. For example, the pitch of the reference frequency of 1000 Hz is 1000 mels (Fig. 6.1); thus a tone with a pitch of 500 mels sounds half as high in pitch as a tone of 1000 mels.

Scaling

Pitch scaling can be done by asking subjects to adjust a variable tone until its pitch appears to be half that of a fixed tone (fractionation) or to adjust a variable tone to a pitch halfway between the pitches of two fixed tones (bisection). As seen in Figure 6.1 the pitch of pure tones does not vary linearly as a function of frequency.

A single numerical estimation method can also be used to measure pitch. The subjects are instructed to assign a number in the range 1 to 100 for the pitch of a single presentation. The scale could be expanded in either direction if required. This procedure is based on the research of Stevens and Greenbaum (1966) on magnitude estimation in matching numbers to loudness.

Intensity

Pitch can affect loudness and vice versa. For example, increasing the intensity of a sound not only increases the loudness, but also there may be a small change in

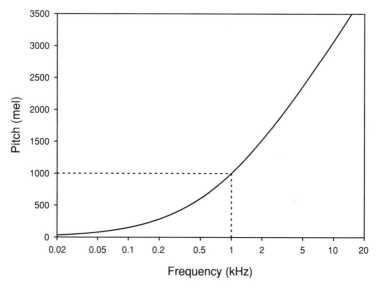

FIGURE 6.1. The pitch in mels versus frequency (Stevens and Volkmann 1940).

pitch. Stevens (1935) noted that increasing the intensity decreased pitch for tones less than 800 Hz and increased pitch for tones above 3000 Hz. Later studies showed the effects highly dependent on the subject, but overall supported the findings of Stevens even though the effects were small (Morgan et al 1951; Cohen 1961; Verschuure and van Meeteren 1975). For frequencies below 2000 Hz there can be a maximum 5% decrease in pitch with an increase in intensity, and a 5% increase in pitch for frequencies above 4000 Hz (Moore 1997). Fortunately, with music, intensity does not influence the pitch of the complex tones of the instruments.

Duration

Recognizing pitch requires a tone of minimal duration. Regardless of frequency, durations of only a few milliseconds are heard as clicks. As the duration is lengthened the click develops a tonal quality that allows some listeners to discriminate among clicks on the basis of click pitch. The minimum period of time for a frequency to be perceived as pitch varies with frequency below 1000 Hz when two to three cycles are needed. Above 1000 Hz a minimum duration of 10 ms is required.

Selectivity

Frequency selectivity is the ability to select separate pitch for one frequency from that at another when the two are sounded together. The ability to select the different harmonics in a complex sound is related to the critical band. The critical band is the frequency region in which there is masking of a pure tone, the loudness of a band of noise of fixed intensity remains constant, and phase differences in amplitude and frequency modulated sounds are detected. It is described in some detail below (see Critical Band and Ratio).

Plomp (1976) found that a harmonic could be heard from a multitone complex only if that harmonic were separated from neighboring harmonics by about one critical band. For a two-tone complex the harmonics could be identified for a separation less than this. Studies by Plomp (1964) and Plomp and Mimpen (1968) have shown that from five to eight harmonics are the maximum number that can be detected from a complex tone. The above studies were important for the development of a multiple-channel implant. They suggested that electrodes need not be spaced closer than a critical band for frequency selectivity, but would need to be closer if phase differences were important in the perception of frequencies within the critical band.

Discrimination

Frequency discrimination is the ability to distinguish a change or difference in frequency when the two tones are played one after the other. It must not be confused with frequency selectivity. Smaller differences in frequency can be detected if the frequencies are played one after the other (discrimination), than if they are to be selected from the sounds presented simultaneously (selectivity). Discrimination is measured as a difference limen (DL). The absolute DL is the smallest detectable change in frequency (Δf). The relative DL is the change relative to the initial frequency ($\Delta f/f$). The DL was measured initially by the modulation method (Shower and Biddulph 1931). This involved modulating frequencies up and down about four times per second. The degree of modulation required to detect a frequency variation is the DL. It is not the more reliable method as the frequency changes are accompanied by loudness fluctuations. The better method is to present two sounds, and to ask the subject to determine which is the higher pitch (a two-alternative forced choice task). The DL is the frequency separation where the subject achieves 75% correct. When measured this way, the results show that the relative DL ($\Delta f/f$) is fairly constant from 50 Hz to 5000 Hz (i.e., it obeys the Weber-Fechner law), and is approximately 0.2% (Moore 1974; Wier et al 1977), but it varies with the duration of the tone. The ability to discriminate frequency is remarkably good with DLs as small as 2 Hz for a 1000-Hz tone.

It was important for understanding the coding of frequency to know whether the detection of a change in frequency could be explained in terms of place or temporal frequency codes. It was also essential that some of the first psychophysical studies on the first University of Melbourne's patient were to determine whether changes in rate or place could be perceived for different stimulus durations required for the perception of vowels and consonants (Tong et al 1982).

With regard to place coding, the tuning curves of cells in the auditory brainstem are too broad to account for the small size of the frequency DL. However, a model was developed by Zwicker (1970) that predicted a change in frequency could be detected whenever the site of neuronal excitation varied by more than a threshold value. The greatest difference between the two patterns would occur on the steeply sloping high-frequency side. The steepness of the tuning curve was shown by Maiwald (1967) to be constant when expressed in units of the critical bandwidth. This predicted that frequency DLs at any frequency should be a constant fraction of the critical bandwidth at that frequency. Nevertheless, this has not been found to be the case. Moore and Glasberg (1989) discovered with tones randomized in intensity that the DLs were lower than the model predicted for frequencies from 500 to 4000 Hz, but were consistent at higher frequencies.

Temporal frequency coding is an alternative explanation for DLs, but the DL at 1000 Hz requires an accuracy of about 2 μs in the neural processing of information. This is unlikely, as there is a jitter in the initiation of nerve impulses with a standard deviation of 100 μs for a 1000-Hz tone (de Boer 1969). It is more likely that both place and temporal coding together are the mechanisms involved, and they may contribute to different degrees over the whole frequency range.

Complex Tones

Sounds composed of more than one sinusoid are complex tones. Ohm's acoustic law states that the ear can separate a complex tone into the individual sinusoids so that the listener can perceive the separate components. The law does not always hold, as the listener may not hear all the components or may hear different ones.

When two tones are presented together, a number of phenomena occur. If they are widely separated in frequency they are heard as two distinct tones on the basis of Ohm's law. If they are very close (e.g., 1000 Hz and 1003 Hz), a tone will be heard that beats at a frequency that is the difference between the two frequencies (3 Hz). Beating is used to tune pianos.

If two tones have frequencies that are not too close, other sounds of low intensity not in the signal can be detected. These are summation or difference tones and are referred to as combination tones. The most common one is the low pitch called the missing fundamental or the residue pitch (Schouten 1940a,b). The pitch persists even when the low frequencies are masked, and this is presumed to arise from the high-frequency end of the cochlea. This can be explained through the physiology of temporal coding, as Rose et al (1967) have recorded the intervals between nerve action potentials for two tones, and found a dominant interval equaling the missing fundamental from the high-frequency neurons.

Other tones not present can also be heard if a probe tone is used to beat with them. These combination tones occur at higher intensities and can be masked at their frequency regions on the basilar membrane. Ones most commonly heard are $f_2 - f_1$, $2f_1 - f_2$ and to a lesser extent others of the class $f_1 - n(f_2 - f_1)$, where f_1 and f_2 are the first and second harmonics respectively. They are produced by a nonlinearity of the middle ear and basilar membrane. The $2f_1 - f_2$, however, can be heard at low intensities and cannot be masked on the basilar membrane location.

Periodicity Detection and Roughness

Normally in recording frequency DLs, place as well as temporal information is transmitted. Spectral information can be removed by amplitude modulation of noise (Harris 1963; Burns and Viemeister 1976, 1981). For normal listeners, non-spectral pitch saturates at about 1000 Hz with DLs in the range 2% to 5% of the stimulating frequency. Amplitude modulated noise has been thought to simulate electrical stimulation. It is, however, not an ideal representation because the pitch percepts for low rates of electrical stimulation are described as being more pitch-like.

Roughness is the result of the interference of two or more tones in the ear (Terhardt 1974a,b). Combination tones and beats are not the only result of the interaction of tones. If they are separated by approximately 40 Hz, they produce the sensation of roughness (Plomp 1976). Roughness is also produced if the amplitude of a tone is sinusoidally modulated. The roughness of an AM tone depends almost entirely on the relative amplitude fluctuation (m) of the sound pressure (P). Thus $m = \Delta P/P$. The slow loudness variations audible for low modulation frequencies change into a "rattle" for faster modulations.

Jitter Detection

Jitter occurs if temporal irregularities are introduced into a sequence of interpulse patterns. Jitter is perceived as roughness and harshness especially at high pulse frequencies. If a temporal irregularity is introduced for two interpulse intervals, then this may be perceived as pitch changes.

Timbre—A Multidimensional Attribute of Complex Tones

Differences in timbre enable people to distinguish between the same note played on a piano and on the flute. Von Helmholtz (1859, 1863) provided the first comprehensive treatise on timbre and its relation to the property of sounds and hearing. He considered it to be due to the harmonic compositions of the sounds. They fuse into a sound as a whole, but their existence is established by their influence on timbre. The timbre depended on the amplitude patterns of the harmonics. It was found, however, that the phase between the harmonics had little effect on timbre. Timbre should be considered as a multidimensional quality, and not like the one-dimensional attributes of loudness and pitch that make up the sensation. Timbre was studied through speech perception where differences in timbre between vowels were correlated with peaks in the amplitude pattern, which Herman coined formants. This was confirmed by experiments by von Helmholtz. It was established that at least the first and second formant frequencies F1 and F2, being the lower two resonance frequencies of the cavities of the vocal tract, were required to characterized each vowel. This was studied later by Peterson and Barney (1952).

Loudness

Loudness is defined as that attribute of auditory sensation in terms of which sounds can be ordered on a scale extending from quiet to loud. It is a subjective sensation, which correlates most with sound intensity, but is also affected by the frequency.

Physical and Physiological Correlates

Sound can be quantified in terms of intensity (watts/m^2) or pressure (N/m^2 or Pa). In psychophysics intensity and pressure are measured as bels or decibels. The decibel is one tenth of a bel. The bel is the logarithm to base 10 of the ratio of the measured intensity to the reference intensity, and was discussed in Chapter 5.

The population of neurons excited is very important in the coding of intensity or its psychophysical correlate loudness. This is reflected in the growth of the field potential recorded from the experimental animal in response to increasing intensities as well as the auditory brainstem responses (ABR) in the human. The height and area of the wave represent the number and rate of fibers firing in response to the stimulus. This was discussed in more detail in Chapter 5.

Rate of nerve firing is also important in the coding of intensity. This is supported by psychophysical studies with simultaneous high- and low-frequency noise (i.e., band-stop noise) (Viemeister 1974; Moore and Raab 1975). The band-stop noise limits the spread of the stimulus to neighboring fibers when the intensity is increased, and so the total number of fibers recruited is restricted. The perception of loudness thus depends on the rate of firing in a narrow range of fibers, and increases were perceived.

Measurement of Loudness

Loudness cannot be measured directly because, like pitch, it is subjective. It can be measured indirectly by matching the loudness of a sound to a standard comparison stimulus (e.g., a 1000-Hz tone). The loudness in phons is the pressure level of the sound in decibels at a frequency of 1000 Hz. As shown in Figure 6.2, however, an equal loudness at another frequency is not at the same pressure level. Alternatively, subjects can rate loudness on a numerical scale (magnitude estimation); for example, it might sound twice as loud as a reference. The reference or 1 sone is the loudness at 40 dB for a frequency of 1000 Hz. A twofold change in loudness is produced by a 10-dB change in level, and so a 50-dB sound will be twice as loud as a 40-dB one.

Threshold

Thresholds (T) to tone pips are lower if the sounds are decreased in intensity to the point where they are not heard (descending method), as reported by Rosenblith and Miller, cited by Hirsh (1952). The ascending T level is approximately 4 dB higher due to the difficulty in first identifying a quiet sound. If the tones are

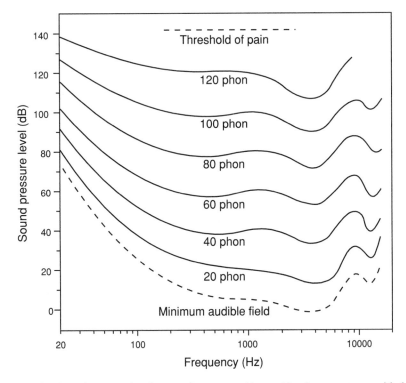

FIGURE 6.2. Sound pressure level versus frequency with equal loudness contours, with the loudness level given in phons for 20-dB increases in sound pressure level. The estimated loudness in sones is also shown. (Modified from the data of Robinson and Dadson, 1956. © Crown Copyright 2002. Reproduced by permission of the Controller of HMSO.)

continuous, the reverse will apply, with the descending T level being higher. This is due to the fact that the continuous sound produces sensory fatigue. For clinical assessment Carhart and Jerger (1959) recommend an ascending technique with short duration tones, and a T level when the sounds are heard repeatedly. In psychophysics, however, a T level is defined as the point where 50% of the responses are positive.

Dynamic Range and Loudness Growth

The growth in loudness can be scaled by estimating its magnitude through asking the subject to assign a number to the loudness or by adjusting the level to compare its loudness with that of another sound (e.g., twice as loud). Stevens (1975) showed that loudness (L) was a power function of intensity (I). The relation is as follows:

$$L = kI^{0.3}$$

where k is a constant that depends on the subject and the units used. Thus a feature of the power function is that when it is plotted in log-log coordinates, it becomes a straight line.

Studies have shown considerable variation across subjects and test procedures (Warren 1970; Warren and Bashford 1980). Another criticism is that the function depends on the relation between the sensation and the number assigned, in particular whether this is linear or logarithmic (Treisman 1964). Loudness scaling models have been used by Zwicker (1958), Zwicker and Scharf (1965), and Stevens (1972) to determine the loudness of complex sounds. As described by Moore (1997), they essentially involve converting the energy in a number of frequency bands into loudness using the Stevens power law. The loudness in each band is summed to produce the total loudness.

The dynamic range is from threshold to the maximum comfortable (MC) level. In measuring the MC level the intensity of the sound is increased until it becomes too loud or uncomfortable. It is then decreased to a level that is just comfortable. Thus the MC is just below the minimum discomfort level.

Difference Limens

The DLs for sound intensity can be determined by (1) modulating the amplitude, (2) making an incremental change in the background intensity, and (3) presenting two pulses and asking the subject which is the louder through a two-alternative forced-choice procedure. With wide band or band pass filtered noise, the ratio of the change in intensity ΔI to the intensity I ($\Delta I/I$), the Weber fraction, is constant from 20 dB to 100 dB above threshold (Miller 1947), thus upholding the Weber law. The data of Riesz (1928) using modulation detection gave values of 1.5 dB at 20 dB SL (Sensation Level), 0.7 dB at 40 dB SL, and 0.3 dB at 80 dB SL. Sensation level is the sound pressure level of a sound relative to the threshold level. The DL was also found to be 0.5 to 1.0 dB by Rodenburg (1977). For pure tones, however, Weber's law is only partially correct, with discrimination better at higher intensity levels.

Temporal Integration

Threshold and loudness depend on duration (Exner 1876), but only for sound durations less than about 200 ms (Moore 1997). The relation is best described by the following expression:

$$(I - I_L) \times t = I_L \times \tau = k$$

where I is the threshold intensity for a tone of duration t, I_L is the threshold intensity for a long duration tone; τ is a constant representing the integration time of the auditory system, and k is a constant.

The integration of energy is undertaken by the nervous system, and helps explain how threshold varies with duration. Plomp and Bouman (1959) found the time course for integration also varied with frequency being 375 ms at 250 Hz and 150 ms at 8000 Hz. Green et al (1957), using signal detection theory, found integration was generally constant for durations from 15 to 150 ms. This indicated there was a limit over which the nervous system could integrate energy. Sheeley and Bilger (1964) also found the plateau occurred at longer durations for low than high frequencies. At a given intensity, loudness increased with duration. Boone (1973) found that for a frequency of 1000 Hz, loudness was related to the total energy of the tone burst. Stephens (1974) also concluded that equal energy meant equal loudness. The mechanisms underlying temporal integration of acoustic stimuli have been modeled by Viemeister (1979), Buus and Florentine (1985), Green and Forrest (1988), Moore et al (1988), Plack and Moore (1990, 1991), and Oxenham and Moore (1994). The first two stages of the models are the auditory filter and compressive nonlinearity representing the processing at the periphery. The third stage is a sliding temporal integrator with a window of 3 to 10 ms representing central neural mechanisms. Thus physical excitation of the cochlea is transformed into neural responses. Finally, the specific loudness is integrated across the cochlear place of excitation to determine the overall loudness of the sound.

Critical Band and Ratio

The concept of the critical band was first suggested by Fletcher (1940). The theory was (1) that in masking with white noise, the only components of the noise that have any masking effect are those frequencies that lie within a narrow band around the frequency of the test tone; and (2) that when the tone is just audible against the noise background, the total acoustic power of the components within the narrow band is the same as that of the pure tone. Fletcher defined this restricted range of frequencies as the critical band. Fletcher's work gave rise to two types of masking experiments: (1) the masking of a pure tone was measured in the presence of a wide band (white) noise, and (2) the masking of pure tones was measured using bands of noise of different width.

Critical Ratio

If Fletcher's hypothesis was correct, the widths of the critical bands could be calculated directly by measuring the masking of pure tones by white noise. This was done by Hawkins and Stevens (1950). Knowing the intensity of the just-masked tone and the intensity of the masking noise, it is then possible to calculate how large a band within the noise contains the same energy as the tone. The width of this band is by definition the masking band. Its width is calculated by taking the ratio of the intensity of the tone to the intensity per cycle of the noise (since a white noise contains all audible frequencies of equal intensity, the intensity per cycle is uniform throughout). Hawkins and Stevens (1950) found that at 1000 Hz, the level per cycle of the masking noise was 18 dB below the level of a 1000-Hz tone at its masked threshold. The 18-dB difference is a ratio and corresponded to an energy ratio of 63:1. This means that the band of frequencies having a total energy equal to that of the 1000-Hz tone would therefore be 63 cycles wide because each cycle contains 1/63 of the energy. This of course assumes that Fletcher's second hypothesis is correct, that the total acoustic power of the components within the narrow band is the same as that of the pure tone.

Hawkins and Stevens (1950) measured the masked thresholds of frequencies from 100 to 900 Hz in the presence of white noise at levels from 20 to 90 dB. They found that the band did not change as a function of the level of the masking noise; however, the band was found to vary with different center frequencies. It should be noted that using the above procedures the critical band was measured as a result of a ratio, and it is more correctly referred to a critical ratio. This must be distinguished from the critical band, where the bandwidth is measured directly.

Critical Band

The bandwidth has been measured directly by narrowband masking. Hamilton (1957) determined the masked thresholds for tones in the presence of bands of noise. It was centered at the frequency of the tone, and the bandwidths were varied. It was found that for a frequency of 800 Hz, the masked threshold increased up to a bandwidth of 145 Hz beyond which the threshold remained constant. Thus the masking band at 800 Hz, was 145 Hz wide. This finding was subsequently confirmed in other types of experiments.

Another method for measuring the critical band is the loudness of complex sounds. Experiments have examined the loudness of noise bands as a function of bandwidth. They showed that the loudness of a band of noise of fixed intensity remained constant until the bandwidth of the noise exceeded the critical band when the loudness increased with width. The bandwidth remained constant if the intensity was increased up to 80 dB.

Just as the energy outside a critical band may not contribute to the masking of the signal within that band, it may also not contribute to the audibility threshold of a complex sound. Gassler (1954) demonstrated this by measuring the threshold of a multitone complex composed of from 1 to 40 sinusoids evenly spaced 10 or 20 Hz apart. The threshold of a single pure tone was first recorded, and then another tone of the same intensity spaced 10 Hz lower was added and the threshold again measured. To reach the threshold the intensity of each tone in the complex could be less than when presented alone. As each tone was added to the complex, the overall sound pressure level at the threshold remained constant up to a critical bandwidth; that is, with the addition of each new tone, the levels of the original tones were decreased. When tones were added beyond the frequency limits of the critical band, the overall sound pressure level of the complex increased. The results indicated that the total energy necessary for a sound to be heard remained constant so long as the energy was confined to a single band.

The masking of a narrow band of noise by two tones (two-tone masking) provides another measure of the critical band. Zwicker (1952) measured the threshold of a narrow band of noise in the presence of two tones, one on either side of the noise. By increasing the difference in frequency between the two tones, the masked threshold for the noise remained unchanged until a critical bandwidth was reached, whereupon the threshold fell sharply and generally continued to fall as the change in frequency was increased further.

Phase sensitivity is another means of measuring the critical band. When the amplitude of a tone is modulated (AM), a three-tone complex is produced with the original tone, the carrier at the center, and a tone on either side (side bands). Similarly when the frequency of a tone is modulated (FM) over a narrow range, a three-tone complex is also produced. The only important difference between the two three-tone complexes concerns the phase relationships among the components. Consequently, any difference in the ear's sensitivity to AM and FM would presumably depend on these phase relations. Zwicker (1952) found that in order for a subject to just hear the difference between a modulated and a pure unmodulated tone, a smaller amount of AM is required than for FM. However, the ear is more sensitive to AM than FM only at low ranges of modulation. As the rate of modulation is increased, the difference in sensitivity to AM and FM disappears. The relevance to the concept of the critical band is that the sensitivity to AM and FM becomes the same when the frequency separation between the side bands of the three-tone spectrum is equal to the critical band. When the separation in the frequencies is greater than the critical band, there is no difference in sensitivity to AM and FM, implying that beyond the critical band, the phase relationships within the complex no longer serve as significant cues, as in the other experiments. The critical band did not vary with changes in the sound pressure level of the stimulus.

If the critical bands measured directly are compared with the results measured indirectly by Fletcher, the so-called critical ratio, it was found the critical band was two and a half times greater than the critical ratio. Fletcher assumed a signal-to-noise ratio of 0 dB. If a signal-to-noise ratio of -4 dB is assumed the differences could be explained.

The critical band may be due to the filtering of frequency either at the periphery or centrally. The effective bandwidth of the filter can be defined by the bandwidth of the half-power point. This is the width of the filter in Hz at a point 3 dB up from the threshold of an auditory neuron to its characteristic frequency or frequency of best response (frequency tuning curve). There is good correspondence between a critical band measured this way in experimental animals, and a band measured by psychophysical means in humans. It is unclear whether the critical band is related to the tuning curve alone or whether lateral suppression is necessary for sharpening it. Lateral suppression has been demonstrated psychophysically by Houtgast (1974), who found edge effects with the nonsimultaneous masking of probe tones.

A relationship was derived from masked audiograms to relate the critical band to frequency and the position of the maximum amplitude of vibration of the basilar membrane to frequency. It was assumed that the frequency range of the critical bands represented equal distances on the basilar membrane (1 mm), and that the critical band increased exponentially from the helicotrema. The frequency-position function had a similar coefficient to the exponential function fitting von Békésy's basilar membrane elasticity data (Greenwood 1961). Newer data

(Greenwood 1990) on cadaver ears and living animal preparations are also predicted by the earlier function. Thus the model is of value in developing cochlear models and spectral transforms of speech.

Furthermore, recent psychophysical masking studies have given results for the critical band that differ from those obtained by Zwicker (1961). The newer estimates have been referred to as equivalent rectangular bands (ERBs) (Moore and Glasberg 1986; Glasberg and Moore 1990). The shape of the auditory filter has been estimated using a psychophysical tuning curve (Patterson and Moore 1986), and the notched noise method of Patterson (1976). The ERB is determined from the change in the threshold of each tone as the amount of noise passing through the filter is reduced as the width of the filter increases. This and other studies have demonstrated that the critical bandwidth decreases below 1000 Hz compared to the data of Zwicker (1961), which showed that it was constant at approximately 100 Hz. The above studies have also shown that the ERB = 0.89 mm along the basilar membrane rather than 1 mm as stated above.

The critical band is an important concept for investigating channel interaction with electrical stimulation of the auditory nerve. The acoustic findings are relevant to place and temporal pitch perception as well as loudness summation. Electrical stimulation too is able to throw light on the underlying neural and perceptual mechanisms for the critical band. Studies should also lead to improved speech processing for cochlear implant patients. For example, there are 14 to 19 critical bands that span the speech frequencies. Thus cochlear implants should at least attempt to provide this many stimulus channels. The critical band is equivalent to a distance of approximately 0.89 to 1.0 mm along the basilar membrane (Clark, Tong et al 1977), thus setting a limit on both the electrode spacing and spread of current.

Musical Acoustics

Definition

Music has been defined by the composer John Cage as "organized sound" (Ford 1993), but this could include communication signals. It has been discussed in some depth by Stainsby (2000). Music has pitch, rhythm, and timbre or tonal color. The sounds can be produced by instruments, by a mixture of instruments and voice, or by voice alone. Music differs from speech in that all aspects are equally important, whereas there is a great deal of redundancy in speech. A cochlear implant speech-processing strategy selects out appropriate signals to transmit through an electroneural bottleneck at the interface between sound and the nervous system, but all aspects of music need to be transmitted for the listener to appreciate the quality and timbre of the sound (Stainsby 2000; Stainsby et al 2002).

In music the succession of pitches constitutes melody. Music also has harmony and counterpoint. Harmony results when more than one pitch is sounded simultaneously, and counterpoint is the combination of multiple unfolding melodies.

Most musical traditions have rules for the combination of notes that sound pleasant (consonance) or unpleasant (dissonance). The control of consonance and dissonance is a primary component of composition.

Pitch

A succession of pitches constitutes melody, and is based around specific steps in pitch that are part of a musical scale. These steps are coded by temporal and place mechanisms. With speech, tonal variations over a wide frequency range are important for providing meaning at a syllabic level in some languages, especially in Southeast Asia. They also provide cross-segmental patterns for pitch-accentuated languages such as Swedish and Japanese. Other languages primarily provide meaning through the formant structure of the speech sounds. Formants are concentrations of frequency energy of importance for intelligibility.

In developing appropriate cochlear implant strategies to reproduce music, it is important to consider theories of pitch perception. One theory is a pattern matching model. This theory proposes that the brain has templates for frequencies that are harmonically related. The fundamental frequency of a complex sound that has harmonics matching the template determines the pitch (Goldstein 1973; Wightman 1973a,b; Terhardt 1974b). This model has a spectral frequency pattern, or temporally coded components.

An alternative theory was based on a spatiotemporal model for the integration of both place and temporal mechanisms (Licklider 1951; Moore 1977; Duifhuis et al 1982; Meddis and Hewitt 1991a,b). The first stage is an analysis of the complex spectrum by the auditory filters. An autocorrelation is performed across the outputs of the filters, and this provides an indication of the most common periodicities across all filters, and thus the overall pitch. The model was extended by including a model of inner hair cell transduction, as an early preprocessing stage (Meddis and Hewitt 1991a,b). De Cheveigne (1998) subtracted the outputs of the separate autocorrelation functions (the cancellation model), and so the candidate pitch is represented by dips in the rectified difference functions instead of peaks. This model is more physiologically plausible and has a greater sensitivity to phase than simple autocorrelation.

Rhythm

Rhythm can be defined as medium scale temporal organization of events at the level of musical notes, as distinct from larger-scale structural organization of phrases and themes (Stainsby 2000). Music is most commonly constructed on a regular metrical grid. The accents, notes, durations, and placement of the beginnings and ends of phrases are constrained by this grid. In contrast, speech has more irregular stressing referred to as prosody. Emphases and accents in speech are placed according to the significance of the words in the sentence. An underlying regular pulse or beat is seen in music, frequently in the range of 40 to 220 beats per minute. A different rhythm is seen with speech, and it can be recognized from other complex sounds by its prosody. The temporal fidelity of the current cochlear implant systems is capable of conveying the rhythm required for music. However, specific processing techniques that interpret musical rhythm may become relevant. It is possible that future specialized strategies will aim to model a particular musical rhythmic structure to enhance reconstruction of a musical signal (Stainsby 2000).

Timbre

Timbre, or tone color, is a complicated perceptual attribute of sound. Timbre, as defined above, is the quality that allows the distinction between two instruments playing the same note and the same loudness. One of the primary characteristics of sound determining the timbre is the shape of the frequency spectrum. This is the amount of energy present at each frequency across the audible range. For musical instruments, as distinct from speech, there is a strong continuity of the frequency spectra over time. Plomp (1970) and Slawson (1985) have suggested that the term *tone color* or *sound color* should be used to refer to a steady-state frequency spectrum as distinct from timbre.

A simple tone or sine wave has a very pure timbre as distinct from a more complex sound with multiple frequencies and a harmonic frequency spectrum. Other types of sounds possess more complicated frequency spectra containing discrete frequency components that do not possess frequencies that are simple numerical frequency ratios. There are other sounds that contain bands of noise in certain frequency regions. The steady-state frequency spectrum is an important component in a musical instrument's timbre. For example, the clarinet has a frequency spectrum that consists almost exclusively of odd-numbered harmonics. Resonances due to the shape or structure of an instrument give rise to spectral formants. The consistency of the formant structure contributes significantly to an instrument's timbre and this formant structure does not change significantly as the notes change, so all the pitches played on a particular instrument have a similar timbre (Rossing 1982). Slawson (1968) showed that when the fundamental frequencies of sounds were varied, the timbre changed less when the absolute frequencies of the formants were held constant than when the relative amplitude of partials remained constant, thus demonstrating the importance of the absolute formant structure in identifying aspects of timbre.

Plomp and Steeneken (1969) investigated the effect phase relationships had on the timbre of complex tones. This was done using synthesized sounds. Research concluded that modification of phase alone could have an effect on a sound's timbre, but it was only of minor importance. The greatest difference was between a sound with all sine harmonics and another with alternating sine and cosine harmonics. This is also not likely to be of significance, as reverberation in common listening environments would blur the relationship between the partials. Plomp (1970) used a multidimensional scaling technique and found three dimensions were adequate to describe the timbre similarity space. Grey (1977) also found three dimensions: rise time, spectral centroid, and spectral flux. The spectral centroid is the frequency band where the greatest spectral energy is located, and reflects whether the sound is bright or dull. Spectral flux refers to the dynamic property of the spectrum and the introduction of more harmonics during the rise time.

Other nonspectral components of timbre are temporal amplitude variation, dynamic changes in the frequency spectrum, and a sound's spatial placement. Of particular interest are the effects of the three stages in the presentation of a sound—namely attack, sustain, and decay. Saldhana and Corso (1964) found that the most significant clue to identifying an instrument was the attack characteristics and then the sustain section. The attack gives a clear indication of the way the instrument is played. Certain dynamic spectral cues are characteristic of classes of instruments; for example, brass instruments have a harmonic frequency spectrum that swells during the onset. [The above section on timbre is a summary of a discussion presented by Stainsby (2000)].

Bilateral Hearing

Bilateral hearing is important for localizing sound in space, hearing in noise, and providing a 3- to 6-dB increase in the intensity or pressure of the sound.

Sound Localization and Lateralization

For most mammals the ability to localize sounds plays an important role in the tracking of prey and avoiding predators. The cues depend on the range of hearing, the size of the head, the shape and mobility of the external ears, and the acoustic spectrum of the sound source. It is important to know not only where a sound comes from but also what is heard. In humans the ability to localize sound is also important socially, and especially in the presence of background noise.

The first experiments on sound localization led to the "duplex" theory put forward by Lord Rayleigh (Strutt 1907). This theory states localization of low-frequency sounds depends on interaural time differences (ITDs) and high-frequency sounds on interaural intensity differences (IIDs). Stevens and Newman (1936) showed the ability to localize pure tone stimuli was greater at frequencies below 1000 Hz and above 4000 Hz. Mills (1958) studied the minimum audible angle (MAA) or the minimum angle of separation that could be detected, and confirmed the loss of acuity in the midfrequency region and found a gradual degradation in performance as the source was moved to the listener's side. The duplex theory was consistent with the degradation of performance in the midfrequency region, but did not explain localization in the midline.

This duplex theory was explained on the grounds that the head reduced the intensity of sound from one side to the other (the head shadow effect) to produce an IID. This process was effective at high frequencies, but only minimally effective at low frequencies. However, there would be ITDs for a transient sound arriving at each ear (Fig. 6.3). If it is assumed that the distance between each ear is 130 to 170 mm (it is smaller in children), and with the velocity of sound in air at 331.3 m/s at 0°C, the time taken would be 0.39 to 0.51 ms. But if a continuous

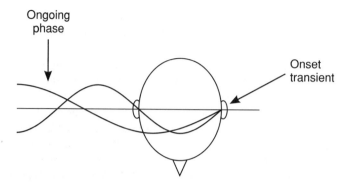

FIGURE 6.3. A diagram of the phase of two sine waves at each ear for an interaural separation that is half the period of one and smaller than the period of another. Note also that the burst of sine waves has onset and ongoing components.

tone is presented, there will be a phase difference between the sine wave at each ear, and this will depend on the interaural distance (Fig. 6.3): A 1200-Hz tone has a wavelength of 276 mm, and this is approximately twice the distance between the ears. In the human, 1200 Hz has been shown as the upper limit for sensitivity to ITDs (Mills 1960; Hershkowitz and Durlach 1969; Yost and Hafter 1987), and is assumed to be due to the fact that there will be ambiguity between information from the first half of the cycle arriving at one ear and information from the second half arriving at the other ear. This limitation on ITDs is also seen under head-phones, so it is due to neural mechanisms and not the interaural distance per se. The ITD for phase is normally from 300 to 400 μs, but a phase delay of 700 to 800 μms was found by Wightman and Kistler (1989) as the upper limit for sounds located to one side of the head. In contrast, the sensitivity to transients such as clicks is greater and can be as small as 10 μs.

The relative importance of the onset transient versus the ongoing phase infor-mation was investigated by Tobias and Schubert (1959). They found that small ongoing phase differences balanced quite large onset or transient disparities. Even for a short burst of 10 ms, the ongoing disparity had the greater effect. The onset disparity was only the greater for clicks.

Apart from IIDs and ITDs, it has been found that sound reflections from the convolutions in the pinna introduce resonances in the sound spectrum of 5000 to 7000 Hz. That is also an important cue for sound localization, especially with one ear. The reflections vary according to the direction of the sound, and this produces a different spectrum of frequencies (Batteau 1967; Gardner and Gardner 1973). Other cues are the change in the spectra with head movement (Thurlow and Runge 1967), visual cues (Gardner 1968), prior knowledge of the stimulus properties (Coleman 1962), and postural variables.

Sound lateralization is the experience of sound moving within the head when signals are presented under headphones; sound localization is when the sound source is externalized, and this occurs when sound is presented from loud speakers in a free field environment. Discrimination of IIDs and ITDs under headphones provides important information on the ability of the auditory system to code these basic signals, but it does not definitively explain the discrimination of positions in space (their azimuth).

Humans can detect a 1-degree separation in signals when they are from directly in front, and 5 to 7 degrees when they are to one side. The location of the signal can be shifted in the head when the signals are presented through headphones or in space, that is, free field, by varying either the IIDs or ITDs. These differences can be traded against each other. The just discriminable level differences (JLDs) can be determined by detecting the change required to move the position in the head or in space.

Studies have shown that for IIDs when the image is in the middle of the head, the JLD is between 0.5 and 1.0 dB except at 1000 Hz when it is higher. As the image is moved off-line, the binaural system is less sensitive to IIDs (Mills 1960; Rowland and Tobias 1967; Hershkowitz and Durlach 1969; Grantham 1984; Yost and Dye 1988). The higher DLs at 1000 Hz may be due to an interaction of two underlying processes for either ITD and IID, where neither process is functioning best. It is important in measuring the IIDs to allow for the effects of binaural summation of loudness, and therefore the overall loudness levels should be randomized.

The data from psychophysical studies on ITDs have shown a JLD of approximately 2 degrees at low frequencies, and it increases with frequency. Above 1200 to 1500 Hz, the binaural system is insensitive to ITDs (Klumpp and Eady 1956; Zwislocki and Feldman 1956; Mills 1960; Sayers 1964; Hershkowitz and Durlach 1969). For some frequencies the 2 degrees of interaural phase shift corresponds to a 10-μs interaural delay. Psychophysical studies have indicated that it is not necessary to stimulate places in the cochlea corresponding to low frequencies to convey ITD information (Henning 1974; Nuetzel and Hafter 1981; Joris and Yin 1992). ITDs can be detected for high frequencies when the signal is amplitude modulated (AM). The interaural processing is from the wave envelope, and has been demonstrated for AM high-pass noise. It requires the modulation rate to be low. Thus ITD information could be used with cochlear implants even though the electrodes are located in the high-frequency or basal region.

Research has been undertaken to develop procedures that allow overall localization performance to be more accurately defined (Whitman et al 1987). This has involved developing models to allow free-field testing conditions under headphones to simulate free-field listening conditions. The binaural system operates over a wide spectrum of frequencies. It is important to understand how the system processes information across the channels. This is critical in daily life, for example, knowing the location of two people talking at the same time at a noisy party.

The physiological aspects of sound localization were discussed in Chapter 5; also see more comprehensive reviews, such as Yost and Gorevitch (1987).

The Precedence Effect

The precedence effect in sound localization occurs when two binaural sounds are presented with a brief delay between them, and its location depends on the directional cues for the first sound. It is an important coding strategy for hearing in noise. When sound arrives in a noisy room the first cues are given greater weight than the subsequent ones from reverberation. The effect has been extensively studied in architectural acoustics (Parkin and Humphreys 1958). Walloch et al (1949) demonstrated that the initial transient had a strong effect on the localization of the sound. The precedence effect occurred when the delay was greater than 500 μs (Blauert 1971). The precedence effect may be a useful measure for the localization of sound in multiple-source sound fields, and could be useful in assessing the performance of bimodal and bilateral cochlear implants.

Binaural Release from Masking

Binaural hearing is important for hearing speech in noise as well as for sound localization (Hirsh 1950; Pollack and Pickett 1958). Improving the recognition of a signal in noise is referred to as the masking level difference and the binaural release from masking (squelch), and has been demonstrated in a number of studies (Licklider 1948; Hirsh 1950; Pollack and Pickett 1958; Carhart 1965; MacKeith and Coles 1971; Peissig and Kollmeier 1997; Litovsky and Colburn 1999). It has been shown for diotic signals in dichotic noise, and for dichotic signals in diotic noise (Licklider 1948). Diotic sounds reaching the ear are the same, and dichotic sounds are different. The ability to hear speech in noise is due to IIDs and ITDs, as discussed above and in Chapter 5. It is also due to the correlation of the signal from each ear compared to the lack of correlation with the noise.

If the same tone and noise are fed to both ears, the threshold of the tone can be reduced by inverting the phase of the tone. The difference between the two thresholds is the binaural masking level difference (MLD). It decreases from about 15 dB at low frequencies (500 Hz) to 2 to 3 dB at 1500 Hz. As it only operates up to 1500 Hz, it is probably related to the ability to compare temporal information between the ears as seen with ITDs. The conditions that improve MLDs also do the same for speech intelligibility in background noise. This is referred to as binaural release from masking or the squelch effect. It was shown that intelligibility was better in antiphasic conditions, that is, when the phase relations of either the noise or signal are opposite (Hirsh 1950).

Loudness Summation

Bilateral summation of loudness occurs at all frequencies (Fletcher and Munson 1933), and is greatest when the stimulus intensity is the same in each ear (Irwin 1965; Marks 1978). The loudness of a binaural stimulus is independent of the frequency separation of the tones (Scharf 1969), and loudness summation is not affected by a hearing loss (Hall and Harvey 1985; Hawkins et al 1987).

Loudness summation has been reviewed by Durlach and Colburn (1978). It is usually assessed by matching the loudness of the binaural sound to a monaural component and adjusting the intensity of one until the loudness is equal. The amount of summation varies between 0 and 12 dB depending on type, level, and interaural differences of the signals. At low levels the monaural signal has to be increased by only about 3 dB to match the loudness of the diotic signal, but it needs to be increased by 12 dB at high levels (Reynolds and Stevens 1960). For broadband stimuli, binaural loudness summation increases with level, but for narrow-band stimuli it first increases and then decreases slightly (Scharf 1968).

Studies have also examined the effects of interaural time and phase as well as frequency differences on loudness summation. With clicks the loudness of the click in one ear compared to the other depends on the delay (Hirsh and Sherrick 1961). In contrast, differences in phase of two tones have no effect on summation (Scharf 1969).

Binaural Creation of Pitch

It is possible to construct stimuli in which pitch is perceived entirely by binaural interaction. This can be experienced, for example, with gaussian noise presented to one ear and the same signal to the other ear except for a narrow-band phase shift at relatively low frequencies (Cramer and Huggins 1958). This is evidence for the importance of fine temporospatial patterns of neural responses for the temporal coding of frequency, as was discussed in Chapter 5.

Electrical Stimulation

A key approach to speech processing with electrical stimulation of the auditory nerve by the University of Melbourne/Bionic Ear Institute was to determine the percepts for simple and complex electrical stimulation, and relate them to the percepts for similar acoustic stimuli. This provided both a rationale and direction for speech-processing strategies with electrical stimulation. It also gave insights into the coding of sound by the brain.

Temporal Information

As was discussed in Chapter 5, the relative importance of the temporal and place coding for different frequencies was not completely clear. In addition, there are situations, for example the coding of speech in noise, in which both place and temporal coding are necessary.

Rate Discrimination

One method of assessing the ability of subjects to process temporal information with electrical stimulation is rate or frequency discrimination. This is measured, as discussed above (see Discrimination) as the just detectable differences the (DLs) in rate of stimulation. Simmons (1966) found in one subject for electrodes in the auditory nerve (AN) there was an absolute DL for rate of stimulation at 100 pulses/s of approximately 20 to 30 pulses/s (20% to 30%), and over 100 pulses/s at 300 pulses/s (greater than 33%). The upper limit was not defined. Michelson (1971) implanted four people with scala tympani electrodes, and they heard pitches up to 12,000 Hz and could distinguish rates of 50 to 100 Hz in the low to middle range. In a later study on these patients in which the stimuli were balanced for loudness, the apparent pitch changed little above 600 Hz. The DLs varied from 1% at 200 Hz to 7% at 900 Hz. Electrophonic hearing could have accounted for these results, as air conduction thresholds for the frequencies 250 to 1000 Hz in one patient varied from 50 to 60 dB.

Electrophonic hearing was seen in a behavioral study on cats (Clark, Kranz et al 1973) in which rate of stimulation could be discriminated up to 2400 pulses/s, but with the destruction of hair cells the upper limit was 800 pulses/s. Subsequently, studies by McAnally and Clark (1991a,b, 1994) and McAnally et al (1993, 1997a–c) have shown that with electrical stimulation of the basal turn, low frequencies can initiate a basilar membrane traveling wave to excite residual hair cells in the low-frequency region. Thus interpreting frequency discrimination with electrical stimulation requires care to exclude electrophonic hearing and to mask the apical region of the cochlea.

Eddington et al (1978) reported that for two people with electrodes in the scala tympani, as the rate of stimulation increased from 80 to 800 pulses/s the pitch became higher, but they did not indicate whether DLs varied with rate. Hochmair-Desoyer et al (1981) found in some patients a monotonic increase in subjective pitch up to 700 Hz or even 1000 Hz. Psychophysical studies on the first patient operated on in Melbourne in 1978, and a second in 1979, showed that the DLs for electrical stimulation, like those in the experimental animal, were better for lower rates (Clark, Nathar et al 1972; Clark, Kranz et al 1973; Williams et al 1976). They ranged from approximately 2% to 6% for stimulus rates of 100 and 200 pulses/s (Tong et al 1982, 1983a,b). The DLs for 100 and 200 pulses/s were higher than the DLs for acoustic stimuli of the same frequencies. The DL for a sound of 200 Hz is about 1%. In general, although rate DLs for electrical stimuli were higher than those for sound, they were more comparable with acoustic stimuli at frequencies up to 200 pulses/s, but not at higher stimulus frequencies. With pulse rates below 200 pulses/s, the pitch produced by electrical stimulation increased with pulse rate. For example, on one electrode the pitch increased from 78 mels at 50 pulses/s to 3000 mels at 200 pulses/s. However, for pulse rates above 200 pulses/s, there was little increase in pitch (Tong et al 1979, 1982, 1983a). As DLs and the scaling of pitch for rate of electrical stimulation up to 200 pulses/s were more comparable with those for sound than at higher rates, this suggested that a rate (time/period) code for frequency could be used to convey low-frequency information in speech using electrical stimulation. This applied in particular to voicing which has a mean frequency of 124 Hz for adult males, 220 Hz for adult females, and 297 Hz for children (Eguchi and Hirsh 1969).

It was also of importance to discover that DLs for rate of stimulation were the same for electrical stimuli on both the apical and basal electrodes, as shown in the study by Tong et al (1983a,b). This was consistent with the results from the behavioral study on the experimental animal (Clark, Nathar et al 1972). The finding suggested that temporal pitch perception for low frequencies was at least partly independent of place pitch perception, and in turn the time/period code independent of the place code for frequency. It also indicated that with speech-processing strategies, voicing, which is low in frequency, could be coded on either apical or basal electrodes.

To establish the variability in rate discrimination across patients, as this might reflect differences in speech perception, a study was undertaken in Melbourne to plot discrimination as a cumulative (d') (Tong and Clark 1985). In a total of six patients, rate discrimination commenced to plateau at approximately 200 pulses/s in two, at 250 pulses/s in two, and at 600 pulses/s in two. Calculating d' to measure the perceptual space was after the method of Braida and Durlach (1970).

Having shown that patients could discriminate rate of stimulation at low frequencies, and that rate of stimulation could be used to provide voicing information, it was desirable to know whether the discriminations and categorizations were due to the perception of pitch. One way of assessing the patients' abilities to perceive pitch for rate of stimulation was to compare the pitch of a test stimulus with a standard stimulus. The results of the study (Tong et al 1983a), illustrated in Figure 6.4, showed the pitch ratio increased rapidly with stimulus rate up to 300 pulses/s, when compared with sound. Above 300 pulses/s the pitch estimate did not change appreciably. The pitch ratios were the same when stimulating apical, middle, and basal electrodes. This study indicated that rate of electrical

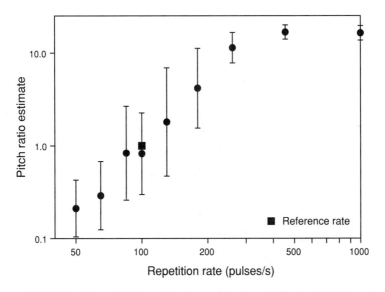

FIGURE 6.4. Ratios of the pitch for a stimulus rate compared with the pitch of a reference stimulus rate of 100 pulses/s [multiple pulses/period (MPP) stimuli] plotted against repetition rate. (Modified from Tong et al, 1983a, with permission from The Acoustical Society of America.)

stimulation is perceived as pitch. Pitch ratio results for rate of stimulation were also consistent with results for rate discrimination, where DLs increased greatly above 300 to 600 pulses/s. This supported the hypothesis that pitch is used in making the discrimination.

An acoustic model of electrical stimulation of the auditory nerve was developed by Blamey et al (1984a,b), to assess speech-processing strategies. First the validity of the model was determined by seeing how well it replicated the psychophysical results for electrical stimulation of the cochlear nerve fibers in the cochlea. The simulation was done with a pseudo-random white noise generator, with the output fed through seven separate band-pass filters with center frequencies corresponding to the electrode sites (Blamey et al 1984a,b). The mean pulse rate relative DLs for the acoustic model were 4% at a rate of 100 pulses/s and 7.4% at 200 pulses/s. For electrical stimulation in the first two patients the DLs were 5.2% and 5.9%, respectively. The curves plotting the pitch ratio versus rate of stimulation showed a plateau at 300 pulses/s as seen with electrical stimulation. Pitch as a function of electrode position and filter frequency was evaluated by measuring the pitch ratios and d' measures of the distances of the perceptual space between stimuli on successive filters. The results for the acoustic model of place of stimulation were similar to those for site of electrical stimulation, as discussed below (see Place Information). Finally, the ability of subjects and patients to judge stimuli differing in pulse rate as well as filter frequency or electrode position was evaluated using multidimensional scaling. The spatial similarity or distance between the combinations of stimuli was measured, and showed good agreement between the acoustic model and multiple-channel electrical stimulation using nonmetric multidimensional scaling (Clark et al 1987). Thus the results for the acoustic model were comparable to those obtained with electrical stimulation on implant patients.

In undertaking studies on the temporal processing of pitch through rate discrimination, it is important to ensure that the stimuli are loudness balanced. Not only may loudness vary with rate, as discussed below (see Neural Population Excited), but a change in current may alter pitch. This was demonstrated in a study by Townshend et al (1987), who found a significant decrease in the perceived pitch with an increased stimulus level. The variation in pitch was less pronounced for a 100 pulses/s than a 200-pulses/s stimulus.

Time-Varying Rate Discrimination Versus Duration

Having shown that electrical stimulation could convey some low-frequency information using a time/period code, the next task was to see if the time/period code could be used to convey segmental and suprasegmental speech information. The ability of patients to discriminate rate of stimulation over durations similar to those of consonants and vowels (segmental information) was first investigated (Tong et al 1982). The results for varying the rate of stimulation over durations of 25, 50, and 100 ms showed that a change in rate from 150 to 240 pulses/s could be well discriminated for the stimuli of longest duration (50 and 100 ms), but not at a short duration of 25 ms, which is comparable in duration to that of a consonant. This is illustrated in the left of Figure 6.5.

From the results in this study it was concluded that the time/period code for frequency would not be appropriate for coding consonants, but would be appropriate for coding frequency changes over the longer durations that occur with suprasegmental speech information, in particular voicing.

Voicing Categorization and Across Electrodes

To confirm that a time/period code was suitable for conveying voicing, it was necessary to see if the sensation obtained with rate of stimulation could be categorized in the speech domain. Furthermore, it was important to confirm that voicing could be transmitted independently of place of stimulation by varying the rate of stimulation on different electrodes, as stimulus rate would be affected by a change in the place of stimulation required to code the formants in vowels and consonants. Studies were undertaken where repetition rates were varied both upward and downward, for a stimulus range of from 60 to 160 pulses/s over a duration of 300 ms. In one study (Tong et al 1983a) six rate trajectories were

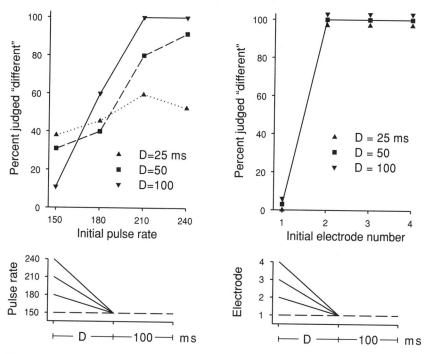

FIGURE 6.5. Rate and place discrimination versus duration. The percents of judgments called different are shown versus pulse rate and stimulus place trajectories. The trajectories were 25, 50, and 100 ms in duration. Left: The initial pulse rates of the trajectory varied from 240, 210, 180, and 150 pulses/s to 150 pulses/s. Right: The initial electrodes of the trajectory varied from electrodes 4, 3, and 2 to the baseline 1 (Tong et al 1982, Psychophysical studies for two multiple-channel cochlear implant patients. *Journal of Acoustical Society of America* **71**: 153–160. Reprinted with permission.)

presented on either the apical, middle, or basal electrodes, and the patients were asked to categorize the stimuli as questions or statements according to whether the sensations were rising or falling in pitch. A typical result for the categorization of voicing using stimuli on these individual electrodes (Fig. 6.6) showed that as the trajectory rose more steeply, the proportions of stimuli judged to be a question reached 100%, and for steeply falling trajectories it approached 0%.

The effects of varying the repetition rate across four electrodes on question-/statement judgments were also studied, and the results were similar to those for variations on a single electrode. Thus the patient was able to categorize questions and statements from the slope of the voicing frequency, while electrode trajectories moved in an apical or basal direction. A two-way analysis of variance on the data showed no significant difference between the performance for the repetition trajectories on individual electrodes or across a number of electrodes. The categorization was comparable with that reported by Fourcin et al (1979) for acoustic stimulation with fundamental frequency trajectories superimposed on the syllable "oh." From these results it was concluded that (1) rate of stimulation was effectively perceived as voicing, (2) voicing was perceived independently of place of stimulation, and (3) voicing was perceived by varying the rate of stimulation across different nerve populations.

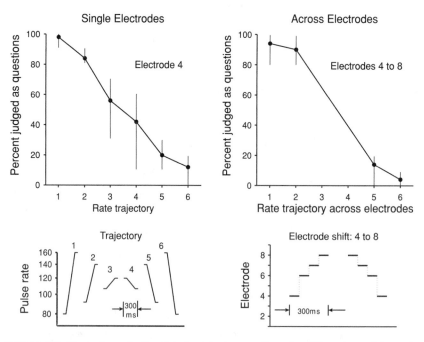

FIGURE 6.6. Proportion of stimuli judged to be questions for different repetition rate trajectories. Left: Judgments for six rate trajectories on a single electrode (rate trajectories shown bottom left). Right: Judgments for rate trajectories as electrode position changes over time (electrode trajectories shown bottom right). Each data point is a mean with the range of results indicated by bar. (Modified from Tong et al, 1983a)

Pitch Matching of Acoustic and Electrical Stimuli

Another way of assessing pitch changes for rate of stimulation was to compare the sensation for electrical stimulation in one ear with that for different sound frequencies in the opposite ear, when some residual hearing was present. This was first done by Bilger et al (1977), who compared the sensations in a patient who had a single-channel implant on one side and some residual hearing in the other ear. They found that the pitch for electrical stimulation below 250 pulses/s could be matched to that for an acoustic tone of the same frequency, but above 250 pulses/s a proportionately higher sound frequency was needed.

Blamey et al (1995) carried out a more extensive pitch matching study on eight subjects using the Cochlear Limited multiple-electrode implant in one ear and a hearing aid in the other. Although the results showed significant variation, five out of eight could match a stimulus of 250 pulses/s to a sound frequency within 20% of this rate, three out of eight at a pulse rate of 500 pulses/s, and two out of eight at a pulse rate of 800 pulses/s (Fig. 6.7). The results from the studies by Bilger et al (1977) and Blamey et al (1995) are consistent with the pitch ratio and frequency DL data summarized above, and help show that rate of electrical stimulation could be used for temporal pitch perception up to frequencies of 250 Hz and even up to 800 Hz. As some subjects (Blamey et al 1995) could match the higher frequencies of 800 pulses/s, there may be a number of individual variables that are important for temporal pitch perception.

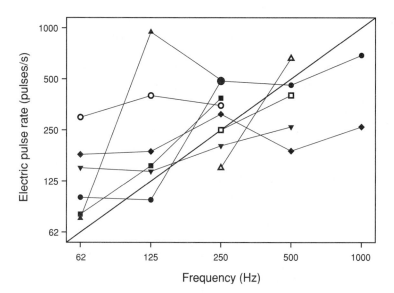

FIGURE 6.7. Pitch matching of sound frequency and rate of electrical stimulation derived from a pitch-matching experiment with eight implantees. The diagonal line shows the equality of pulse rate and frequency. (Blamey et al, 1995, Pitch matching of electric and acoustic stimuli. *Annals of Otology, Rhinology and Laryngology* **104**; Supplement 166: 220–222. Reprinted with permission.)

Amplitude Modulation and Pitch Perception

It was thought that the pitch experienced for rate of electrical stimulation would be similar to that with amplitude modulated white noise (Evans 1978). A weak pitch was perceived with modulated white noise corresponding to the modulation frequency up to approximately 500 Hz (Burns and Viemeister 1981). The sensation was considered to be pitch, as subjects were able to identify simple melodies and musical intervals. With amplitude-modulated electrical stimuli at a comfortable listening level, the modulation frequency could be readily detected up to 100 Hz. Above this frequency the detectability fell off rapidly (Shannon 1992; Busby et al 1993b). The pitch perceived with amplitude-modulated electrical stimuli was studied further by varying the modulation depth, and comparing the resulting pitch with that of an unmodulated stimulus. The effects of different modulation and carrier frequencies were examined (McKay et al 1995b). For modulation frequencies in the range of 80 to 300 pulses/s, the pitch fell exponentially from a value close to the carrier rate to a value toward the modulation frequency, as the modulation depth increased. The pitch could be predicted on the basis of the weighted average of the two neural firing rates in the stimulated population, with the weightings proportional to the respective numbers of neurons firing at each frequency. The population of neurons excited for currents at different modulation depths was estimated indirectly from a masking study showing the relationship between loudness and the area of the basilar membrane excited (Cacace and Margolis 1985) and other studies demonstrating the relationship between current level and loudness (Eddington et al 1978). The model could predict the findings for carrier rates up to 700 Hz. These findings support the hypothesis that a time/period code for frequency depends on a weighting of the interspike intervals in a population of neurons. The weighting of intervals would be a separate mechanism from a coincidence detection model of temporal coding proposed by Carney (1994).

Neither these psychophysical studies nor other mathematical modeling studies (Irlicht and Clark 1995; Bruce et al 1999; Burkitt and Clark 2000) clearly show whether the temporal coding mechanisms used by the brain for electrical stimulation are similar to those for amplitude modulated white noise. However, it must be emphasized that the pitch perceived for unmodulated low rates of electrical stimulation appears as a stronger pitch sensation than for modulated white noise.

Gap Detection

The ability to detect the duration of gaps between sounds is another aspect of the temporal coding of information of importance to speech understanding. This ability is also related to DLs for duration. The detection of gaps could be used for the recognition of consonants, voice-onset time, and the overall segmentation of speech. In normal-hearing subjects, gap detection thresholds varying from 3 to 10 ms have been recorded (Penner 1977; Fitzgibbons and Wightman 1982; Florentine and Buus 1984; Fitzgibbons and Gordon-Salant 1987; Hall and Grose 1997). If tones rather than noise were used, there was a relationship between performance and the frequency of the stimuli. Normal-hearing subjects could detect smaller gaps for high frequencies due to the shorter ringing time of the basilar membrane at these frequencies (Green 1973; Shailer and Moore 1983). With some hearing-impaired subjects, there was degradation in performance due possibly to central neural degeneration and tone decay. Gap detection thresholds for normal-hearing subjects were compared with those for electrical stimulation of the auditory nerve. This would also help show whether neural degeneration following a hearing loss affected this form of temporal processing. The duration DLs in two postlinguistically deaf subjects were 3 and 6 ms, and the gap detection thresholds were 2 and 17 ms (Tong et al 1988). They were thus almost identical to those for sound. The gap detection thresholds for acoustic and electrical stimulation both became poorer near hearing thresholds, and were both on the order of 20 to 80 ms (Penner 1977; Florentine and Buus 1984; Fitzgibbons and Gordon-Salant 1987; Moore and Glasberg 1988; Preece and Tyler 1989; Shannon 1989a; Hall and Grose 1997). The similarity of the results suggested that the differences in the temporal processing of information between the two types of stimuli were not critical. Furthermore, the research of Moore and Glasberg (1988) showed that fluctuations in noise levels can limit gap detection thresholds when "dips" in the noise may be confused with the gap. This was more apparent in the three single-channel implant patients investigated than in the subjects with normal hearing. It was assumed that this was due to the limited dynamic range of electrical stimulation, the compression of the signal, and the signal-to-noise ratio.

A study by Shannon (1989a) compared gap detection in normal-hearing subjects and patients receiving pulsatile stimulation from the Nucleus device or sinusoids from the Symbion implant. The data from most patients were well fitted by a trading relationship between the duration of the gap and the square of the stimulus intensity, indicating energy detection. As gap detection may be a speech cue when stimulating two electrodes, a study by Hanekom and Shannon (1998) found that thresholds were a function of the spatial separation of the electrode pairs for the two stimuli. Lower gap thresholds were found when two electrode pairs were closely spaced, and gap thresholds increased as the separation increased. The sharpness of tuning was generally better for patients with better speech perception.

Temporal Information: Prelinguistically Deaf

Exposure to sound in early life is thought necessary for developing the neural connectivity for temporal and place coding of frequency. This coding underlies speech recognition. Thus for children who are prelinguistically deaf there is a need to know when and how best to present electrical stimuli. As discussed above, a number of studies have examined the limitations of postlinguistically deaf subjects in discriminating changes in rate of stimulation, and these studies required repeating in prelinguistically children.

A study was done initially for rate identification, duration DL, gap detection, and numerosity on two adults and one child who were prelinguistically deaf. The results were compared with those for people with postlinguistic deafness (Clark, Blamey et al 1987; Clark, Busby et al 1987; Tong et al 1988). The subjects were two adults aged 23 and 24 years (Pre-1 and Pre-2, respectively), and one child aged 14 years (Pre-3). Pre-1 had meningitis at age 15 to 16 months and the implant at age 25 years. Pre-2 was congenitally deaf and had the implant at age 24 years. The child (Pre-3) had meningitis at age 16 months and the operation at 14 years. The results for rate identification at 100 and 200 pulses/s, when measured as d' values on all three subjects, were initially worse than for two postlinguistically deaf adults. In addition, speech perception scores were poorer than for the average postlinguistically deaf adults using formant extraction speech-processing strategies. This applied, in particular, to the perception of consonants. The child, however, showed an improvement in rate discrimination with training over time, which was associated with an improvement in speech perception.

The temporal processing of on/off patterns was also investigated in these subjects by Tong et al (1988) using duration DLs, gap detection, and numerosity. All test results were worse for the two prelinguistically deaf adults compared with a control group of two postlinguistically deaf subjects. The duration DLs were 9 ms and 18 ms compared to 3 ms and 6 ms for the postlinguistic subjects. The gap detection thresholds were 64 ms and 200 ms at 7 months, compared to 2 ms and 17 ms for the two postlinguistic subjects. A pulse train at 1000 pulses/s was used to provide more accurate sampling of information. However, after 33 months the gap detection thresholds had improved to 14 ms and 95 ms. With numerosity, detection occurred at a rate of 3/s for all subjects, compared to 5/s for the postlinguistic group. The child (Pre-3) had results similar to those of the postlinguistic group for duration DLs (6 ms) and gap detection (3 ms), and numerosity improved from 1 month to 18 months. Although there was better temporal processing for the prelinguistic 14-year-old compared to the prelinguistic adults, there was little evidence of better vowel or consonant identification. However male–female speaker and one–two syllable identification improved. It was not clear if the poorer speech perception scores overall for the prelinguistic group were due to inadequate rate or place identification. A multidimensional analysis of the vowel data, however, showed a one-dimensional solution. This was interpreted as vowel length, and thus indicated that intensity and not temporal or place frequency information was responsible. Furthermore, a clustering analysis indicated a high degree of perceptual confusion between consonants. This suggested that neither rate nor electrode place was being used for identification.

In a subsequent and larger study on 10 early-deafened people ranging in age at implantation from 5 to 23 years, seven had rate DLs (4% to 36% for rates from 50 to 200 pulses/s) that were similar to those for late-deafened people (Busby et al 1992). Six of these had meningitis at ages varying from 1 year and 3 months to 3 years and 9 months. However, the remaining three people who were congenitally deaf with Usher's syndrome (operations at 16.4 years, 20.4 years, and 23.1 years) had larger DLs than the other seven in the group. At 200 pulse/s the relative DLs were 73%, 72%, and 105%, respectively.

The duration DLs ranged from 3 ms to 61 ms, and were much greater for the three congenitally deaf children with Usher's syndrome. One of the children with meningitis had a limen of 61 ms, but this was reduced to 8 ms after 30 months. The gap detection thresholds were 5 ms or less for all except two who had Usher's syndrome (16.4 years and 23.1 years), and were much better than those for the prelinguistically deaf people implanted in their early twenties discussed in the study by Tong et al (1988). With numerosity, most subjects underestimated rates of 3 to 4 pulses/s.

Speech is a dynamic stimulus, which requires coding not only steady-state stimuli but also those with time-varying rate and place of stimulation. A study on these complex stimuli on postlinguistically deaf adults by Tong et al (1982), discussed above (see Time-Varying Rate Discrimination Versus Duration), showed the limitations of using rate for coding a change in frequency over the short durations of 20 ms required for consonant recognition. As the processing of rate-varying stimuli could be affected by prior exposure to sound and the maturation of the nervous system, a similar study to that of Tong et al (1982) was undertaken to compare performance between early- and late-deafened patients (Busby et al 1993a). There were four early-deafened patients who lost hearing at 1 to 3 years of age and were implanted at 5 to 14 years of age. The four postlinguistically deaf patients were all deafened in adult life. The extent to which the stimulus rate changed (e.g., from 130 to 170 pulse/s) and the duration of the changes were studied. It was found that three of the four late-deafened patients were more successful than the early-deafened patients in discriminating repetition rate trajectories. Comparisons were made between the psychophysical results for subjects and their speech perception for single and multiple-electrode speech-processing strategies. The multiple-electrode speech-processing strategy coded the fundamental frequency (F0) as repetition rate, and the first and second formant frequencies (F1 and F2) as two time-varying electrode positions. Two temporally nonoverlapping current pulses, of current levels related to the amplitudes of the two formants (A1 and A2), were delivered in quick succession to the two electrode positions (F1 and F2) within each stimulus period (1/F0). The single-electrode strategy coded F0 as the repetition rate to a time-invariant electrode position. Two temporally nonoverlapping pulses corresponding to A1 and A2 in the multiple-electrode strategy were delivered in quick succession to the same single electrode position within each stimulus period (1/F0). This strategy was a single-electrode version of the above F0/F1/F2 multiple-electrode strategy. Speech perception performance in the single-electrode strategy was closely related to performance in repetition rate discrimination. This suggested that either F1 and F2 rates or the amplitude envelope cues from the F1 and F2 filters could be partly processed along a single channel to improve speech understanding. The difficulty, however, was the limited capacity of electrical stimulation in coding rate. As will be discussed below, it was shown that there was a significant improvement in speech perception performance from the single- to multiple-electrode strategy that was consistent with successful electrode place discrimination. The results of the above studies suggest that early exposure to sound is necessary for all aspects of temporal processing. Some learning of the discrimination of duration and gap detection was possible. The relationship between temporal processing skills and speech perception has not been fully clarified.

Place Information

Initial studies on a human volunteer by Simmons (1966) with six electrodes placed directly into the auditory nerve demonstrated different pitch-like sensations that were related to the electrode site and to whether the stimulus was monopolar or bipolar. The subject described some sensations as like "ping" and others as dull like "tapping on a door." This result was consistent with the place coding of frequency.

Place pitch perception needed to be established more definitively in subsequent cochlear implant studies on patients, as the physiological and behavioral research on the experimental animal had shown the limitations of using a time/period code on individual electrodes to convey the middle to high frequencies of speech (Clark, Nathar et al 1972; Clark, Kranz et al 1973), and that place coding of frequency was possible as electrical current could be adequately localized with bipolar or common ground stimulation in the scala tympani (Merzenich 1975; Black and Clark 1978, 1980; Black et al 1981). It has subsequently been shown that monopolar stimulation with the electrodes placed close to the spiral ganglion cells provides localized excitation of cochlear nerves.

As discussed in Chapter 5, bipolar stimulation is the passage of current between neighboring electrodes, common ground stimulation between an active electrode with the others all connected together electronically, and monopolar stimulation normally between on active electrode and a distant ground outside the cochlea.

Electrode Discrimination and Scaling of Place Pitch

With multiple-electrode cochlear implants it was first essential to know whether the electrical current could be localized to separate groups of auditory nerve fibers to allow the place coding of frequency and presumably place pitch perception. Studies were required to explore the effect of mode of stimulation and electrode geometry on current spread. Simmons (1966) found that stimuli on each of six electrodes in the modiolus produced a pitch sensation, which varied according to the position of the stimulating electrode. Pialoux et al (1976) after drilling directly into different sites in the cochlea reported that the sounds perceived after stimulating electrodes in succession depended on the topography of the electrode. Studies by Clark, Tong et al (1978) and Tong et al (1979, 1980b, 1982) showed that, with common ground stimulation on banded electrodes, pitch estimates varied sequentially along the array and around the basal turn of the cochlea. The loudness of the stimuli was balanced to prevent loudness being used as a cue. Eddington et al (1978) reported first on a patient (C.R.) of William House (one of the few from the Los Angeles group with multiple electrodes) in whom all electrode pairs separated by two or more inactive pairs could be detected nearly 100% of the time. A second patient deaf from birth had difficulty in distinguishing pitch, but a third had tonotopic ordering of pitch according to the place of the electrode.

Burian et al (1979) referred to stimulation on various channels producing tones differing in pitch.

One of the most important findings in the study by Clark, Tong et al (1978) and Tong et al (1979, 1980b, 1982) was that the patient described constant rate stimuli on different electrodes as sharp or dull, according to whether they excited higher or lower frequency regions of the cochlea, respectively. The other most important finding was that stimulating each electrode produced different vowel or spectral colors. The vowels perceived were similar to the vowels heard when a similar region of the cochlea would have been excited by an acoustic formant in a normally hearing subject. This was the rationale underlying the University of Melbourne's formant extracting speech-processing strategies (see Chapter 7).

The spectral quality or timbre perceived as sharp or dull was used to rank the electrodes with respect to each other on the initial patient and on a second patient (Tong et al 1982). A typical result is shown in Figure 6.8. The ordering of the columns in the data matrix gives an approximate ordering of the electrodes with respect to sharpness. The electrodes were numbered from 9 to 0 in a basal to apical direction. A fixed pulse rate of 100 pulses/s was used. Each standard electrode was paired once with itself and the other seven comparison electrodes, and the patient was to identify whether the comparison was duller or sharper than the standard. Each cell of the matrix contains an S if the comparison was sharper than the standard, and a D if the comparison was the same or duller than the standard. The dull sensation on electrode 9 was probably due to the fact that it was outside the round window and the current flowed outside the cochlea to fibers conducting low-frequency information. As can be seen, there was good ranking of place pitch in the sharp–dull domain. This ranking of pitch was also seen in

Standard stimulus (electrode no.)

		1	2	3	4	6	7	8
	1	=	D	D	D	D	D	D
	2	S	=	S	S	D	D	D
Comparison stimulus	3	S	S	=	D	D	D	D
	4	S	S	S	=	S	D	D
	6	S	S	D	S	=	D	D
	7	S	S	S	S	S	=	D
	8	S	S	S	S	S	S	=

FIGURE 6.8. The scaling of place pitch. A comparison of the timbre described as sharp (S) or dull (D) with a standard stimulus for the nine electrodes in use for the first patient MC-1. Electrode 9 was extracochlear, and the current flow was probably extracochlear directly to the auditory nerve to produce a dull rather than sharp sensation (Tong et al 1982, Psychophysical studies for two multiple channel implant patients. *Journal of the Acoustical Society of America* **71**: 153–160. Reprinted with permission.).

the second and subsequent Melbourne research patients. The ability to rank pitch is used routinely in the clinical setting to produce a MAP of the patient's speech-processing strategy. It should also be noted to that in the acoustic modeling studies of electrical stimulation by Blamey et al (1984a,b), reported above (see Rate Discrimination); that the perceptual space between the model filter frequencies was very similar to those obtained with corresponding electrode stimulus sites in patients. The representation of formant and spectral information in speech through place pitch and electrical stimulation of separate groups of nerve fibers was a key principle underlying the University of Melbourne's initial (F0/F2) and subsequent formant vocoder strategies (F0/F1/F2, Multipeak) (Clark, Tong et al 1978; Tong et al 1979, 1980a; Dowell et al 1990) as well as the fixed filter (channel vocoder) schemes of Laird (1979), Eddington (1980), Chouard et al (1983), Merzenich et al (1984), Clark (1987), McKay et al (1991), and Wilson et al (1992, 1993). Speech energy at different frequency regions is transformed into electrical stimulation of appropriate groups of residual nerve fibers.

The speech-processing strategies, to be discussed in more detail in Chapter 7, differ in how spectral information is presented to the electrodes, but nevertheless there is a dynamic spatial pattern of neural excitation representing speech for each strategy. Understanding how the discrimination of pitch varied across patients and the relationship between the discrimination of the percepts for the each electrode and speech understanding is important for improving speech-processing strategies.

Because of the importance of place coding, there was a need to carry out studies in more patients to determine their ability to discriminate place of stimulation, depending on the separation of the electrodes for each bipolar pulse (spatial extent) and the number of electrodes between each pair (spatial separation), as illustrated in Figure 6.9. This would show the effect of overlap in stimulated populations of auditory nerves on discrimination. It was also necessary to determine how this varied within and across patients, as the variability of these skills would provide a basis for comparison with speech-processing performance. The study on seven patients (Tong and Clark 1985) determined their ability to discriminate and identify electrodes using the perceptual sensitivity index (d'). d' was calculated from signal detection theory (Green and Swets 1966). The spatial extent of the bipolar pulses and the pulse rate were fixed. It was found that discrimination improved as the spatial separation of the electrodes increased from 0.75 mm to 1.5 mm, but there was no further increase at 2.5 mm. This indicates the minimum overlap of neural populations stimulated for perceptual discrimination. There was no effect of the spatial extent on place discrimination.

The relationship between the population of neurons stimulated and the discrimination of different electrode spatial separations was also studied using masking techniques (Lim et al 1986, 1989a). This helped determine the distribution of neural excitation for one and two electrode stimuli. This is discussed below (see Channel Interaction).

The above studies demonstrated that banded and ball electrodes could produce localized stimulation of auditory nerve fibers and thus place pitch. The electro-physiological studies had shown this depended on the position of the electrodes

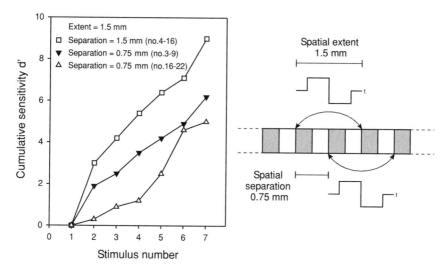

FIGURE 6.9. Left: The cumulative d' curves for the identification of seven bipolar electrode pairs for patients in the study by Tong and Clark (1985, Absolute identification of electric pulse rates and electrode positions by cochlear implant patients. *Journal of the Acoustical Society of America* **77**: 1881–1888. Reprinted with permission.). There was a fixed spatial extent of 1.5 mm, and the curves were for spatial separations of 0.75 and 1.5 mm. Right: A diagram of the spatial extent and spatial separation of electrodes.

within the scala tympani, and the current flow. The direction and spread of current depended on the electrode configuration or mode of stimulation. As discussed above and in Chapter 5, the three main modes of stimulation were bipolar, common ground, and monopolar.

A study on the estimation of pitch for three modes of stimulation using the Nucleus 22 banded array was undertaken by Busby et al (1994) on nine postlinguistically deaf patients. The modes of stimulation were bipolar, common ground, and monopolar. Monopolar stimulation was really pseudo-monopolar as it was between the active electrode and the most basal one, rather than a more distant extracochlear ground. In some patients the return electrode lay external to the cochlea. The Nucleus 22 system did not have the capacity for true monopolar stimulation between an active electrode and an extracochlear one in the surrounding tissue. As with all pitch-scaling experiments, the stimuli were balanced for loudness. The threshold (T) and MC listening levels were measured as discussed above.

The data were analyzed to determine the interaction effects and which electrodes contributed to these. The general pattern of pitch estimations across electrodes for each patient was consistent with the tonotopic organization of the cochlea. It was initially thought that current from monopolar stimulation would be too widespread for place coding; however, the study by Busby et al (1994) demonstrated it could be localized. There was a marked reversal of pitch ordering for electrodes in the middle of the array with common ground stimulation for three of the nine patients. This was assumed to be due to the effect of the current flow with this mode of stimulation associated with a loss of neurons or pathology in the cochlea (Busby et al 1994). This is supported by the finding that there was a trend for T and MC levels to be highest in the middle of the array with common ground stimulation. There was a reduced range in pitch for monopolar simulation when the return electrode was in the cochlea. This research thus showed that pitch or timbre may not be ranked consistently from basal to apical regions in some patients, especially for certain modes of stimulation. In this study patients were shown not only to be able to assign pitch labels for different sites of electrode stimulation, but also to discriminate one from the other. The data indicated that when switching modes of stimulation, the frequency-to-electrode mapping in speech processors should be reviewed because of variations in pitch estimation for different modes. There did not appear to be any advantage in using bipolar versus common ground stimulation for half the subjects. A study by Nelson et al (1995) using the Nucleus banded array also examined the variation in patients' abilities to distinguish the pitch (sharpness) with multiple-electrode stimulation. They used only bipolar stimulation and usually with electrodes separated by an inactive intervening electrode (BP + 1). The study was on 14 subjects. The ranking matrix for pitch was transformed into perceptual sensitivity units of d'/mm of the distance between comparison electrodes. There was considerable variability in abilities in ranking place pitch from spatial separations as little as 0.75 mm to over 13 mm. This indicated that the maximal spatial separation of 2.5 mm found by Tong and Clark (1985) was too low for some patients, and could reflect large differences in the ganglion cell population. The timbre (pitch) was generally ordered from sharp to dull in a basal to apical direction; however, the ability to rank the electrodes was sometimes better in the basal and at other times in the apical half of the array. In some there were reversals in place-pitch ordering for electrodes as was seen in the study by Busby et al (1994) for common ground stimulation.

In the initial development of cochlear implants it had been assumed that with multiple-electrode stimulation one's ability to perceive spectral speech cues would be related to one's ability to differentiate electrodes on the basis of place pitch (Clark 1969). In a study on vowel perception in children (Busby et al 1984), a multidimensional scaling procedure demonstrated that they were using three perceptual dimensions to recognize the vowels. These were interpreted as first and second formant frequencies and vowel length. The recognition of formant frequencies correlated with the ability to discriminate electrodes on the basis of place pitch (Tong et al 1980b). The importance of pitch ranking in the perception of consonants was evaluated by Nelson et al (1995) on 10 subjects. The consonant feature for place of articulation was examined, in particular, as cochlear implant psychophysical research had suggested the relationship of place pitch to the perception of plosives (Blamey et al 1987; Van Tasell et al 1987, 1992). However, in the study by Nelson et al (1995) there was only limited additional information transmitted on place information with the Nucleus 22 Multipeak-MSP strategy (Dowell et al 1990). This suggested that the extraction of spectral information, and its presentation on a place-coding basis for consonant place recognition, was not correlated with pitch ranking for this particular strategy. This contention was supported by the study by Zwolan et al (1997), who found no significant relationship between electrode discrimination and speech recognition in 11 Multipeak-MSP users. When only discriminable electrodes were included in the MAP in nine subjects, five showed improvements in at least one speech perception measure. In contrast, Collins et al (1997) found a significant correlation between pitch ranking and speech perception for 11 Nucleus Multipeak-MSP patients. Thus the pitch ranking could reflect the transmission of other speech features not examined in the study by Nelson et al (1995). Dorman et al (1990) found above average Ineraid users perceived a wider range of pitch than subjects who were classified as having poor speech perception. In addition, with a comparison of the Multipeak-MSP and SPEAK Spectra-22 systems, there was improvement in the perception of the place speech feature for the SPEAK strategy (McKay and McDermott 1993), but it was not clear if this was due to better spectral information or other cues.

For speech understanding, people also need to recognize place of stimulation in spite of loudness differences. A study by Henry et al (1996, 1997) investigated the relationship between speech perception and electrode discrimination using stimuli that varied in loudness in eight patients using the Nucleus SPEAK Spectra 22 system. An analysis of the data showed that speech perception and electrode discrimination were correlated in the apical and middle regions of the basal turn of the cochlea, but not the basal region. In an extension of this study Henry et al (2000) used the Speech Intelligibility Index (SII) (French and Steinberg 1947) to also measure the amount of speech information in the five frequency bands (170–570, 570–1170, 1170–1768, 1768–2680, and 2680–5744 Hz) by 15 users of the Nucleus SPEAK Spectra-22 system; the results are discussed in Chapter 7. There was a significant correlation between electrode discrimination ability for intensity variations over 20% or more of the dynamic range, and the amount of speech information received in the frequency region 170 to 2680 Hz. There was no correlation in the region 2680 to 5744 Hz.

The electrode discrimination at different stimulus levels was studied further by McKay et al (1999) for the Nucleus banded electrode array, at 40%, and 70% of the dynamic range, and close to the maximum comfortable level. The results only showed a small deterioration in discrimination with a reduced current level. Similar findings were obtained by Pfingst et al (1999), and in both studies the effects were variable across electrodes and patients.

It is assumed that improvements in speech perception will be achieved with more channels of stimulation, and that this will require the array to have more electrodes and to lie close to the modiolus, and not around the outer wall or in the center of the scala tympani. With this aim, research in the Human Communication Research Center (HCRC) at the University of Melbourne/Bionic Ear Institute in 1989 commenced to develop a precurved array that would curl and lie close to the spiral ganglion cells in the modiolus. The bioengineering development leading to the Contour array manufactured by Cochlear Limited is described in Chapter 8. With the first prototype of the precurved array, it was important to determine if the discrimination of electrodes was better with the electrodes closer to the inner wall. In the study by Cohen et al (2001), the discrimination of electrode place was made with and without loudness jitter. The electrode DLs decreased with reduced radial distance for patients 2 and 3, especially in the presence of loudness jitter as illustrated in Figure 6.10.

A masking study was also undertaken to evaluate the spread of current to the cochlear nerve fibers (Fig. 6.11). A masking stimulus was followed by the probe stimulus at different electrode sites. The intensity of the probe required for masking determined the spread of current to that site, and enabled a masking curve to be plotted. An example of the results are shown for patient 1 for the basal electrode 10, which was at some distance from the inner wall, and the masking curve is broad. But where the apical electrode 18 lies very close to the inner wall, the masking curve is much narrower, indicating a discrete area of neural stimulation. These initial psychophysical studies provide the basis for developing and evaluating speech-processing strategies with increased representation of spectral information through higher density arrays lying close to the spiral ganglion cells.

Time-Varying Place-Pitch Discrimination and Stimulus Duration

The study by Tong et al (1982) referred to above examined pitch scaling for steady-state stimuli with a duration of at least 200 to 300 ms, which is comparable to a vowel. Speech, however, is a dynamic stimulus, and for consonants, frequencies change rapidly over a duration of approximately 20 ms. For this reason, the discrimination of sensations for time-varying electrode place (change in site of stimulation and duration) was determined (Tong et al 1982). The test stimulus had a period over which the stimulus changed from an initial electrode to the reference electrode, and this test stimulus was compared with the reference. A constant rate was used for both electrodes. At the right of Fig. 6.5 the percentage

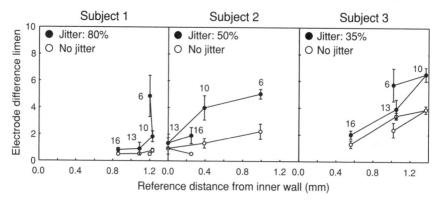

FIGURE 6.10. Electrode discrimination (difference limens) versus radial distance from the inner wall of the cochlea in three cochlear implant subjects both with and without loudness jitter (Reprinted from *Hearing Research* **153**, Cohen et al, Psychophysics of a prototype perimodiolar cochlear implant electrode array, pp 63–74, © 2001, with permission from Elsevier Science.).

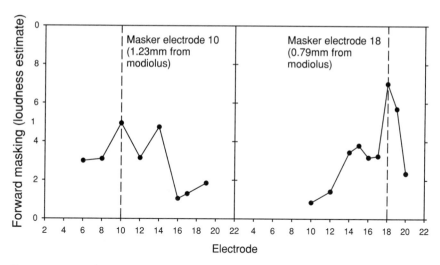

FIGURE 6.11. Forward masking with the precurved array (BP + 1) illustrating the sharper tuning with the electrode closer to the modiolus in patient 1 (Reprinted from *Hearing Research* **153**, Cohen et al, Psychophysics of a prototype perimodiolar cochlear implant electrode array, pp 63–74, © 2001, with permission from Elsevier Science.).

of judgments of difference are plotted for test stimuli whose initial electrode varied from 4 down to the reference that was electrode 1. From this figure it can be seen that a shift in the place of stimulation across two or more electrodes could be discriminated 100% of the time for durations of 25, 50, and 100 ms. This finding indicated that place of stimulation was appropriate for coding the formant frequency transitions present in consonants.

Furthermore, the patient described the sounds for stimuli with time-invariant electrode position toward the higher frequency region as "berp," and those for a time-varying shift toward the lower frequency region as "jerp." This was consistent with the fact that "berp" is characterized by a steady or rising F2, and "jerp" by a falling F2 transition.

Temporal and Place-Pitch Perception

Having found that varying place of stimulation from base to apex resulted in a pitch percept described as sharp to dull, and varying rate of stimulation from high to low resulted in a pitch percept from high to low, the next task was to determine whether these two percepts would be best described as percepts along one or two perceptual dimensions. The perceptual dimensions for rate and place pitch were studied (Tong et al 1983a) by asking the patient to say of three combinations of place and rate of stimulation which two were least similar and which two were most similar. The data on dissimilarity values were analyzed by a multidimensional scaling procedure (Kruskal 1964), which showed how many perceptual dimensions were used by the subjects to complete the task. The results in Figure 6.12 show that a two-dimensional representation provided the best solution. This indicated a low degree of interaction between electrode position and repetition

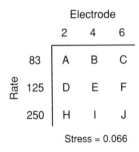

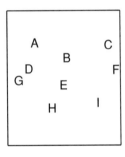

FIGURE 6.12. A two-dimensional configuration representing the dissimilarities among the nine possible combinations of three electrode positions and three repetition rates obtained by multidimensional scaling. Left: the stimulus matrix for rate and place with corresponding alphabetical symbols. Right: perceptual space for these configurations. The two-dimension solution had a stress value of 0.066, which was much lower than the stress of a one-dimensional solution at 0.227 (Reprinted with permission from Tong et al. Psychophysical studies evaluating the feasibility of a speech processing strategy for a multiple-channel cochlear implant, *Journal of the Acoustical Society of America* **74**(1): 73–80. © 1983, Acoustical Society of America.)

rate in their effects on the hearing sensation. It was concluded that temporal and place information provided two components to the pitch of an electrical stimulus.

The data for direct electrical stimulation of the cochlear nerve fibers in implant patients are consistent with the findings for an acoustic model of electrical stimulation tested on normal-hearing subjects (see Rate Discrimination, above). Multidimensional scaling showed good agreement between the model and electrical stimulation and that rate and place of stimulation produced perceptually different sensations (Clark 1987).

Channel Interaction

The initial studies on the University of Melbourne's physiologically based speech processor (Laird 1979; Clark 1987) demonstrated that with simultaneous stimulation there were unpredictable variations in loudness due to the simultaneous overlap of the electrical fields. This interaction was due to the direct effects of electrical fields. It became an important principle in multiple-electrode cochlear implants that speech information should not be presented simultaneously to separate electrodes. The Nucleus F0/F1/F2 speech processor and receiver-stimulator were designed with a 0.4- 0.8-ms separation between any two pulses to avoid this interaction of electrical fields. Nevertheless, neural interactions could occur with time separations as the neural metabolic activity has a longer time course.

After it was found that place pitch for separate electrodes could be scaled from sharp to dull, and that time-varying electrode place could be discriminated for short durations, it was necessary to find out the channel interaction arising from the spatial spread of current. This was investigated using forward masking to determine the population of auditory neurons excited (Lim et al 1986, 1989a). The forward masking was the result of suppression of neural activity (probably inhibition) as the probe followed the masking stimulus by approximately 5.9 ms.

A four-interval forced-choice technique was used (Bilger et al 1977) for the study. A bipolar masker stimulus was applied to one electrode, and the current levels on the others, which were used as probes, were increased until masking occurred. As a result the spread and shape of auditory neural excitation was graphed.

It was found that the spread of current around the masker electrode was asymmetrical with masking decreasing rapidly in the apical direction and more gradually in the basal direction (Lim et al 1986, 1989a) (Fig. 6.13). As the stimulus current increased, the degree of masking toward the basilar region increased. If the spatial extent of the bipolar masking stimulus increased from 1.5 to 4.5 mm (i.e., the distance between the basal and apical member of the bipolar electrode pair), there was also an increase in the masking toward the basal region. Furthermore, with a small spatial extent (1.5 mm), the more basal electrode pairs (higher threshold and smaller dynamic range) produced broader masking patterns than did the more apical electrode pairs (lower threshold and wider dynamic range). This helped confirm the initial finding that there was more current spread in the lower basal region. This could have been due to a greater area of the scala basalward and more short-circuiting of current.

In addition, the masking patterns for loudness-balanced stimuli on two pairs of electrodes were plotted (Fig. 6.14). The stimulus pulses were nonoverlapping in time. The two-electrode masking patterns were similar to the maximum masked level for each of the separate masker stimuli. When the two overlapped, there

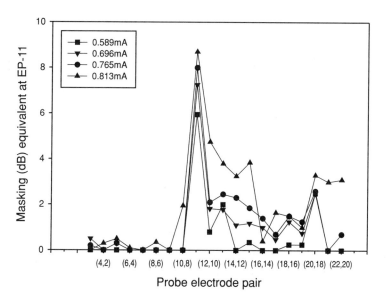

FIGURE 6.13. The masking curves for a single-electrode bipolar pair at masker levels of 0.589, 0.696, 0.765, and 0.813 mA. The electrodes are numbered from low to high according to whether they are in the low- or high-frequency regions of the basal cochlear turn (Reprinted with permission from Lim et al. Forward masking patterns produced by intracochlear electrical stimulation of one and two electrode pairs in the human cochlea, *Journal of the Acoustical Society of America* **86**(3), pp 971–980. © 1989. Acoustical Society of America.).

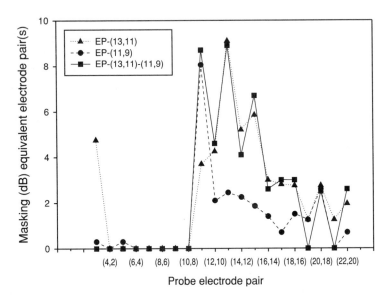

FIGURE 6.14. The masking curves for two-electrode bipolar pairs presented with nonoverlapping pulses, and the masking curves for the electrodes stimulated individually (Reprinted with permission from Lim et al, Forward masking patterns produced by intracochlear electrical stimulation of one and two electrode pairs in the human cochlea, *Journal of the Acoustical Society of America* **86**(3), pp 971–980. © 1989, Acoustical Society of America.).

was little summation and no suppression of excitation in the overlapped region to produce a sharpening of the two populations excited (Fig. 6.14). Nevertheless, the two maxima seen in the vicinity of each electrode pair would form the neurophysiological substrate for the two-component hearing sensation produced by electrical stimulation at two electrode positions. It was also of interest that for a stimulus with wide extent on a bipolar pair, located at the center of the array, there were two maxima. The more apical maximum was probably due to the spread of current to the modiolus exciting fibers coursing from the more apical region. This detailed study was on one patient and there was a need to establish the generality of the results.

Another question arising from the psychophysical research of Tong and Clark (1985) and Nelson et al (1995) was the degree to which current from different electrodes overlap for effective speech understanding. Hanekom and Shannon (1996) investigated the relation between electrode spatial separation or perceptual distance and speech perception on three subjects. Electrode separation was an average of the interelectrode separations. Perceptual distance between electrodes was quantified by d'. The experiments were done by varying the distribution of the outputs of the 20 band-pass filters to seven electrode pairs in patients using the Nucleus SPEAK Spectra-22 system. The results showed in general the larger the spacing or the greater the perceptual distance between the electrodes the better the speech perception, especially for vowels. However, the relationship was not so well seen for consonants where the speech cues are more complex.

Two-Component Place-Pitch Perception

The above research showed that place pitch could be perceived when stimulating individual electrodes, and place and rate pitch were two different percepts. These perceptual processes underlay the University of Melbourne inaugural speech-processing strategy where the midfrequency range (F2) was encoded as place pitch on separate electrodes and voicing as stimulus rate. To further improve the amount of speech information on a place-encoding basis, it was next important to know if stimulating a second electrode nonsimultaneously for frequencies in the low (F1) or high (F3) range could produce a two-component sensation as might be expected for two formant sounds. The perceptual dissimilarities among 10 two-electrode stimuli shown in Figure 6.15 were estimated by triadic comparisons (Plomp 1976). A matrix of dissimilarities was constructed, as shown in the figure, and the matrix analyzed by nonmetric multidimensional scaling for spatial separation of the stimuli. A two-dimensional solution was best and indicated that two percepts were obtained. This finding (Tong et al 1983b) formed the basis for the first speech-processing improvement (F0/F1/F2) where F1 as well as F2 was coded on a place-coding basis (see Chapter 7).

The neural processing required for perceiving two place-pitch components was examined in more detail on four postlinguistically deaf implant subjects by McKay et al (1996). The aim of the study was to determine whether an overlap in the excited neural populations due to a reduced spatial separation between the two pairs of electrodes and an increased distance between the two electrodes in each pair (spatial extent), as well as time delays between the two stimuli, would affect the results. As with the study by Tong et al (1983b), the perceptual space was determined by nonmetric multidimensional scaling techniques. There were variations in the subjects' performance; however, two perceptual dimensions were observed—whether the spatial extent of each electrode was widened from 1.5 mm to 3.75 mm, or the interelectrode timing increased from 0.63 ms to 2 ms. It was hypothesized that a well-ordered perceptual space would correspond to a similarly

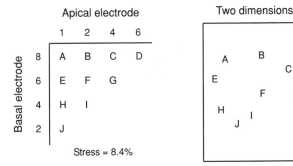

FIGURE 6.15. A two-dimensional configuration representing the dissimilarities among the 10 two-electrode stimuli. Left: The spatial configuration obtained by multidimensional scaling. Right: The stimulus matrix. (Reprinted with permission from Tong et al, Two-component hearing sensations produced by two-electrode stimulation in the cochlea of a deaf patient. *Science* **219**: 993–994, © 1983 American Association for the Advancement of Science.)

ordered vowel space with the two electrode positions corresponding to the first two vowel formants. However, there was only fair correlation between the ordering of perceptual space and vowel recognition. This suggests that spatial discrimination was at fault and/or that temporal coding was required.

Dual-Electrode Intermediate Place-Pitch Percepts

For future speech-processing strategies, it was necessary to know not only if two electrode stimulation could produce two pitch components, but also if a greater range of pitch could be achieved by altering the balance in intensities on two neighboring electrodes. Was there a mechanism in the brain for determining spectral pitch through the weighting given to the population of neurons excited? The first evidence for this came when electrical stimulation on two electrodes produced a vowel color differing from the vowel colors for the two stimulated separately (Tong et al 1979). This was thought to be due to a central averaging process. This phenomenon was subsequently explored by Townshend et al (1987), who showed that when two widely separated electrodes were stimulated, changing the current ratio on the electrodes caused the pitch to shift smoothly between the two. The subjects did not report a two-component sensation. The difference between the findings of Tong et al (1979) and Townshend et al (1987) on intermediate place pitch and those of Tong et al (1983b) on a two-component sensation for comparable separations could have been due to the nature of the task.

Furthermore, intermediate pitch from dual-electrode stimulation was demonstrated by McDermott and McKay (1994); the stimuli were nonsimultaneous with a 0.4-ms time delay. They also found that for some patients the electrode separation needed to be increased to 3 mm to produce an intermediate pitch. Whether this can be explained on the basis of two spatially distinct populations of neurons is not clear, as the masking studies of Lim et al (1986, 1989a) showed two peaks for electrode separations of 1.5 mm. The reconciliation of two-component and intermediate pitch is probably through mechanisms (external and internal spectral patterns) that underlie musical perception discussed above (see Musical Acoustics) and below (see Musical Perception). A spectrum of frequencies determines timbre of an instrument or voice, but component frequencies can be selected in a psychophysical task. Producing a variable pitch, by varying the current on two electrodes, was referred to as "virtual" channels by Wilson, Lawson et al (1994). Dual-electrode stimulation with weighted current could be useful for people with a small number of functional electrodes in "bridging the gap" to produce better speech recognition.

Place Information: Prelinguistically Deaf

The initial results on the first University of Melbourne's patients, who were postlinguistically deaf, demonstrated how the use of the place coding of pitch for multiple-electrode speech-processing strategies (Clark, Tong et al 1981a,b) could lead to open-set speech understanding. It was therefore crucial in the management of prelinguistically deaf people to know if place as well as rate pitch was perceived in the absence of exposure to sound.

An initial study at the University of Melbourne on two adults and one 14-year-old showed that electrode position identification as well as temporal perception was worse for the prelinguistically deaf compared to the postlinguistic controls (Tong et al 1988). The poor speech for this group of prelinguistically deaf people was most probably due to their inability to use place and temporal information. In a larger study on 10 early-deafened people (Busby et al 1992), it was found that for electrode position identification on three children, performance was much poorer than for postlinguistically deaf adults. This may have been due to the difficulty of the task or to the fact that the subjects did not have their implant until they were 10 to 16 years of age. The difference limens for electrode position were one to three electrodes for all but three subjects where limens of six to 10 electrodes were recorded. It was also essential to determine how time-varying place of stimulation was perceived. Did this change with age at implantation and duration of exposure to sound? The discrimination of electrode place as well as rate trajectories was examined by Busby et al (1993a). The results showed in general that the four postlinguistically deaf adults and one of the prelinguistically deaf children (ED2) were more successful than the other three prelinguistically deaf children in discriminating electrode place trajectories, and in speech perception using the multiple-electrode strategy. A second prelinguistically child (ED4) also performed at levels above chance. The electrode place trajectories were across two, four, and six electrodes for durations of 48 ms and 96 ms. The child (ED2) was older than the others when the hearing loss occurred (3 years) and was younger at age at implantation (5 years). Later clinical studies showed that open-set speech perception was significantly better the younger the child (Clark 1997, 1999) (see Chapter 12). It was therefore important to see whether improved speech results in younger children were associated with better electrode discrimination and place pitch estimation.

The DLs for electrode discrimination were measured in 16 prelinguistically deaf patients at apical, middle, and basal positions on the array (Busby and Clark 2000a). With one exception (age 20) the patients were all implanted under the age of 18 years. Only three children had meningitis, and the rest were congenitally deaf. Stimulus current levels were randomized to reduce loudness cues. The correlation between the DLs for electrode discrimination and auditory deprivation and auditory experience were assessed. Auditory deprivation depended on the age at onset of deafness, the duration of deafness, and the age at implantation. The auditory experience depended on the duration of implant use and any hearing prior to surgery. The correlation between electrode place DLs and these variables as well as speech perception using closed and open-set tests was determined.

The average DLs for the three positions on the array were less than two electrodes for 75% of patients (Fig. 6.16). For the remaining 25% the average DLs were between two and 6.5 electrodes. There were significant variations in DLs across positions in 69%. The average DLs for the electrodes correlated positively with the duration of auditory deprivation. There were larger DLs for people im-

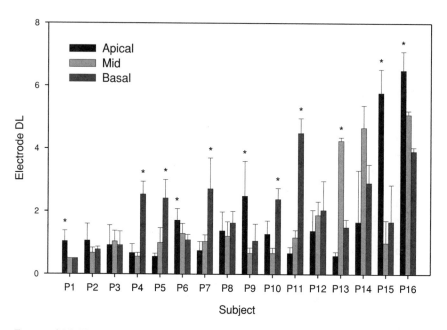

FIGURE 6.16. The electrode difference limens from the 16 prelinguistically deaf patients for the apical, middle, and basal electrodes. The standard deviations are shown. The asterisks show a significant difference for the electrode positions (Busby and Clark 2000, Electrode discrimination by early-deafened subjects using Cochlear Limited multiple-electrode cochlear implant. *Ear and Hearing* **21**: 291–304. Reprinted with permission from Lippincott Williams & Wilkins.).

planted at a later age, and with a longer duration of deafness. The DLs at the apical position correlated negatively with closed-set perception scores, with higher scores for patients with smaller DLs. But this was not the case with open-set tests. The speech scores also correlated negatively with auditory deprivation.

The above study thus showed that prelinguistically deaf patients were generally successful in making electrode discrimination, although the performance varied along the array for more than half the subjects. Electrode discrimination was affected by auditory deprivation, and both electrode discrimination and auditory deprivation affected closed-set speech perception.

The findings from the above study support the view that during the critical phase in maturation "plasticity" (see Chapter 5), sound exposure or electrical stimulation is required for the neural connections to achieve place coding of frequency, and this is important for the best speech perception.

Although loudness was balanced for the electrode DL study, the task was not the same as assigning a pitch to a stimulus, and scaling pitch along the array. For this reason, a pitch estimation study was also undertaken by Busby and Clark (2000b). The study was on 18 prelinguistically deaf subjects. The group was the same as for the discrimination study but with two additional children—one with rubella and one with meningitis. Pitch estimation was assessed, versus the duration of sound deprivation prior to implantation, and its effect on speech perception determined. Speech perception was measured using closed-set monosyllabic words and open-set Bamford-Kowal-Bench (BKB) sentences (Bench and Bamford 1979). The main finding of the study was that a tonotopic ordering of pitch for electrode stimulation was present for 56% of the children as shown by the large slopes of the function accompanied by the small standard deviations of the estimates (Fig. 6.17). A consistent but deviant tonotopic order of pitch was found for 22% of children, as shown by the small slopes of the estimate function. There was essentially no pitch order for the remaining 22% where there were large standard deviations for the estimates. Some of the between-subject differences in pitch estimation were related to the duration of auditory deprivation prior to implantation, with the poorest results in those not able to order pitch. These had a longer duration of deafness and were implanted at a later age. These children also had the lowest scores on the BKB sentence test.

The results suggest that developmental plasticity in establishing the neural connectivity for place discrimination is an important factor for speech perception. It is probably not the only factor, as language has also been show to be important.

Loudness

Loudness, as discussed earlier in the chapter, is the percept that can be ordered from quiet to loud, and with electrical stimulation is correlated with the spread of current and flow of charge across the nerve membrane.

Threshold and Maximum Comfortable Level

Thresholds for electrical stimulation depend on how the procedure is carried out, the stimulus parameters, electrode geometry, and neural population. The T and

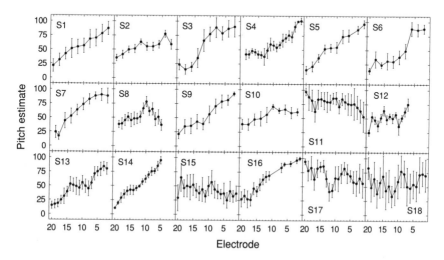

FIGURE 6.17. Pitch estimates from the 18 patients for stimulating the electrodes on the array. The electrodes are numbered from 1 to 21 in an apical to basal direction (Busby and Clark 2000, Pitch estimation by early deafened subjects using a multiple electrode implant. *Journal of the Acoustical Society of America* **107**: 547–558. Reprinted with permission.).

MC levels are determined for electrical stimulation in the same way as for sound (as discussed earlier in the chapter). However, there can be variable test-retest reliability; for example, Simmons (1966) found with an intraneural array a difference of up to 10 dB in μA not unusual for monopolar trains of sine waves.

As discussed in Chapter 4 and 5, the threshold of the neuron depends on the transfer of charge across the nerve membrane. When a voltage is applied, the transfer of charge depends on the biophysical properties of the nerve membrane that acts like a leaky capacitor with a time constant determined by its capacitance. A minimum amount of electrical charge [i.e., charge (Q) is current (I) over time (t)] must pass through the nerve membrane to produce an action potential (Katz 1966). If the pulse is shortened, the intensity must be increased. The relationship between intensity and pulse duration can be plotted as a strength duration curve. This was demonstrated for the auditory nerve in the experimental animal by Black and Clark (1977) and Clark, Black et al (1977), and was illustrated in Chapter 5.

A similar relationship between the current amplitude and pulse width in producing threshold was demonstrated for a patient by Eddington et al (1978). It was also shown that a patient's threshold could be lowered if an initial subthreshold pulse were delivered within 0.5 ms of a second pulse. This was again due to the biophysical properties of the nerve membrane, leading to a summation in time of the depolarizing potentials. Just how close the second pulse needed to be for temporal summation depended on the biophysical properties of the membrane. Furthermore, the thresholds were lower for high rates of stimulation (Eddington et al 1978). When stimulating at higher rates with pulses of fixed width, more charge was delivered to the nerve membrane. Thus the threshold was lower or the response stronger. However, the studies to be described below (see Neural Population Excited) suggest that the threshold, dynamic range, and loudness were not simply related to total charge, but could depend also on the pulse rate, interpulse interval, and burst duration.

The effect of varying the time between the phases of a biphasic stimulus on thresholds was studied at the University of Melbourne by Johnson (1980) on the behavioral experimental animal. This was undertaken to determine the effects of the positive and negative phases of a stimulus on the nerve membrane. It was carried out for pulses of 100 μs/phase at 200 pulses/s. The results showed that as the interphase gap was increased the threshold decreased rapidly until approximately 800 μs (Fig. 6.18). This suggests there is an integration time at the nerve membrane or higher centers for threshold.

The T and MC levels depend first on the stimulus parameters, the properties of the neural membrane, and the population of neurons excited. These in turn create the strength of the sensation, in this case loudness, through an increase in the population of neurons excited, or the firing rate. The population at a comfortable loudness was studied for the straight but flexible banded array using psychophysical masking procedures (Tong et al 1987). The results showed the spread of excitation to roll off at 3 to 9 dB/mm for a bipolar pair of electrodes in the scala tympani (electrodes separated by 1.5 mm).

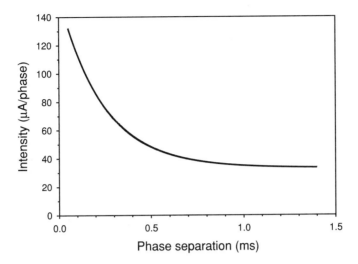

FIGURE 6.18. Behavioral thresholds for intensity as a function of the phase separation in a cat (analysis of the data of Johnson 1980).

A study was also undertaken on nine postlinguistically deaf adults with the Nucleus straight flexible array to determine how the T and MC levels varied for different modes of stimulation (bipolar, common ground, and monopolar where the most basal electrode was the return path) and across patients (Busby et al 1994). The study showed that the current levels for T and MC levels were highest for bipolar and lowest for monopolar stimuli. This is consistent with the hypothesis that the T and MC levels depend on the population of fibers stimulated, as the spread of current is greatest for monopolar, and least for bipolar stimulation. For common ground stimulation there was a trend for T and MC levels to be highest in the middle of the array. This may have been due to the fact that half the current from the active electrode passes to the apical and the other half to the basal end of the turn. This would spread the current with less concentrated stimulation of the nerves. With the monopolar stimulation T and MC levels increased from the apical to basal ends. This was probably due to the fact that the more basal region is larger with the electrode further from the ganglion cells, and there is often more fibrous tissue and bone near the round window affecting the spread of current. There was no consistent pattern for bipolar stimulation.

A precurved electrode array (see Chapter 8) was developed not only to improve current localization, as discussed above, but also to lower the T and MC levels for better power consumption and battery life. In an initial study with the prototype precurved array on three patients, T and MC levels were also plotted as a function of the distance of the array from the inner wall. With each array the distance varied sequentially from the basal (B) to the apical (A) end, as shown in Figure 6.19. An inset of the electrode locations derived from x-rays is included for reference. In patient 1, as the electrode got closer to the inner wall there was

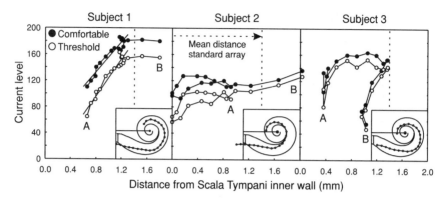

FIGURE 6.19. Thresholds and maximum comfortable levels with distance for three patients implanted with the prototype precurved array. A, apical end of the basal turn; B, basal end of the basal turn. Reprinted from *Hearing Research* **153**, Cohen et al, Psychophysics of a prototype perimodiolar cochlear implant electrode array, pp 63–74, © 2001, with permission from Elsevier Science.

a progressive decrease in T and MC levels. In patient 2, the electrode moved close to the inner wall in the middle to apical third, and this is where the T and MC levels were lowest. In patient 3, the electrode was moderately close to the inner wall in the basal third, then moved away from it where the levels were high, and returned to be close to it in the apical third where there were low T and MC levels.

In a study on 35 patients with the Nucleus straight but flexible banded array, and on seven patients with the Cochlear Limited version of the University of Melbourne's precurved array Contour, the population means for T and MC levels were lower for the Contour (Fig. 6.20). An analysis of thresholds for 680 electrodes on the Nucleus straight array and 70 for the Contour array also showed a bimodal population and two distinct peaks for the two arrays. The dynamic ranges were the same. The T and MC levels were plotted against radial distance from the modiolus. As with the initial study, there was good correlation between the average T levels and radial distance, a finding that would be expected. Furthermore, the average MC levels also mirrored the T levels. However, there was considerable individual variation for T and MC levels versus radial distance in a group of 19 patients (Saunders et al, 2002).

The behavioral T and MC levels were compared with the compound action potentials (CAPs) (field potentials) from the auditory nerve recorded by neural response telemetry (NRT) for eight Melbourne and 156 European patients (Dillier 1998, Dillier et al 2000, 2002). The NRT could be very important clinically for evaluating T and MC levels in young children, as behavioral testing at this age can be imprecise. The NRT levels at MC listening levels were determined. The T levels were also analyzed as a function of electrode distance from the modiolus. The results showed the NRT thresholds mirrored the T and MC levels, and the radial distance of the electrode from the spiral ganglion cells. The results for one patient (M2) from the study by Saunders et al (2002) are shown in Figure 6.21.

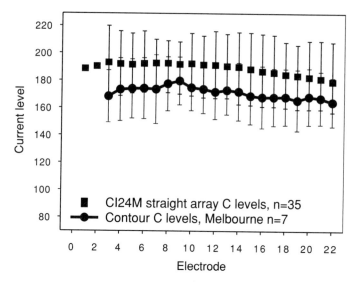

FIGURE 6.20. Threshold levels for the straight and Contour arrays (E. Saunders, personal communication).

Loudness Growth Functions

Variations in the amplitude of the speech signal, as well the component frequencies, are essential for speech recognition. It is therefore important to know the effects of stimulus parameters on loudness from the T to MC levels and the range of parameters that could be used to convey variations in sound intensity.

The normal-hearing ear can respond comfortably to an intensity range of 120 dB at a frequency of 1000 Hz. In contrast, an initial study by Simmons (1966) found the range from threshold to uncomfortably loud to vary from 15 to 25 dB for intraneural stimulation of the auditory nerve at 50 pulses/s. The loudness growth or dynamic range for stimulation at 200 pulses/s with the University of Melbourne/Nucleus banded electrode array in the scala tympani was found to vary from 5 to 10 dB (Clark, Tong et al 1978; Tong et al 1979). Figure 6.22 shows a steep linear function between the logarithm of current amplitude, plotted in decibels, versus loudness. The dynamic range for electrical stimulation was much smaller than the 30- to 40-dB range for speech sounds or the 120-dB range for sound.

A linear relation was observed between loudness in decibels and current amplitude by Eddington et al (1978). For a constant amplitude and frequency an increase in the pulse width produced an increase in charge leading to an increase in loudness. This would be expected from the properties of the neural membrane as demonstrated with the strength-duration curve. In addition, when the frequency of stimulation with biphasic pulses was lowered, the amplitude had to be increased to maintain equal loudness, again indicating the importance of charge delivered per unit time. However, the relationship between charge and loudness is complex; for example, the amount of charge delivered at short pulse durations may need

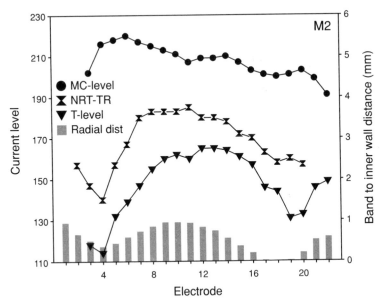

FIGURE 6.21. Behavioral (T) and NRT thresholds, maximum comfortable listening (MC) levels, and electrode radial distance from the spiral ganglion cells in the modiolus (E. Saunders, personal communication).

to be disproportionately increased with increased current amplitude. Other parameters including rate, amplitude, duration, and pulse period may also code for loudness. For example, Shannon (1989b) modeled threshold data or sensory magnitude with power law functions of the instantaneous current amplitude, and found good agreement. The effect of other parameters is discussed below.

The modeling studies of Rubinstein et al (1999) showed that a high stimulus rate of 5000 pulses/s would lead to independent spontaneous activity in neurons. Physiological research demonstrated that this high rate produced a greater dynamic range for the response rate in individual units (Litvak 2002; Runge-Samuelson 2002). These findings were confirmed by Hong et al (submitted) in a study with the Clarion CII device on 28 patients presented with sinusoidal stimuli with various levels of superimposed pulses at 5000 pulses/s. The mean of the largest increase in dynamic range for each subject was 6.6 dB.

Difference Limens

To determine how to convert variations in speech sound energy to electrical current and charge, it was necessary to know not only the T and MC levels and as a consequence the dynamic range for electrical stimulation, but also the number of discriminable current levels in the range. The DL for sound intensity varied according to the overall sound level as discussed above (see Difference Limens), and was 1.5 dB at 20 dB SL, 0.7 dB at 40 dB SL, and 0.3 dB at 80 dB SL. The DLs also varied to a limited extent with frequency.

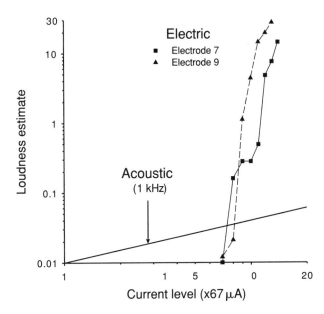

FIGURE 6.22. Loudness growth functions for two Nucleus banded electrodes in the scala tympani at 200 pulses/s for the first Melbourne patient, compared with the loudness growth for sound. (Reprinted with permission from Tong et al, 1979. A preliminary report on a multiple-channel cochlear implant operation. *Journal of Laryngology and Otology* **93**(7): 679–695.)

Patients' discrimination of electrical current was studied by Simmons (1966), Douek et al (1977), Eddington et al (1978), Fourcin et al (1979), Aran (1981), House and Edgerton (1982), Dillier et al (1983), Hochmair and Hochmair-Desoyer (1983), Shannon (1983a,b), Tong et al (1988), and Nelson et al (1995). The just discriminable differences in electrical current varied from 1% to 8% of the dynamic range. The just discriminable steps for sound can vary from 0.3 to 1.5 dB, and with a dynamic range of 100 dB the DLs are from 0.3% to 1.5% of the range. The steps for electrical stimulation are larger but still of the same order as acoustic stimulation. Thus electrical current can be used to convey variations in acoustic intensity information in speech.

A study by Nelson et al (1995) examined in more detail the ability of implantees to discriminate intensity over the dynamic range. Their research was on seven postlinguistically deaf adults and one prelinguistically deaf adult using the Nucleus system and banded electrode array. The relative DLs or Weber fractions measured in decibels [$10 \log (\Delta I / I)$] decreased as a power function of stimulus intensity relative to absolute threshold, which had also been reported for acoustic stimuli. The exponents of the power function, however, were one order of magnitude larger than for sound. The 10-to-1 difference in the Weber function between acoustic and electrical stimulation was presumed due to the ability of the cochlea to compress intensity. The Weber fractions were the same for the postlinguistically deaf patients and one prelinguistically deaf patient. The differences for electrical stimulation across patients in part would be due to the neural population, and cochlear pathology affecting current spread and/or electrode geometry. The cumulative number of discriminable steps in intensity across the dynamic range varied from approximately 7 to 45. The data from the human studies were consistent with the findings of Pfingst et al (1983) on monkeys. They showed that with sinusoidal electrical stimuli there was a progressive improvement in the DL with stimulus level.

The neural population, in particular the presence of residual peripheral processes (dendrites) as well as spiral ganglion cells, could account for variations in the intensity DLs. Studies by Stypulkowski and van den Honert (1984) and Javel et al (1987) showed that at low stimulus levels a long latency response with a gradual growth in amplitude occurred, assumed to be due to stimulation of peripheral processes. In contrast, a response at higher stimulus levels had a shorter latency and a steeper amplitude growth rate, assumed to be from the spiral ganglion cells. It would be expected that with the more gradual growth in the response from the peripheral processes, more discriminable steps in intensity could be processed. The firing statistics of the neurons as well as current flow would affect the population of neurons excited, as demonstrated by the model of Bruce et al (1998), and account for individual variability.

The importance of electrode geometry and current spread is emphasized in a study by Cohen et al (2001). This study was undertaken on three patients fitted with a perimodiolar electrode array that varied in its proximity to the modiolus, but lay on average about halfway between the inner and outer cochlear walls, as distinct from the straight flexible array that lies close to the outer wall. It was found that current level discrimination, normalized with respect to dynamic range, improved with decreasing distance of the electrode from the modiolus in two of the three subjects.

From the above data it can be concluded that if the number of discriminable steps that can be used with electrical stimulation over the dynamic range were 45, this would be more than half the 60 discriminable steps for speech sounds over a 30-dB range. The data thus indicate that if the amplitude variations of speech are also compressed into a narrower dynamic range, there would be enough loudness steps for current to convey essential loudness information in a majority of patients.

Neural Population Excited

The coding of intensity and its perception as loudness is due to the population of nerves excited and/or their firing rate. Other parameters may also be involved as discussed below. Data have shown a direct relationship between both sound intensity and electrical current level and the amplitude of evoked potentials in both the experimental animal and the human. Dobie and Kimm (1980) and Charlet de Sauvage et al (1983) found the amplitude of the field potential increased linearly with the current level in monkeys and guinea pigs, respectively. Smith et al (1994) also discovered that there was good correlation between the electrically evoked brainstem response (EABR) and the behavioral threshold. This was also found for implant patients where the NRT thresholds were related to the T and MC levels, as shown in Figure 6.21. These and other data support the hypothesis that the population of neurons excited is important in coding intensity.

Mean Stimulus Rate

It is also known that rate of individual nerve firing in the auditory system as in other systems increases with the strength of the stimulus (Kiang et al 1965). This is discussed in Chapter 5.

With cochlear implant patients the threshold was shown by Eddington et al (1978) to decrease with an increase in pulse rate (i.e., become louder), and this was attributed to an increase in total charge. The relationship between loudness and charge is complex, and other parameters such as pulse rate could be factors. For the above reasons a study was undertaken by Tong et al (1983a) to help determine whether rate of stimulation as well as charge affected loudness. The loudness of two stimuli, one with a single pulse per period and another with multiple pulses per period were compared for different overall stimulus rates. The pulses had fixed current levels and the multiple pulses were at a rate of 1000 pulses/s. With multiple-pulse-per-period stimuli, as the overall pulse rate increased the number of pulses per period were reduced to keep the total number of pulses and total charge constant (Fig. 6.23). The results showed loudness increased with rate of stimulation when using a single pulse per period (Fig. 6.23). On the other hand, with multiple-pulse-per-period stimuli, as the overall periodic rate increased there was a small fall in loudness on two electrodes, and a very slight increase on another (Fig. 6.23). This supported the importance of total charge in coding intensity, but did not exclude other parameters.

Busby and Clark (1997) extended the multiple-pulse-per-period study on loudness by Tong et al (1983a) to six postlinguistically deaf adults, and eight prelinguistically deaf adults and children. Bipolar stimuli (200 μs/phase) were used. Loudness estimates for the single-pulse-per-period stimuli increased with stimulus rate for all postlinguistically deaf patients. Loudness estimates for the multiple-pulse-per-period stimuli decreased with increases in the pulse rate for three of the six postlinguistically deaf subjects. For the others loudness marginally increased with pulse rate. In addition, pitch estimates increased with both the single-pulse-per-period and multiple-pulse-per-period stimuli for all six of the postlinguistically deaf people. In those patients in whom the loudness was constant across the rate of stimulation, it was assumed that the total charge was the prime coding mechanism. In contrast, in those in whom there was a decrease in the loudness with an overall increase in the pulse rate for multiple-pulse-per-period stimuli, this could have been due to the operation of a short-term integration mechanism. As the rate increased the number of pulses in the first half of the cycle was reduced, with a reduced integration of energy, thus leading to a reduction in loudness (Busby and Clark 1997). This applied to pulse intervals and not average pulse rate.

Studies by Shannon (1985, 1989b), Moon et al (1993), and Pfingst and Morris

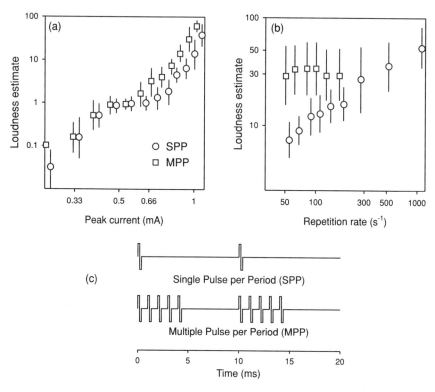

FIGURE 6.23. (a) Loudness versus peak electric current, (b) loudness versus mean stimulus rate, and (c) SPP and MPP pulse patterns for a repetition rate of 100 pulses/s. (Reprinted from Tong et al 1983a with permission from The Acoustical Society of America.)

(1993) showed that the threshold decreased monotonically with an increase in stimulus rate for short pulse durations (less than 500 μs/phase), but was bowl-shaped for pulse durations of 2 ms. The data in the study by Pfingst et al (1996) also suggested that pulse separation (interpulse interval) contributed significantly to the threshold versus rate curves as distinct from pulse rate per se or number of stimulus pulses. The results were consistent with an integrating mechanism up to 5 ms in duration. The nonmonotonic rate functions with long duration pulses (2 ms) are unexplained, but could have been due to greater inhibition with the longer duration pulse. Furthermore, Shannon (1985, 1989b), Pfingst and Morris (1993), and Busby and Clark (1997) showed there was a reduction in loudness when the interval between stimuli was reduced to less than 1 ms. This is at variance with the data from McKay and McDermott (1996).

The data referred to above show that charge per phase and total charge is a key parameter for loudness, but this relationship is not linear for short pulse widths where amplitude makes a greater contribution. Pulse rate affects loudness through varying the total charge, but this depends on the pulse width. With a long pulse width, neural mechanisms such as accommodation may be involved to reduce excitation. The pulse intervals are also important, as they relate to how well the nerves can integrate the energy. Thus charge is the prime parameter, and the time for integration of this energy is the other parameter.

Thus the transmission of the electrical energy depends not only on the amplitude and pulse width, but also on the interpulse interval, the number of stimuli, and the duration of the burst. Their effects vary with the properties and extent of the neural population stimulated, as hypothesized by Busby and Clark (1997). Research was undertaken by McKay and McDermott (1998) to examine the effect of interpulse intervals on loudness and threshold for eight postlinguistically deaf adults. It is difficult to determine the effects of interpulse intervals alone, as they covary with pulse rate, the number of stimuli, and the duration of the burst when keeping the charge constant. Equal loudness contours and thresholds were obtained for two bursts of pulses with their interpulse intervals varied in a fixed repetition period of 4 and 20 ms. The results showed that as the interval increased, there was a monotonic reduction of the current for threshold as seen in the above studies for short duration stimuli. The reduction for the maximal comfortable level was smaller, leading to a steadily increasing dynamic range.

A model to explain loudness for pulse interval and rate of electrical stimulation was developed by McKay et al (1994) from a similar one for the perception of loudness to sound (Viemeister 1979; Buus and Florentine 1985; Green and Forrest 1988; Moore et al 1988; Plack and Moore 1990, 1991). It was postulated that the time course for the peripheral integration was due to the absolute and relative refractory periods of nerves. But as the absolute and relative refractory periods are short (0.5 and 0.2 ms; see Chapter 5), its duration would also require inhibition to account for the data. The model predicted the average spike probability data for the different interpulse intervals, as outlined in the study by McKay et al (1994). With an increase in the number of pulses there would be an increase in the number of pulses in an integration window, but this would be offset by a decrease in excitation per pulse due to a falling firing probability. The relative excitation per pulse or stage 1 was used as the input to stage 2, the temporal integrator modified from Oxenham and Moore (1994). This transformed excitation into specific loudness. The third and final stage was an equal loudness decision criterion based on an equal output from stage 2. There was a spatial integration of specific loudness to obtain the overall loudness. An average integration time of 6.8 ms was obtained for eight implant patients, and was consistent with the stage 3 temporal integration for acoustic stimuli discussed above (see Temporal Integration). The above studies suggest that the interpulse interval is one mechanism affecting loudness, but they did not indicate the relationship between this and pulse rate or number. Thus when successive current pulses stimulate the same cochlear place, loudness is determined by the total neural excitation within a time window of about 7 ms.

Loudness Summation

With sound, the loudness in sones of two or more frequencies sum completely if the frequencies are separated by more than one critical bandwidth. If not, it depends on first summing the energy of the sounds and then determining the relationship between the change in intensity and the resulting loudness. In cochlear implant speech processing with multiple-electrode stimulation, it is important to predict the manner in which loudness is summed to produce a more natural and intelligible speech sound.

Summation of loudness for electrical stimuli was first studied in two implant subjects for bipolar stimulation with two pairs of Nucleus banded electrodes activated nonsimultaneously (Tong et al 1987; Lim et al 1989b). Loudness summation increased for spatial separations up to 3 mm. The summation was assumed to be due to an integration of the effects of current spread, but without suppression of neural excitation due to the refractory properties of nerve fibers, as one pulse followed the other by 0.7 ms. This is just outside the absolute and relative refractory periods, as discussed in Chapter 5. The result was consistent with the critical band concept for loudness summation and it occurs over approximately 0.89 to 1.0-mm segments of the basilar membrane (Clark, Tong et al 1977). However, loudness summation was not affected by electrode separations greater than 0.75 mm in a study by McKay et al (1995a) on four subjects also using bipolar stimulation with the Nucleus electrode where one pulse followed the other by 0.48 ms. The discrepancy between the above two results could depend on the considerable intrasubject differences in current flow and neural populations demonstrated by place pitch and current threshold variations (Busby et al 1994). However, a further study by McKay et al (2001) on 12 postlinguistically deaf adults confirmed no change in the summation of loudness for only 0.75 mm, and only for a rate of 1000 pulses/s. Significant summation was also seen for bipolar and not monopolar stimulation.

The difference between the results of Tong et al (1987), Lim et al (1989b), and McKay et al (2001) is most likely to be due to the fact that McKay et al (1995a, 2001) used a time delay between the pulses of 0.48 ms, which would have meant stimulating an overlapping population within the refractory period of the first pulse, and impaired loudness summation for that population. Loudness summation would more likely have occurred with a time separation of 0.7 ms as used by Tong et al (1987) and Lim et al (1989b), where the overlapping fields would have contributed to a summation of energy within a critical band. It is hypothesized the total energy in each critical band depends on the temporal integration time within that band as well as the current spread, that is, the gradient of the population of the neurons excited.

Intensity Information: Prelinguistically Deaf

Two prelinguistically deaf adults were first studied by Tong et al (1988) to determine their ability to identify changes in current level, when compared to that for postlinguistically deaf subjects. The DLs in decibels were calculated from a perceptual sensitivity index (d') between two stimuli. This was found to be 0.4 dB for both patients, and corresponded to 9.5% and 14.3% of the dynamic range. These were similar to the difference limens of 4% and 8% for postlinguistically deaf adults reported by Shannon (1983a).

In a study on 18 early deafened or prelinguistic subjects, all experienced increased loudness with current level (Busby and Clark 2000a). There were, however, considerable differences in dynamic range that varied from 1.21 dB to 10.96 dB. So although they had an ability to discriminate loudness, there could have been limitations in the perception of loudness variations for optimal speech perception.

In addition, Busby and Clark (1997) compared loudness estimates for single-pulse-per-period (SPP) and multiple-pulse-per-period (MPP) stimuli as discussed above for six postlinguistically deaf adults and eight prelinguistically deaf adults and children. With MPP stimuli in the prelinguistically deaf subjects the loudness estimates fell in six of eight and remained flat in the other two. The pitch estimates rose in five and decreased in the others. The different proportion of postlinguistically deaf subjects with an increase in loudness may indicate different integration of information.

Musical Perception

Musical perception through cochlear implants has been well reviewed by Stainsby (2000). It has been studied by presenting controlled psychophysical stimuli to the implants.

Single-Channel Processor

The perception of pitch through the recognition of musical melodies was investigated by Moore and Rosen (1979), with single-electrode stimulation (Fourcin et al 1979). In the study there were 20 normal-hearing subjects and one single-channel cochlear implant user. The normal-hearing subjects were asked to recognize melodies in five conditions that minimized or distorted acoustic temporal and place pitch cues. The conditions were high-pass filtered pulse trains (the low cut was 2000 or 4000 Hz), sine waves (compressed 1:1.3), expanded octave ratios (1:4), frequency held constant, and melody coded in an amplitude scale. The results showed that melodic pitch was best recognized with high-pass filtered pulse trains (cutoff was 2000 or 4000 Hz). Distorted frequency ratios impaired the recognition of melodies, and it was absent with only amplitude cues. Melodies were presented to the single-channel implant subject as pulse trains with different rates, and the tunes were all identified. This showed musical pitch for melodies can be represented in the temporal pattern of pulses on a single electrode, without additional place pitch information. It must be emphasized that variations in pitch perception cannot be conveyed with single-electrode stimulation above 800 pulses/s.

Multiple-Channel Processors

Gfeller and Lansing (1991, 1992) measured the melodic and rhythmic perceptual abilities of ten Nucleus and eight Ineraid adult users with the Primary Measures of Music Audiation (PMMA) tests (Gordon 1979). The Nucleus subjects used the WSP-III processor (Wearable Speech Processor), which presented voicing as rate of stimulation and the first and second formants as place of stimulation, and the Ineraid subjects used a fixed filter processor with analog stimuli (see Chapter 7 for details). The PMMA is a standardized test of tonal and rhythmic patterns. The subject's perception of timbre was measured using the Musical Instrument Quality Rating. The melodic tests had no differences in the rhythmic patterns but the pitches varied. The rhythmic tests had notes that varied in duration and intensity, but were presented at the same frequency. Overall the subjects performed better on the rhythmic than on the melodic tests. A more detailed assessment was made by Gfeller and Lansing (1992) on 17 Nucleus (F0/F1/F2) and 17 Ineraid users. The results again showed they performed better on the rhythmic than on the melodic tests. There was no significant difference between the performances of listeners with the two devices. The results demonstrated that the rhythmic elements of music were more accessible to cochlear implant users than the melodic elements.

The perception of musical sounds with the Ineraid implant was also investigated by Dorman et al (1991). The majority was unable to determine if an ascending or descending scale was played. Open-set identification of five melodies was poor. This improved when the melodies were presented as a closed set. When asked to identify five instruments in an open-set, a mean score of 30% resulted, and this increased to 77% for a closed set.

Additional studies on the perception of musical pitch with the Nucleus CI 22 system were undertaken on 17 subjects by Pijl and Schwarz (1995b). For open sets of 30 rhythmically intact melodies, a mean score of 44% was obtained. Three were presented with tunes without rhythm, and they had different pulse rates and electrode positions. Performance was superior for the lower pulse rates and the more apical electrodes. Some melodic recognition was still possible for rates up to 600 to 800 pulses/s. Music interval identification was also tested with the intervals represented as frequency ratios between different pulse trains. For the lower pulse rates, subjects assigned similar ratios to the musical intervals as for acoustic stimuli. The authors concluded that musical pitch could be conveyed by the rate of stimulation alone, at least for the frequencies in the lower half of the range common in music. Further investigations of musical pitch by Pijl and Schwarz (1995a) examined the ability of Nucleus CI 22 patients to produce musical intervals by controlling stimulation pulse rate. Target intervals consisted of a minor third (three semitones), a perfect fourth (five semitones), and a perfect fifth (seven semitones). The ability of three patients to select these intervals was similar to that of listeners with normal hearing. In a second study the ability of two implant subjects to produce the same intervals at transposed frequencies was assessed. Accuracy was best at low rather than high frequencies, and similar to the upper limit of 800 to 1000 Hz for acoustic temporally mediated pitches, reported by Burns and Viemeister (1976).

A comparison of the musical perception abilities of 17 adults who used the Nucleus 22 F0/F1/F2 WSP-III and Multipeak-MSP systems alternatively, was made by Gfeller et al (1998). No differences were observed for the PMMA test battery. The Multipeak strategy coded additional high-frequency information through fixed filters in the three frequency bands of 2.0 to 2.8 kHz, 2.8 to 4.0 kHz, and >4.0 kHz. Although this information was useful for speech perception, it was no better for the representation of musical pitch.

The ability to recognize music through the Multipeak-MSP and SPEAK Spectra-22 systems was evaluated by Fujita and Ito (1999). Seven subjects had the Multipeak and one had the SPEAK strategy. They were asked to identify nursery rhymes played with and without speech, discriminate between instrumental melodies played with identical rhythms, and discriminate different instrumental sounds playing the same scale. The subjects were able to identify the nursery rhymes with words 39% of the time in an open set, and 53% of the time in a closed set. However, only two were able to identify the rhymes when played only with an instrument. Although the subjects could not identify melodies played with identical rhythms, they could distinguish instruments playing the same scale. The one subject with the SPEAK strategy showed the best performance in recognizing songs with vocal and instrumental accompaniment and the poorest performance on melodic interval recognition. It was thought that this was due to the relatively constant rate of stimulation with the SPEAK as opposed to variations in rate for the fundamental frequency (F0) with the Multipeak strategy. The authors considered that good spectral information was required for the identification of speech or instrumental colors. Straight pitch perception possibly from temporal processing could explain the generally poor performance with implant for the recognition of melodies. However, in a questionnaire developed by Skinner et al (1994), a clear majority (84%) of subjects reported that music sounded better with the SPEAK strategy than with the Multipeak.

In a study by McDermott and McKay (1997) the effect of changing the rate and place of electrical stimulation on the judgments of musical interval size was evaluated on a multiple-electrode implant patient who was also a piano tuner by profession. First, place pitch was found to dominate over rate of stimulation in pitch judgments. Second, the subject could also make pitch judgments for sinusoidal amplitude-modulated stimuli of electrical pulses at carrier rates of 1200. With these stimuli pitch judgments could be made over one octave in the 100- to 200-Hz range.

A study of music perception was undertaken on the Med El device by Schulz and Kerber (1994), but the processing strategy was not specified. Eight subjects were evaluated; half were prelinguistically deaf. A test battery was constructed where subjects were asked to rate musical instruments and rhythm patterns, discriminate tones sequences, and recognize rhythms, instruments, and tones. The implant subjects performed worse on all tests except the reproduction of rhythm. This indicated that the implant provided the correct coarse temporal and intensity information. It may also mean that the subjects developed good rhythm skills to compensate for their lack of ability to recognize tonal sequences.

The perception of musical timbre by the Clarion cochlear implant patients was investigated by Gfeller et al (1998). The 28 implantees used the continuous interleaved sampler (CIS) strategy, and there were 41 normal-hearing subjects. The timbre of four instruments—clarinet, piano, violin, and trumpet—was investigated. Timbre was appraised for "likability," and the instruments identified from closed sets. The violin and trumpet were rated as more likable by the hearing listeners than by the implant subjects, and the normal-hearing subjects were able to identify all instruments with greater accuracy than the implant users.

As the perception of the frequency spectrum of a musical instrument is responsible for its timbre, the steady-state physical spectra of a number of instruments and vowels were correlated with their perception (as represented by their internal spectra) through electrical stimulation with a Spectra-22 speech processor (Stainsby 2000; Stainsby, McDermott et al. 2002). The internal spectrum of the sounds was determined by forward masking. The closest relation to the physical spectrum was found for normal hearing listeners. However, some implant subjects had correlations that were almost as good as those of the normally hearing listeners. Although there was a relationship between the internal spectrum and the ability to discriminate stimuli, a similar relationship between internal spectrum and the ability to identify instruments was not found. This was attributed to the lack of temporal cues, such as attack and decay transients, in these steady-state stimuli.

Bimodal Stimulation

Bimodal stimulation was defined by Clark et al (1991) as electrical stimulation in one ear with a cochlear implant, and the presentation of sound to the other ear with residual hearing through a hearing aid. Bimodal psychophysical studies were necessary to see how best to develop bimodal speech processing. Bimodal psychophysical and speech perception studies commenced in the Australian Research Council's HCRC at the University of Melbourne/Bionic Ear Institute in 1989. A study was first undertaken on eight subjects at the HCRC (Blamey et al 1995), and then on 13 subjects both in Melbourne and the Denver Ear Institute (Blamey et al 1996). The research provided basic data on loudness growth and pitch matching in the two ears for the hearing aid and cochlear implant.

Loudness Growth Function and Summation

The optimal fitting of a hearing aid and implant requires setting the loudness of the combined signal appropriately. Not only is it necessary to have loudness in a comfortable range and smooth variation in loudness with speech, but the balance in loudness between the ears should permit differential loudness cues for sound localization. With the hearing aid it is not enough to set the gain and maximum power output, as many hearing-impaired listeners have an abnormal loudness growth (change in subjective loudness with increasing intensity) (Pohlman and Kranz 1924; Fowler 1928; Galanter and Messick 1961; Pascoe 1978, 1988). The gain required for an appropriate loudness, therefore, would not be uniform over the dynamic range. Studies have suggested that loudness growth measures could be the most effective way of fitting hearing aids (Galanter and Messick 1961;

Pascoe 1978, 1988; Cox et al 1997). With a cochlear implant the loudness growth between T and MC levels could also vary on each electrode, and the correct conversion of current to loudness at intermediate levels should be achieved for a smooth variation in loudness across electrodes.

The summation of loudness for electrical stimulation in one ear and hearing with an aid in the other ear was studied by Blamey et al (2000). In the study on nine patients, the loudness for monaural and bimodal stimulation was reported in six categories from too soft to too loud according to the method of loudness growth in half-octave bands (Allen et al 1990). When the standard fitting procedures were used for cochlear implants and hearing aids, there was a complex pattern of loudness differences between the ears. Loudness summation occurred for some of the subjects. The summation was greatest when the acoustic and electrical signals were of equal loudness, but there was individual variability. Thus the study indicated that standard monaural fitting procedures are unlikely to be successful, and the output levels of both devices will need to be reduced relative to a monaural fitting to compensate for the binaural summation of loudness.

Pitch Matching

The rate and place pitch for electrical stimulation was compared with the pitch for acoustic stimulation of the opposite ear (Blamey et al 1995). As discussed in the section on temporal pitch perception, five of eight subjects could match a stimulus of 250 pulses/s to a sound frequency within 20% of this rate, three of eight at a pulse rate of 500 pulses/s, and two of eight at a pulse rate of 800 pulses/s (Fig. 6.7). However, the match was better when the expected frequency for the site of stimulation matched the stimulus rate. Nevertheless, rate pitch could be matched regardless of the site of the stimulus electrode. For place pitch the electrode positions that matched pure tones were more basal than predicted from the characteristic frequency coordinates of the basilar membrane in the human cochlea (Fig. 6.24). In other words, the perceived pitch of electrical stimuli was lower than that predicted from the positions of electrodes in the cochlea and from physiological studies of frequency versus place in the normal ear. It could help to explain why cochlear implants have been so successful for speech recognition despite the fact that they can use only approximately two thirds of the length of the cochlea. This finding was most probably due to the passage of auditory nerve fibers from the more apical region (lower frequency region of the cochlea) past the region of the basilar membrane responding to a higher frequency traveling wave. Furthermore, the spiral ganglion cells in Rosenthal's canal for the whole frequency scale span 1.875 turns, but the basilar membrane filtering those frequencies covers 2.625 turns (Bredberg 1968). The spread of current from the electrodes would thus excite the fibers from the lower frequency regions. There was no need to stimulate the low frequencies on a place-coding basis, as temporal coding was shown to be more important for these frequencies. Furthermore, as rate of stimulation corresponded with the pitch of sound for the voicing frequen-

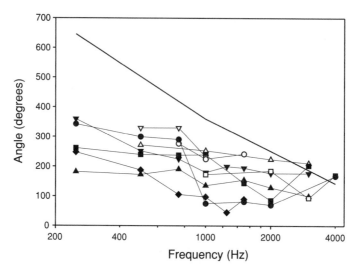

FIGURE 6.24. Frequency versus position in cochlea from pitch-matching experiments with nine cochlear implant users compared to normal listeners (diagonal bold line). Pulse rate was 250 pulses/s. Symbols indicate different subjects (Reprinted with permission from Blamey et al 1995. Pitch matching of electric and acoustic stimuli. *Annals of Otology, Rhinology and Laryngology* **104**: 220–222.).

cies (Tong et al 1979, 1980a), there was no need to transform the rate of stimulation to achieve the best quality sound.

Bilateral Stimulation

In 1989 research commenced in the HCRC at the University of Melbourne/Bionic Ear Institute to explore the use of bilateral cochlear implants in two cochlear implant patients. The aim of implanting two ears rather than one was predominantly to achieve the transmission of more information, and better performance in noise. A series of psychophysical studies investigating the basic characteristics of binaural stimulation were completed on these patients. A summary of the main results from this work from the HCRC and more recent studies at the Cooperative Research Centers for Cochlear Implant Speech and Hearing Research and Cochlear Implant and Hearing Aid Innovation is presented below. The results are useful for developing binaural speech processing that maximizes performance in everyday life. The benefits of bilateral implants should be compared with the alternative approach of improving performance in noise via the use of multiple microphones with algorithms such as adaptive beam-forming. The speech processing for bilateral implants is discussed in Chapter 7.

Fusion and Lateralization

Psychophysical studies on binaural stimulation initially examined whether stimulation of each ear could lead to a fused single auditory image, and to what extent

IIDs or the time of arrival of a stimulus between each ear (ITDs) could shift the image (van Hoesel and Clark 1995). Could implant patients experience a sound image inside the head as did normal-hearing people presented with acoustic clicks under headphones?

A preliminary study showed that if corresponding sites in the cochleae were stimulated, one image would be experienced. This was consistent with the physiological findings of Merzenich and Reid (1974) and Semple and Aitkin (1979) that the frequency tuning curves for units in the inferior colliculus (IC) were approximately the same for contralateral and ipsilateral excitation. In the psychophysical studies the image appeared to originate from different spatial locations in the head or in the ears for the right combination of parameters (van Hoesel et al 1990). If the stimulating electrodes did not excite corresponding frequency areas of the cochlea, or if the IIDs were large, the initial patient described two images.

Before undertaking psychophysical studies it is desirable to compare the position of the electrodes, as they may be offset in the cochlea with respect to each other. An excellent x-ray for measuring any offset is the modified Stenver's view (the cochlear view) described by Marsh et al (1993). The angle of insertion can be accurately determined, as described in Chapter 8. In a temporal bone study the standard deviation of the mean error for the estimated insertion angle was 3.7 degrees. The upper bound was assessed to be 5 degrees (van Hoesel 1998). That is less than 0.5 electrode width in the basal turn of the cochlea.

Interaural Intensity Differences and Lateralization

Studies to determine the effects of IIDs and ITDs on the lateralization of the image commenced on one patient in 1989 and then a second in 1992. The stimulating pair of electrodes were matched for place pitch using the x-ray procedure discussed above. The study on IIDs aimed to determine the range of amplitude differences required to move the image from one side of the head to the other and assess the just noticeable difference (JND) in interaural loudness that could be detected. The loudness was balanced by presenting stimuli to each ear, and adjusting the loudness of one until they were equal. Amplitude ratios were selected so that for a zero interaural time delay the image moved from the left to right of the head. A range of amplitude differences and interaural time delays up to 16 ms were examined. The biphasic pulses were presented at a rate of 200 pulses/s and in bursts of 300 ms. The patients were asked to indicate on a computer screen, representing the centerline between the ears, the positions and widths (spread or compactness) of the images heard. The results were plotted as a matrix with interaural amplitude ratios varying across each column, and the time differences across the rows. The patient indicated the position and width of the images heard. The width of the histogram reflected the variability of the position and the spread of the image. The percentage of presentations with a split image was noted (van Hoesel et al 1990, 1993; van Hoesel and Clark 1995, 1997). The results are summarized in Figure 6.25 for the first patient. The graph is for an amplitude ratio and plots the lateral position (percent to the right) versus interaural time

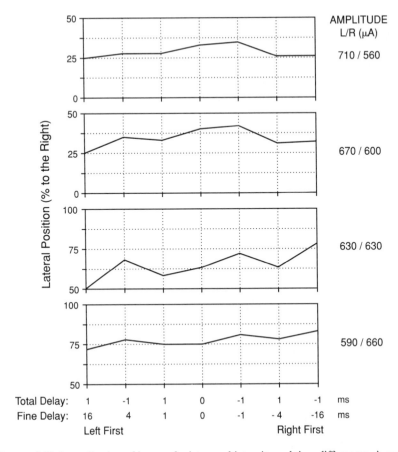

FIGURE 6.25. Lateralization of images for interaural intensity and time differences. A graph of the relative position of the percept to the right for amplitude ratios and time delays (Reprinted with permission from van Hoesel et al 1993. Psychophysical and speech perception studies: a case report on a binaural cochlear implant subject. *Journal of the Acoustical Society of America* **94**: 3178–3189.).

difference. From this it can be seen that for an intensity difference of 670 μA (left) and 600 μA (right) and 630 μA (left) and 630 μA (right), the percept moves from left to right. Although the data do not allow a JND for loudness to be calculated, they do show that an interaural current amplitude difference of 70 μA clearly moved the percept from the left to the right. These current level differences can be equated to acoustic differences through the sound to current mapping for the Nucleus device, and indicate that a perceptible interaural loudness level difference would be a few decibels, and thus quite similar to sound, which as discussed above (see Sound Localization and Lateralization) is between 0.5 and 1.0 dB. In contrast, ITDs had a smaller effect for the burst onset over the range from 0 to 16 ms, and time-intensity trading was weak.

In a combined study on five patients by the Cooperative Research Center (CRC) for Cochlear Implant and Hearing Aid Innovation the IIDs were determined using a two-alternative forced-choice procedure with pitch-matched stimuli and a threshold of 75% correct (van Hoesel and Tyler, submitted). The IIDs varied from less than a 0.17-dB level to 0.68 dB. A 0.17-dB current level was one current level step in the output of the Nucleus receiver-stimulator, and it equated to approximately to 1 dB in sound intensity. Thus the ILD s varied from 1 to 4 dB.

Interaural Time Delay and Lateralization

In an initial study on the first patient (van Hoesel et al 1990), ITDs over the range of 0 to 4 ms were applied to the entire waveform. The ITDs measured the difference between the entire stimulus pulse trains at both sides. Thus the onset or transient of the envelope as well as the ongoing fine structure (analogous to the phase of acoustic signals) were assessed jointly. Systematic shifts in lateralization corresponding to the time delays were not observed. A JND time of approximately 1.5 ms for ITDs was measured. This was significantly worse than the physiological range discussed above. The poor perception of ITDs to electrical stimulation was seen in an additional study that included the first two patients (van Hoesel and Clark 1995). The results showed ITDs produced only a small percentage shift in the lateralized position over a range from -1 to $+1$ ms (van Hoesel and Clark 1995). For both patients the effect of time became smaller when the interaural intensity cues were strong. Thus there could have been some time-intensity trading.

Research was necessary to see why there was poor discrimination of ITDs. A number of electrical parameters were varied, including site in the cochlea and rate of stimulation (van Hoesel and Clark 1997). It was not expected that the decreased sensitivity to ITDs was the result of basal stimulation by the implants, since data on hearing listeners point to good sensitivity to time delays with high-frequency amplitude-modulated signals containing no low-frequency spectral components. It was, however, postulated that the limited effects of time delays could have resulted from inaccurate place matching on the two sides.

To determine the effect of an offset on the placement of the electrode array on the ITD, a pitch magnitude estimation was first undertaken on electrodes on both sides to ensure that the electrodes were matched. The results agreed to within less than half an electrode band spacing as measured by x-ray. This validated the pitch magnitude estimation procedure as complementary to good x-ray data.

None of the electrode combinations gave rise to JND measurements of less than about 0.5 to 1 ms for either patient. For the optimal place-matched condition the JND was the same as when the place offset between the two sides increased to about 4 to 8 electrodes (3 to 6 mm) as seen in Figure 6.26. It appeared, therefore, that the relatively poor discrimination of interaural time delays with cochlear implants should not be subscribed to place mismatching. The ITDs for rates of stimulation, which varied from 50 to 300 pulses/s, were studied. The results showed greatly increased JND values above 200 to 300 pulses/s, the region in which monaural rate perception typically starts to deteriorate (van Hoesel and Clark 1997). Again, none of the rates tested showed interaural time delay JNDs much less than around 0.5 ms.

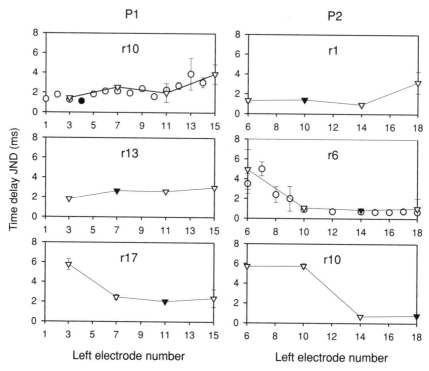

FIGURE 6.26. Interaural time delay just noticeable differences (JNDs) as a function of stimulation on the left side with the place of stimulation held fixed on the right side. Each graph is for a fixed place on the right side (r1, r6, r10, r13, r17). Graphs in the left column are for subject P1; those on the right are for P2. Solid symbols indicate bilateral best matched place pair. Circular symbols are data from the first place experiment, triangular symbols are from the second (Reprinted with permission from van Hoesel and Clark 1997. Psychophysical studies with two binaural cochlear implant subjects. *Journal of the Acoustical Society of America* **102**: 495–507.).

The bilateral coding of sound was further studied on a patient who came to the CRC for Cochlear Implant and Hearing Aid Innovation from Manchester, UK (van Hoesel et al 2002). The JND for detecting the time of arrival of sound at each ear was plotted versus rate of stimulation. Results showed that there was a JND of 400 μs for continuously presented electrical stimuli at rates up to 200 pulses/s. Above that rate it climbed to 1600 μs. The results with sound were similar to those of electrical stimulation up to 200 pulses/s, but markedly better at higher frequencies.

In a study on five patients by the CRC, the ITDs were also determined (van Hoesel and Tyler, submitted). The best scores for the individual patients varied from 100 to 200 μs with a mean of approximately 120 μs. These ITDs were at 50 pulses/s, and were comparable to those for sound. They were also affected by rate of stimulation above 200 pulses/s as in the studies by van Hoesel and Clark (1997) and van Hoesel et al (2002).

The poor ITD performance of the first two Melbourne patients in particular may be due to the long-term effects of deafness on the brain centers responsible for the coding of ITDs, and in particular the medial superior olive (MSO). Its role in sound localization was discussed in Chapters 1 and 5. The brainstem of the first patient, who died 13 years after his first implant, was sectioned, and there was a significant loss in the number and size of the neurons in this center (this was discussed in Chapter 2).

Localization

The above studies were undertaken with the signals fed to the microphones or directly to the speech processor, and are similar to sound lateralization experiments in hearing subjects where the sound is heard in the head. They are different from localizing sound in space. For this reason a subject from Manchester, UK, who visited Melbourne (van Hoesel et al 2002) was tested with a half circle of speakers separated by 18 degrees at 1.5 m to determine the effect of both interaural intensity and temporal cues, and both occurred together. The results showed the average root mean square (RMS) errors (Hartmann 1983) for left and right stimulation to be 81 degrees [standard deviation (SD) 10 degrees] and 85 degrees (SD 17 degrees) respectively, but when both were stimulated together there was a dramatic improvement in the ability to localize sound with the errors falling to 16 degrees (SD 18 degrees). In a study on sound localization with bilateral stimulation on five patients in Iowa by the CRC for Cochlear Implant and Hearing Aid Innovation, chance levels were obtained for monaural stimulation, and with binaural hearing the RMS was a directional sensitivity of less than 10 degrees (van Hoesel and Tyler, submitted). This result is not as good as for normal hearing, where an angular separation of 1 degree can be detected from directly in front and 7 degrees to one side.

Binaural Summation of Loudness

Binaural summation of loudness was studied for matched and unmatched place of stimulation (van Hoesel and Clark 1997). A fixed electrode was selected on one side and combined with both a matched and an unmatched place on the other side. For one of the two patients, both matched and unmatched places produced an overall loudness sensation approximately twice as high as the loudness of either side alone for all five intensity levels. This was also true for the second patient for the matched place, but not for the unmatched place combination. It is possible, however, that this second patient, who at times seemed to have difficulty separating loudness and pitch effects, may have allowed pitch effects to influence his loudness estimates. This emphasizes the importance of pitch matching in these studies. Alternatively, the patient may have listened to just one side or the other rather than to the overall loudness sensation for the unmatched place condition, where two images rather than one fused image may have been heard. The finer details of how loudness is combined from the two sides is perhaps yet to be determined from a larger population of bilateral implant patients, from which it may be possible to determine whether the amount of summation is indeed independent of overall level as it is with narrow band signals for normal hearing, or whether it varies with level as it does with wide-band noise with normal hearing.

The summation of loudness needs to be taken into consideration in binaural studies, as the increased loudness may lead to results that appear better as a result of the summation. Thus the loudness should be balanced. Furthermore, the automatic gain control (AGC) in the speech processor may affect the findings. If a study is carried out with spatially separated noise, the signal may be presented to one ear at a level at which the AGC compresses the sound and so removes some of the cues for binaural hearing.

Central Masking

With central masking, elevated hearing thresholds in one ear occur when a sound of similar spectral content is presented in the contralateral ear at the same time (Zwislocki 1978). To investigate this phenomenon for electrical stimulation, a fixed place was again selected on the right side. The threshold level was measured both with and without the contralateral masker. The stimuli at each place were balanced for loudness at 80% of the dynamic range. The difference between the two measurements is the amount of central masking. The results showed an increase in threshold of around 20% to 40% of the dynamic range for the masked compared to the monaural case, but no place-dependent effects. The mechanism of central masking with normal hearing is not yet well understood (van Hoesel and Clark 1997).

Dichotic Rate Presentation

Binaural diotic rate JNDs were compared with monaural JNDs to see if the patients could combine the information. This was shown to be the case for loudness and frequency discrimination in people with normal hearing by Jesteadt et al (1977). In addition, rate JNDs were measured with fixed rates on the opposite side (dichotic stimulation). With normal hearing this produces beats. There was, however, no difference in the JNDs for binaural and monaural stimulation (van Hoesel and Clark 1997), which may have been due to the fact that deterministic stimulation does not produce the fine temporospatial patterns of excitation needed for producing beats in normal hearing subjects. It could also have been due to the poor temporal processing of interaural time information seen with these two subjects. This was probably due to the loss of neurons in the MSO seen in the histopathological study. With the dichotic presentation of stimuli there were no beats experienced.

In another study on four patients in which coincident signals and noise were presented, there was no difference for the binaural stimuli. These people had better ITDs for the low stimulus rates of 50 pulses/s than the first two. This suggests that improvements require advances in fine temporal processing.

References

Allen, J. B., J. L. Hall and P. S. Jeng. 1990. Loudness growth in 1/2-octave bands (LGOB)—a procedure for the assessment of loudness. Journal of the Acoustical Society of America 88: 745–753.

Aran, J. M. 1981. Electrical stimulation of the auditory system and tinnitus control. Journal of Laryngology and Otology 95 (suppl 4): 153–161.

Batteau, D. W. 1967. The role of the pinna in human localization. Proceedings of the Royal Society 168: 158–180.

Bench, R. J. and J. Bamford. 1979. Speech-hearing tests and the spoken language of hearing-impaired children. London, Academic Press.

Bilger, R. C., F. O. Black and N. T. Hopkinson. 1977. Evaluation of subjects presently fitted with implanted auditory prostheses. Annals of Otology, Rhinology and Laryngology 86(suppl 38): 1–176.

Black, R. C. and G. M. Clark. 1977. Electrical transmission line properties of the cat cochlea. Proceedings of the Australian Physiological and Pharmacological Society 8: 137.

Black, R. C. and G. M. Clark. 1978. Electrical network properties and distribution of potentials in the cat cochlea. Proceedings of the Australian Physiological and Pharmacological Society 9: 71P.

Black, R. C. and G. M. Clark. 1980. Differential electrical excitation of the auditory nerve. Journal of the Acoustical Society of America 67(3): 868–874.

Black, R. C., G. M. Clark and J. F. Patrick. 1981. Current distribution measurements within the human cochlea. IEEE Transactions on Biomedical Engineering 28: 721–724.

Blamey, P. J., G. J. Dooley, E. S. Parisi and G. M. Clark. 1996. Pitch comparisons of acoustically and electrically evoked auditory sensations. Hearing Research 99(1–2): 139–150.

Blamey, P. J., R. C. Dowell, A. M. Brown, G. M. Clark and P. M. Seligman. 1987. Vowel and consonant recognition of cochlear implant patients using formant-estimating speech processors. Journal of the Acoustical Society of America 82: 48–57.

Blamey, P. J., R. C. Dowell, Y. C. Tong, A. M. Brown, S. M. Luscombe and G. M. Clark. 1984a. Speech processing studies using an acoustic model of a multiple-channel cochlear implant. Journal of the Acoustical Society of America 76: 104–110.

Blamey, P. J., R. C. Dowell, Y. C. Tong and G. M. Clark. 1984b. An acoustic model of a multiple-channel cochlear implant. Journal of the Acoustical Society of America 76: 97–103.

Blamey, P. J., C. J. James, G. J. Dooley and E. S. Parisi. 2000. Monaural and binaural loudness measures in cochlear implant users with contralateral residual hearing. Ear and Hearing 21: 6–17.

Blamey, P. J., E. Parisi and G. M. Clark. 1995. Pitch matching of electric and acoustic stimuli. Annals of Otology, Rhinology and Laryngology 104: 220–222.

Blauert, J. 1971. Localization and the law of the first wavefront in the median plane. Journal of the Acoustical Society of America 50: 466–470.

Boone, M. M. 1973. Loudness measurements on pure tone and broad band impulsive sounds. Acustica 29: 198–204.

Braida, L. D. and N. I. Durlach. 1970. Intensity perception. II. Resolution in one-interval paradigms. Journal of the Acoustical Society of America 51: 483–502.

Bredberg, G. 1968. Cellular pattern and nerve supply of the human organ of Corti. Acta Oto-Laryngologica-supplement 236: 1–138.

Bruce, I. C., L. S. Irlicht and G. M. Clark. 1998. A mathematical analysis of spatiotemporal summation of auditory nerve firings. Information Sciences: Applications 111: 303–334.

Bruce, I. C., M. W. White, L. S. Irlicht, S. J. O'Leary and G. M. Clark. 1999. The effects of stochastic neural activity in a model predicting intensity perception with cochlear implants: low-rate stimulation. IEEE Transactions on Biomedical Engineering 46: 1393–1404.

Burian, K., E. Hochmair and I. Hochmair-Desoyer. 1979. Designing of and experience with multichannel cochlear implants. Acta Oto-Laryngologica 87: 190–195.

Burkitt, A. N. and G. M. Clark. 2000. Analysis of synchronization in the response of neurons to noisy periodic synaptic input. Neurocomputing 32–33: 67–75.

Burns, E. M. and N. F. Viemeister. 1976. Nonspectral pitch. Journal of the Acoustical Society of America 60: 863–869.

Burns, E. M. and N. F. Viemeister. 1981. Played-again SAM: further observations on the pitch of amplitude-modulated noise. Journal of the Acoustical Society of America 70: 1655–1660.

Busby, P. A. and G. M. Clark. 1997. Pitch and loudness estimation for single and multiple pulse per period electric pulse rates by cochlear implant patients. Journal of the Acoustical Society of America 101: 1687–1695.

Busby, P. A. and G. M. Clark. 2000a. Electrode discrimination by early-deafened subjects using the Cochlear Limited multiple-electrode cochlear implant. Ear and Hearing 21: 291–304.

Busby, P. A. and G. M. Clark. 2000b. Pitch estimation by early-deafened subjects using a multiple-electrode cochlear implant. Journal of the Acoustical Society of America 107: 547–558.

Busby, P. A., Y. C. Tong and G. M. Clark. 1984. Underlying dimensions and individual differences in auditory, visual and auditory-visual vowel perception by hearing impaired children. Journal of the Acoustical Society of America 75: 1858–1865.

Busby, P. A., Y. C. Tong and G. M. Clark. 1992. Psychophysical studies using a multiple-electrode cochlear implant in patients who were deafened early in life. Audiology 31(2): 95–111.

Busby, P. A., Y. C. Tong and G. M. Clark. 1993a. Electrode position, repetition rate, and speech perception by early- and late-deafened cochlear implant patients. Journal of the Acoustical Society of America 93: 1058–1067.

Busby, P. A., Y. C. Tong and G. M. Clark. 1993b. The perception of temporal modulations by cochlear implant patients. Journal of the Acoustical Society of America 94: 124–131.

Busby, P. A., L. A. Whitford, P. J. Blamey, L. M. Richardson and G. M. Clark. 1994. Pitch perception for different modes of stimulation using the Cochlear multiple-electrode prosthesis. Journal of the Acoustical Society of America 95: 2658–2669.

Buus, S. and M. Florentine. 1985. Gap detection in normal and impaired listeners: the effect of level and frequency. In: Michelsen, A., ed. Time resolution in auditory systems. Berlin, Springer-Verlag 159–179.

Cacace, A. T. and R. H. Margolis. 1985. On the loudness of complex stimuli and its relationship to cochlear excitation. Journal of the Acoustical Society of America 78: 1568–1573.

Carhart, R. 1965. Monaural and binaural discrimination against competing sentences. International Audiology 4: 5–10.

Carhart, R. and J. F. Jerger. 1959. Preferred method for clinical determination of pure-tone thresholds. Journal of Speech and Hearing Disorders 24(4): 330–345.

Carney, L. H. 1994. Spatiotemporal encoding of sound level: models for normal encoding and recruitment of loudness. Hearing Research 76: 31–44.

Charlet de Sauvage, R., Y. Cazals and J.-P. Erre. 1983. Acoustically derived auditory nerve action potential evoked by electrical stimulation: an estimation of the waveform of single unit contribution. Journal of the Acoustical Society of America 73: 616–627.

Chouard, C. H., C. Fugain, B. Meyer and H. Lacombe. 1983. Long-term results of the multichannel cochlear implant. Annals of the New York Academy of Sciences 405: 387–411.

Clark, G. M. 1969. Middle ear and neural mechanisms in hearing and the management of deafness. PhD thesis.

Clark, G. M. 1987a. The University of Melbourne–Nucleus multi-electrode cochlear implant. Advances in oto-rhino-laryngology. Volume 38. Basel, Karger.

Clark, G. M. 1997. Auditory central nervous system plasticity: application to cochlear implantation. In: Clark, G. M., ed. Cochlear implants. XVI World Congress of Otorhinolaryngology Head and Neck Surgery. Bologna, Monduzzi Editore: 19–23.

Clark, G. M. 1999. Cochlear implants in the third millennium. American Journal of Otology 20: 4–8.

Clark, G. M., R. C. Black, D. J. Dewhurst, I. C. Forster, J. F. Patrick and Y. C. Tong. 1977. A multiple-electrode hearing prosthesis for cochlear implantation in deaf patients. Medical Progress through Technology 5: 127–140.

Clark, G. M., P. J. Blamey, P. A. Busby, et al. 1987b. A multiple-electrode intracochlear implant for children. Archives of Otolaryngology 113: 825–828.

Clark, G. M., P. A. Busby, S. A. Roberts, et al. 1987. Preliminary results for the Cochlear Corporation multi-electrode intracochlear implants on six prelingually deaf patients. American Journal of Otology 8: 234–239.

Clark, G. M., G. J. Dooley and P. J. Blamey. 1991. Combined electrical and acoustical stimulation using a bimodal speech processor. American Pediatric Otolaryngological Society Meeting, Hawaii.

Clark, G. M., H. G. Kranz and H. Minas. 1973. Behavioral thresholds in the cat to frequency modulated sound and electrical stimulation of the auditory nerve. Experimental Neurology 41: 190–200.

Clark, G. M., J. M. Nathar, H. G. Kranz and J. S. Maritz. 1972. A behavioral study on electrical stimulation of the cochlea and central auditory pathways of the cat. Experimental Neurology 36: 350–361.

Clark, G. M., Y. C. Tong, Q. R. Bailey, et al. 1978. A multiple-electrode cochlear implant. Journal of the Oto-Laryngological Society of Australia 4: 208–212.

Clark, G. M., Y. C. Tong, R. C. Black, I. C. Forster, J. F. Patrick and D. J. Dewhurst. 1977. A multiple electrode cochlear implant. Journal of Laryngology and Otology 91: 935–945.

Clark, G. M., Y. C. Tong and L. F. Martin. 1981a. A multiple-channel cochlear implant. An evaluation using open-set CID sentences. Laryngoscope 91: 628–634.

Clark, G. M., Y. C. Tong, L. F. Martin and P. A. Busby. 1981b. A multiple-channel cochlear implant. An evaluation using an open-set word test. Acta Oto-Laryngologica 91: 173–175.

Cohen, A. 1961. Further investigation of the effects of intensity upon the pitch of pure tones. Journal of the Acoustical Society of America 33: 1363–1376.

Cohen, L. T., E. Saunders and G. M. Clark. 2001. Psychophysics of a prototype perimodiolar cochlear implant electrode array. Hearing Research 155: 63–81.

Coleman, P. D. 1962. Failure to localize the source distance of an unfamiliar sound. Journal of the Acoustical Society of America 34: 345–346.

Collins, L. M., T. A. Zwolan and G. H. Wakefield. 1997. Comparison of electrode discrimination, pitch ranking, and pitch scaling data in postlinguistically deafened cochlear implant subjects. Journal of the Acoustical Society of America 101: 440–455.

Cox, R. M., G. C. Alexander, I. M. Taylor and G. A. Gray. 1997. The contour test of loudness perception. Ear and Hearing 18: 388–400.

Cramer, E. M. and W. H. Huggins. 1958. Creation of pitch through binaural interaction. Journal of the Acoustical Society of America 30: 413–417.

de Boer, E. 1969. Encoding of frequency information in the discharge pattern of auditory nerve fibres. International Audiology 8: 547–556.

De Cheveigne, A. 1998. Cancellation model of pitch perception. Journal of the Acoustical Society of America 103: 1261–1271.

Dillier, N. 1998. Intracochlear recordings of electrically evoked compound action potentials. In: Barber, C., ed. Programme and abstracts of First international symposium and workshop on objective measures in cochlear implantation. Nottingham UK: 54.

Dillier, N., W. K. Lai, B. Almquist, et al. 2002. Measurement of the electrically evoked compound action potential (ECAP) via a neural response telemetry system. Annals of Otology Rhinology and Laryngology 111(5 pt 1): 407–414.

Dillier, N., W. K. Lai, D. Cafarelli-Dees and E. von Wallenberg. 2000. Post-operative neural response telemetry findings in adults submitted for publication. Neural esponse telemetry: results from a European field trial, 12th AAA Convention, Chicago: 137.

Dillier, N., T. Spillman and J. Guntensperger. 1983. Computerized testing of signal encoding strategies with round window implants. In: Parkins, C. and J. Anderson, eds. Cochlear prostheses: an International Symposium New York, New York Academy of Sciences: 360–369.

Dobie, R. A. and J. Kimm. 1980. Brainstem responses to electrical stimulation of the cochlear. Archives of Otolaryngology 106: 573–577.

Dorman, M. F., K. Basham, G. McCandless and H. Dove. 1991. Speech understanding and music appreciation with the Ineraid cochlear implant. Hearing Journal 44: 34–37.

Dorman, M. F., L. Smith, G. McCandless, G. Dunnavant, J. Parkin and K. Dankowski. 1990. Pitch scaling and speech understanding by patients who use the Ineraid cochlear implant. Ear and Hearing 11: 310–315.

Douek, E., A. J. Fourcin, B. C. J. Moore and G. P. Clarke. 1977. A new approach to the cochlear implant. Proceedings of the Royal Society of Medicine 70: 379–383.

Dowell, R. C., L. A. Whitford, P. M. Seligman, B. K.-H. Franz and G. M. Clark. 1990. Preliminary results with a miniature speech processor for the 22-electrode/Cochlear hearing prosthesis. In: Sacristan, T., ed. Otorhinolaryngology, head and neck surgery. Amsterdam, Kugler and Ghedini: 1167–1173.

Duifhuis, H., L. F. Willems and R. J. Sluyter. 1982. Measurement of pitch in speech: an implementation of Goldstein's theory of pitch perception. Journal of the Acoustical Society of America 71: 1568–1580.

Durlach, N. I. and H. S. Colburn. 1978. Binaural phenomena. In: Carterette, E. C. and M. P. Friedman, eds. Handbook of perception, vol. 4. New York, Academic Press: 365–466.

Eddington, D. K. 1980. Speech discrimination in deaf subjects with cochlear implants. Journal of the Acoustical Society of America 68: 885–891.

Eddington, D. K., W. H. Dobelle and D. E. Brackmann. 1978. Auditory prostheses research with multiple channel intracochlear stimulation in man. Annals of Otology 87: 1–39.

Eguchi, S. and I. J. Hirsh. 1969. Development of speech sounds in children. Acta Oto-Laryngologica 257: 1–51.

Evans, E. F. 1978. Peripheral auditory processing in normal and abnormal ears: physiological considerations for attempts to compensate for auditory deficits by acoustic and electrical prostheses. Scandinavian Audiology Supplementum 6: 10–46.

Exner, S. 1876. Zur lehre von den gehorsempfindungen. Pflugers Archiv 13: 228–253.

Fitzgibbons, P. J. and S. Gordon-Salant. 1987. Temporal gap resolution in listeners with high-frequency sensorineural hearing loss. Journal of the Acoustical Society of America 81: 133–137.

Fitzgibbons, P. J. and F. L. Wightman. 1982. Gap detection in normal and hearing-impaired listeners. Journal of the Acoustical Society of America 72: 761–765.

Fletcher, H. 1940. Auditory patterns. Review of Modern Physics 12: 47–65.

Fletcher, H. F. and W. A. Munson. 1933. Loudness, its definition, measurement, and calculation. Journal of the Acoustical Society of America 5: 82–108.

Florentine, M. and S. Buus. 1984. Temporal gap detection in sensorineural and simulated hearing impairments. Journal of Speech and Hearing Research 27: 449–455.

Ford, A. 1993. Composer to composer: conversations about contemporary music. St Leonards, Australia, Allen and Unwin.

Fourcin, A. J., S. M. Rosen, B. C. Moore, et al. 1979. External electrical stimulation of the cochlea: clinical, psychophysical, speech-perceptual and histological findings. British Journal of Audiology 13(3): 85–107.

Fowler, E. P. 1928. Marked deafened areas in normal ears. Archives of Otolaryngology 8: 151–155.

French, N. R. and J. C. Steinberg. 1947. Factors governing the intelligibility of speech sounds. Journal of the Acoustical Society of America 19: 90–119.

Fujita, S. and J. Ito. 1999. Ability of nucleus cochlear implantees to recognize music. Annals of Otology, Rhinology and Laryngology 108: 634–640.

Galanter, E. and S. Messick. 1961. The relation between category and magnitude scales of loudness. Psychological Review 68: 363–372.

Gardner, M. B. 1968. Proximity image effect in sound localization. Journal of the Acoustical Society of America 43: 163.

Gardner, M. B. and R. S. Gardner. 1973. Problem of localization in the median plan: effect of pinnae cavity occlusion. Journal of the Acoustical Society of America 53: 400–408.

Gassler, G. 1954. Uber die horschwelle fur schallereignisse mit verschieden breitem frequenzspektrum. Acustica 4: 408–414.

Gfeller, K., J. F. Knutson, G. Woodworth, S. Witt and B. DeBus. 1998. Timbral recognition and appraisal by adult cochlear implant users and normal-hearing adults. Journal of the American Academy of Audiology 9: 1–19.

Gfeller, K. and C. R. Lansing. 1991. Melodic, rhythmic, and timbral perception of adult cochlear implant users. Journal of Speech and Hearing Research 34: 916–920.

Gfeller, K. and C. R. Lansing. 1992. Musical perception of cochlear implant users as measured by the primary measures of music audiation: an item analysis. Journal of Music Therapy 29: 18–39.

Glasberg, B. R. and B. C. Moore. 1990. Derivation of auditory filter shapes from notched-noise data. Hearing Research 47: 103–138.

Goldstein, J. L. 1973. An optimum processor theory for the central formation of the pitch of complex tones. Journal of the Acoustical Society of America 54: 1496–1516.

Gordon, E. E. 1979. Primary measures of music audiation. Chicago, G.I.A. Publications.

Grantham, D. W. 1984. Interaural intensity discrimination: insensitivity at 1000 Hz. Journal of the Acoustical Society of America 75: 1191–1194.

Green, D. M. 1973. Temporal acuity as a function of frequency. Journal of the Acoustical Society of America 54: 373–379.

Green, D. M., T. G. Birdsall and W. P. Tanner. 1957. Signal detection as a function of signal intensity and duration. Journal of the Acoustical Society of America 29: 523–531.

Green, D. M. and T. G. Forrest. 1988. Detection of amplitude modulation and gaps in noise. In: Duifhuis, H., ed. Basic issues in hearing. New York, Academic Press.

Green, D. M. and J. A. Swets. 1966. Signal detection theory and psychophysics. New York, John Wiley and Sons.

Greenwood, D. D. 1961. Critical bandwidth and the frequency coordinates of the basilar membrane. Journal of the Acoustical Society of America 33: 1344–1356.

Greenwood, D. D. 1990. A cochlear frequency-position function for several species—29 years later. Journal of the Acoustical Society of America 87: 2592–2605.

Grey, J. M. 1977. Multidimensional scaling of musical timbres. Journal of the Acoustical Society of America 61: 1270–1277.

Hall, J. W. and J. H. Grose. 1997. The relation between gap detection, loudness, and loudness growth in noise-masked normal-hearing listeners. Journal of the Acoustical Society of America 101: 1044–1049.

Hall, J. W. and A. D. G. Harvey. 1985. Diotic loudness summation in normal and impaired hearing. Journal of Speech and Hearing Research 28: 445–448.

Hamilton, P. M. 1957. Noise masked thresholds as a function of tonal duration and masking noise bandwidth. Journal of the Acoustical Society of America 29: 506–511.

Hanekom, J. J. and R. V. Shannon. 1996. Place pitch discrimination and speech recognition in cochlear implant users 43: 27–40.

Hanekom, J. J. and R. V. Shannon. 1998. Gap detection as a measure of electrode interaction in cochlear implants. Journal of the Acoustical Society of America 104: 2372–2384.

Harris, G. G. 1963. Periodicity perception by using gated noise. Journal of the Acoustical Society of America 35: 1229–1233.

Hartmann, W. M. 1983. Localization of sound in rooms. Journal of the Acoustical Society of America 74: 1380–1391.

Hawkins, D. B., R. A. Prosek, B. E. Walden and A. A. Montgomery. 1987. Binaural loudness summation in the hearing impaired. Journal of Speech and Hearing Research 30: 37–43.

Hawkins, J. E. and S. S. Stevens. 1950. The masking of pure tones and of speech by white noise. Journal of the Acoustical Society of America 22: 6–13.

Henning, D. B. 1974. Detectability of interaural delay in high-frequency complex waveforms. Journal of the Acoustical Society of America 55: 84–90.

Henry, B. A., C. M. McKay, H. J. McDermott and G. M. Clark. 1996. The importance of different frequency bands to the speech perception of cochlear implantees. Australian Journal of Audiology 17(suppl): 44–45.

Henry, B. A., C. M. McKay, H. J. McDermott and G. M. Clark. 1997. Speech cues for cochlear implantees: spectral discrimination. In: Clark, G. M., ed. Cochlear implants. XVI World Congress of Otorhinolaryngology Head and Neck Surgery. Bologna, Monduzzi Editore: 89–93.

Henry, B. A., C. M. McKay, H. J. McDermott and G. M. Clark. 2000. The relationship between speech perception and electrode discrimination in cochlear implantees. Journal of the Acoustical Society of America 108: 1269–1280.

Hershkowitz, R. M. and N. I. Durlach. 1969. Interaural time and amplitude JNDs for a 500-Hz tone. Journal of the Acoustical Society of America 46: 1464–1467.

Hirsh, I. J. 1950. The relation between localization and intelligibility. Journal of the Acoustical Society of America 22: 196–200.

Hirsh, I. J. 1952. The measurement of hearing. New York, McGraw-Hill.

Hirsh, I. J. and C. E. Sherrick. 1961. Perceived order in different sense modalities. Journal of Experimental Psychology 62: 423–432.

Hochmair, E. S. and I. J. Hochmair-Desoyer. 1983. Percepts elicited by different speech coding strategies. Annals of the New York Academy of Sciences 405: 268–279.

Hochmair-Desoyer, I. J., E. S. Hochmair and K. Burian. 1981. Four years of experience with cochlear prostheses. Medical Progress Technology 8: 107–119.

Hong, R. S., J. T. Rubinstein, D. Wehner and D. Horn. Submitted. Dynamic range enhancement for cochlear implants. Otology and Neuro-Otology.

House, W. F. and B. J. Edgerton. 1982. A multiple-electrode cochlear implant. Annals of Otology, Rhinology and Laryngology 91: 104–116.

Houtgast, T. 1974. Lateral suppression and loudness reduction of a tone in noise. Acustica 30: 214–221.

Irlicht, L. S. and G. M. Clark. 1995. Control strategies for neurons modeled by self-exciting point processes. Journal of the Acoustical Society of America 98: 2927.

Irwin, R. J. 1965. Binaural summation of thermal noises of equal and unequal power in each ear. American Journal of Psychology 78: 57–65.

Javel, E., Y. C. Tong, R. K. Shepherd and G. M. Clark. 1987. Responses of cat auditory fibres to biphasic electrical current pulses. Annals of Otology, Rhinology and Laryngology 96: 26–30.

Jesteadt, W., C. C. Wier and D. M. Green. 1977. Intensity discrimination as a function of frequency and sensation level. Journal of the Acoustical Society of America 61: 169–177.

Johnson, P. M. 1980. A behavioural study of electrical stimulation of the auditory nerve of the cat: with application to the design of a cochlear implant prosthesis for sensory deafness. Master of Surgery dissertation. University of Leeds.

Joris, P. X. and T. C. T. Yin. 1992. Responses to amplitude-modulated tones in the auditory nerve of the cat. Journal of the Acoustical Society of America 91: 215–232.

Katz, B. 1966. Nerve, muscle and synapse. New York, McGraw-Hill.

Kiang, N. Y.-S., R. F. Pfeiffer and W. B. Warr. 1965. Stimulus coding in the cochlear nucleus. Annals of Otology, Rhinology and Laryngology 74: 2–23.

Klumpp, R. G. and H. R. Eady. 1956. Some measurements of interaural time difference thresholds. Journal of the Acoustical Society of America 28: 859–860.

Kruskal, J. B. 1964. Multidimensional scaling by optimizing goodness of fit to a nonmetric hypothesis. Psychometrics 29: 1–27.

Laird, R. K. 1979. The bioengineering development of a sound encoder for an implantable hearing prosthesis for the profoundly deaf. Master of engineering science thesis. University of Melbourne.

Licklider, J. C. R. 1948. The influence of interaural phase upon the masking of speech by white noise. Journal of the Acoustical Society of America 20: 150–159.

Licklider, J. C. R. 1951. A duplex theory of pitch perception. Experientia 7: 128–133.

Lim, H. H., Y. C. Tong and G. M. Clark. 1989a. Forward masking patterns produced by intracochlear electrical stimulation of one and two electrode pairs in the human cochlea. Journal of the Acoustical Society of America 86(3): 971–980.

Lim, H. H., Y. C. Tong, G. M. Clark and P. A. Busby. 1986. Forward masking, studies on a multichannel cochlear implant patient. Journal of the Acoustical Society of America 80(suppl 1): S29.

Lim, H. H., Y. C. Tong, R. D. Hollow, A. Vandali and R. van Hoesel. 1989b. Loudness summation study on multielectrode pair stimulation. Journal of the Acoustical Society of America 86(suppl 1): S81–S82.

Litovsky, R. Y. and H. S. Colburn. 1999. The precedence effect. Journal of the Acoustical Society of America 106: 1633–1654.

Litvak, L. 2002. Towards a better speech processor for cochlear implants: auditory nerve responses to high rate electric pulse trains. PhD thesis. Massachesetts Institute of Technology.

MacKeith, N. W. and R. R. Coles. 1971. Binaural advantages in hearing of speech. Journal of Laryngology and Otology 85: 213–232.

Maiwald, D. 1967. Die berechnung von modulationsschwellen mit hilfe eines funktionsschemas. Acustica 18: 193–207.

Marks, L. E. 1978. Binaural summation of the loudness of pure tones. Journal of the Acoustical Society of America 64: 107–113.

Marsh, M. A., J. Xu, P. J. Blamey, et al. 1993. Radiologic evaluation of multichannel intracochlear implant insertion depth [published erratum appears in Am J Otol 1993 Nov; 14(6):627]. American Journal of Otology 14(4): 386–391.

McAnally, K. I., M. Brown and G. M. Clark. 1997a. Acoustic and electric forward-masking of the auditory nerve compound action potential: evidence for linearity of electro-mechanical transduction. Hearing Research 106: 137–145.

McAnally, K. I., M. Brown and G. M. Clark. 1997b. Comparison of current waveforms for the electrical stimulation of residual low frequency hearing. Acta Oto-Laryngologica 117: 831–835.

McAnally, K. I., M. Brown and G. M. Clark. 1997c. Estimating mechanical responses to pulsatile electrical stimulation of the cochlea. Hearing Research 106: 146–153.

McAnally, K. I. and G. M. Clark. 1991a. Electrophonic excitation from electrical stimulation of the hearing-impaired cochlea of the cat. Journal of the Acoustical Society of America 90: 2291.

McAnally, K. I. and G. M. Clark. 1991b. The origin of auditory compound action potentials induced by electrical stimulation of the cat cochlea. Proceedings of the Australian Physiological and Pharmacological Society 22: 178P.

McAnally, K. I. and G. M. Clark. 1994. Stimulation of residual hearing in the cat by pulsatile electrical stimulation of the cochlea. Acta Oto-Laryngologica 114(4): 366–372.

McAnally, K. I., G. M. Clark and J. Syka. 1993. Hair cell mediated responses of the auditory nerve to sinusoidal electrical stimulation of the cochlea in the cat. Hearing Research 67: 55–68.

McDermott, H. J. and C. M. McKay. 1994. Pitch ranking with nonsimultaneous dual-electrode electrical stimulation of the cochlea. Journal of the Acoustical Society of America 96: 155–162.

McDermott, H. J. and C. M. McKay. 1997. Musical pitch perception with electrical stimulation of the cochlea. Journal of the Acoustical Society of America 101: 1622–1631.

McKay, C. M. and H. J. McDermott. 1993. Perceptual performance of subjects with cochlear implants using the spectral maxima sound processor (SMSP) and the mini speech processor (MSP). Ear and Hearing 14: 350–367.

McKay, C. M. and H. J. McDermott. 1996. The effect on pitch and loudness of major interpulse intervals within modulated current pulse trains in cochlear implantees. Journal of the Acoustical Society of America 99: 2584.

McKay, C. M. and H. J. McDermott. 1998. Loudness perception with pulsatile electrical stimulation: the effect of interpulse intervals. Journal of the Acoustical Society of America 104: 1061–1074.

McKay, C. M., H. J. McDermott and G. M. Clark. 1991. Preliminary results with a six spectral maxima speech processor for the University of Melbourne/Nucleus multiple electrode cochlear implant. Journal of the Oto-Laryngological Society of Australia 6: 354–359.

McKay, C. M., H. J. McDermott and G. M. Clark. 1994. Pitch percepts associated with amplitude-modulated current pulse trains in cochlear implantees. Journal of the Acoustical Society of America 96: 2664–2673.

McKay, C. M., H. J. McDermott and G. M. Clark. 1995a. Loudness summation for two channels of stimulation in cochlear implantees: effects of spatial and temporal separation. Annals of Otology, Rhinology and Laryngology 104 (suppl 166): 230–233.

McKay, C. M., H. J. McDermott and G. M. Clark. 1995b. Pitch matching of amplitude-modulated current pulse trains by cochlear implantees: the effect of modulation depth. Journal of the Acoustical Society of America 97: 1777–1785.

McKay, C. M., H. J. McDermott and G. M. Clark. 1996. The perceptual dimensions of single-electrode and non-simultaneous dual-electrode stimuli in cochlear implantees. Journal of the Acoustical Society of America 99: 1079–1090.

McKay, C. M., A. O'Brien and C. J. James. 1999. Effect of current level on electrode discrimination in electrical stimulation. Hearing Research 136: 159–164.

McKay, C. M., M. D. Remine and H. J. McDermott. 2001. Loudness summation for pulsatile electrical stimulation of the cochlea: effects of rate, electrode separation, level, and mode of stimulation. Journal of the Acoustical Society of America 110: 1514–1524.

Meddis, R. and M. J. Hewitt. 1991a. Virtual pitch and phase sensitivity of a computer model of the auditory periphery. I: Pitch identification. Journal of the Acoustical Society of America 89: 2866–2882.

Meddis, R. and M. J. Hewitt. 1991b. Virtual pitch and phase sensitivity of a computer model of the auditory periphery. II: Phase sensitivity. Journal of the Acoustical Society of America 89: 2883–2894.

Merzenich, M. M. 1975. Studies on electrical stimulation of the auditory nerve in animals and man: cochlear implants: In: Tower, D. B. ed. The nervous system vol. 3, Human communication and its disorders. New York, Raven Press. 537–548.

Merzenich, M., C. Byers and M. White. 1984. Scala tympani electrode arrays. Fifth quarterly progress report. NIH contract NO1-NS9-2353: 1–11.

Merzenich, M. M. and M. D. Reid. 1974. Representation of the cochlea within the inferior colliculus. Brain Research 77: 397–415.

Michelson, R. P. 1971. Electrical stimulation of the human cochlea—a preliminary report. Archives of Otolaryngology 93: 317–323.

Miller, G. A. 1947. Sensitivity to changes in the intensity of white noise and its relation to asking and loudness. Journal of the Acoustical Society of America 191: 609–619.

Mills, A. 1958. On the minimum audible angle. Journal of the Acoustical Society of America 30: 237–246.

Mills, A. W. 1960. Lateralization of high-frequency tones. Journal of the Acoustical Society of America 60: 410–417.

Moon, A. K., T. A. Zwolan and B. E. Pfingst. 1993. Effects of phase duration on detection of electrical stimulation of the human cochlea. Hearing Research 67: 166–178.

Moore, B. C. J. 1974. Relation between the critical bandwidth and the frequency-difference limen. Journal of the Acoustical Society of America 55: 359.

Moore, B. C. J. 1977. Effects of relative phase of the components on the pitch of three-component complex tones. In: Evans, E. F., and J. P. Wilson, eds. Psychophysics and physiology of hearing. London, Academic Press: 349–358.

Moore, B. C. J. 1997. An introduction to the psychology of hearing. San Diego, Academic Press.

Moore, B. C. J. and B. R. Glasberg. 1986. The role of frequency selectivity in the perception of loudness, pitch and time. In: Moore, B. C. J., ed. Frequency selectivity in hearing. London, Academic: 251–308.

Moore, B. C. J. and B. R. Glasberg. 1988. Gap detection with sinusoids and noise in normal, impaired, and electrically stimulated ears. Journal of the Acoustical Society of America 83: 1093–1101.

Moore, B. C. J. and B. R. Glasberg. 1989. Mechanisms underlying the frequency discrimination of pulsed tones and the detection of frequency modulation. Journal of the Acoustical Society of America 86: 1722–1732.

Moore, B. C. J., B. R. Glasberg, C. J. Plack and A. K. Biswas. 1988. The shape of the ear's temporal window. Journal of the Acoustical Society of America 83: 1102–1116.

Moore, B. C. J. and D. H. Raab. 1975. Intensity discrimination for noise bursts in the presence of a continuous, bandstop background: effects of level, width of the bandstop, and duration. Journal of the Acoustical Society of America 57: 534–539.

Moore, B. C. J. and S. M. Rosen. 1979. Tune recognition with reduced pitch and interval information. Quarterly Journal of Experimental Psychology 31: 229–240.

Morgan, C. T., W. R. Garner and R. Galambos. 1951. Pitch and intensity. Journal of the Acoustical Society of America 23: 658–663.

Nelson, D. A., D. J. Van Tasell, A. C. Schroder et al. 1995. Electrode ranking of "place pitch" and speech recognition in electrical hearing. Journal of the Acoustical Society of America. 98: 1987–1999.

Nuetzel, J. M. and E. R. Hafter. 1981. Discrimination of interaural delays in complex waveforms: spectral effects. Journal of the Acoustical Society of America 69: 1112–1118.

Oxenham, A. J. and B. C. J. Moore. 1994. Modeling the additivity of nonsimultaneous masking. Hearing Research 80: 105–118.

Parkin, P. H. and H. R. Humphreys. 1958. Acoustics noise and buildings. London, Faber.

Pascoe, D. P. 1978. An approach to hearing aid selection. Hearing Instruments 29: 12–16.

Pascoe, D. P. 1988. Clinical measurements of the auditory dynamic range and their relation to formulas for hearing aid gain. In: Jensen, J. H., ed. Hearing aid fitting: theoretical and practical views. 13th Danavox symposium. Copenhagen, Sougaard Jensen: 129–151.

Patterson, R. D. 1976. Auditory filter shapes derived with noise stimuli. Journal of the Acoustical Society of America 59: 640–654.

Patterson, R. D. and B. C. J. Moore. 1986. Auditory filters and excitation patterns as representations of frequency resolution. In: Moore, B. C. J., ed. Frequency selectivity in hearing. London, Academic: 123–177.

Peissig, J. and B. Kollmeier. 1997. Directivity of binaural noise reduction in spatial multiple noise-source arrangements for normal hearing and impaired listeners. Journal of the Acoustical Society of America 105: 1660–1670.

Penner, M. J. 1977. Detection of temporal gaps in noise as a measure of the decay of auditory sensation. Journal of the Acoustical Society of America 61: 552–557.

Peterson, G. E. and H. L. Barney. 1952. Control methods used in a study of the vowels. Journal of the Acoustical Society of America 24: 175–184.

Pfingst, B. E., P. A. Burnett and D. Sutton. 1983. Intensity discrimination with cochlear implants. Journal of the Acoustical Society of America 73(4): 1283–1292.

Pfingst, B. E., L. A. Holloway and S. A. Razzaque. 1996. Effects of pulse separation on detection thresholds for electrical stimulation of the human cochlea. Hearing Research 98(1–2): 77–92.

Pfingst, B. E., L. A. Holloway, T. A. Zwolan and L. M. Collins. 1999. Effects of stimulus level on electrode-place discrimination in human subjects with cochlear implants. Hearing Research 134: 105–15.

Pfingst, B. E. and D. J. Morris. 1993. Stimulus features affecting psychophysical detection thresholds for electrical stimulation of the cochlea II: frequency and interpulse interval. Journal of the Acoustical Society of America 94: 1287–1294.

Pialoux, P., C. H. Chouard and P. MacLeod. 1976. Physiological and clinical aspects of the rehabilitation of total deafness by implantation of multiple intracochlear electrodes. Acta Oto-Laryngologica 81: 436–441.

Pijl, S. and D. W. F. Schwarz. 1995a. Intonation of musical intervals by musical intervals by deaf subjects stimulated with single bipolar cochlear implant electrodes. Hearing Research 89: 203–211.

Pijl, S. and D. W. F. Schwarz. 1995b. Melody recognition and musical interval perception by deaf subjects stimulated with electrical pulse trains through single cochlear implant electrodes. Journal of the Acoustical Society of America 98: 886–895.

Plack, C. J. and B. C. J. Moore. 1990. Temporal window shape as a function of frequency and level. Journal of the Acoustical Society of America 87: 2178–2187.

Plack, C. J. and B. C. J. Moore. 1991. Decrement detection in normal and impaired ears. Journal of the Acoustical Society of America 90: 3069–3076.

Plomp, R. 1964. The ear as a frequency analyzer. Journal of the Acoustical Society of America 36: 1628–1636.

Plomp, R. 1970. Timbre as a multidimensional attribute of complex tones. In: Plomp, R. and G. F. Smoorenburg eds. Frequency analysis and periodicity detection in hearing. Leiden, Sijthoff: 397–414.

Plomp, R. 1976. Aspects of tone sensation: a psychophysical study. London, Academic Press.

Plomp, R. and M. A. Bouman. 1959. Relation between hearing threshold and duration for tone pulses. Journal of the Acoustical Society of America 31: 749–758.

Plomp, R. and A. M. Mimpen. 1968. The ear as a frequency analyzer. II. Journal of the Acoustical Society of America 43: 764–767.

Plomp, R. and H. J. M. Steeneken. 1969. Effect of phase on the timbre of complex tones. Journal of the Acoustical Society of America 46: 409–421.

Pohlman, A. G. and R. W. Kranz. 1924. Binaural minimum audition in a subject with ranges of deficient acuity. Proceedings of the Society for Experimental Biology and Medicine 20: 335–337.

Pollack, I. and J. M. Pickett. 1958. Stereophonic listening and speech intelligibility against voice babble. Journal of the Acoustical Society of America 30: 131–133.

Preece, J. P. and R. S. Tyler. 1989. Temporal-gap detection by cochlear prosthesis users. Journal of Speech and Hearing Research 32: 849–856.

Reynolds, G. S. and S. S. Stevens. 1960. Binaural summation of loudness. Journal of the Acoustical Society of America 32: 1337–1344.

Riesz, R. R. 1928. Differential intensity sensitivity of the ear for pure tones. Physical Review 31: 867–875.

Robinson, D. W. and R. S. Dadson. 1956. A re-determination of the equal-loudness relations for pure tones. British Journal of Applied Physiology 7: 166–181.

Rodenburg, M. 1977. Investigation of temporal effects with amplitude modulated signals. In: Evans, E. F., ed. Psychophysics and physiology of hearing. London, Academic Press.

Rose, J. E., J. F. Brugge, D. J. Anderson and J. E. Hind. 1967. Phase-locked response to low-frequency tones in single auditory nerve fibers of the squirrel monkey. Journal of Neurophysiology 30: 769–93.

Rossing, T. D. 1982. The science of sound. Reading, MA, Addison-Wesley.

Rowland, R. C. and J. V. Tobias. 1967. Interaural intensity difference limen. Journal of Speech and Hearing Research 10: 745–756.

Rubinstein, J. T., B. S. Wilson, C. C. Finley and P. J. Abbas. 1999. Pseudospontaneous activity: stochastic independence of auditory nerve fibers with electrical stimulation. Hearing Research 127: 108–118.

Runge-Samuelson, C. 2002. Response of the auditory nerve to sinusoidal electrical stimulation: effects of high-rate pulse trains. PhD thesis. University of Iowa.

Saldhana, E. L. and J. F. Corso. 1964. Timbre cues and the identification of musical instruments. Journal of the Acoustical Society of America 36: 2021–2026.

Saunders, E., L. Cohen, A. Aschendorff, et al. 2002. Threshold, comfortable level and impedance changes as a function of electrode-modiolar distance. Ear and Hearing 23(1 suppl): 28S–40S.

Sayers, B. M. 1964. Acoustic-image lateralization judgements with binaural tones. Journal of the Acoustical Society of America 36: 923–926.

Scharf, B. 1968. Binaural loudness summation as a function of band-width. Sixth International Congress of Acoustics Tokyo: A3–5, 25–28.

Scharf, B. 1969. Dichotic summation of loudness. Journal of the Acoustical Society of America 45: 1193–1205.

Schouten, J. F. 1940a. The residue and the mechanism of hearing. Proceedings of the Koninklijke Nederlandse Akademie van Wetenschappen 43: 991–999.

Schouten, J. F. 1940b. The residue, a new component in subjective sound analysis. Proceedings of the Koninklijke Nederlandse Akademie van Wetenschappen 43: 356–365.

Schulz, E. and M. Kerber. 1994. Music perception with the MED-EL implants. In: Hochmair-Desoyer, I. J. and E. S. Hochmair, eds. Advances in cochlear implants. Manz: Vienna, 326–332.

Semple, M. N. and L. M. Aitkin. 1979. Representation of sound frequency and laterality by units in central nucleus of cat inferior colliculus. Journal of Neurophysiology 42: 1626–1639.

Shailer, M. J. and B. C. J. Moore. 1983. Gap detection as a function of frequency, bandwidth, and level. Journal of the Acoustical Society of America 74: 467–473.

Shannon, R. V. 1983a. Multichannel electrical stimulation of the auditory nerve in man. I. Basic psychophysics. Hearing Research 11: 157–189.

Shannon, R. V. 1983b. Multichannel electrical stimulation of the auditory nerve in man. II. Channel interaction. Hearing Research 12: 1–16.

Shannon, R. V. 1985. Threshold and loudness functions for pulsatile stimulation of cochlear implants. Hearing Research 18: 135–143.

Shannon, R. V. 1989a. Detection of gaps in sinusoids and pulse trains by patients with cochlear implants. Journal of the Acoustical Society of America 85: 2587–2592.

Shannon, R. V. 1989b. A model of threshold for pulsatile electrical stimulation of cochlear implants. Hearing Research 40(3): 197–204.

Shannon, R. V. 1992. Temporal modulation transfer functions in patients with cochlear implants. Journal of the Acoustical Society of America 91: 2156–2164.

Sheeley, E. C. and R. C. Bilger. 1964. Temporal integration as a function of frequency. Journal of the Acoustical Society of America 36: 1850–1857.

Shower, E. G. and R. Biddulph. 1931. Differential pitch sensitivity of the ear. Journal of the Acoustical Society of America 2: 275–287.

Simmons, F. B. 1966. Electrical stimulation of the auditory nerve in man. Archives of Otolaryngology 84: 2–54.

Skinner, M. W., G. M. Clark, L. A. Whitford, et al. 1994. Evaluation of a new spectral peak coding strategy for the Nucleus 22 channels cochlear implant system. American Journal of Otology 15: 15–27.

Slawson, A. W. 1968. Vowel quality and musical timbre as functions of spectrum envelope and fundamental frequency. Journal of the Acoustical Society of America 43: 97–101.

Slawson, W. 1985. Sound colour. Berkeley, University of California Press.

Smith, D. W., C. C. Finley, C. van den Honert, V. B. Olszyk and K. E. Konrad. 1994. Behavioral and electrophysiological responses to electrical stimulation in the cat. I. Absolute thresholds. Hearing Research 81: 1–10.

Stainsby, T. H. 2000. The perception of musical sounds with cochlear implants. PhD thesis. University of Melbourne.

Stainsby, T., H. McDermott, C. McKay and G. M. Clark. 2002. Musical timbre perception with cochlear implants: investigations using forward masking. 7th International Conference on Music Perception and Cognition, Sydney, July.

Stephens, S. D. G. 1974. Methodological factors influencing loudness of short duration sounds. Journal of Sound Vibration 37: 235–246.

Stevens, S. S. 1935. The relation of pitch to intensity. Journal of the Acoustical Society of America 6: 150–154.

Stevens, S. S. 1972. Perceived level of noise by Mark VII and decibels (E). Journal of the Acoustical Society of America 51: 575–601.

Stevens, S. S. 1975. Psychophysics. New York, John Wiley and Sons.

Stevens, S. S. and H. B. Greenbaum. 1966. Regression effect in psychophysical measurement. Perceptive Psychophysics 1: 439–446.

Stevens, S. S. and E. B. Newman. 1936. The localization of actual sources of sound. American Journal of Psychology 48: 297–306.

Stevens, S. S. and J. Volkmann. 1940. The relation of pitch to frequency: a revised scale. American Journal of Physiology 53(3): 329–353.

Strutt, J. W. 1907. The theory of sound. Proceedings of the Royal Society of London 79: 399.

Stypulkowski, P. H. and C. van den Honert. 1984. Physiological properties of the electrically stimulated auditory nerve. I. Compound action potential recordings. Hearing Research 14: 205–223.

Terhardt, E. 1974a. On the perception of periodic sound fluctuations (roughness). Acustica 30: 201–213.

Terhardt, E. 1974b. Pitch, consonance and harmony. Journal of the Acoustical Society of America 55: 1061–1069.

Thurlow, W. R. and P. S. Runge. 1967. Effect of induced head movements on localization of direction of sounds. Journal of the Acoustical Society of America 42: 480–488.

Tobias, J. V. and E. D. Schubert. 1959. Effective onset duration of auditory stimuli. The Journal of the Acoustical Society of America 31: 1595–1605.

Tong, Y. C., R. C. Black, G. M. Clark, et al. 1979. A preliminary report on a multiple-channel cochlear implant operation. Journal of Laryngology and Otology 93: 679–695.

Tong, Y. C., P. J. Blamey, R. C. Dowell and G. M. Clark. 1983a. Psychophysical studies evaluating the feasibility of a speech processing strategy for a multiple-channel cochlear implant. Journal of the Acoustical Society of America 74: 73–80.

Tong, Y. C., P. A. Busby and G. M. Clark. 1988. Perceptual studies on cochlear implant patients with early onset of profound hearing impairment prior to normal development of auditory, speech, and language skills. Journal of the Acoustical Society of America 84: 951–962.

Tong, Y. C. and G. M. Clark. 1982. Percepts produced by electrical stimulation of the human cochlea. Proceedings of the Australian Physiological and Pharmacological Society 13: 150P.

Tong, Y. C. and G. M. Clark. 1985. Absolute identification of electric pulse rates and electrode positions by cochlear implant patients. Journal of the Acoustical Society of America 77: 1881–1888.

Tong, Y. C., G. M. Clark, P. J. Blamey, P. A. Busby and R. C. Dowell. 1982. Psychophysical studies for two multiple-channel cochlear implant patients. Journal of the Acoustical Society of America 71: 153–160.

Tong, Y. C., G. M. Clark and H. H. Lim. 1987. Estimation of the effective spread of neural excitation produced by a bipolar pair of scala tympani electrodes. Annals of Otology, Rhinology and Laryngology 96(suppl 128): 37–38.

Tong, Y. C., G. M. Clark, P. M. Seligman and J. F. Patrick. 1980a. Speech processing for a multiple-electrode cochlear implant hearing prosthesis. Journal of the Acoustical Society of America 68: 1897–1899.

Tong, Y. C., R. C. Dowell, P. J. Blamey and G. M. Clark. 1983b. Two-component hearing sensations produced by two-electrode stimulation in the cochlea of a deaf patient. Science 219: 993–994.

Tong, Y. C., J. B. Millar, G. M. Clark, L. F. Martin, P. A. Busby and J. F. Patrick. 1980b. Psychophysical and speech perception studies on two multiple-channel cochlear implant patients. Journal of Laryngology and Otology 94: 1241–1256.

Townshend, B., N. E. Cotter, D. Van Compernolle and R. L. White. 1987. Pitch perception by cochlear implant subjects. Journal of the Acoustical Society of America 82(1): 106–115.

Treisman, M. 1964. Sensory scaling and the psychophysical law. Quarterly Journal of Experimental Psychology 16: 11–22.

van Hoesel, R. J. M. 1998. Bilateral electrical stimulation with multi-channel cochlear implants. PhD thesis. University of Melbourne.

van Hoesel, R. J. M. and G. M. Clark. 1995. Fusion and lateralization study with two binaural cochlear implant patients. Annals of Otology, Rhinology and Laryngology 104(suppl 166): 233–235.

van Hoesel, R. J. M. and G. M. Clark. 1997. Psychophysical studies with two binaural cochlear implant subjects. Journal of the Acoustical Society of America 102: 495–507.

van Hoesel, R., R. Ramsden and M. O'Driscoll. 2002. Sound-direction identification, interaural time delay discrimination and speech intelligibility advantages in noise for a bilateral cochlear implant user. Ear and Hearing 23(2): 137–149.

van Hoesel, R. J. M., Y. C. Tong, R. D. Hollow and G. M. Clark. 1993. Psychophysical and speech perception studies: a case report on a binaural cochlear implant subject. Journal of the Acoustical Society of America 94: 3178–3189.

van Hoesel, R. J. M., Y. C. Tong, R. D. Hollow, J. Huigen and G. M. Clark. 1990. Preliminary studies on a bilateral cochlear implant user. Journal of the Acoustical Society of America 88(suppl 1): S193.

van Hoesel, R. and R. Tyler. Submitted. Speech perception, localization, and lateralization with bilateral cochlear implant users. Journal of the Acoustical Society of America.

van Tasell, D. J., D. G. Greenfield, J. J. Logemann and D. A. Nelson. 1992. Temporal cues for consonant recognition: training, talker generalization, and use in evaluation of cochlear implants. Journal of the Acoustical Society of America 92: 1247–1257.

van Tasell, D. J., S. D. Soli and V. M. Kirby. 1987. Speech waveform envelope cues for consonant recognition. Journal of the Acoustical Society of America 82(4): 1152–1161.

Verschuure, J. and A. A. van Meeteren. 1975. The effect of intensity on pitch. Acustica 32: 33–44.

Viemeister, N. F. 1974. Intensity discrimination of noise in the presence of band-reject noise. Journal of the Acoustical Society of America 56: 1594–1600.

Viemeister, N. F. 1979. Temporal modulation transfer functions based on modulation thresholds. Journal of the Acoustical Society of America 66: 1364–1380.

von Helmholtz, H. L. F. 1859. Ueber die Klangfarbe der Vocale. Annals of Physical Chemistry 18: 280–290.

von Helmholtz, H. L. F. 1863. Die Lehre von den tonempfindungen als physiologische grundlage fur die theorie der musik. Braunschweig, F Vieweg and Sohn.

Walloch, H., E. B. Newman and M. R. Rosenzweig. 1949. The precedence effect in sound localization. American Journal of Psychology 52: 315–336.

Warren, R. M. 1970. Elimination of biases in loudness judgments for tones. Journal of the Acoustical Society of America 48: 1397–1413.

Warren, R. M. and J. A. Bashford. 1980. Broadband repetition switch: spectral dominance or pitch averaging? Journal of the Acoustical Society of America 84: 2058–2062.

Whitman, F. L., D. J. Kistler and M. E. Perkins. 1987. A new approach to the study of human sound localization. In: Yost, W. A. and G. Gorevitch, eds. Directional hearing. New York, Springer-Verlag: 26–48.

Wier, C. C., W. Jesteadt and D. M. Green. 1977. Frequency discrimination as a function of frequency and sensation level. Journal of the Acoustical Society of America 61: 178–184.

Wightman, F. L. 1973a. The pattern transformation model of pitch. Journal of the Acoustical Society of America 54: 407–416.

Wightman, F. L. 1973b. Pitch and stimulus fine structure. Journal of the Acoustical Society of America 54: 397–406.

Wightman, F. L. and D. J. Kistler. 1989. Headphone simulation of free-field listening I. Stimulus synthesis. Journal of the Acoustical Society of America 85: 858–867.

Williams, A. J., G. M. Clark and G. V. Stanley. 1976. Pitch discrimination in the cat through electrical stimulation of the terminal auditory nerve fibers. Physiological Psychology 4: 23–27.

Wilson, B. S., D. T. Lawson, M. Zerbi and C. C. Finley. 1992. Speech processors for auditory prostheses. Twelfth quarterly progress report, April 1992. NIH contract N01-DC-9-2401. Research Triangle Institute.

Wilson, B. S., D. T. Lawson, M. Zerbi and C. C. Finley. 1993. Speech processors for auditory prostheses. Fifth quarterly progress report, Oct 1993. NIH contract N01-DC-2-2401. Research Triangle Institute.

Wilson, B. S., D. T. Lawson, M. Zerbi and C. C. Finley. 1994. Recent developments with the CIS strategies. In: Hochmair-Desoyer, I. J. and E. S. Hochmair, eds. Advances in Cochlear Implants, Proceedings of the Third International Cochlear Implant Conference, Innsbruck, Austria, April 1993.

Yost, W. A. and R. H. Dye. 1988. Discrimination of interaural differences of level as a function of frequency. Journal of the Acoustical Society of America 83: 1846–1851.

Yost, W. A. and G. Gorevitch. 1987. Directional hearing. New York, Springer-Verlag.

Yost, W. and W. Hafter 1987. Lateralization. In: Yost, W. and G. Gourevitch, eds. Directional hearing. New York, Springer.

Zwicker, E. 1952. Die grenzen der horbarkeit der amplitudenmodulation und der frequenzmodulation eines tones. Acustica 2: 125–133.

Zwicker, E. 1958. Uber psychologische und methodische Grundlagen der Lautheit. Acustica 8: 237–258.

Zwicker, E. 1961. Subdivision of the audible frequency range into critical bands (Frequenzgruppen). Journal of the Acoustical Society of America 33: 248.

Zwicker, E. 1970. Masking and psychological excitation as consequences of the ear's frequency analysis. In: Plomp, R., and G. Smoorenburg, eds. Frequency analysis and periodicity detection in hearing. Leiden, Sijthoff: 376–394.

Zwicker, E. and B. Scharf. 1965. A model of loudness summation. Psychological Review 72: 3–26.

Zwislocki, J. J. 1978. Masking: experimental and theoretical aspects of simultaneous, forward, backward, and central masking. In: Carterette, E. C. and M. P. Friedman, eds. Handbook of perception. Volume 4. New York, Academic Press: 283–336.

Zwislocki, J. and R. S. Feldman. 1956. Just noticeable dichotic phase difference. Journal of the Acoustical Society of America 28: 152–153.

Zwolan, T., L. Collins and G. Wakefield. 1997. Electrode discrimination and speech recognition in postlinguistically deafened adult cochlear implant subjects. Journal of the Acoustical Society of America 102: 3673–3685.

7
Speech (Sound) Processing

Acoustic

Human communication is achieved when thought is transformed through language into speech. The sounds of speech are initiated by activity in the central nervous system, and this stimulates the respiratory and vocal tracts. Speech sounds are usually produced through the expulsion of air through the larynx. Voiced sounds are due to the vibration of the vocal cords, allowing puffs of air to be transmitted to the vocal tract. With unvoiced speech the larynx is held open, and a turbulent flow of air at a point of constriction creates the basic sound. The vocal tract acts as a multiresonant filter for the transmitted sound. Resonances that result are referred to as formants. Communication is completed when the speech sounds are received by the auditory system of another person and translated back into thought. Speech sounds are important, but not essential for communication. Visual signals such as sign language of the deaf or signed English can also be used, as well as the written word.

Speech is a complex acoustic signal, and information is transmitted to the brain at a rapid rate. For example, three to seven syllables are uttered per second during a conversation (Pickett 1980). The phonemes are discrete perceptual units consisting of complex sound elements and are coded into patterns of neural discharges in the lower auditory brain centers for decoding into speech by the higher auditory centers. Speech understanding by cochlear implant patients requires rapid processing of the speech signals into patterns of electrical stimulation in the auditory nerve in the cochlea that partially reproduce those for normal hearing.

A knowledge of speech science is important for developing cochlear implant speech processing strategies. There are many excellent references, such as Fant (1973), Flanagan and Rabiner (1973), Ainsworth (1976, 1992), O'Shaughnessy (1987), Ainsworth (1992), and Ladefoged (1993). This chapter discusses issues that are especially relevant to cochlear implants.

Articulators and Vocal Tract Shape

The articulators that shape the cavities to create resonances (formants) in the speech signal are part of the oropharynx. They can be classified for consonants into those for place and manner of articulation. The articulators for place of articulation in the following words can be classified as (1) bilabial, with the two lips together for "bee"; (2) labiodental, with the lower lip behind the upper front teeth for "fee"; (3) dental, with the tip of the tongue or blade next to the upper front teeth for "thee"; (4) alveolar, with the tongue tip or blade close to or against the upper alveolar ridge for "dee"; (5) retroflex, with the tip of the tongue close to the back of the alveolar ridge for "re"; (6) palato-alveolar, with the tongue blade at the back of the alveolar ridge for "she"; and (7) velar, with the back of the tongue against the soft palate for "key."

The manner of articulation may be (1) voiced or unvoiced, providing discrimination between "do" and "to"; (2) nasal, where the soft palate is down and the air passes out through the nose for the sounds "me," "knee," and "ing"; (3) oral stop (plosive), where the air stream is completely obstructed at a point in the vocal tract as in "bet," "debt," and "get"; (4) fricative, due to the close approximation of two articulators, so that the air stream is partially obstructed but turbulent (the higher pitch sounds with more acoustic energy as in "see" are sibilants, and others are fricative nonsibilants); (5) lateral sounds, where obstruction of the air stream occurs at a point along the center of the oral tract with incomplete closure between one or both sides of the tongue and the roof of the mouth as in "lee"; and (6) affricate, where the sound is a combination of plosive and fricative as in "cheer."

The complex speech sounds produced by the articulators referred to above have specific acoustic features, and can be classified accordingly. Their features are processed by the central auditory pathways, and are discussed below.

Speech Analysis

Speech is a complex signal, and its analysis provides an understanding of the cues of importance in perception. Representing these cues with electrical stimulation helps provide optimal speech processing for profoundly deaf people with cochlear implants.

Band-Pass Filtering

Speech consists of the fundamental or voicing frequency and the harmonic frequencies produced by the resonant cavities. An effective method of analyzing speech is to pass the signal through a bank of band-pass filters. The pattern of energy in each filter or the relative strength of the frequency components is the speech spectrum. The speech spectrum over a period of time can be represented graphically as a spectrogram, as illustrated in Figure 7.1.

Each filter represents the energy of speech across a certain frequency band.

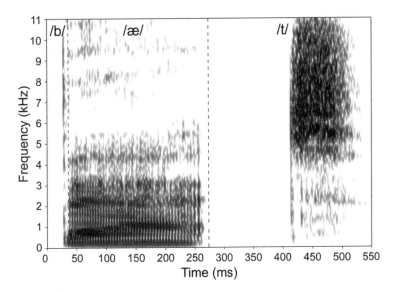

FIGURE 7.1. A speech spectrograph of the word "bat" showing frequency versus time with the intensity of the sound increasing from light to dark. The locations of the three phonemes /b/, /æ/, and /t/, are shown by the dashed lines.

The center frequency and the slope of the filter on the high- and low-frequency side can be varied, and thus change the extent of the filtering over the frequency range. The ideal filter should transmit only the frequencies within the pass band. Their amplification should be the same, and there should be no change in the phase relations of the frequencies within the pass band. It is usual to consider the upper and lower ends of the band pass separately. A low-pass (high-cut) filter determines the upper limit of the band pass. A high-pass (low-cut) filter determines the lower limit of the pass band. The simplest and most fundamental type of filter is the RC circuit that has a resistor (R) and capacitor (C) in parallel. The gain and phase shift of this filter falls short of the ideal. For this reason digital filters or active filters incorporating operational amplifiers are used. Filters are discussed further in Chapter 8, and reference can be made to an appropriate text such as Dewhurst (1976), Brophy (1977), and O'Shaughnessy (1987).

Spectrogram

The sound spectrogram, referred to above, is produced by a spectrograph that was first developed at the Bell Telephone Laboratories to analyze and synthesize speech. It plots the amplitude of the signal from variable filters over time. The electromagnetic data were originally displayed as a paper readout, and now digitally on a computer. It is an approximation to the continuous spectrum at a number of instances in time. An example is shown in Figure 7.1 for the word "bat." The intensity of the signal at each frequency is indicated by the depth of the shading. It is possible to see the concentrations of energy that indicate formant frequencies. The voicing frequency is represented by the periodic striations resulting from the repetitive puffs of air produced by the larynx. The spectrogram is helpful in understanding how speech can be analyzed by cochlear implant speech processors.

Formants

When speech is filtered into its spectral components, there are peaks of energy at certain frequencies called formants. They are, as referred to above, resonances produced by variations in the dimensions of the pharynx during articulation. Speech can have many well-defined formants, but the first (F1) and second (F2) formants are the most important for the identification of vowels. The F1 frequency range for American English and Swedish is from approximately 250 to 700 Hz, and the F2 range is from 700 to 2300 Hz (Peterson and Barney 1952; Fant 1959).

Fourier Analysis—Fast Fourier Transform

The most obvious way to view sound is in the time domain (Fig. 7.2). The amplitude of the waves can be inspected over instants in time. However, with a complex wave it is difficult to determine the constituent frequencies. This can be done using a Fourier analysis. First, the oscillations of sound can be expressed mathematically as both sine and cosine terms. Second, the constituent frequencies are those that are determined by the analysis to correlate with the sine and cosine waves. The waves can be defined by either sine or cosine functions, and the difference between the two provide the phase information.

An example of the Fourier transforms of two composite waves from two sine waves of the same frequency and amplitude, but with a 45-degree phase shift, are shown in Figure 7.3. From this it can be seen that the composite waves are different even though the components have the same frequencies and amplitudes. This difference is reflected in the pattern of interspike intervals in the neural

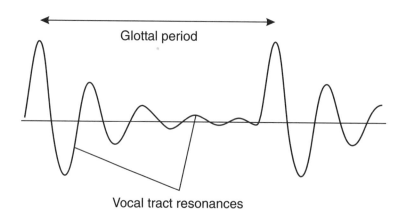

FIGURE 7.2. A diagram of the speech waveform for a vowel showing the glottal period and the vocal tract resonance.

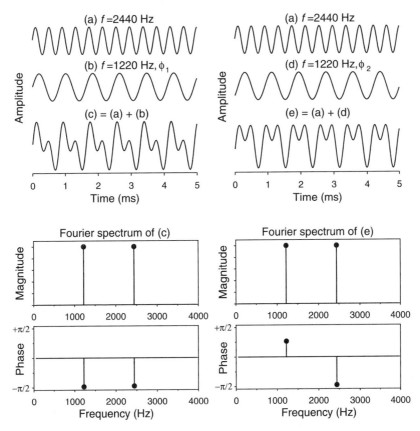

FIGURE 7.3. Top: The superposition of sine waves of frequency (a) 2440 Hz, and (b and d) 1220 Hz, resulting in waveforms (c) and (e). The phase difference between waves (b) and (d) is $\phi_1 - \phi_2 = 45$ degrees. Bottom: The Fourier spectra of waves (c) and (e) showing the magnitude and phase of each frequency component.

activity, as shown in the study by Rose et al (1969). If a Fourier analysis of the speech signal is carried out, the frequency spectrum is well preserved but the phase information is lost. The fine time information is probably important for the encoding of frequency and should be preserved in new speech-processing strategies.

There are other methods of analyzing speech such as models of cochlear function and linear predictive coding, but they will not be discussed as they are not yet used routinely in speech-processing strategies for cochlear implants.

Speech Perception and Production

Speech conveys meaning (semantics) through words connected as sentences based on grammatical rules (syntax). The words and grammatical rules are the same for spoken and written speech, and constitute language. The perception of words and, to a lesser extent, sentences is through a sequence of perceptual units. The units making up words are referred to as phonemes. The number varies according to the dialect of the speaker and the system of the linguist doing the classification.

The phonemes are discrete and successive units as illustrated in Figure 7.4 (Fant 1973). For example, the word "pat" is defined by the phonemes /p/, /æ/, and /t/ following each other. On the other hand, the sounds that make up each phoneme (the phonetic representation) may vary within its boundary or overlap the boundary in time to a neighboring phoneme. For example, the consonant /p/ is represented first by a period of silence (occlusion) followed by a burst of noise. The noise burst may extend beyond the time when the vowel is initiated. Furthermore, a consonant influences the acoustic representation of the following vowel and vice versa. This effect of phonemes on each other is called coarticulation.

Vowels

Vowels are voiced sounds that are more constant in frequency and amplitude and of longer duration than consonants. They are referred to as monophthongs, as distinct from diphthongs where the shape of the vocal tract changes from one configuration to another during articulation. The duration of a short vowel varies from 138 to 262 ms and a long vowel from 188 to 410 ms (Lehiste and Peterson 1959). They are more intense than consonants. The intensity of vowels with respect to /θ/ as in "thin" varies from 29 to 22 dB, and consonants from 21 to 7 dB (Fry 1979). The vowels for English are listed in Table 7.1. They are produced by the resonances in the larynx and pharynx, emphasizing frequencies in the sounds induced by laryngeal vibrations. An example of the raw acoustic waveform for a vowel (Fig. 7.2) shows the glottal period and the influence of the vocal tract resonances. The main articulator for vowel sounds is the tongue. As there is always a gap between the tongue and the palate, the air stream is relatively unobstructed. In normal conversation the glottal period due to the vibrations of the vocal cords is changing, for example, when voiced pitch is used for stressing syllables.

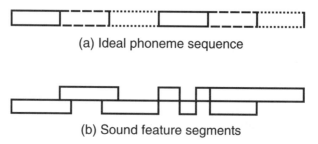

(a) Ideal phoneme sequence

(b) Sound feature segments

FIGURE 7.4. An illustration of the perceptual and acoustic elements of speech (Fant 1973). Reprinted with permission from Fant G., 1973. Speech sounds and features. Cambridge, Mass., MIT Press.

TABLE 7.1. Monophthongs (vowels) and diphthongs.

Phoneme	Class	Example
/i/	Front vowels	beet
/ɪ/		bit
/ɛ/		bet
/æ/		bat
/ɑ/	Back vowels	cart
/ɒ/		rod
/ɔ/		cord
/ʊ/		would
/u/		rude
/ɜ/	Middle vowels	dirt
/ʌ/		hut
/ə/		the
/ɛi/	Diphthongs	pay
/ɑi/		high
/əʊ/		road
/ɑʊ/		cow
/iə/		hear
/uə/		endure
/ɔə/		sore
/ɔi/		boy
/ɛə/		care

Based on Ainsworth (1976).

The vowel sounds vary according to the position of the highest point of the tongue and the position of the lips. For the front vowels in the words "heed" /i/, "hid" /ɪ/, "head" /ɛ/, and "had" /æ/, the position of the highest point of the tongue is in the front of the mouth, as illustrated in Figure 7.5. The height of the tongue varies from high to low for the above vowels.

With the middle vowels in the words "the" /ə/, "hurt" /ɜ/, and "hut" /ʌ/, the tongue is elevated at the center of the palate. For the back vowels in the words "heart" /ɑ/, "hod" /ɒ/, "hoard" /ɔ/, "hood" /ʊ/, and "who'd" /u/, the tongue is close to the posterior wall of the oropharynx. The height of the tongue for these vowels also varies as for the front vowels. Vowels are also classified according to whether the lips are more open or closed.

The front vowels tend to have higher second formants than the back vowels, due to the fact that the front resonating chamber is smaller. The frequency of the first formant is more influenced by the height of the tongue. A higher tongue results in a lower first formant and vice versa. Hence /a/ has the highest first formant. The closed lips of /u/ and other back vowels have the effect of reducing the second formant further than it would be normally. This is an oversimplification, as each formant depends on the entire shape of the vocal tract (Dunn 1950; Fant 1973).

As stated above, the diphthongs are produced in a manner similar to that of vowels, with the vocal tract moving from one position to another over time. For example, the diphthong /ɑi/ as in the word "high" is produced by changing from

Front   Middle   Back

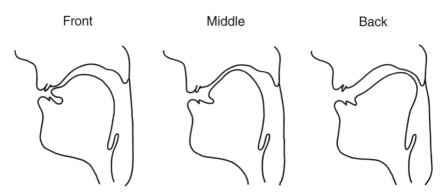

FIGURE 7.5. A diagram of the cross section of the mouth showing the placement of the tongue for front, middle, and back vowels. The height of the tongue can vary at each site. Note also the lips and jaws can be either more closed or open (From A Course in Phonetics 3E HSIE 3rd edition by Ladefoged, © 1993. Reprinted with permission of Heinle, a division of Thomson Learning: www.thomsonrights.com. Fax 800 730-2215.).

an open-back shape near /ɑ/ to a closed-front one near /i/. The diphthongs vary for different dialects of English and other languages.

Vowels are recognized primarily on the basis of the first and second formant frequencies. This was demonstrated by Pols et al (1969), who presented a set of vowels to listeners to determine which were most similar, and found that two dimensions (i.e., two formants) explained 70% of the variance. The range of the two-formant frequencies required for the perception of different vowels has been studied using natural and synthetic sounds, and plotting a two-dimensional space for the perception of the vowels (Peterson and Barney 1952; Ainsworth and Millar 1972). This is shown in Figure 7.6. The center of each space corresponded to the mean F1-F2 positions for naturally occurring vowels (Peterson and Barney 1952). Variation in the data was due, in particular, to the higher formant frequencies in women and children as a result of their smaller vocal cavities. The relative amplitudes of the formants were not found to be critical for recognition. The synthetic study (Ainsworth and Millar 1972) showed F2 could be reduced to 28 dB below the F1 before the vowel changed identity. This supported a peak-picking mechanism rather than one using a weighted average of the formant amplitudes. If the F2 amplitude was markedly reduced, the front vowel shifted to a back vowel.

In contrast, single-formant synthetic vowels were perceived as front as well as back vowels. For example, the vowels with frequencies at 720, 2160, and 3000 Hz were identified as /ɔ/, /ɛ/, and /i/ (Delattre et al 1952). It was originally thought by the research group at the University of Melbourne's Department of Otolaryngology that stimulating single electrodes at different frequency locations in the cochlea could represent single formant vowels and be used as a speech-processing strategy (Tong et al 1979). However, this would have required a special algorithm to determine the formant to be represented by the vowels and consonants, and the electrode that would convey the percept. Instead the F2 was extracted, and coded on a frequency place basis as a more straightforward alternative.

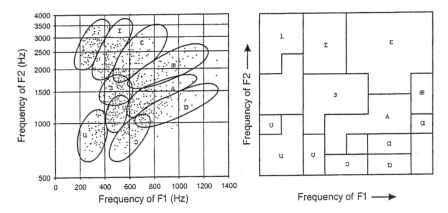

FIGURE 7.6. A two-dimensional space for the range of F1 and F2 frequencies for the recognition of vowels. Left: Natural vowels (Peterson and Barney 1952). Right: Synthetic vowels (Reprinted from Ainsworth 1976 with permission.).

The vowel perceived is affected by the neighboring consonants as well as the tempo and stress of the sentence. If these elements are factored out, speech studies have shown the perceived vowels with similar F2 frequencies can be altered by duration. For example, the short vowel /ɪ/ becomes /i/, and /ɒ/ becomes /ɔ/ (Ainsworth 1976). Thus it was important for cochlear implant speech-processing strategies to provide clear information on duration as well as the second and subsequently the first formant peaks for good speech understanding. The effect of duration was seen in the initial speech-processing studies on electrical stimulation, and confusions between the long (/ɔ/, /ɑ/, /i/) and short (/ɒ/, /ʌ/, /ɪ/) vowels with the inaugural University of Melbourne F0/F2 speech-processing strategy were rare (Tong et al 1979). The steady-state frequency cues are responsible not only for vowel identification but also for the direction and rate of change in formants (Lindblom and Studdert-Kennedy 1967). This is due to the fact the articulators may not reach the required position before the next consonant is initiated (coarticulatory effects). Cochlear implant speech processors and implants should provide this time-varying information.

It was also of importance for cochlear implant speech processing to know what the tolerated variability in F1 and F2 frequencies for the correct identification of individual vowels would be, given that the voicing frequency was constant. A study by Ainsworth and Millar (1972) using synthetic vowels found a standard deviation of 110 Hz for F1 and 183Hz for F2 from the centers of the vowel areas. This information was relevant to the optimal spacing of electrodes for the place coding of the formant frequencies.

Finally, the difference limens (DLs) for the sound parameters of vowels have been determined to relate perceptual discrimination to vowel identification (Flanagan and Saslow 1958). The DL for fundamental frequency was 0.3% to 0.5%. This is the same as the DL for pure tones (Nordmark 1968). Flanagan (1957) found the DL for formant frequencies depended on their proximity, but was on the order of 3% to 5% of the frequency of both F1 and F2. This was similar to the accuracy of vowel identification (Ainsworth and Millar 1972), and suggested a similar mechanism was involved. The DL for the amplitude of the F2 of the synthesized vowel /æ/ was 3 dB, which is the same order of magnitude as the DL for the amplitude of pure tones (Flanagan 1957). This corresponded in relative terms to about 40% of the second formant amplitude (Ainsworth 1976).

From the above studies it can be seen that the discrimination of the fundamental frequency measured as a percentage is an order of magnitude better than the formant frequency, and the discrimination of the formant amplitude was poor (Ainsworth 1976). This information was taken into consideration in the design of the University of Melbourne multiple-channel implant, where there was more provision for changes in the rate of stimulation than amplitude (Clark et al 1977, 1978).

Vowels also have timbre and this was investigated by Plomp (1975, 1976), who applied a multidimensional analysis to the steady-state frequency spectra of Dutch vowels, aiming to determine the number of dimensions that could adequately describe differences between vowel timbres. It was found that much of the distance in sample space could be accounted for with only two dimensions, which were related to the strength of the first two formants.

Consonants

Consonants are speech segments like vowels, but they convey up to 80% of the information required for intelligibility. They are usually of shorter duration, can vary more rapidly over time, and have more cues for their identification. In contrast, the transitions of consonants can be as short as 5 to 8 ms (Lehiste and Peterson 1961). The consonants in English are listed in Table 7.2, with their international phonetic alphabetical transcripts.

Classifications

Consonants can be divided into five groups according to their method of articulation, which in turn reflects their acoustic features. The groups are nasals, semivowels, plosives, fricatives, and affricates. The nasals are / m, n, ŋ /, the semivowels / w, j, r, l /; the plosives / p, t, k, b, d, g /, the fricatives / s, ʃ, f, θ, z, ʒ, v, ð /, and the affricates / tʃ, dʒ /.

With the nasals the sound is voiced with the soft palate lowered, so that the nasal cavity forms an open resonating tube and the oral cavity becomes a closed resonating side arm. The length of the side arm determines the frequencies of the sound. If it is closed at the lips it is an /m/, with the tongue at the roof of the mouth an /n/, and with the tongue against the soft palate an /ŋ/. With the semivowels (the glides /w/ and /j/) the articulators are initially in the position of a vowel, and then they move rapidly to a following vowel. The semivowel liquids /r/ and /l/ can be held in continuous articulation. With the plosives or stop con-

TABLE 7.2. Consonants.

Phoneme	Class	Example
/b/	Voiced plosives	bat
/d/		dog
/g/		get
/p/	Voiceless plosives	pig
/t/		tell
/k/		kick
/m/	Nasals	man
/n/		null
/ŋ/		sing
/w/	Semivowels (glides)	well
/j/		you
/r/	Semivowels (liquids)	ran
/l/		let
/h/	Voiceless fricatives	hat
/f/		fix
/θ/		thick
/s/		sat
/ʃ/		ship
/v/	Voiced fricatives	van
/ð/		this
/z/		zoo
/ʒ/		azure
/dʒ/	Affricates	joke
/tʃ/		chew

Based on Ainsworth (1976).

sonants, the vocal tract is closed at some point for a short period of time. The pressure builds up and then it is suddenly released. They are distinguished by the place of closure (bilabial, alveolar, and glottal), and the manner of the release (voiced or unvoiced). With the fricatives a narrow constriction is produced in the vocal tract, and air is forced past this to create a turbulent sound that can be voiced or unvoiced. With an affricate, a stop consonant release is followed by frication to form a single consonant.

Consonants have been classified by Miller and Nicely (1955) into five categories based on acoustic as well as articulatory features. These have proved very useful in evaluating cochlear implant speech-processing strategies to see how well phonetic information is transmitted and perceived (Dowell et al 1982). The categories are (1) voicing with /b, d, g, v, ð, z, ʒ, m, n/ voiced and /p, t, k, f, θ, s, ʃ/ unvoiced; (2) nasality with /m, n, ŋ/ having nasal resonance; (3) frication with /f, θ, s, ʃ, v, ð, z, ʒ/, where there is a turbulent flow of air past a constriction as compared with a complete obstruction to the air flow as in stops or nasals b, d, g, p, t, k, m, n/; (4) duration, with /s, ʃ, z, ʒ/ having a longer duration than the others; and (5) place, with /p, b, f, v, m/ classed as front, /t, d, θ, s, ð, z, n/ as middle, and /k, g, ʃ, ʒ/ as back according to the articulator placement.

Acoustic Cues

Studies have been undertaken to determine the specific acoustic cues or their combination that underlie the perception of consonant speech features. The studies manipulate the acoustic elements in natural speech, or create speech synthetically. This determines their effect on intelligibility. This approach can also be used to determine the information transmitted by electrical stimulation with cochlear implants. For example, if the fine temporal cues of speech are removed and the speech wave used to modulate noise, normal-hearing listeners still have cues available. In particular, these are envelope cues, conveying mostly information about phoneme duration, manner, and voicing.

It has been shown that the voice-onset time (VOT) distinguishes the voiced from unvoiced stop consonants (Liberman et al 1958; Lisker and Abramson 1967). The VOT is the time between the release of the vocal tract closure and the commencement of voicing. With voiced stops there is virtually no delay, but with voiceless stops it is on the order of 50 ms or more. The length of the vowel is also a vital cue for final consonant voicing.

The most important cue for the recognition of the voiceless stop is the frequency of the noise burst at the moment of release (Delattre et al. 1952; Liberman et al. 1956; Stevens and Blumstein 1981; Kewley-Port and Pisoni 1983). The burst of noise varies in duration from 5 to 50 ms. A /t/ is heard if it is above 3000 Hz, a /k/ if it is equal to F2, and a /p/ at other frequencies depending on the vowel (Delattre et al 1952). With voiced stops not only is the frequency of the burst important, but also the direction of the second formant transition (Liberman et al 1954). These too are short and in the range 5 to 50 ms.

The importance of F2 is illustrated for the syllables /ba/, /da/, and /ga/ in Figure 7.7. This shows that they can be distinguished on the basis of the F2 transitions. The formants or vocal tract resonances can be observed in the waveform, but determining the actual formant frequencies is difficult. As illustrated in Figure 7.7, a rising second formant is identified as /b/ and a falling one as /d/ or /g/. The rate of fall may distinguish the two as well as the locus of frequency

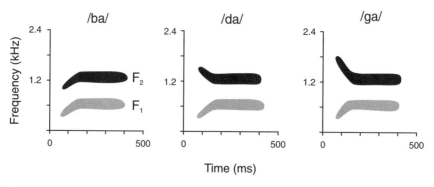

FIGURE 7.7. The first (F1) and second formants (F2) for the syllables /ba/, /da/, and /ga/.

from which the transition has arisen. The locus varies whether there is a following front or back vowel. There may also be more cues in the vowel sound as to the identity of a final rather than initial consonant (Wang 1959). As with vowels the cues for the perception of consonants adapt to or are influenced by the fundamental frequency of the speaker (Fourcin 1968), as well as to the rate of speaking (Summerfield and Haggard 1972).

Nasals have their strongest formant frequency quite low (approximately 300 Hz), because the open nasal tract is longer than the oral tract. The closed arm or oral cavity is a second resonating chamber, and there are other resonances (poles) in the 800- to 2300-Hz range (Fujimura 1962). There is also an antiresonance (zero) at 600 Hz for /m/. They can be distinguished from each other by the position of the poles and zeros in the steady part of the sound, or the formant transitions toward or away from the vowel (Ainsworth 1976).

Semivowels referred to as glides can be distinguished from other sounds by the duration of the formant transitions. For example, Liberman et al (1956) showed with the vowel /u/ that if the formant transitions were shortened to 150 ms, it was perceived as a /w/, and if they were shortened to 50 ms, it was a /b/. Studies also demonstrated that it was the duration rather than the rate of change in formant frequencies that was important. It was also shown by O'Connor et al (1957) that the main cue for the perceived semivowel was the initial frequency of the second formant, but it also depended on the following vowel. The other semivowels (the liquids /r/ and /l/) can be distinguished by formant location especially for /r/, which has a very low third formant (F3).

The voiced fricatives can be distinguished from the unvoiced ones as they have more energy in the low-frequency part of the speech spectrum. Furthermore, voicing striations (periodicity) can be seen in the spectrograms. They can be distinguished from each other by the spectrum of the frication and the transitions of the formants. The spectrum of the noise is the most important cue for the phonemes /ʃ/ and /s/ and their voiced equivalents /ʒ/ and /z/. With /s/ the frication energy is almost always above 3500 Hz, but the lowest frequency of frication energy for /ʃ/ is between 1600 and 2500 Hz and the amplitude pattern is weighted to the bottom of the spectrum (Strevens 1960).

Studies with synthesized sounds have found that consonant recognition is associated with categorical perception (Liberman et al 1957). With categorical perception, a perceived consonant changes to another over a short range of parameters. Discrimination between consonants is worst at this categorical boundary (Ainsworth 1976).

Prosody

The greater proportion of speech information is carried by the phonetic segments; vowels, diphthongs, and consonants. Additional information is also transmitted through suprasegmental or prosodic features; stress, tempo and rhythm, and intonation. Stress on the syllables in a word distinguishes alternative meanings, and stress on words in a sentence indicates the most important words. Tempo is the rate at which an utterance is spoken and rhythm the pattern of time intervals between the stressed syllables. Finally, intonation depends on the fundamental or voicing frequency. There are a number of intonation patterns, but a rising one is usually interpreted as a question and a flat or falling one a statement. The patterns were also described by Halliday (1963) as either rising, or falling, or combinations of rising and falling with low, middle, and high terminations. The voicing frequency is on average lower in males (120 Hz) than in females (225 Hz) and in turn lower than that of children (approximately 265 Hz) (Fry 1979). The range in the fundamental frequency is from 60 to 500 Hz.

Tonal Language

Changes in tone convey meaning in certain languages, whereas in others (e.g., English) they provide prosody as discussed above. Mandarin is a good example in which there are three tones that involve gliding movements. They can be described through reference to five frequency points within the normal pitch range of a speaker's voice, that is, low, half-low, middle, half-high, and high (Table 7.3). This system of classification was described by Chao (1930). The contour tone is then a movement from one of these points to another. A vertical line as shown in (Table 7.3) represents the normal range of a speaker's voice, and the graph to the left the pitch contour. Thus, for example, tone 1 is a high-level tone remaining at pitch level 5, and tone 2 is a rising one going from pitch level 3 to pitch level 5.

It was important to know that cochlear implant speech processors would be able to help in understanding all languages including the tonal ones. The University of Melbourne F0/F2 and subsequent Nucleus strategies provided these changes for speech perception as was first demonstrated by Xu et al (1987) for Mandarin. It has been studied for Cantonese with the Nucleus Spectral Maxima Speech Processor (SPEAK) and advanced combination encoder (ACE) strategies (Barry et al 2002a,b), and is discussed in the results.

Words and Sentences

Words convey meaning (semantic information), and they are assembled according to the rules (syntax) of the language as phrases or sentences. These grammatical rules are referred to as syntax. Meaning can be conveyed even if the subject cannot recognize all the words in a sentence. They are helped by their knowledge of

TABLE 7.3. Tonal contrasts in Mandarin.

Tone number	Description	Pitch	Tone letter	Example	Gloss
1	High level	5-5	˥	mā	mother
2	High rising	3–5	˦	má	hemp
3	Low falling rising	2-1-4	˧	mǎ	horse
4	High falling	5-1	˥	mà	scold

Based on Ladefoged (1975).

grammar and the context of the conversation. People vary in these top-down processing skills that enable them to understand degraded speech signals. This is the reason tests that score words recognized in sentences [e.g., Central Institute for the Deaf (CID) Words in Everyday Sentences] are easier with higher scores than phonetically balanced (PB) monosyllabic words.

Binaural Hearing

Binaural hearing is important for locating a speaker in space and for hearing speech in noise (Hirsh 1950; Pollack and Pickett 1958). There is also an increase in acuity from loudness summation. Improving the recognition of a signal in noise is referred to as the binaural release from masking or masking level difference, and has been demonstrated in a number of studies (Licklider 1948; Hirsh 1950; Pollack and Pickett 1958; Carhart 1965; MacKeith and Coles 1971; Peissig and Kollmeier 1997; Litovsky and Colburn 1999). It has been shown for diotic signals in dichotic noise, and for dichotic signals in diotic noise (Licklider 1948). Diotic sounds reaching the ear are the same, and dichotic sounds are different.

The ability to localize sound and hear speech in noise with bilateral hearing is achieved by the coding of interaural time delays and intensity differences between the two ears, as was discussed in Chapter 5. The localization of sound can also occur with a single ear and is due to the introduction of harmonics by reflections from the folds in the pinna. These differ depending on whether the sound is from the front or back.

Binaural release from masking (Licklider 1948), also referred to as the squelch effect, has been thought to be due to the fact that the noise signals in each ear are uncorrelated and the speech signals are correlated or vice versa. The neuro-physiological mechanisms underlying the phenomenon are not clear, but could be due to the fact that the phase locking of a signal is robust in noise, and also due to the cross correlation model of Jeffress (1948) for binaural interaction as shown in Figure 5.16 in Chapter 5.

Acoustic Models of Cochlear Implant Speech-Processing Strategies

Channel Vocoders and Fixed Filters

Systems for analyzing, transmitting, and synthesizing speech, such as vocoders, have provided a basis for the subsequent development of cochlear implant speech processors. In telephone or communications engineering, it was learned that speech information could be compressed and transmitted to a receiver, where it could be reassembled without much degradation of the signal This compression allowed speech to be transmitted more efficiently and economically.

One system for compressing information was the channel vocoder arising from work by Dudley (1939) (Fig. 7.8). This vocoder consisted of a set of band-pass

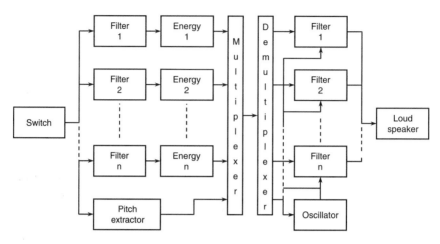

FIGURE 7.8. A block diagram of the channel vocoder (Dudley 1939).

filters covering the frequency range of speech. The amplitude of the signal in each of the filter bands was measured continuously, as was the fundamental frequency of speech (F0). A decision was made as to whether the sound was voiced or unvoiced. The amplitudes of the signals from the band-pass filters and F0 were multiplexed (mixed) and transmitted to the receiver. In the receiver the parameters were de-multiplexed (separated), and the amplitude parameters used to control the gains of a similar set of band-pass filters, which were excited by a pulse train controlled by F0 or a noise source. The number of filter bands or channels required for intelligible speech was 10 to 20. Hill et al (1968) reported good consonant and vowel recognition for eight channels of spectral information. However, the speech was not perfectly intelligible. This was largely due to the difficulty in some circumstances of making a decision between voiced and unvoiced signals, for example with the mixed excitation of the vocal tract in the case of a voiced fricative. This was especially difficult for female voices, and in a noisy reverberant room with competing speech. It was also difficult to track the fundamental frequency. To avoid this problem, other devices such as the voice-excited vocoder were developed (Schroeder and David 1960). With this system the low-frequency part of the spectrum was transmitted intact and used to excite the filters. This improved the quality of speech, but a greater bandwidth was needed for transmission (Ainsworth 1976).

With the channel vocoder the band-pass filter width needed only to be as wide as the critical band for optimal speech transmission. As there are 14 to 19 critical bands over the speech frequency range, this was also thought to be the limit for the number of electrodes required for the multiple-electrode cochlear implant speech-processing strategies (Clark, Black et al 1977; Clark, Tong et al 1977).

It must be pointed out that the fixed filter schemes for electrically stimulating the cochlear nerve described below are not strictly channel vocoders, as they usually do not specifically extract the voicing frequency. The fundamental frequency is a product of the outputs from the filters. The optimal number of filters for channel vocoder and fixed filter strategies is discussed below.

Formant Vocoders

As discussed above, formants are concentrations of frequency energy of importance for intelligibility. The formant vocoder was an advance in communication systems as the data rate was much less than when transmitting an unprocessed speech signal. Speech was first analyzed with a voiced/unvoiced detector, and a pitch detector measured the frequency of the glottal openings. Information about the frequencies and amplitudes of the formants was obtained from a bank of band-pass filters and envelope detectors.

Formant vocoders were first developed by Lawrence (1953) [Parametric Artificial Talker (PAT)], and Fant and Martony (1962) [Orator Verbis Electris (OVE II)]. These were adequate for the transmission of vowels, but for some consonants the resonances (poles) and antiresonances (zeros) needed to be specified. This formant vocoder required less bandwidth than the channel vocoder, but was not used in communication systems because of the complexity of the circuitry. However, it has been very useful for studies on speech perception (Ainsworth 1976). The design of a formant synthesizer is illustrated in Figure 7.9.

The formant vocoder became the basis for the formant speech-processing strategies used with multiple-electrode stimulation discussed below, and in Chapter 8.

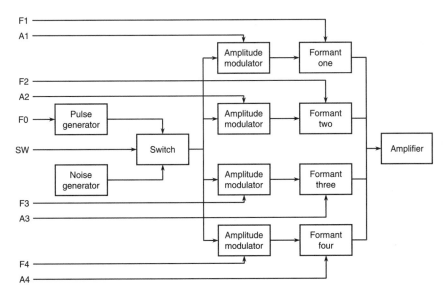

FIGURE 7.9. Block diagram of a parallel-coupled formant synthesizer (Reprinted from Ainsworth 1976 with permission.).

Acoustic Representation of Electrical Stimulation

An acoustic model to evaluate multiple-electrode speech-processing strategies was developed by Blamey et al (1984a,b). The model used a pseudo-random white noise generator with the output fed through seven separate band-pass filters with center frequencies corresponding to the electrode sites (Blamey et al 1984a,b). This model was first evaluated psychophysically for pulse rate difference limens; pitch scaling for stimuli differing in pulse rate; pitch scaling and categorization of stimuli differing in filter frequency (equivalent to electrode position); and similarity judgments of stimuli differing in pulse rate as well as filter frequency (electrode position). The results for the acoustic model were comparable to those obtained with electrical stimulation on implant patients, and were discussed in Chapter 6.

Having established that the acoustic model gave similar psychophysical results to those for multiple-channel electrical stimulation, a further study was undertaken to see if similar results could be obtained for an acoustic model of the fundamental (F0) and second formant (F2) speech processor to those for electrical stimulation with the same strategy in the Nucleus multiple-electrode system. Speech perception tests were administered in hearing alone, speech reading alone, and speech reading plus hearing conditions. The scores for the first University of Melbourne cochlear implant patient, and the three normal-hearing subjects using the acoustic model are shown in Table 7.4 (Blamey et al 1984a,b). There was good correspondence between the speech tests (male–female speaker, question–statement, vowels, final and initial consonants, AB (Arthur Boothroyd) words and phonemes (Boothroyd 1968), and CID sentences (Davis and Silverman 1970), for a multiple-channel cochlear implant patient using the F0/F2 speech processor and subjects using the F0/F2 model (Clark 1987). The acoustic model and cochlear implant performances were also compared on the basis of the percentage information transferred for each speech feature on a 12-consonant test (Table 7.5). The consonants were /b, p, m, v, f, d, t, n, z, s, g, k/. The speech features were voicing, nasality, affrication, duration, and place. The results for the first implant

TABLE 7.4. Speech perception test scores (%) for the F0/F2 cochlear implant speech processor and an acoustic model (mean scores for three subjects).

Test	Implant patient hearing alone	Subjects with acoustic model
Male–female speaker	24	17
Question statement	19	16
Vowels	29	23
Final consonant	30	29
Initial consonant	21	28
AB words	2	0
Phonemes	12	7
CID sentences	7	4

CID, Central Institute for the Deaf; HA, hearing or electrical stimulation alone. Based on Blamey et al (1984a,b) and Clark (1987).

TABLE 7.5. Information transmission (%) for the F0/F2 cochlear implant speech processor and an acoustic model.

Feature	Electrical stimulation alone	Acoustic model
Voicing	36	39
Nasality	25	68
Affrication	38	31
Duration	79	85
Place	25	15
Overall	46	45

Based on Clark (1987).

patient and the average results for the three subjects are shown in Table 7.5. It can be seen that the trends are very similar for both electrical stimulation and the acoustic model except for better transmission of nasality with the acoustic model. This could have been due to the fact nasal sounds are distinguished from other speech sounds by a strong formant at about 200 to 300 Hz, and the information from F2 would not have provided this spectral detail. It could also have been due to better representation of the antiresonances (zeros) with the acoustic model.

As the acoustic model on normal-hearing subjects proved to be good at reproducing the speech and speech feature results for the F0/F2 speech-processing strategy, a study using the acoustic model was undertaken to determine whether an F0/F1/F2 speech-processing strategy (F1, first formant) would give better results than the F0/F2 processor, and to what extent additional information due to F1 would be transmitted. An additional strategy was also evaluated in which F2 was coded as rate of stimulation. A confusion study on the 11 Australian English vowels was carried out on six subjects to determine the information transmission for the vowels grouped according to duration, F1 and F2. The results in Table 7.6 show there was a small increase in the total information transmitted for the F0/F1/F2 strategy. The F0/F1/F2 strategy was the only one that transmitted a large proportion of the F1 information. A much greater proportion of the F2 information was transmitted when coded as filter frequency rather than as pulse rate.

With consonants the acoustic model of the F0/F1/F2 speech processor led to better transmission of the voicing, nasality, affrication, duration and amplitude envelope features, but not place of articulation than for the F0/F2 strategy (Table 7.7). The F2 (rate) had poorer results than F0/F2 for place of articulation and high F2. The addition of F1 would provide the low-frequency information for identifying voicing through the VOT and a rising F1, as well as the essential cues for nasality. Further cues for duration and amplitude envelope would be provided by the greater energy in F1. Amplitude envelope information (Blamey et al 1985) improved significantly as well, and as a result so did information on manner of articulation. The speech-processing strategies were also compared for connected discourse using the speech-tracking test. The F0/F1/F2 strategy was superior to the others, and the F0/F2 strategy superior to the F2 (rate) strategy.

TABLE 7.6. Acoustic model: comparison of speech-processing strategies—information transmission for vowels.

	F2 (rate) (%)	F0/F2 (%)	F0/F1/F2 (%)
Total	34	56	72
Duration	83	85	95
F1 grouping	12	27	81
F2 grouping	25	68	55

Based on Blamey et al (1985).

Having shown with an acoustic model of multiple-electrode stimulation that the F0/F1/F2 speech-processing strategy was better than the F0/F2 strategy, it was implemented as the Nucleus F0/F1/F2 WSP-III system, and a clinical study carried out to determine if the same would apply to cochlear implant patients. This would not only be a good test of the predictive value of the model, but more importantly further support the rationale for the multiple-electrode speech-processing strategy, which still had not been completely proven over single-channel devices at this time.

In comparing the results for the F0/F2 and F0/F1/F2 cochlear implant speech processors, it was considered important to analyze the information received by the patient and whether the information transmitted was consistent with the type of speech-processing strategy used. The percentage information transmitted for vowels and consonants was determined for a group of 13 patients with the F0/F2 processor and seven patients with the F0/F1/F2 processor. For vowels the scores were 51% (F0/F2) and 64% (F0/F1/F2). For consonants the scores were 36% (F0/F2) and 50% (F0/F1/F2). The information transmitted for duration, F1, and F2 was greater for the F0/F1/F2 strategy. From Table 7.8 it can be seen that for consonants, information transmission was also better for the F0/F1/F2 speech processor compared to the F0/F2 processor for all speech features (Clark 1987). The information transmission was calculated from a confusion study on the consonants /p, t, k, b, d, g, m, n, s, z, v, f/. The information transmission was for the features of Miller and Nicely (1955), and an additional two features, the amplitude envelope and high F2. The amplitude envelope feature classified the consonants into four groups, as shown in Figure 7.10. These groups were easily recognized visually from the traces of the amplitude envelopes produced by the real-time speech processor. The high F2 feature refers to the output of the speech processor's F2 frequency extraction circuit during the burst for the stops /t/ and /k/ or during the frication noise of /s/ and /z/. /f/ and /g/ did not give rise to the feature because the amplitude of the signal was too low during the period the F2 frequency was high. Thus the F2 feature was a binary grouping with /t, k, s, z/ in one group and the remainder of the consonants in the other (Blamey et al 1985).

As the results for information transmission for vowels and consonants for multiple-electrode stimulation were similar to those obtained for the acoustic model, it confirmed the predictive value of the acoustic model. The features for acoustic

TABLE 7.7. Acoustic model: comparison of F2 (rate), F0/F2, and F0/F1/F2 speech-processing strategies—information transmission for consonants.

	F2 (rate) (%)	F0/F2 (%)	F0/F1/F2 (%)
Total	37	43	49
Voicing	35	. 34	50
Nasality	86	84	98
Affrication	31	32	40
Duration	62	71	81
Place	19	28	28
Amplitude envelope	47	46	61
High F2	48	68	64

and electrical stimulation could be compared to the speech perception data presented below and in Chapter 12.

Speech Cues

The importance of different speech cues in perception can be examined by presenting natural or synthetic speech without these cues. This helps determine how important they are for reproducing with electrical stimulation. The acoustic representation of electrical stimulation can also help in optimizing fixed-filter speech-processing strategies as well as formant processors.

The importance of speech wave envelope cues can be studied by using them to modulate noise, thus separating them from spectral and fine temporal information. The fine temporal information is, for example, phase and frequency modulation. The envelope cues convey information mostly about phoneme duration, voicing, and manner. Rosen (1989) transformed speech wave envelopes into "signal-correlated noise," as described by Schroeder (1968). This was equivalent to multiplying the envelopes by white noise, resulting in a signal with an instantaneous amplitude identical to that of the original signal, but with a frequency spectrum that was white. It was found that manner distinctions were present for sampling rates down to 20 Hz. Thus the cues from the amplitude envelope, as shown in Figure 7.10, could be defined at these low frequencies. Voicing was best with unfiltered speech or when filtered with a cut at 2000 Hz. Place recognition was poor. Similar information transmission to "signal-correlated noise" was obtained for the single-electrode cochlear implant (3M, Los Angeles) (Van Tasell et al 1987). With this system, as discussed below, the speech signal was filtered over the frequency range of 200 to 4000 Hz, and the output modulated a 16,000-Hz carrier wave. At 16,000-Hz there would be no fine time structure in neural firing, and the information would be from the amplitude variations.

Cues for consonant recognition are not only from frequency spectra (provided by multiple-electrode stimulation) but also from the fine time variations in the amplitude envelopes. These variations were studied with speech processors based on an acoustic model of electrical stimulation (Blamey et al 1985, 1987). The

TABLE 7.8. Consonant speech features for the F0/F2 and F0/F1/F2 speech-processing strategies.

	Voicing	Nasality	Affrication	Place	Amplitude envelope	High F2
F0/F2 (A0) ($n = 13$)	33	38	36	20	36	36
F0/F1/F2 ($n = 7$)	56	49	45	35	54	48
% increase	70	29	25	75	50	33

A0 is the amplitude of the whole speech wave envelope. Based on Clark (1987).

groups of consonants on the basis of the envelope variations were unvoiced stops or plosives, unvoiced fricatives, voiced fricatives and stops together, and nasals (Fig. 7.10). Within these groups, the distinctions of place of articulation must also be made with other coding mechanisms. The amplitude envelope cues are available for cochlear implant patients (Blamey et al 1987; Dorman et al 1990). They may be especially important for those who have poor electrode place identification, and so do not receive the spectral shape of speech. Research also suggested these cues might be used by those with hearing aids (Van Tasell et al 1987).

Studies by Erber (1972) and Van Tasell et al (1987, 1992) have shown that an essential cue for consonant place perception is the distribution of speech energy across frequency. Acoustically this is represented by both place coding and the fine temporal coding of frequency in the frequency bands. With the present methods of electrical stimulation, as was discussed in Chapter 6, the temporal resolution is very limited. Consequently, with the cochlear implant the coding of place of stimulation becomes the primary cue. However, as was discussed in Chapter 6, the correlation between electrode place discrimination and the place speech feature recognition is not as good as expected.

Channel Numbers

The number of stimulus channels required to transmit speech information is important for understanding how to optimize multiple-electrode stimulation. Shannon et al (1995) and Turner et al (1995) used acoustic models to study, in particular, the speech information transmitted by fixed filter speech-processing schemes, to assess the optimal number of filters to be used as well as the number of electrodes to be stimulated. The research first studied the effects of modulating high-pass and low-pass noise, divided at 1500 Hz, with the speech wave envelope. This showed almost 100% recognition of voicing and manner cues, but the two channels provided only limited speech understanding. Information transmission analysis showed that the addition of a third and fourth band improved place of articulation. Shannon et al (1995) found that with a four-channel processor normal-hearing listeners could obtain near-normal speech recognition in quiet listening conditions. This suggested to the authors that only four channels may be required for good speech recognition with a cochlear implant. Furthermore, in a study in normal-hearing listeners by Dorman et al (1997), in which the amplitudes of the center frequencies of increasing numbers of filters were used to represent speech, it was found that four filters would provide greater than 90% speech

[vowel   ] [consonant] [vowel   ]

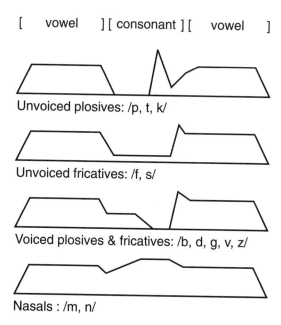

Unvoiced plosives: /p, t, k/

Unvoiced fricatives: /f, s/

Voiced plosives & fricatives: /b, d, g, v, z/

Nasals : /m, n/

FIGURE 7.10. Schematic diagrams of the amplitude envelopes for the grouping of consonants from inspection of the outputs of speech processors using an acoustic model of electrical stimulation (Reprinted with permission from Blamey et al 1985. A comparison of three speech coding strategies using an acoustic model of cochlear implant. *Journal of the Acoustical Society of America* **77**: 209–217.).

perception accuracy in quiet. The data indicate that speech understanding in quiet is in part due to a fluctuating spatially distributed pattern of neural responses to amplitude variations in the speech signal. The study did not address the importance of the fine temporal or frequency information in each channel for both naturalness and intelligibility especially in noise.

The interaction of the limited spectral channels and associated temporal envelope cues was studied for four filtered bands of speech by Shannon et al (1998). The envelope from each speech frequency band modulated a band-limited noise. It was found that significant variation in the cutoff frequencies for the bands, or an overlap in the bands that would simulate current interaction with a cochlear implant, produced only limited deterioration in speech recognition. However, it was essential for the temporal envelope cues to be those derived from the same frequency band as the noise being modulated.

In a study by Fu and Shannon (1999) the temporal envelopes from 4, 8, and 16 band-pass filters were used to modulate noise bands shifted in frequency relative to the tonotopic representation of spectral envelope information. It was found that the frequency of the bandwidth and envelope cues did not interact, and were therefore independent in their effect on intelligibility for a shift equivalent to 3 mm along the basilar membrane, that is, a frequency shift of 40% to 60%.

The temporal information from the amplitude-modulated speech wave in the presence of reduced spectral information was studied by varying the low-pass cutoffs (Shannon et al 1998, 1999). No change was observed in vowel, consonant, or sentence recognition for low-pass filter cutoffs above 50 Hz. It was only when the envelope fluctuations between 20 and 50 Hz were removed that a marked reduction in phoneme discrimination occurred. This indicated that in the previous studies of Blamey et al (1987) and Van Tasell et al (1987) on the importance of amplitude envelope patterns for consonant recognition, only a frequency resolution below 50 Hz was required. The data also indicated the upper frequency limit required to refresh the neural patterns for the recognition of vowel spectral information. For cochlear implants the data help determine the rate of stimulation required to represent the amplitude variations in speech and the update rate of information by the hardware.

Speech in Noise

A study was undertaken by Dorman et al (1998) to investigate the number of filter bands for speech perception in noise using a model of the Med El Combi-40 implementing the continuous interleaved sampler (CIS) speech-processing strategy (Hochmair and Hochmair-Desoyer 1983) on normal-hearing listeners. The current outputs of the filters with center frequencies distributed on a logarithmic scale from approximately 160 to 5200 Hz were used. The results showed that at $+2$ dB signal-to-noise ratio (SNR), the maximum speech recognition was achieved with 12 stimulus channels, and at -2 dB SNR the performance maximum occurred with 20 channels of stimulation. For the same strategy the maximum performance in quiet was with five channels. The results suggest the importance of having adequate stimulus channels for electrical stimulation particularly in noise. This is supported by the evidence obtained earlier for cochlear implants with the F0/F1/F2 compared to the F0/F2 strategy (Dowell et al 1987a).

Channel Selection

With electrical stimulation it is also important to determine the frequency-to-electrode mapping. In what frequency region of the cochlea should the electrodes be concentrated, and how should they be spaced? The contributions of frequencies to speech understanding were initially investigated by Fletcher and Steinberg (1929), who found that 1500 Hz was the frequency around which low- and high-frequency contributions to speech recognition were equal.

A key to the analysis of the contribution of different frequencies to speech understanding is the Speech Intelligibility Index (SII) theory that was developed by Fletcher and Steinberg (1929) and French and Steinberg (1947). It has important application to the assessment of hearing loss and the optimization of cochlear implant speech-processing strategies. It is a measure of the amount of information in the speech signal available to the listener. It is defined by the following equation:

$$SII = \sum_{i=1}^{n} I_i \times W_i$$

where n is the number of frequency bands, and I_i and W_i are the values associated with the frequency band (i) of the importance function (I) representing the relative contribution of different frequency bands to speech perception, and the audibility function W representing the effective proportion of the dynamic range audible within each band. SII has been used by a number of researchers to determine the speech perception of listeners with a sensorineural hearing loss (Skinner et al 1982; Dirks et al 1986; Pavlovic et al 1986).

Electrical Stimulation: Principles

Processing speech for electrical stimulation of the cochlear nerve should ideally present the information used by people with normal hearing, and their neural pathways are interconnected to process the information. An adjustment of neural connectivity occurs in young children after exposure to speech to facilitate the processing.

In presenting speech to the central auditory pathways by electrical stimulation of the cochlear nerve, the normal transduction mechanisms in the intact inner ear are bypassed. Physiological and psychophysical studies (see relevant chapters) have shown the limitations of reproducing the coding of speech frequencies and intensities through electrical stimulation. This created an electroneural bottleneck between the world of sound and the central auditory nervous system, as was discussed in more detail in Chapter 5. Solutions to this problem were to analyze the most important speech information and optimize its transmission through the bottleneck. Nevertheless, this required transmitting the information by attempting to reproduce the coding sound. Cochlear implant speech processing had to use a multiple-electrode implant to transmit sufficient information through the bottleneck (Fig. 7.11). Speech perception has been achieved with studies using electrical stimulation as discussed below, and helped through the acoustic model studies of electrical stimulation discussed above.

The perception of speech incorporates both bottom-up and top-down processing of information. Bottom-up is the transmission of perceived sound and its features up the brain central pathways. Top-down is the anticipation of words and syntax/semantic influences applied by knowledge of the context and the language. The bottom-up processing codes the complex sounds or elements of speech in the central auditory pathways. There is a complex pattern of neural activity underlying speech perception consisting of (1) time-varying changes in the number of neurons firing in spatially distributed groups at different intensities, and (2) fine temporal activity within and across groups. The fine temporal component in the pattern is supported by the study of Remez et al (1981). In this study time varying patterns of sine waves were produced to represent the center frequency of the one to three formants in speech every 15 ms, as well as their amplitudes. In the signal

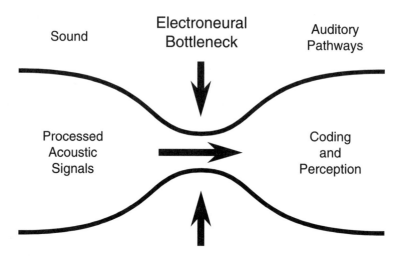

FIGURE 7.11. A diagram showing how the cochlear implant acts as an electroneural bottleneck between sound and the coding mechanisms in the central auditory pathways.

there were no formant frequency transitions, and no fundamental frequency changes. With three frequencies most words were recognized, but the signal was not speech-like. In contrast, top-down processing is achieved through processes in the primary auditory cortex, association areas, and other cognitive centers.

Channel Numbers

The number of stimulus channels required to transmit speech information has been evaluated with acoustic models as referred to above, but ultimately requires validation with electrical stimulation on cochlear implant patients. The Nucleus formant processors extracted peaks of frequency energy, and there was a need to vary their position along the array. Furthermore, as distinct from fixed-filter electrical stimulation the Nucleus F0/F2, F0/F1/F2 (Clark, Tong et al 1978; Tong et al 1979, 1980; Clark and Tong 1981) and Multipeak (Dowell et al 1990) strategies presented the voicing (F0) frequency at each electrode. The F0/F2 strategy extracted the second formant frequency (F2) and coded this as place of stimulation, the fundamental (F0) as rate of stimulation, and the amplitude of F2 as the current level (A2) (Clark, Tong et al 1978). The F0/F1/F2 coded the first formant (F1) as place of stimulation as well. The Multipeak is a misnomer, as it extracted not only the F1 and F2 peaks, but also the energy in fixed filters in the bands (2000–2800 Hz; 2800–4000 Hz, and >4000 Hz), together with voicing as rate of stimulation.

Holmes et al (1987) found that open-set word recognition and continuous discourse tracking results for the Nucleus F0/F1/F2 speech processor increased using up to 15 active electrodes. The correlation between electrode number and open-set CID word-in-sentence scores was examined statistically for a combined group of patients at the University of Melbourne clinic with the F0/F2 ($n = 16$) and F0/F1/F2 ($n = 48$) speech processors (Blamey et al 1992). The minimum number of electrodes was 9 and the maximum 21. There was a positive correlation between speech perception and the number of electrodes in use. The regression analysis showed the difference between 9 and 21 electrodes (12 electrodes) accounted for a 24% increase in score (i.e., 2% per electrode). Thus the additional electrodes would be of marked benefit to the patient.

The number of stimulus channels for a fixed-filter (modified channel vocoder) was examined by Dorman et al (1989). They used a ball electrode array with analog and monopolar stimulation. They compared consonant recognition for 1, 2, 3, and 4 channels of stimulation. In the initial group of six subjects the mean scores were channel 1, 23%; channels 1, 2, 27%, channels 1, 2, 3, 49%; and channels 1, 2, 3, 4, 55%. A similar trend was seen for other subjects in that a low- and high-frequency channel provided the most information. It was unclear from this study the nature of the information of importance for consonant recognition. It suggested, however, that either temporal information or noise in specific frequency bands is quite crucial.

With the Nucleus SPEAK strategy as distinct from the F0/F2, F0/F1/F2, and Multipeak strategies, six or eight spectral maxima rather than frequency peaks were selected from a bank of 20 band-pass filters (McKay et al 1991). A continuous stimulus rate was used on each channel, so there was no fine temporal information transmitted on each channel. In a study by Fishman et al (1997) using the SPEAK strategy, the outputs of varying numbers of adjacent filters were used to stimulate electrodes ranging in number from 1 to 20. Performance increased dramatically up to four electrodes, but no difference was seen for 7, 10, or 20 electrodes. The findings were similar to the optimal number required for the fixed filter strategies typified by the CIS (Wilson et al 1992; Battmer et al 1994).

This does not mean that only four electrodes are required for adequate performance of the SPEAK or formant vocoder strategies. It was shown that 20 rather than eight electrodes gave improved performance with the Multipeak and SPEAK strategies especially in noise (Blamey et al 1992). They also give the ability to select the electrodes transmitting the most useful information as well as allowing the pattern of electrodes to be altered in the presence of reduced neural populations in certain regions of the cochlea or spread of current to the facial nerve (McKay et al 1994).

Channel Selection

With electrical stimulation it is also important to determine the frequency-to-electrode mapping. In a study with the Nucleus F0/F1/F2 strategy, Kileny et al (1992) showed there was no significant difference in speech recognition between a processor that used a full 20 electrode array and one that used only the basal 10 electrodes. This may in part be explained by the finding of Blamey et al (1995) that electrodes in the basal turn stimulated neurons with lower pitches than the place of excitation, and these would be in the speech frequency range. The effects of plasticity occurring after 1 month could also have been a factor, as discussed in Chapters 5 and 11. Geier and Norton (1992) found using the Nucleus F0/F1/F2 strategy that the removal of the five most apical electrodes gave reduced speech recognition in three of five subjects. Collins et al (1994) also found in some patients that if electrodes that were not discriminated were removed from the mapping, there was an improvement in speech scores. Hanekom and Shannon (1996) discovered with the Nucleus SPEAK Spectra-22 system that speech recognition was a function of which seven electrodes were selected. Lawson et al (1996) found a larger difference between two different selections of six electrodes than between six and 20 electrodes.

The above studies on frequency-to-electrode mapping or channel selection suggest that the distribution of stimulus channels along the cochlea is likely to be of greatest importance in the low- to midfrequency range for both vowels and consonants. This applies to the F1 (frequency range is from approximately 250 to 700 Hz) and the F2 (range from 700 to 2300 Hz) frequencies. However, the number of channels and the tolerable overlap in electrical fields still has to be determined.

In a further study on cochlear implant patients (Henry et al 2000) the SII (see Channel Selection, above) was used to measure the amount of speech information in five frequency bands (170–570 Hz, 570–1170 Hz, 1170–1768 Hz, 1768–2680 Hz, and 2680–5744 Hz) by 15 users of the Nucleus SPEAK Spectra-22 system. Random variations in loudness were introduced into the signal to make the test more difficult and more like everyday conditions. Relative to normal-hearing subjects, speech information was significantly more reduced in the four frequency regions between 170 and 2680 Hz than in the region 2680 to 5744 Hz. There was also a significant correlation between electrode discrimination ability and the amount of speech information received in the regions between 170 and 2680 Hz for intensity variations over 20% or more of the dynamic range. There was no correlation in the region 2680 to 5744 Hz. The results indicated that speech information in the low- to midfrequency regions of the cochlea is most critical for implant patients, and their recognition of speech correlated with electrode discrimination in this region. Fine spectral discrimination may be more important in the vowel formant regions than in the higher frequency regions. This study emphasizes that it is important to select the outputs of the frequency bands carrying the greatest amount of information for stimulating electrodes.

Speech in Noise

Speech is corrupted by noise in part because of the voicing decision required for channel and formant vocoders. It was also shown by Miller and Nicely (1955) that the consonant place feature was most affected by noise while voicing and nasality were the least corrupted. With the cochlear implant the effect of noise on the voicing decision was seen when comparing the Nucleus F0/F2 and University of Utah (Ineraid) strategies. The Nucleus strategy used a voicing detector, and the University of Utah had a four fixed-filter scheme but no voicing decision. The open-set speech perception results were similar, but in multispeaker babble there was a trend for degradation in the performance of the F0/F2 processor at a 10-dB SNR compared with the Utah system (Gantz et al 1987). It was only when F1 was coded as well as F2 in the Nucleus system that the results were the same (Cohen et al 1993). The addition of F1 allowed VOT and F1 transition cues to be used by the brain, rather than have an algorithm make the decision. When the F0/F2 and F0/F1/F2 processors were compared in five patients with four-choice spondee words and competing four-speaker babble, the results were significantly better for the F0/F1/F2 processor at 0-dB and 10-dB SNRs (Dowell et al 1987b).

The number of stimulus channels is important for speech perception in noise. Blamey et al (1992) have shown that 20 rather than eight banded electrodes provide improved speech processing for both the Nucleus Multipeak and SPEAK strategies in quiet and especially in noise. Thus more electrodes or channels of stimulation provide the additional spectral information to assist with speech recognition in adverse noisy conditions.

Speech-Processing Strategies

Speech-processing strategies for electrical stimulation have originated to a certain extent from the auditory neurophysiological, psychophysical, and speech sciences. The evaluation of speech-processing schemes has provided an understanding of how responses to electrical simulation differ from those to sound. This has led to not only effective speech recognition, but also a better knowledge of the sciences that gave birth to this discipline.

Single-Channel (Electrode) Strategies

Single-channel systems were more frequently explored initially as they were simpler to engineer, and there was initially insufficient evidence that multiple-channel stimulation would allow the additional speech information to pass through the electroneural bottleneck for speech understanding. It was thought, as discussed in Chapter 1, that 10 or even more electrodes in the inner ear would be inadequate to replace the 10,000 or more auditory nerve fibers normally transmitting information on the speech frequencies. In the 1960s and 1970s it was not clear to what extent the time/period (volley) rather than the place theory was important in the coding of speech frequencies. The debate was heightened by the key study of Rose et al (1967), which even showed some phase locking of cochlear nerve responses at 5000 Hz.

Minimal Preprocessing of the Acoustic Signal

Initially some thought that a single-channel strategy should present as much information to the brain as possible, even though a great deal was in a form that was not usable (Chouard et al 1985). It was assumed that the brain would find the important information for hearing speech. The single-channel (electrode) implant system developed in Los Angeles (House et al 1981) and commercialized by 3M embodied this principle. It did so by filtering the signal over the frequency range of 200 to 4000 Hz, and providing nonlinear modulation of a 16,000-Hz carrier wave with the output.

This simple strategy provided the patient with information about not only the boundaries for speech events such as syllables, words, phrases, and sentences, but also the stress for words or syllables where extra vocal energy was applied. It also enabled the rapid intensity changes in plosives to be coded (amplitude envelop variations) (Blamey et al 1987) and vowel durations to be discriminated. Intensity and coarse temporal cues permitted the discrimination of voiced from unvoiced speech, and low first formant from high first formant information. A low F1 frequency has more energy. There was, however, insufficient information to discriminate formants and their transitions and other important segmental information. This was reflected in the fact that no open-set speech was obtained for electrical stimulation alone, but closed-set consonant and vowel recognition could be achieved in some of the patients.

Preprocessing of the Acoustic Signal

To improve the performance of the Los Angeles/3M system, an optimized version was developed by Edgerton and Brimacombe (1984) that emphasized the mid- and high-frequency cues, and reduced low-frequency masking effects. The masking of neural excitation with depth of modulation was demonstrated in the psychophysical research of McKay et al (1993). This single-electrode system gave improved recognition of a limited set of plosives /p/, /k/, /b/, /g/, and fricatives /s/, /f/, /ʃ/, /v/, but not the nasals /m/, /n/, and liquids /r/, /l/ (semivowels) when compared to the standard Los Angeles/3M system (Edgerton 1985). The improved results were probably due to better representation of the energy of the noise bursts, envelope timing, and low-frequency periodicity. With the standard Los Angeles/3M system, stops, fricatives, and sibilants were confused across and within classes (Edgerton 1985). The improved speech feature recognition was reflected in better CID spondee (closed-set) word results. Two of three subjects scored 9 out of a set of 36 words.

Some preprocessing of speech was utilized by the system developed in Vienna (Hochmair et al 1979). With their best strategy, there was gain compression, followed by frequency equalization from 100 to 4000 Hz, and the stimulus was mapped onto an equal loudness contour at a comfortable level. This helped ensure that the energy in the low rates of modulation did not mask the higher rate stimuli. Although four electrodes were implanted, only the electrode where the best performance was achieved was stimulated with bipolar pulses (Burian et al 1984).

Some patients with the Vienna system were reported to get significant open-set scores for words and sentences for electrical stimulation alone (Hochmair-Desoyer et al 1980, 1981), but open-set speech recognition was not found in a controlled study in which this device was compared with the Los Angeles/3M single-channel and the University of Utah/Salt Lake city (Ineraid) and University of Melbourne (Nucleus) multiple-channel devices (Gantz et al 1987).

Tyler et al (1989), however, found that for better Vienna/3M patients the word-in-sentence scores were on average half those for the better multiple-channel patients using the Ineraid (Symbion) four-channel fixed-filter and Nucleus F0/F2 strategy. The vowel recognition scores were also half those of the multiple-channel patients, and the consonants scores were slightly lower. This suggested that the spectral information from frequency place coding was especially important for coding vowels. Later multiple-channel strategies (F0/F1/F2, Multipeak, CIS, and SPEAK), which provided more formant, spectral, amplitude, and temporal information, not only further improved vowel recognition but also greatly increased consonant recognition and in turn speech perception.

Speech was also preprocessed by the single-channel system developed in London (Fourcin et al 1979). This stimulated a single extracochlear electrode with a pulsatile current source triggered by a voicing detector. With this system the signal retained information about the precise timing of glottal closure, and fine details of the temporal aspects of phonation. It was found that patients could reliably detect small intonation variations, and when combined with a visual signal the information on voicing improved scores on closed-sets of consonants.

Multiple-Channel Strategies: Fixed-Filter Schemes

Multiple-channel strategies were developed on the one hand to reproduce the neurophysiological responses to sound and on the other hand to select information from speech in ways that were similar to vocoders. With channel vocoders speech could be filtered into a number of channels, and reconstituted without significant degradation as discussed above (see Channel Vocoders and fixed filters). Initially the fixed-filter schemes did not make voiced/voiceless decisions as occurs with a channel vocoder, or especially address the issue of how to present the information through the electroneural bottleneck.

Cochlear and Neural Models

Prior to developing the University of Melbourne's inaugural formant or cue extraction strategy in 1978, a fixed-filter strategy, which modeled the physiology of the cochlea and the neural coding of sound, was tested (Laird 1979). This strategy had band-pass filters to approximate the frequency selectivity of auditory neurons, delay mechanisms to mimic basilar membrane delays, and stochastic pulsing for maintaining the fine time structure of responses, and a wide dynamic range. With this fixed-filter strategy unsatisfactory results were obtained due to simultaneous stimulation of electrodes leading to channel interaction (Laird 1979; Clark 1987). The summation of the overlapping electrical fields could not be easily determined, and as a result unpredictable variations in loudness occurred. This led to the important principle in cochlear implant speech processing of presenting electrical stimuli nonsimultaneously.

Fixed Filter and Simultaneous Analog Stimulation

Other fixed-filter speech-processing strategies were not based on cochlear models, but the outputs of the filters stimulated separate electrodes opposite appropriate frequency sites in the cochlea. They were similar in concept to the channel vocoders used, as discussed above, but there was no voicing decision. One of the first of these fixed-filter strategies was evaluated by the University of Utah in Salt Lake City (Eddington 1980, 1983), and subsequently manufactured and marketed by Symbion and then by Smith and Nephew as the Ineraid device.

The Ineraid system presented the outputs of four fixed filters by simultaneous monopolar analog stimulation between the electrodes in the cochlea and a remote reference. Thus it was a simultaneous analog system (SAS). Compression of the amplitude variations in speech to bring them within the dynamic range of electrical stimulation was achieved with a variable gain amplifier operating in compression mode. It could thus be referred to as a compressed analog (CA) scheme. To avoid destructive channel interaction with simultaneous stimulation the electrodes (channels) needed to be well separated spatially so the voltage fields did not overlap unnecessarily. This limited the number of electrodes that could be used. Six electrodes were spaced at 4-mm intervals along an array 22 mm in length. In most patients only the apical four electrodes were excited. The center frequencies of the filters for these electrodes were 500, 1000, 2000, and 3400 Hz (Dorman et al 1989).

A study with the Symbion/Ineraid four-fixed-filter strategy examined the median score for open sets of CID sentences with electrical stimulation alone and vowel and consonant recognition. The mean word-in-sentence score was 45% (range 0–100%) (Dorman et al 1989). With vowel recognition the errors were mainly limited to the vowels with the most similar formant frequencies. This could be attributed to the limited number of electrodes used, resulting in a large overlap in neurons excited by these frequencies. With closed sets of consonants, manner and voicing were well recognized. With these features temporal information from the fundamental frequency and the amplitude wave envelope is important, and would explain the satisfactory results for only four electrodes. The patients with the better scores had more recognition of stop consonant place of articulation, and improved discrimination between /s/ and /ʃ/, suggesting more information received from the middle to high frequencies (Dorman 1993).

It is unlikely that the analog stimulation provided by the Ineraid SAS transmitted any additional temporal information over pulsatile stimulation. Analog electrical stimulation, as used in the system described above, was found by neurophysiologists in the 1940s and 1950s to be less suitable than electrical pulses for stimulating the nervous system. Neurons integrate current to produce an action potential regardless of the type of stimulation, and current can be more precisely controlled with a pulse. A preliminary study (Clark 1969) to compare analog and pulsatile stimuli and their effects on synchrony of firing showed little difference. A more detailed evaluation (Hartmann et al 1984a, b) of the effects of biphasic pulses and sinusoidal current waveforms also showed no significant differences in the temporal properties of the responses, although there were differences in synchrony of responses depending on pulse width and frequency.

Advanced Bionics Clarion developed a device referred to as SAS that provided simultaneous analog stimulation. It was similar to the strategy developed by the

University of Utah in Salt Lake City (Eddington 1980, 1983). The main difference from the CA scheme was that it had automatic gain control (AGC) with longer attack and release times, as well as a lower compression ratio making for reduced spectral distortion. It was subsequently used with eight filters in the Clarion processor (Battmer et al 1994). The Clarion SAS electrode array arose from the research at the University of San Francisco (Merzenich et al 1984) as discussed below.

Fixed Filter and Simultaneous Pulsatile Stimulation

A second fixed-filter (channel vocoder) system was developed at the University of San Francisco (Merzenich et al 1984). However, six and later eight electrodes were used for bipolar pulsatile stimulation. This allowed more controlled stimulation for the reasons stated above. The electrodes were embedded in a molded array with the electrodes placed in the vicinity of the peripheral processes of the cochlear nerve fibers in the cochlea. Further information on its design is provided in Chapter 8 and the biological aspects in Chapters 3 and 5. This system was implemented as the Storz (MiniMed) device. Initial results were published for one patient. She obtained 28% discrimination of spondee words (a closed-set test). With CID word in sentences, there was no open-set word recognition with electrical stimulation alone, but the speech reading improved from 32% to 78% with the addition of electrical stimulation.

Fixed Filter with Constant Rate of Stimulation

The first strategy to use a constant rate of stimulation on each electrode was described by Chouard et al (1984, 1985). A bank of 12 filters and biphasic asymmetrical pulses were used at a constant rate of 300 pulses/s unless the filter frequency was less than 300 Hz. Pulse duration coded intensity changes. The authors aimed to transmit all possible information to patients without selecting information or features. This strategy was developed commercially by Bertin St. as the Chorimac-8 and -12. The results (Fugain et al 1984) showed vowels were well recognized. With consonants, voicing was well differentiated (90%).

Unfortunately, the Miller and Nicely (1955) classification of features was not used to make comparisons with other strategies possible. However, different fricatives could be distinguished, but place information was poorly transmitted. Standardized open sets of words were not used and so it was not clear to what extent open-set speech could be recognized (Chouard et al 1985).

Interleaved Pulse (IP) Strategy

A fixed-filter system was developed in which the outputs from a number of band-pass filters were presented to electrodes nonsimultaneously as interleaved pulses (IPs) (Wilson et al 1988). This was undertaken to reduce channel interaction. This had been established as an important principle in speech processing (Clark 1987). The principle was discovered from speech processing research using a cochlear model (Laird 1979), and applied to the Nucleus F0/F1/F2 system (Dowell et al 1987b). The fixed-filter IP strategy was compared with CA processors in eight subjects. The University of California at San Francisco (UCSF)/Storz electrode array was used. Half the patients had better speech perception using the IP processor than the CA. It was hypothesized that the reduced temporal overlap with interleaved pulses from the IP processor benefited those with poor nerve survival. This could be due to neurons in depleted populations having poorer temporal integration, and thus being affected by simultaneous stimulation.

The IP speech-processing strategies were fixed-filter schemes with two, four, and six stimulus channels. Performance improved as the number of channels increased from two to six. The specific selection of filter outputs according to the dynamics of the speech signal, as occurred with the University of Melbourne/Nucleus formant extraction (F0/F2, F0/F1/F2, Multipeak) or spectral maxima (SPEAK) systems, was not reported by Wilson et al (1988). In addition, in one patient intensively studied, voicing performance was better when it was explicitly coded through a channel vocoder (voiced/voiceless decision).

Continuous Interleaved Sampler (CIS)

The CIS strategy evolved from the above fixed-filter scheme that used IPs to avoid channel interaction. It was developed because patients' perceptual boundaries between voiced and unvoiced sounds seemed unnatural with the IP scheme (Wilson 2000). It was considered that a higher pulse rate should be used to provide a better representation of the voicing information. The rate needed to be greater than twice the cutoff frequency of the low-pass filters to avoid aliasing effects in the pattern of stimulation of nerve fibers (Rabiner and Schafer 1978; Wilson 1997). The waveform envelopes from the band-pass filters modulated a high-pulse rate train (Wilson et al 1992). It used biphasic pulses rather than analog stimulation as occurred with the CA scheme. In contrast to the F0/F2, F0/F1/F2, and F0/F1/F2 with high-frequency fixed-filter (Multipeak) strategies, but in line with the SPEAK and ACE strategy, there were no voiced/unvoiced distinctions made, and thus no explicit representation of voicing as rate of stimulation.

The outputs of six or more filters were sampled and used to stimulate the same number of electrodes on a place-coding basis. Various studies were undertaken to optimize the number of filters and stimulus rate (Wilson et al 1992, 1993). Lawson et al (1996) and Wilson (1997) found that as the number of electrodes with the CIS strategy was increased up to seven, there was an improvement in speech perception, but not above this number. This is consistent with the acoustic modeling studies (see Acoustic Representation of Electrical Stimulation, above) that showed that there was no significant improvement in speech recognition with more than six bands of filtered noise modulated by the filter amplitudes for this type of fixed-filter strategy. This does not apply to the Nucleus formant extraction strategies. In the presence of noise more channels are required to distinguish meaningful from random activity in the auditory nerve (AN).

The Advanced Bionics Clarion processor with CIS strategy was implemented with eight band-pass channels coding frequencies ranging from 250 to 5500 Hz. The spectral information was presented at a constant stimulus rate between 833 and 1111 pulses/s per channel for bipolar or monopolar stimulus modes (Battmer et al 1994). It is, however, still not clear up to what rates the auditory nervous pathways can handle the increased information from higher stimulus rates. For example, it was shown that there was a marked decrement in the response of units in the anteroventral cochlear nucleus of the cat when stimulus rates reached 800 pulses/s (Buden et al 1996). As discussed in Chapter 5, data from intracellular recordings from the globular bushy cells in the cochlear nucleus showed they could not convey temporal information at rates greater than about 1200 pulses/s (Paolini and Clark 1997). For this reason very high stimulus rates (greater than 1200 pulses/s) appear to have little value and could damage neural fibers and ganglion cells.

An analysis of results for CIS using the Clarion system was undertaken by Schindler et al (1995). In a group of 73 patients the mean open-set CID sentence score for electrical stimulation alone was 58% six months after implantation. A study by Kessler et al (1995) reported the mean CID sentence score for the first 64 patients implanted with the Clarion device to be 60% six months postoperatively. Kessler et al also reported a bimodal distribution in results with a significant number of poorer performers. It is also of interest to examine the differences in information transmitted for the CA and CIS strategies. A study on seven patients referred to by Dorman (1993) showed better transmission for nasality, frication, and place features for CIS. Nasality and place improvement could have been due to the better transmission of amplitude envelope cues (Blamey et al 1987), and frication due to its better coding by higher rates of stimulation (Grayden and Clark 2000). In a study by Doyle et al (1995), both CA and CIS users scored 50% for closed sets of consonants. However, information transmission analysis showed the best scores for CA were duration, 50%; place of articulation, 29%; manner, 28%; and nasality, 27%. In contrast, for CIS the best features were voicing, 41%; place of articulation, 40%; and duration, 37%.

Multiple-Electrode Strategies: Formant and Spectral Cue Extraction

The University of Melbourne in 1978 first evaluated a speech-processing scheme based on a cochlear and neural model. As discussed above, due to unpredictable variations in loudness from simultaneous stimulation, this scheme was not investigated further. In the same year a scheme based on a formant vocoder was explored in preference to a fixed-filter or channel vocoder. This was because the patient described vowel percepts that were similar to those experienced by normal-hearing subjects when a single formant vowel excited a similar area of the cochlea to that of a normal-hearing person (Delattre et al 1952). Thus the approach became one of preprocessing the signal to present the formant of most importance for speech understanding through multiple electrodes. This was also based on the assumption that presenting the whole signal through the narrow electroneural bottleneck (demonstrated by the physiological and psychophysical studies) using fixed filters could mask or restrict the usable information. A multiple-channel strategy that extracted formants and spectral cues was developed specifically to optimize the information transmitted.

The next approach at the University of Melbourne in developing the initial formant-extraction strategy was to match the psychophysical or speech percept to the pattern of electrical stimulation, as there were difficulties in replicating the coding of sound as discussed above. This was to be the approach until there was a better understanding of how electrical stimulation reproduced the coding of sound. In other words, the psychophysical findings on pitch and loudness were applied to the development and modification of the formant speech-processing strategies.

Fundamental (F0) and Second Formant (F2) Extraction (F0/F2)

With the inaugural formant extraction strategy developed in 1978, the second formant frequency (F2) was extracted and presented as place of stimulation, the fundamental or voicing frequency (F0) as rate of stimulation on individual electrodes, and the amplitude of F2 as the current level (A2) (Clark, Tong et al 1978, 1981a; Tong et al 1979, 1980; Clark and Tong 1981). Information about the fundamental frequency and the presence or absence of voicing was presented by modulating the frequency (pulse rate) on each electrode with the pulse rate proportional to the acoustic fundamental frequency. With voiceless sounds a random electric stimulus pattern was used, as this was described as rough and noise-like. As discussed above, the voicing (fundamental) frequency (coded as stimulus rate) provided linguistic information about the stress and intonation of the speech message, and the voiced/voiceless distinction was therefore one of the important features for the recognition of speech. Sounds were considered unvoiced if the energy of the voicing frequency was low in comparison to the second formant.

The coding of speech by this strategy was based on the studies described in Chapter 6. For example, the F2 frequencies were coded as place of stimulation not only because rate of stimulation could not be discriminated at these high rates, but also because the frequency glides seen in consonants could be perceived over durations that were the same as those of consonants (i.e., on the order of 20 ms). Rate of stimulation could not be adequately perceived over this duration. Rate of stimulation, however, was effective for coding the slower changes seen with F0.

The first clue to developing this strategy, came as discussed above, when it was observed that electrical stimulation at individual sites within the cochlea produced vowel-like sounds, and that the vowel sounds corresponded to the single-formant vowels (Delattre et al 1952).

The inaugural F0/F2 strategy was first evaluated on two patients using a laboratory-based speech processor. The CID sentence test showed the patients obtained marked improvements in communication (188% and 386%) when using electrical stimulation in combination with speech reading, compared to speech reading alone (Clark, Tong et al 1981a). For electrical stimulation alone, the average score for a closed set of six vowels was 77% (Tong et al 1980; Clark and Tong 1982) and for a set of 12 consonants 34% (Tong et al 1980). The average score for open sets of words (scored as words) was 8% when presented by live voice, and 5% when presented using prerecorded test materials (Clark, Tong et al 1981b). Similarly, scores on CID sentences (scored as key words) were 35% for live voice and 11% when prerecorded (Clark, Tong et al 1981a).

A study was undertaken on the first two patients using the University of Melbourne's laboratory speech processor for the consonants /b, p, d, t, g, k, m, n/ to determine how effectively speech features were transmitted by electrical stimulation, and how this was affected by speech reading (Clark, Tong et al 1981c). The results showed that for electrical stimulation alone, voicing and manner distinctions were better than for speech reading alone. This confirmed that electrical stimulation was giving voicing information not visible on the lips. There was also a small further improvement in the recognition of these features when electrical stimulation was combined with speech reading. The place distinctions were not as well recognized for electrical stimulation as with speech reading, but the two combined to give better results. Thus F2 provided additional information on place of articulation.

A further study was undertaken on the first patient to determine the transmission of speech information once the F0/F2 speech processor had been implemented as the University of Melbourne's hard-wired portable speech processor rather than a software algorithm. A larger set of 12 consonants was used (b, p, d, t, g, k, m, n, v, f, z, s), and information transmission analyses carried out for voicing, nasality, affrication, duration, and place (Dowell et al 1982). The results for voicing and manner were similar to those in the earlier study by Clark, Tong et al (1981c). Frication was similar for electrical stimulation and speech reading, but when the two were combined there was a marked improvement. The duration cues provided by electrical stimulation were much better than with speech reading, and they too combined to give high scores. Cues to distinguish manner and affrication were all provided by the additional high-frequency F2 information. Place of articulation scores for electrical stimulation of 25% correct still required improvement. As there are multiple cues for place of articulation (burst frequency, frequency transitions, and amplitude wave envelope), it was not clear which were being provided by the F0/F2 strategy.

As there was still some uncertainty about just what additional information would be provided by coding F2 as place of stimulation, the transmission of speech information for 12 consonants using the F0/F2 strategy and one that provided single-channel stimulation for F0 were compared on the one patient (MC-1) using the University of Melbourne's portable speech processor (Clark, Tong et al 1984). The results for electrical stimulation alone showed that the addition of F2 as well as the voicing frequency resulted in improved frication, duration, and place information.

The F0/F2 strategy was implemented by Cochlear Proprietary Limited in the Nucleus WSP-II wearable speech processor. This was initially tested for the U.S.

Food and Drug Administration (FDA) on 40 postlinguistically deaf adults from nine centers worldwide (see Chapter 1). Three months postimplantation the patients had obtained a mean CID sentence score of 87% (range 45–100%) for speech reading plus electrical stimulation, compared to a score of 52% (range 15–85%) for speech reading alone. In a subgroup of 23 patients the mean CID sentence scores for electrical stimulation alone rose from 16% (range 0–58%) at 3 months postimplantation to 40% (range 0–86%) at 12 months (Dowell et al 1986a,b). The F0/F2 WSP-II was approved by the FDA in October 1985 for use in postlinguistically deaf adults as safe and effective and able to provide speech perception with the aid of speech reading and some open-set speech understanding with electrical stimulation alone.

Fundamental, First and Second Formant Frequencies

Further research at the University of Melbourne aimed, in particular, at improving the recognition of consonants because of their importance for speech intelligibility. To achieve this goal, additional spectral energy (first formant, F1) was extracted and presented on a place-coding basis. This was supported by the psychophysical study that showed that stimuli presented to two electrodes could be perceived as a two-component sensation (Tong et al 1983b). The anticipated improvement expected in providing F1 as well as F2 information was seen in the acoustic model studies of electrical stimulation on normal-hearing individuals discussed above (see Acoustic Representation of Electrical Stimulation) (Blamey et al 1984a,b, 1985). The information transmission analysis for F0/F2 and F0/F1/F2 strategies using the acoustic model (Blamey et al 1985) showed improved speech perception scores with the addition of F1 information.

To overcome the problems of channel interaction, first demonstrated in the University of Melbourne's physiological speech-processing strategy in 1978 (Laird 1979), nonsimultaneous (pulse separation of 0.7 ms), sequential pulsatile stimulation at two different sites within the cochlea was used to provide F1 and F2 information. F0 was coded as rate of stimulation as with the inaugural F0/F2 strategy.

A comparison was made in Melbourne of the F0/F2 WSP-II system used on 13 postlinguistically deaf adults, and the F0/F1/F2 WSP-III system on nine patients (Dowell et al 1987b). The results for electrical stimulation alone were recorded 3 months postoperatively. The average open-set CID sentence score for electrical stimulation alone increased from 16% to 35%. Blamey et al (1987) reported a mean vowel recognition score of 49% and for consonants 37%. Vowel recognition could be accounted for largely due to the place coding of the F1 and F2 frequencies (Blamey and Clark 1990). The F0/F1/F2 WSP-III speech processor was approved by the FDA in May 1986 for use in postlinguistically deaf adults. The findings were also supported by Gantz et al. (1988); Tyler and Lowder (1992), and Hollow et al (1995). Hollow et al reported a mean CID word-in-sentence score of 38.5% ($n = 32$) for the F0/F1/F2 WSP-III system.

The improved F0/F1/F2 WSP-III speech scores (approximately 120%) were related to the better information transmission for consonant features. The speech feature scores for the two-formant strategies F0/F2 and F0/F1/F2 are shown in Table 7.8. From this it can be seen that the addition of F1 on a place-coding basis improved the percentage of voicing (70%), nasality (29%), affrication (25%), place (75%), amplitude envelope information (50%), and high F2 (33%). These findings were supported by Tye-Murray et al (1992), who showed that the features for amplitude envelope, nasality, frication, and voicing were relatively well transmitted, but not the place feature. Blamey et al (1987) reported a mean vowel recognition score of 49% and consonant score of 37%. This consonant score in particular was still low in spite of the percentage improvements in the transmission of speech features. The percentage improvement in speech score with the F0/F1/F2 was assumed to be due to additive or multiplicative effects from the transmission of the speech features, and not a reflection of the vowels and consonants alone.

The formant-based F0/F1/F2 WSP-III system was compared with the Symbion Ineraid device (Cohen et al 1993). The Ineraid device presented the outputs of four fixed filters to the cochlear nerve by simultaneous monopolar stimulation. The Ineraid did not have the preprocessing of speech seen with the F0/F1/F2 WSP-III device to help get useful information through the electroneural bottleneck. The processors were compared for prosody, phoneme, spondee, and open-set speech recognition. There was no significant difference between the F0/F1/F2 WSP-III and Ineraid systems. The data suggest that the two systems provided different types and degrees of speech information.

Fundamental, First and Second Formant Frequencies and High-Frequency Fixed-Filter Outputs

The mean open-set CID word-in-sentence score for electrical stimulation alone increased from 16% ($n = 13$) with the F0/F2 WSP-II system to 35% ($n = 9$) with the F0/F1/F2 WSP-III system (Dowell et al 1987b). This was still well below the ideal. It was assumed, however, that better speech perception would occur if there was improved identification of the place speech feature (only 35% for electrical stimulation alone with the F0/F1/F2 strategy). The place of articulation feature is important for consonant recognition, and in turn for speech understanding. The research at the HCRC at the University of Melbourne/Bionic Ear Institute set out to provide more high-frequency (third formant, F3) information for the place feature. The high-frequency spectral information was extracted to provide additional high-frequency cues to improve consonant perception and speech understanding in quiet as well as in noise.

As a result, a strategy was developed where the outputs of fixed filters in three frequency bands (2000–2800 Hz, 2800–4000 Hz, and >4000 Hz) were presented as well as the first two formants on a place-coding basis, together with voicing as rate of stimulation. This became the Multipeak speech-processing strategy. It is a misnomer, as the high-frequency information was not peaks of energy but the outputs from fixed filters. It was thus a hybrid scheme between formant extraction and fixed filter. The strategy was implemented in a speech processor named the Nucleus Miniature Speech Processor (MSP). The Multipeak-MSP system was approved by the FDA on October 11, 1989, for use in postlinguistically deaf adults.

A study by Dowell et al (1990) was undertaken to compare a group of four experienced subjects who used the WSP-III speech processor with the F0/F1/F2 speech-processing strategy, and four who used the newer MSP speech processor and Multipeak strategy. The patients were not selected using any special criteria. The results showed that for open-set Bench-Kowal-Bamford (BKB) sentences there was a statistically significant improvement in quiet from 54% to 88%. The differences in results became greater with lower SNRs. The improvement was also observed by Skinner et al (1991), Cohen et al (1993), Hollow et al (1995), and Parkinson et al (1996). Skinner et al (1991) found the open-set monosyllabic scores improved from 14% to 29%, and Hollow et al (1995) found that the open-set word-in-sentence scores went from 38.% ($n = 32$) to 59.1% ($n = 27$).

The information transmitted for vowels and consonants with the F0/F1/F2 and Multipeak strategies was compared in four subjects. With vowels the information transmitted for F1 and F2 increased with the Multipeak strategy, and the identification scores went from 80% to 88% (Dowell et al 1990, 1993; Dowell 1991). The information for consonants increased for voicing from 62% to 79% (a 27% increase), nasality from 63% to 95% (a 51% increase), frication from 54% to 81% (a 50% increase), place of articulation from 25% to 32% (a 28% increase), and the identification scores for consonants as a whole went from 48% to 63% (a 31% increase) (Dowell 1991). There was a 10% increase in intelligibility for vowels, and an overall 31% increase for consonants. This was associated with an increase from 33% to 46% in open-set consonant-nucleus-consonant (CNC) word scores (a 39% increase). The improved vowel and consonant scores were not seen in a study by Parkinson et al (1996), but they demonstrated that open-set speech recognition was significantly higher. As with the improvements from the F0/F2 to F0/F1/F2 strategies, the data suggest that the speech features have complex additive or multiplicative effects on speech recognition as a whole.

In the above comparison of the F0/F1/F2 and Multipeak strategies, the addition of the high-frequency spectral information from the fixed filters could have assisted in the identification of voicing by providing temporal information in the high-frequency fibers as well as the lower ones. This is supported by the psychophysical studies of Tong et al (1983a), who showed that an implant patient could categorize questions and statements while electrode trajectories moved in an apical or basal direction. The poor performance seen in noise for both strategies (Fig. 7.12) was assumed to be due to the limitation of using a voiced/voiceless decision. The improved results for nasals was most likely due to the fixed filters in the ranges 2000 to 2800 Hz and 2800 to 4000 Hz, providing the poles and zeros for the four formants necessary for the identification of /ŋ/ as well as the frequency transitions (Liberman et al 1954; Fujimura 1962). The additional high-frequency information was essential for distinguishing fricatives, as their noise frequencies vary considerably from below 1200 Hz to as high as 7000 Hz (Strevens 1960) (see Consonants, above). This additional information was also important in recognizing fricatives in noise, although for multispeaker babble there is

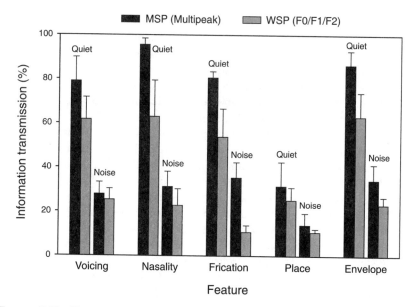

FIGURE 7.12. The mean voicing, nasality, frication, place, and amplitude envelope information transmission for consonants with the F0/F1/F2 WSP and Multipeak-MSP systems in quiet and noise (+5 dB signal to noise) for four patients (Reprinted with permission from Dowell 1991. Speech perception in noise for multichannel cochlear implant users. PhD thesis. The University of Melbourne.).

greater energy in the mid- to low-frequency range. The transmission of place of articulation information was low for both strategies, and the small improvement with Multipeak was assumed to be due to more high-frequency energy required for recognizing plosives. As the amplitude envelope cues were well transmitted, this would have partly contributed to the recognition of place.

The Multipeak strategy was also compared with the Symbion/Ineraid device in the study by Cohen et al (1993). They found a significant difference between the Nucleus Multipeak-MSP and Symbion/Ineraid systems, particularly for the perception of open-set speech presented by electrical stimulation alone. There was a 75% score with the Multipeak-MSP system and only a 42% with the Symbion/Ineraid system. Both speech-processing strategies presented information along approximately the same number of channels (five for Multipeak and six for Ineraid). Although the Ineraid strategy did not use a voicing decision the significantly better results with Multipeak would not have been due to that alone, but also to the selection of formants and presentation of the energy peaks over a range of frequency regions in the cochlea.

Spectral Maxima Sound Processor (SPEAK)

The research with the F0/F1/F2 and hybrid formant and high-frequency fixed-filter strategy (Multipeak) showed that the recognition of place of articulation was considerably less than that for other features. For this reason, studies were undertaken to compare the extraction of three, four, and six frequency peaks to provide additional cues. This was done using the outputs of 16 band-pass filters, and the opportunity was also taken to compare the presentation of information with and without voicing as rate of stimulation, by also using a constant rate of stimulation for the coding of the frequency energy peaks. The research was carried out in 1989 at the University of Melbourne / Bionic Ear Institute. In addition, the extraction of two formants with the F0/F1/F2 speech-processing strategy was compared with fixed-filter schemes that allowed the extraction of three, four, and six peaks of spectral energy. The selection of more peaks was expected to provide a better representation of the place feature for speech articulation (Tong et al 1989, 1991). Two versions of the strategy that picked four spectral peaks were used, one in which F0 was specifically extracted and coded as rate of stimulation, with random stimulation for unvoiced speech, and the other strategy where constant stimulus rates of 125 pulses/s and 166 pulses/s were used on all electrodes to reduce channel interaction. The peaks in the voltage outputs of the filters were used to stimulate appropriate electrodes on a place-coding basis. The perception of vowels and consonants was significantly better for both peak-picking filter bank schemes compared to the F0/F1/F2 WSP-III system, and the perception of consonant duration, nasality, and place improved.

Tong et al (1990) made a comparison between the Multipeak-MSP system and a filter bank strategy that selected the four highest spectral peaks and coded these on a place basis. Electrical stimulation occurred at a constant rate of 166 Hz. This strategy was also implemented using a Motorola DSP56001 digital signal processor (DSP). The mean results for vowels were as follows: Multipeak-MSP, 76%; fixed-filter DSP, 84%. The results for consonants were as follows: Multipeak-MSP, 66%; fixed-filter DSP, 81%. The improved results obtained for the fixed-filter DSP processor extracting four spectral peaks and presenting the energy as the constant rate of stimulation suggested that this type of strategy could lead to better speech results than the Multipeak-MSP. This study indicated the importance of selecting spectral peaks and their representation as place coding using 22 electrodes.

To improve the strategy further, a decision was required whether to have a strategy that presented six spectral peaks or six spectral maxima. As preliminary investigations did not show six peaks made a significant difference over four peaks, it was decided to proceed with a strategy that extracted six spectral maxima instead. The voltage outputs of the filters were also presented nonsimultaneously with nonoverlapping pulses at a constant rate of stimulation (166 pulses/s), as this had been used with some of the peak picking strategies referred to above, to minimize channel interaction.

The strategy called the spectral maxima sound processing (SMSP) scheme was implemented in 1990 on an initial patient using an NEC filter bank chip (D7763). The strategy estimated the spectrum of the speech with a bank of 16 band-pass filters. The first eight had center frequencies distributed linearly over the range of 280 to 1780 Hz, and the remaining eight were logarithmically spaced up to 6000 Hz. When tested on an initial patient it was found to give substantial benefit.

For this reason in 1990 a pilot study was carried out on two other patients who had been using the F0/F1/F2-MSP system (McKay et al 1991). The consonant scores for the two patients with the F0/F1/F2-MSP system were 20% and 16%, and for the SMSP-DSP 43% and 39%. The open-set CNC word scores (scored as words) were 9% and 1% for the F0/F1/F2-MSP system, and 21% and 16% for SMSP-DSP. The open-set CID sentence scores (scored as key words) were 53% and 56% for the F0/F1/F2-MSP system and 80% and 88% for SMSP-DSP. The Multipeak-MSP was evaluated on one of these patients, and the results for electrical stimulation alone were CNC words 3%, and CID sentences 41%.

The SMSP system was then assessed on four patients who had been using the Multipeak-MSP system. The average scores for closed sets of vowels and consonants and open sets of CNC words and words in sentences improved for the SMSP system (McKay et al 1992). In view of the above improvements, the SMSP strategy was implemented by Cochlear Limited as SPEAK. SPEAK (McDermott et al 1992) was implemented in a processor referred to as Spectra-22. SPEAK Spectra-22 (Seligman and McDermott 1995) differed from SMSP and its implementation with analog circuitry in being able to select six or more spectral maxima from 20 rather than 16 filters. A constant stimulus rate that varied adaptively from 180 to 300 pulses/s was used. The description of the above research is from Clark et al (1996).

A multicenter comparison of the SPEAK Spectra-22 and Multipeak-MSP systems was undertaken to establish the benefits of the SPEAK Spectra-22 system (Skinner et al 1994). The field trial was on 63 postlinguistically and profoundly deaf adults at eight centers in Australia, North America, and the UK. The mean scores for vowels, consonants, CNC words, and words in City University of New York (CUNY) and Speech Intelligibility Test (SIT) sentences in quiet were all significantly better for SPEAK. The mean score for words in sentences was 76% for SPEAK Spectra-22 and 67% for Multipeak-MSP. SPEAK performed particularly well in noise. SPEAK Spectra-22 was approved by the FDA for postlinguistically deaf adults on March 30, 1994. In another set of data presented to the FDA in January 1996, a mean open-set CID sentence score of 71% was obtained for the SPEAK strategy on 51 consecutive patients 2 weeks to 6 months after the start-up time. With the CIS strategy (Research Triangle) implemented on the Clarion system (Advanced Bionics), there was a mean open-set CID sentence score of 60% for 64 patients (Kessler et al 1995) 6 months postoperative, as discussed above. The CIS strategy used six fixed filters and stimulated at a rate of 800 pulses/s.

The speech information transmitted for closed sets of vowels and consonants for SPEAK Spectra-22 (McKay and McDermott 1993) was compared to Multipeak-MSP. Vowel and consonant confusion data from five subjects converted from Multipeak-MSP to SPEAK were analyzed. There was an improvement for F1 and F2 in vowels. With consonants there was an increase in the transfer of information for all speech features except consonant voicing, with consonant place and manner of articulation showing the largest improvements. The mean scores on five patients for voicing were Multipeak-MSP 94% and SMSP 93%, for manner Multipeak-MSP 88% and SMSP 92% (5% increase), and for place Multipeak-MSP 71% and SMSP 82% (15% increase). The improved coding of place of articulation produced a significant but not large increase on the word-in-sentence recognition scores (from 67% to 76%) in the study by Skinner, et al (1994).

The differences in information presented to the nervous system with the Multipeak-MSP, SPEAK Spectra-22, and CIS strategies can be seen in the outputs to the electrodes for different words, plotted as electrodograms; the word "choice" is shown in Figure 7.13. From this it can be seen that with SPEAK there is better representation of the consonant transitions from the affricate (tʃ) to the diphthong (ɔi), so more spectral information appears to be presented on a place-coding basis. This finding was supported by the confusion data for the diphthongs, where the greatest improvements for SPEAK were with /ɛɪ/ and /ɑɪ/. Both diphthongs have rising second and falling first formants. The better representation was probably due a greater overlap in the electrodes stimulated and the higher rate. As CIS gave similar results to SPEAK, the differences in the electrodograms suggest that more temporal rather than spectral information has been transmitted with CIS.

The voicing information transmitted by Multipeak compared with SPEAK as well as CIS was coded by a different mechanism. As discussed in Chapter 8, voicing with the Multipeak strategy was extracted with a zero crossing detector, and coded on each electrode as rate of stimulation. In addition a voicing decision

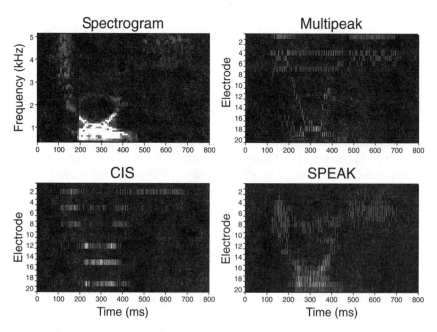

FIGURE 7.13. Spectrogram for the word "choice" and the electrode representations (electrodograms) for this word using the Multipeak, continuous interleaved sampler (CIS), and SPEAK strategies.

was made, and an unvoiced sound coded as an aperiodic stimulus at higher rate. With SPEAK and CIS there was no voicing decision, and F0 was coded through amplitude modulating the output of some of the lower frequency channels. The mean results for combined male and female speakers in identifying intonation patterns was the same for Multipeak and SPEAK. However, the Multipeak was better for males and SPEAK for females. This suggested first that rate of stimulation was better for conveying voicing for males. As the higher F0 of females was not represented in the amplitude modulation of the output of SPEAK, the better result could have been due to a small change in the formant frequencies from the harmonic structure of the voicing frequency being better represented spatially. The improvement in information on place of articulation could have resulted from a better representation of spectral shape resulting from a more normal mapping of frequency to electrodes. This is seen in the electrodograms in Figure 7.13. Throckmorton and Collins (1999) also found a positive correlation between electrode discrimination and speech perception in seven SPEAK Spectra-22 patients.

Spectral Maxima Speech (Sound) Processor at High Rates (ACE)

The SPEAK strategy with six spectral maxima at a rate of 250 pulses/s gave comparable or better results than the six filter CIS strategy at a rate of 800 pulses/s. Thus an advantageous spectral pattern from SPEAK could have been counterbalanced by additional timing information from a higher rate of stimulation with CIS. It was important to see if an increase in stimulus rate would give improved results with SPEAK. A flexible processor was implemented on the Nucleus-24 system that would allow the presentation of SPEAK at different rates and vary the number of stimulus channels. This was the Advanced Combination Encoder (ACE).

A study commenced in the (CRC) for Cochlear Implant Speech and Hearing Research at the University of Melbourne/Bionic Ear Institute on the effects of low (250 pulses/s) and high (800 and 1600 pulses/s) rates of stimulation on five subjects (Fig. 7.14). The mean results for CUNY sentences for the lowest SNR ratio (Vandali et al 2000) showed the performance for the highest rate was significantly the poorest. However, the scores varied in the five individuals. Subject 1 performed best at 807 pulses/s, subject 4 was poorest at 807 pulses/s, and 5 poorest at 1615 pulses/s. There was thus significant intersubject variability for SPEAK at different rates. The physiological limitations of using high rates have been discussed in association with the CIS strategy and in Chapter 5.

With electrical stimulation Fishman et al (1997) varied the number of electrodes in use. The outputs of adjacent filters were directed to a single electrode, allowing the number of stimulus channels to be reduced. They found no increase in speech perception was seen when the number of electrodes was increased from four to 20. However, the study was undertaken in quiet conditions, and a greater number of electrodes would be expected for hearing speech in noise. The advantage in having 22 electrodes was discussed above (see Channel Numbers). They are

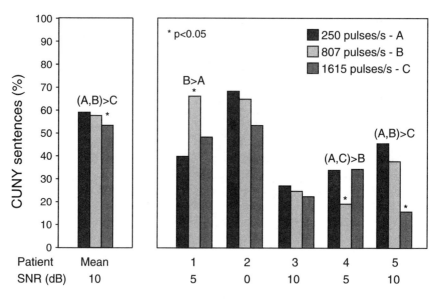

FIGURE 7.14. The effect of rate of stimulation alone on the word in sentence scores for five patients using the SPEAK strategy (Reprinted with permission from Vandali et al 2000).

important in being able to select the optimal placement of the electrodes and allow for variations in the spiral ganglion cell density and cochlear pathology.

The ACE strategy was evaluated in a larger study on 62 postlinguistically deaf adults who were users of SPEAK at 21 centers in the United States (Arndt et al 1999). ACE was compared with SPEAK and CIS. The rate and number of channels were optimized for ACE and CIS. The rates were most frequently 720 pulses/s and 1800 pulses/s for ACE, and 900 pulses/s and 1800 pulses/s for CIS. The number of channels was varied from six to 20, depending on the optimal performance of each subject. Mean HINT (Nilsson et al 1994) sentence scores in quiet were 64.2% for SPEAK, 66.0% for CIS, and 72.3% for ACE. The ACE mean was significantly higher than the CIS mean ($p < .05$), but not significantly different from SPEAK. The mean CUNY sentence recognition at an SNR of 10 dB was significantly better for ACE (71.0%) than for both CIS (65.3%) and SPEAK (63.1%). In addition, the optimal strategy varied greatly from subject to subject as did the best set of stimulus parameters. Overall 61% preferred ACE, 23% SPEAK, and 8% CIS. The strategy preference correlated highly with speech recognition. Furthermore, one third of the subjects used different strategies for different listening conditions.

In a subsequent study (Skinner et al, 2002), 12 new patients were given SPEAK, ACE, and CIS in different orders, after each strategy was adjusted to suit the patient. The results were consistent with those of Arndt et al (1999), as 58% preferred ACE, 25% SPEAK, and 17% CIS. Six of the 12 patients had higher CUNY sentence scores for one strategy rather than for either one or two of the others. There was also a strong correlation between the preferred strategy and the performance on speech recognition.

Transient Emphasis Speech Processor

An alternative speech-processing strategy was investigated by Vandali et al (1995) and Vandali (2001), in which SPEAK had important transient cues, especially for the recognition of plosive consonants, which were identified and given emphasis. The amplitude, frequency, and duration of these segments were probably not adequately sampled by the standard SPEAK strategy. However, their perception could have been obscured by temporal and spatial masking. So emphasizing amplitude or frequency transitions in speech formants provided additional information to pass through the electroneural bottleneck (Vandali et al 1995). This is illustrated in Figure 7.15. It shows the energy output from the speech filters for four electrodes over short durations, and the amplification of these features by the transient emphasis spectral maxima (TESM) speech processor. There was some support for this concept from the acoustic study by Kennedy et al (1998) on hearing-impaired listeners where an increase in the intensity of consonants in relation to vowels improved the perception of some consonants in a vowel-consonant environment. However, an improvement for voiceless stops was not seen by Sammeth et al (1999).

The algorithm in TESM, used in conjunction with the SMSP strategy (developed as SPEAK), produced additional gain during periods of rapid rise in the envelope signal of each band. These periods corresponded to the noise burst in consonants and the onset of vowel formants. It was first evaluated on four subjects at a +5 dB SNR and compared with SMSP (Fig. 7.15). There was a significant

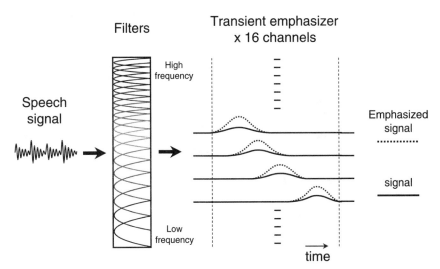

FIGURE 7.15. Transient emphasis strategy (TESM) showing the amplification of the outputs from four speech filters over a short duration (Vandali et al 1995; Clark 2001). Reprinted with permission Clark GM, 2001. Editorial Cochlear implants: climbing new mountains. The Graham Fraser Memorial Lecture 2001. *Cochlear Implants International* 2(2): 75–79.

improvement for TESM in these four patients for open-set word-in-sentence perception in noise, but not for words, consonants, or vowels in quiet (Vandali et al 1995).

A similar strategy was developed by Geurts and Wouters (1999) for the CIS strategy on the LAURA (University of Antwerp) cochlear implant. It also used a multiband gain control, but filtered the signal into fast and slow envelopes. Gain was applied when the fast exceeded the slow envelope. However, when compared with the standard CIS there was some improvement with closed sets, but not open sets, of consonants. As both the prototype TESM and enhanced envelope CIS could have been overemphasizing the onset of long-duration cues such as vowel formants, the TESM was modified (Vandali 2001) to place more emphasis on the rapid changes accompanying short duration signals (5 to 50 ms). A study on eight Nucleus 22 patients found that the CNC open-set word test scores (Fig. 7.16) increased significantly from 53.6% for SMSP to 61.3% for TESM, the open-set sentence scores in multispeaker noise from 64.9% for SMSP to 70.6% for TESM, the consonant scores from 75.1% for SMSP to 80.6% for TESM, and the vowel scores from 83.1% for SMSP to 85.7% for TESM (Vandali 2001). The additional information can be seen in the representation of the patterns of electrode stimulation for the word "mit" shown in the spectrogram in Figure 7.17. From this it can be seen that the stimuli for the F2 transition from /m/ to the vowel /ɪ/ are higher in intensity, as is the energy in the noise burst for the final plosive /t/.

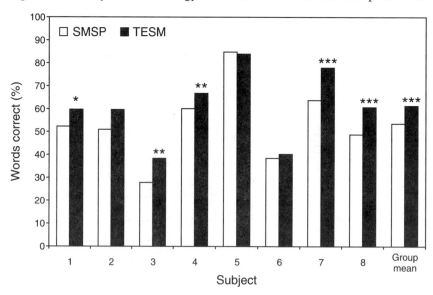

FIGURE 7.16. Open-set consonant-nucleus-consonant (CNC) word scores for electrical stimulation alone in quiet for the standard spectral maxima speech processor (SMSP/SPEAK) and modified transient emphasis speech processor (TESM). Statistically significant differences are indicated by * at the 5% level, ** at the 1% level, and *** at the 0.1% level. (Vandali et al 1995; Vandali 2001). Reprinted with permission from Vandali et al 1995. Multichannel cochlear implant speech processing: further variations of the Spectral Maxima sound processor strategy. *Annals of Otology, Rhinology and Laryngology* **104** (Suppl 166): 378–381.

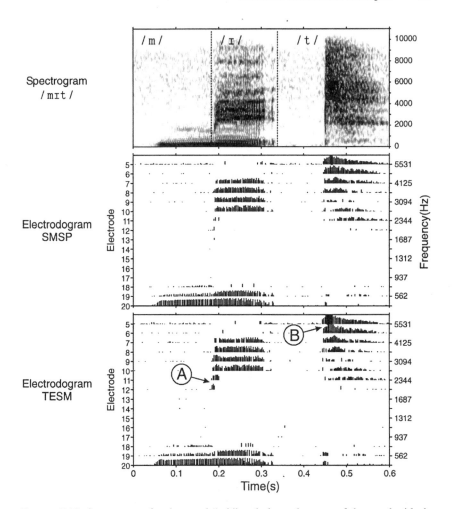

FIGURE 7.17. Spectrogram for the word "mit" and electrodograms of the word with the SMSP and TESM speech processing strategies. A, higher intensity of F2 transition from /m/ to the vowel /ɪ/; B, higher intensity in the noise burst for the final plosive /t/. (Reprinted with permission from Vandali 2001. Emphasis of short-duration acoustic speech cues for cochlear implant users. *Journal of the Acoustical Society of America* 109: 2049–2061.)

These data indicate the importance of representing the spectral and temporal information over short durations in future speech-coding strategies.

Differential Rate Speech Processor

As the studies with ACE reported above showed patient variability, with some performing best at a low rate of 250 pulses/s and others at 800 pulses/s, research was carried out to understand why this occurred, and whether this knowledge could lead to a more advanced strategy (Fig. 7.14). First, research was undertaken to see how rate affected the recognition of phonemes. If rate of stimulation had different effects on speech features, this could account for the variation in speech scores. This was done by constructing a consonant confusion matrix for the 24 Australian English consonants arranged into distinctive feature groups for stimulus rates of 250, 807, and 1615 pulses/s (Grayden and Clark 2000, 2001; Clark 2001).

First the data were examined for an overall difference between the patterns of errors for the low- and high-stimulus rates. Log-linear modeling revealed there was a significant difference in the patterns for four out of five subjects. It was then necessary to see if there was a difference in the pattern of errors for various types of phonemes. The phonemes were divided into their distinctive feature categories (Miller and Nicely 1955; Singh 1968; Chomsky and Halle 1968): nasal, continuant, voicing, sibilant, duration, anterior, coronal, high, back, and distributed. This classification is a variant on the ones discussed above. It was chosen because of its close relationship to speech sounds. The manner of articulation features are as follows: nasal—the oral tract is closed and air flows through the nose; continuant—airflow is blocked at any point in the vocal tract; voiced—there is vibration of the vocal cords; frication—air is forced through a narrow aperture creating noise; strident—considerable noise is produced; sibilant—considerable high-frequency noise is produced; and duration—there are long duration sibilant fricatives. Place of articulation features are, for example, as follows: anterior—obstruction anterior to location for /ʃ/; coronal—tongue blade raised above neutral position /ə/; high—tongue body raised above the neutral position; back—tongue retracted to the back of the mouth; and distributed—relatively long constriction along the vocal tract.

An information transmission analysis was carried out, which showed that there was a trend for manner of articulation to be better perceived for high rates, and place of articulation for low rates, as illustrated in Figure 7.18 for 250 and 1500 pulses/s. Better manner of articulation could be expected for sibilants at high rates of stimulation, as they cause the nerves to fire in a random fashion (Fig. 7.19). With the other manner features (nasal, continuants, and voicing), higher rates of stimulation more accurately represent the speech envelope.

Place of articulation, however, was better perceived with a low rate of stimulation. Studies by Bruce et al (2000) suggest that at a high rate the response patterns at the edge of a population of excited fibers would lead to a poorer transmission of place information for multiple stimuli, due to a less clear-cut distinction between excited and nonexcited fibers. In summary, the phonetic analysis demonstrated that for high rates of stimulation manner of articulation was better perceived, and for low rates of stimulation place of articulation was better perceived.

A speech-processing strategy that provides manner of articulation at high rates of stimulation and place of articulation at low rates has been developed and is illustrated in Figure 7.20. This differential rate speech processor (DRSP) selects place information, which is usually within the low-frequency range, and presents it at low rates of stimulation. Manner of articulation, which is usually in the higher frequency range, is presented at a high rate of stimulation. This strategy is currently being evaluated, and is discussed in Chapter 14.

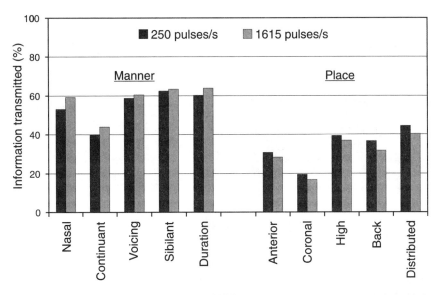

FIGURE 7.18. Information transmission analysis for consonants divided into their distinctive feature categories for the Nucleus SPEAK strategy at 250 and 1615 pulses/s (Grayden and Clark 2000; Clark 2001). Reprinted with permission from Clark G. M. 2001. Editorial Cochlear implants: climbing new mountains. The Graham Fraser Memorial Lecture 2001. *Cochlear Implants International* **2**(2): 75–97.

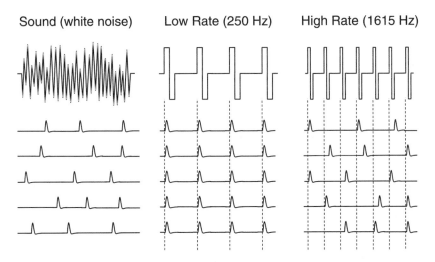

FIGURE 7.19. Manner of articulation—sibilants. The neural response patterns for sound (white noise), low rates of electrical stimulation (250 pulses/s), and high rates of electrical stimulation (1615 pulses/s) (Grayden and Clark 2000; Clark 2001). Reprinted with permission from Clark G. M. 2001. Editorial Cochlear implants: climbing new mountains. The Graham Fraser Memorial Lecture 2001. *Cochlear Implants International* **2**(2): 75–97.

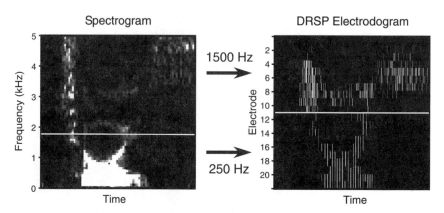

FIGURE 7.20. Differential rate speech processing (DRSP). The low frequencies, which predominantly represent place information, are coded on electrodes in the upper basal turn at low stimulus rates. The high frequencies, which predominantly represent manner, are coded on lower basal electrodes at high stimulus rates (Grayden and Clark 2000; Clark 2001). Reprinted with permission from Clark G.M. 2001. Editorial Cochlear implants: climbing new mountains. The Graham Fraser Memorial Lecture 2001. *Cochlear Implants International* **2**(2): 75–97.

Adaptive Dynamic Range Optimization

Adaptive dynamic range optimization (ADRO) is a mathematical routine that fits the dynamic range for sound intensities in each frequency band into the dynamic range for each electrode (Martin et al. 2000a,b). It arose out of the need in research on bimodal stimulation (i.e., hearing aid in one ear and implant in the other) to ensure that the loudness range with each device was comparable. The dynamic range is from the threshold (T) of hearing to the maximum comfortable (MC) level. The mathematical algorithm is a set of rules to control the output level. An audibility rule specifies that the output level should be greater than a fixed level between T and MC at least 70% of the time. The discomfort rule specifies that the output level should be below MC at least 90% of the time. It operates so that the acoustic input to the speech processor would be mapped to higher stimulus levels on the electrodes, especially at low speech intensities, than with the standard SPEAK strategy. This deficiency with SPEAK is emphasized by reduced speech perception scores with the Nucleus 22 system at low intensity levels (Muller-Deile et al 1995; Skinner et al 1997). It was anticipated that ADRO would improve speech perception at low signal levels.

It was also expected that ADRO would improve the recognition of speech in noise. This is illustrated in Figure 7.21, showing its mode of operation. With speech in the presence of noise (top row), the preemphasis and automatic gain control (AGC) in the standard speech processor reduces the intensity range of all the frequencies (middle). As a result there is a limited dynamic range available for electrical stimulation on many of the electrodes (bottom). In contrast, with

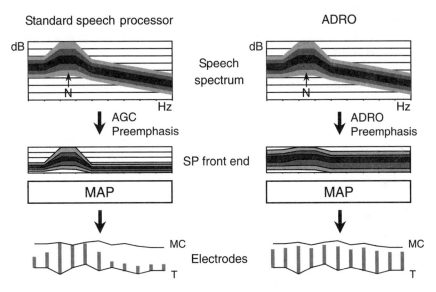

FIGURE 7.21. A diagram showing the frequency to electrode mapping, in the presence of noise (N), for the present cochlear implant speech processor (left column), and speech processor with adaptive dynamic range optimization (ADRO) (right column). Top row: The intensities of speech frequencies with added noise (N). Middle row: The intensity outputs of the speech processors after preemphasis and automatic gain control or ADRO. Bottom row: The relative current levels mapped to electrodes on the array. T, threshold; MC, maximum comfortable level (Martin et al 2000a,b; Clark 2001). Reprinted with permission from Clark G. M. 2001. Editorial Cochlear implants: climbing new mountains. The Graham Fraser Memorial Lecture 2001. *Cochlear Implants International* **2**(2): 75–97.

ADRO as shown on the right of Figure 7.21, there is less compression of the speech frequencies, and a greater dynamic range on all stimulating electrodes.

A preliminary study was undertaken at the University of Melbourne/Bionic Ear Institute on nine postlinguistically deaf subjects who used the Nucleus 24 system and SPEAK strategy. SPEAK with ADRO was compared with the standard SPEAK for speech at different loudness levels in quiet and in background noise (eight-talker babble) (Martin et al 1999, 2000a,b). It was found in quiet that at 50 dB there was a significant 16% improvement in open-set word-in-sentence scores, and at 60 dB there was a 9.5 % improvement in CNC word scores. There was, however, no difference between ADRO and standard SPEAK in the presence of multitalker babble at SNRs of 10 and 15 dB. This suggests a clearer spectral pattern is insufficient for perception in noise, and that phase information across channels needs to be transmitted as well.

Dual Microphones

Dual microphones and an adaptive beam former have been used to improve the recognition of speech in noise for people with cochlear implants. A study by

Peterson et al (1990) showed its value for people with hearing aids. They found an intelligibility gain of 9.5 dB for an adaptive filter of 10 ms with room-filtered white noise under living-room conditions.

The principles underlying the Griffiths/Jim adaptive beam former that was tested with a cochlear implant speech processor is illustrated in Figure 7.22 (van

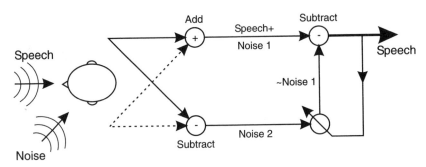

FIGURE 7.22. A diagram of the principle of operation of the Griffiths/Jim adaptive beam former (Reprinted with permission from van Hoesel and Clark 1995a).

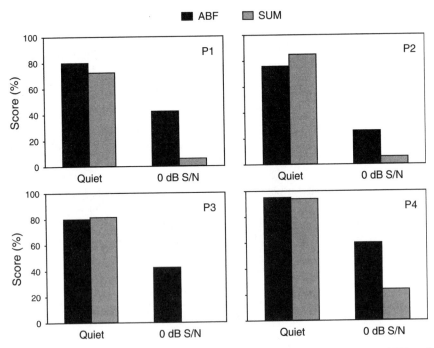

FIGURE 7.23. The open-set sentence test results for the adaptive beam former (ABF) and simple summation of a two microphone broadside array (SUM) for four unilateral cochlear implant users in quiet and in noise at a signal-to-noise (S/N) ratio of 0 dB. (Reprinted from Tong et al 1983 with permission from The Acoustical Society of America.)

Hoesel and Clark 1995a). When speech, for example, comes from directly in front and noise from the side, the signals from both microphones are sent to an adder and a subtractor. The output from the adder contains speech plus added noise. The output from subtractor has removed speech and has subtracted noise. The two signals are then subtracted and an adaptive filter is used to adjust noise to approximately zero with the result that the output is relatively free of noise.

The Griffiths/Jim adaptive beam former was implemented for two microphones as the front end to a SPEAK strategy. With this arrangement, the processor effectively used the two microphones to form a beam directly in front of the patients, and attempts to reject sounds not falling within it. The beam is a region in space that is shaped like a beam of light.

A study on four patients tested speech perception at 0 dB SNR with the signal directly in front of the patients, and the noise at 90 degrees to the left (van Hoesel and Clark 1995a). The results in Figure 7.23 showed dramatic improvements in noise for the adaptive beam-forming (ABF) strategy when compared to a strategy that simply added the two microphone signals together (SUM). There was a mean open-set sentence test score of 43% for the beam former, and 9% for the control at a very difficult (0 dB) SNR. All of the patients showed significant benefits. An analysis of variance showed a significant difference between the ABF and SUM strategies in noise but not in quiet. Further development is required to make this beam former more robust in multispeaker and reverberant conditions.

The characteristics of the ABF system were explored using a Knowles Electronic Manikin for Acoustic Research (KEMAR) in different environments. Figure 7.24 shows that with two microphones, the signal-to-noise advantage decreased from 20 dB in a near-anechoic situation to only 3 dB in a concrete stairwell. This indicates that to accommodate a wider range of real-world situations for patients, the use of four microphones with adaptive beam formers is highly desirable.

In the above type of studies, standardization is important and should be carried out on a KEMAR manikin, under specified direct-to-reverberant power ratios, at defined distances. The number of noise sources and the placement should be the same across tests, and for adaptation the rate at which the noise source is switched between two loud speakers should be defined.

A fixed beam-forming strategy was successfully implemented by Soede et al (1993) using five microphones, and they found a 7.5-dB signal-to-noise improvement in a diffuse field. However, five microphones were used, and this arrangement is not presently suitable for patients.

Bimodal Speech Processing

Bimodal speech processing uses electrical stimulation with an implant in one ear and acoustic stimulation with a hearing aid in the other ear. Research on bimodal stimulation commenced in the HCRC at the University of Melbourne/Bionic Ear Institute in 1989.

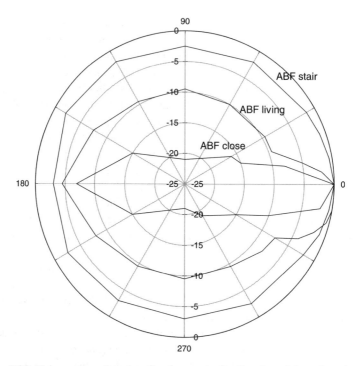

FIGURE 7.24. Noise suppression plots for changes in the direction of the noise when using the adaptive beam former (ABF) on a Knowles Electronic Manikin for Acoustic Research (KEMAR). The plots show the output as the angle of incidence of the noise source (in the absence of target speech) changes. The 0-dB, 0-degree reference condition is when noise is presented at 70 dB sound pressure level (SPL) directly in front of the manikin. There were three test environments: (1) close to the manikin (approaching anechoic), (2) living room (only slightly reverberant), and (3) stairwell (highly reverberant). Positive rotation is to the left from the perspective of the manikin. (Reprinted with permission from van Hoesel and Clark. Evaluation of a portable two-microphone adaptive beam forming speech processor with cochlear implant patients. *Journal of the Acoustical Society of America* **97**(4), pp 2498–2503. © 1995, Acoustical Society of America.)

The results of bimodal stimulation were first reported by Blamey (1990), Clark, Dooley et al (1991), and Dooley et al (1993), the device being referred to as the Combionic Aid. The bench-top system allowed the filter in the acoustic section to be controlled from the implant, as indicated in Figure 7.25. The filter center frequency, bandwidth, and attenuation were programmable. The Nucleus Multi-peak strategy was used in combination with a hearing aid with two acoustic strategies referred to as frequency response tailoring and peak sharpening. With frequency response tailoring the output was similar to that of a well-fitting hearing aid, with the ideal gain calculated from the person's audiogram using the National

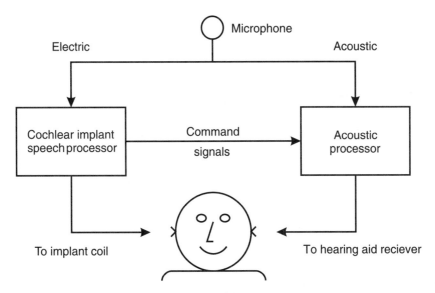

FIGURE 7.25. A diagram of bimodal speech processing showing separate devices and "intelligent" processing of signals from a single microphone.

Acoustic Laboratories rule. With peak sharpening, the ideal gain was calculated for the unvoiced segments of speech, and for the voiced segments the formants were tracked and the peaks emphasized. The schemes were evaluated for a test battery containing words, sentences, consonants, and vowels (Dooley et al 1993). The results were for sound presented to a single microphone and in quiet. The results showed better scores for the implant than the hearing aid strategies for all tests. The mean binaural scores for bimodal stimulation were higher than the Multipeak implant scores alone.

Further studies with this strategy were important because results obtained with advances in cochlear implant speech processing were on average better than for people with severe-to-profound losses using a hearing aid (Skinner et al 1994). Consequently, people with some useful hearing in the opposite ear are now suitable for an implant. This offers the possibility of combining the percepts from bimodal stimulation.

The BKB (Bench and Bamford 1979) word-in-sentence results for monaural and bimodal speech processing in 12 patients are shown in Figure 7.26 (Armstrong et al 1997). The results were obtained in quiet and with competing noise at +10 dB SNR where the noise and signal were spatially separated. The noise data were combined results, with the noise in front and on each side. The only significant difference between monaural and bimodal speech processing occurred with competing noise, and was better for the bimodal stimulation.

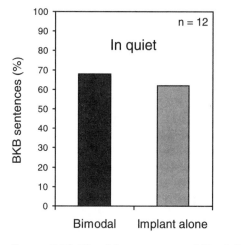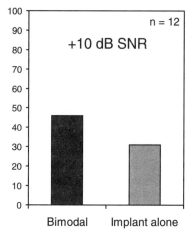

FIGURE 7.26. Bimodal versus monaural Bench-Kowal-Bamford (BKB) sentence results: (left) in quiet, and (right) with + 10 dB signal-to-noise ratio (SNR) (Reprinted with permission from Armstrong et al 1997).

Bilateral Speech Processing

The development of bilateral speech processing first required psychophysical research to determine the parameters to use, and this was discussed in Chapter 6. The speech processing for bilateral implants as well as with bimodal stimulation can be divided into three main approaches: (1) two independent processors (with two microphones), (2) one microphone with information distributed across both ears, and (3) two microphones with "central intelligence" to enhance binaural differences or reduce noise (van Hoesel and Clark 1995b, 1997, 1999).

Preliminary reports were made of bilateral electrical stimulation by Balkany et al (1988), who compared a single-channel implant in one ear and a multiple-channel implant in the other; by Pelizzone et al (1990), who discuss predominantly electrically evoked brainstem potentials with a binaural implant, and briefly mention subjective effects; and by Green et al (1992), who examined one of 12 subjects with bilateral implants, but did not evaluate speech perception.

Bilateral speech processing studies commenced at the HCRC in 1989. The study on the first bilateral implant patient was on the use of two independent processors (van Hoesel et al 1990, 1993). Each strategy was F0/F1/F2 with the MSP processor. The multispeaker babble was presented through the same loud speaker as the speech, and the dual independent binaural stimuli were adjusted so they were comfortably loud to avoid loudness summation. When the data were grouped for pairs of comparisons, the binaural stimuli were better than for stimulation on one side. The binaural results in quiet were on average 10% better than the best monaural side (Fig. 7.27). At a 10 dB SNR they were 30% better, and at a 5 dB SNR they were almost 50% better (van Hoesel et al 1990, 1993; van Hoesel and Clark 1993).

In another independent two-microphone strategy (van Hoesel et al 1993), both

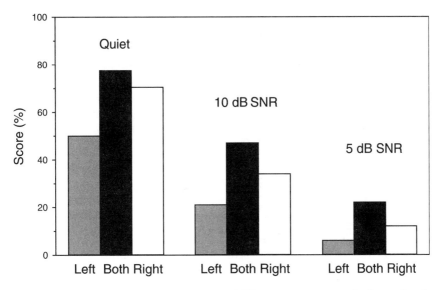

FIGURE 7.27. Binaural and monaural open-set BKB sentence scores on the first patient in quiet, and at S/N ratios of 5 and 10 dB. Dual independent F0/F1/F2 speech processors were used for binaural stimulation (Reprinted with permission from van Hoesel et al 1993. Psychophysical and speech perception studies: a case report on a binaural cochlear implant subject. *Journal of the Acoustical Society of America* **94**: 3178–3189.).

the Multipeak-MSP and the initial SMSP, a 16-filter-band precursor of SPEAK, were used. With SMSP the largest six spectral components were presented to the implant on each side. The results for the bilateral performance in quiet were similar to the best unilateral side, but in noise the binaural strategy performed considerably better.

With the single microphone, fixed-filter strategy evaluated on the same patient, the signal was analyzed into 12 spectral bands. These 12 bands and stimulus channels were presented to the patient in an interleaved fashion, six channels to the left and six to the right implant. Thus complementary spectral information was presented to the patient. Both monaural and binaural tests (in quiet) using SIT sentences were performed. The results showed only relatively small increases in performance for the binaural condition compared to the better monaural side alone. The lack of improvement in results for the binaural interleaved pulses could have been due to the fact that there is a limit on the number of sites of stimulation needed. This is supported by acoustic studies in which the formants have been divided between the ears and not led to significantly better results. Further work is needed to determine the possible benefits of presenting complementary information to the two sides.

With the above preliminary findings, using two independent processors the speech and noise were presented through the same loudspeaker. The results thus suggested the good interaural loudness discrimination in this patient was responsible for the better speech scores, operating in some way through a binaural release from masking (van Hoesel et al 1990, 1993). However, they could have been due to the method used to loudness balance the speech.

The above study was not designed to determine the effect of the head shadow where the noise and sound sources need to be spatially separated. It was assumed, however, that the good interaural loudness discrimination, as seen in the second patient, would lead to improved speech perception in noise tested in a free field. This was shown to be the case for a second patient (van Hoesel and Clark 1999) who had the SPEAK strategies on Spectra-22 drive each Nucleus CI-22 implant. The test setup with the separation of noise and signal is illustrated in Figure 7.28. The performance of the left ear (60%) was better than that of the right (20%). The results in Figure 7.29 show that when the signal is on the side of the better ear and the noise on the opposite side, there is no binaural advantage. On the other hand, when the noise is on the side of the better ear and the signal on the worse side, there is an advantage with the binaural hearing. It suggests that the ear attended to is the one with the better SNR even if the performance is poorer. The SNR is presumably due to the head shadow effect of the noise from the left side.

In another study in Melbourne on a patient from Manchester, UK, the SPEAK

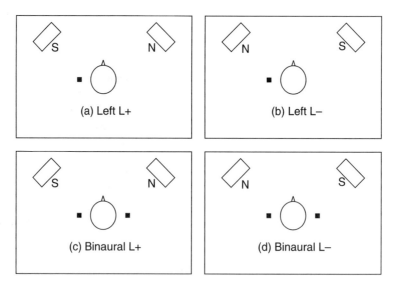

FIGURE 7.28. Test configuration to evaluate better monaural side (left) and binaural listening with 10 cm spatially separated signal (S) and noise (N) sources. (a) Left monaural test configuration for signal on the left, noise on the right (L+). (b) Left monaural test configuration for signal on the right, noise on the left (L−). (c) Binaural test configuration for signal on the left, noise on the right (L+). (d) Binaural test configuration for signal on the right, noise on the left (L−). (Reprinted with permission from van Hoesel and Clark 1999. Speech results with a bilateral multi-channel cochlear implant subject for spatially separated signal and noise. *Australian Journal of Audiology* **21**: 23–28.)

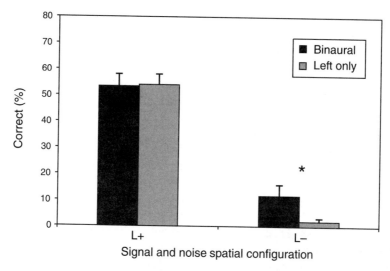

FIGURE 7.29. The average results from five test lists of City University of New York (CUNY) word in sentences presented at 70 dB, in the presence of multitalker babble at 65 dB. Two test configurations are illustrated: signal on the left with noise on the right (L+) and signal on the right with noise on the left (L−). The asterisk indicates a statistically significant difference between monaural and binaural results. (Reprinted with permission from van Hoesel and Clark 1999. Speech results with a bilateral multi-channel cochlear implant subject for spatially separated signal and noise. *Australian Journal of Audiology* **21**: 23–28.)

strategy selected the maximal outputs from eight out of 17 filters and was used with a Nucleus behind the ear (ESprit processor) (van Hoesel et al 2002). Both low (250 pulses/s) and high (1800 pulses/s) rates of stimulation were used on each electrode. The tests were conducted in a mildly reverberant room with the direct-to-reverberant ratio greater than 13 dB. The speech perception for this subject was evaluated in the presence of multispeaker babble at an SNR of 5 dB. For the tests on spatially separated noise the speech came from in front and the noise at 90 degrees to the left or right. The data showed that if speech was presented from the front and noise was presented to the right implanted ear, the results were significantly worse for the right ear, as it was the side receiving the noise. The better results for the left ear were due to the head shadow effect shielding the left ear from the noise on the right. There was no significant advantage when both implants were used together. There was also no advantage when using high rather than low rates of stimulation. It should also be noted that with the coincident noise and speech there was no advantage with binaural stimulation, indicating that the preliminary findings on release from masking may have been due to loudness balancing effects.

A study by the CRC for Cochlear Implant and Hearing Aid Innovation in Melbourne on four patients at Iowa also examined the effect of spatially separated noise on speech perception (van Hoesel and Tyler, submitted). The signal and noise were first presented together from directly in front, and speech scores were determined for monaural and binaural hearing. The noise was then moved opposite the left or right ear and the SNRs adjusted to return the speech perception scores to those for the frontal presentation. The results showed that there was a very significant head shadow effect of 4 to 5 dB ($p < .001$). There was a possible squelch effect of 1 to 2 dB ($p = .04$). However, this was seen only when the noise was on the left but not the right, and therefore not a strong effect, if at all (van Hoesel and Tyler, submitted).

As the psychophysical results from the first two University of Melbourne patients and the Manchester subject showed the limitations of using interaural time differences, future bilateral speech-processing research may need to especially use amplitude and place information rather than detailed interaural timing. In addition, the use of "intelligent" processing with one and two microphones, as discussed above (see Dual Microphones), should be incorporated as well. The presentation of complementary information on the two sides to provide optimal fusion of all the information still requires further investigation. The advantages of the head shadow effect in overall listening conditions require consideration in each person's situation, and should be evaluated before bilateral implants are routinely used.

References

Ainsworth, A. 1992. Advances in speech hearing and language processing. London, JAI Press.

Ainsworth, W. A. 1976. Mechanisms of speech recognition. Oxford, Pergamon Press: 42.

Ainsworth, W. A. and J. B. Millar. 1972. The effect of relative formant amplitude on the perceived identity of synthetic vowels. Language and Speech 15: 328.

Armstrong, M., P. Pegg, C. James and P. J. Blamey. 1997. Speech perception in noise with implant and hearing aid. American Journal of Otology 18: S140–S141.

Arndt, P., S. Staller, J. Arcaroli, A. Hines and K. Ebinger. 1999. Within-subject comparison of advanced coding strategies in the Nucleus 24 cochlear implant. Cochlear Corporation Report.

Balkany, T., W. Bogess and B. Dinner. 1988. Binaural cochlear implantation: comparison of 3M/House and Nucleus 22 devices with evidence of sensory integration. Laryngoscope 98: 1040–1043.

Barry, J. G., P. J. Blamey and L. F. A. Martin. 2002a. A multidimensional scaling analysis of tone discrimination ability in Cantonese-speaking children using a cochlear implant. Clinical Linguistics and Phonetics 16: 101–113.

Barry, J. G., P. J. Blamey, L. F. A. Martin, et al. 2002b. Tone discrimination in Cantonese-speaking children with cochlear implants. Clinical Linguistics and Phonetics 16: 79–99.

Battmer, R.-D., D. Gnadeberg, D. J. Allum-Mecklenberg and T. Lenarz. 1994. Matched-pair comparisons for adults using the Clarion or Nucleus devices. Annals of Otology, Rhinology and Laryngology 104: 251–254.

Bench, R. J. and J. Bamford. 1979. Speech-hearing tests and the spoken language of hearing-impaired children. London, Academic Press.

Blamey, P. J. 1990. Multimodal stimulation for speech perception. In: Rowe, M. and L. Aitkin, eds. Information processing in mammalian auditory and tactile systems. New York, Wiley-Liss: 267–280.

Blamey, P. J. and G. M. Clark. 1990. Place coding of vowel formants for cochlear implant patients. British Journal of Audiology 88: 667–673.

Blamey, P. J., R. C. Dowell, A. M. Brown, G. M. Clark and P. M. Seligman. 1987. Vowel and consonant recognition of cochlear implant patients using formant-estimating speech processors. Journal of the Acoustical Society of America 82: 48–57.

Blamey, P. J., R. C. Dowell, Y. C. Tong, A. M. Brown, S. M. Luscombe and G. M. Clark. 1984a. Speech processing studies using an acoustic model of a multiple-channel cochlear implant. Journal of the Acoustical Society of America 76: 104–110.

Blamey, P. J., R. C. Dowell, Y. C. Tong and G. M. Clark. 1984b. An acoustic model of a multiple-channel cochlear implant. Journal of the Acoustical Society of America 76: 97–103.

Blamey, P. J., L. F. Martin and G. M. Clark. 1985. A comparison of three speech coding strategies using an acoustic model of a cochlear implant. Journal of the Acoustical Society of America 77: 209–217.

Blamey, P. J., E. Parisi and G. M. Clark. 1995. Pitch matching of electric and acoustic stimuli. Annals of Otology, Rhinology and Laryngology 104: 220–222.

Blamey, P. J., B. C. Pyman, M. Gordon, et al. 1992. Factors predicting postoperative sentence scores in postlinguistically deaf adult cochlear implant patients. Annals of Otology, Rhinology and Laryngology 101: 342–348.

Boothroyd, A. 1968. Developments in speech audiometry. Sound 2: 3–10.

Brophy, J. T. 1977. Basic electronics for scientists. 3rd edition. Tokyo, McGraw Kogak-usha.

Bruce, I. C., L. S. Irlicht, M. W. White, S. J. O'Leary and G. M. Clark. 2000. Renewal-process approximation of a stochastic threshold model for electrical neural stimulation. Journal of Computational Neuroscience 9: 119–132.

Buden, S. V., M. Brown, A. G. Paolini and G. M. Clark. 1996. Temporal and entrainment response properties of cochlear nucleus neurons to intra cochleal electrical stimulation in the cat. Proceedings of the Australian Neuroscience Society 7: 104.

Burian, K., B. Eisenwort, E. Hochmair and I. Hochmair-Desoyer. 1984. Clinical experience with the "Vienna cochlear implant." In Keidel, W. D. and P. Finkenzeller, eds. Cochlear implants in clinical use. Advances in audiology. Basel, Karger 2: 19–29.

Carhart, R. 1965. Monaural and binaural discrimination against competing sentences. International Audiology 4: 5–10.

Chao, Y. R. 1930. A system of tone letters. Le Maître Phonétique 45: 24–27.

Chomsky, N. and M. Halle. 1968. The sound pattern of English. New York, Harper and Row.

Chouard, C. H., C. Fugain, B. Meyer and F. Chabolle. 1985. The Chorimac-12. A multi-channel cochlear implant for total deafness. Description and clinical results. Acta Oto-Rhino-Laryngologica Belgica 39(4): 735–748.

Chouard, C. H., B. Meyer, F. Chabolle, N. Alcaras and D. Gegu. 1984. Sound signal processing. Acta Oto-Laryngologica-supplement 411: 95–104.

Clark, G. M. 1969. Responses of cells in the superior olivary complex of the cat to electrical stimulation of the auditory nerve. Experimental Neurology 24: 124–136.

Clark, G. M. 1986. The University of Melbourne/Cochlear Corporation (Nucleus) program. Otolaryngologic Clinics of North America 19: 329–354.

Clark, G. M. 1987. The University of Melbourne–Nucleus multi-electrode cochlear implant. Advances in Oto-Rhino-Laryngology. Volume 38. Basel, Karger.

Clark, G. M. 2001. Editorial Cochlear implants: climbing new mountains. The Graham Fraser Memorial Lecture 2001. Cochlear Implants International 2(2): 75–97.

Clark, G. M., R. C. Black, D. J. Dewhurst, I. C. Forster, J. F. Patrick and Y. C. Tong. 1977. A multiple-electrode hearing prosthesis for cochlear implantation in deaf patients. Medical Progress Through Technology 5: 127–140.

Clark, G. M., R. C. Black, I. C. Forster, J. F. Patrick and Y. C. Tong. 1978. Design criteria of a multiple-electrode cochlear implant hearing prosthesis. Journal of the Acoustical Society of America 63: 631–633.

Clark, G. M., G. J. Dooley and P. J. Blamey. 1991. Combined electrical and acoustical stimulation using a bimodal speech processor. American Pediatric Otolaryngological Society Meeting, Hawaii, May.

Clark, G. M., R. C. Dowell, R. S. C. Cowan, B. C. Pyman and R. L. Webb. 1996. Multicenter evaluations of speech perception in adults and children with the Nucleus (Cochlear) 22-channel cochlear implant. In: Portmann, M., ed. Transplants and implants in otology III. Amsterdam, Kugler: 353–363.

Clark, G. M. and Y. C. Tong. 1981. Multiple-electrode cochlear implant for profound or total hearing loss: a review. Medical Journal of Australia 1: 428–429.

Clark, G. M. and Y. C. Tong. 1982. A multiple-channel cochlear implant. A summary of results for two patients. Archives of Otolaryngology 108: 214–217.

Clark, G. M., Y. C. Tong, Q. R. Bailey, et al. 1978. A multiple-electrode cochlear implant. Journal of the Oto-Laryngological Society of Australia 4: 208–212.

Clark, G. M., Y. C. Tong, R. C. Black, I. C. Forster, J. F. Patrick and D. J. Dewhurst. 1977. A multiple electrode cochlear implant. Journal of Laryngology and Otology 91: 935–945.

Clark, G. M., Y. C. Tong and R. C. Dowell. 1984. Comparison of two cochlear implant speech-processing strategies. Annals of Otology, Rhinology and Laryngology 93: 127–131.

Clark, G. M., Y. C. Tong and L. F. Martin. 1981a. A multiple-channel cochlear implant. An evaluation using open-set CID sentences. Laryngoscope 91: 628–634.

Clark, G. M., Y. C. Tong, L. F. Martin and P. A. Busby. 1981b. A multiple-channel cochlear implant. An evaluation using an open-set word test. Acta Oto-Laryngologica 91: 173–175.

Clark, G. M., Y. C. Tong, L. F. A. Martin, et al. 1981c. A multiple-channel cochlear implant: an evaluation using nonsense syllables. Annals of Otology, Rhinology and Laryngology 90: 227–230.

Cohen, N. L., S. B. Waltzman and S. G. Fisher. 1993. A prospective, randomised study of cochlear implants. New England Journal of Medicine 328: 233–282.

Collins, L., T. Zwolan, J. O'Neill and G. Wakefield. 1994. Analysis of electrode pair confusions and implications for speech recognition in cochlear implant listeners. In: Popelka, G.R, ed. Abstracts of the Seventeenth Midwinter Meeting of the Association for Research in Otolaryngology, St. Petersburg, FL: 161.

Davis, H. and S. R. Silverman. 1978. Hearing and deafness, 4th edition. New York, Holt, Rinehart and Winston.

Delattre, P., A. M. Liberman and F. S. Cooper. 1952. An experimental study of the acoustic determinants of vowel color observations on one and two formant vowels synthesised from spectrographic patterns. Word 8: 195–210.

Dewhurst, D. J. 1976. An introduction of biomedical instrumentation. London, Pergamon Press.

Dirks, D. D., T. S. Bell, R. N. Rossman and G. E. Kincaid. 1986. Articulation index predictions of contextually dependent works. Journal of the Acoustical Society of America 80: 82–92.

Dooley, G. J., P. J. Blamey, P. M. Seligman, et al. 1993. Combined electrical and acoustical stimulation using a bimodal prosthesis. Archives of Otolaryngology—Head and Neck Surgery 119: 55–60.

Dorman, M. F. 1993. Speech perception by adults. In: Tyler, R. S., ed. Cochlear implants. Audiological foundations. San Diego, Singular: 145–190.

Dorman, M. F., K. Dankowski and G. McCandless. 1989. Consonant recognition as a function of the number of channels of stimulation by patients who use the Symbion cochlear implant. Ear and Hearing 10: 288–291.

Dorman, M. F., P. C. Loizou, J. Fitzke and Z. Tu. 1998. The recognition of sentences in noise by normal-hearing listeners using simulations of cochlear-implant signal processors with 6–20 channels. Journal of the Acoustical Society of America 104: 3583–3585.

Dorman, M. F., P. C. Loizou and D. Rainey. 1997. Speech intelligibility as a function of the number of channels of stimulation for signal processors using sine-wave and noise-band outputs. Journal of the Acoustical Society of America 102: 2403–2411.

Dorman, M. F., S. Soli, K. Dankowski, L. M. Smith, G. McCandless and J. Parkin. 1990. Acoustic cues for consonant identification by patients who use the Ineraid cochlear implant. Journal of the Acoustical Society of America 88(5): 2074–2079.

Dowell, R. C. 1991. Speech perception in noise for multichannel cochlear implant users. PhD thesis. University of Melbourne.

Dowell, R. C., G. M. Clark, P. M. Seligman, P. J. Blamey, A. M. Brown and Y. C. Tong. 1986a. Perception of connected speech without lipreading, using a multichannel hearing prosthesis. Acta Oto-Laryngologica 102: 7–11.

Dowell, R. C., M. A. Marsh, R. D. Hollow, et al. 1993. Clinical comparison of open-set speech perception with the MSP and WSPIII speech processors and preliminary results for the new SPEAK processor. Abstracts of Third International Cochlear Implant Conference, Innsbruck: 14.6.

Dowell, R. C., L. F. Martin, Y. C. Tong, G. M. Clark, P. M. Seligman and J. F. Patrick. 1982. A 12-consonant confusion study on a multiple-channel cochlear implant patient. Journal of Speech and Hearing Research 25: 509–516.

Dowell, R. C., D. J. Mecklenburg and G. M. Clark. 1986b. Speech recognition for 40 patients receiving multichannel cochlear implants. Archives of Otolaryngology 112: 1054–1059.

Dowell, R. C., J. F. Patrick, P. J. Blamey, P. M. Seligman, D. K. Money and G. M. Clark. 1987a. Signal processing in quiet and noise. In: Banfai, P., ed. Cochlear implant: current situation. Proceedings of the International Cochlear Implant Symposium, September 7–12, 1987. Duren, West Germany. 495–498.

Dowell, R. C., P. M. Seligman, P. J. Blamey and G. M. Clark. 1987b. Speech perception using a two-formant 22-electrode cochlear prosthesis in quiet and in noise. Acta Oto-Laryngologica 104(5–6): 439–446.

Dowell, R. C., L. A. Whitford, P. M. Seligman, B. K.-H. Franz and G. M. Clark. 1990. Preliminary results with a miniature speech processor for the 22-electrode/Cochlear hearing prosthesis. In: Sacristan, T., ed. Otorhinolaryngology, head and neck surgery. Amsterdam, Kugler and Ghedini: 1167–1173.

Doyle, K. J., D. Mills, J. Larky, D. Kessler, W. M. Luxford and R. A. Schindler. 1995. Consonant perception by users of Nucleus and Clarion multichannel cochlear implants. American Journal of Otology 16: 676–681.

Dudley, H. 1939. The vocoder. Bell Labs Record 17: 122.

Dunn, H. K. 1950. The calculation of vowel resonances, and an electrical vocal tract. Journal of the Acoustical Society of America 22(6): 740–753.

Eddington, D. K. 1980. Speech discrimination in deaf subjects with cochlear implants. Journal of the Acoustical Society of America 68: 885–891.

Eddington, D. K. 1983. Speech recognition in deaf subjects with multichannel intracochlear electrodes. Annals of the New York Academy of Science 405: 241–258.

Edgerton, B. J. 1985. Implications of optimized single-channel cochlear implants. Hearing Journal 38: 17–20.

Edgerton, B. J. and J. A. Brimacombe. 1984. Effects of signal processing by the House-3M cochlear implant on consonant perception. Acta Oto-Laryngologica-supplement 411: 115–123.

Erber, N. P. 1972. Auditory, visual, and auditory-visual speech recognition of consonants by children with normal and impaired hearing. Journal of Speech and Hearing Research 15: 413–422.

Fant, G. 1959. Acoustic analysis and synthesis of speech with applications to Swedish. Ericsson Technics 15: 3–108.

Fant, G. 1973. Speech sounds and features. Cambridge, MIT Press.

Fant, G. and J. Martony. 1962. Instrumentation for parametric synthesis (OVE II). Stockholm, Speech transmission lab QPR: 18.

Fishman, K. E., R. V. Shannon and W. H. Slattery. 1997. Speech recognition as a function of electrodes used in the SPEAK cochlear implant processor. Journal of Speech, Language and Hearing Research 40: 1201–1215.

Flanagan, J. L. 1957. Difference limen for formant amplitude. Journal of Speech and Hearing Disorders 22: 205.

Flanagan, J. L. and L. R. Rabiner. 1973. Speech synthesis. Stroudsburg, Hutchinson and Ross.

Flanagan, J. L. and M. G. Saslow. 1958. Pitch discrimination for synthetic vowels. Journal of the Acoustical Society of America 30(5): 435–442.

Fletcher, H. and J. C. Steinberg. 1929. Articulation testing methods. Bell Systems Technology Journal 8: 806–854.

Fourcin, A. J. 1968. Speech source inference. IEEE Transactions of Audio Electroacoustics AU-16: 65.

Fourcin, A. J., S. M. Rosen, B. C. Moore, et al. 1979. External electrical stimulation of the cochlea: clinical, psychophysical, speech-perceptual and histological findings. British Journal of Audiology 13(3): 85–107.

French, N. R. and J. C. Steinberg. 1947. Factors governing the intelligibility of speech sounds. Journal of the Acoustical Society of America 19: 90–119.

Fry, D. B. 1979. The physics of speech. Cambridge textbooks in linguistics. Cambridge, Cambridge University Press.

Fu, Q.-J. and R. V. Shannon. 1999. Recognition of spectrally degraded and frequency-shifted vowels in acoustic and electric hearing. Journal of the Acoustical Society of America 105: 1889–1900.

Fugain, C., B. Meyer, F. Chabolle and C. H. Chouard. 1984. Clinical results of the French multichannel cochlear implant. Acta Oto-Laryngologica-supplement 411: 237–246.

Fujimura, O. 1962. Analysis of nasal consonants. Journal of the Acoustical Society of America 34: 1865–1875.

Gantz, B. J., B. F. McCabe, R. S. Tyler and J. P. Preece. 1987. Evaluation of four cochlear implant designs. Annals of Otology, Rhinology and Laryngology 96: 145–147.

Gantz, B. J., R. S. Tyler, J. F. Knutson, et al. 1988. Evaluation of five different cochlear implant designs: audiologic assessment and predictors of performance. Laryngoscope 98: 1100–1106.

Geier, L. L. and S. J. Norton. 1992. The effects of limiting the number of Nucleus 22 cochlear implant electrodes programmed on speech perception. Ear and Hearing 13: 340–348.

Geurts, L. and J. Wouters. 1999. Enhancing the speech envelope of continuous interleaved sampling processors for cochlear implants. Journal of the Acoustical Society of America 105: 2476–2484.

Grayden, D. B. and G. M. Clark. 2000. The effect of rate stimulation of the auditory nerve on phoneme recognition. In: Barlow, M., ed. Proceedings of the Eighth Australian International Conference on Speech Science and Technology. Canberra, Australian Speech Science and Technology Association: 356–361.

Grayden, D. B. and G. M. Clark. 2001. Improved sound processor for cochlear implants. International patent application no. PCT/AU00/01038.

Green, J. D., D. M. Mills, B. A. Bell, W. M. Luxford and L. L. Tonokawa. 1992. Binaural cochlear implants. American Journal of Otology 13(6): 502–506.

Halliday, M. A. K. 1963. The tones of English. Archivum Linguisticum 15: 1.

Hanekom, J. J. and R. V. Shannon. 1996. Place pitch discrimination and speech recognition in cochlear implant users 43: 27–40.

Hartmann, R., G. Topp and R. Klinke. 1984a. Discharge patterns of cat primary auditory fibres with electrical stimulation of the cochlea. Hearing Research 13: 47–62.

Hartmann, R., G. Topp and R. Klinke. 1984b. Electrical stimulation of the cat cochlea—discharge pattern of single auditory fibres. Advances in Audiology 1: 18–29.

Henry, B. A., C. M. McKay, H. J. McDermott and G. M. Clark. 2000. The relationship between speech perception and electrode discrimination in cochlear implantees. Journal of the Acoustical Society of America 108: 1269–1280.

Hill, F. J., L. P. McRae and R. P. McClellan. 1968. Speech recognition as a function of channel capacity in a discrete set of channels. Journal of the Acoustical Society of America 44: 13–18.

Hirsh, I. J. 1950. The relation between localization and intelligibility. Journal of the Acoustical Society of America 22: 196–200.

Hochmair, E. S. and I. J. Hochmair-Desoyer. 1983. Percepts elicited by different speech coding strategies. Annals of the New York Academy of Sciences 405: 268–279.

Hochmair, E. S., I. J. Hochmair-Desoyer and K. Burian. 1979. Investigations towards an artificial cochlea. International Journal of Artificial Organs 2(5): 255–261.

Hochmair-Desoyer, I. J., E. S. Hochmair and K. Burian. 1981. Four years of experience with cochlear prostheses. Medical Progress Technology 8: 107–119.

Hochmair-Desoyer, I. J., E. S. Hochmair, R. E. Fischer and K. Burian. 1980. Cochlear prostheses in use: recent speech comprehension results. Archives of Otorhinolaryngology 229: 81–98.

Hollow, R. D., R. C. Dowell, R. S. C. Cowan, M. C. Skok, B. C. Pyman and G. M. Clark. 1995. Continuing improvements in speech processing for adult cochlear implant patients. Annals of Otology, Rhinology and Laryngology 104(suppl 166): 292–294.

Holmes, A. E., F. J. Kemler and G. E. Merwin. 1987. The effects of varying the number of cochlear implant electrodes on speech perception. American Journal of Otology 8: 240–246.

House, W. F., K. I. Berliner and L. S. Eisenberg. 1981. The cochlear implant: 1980 update. Acta Oto-Laryngologica 91: 457–462.

Jeffress, L. A. 1948. A place theory of sound localization. Physiological Psychology 41: 35–39.

Kennedy, E., H. Levitt, A. C. Neuman and W. Weiss. 1998. Consonants-vowel intensity ratios for maximizing consonant recognition by hearing-impaired listeners. Journal of the Acoustical Society of America 103: 1098–1114.

Kessler, D. K., G. E. Loeb and M. J. Barker. 1995. Distribution of speech recognition results with the Clarion cochlear prosthesis. Annals of Otology, Rhinology and Laryngology 104: 283–285.

Kewley-Port, D. and D. B. Pisoni. 1983. Perception of static and dynamic acoustic cues to place of articulation in initial stop consonants. Journal of the Acoustical Society of America 73: 1779–1793.

Kileny, P. R., T. A. Zwolan, S. Zimmerman-Phillips and J. L. Kemink. 1992. A comparison of round-window and transtympanic promontory electric stimulation in cochlear implant candidates. Ear and Hearing 13: 294–299.

Ladefoged, P. 1975. A course in phonetics. New York, Harcourt Brace Jovanovich.

Ladefoged, P. 1993. A course in phonetics. 3rd ed. New York, Harcourt Brace Jovanovich.

Laird, R. K. 1979. The bioengineering development of a sound encoder for an implantable hearing prosthesis for the profoundly deaf. Master of engineering science thesis. University of Melbourne.

Lawrence, W. 1953. The synthesis of speech from signals which have a low information rate. In: Communication theory. London, Butterworth: 460.

Lawson, D. T., B. S. Wilson, M. Zerbi and C. C. Finley. 1996. Speech processors for auditory prostheses. Third quarterly progress report. NIH contract No. 1-DC-5-2103.

Lehiste, I. and G. E. Peterson. 1959. Vowel amplitude and phonemic stress in American English. Journal of the Acoustical Society of America 31: 428–435.

Lehiste, I. and G. E. Peterson. 1961. Transitions, glides and diphthongs. Journal of the Acoustical Society of America 33: 268–277.

Liberman, A. M., P. C. Delattre and F. S. Cooper. 1958. Some cues for the distinction between voiced and voiceless stops in initial position. Language and Speech 1: 153.

Liberman, A. M., P. C. Delattre, F. S. Cooper and L. J. Gerstman. 1954. The role of consonant-vowel transitions in the perception of the stop and nasal consonants. Psychological Monographs 8: 1–13.

Liberman, A. M., P. C. Delattre, L. J. Gerstman. and F. S. Cooper. 1956. Tempo of frequency change as a cue for distinguishing classes of speech sounds. Journal of Experimental Psychology 52: 127.

Liberman, A. M., K. S. Harris, H. S. Hoffman and B. C. Griffith. 1957. The discrimination of speech sounds within and across phoneme boundaries. Journal of Experimental Psychology 54: 358.

Licklider, J. C. R. 1948. The influence of interaural phase upon the masking of speech by white noise. Journal of the Acoustical Society of America 20: 150–159.

Lindblom, B. E. F. and M. Studdert-Kennedy. 1967. On the role of formant transitions in vowel recognition. Journal of the Acoustical Society of America 42: 830–843.

Lisker, L. and A. S. Abramson. 1967. Some effects of context on voice onset time in English stops. Language and Speech 10: 1.

Litovsky, R. Y. and H. S. Colburn. 1999. The precedence effect. Journal of the Acoustical Society of America 106: 1633–1654.

MacKeith, N. W. and R. R. Coles. 1971. Binaural advantages in hearing of speech. Journal of Laryngology and Otology 85: 213–232.

Martin, L. F. A., P. J. Blamey, C. James, K. L. Galvin and D. MacFarlane. 2000a. Adaptive range of optimisation for hearing aids. In: Barlow, M., ed. Proceedings of the Eighth

Australian International Conference on Speech Science and Technology. Canberra, Australian Speech Science and Technology Association: 373–378.

Martin, L. F. A., C. James, P. J. Blamey, et al. 2000b. Adaptive dynamic range optimisation for cochlear implants. Australian Journal of Audiology 22(suppl): 64.

Martin, L. F. A., C. James, P. J. Blamey, B. Swanson, Y. Just and D. S. MacFarlane. 1999. Adaptive dynamic range optimisation; pre-processing for cochlear implants. 1999 Conference on Implantable Auditory Prostheses, Asilomar Conference Center, California: 127.

McDermott, H. J., C. M. McKay and A. Vandali. 1992. A new portable sound processor for the University of Melbourne/Nucleus Limited multi-electrode cochlear implant. Journal of the Acoustical Society of America 91: 3367–3371.

McKay, C. M. and H. J. McDermott. 1993. Perceptual performance of subjects with cochlear implants using the spectral maxima sound processor (SMSP) and the mini speech processor (MSP). Ear and Hearing 14: 350–367.

McKay, C. M., H. J. McDermott and G. M. Clark. 1991. Preliminary results with a six spectral maxima speech processor for the University of Melbourne/Nucleus multiple electrode cochlear implant. Journal of the Oto-Laryngological Society of Australia 6: 354–359.

McKay, C. M., H. J. McDermott and G. M. Clark. 1993. Temporal pitch coding for cochlear implantees: the effects of carrier rate and amplitude-modulation of pulsatile electrical stimuli. Journal of the Acoustical Society of America 93: 2333.

McKay, C. M., H. J. McDermott and G. M. Clark. 1994. The beneficial use of channel interactions for the improvement of speech perception for multichannel cochlear implants. Australian Journal of Audiology 15(2 suppl): 20–21.

McKay, C. M., H. J. McDermott, A. Vandali and G. M. Clark. 1992. A comparison of speech perception of cochlear implantees using the Spectral Maxima Sound Processor (SMSP) and the MSP (Multipeak) processor. Acta Oto-Laryngologica 112: 752–761.

Merzenich, M., C. Byers and M. White. 1984. Scala tympani electrode arrays. Fifth quarterly progress report. NIH contract NO1-NS9-2353: 1–11.

Miller, G. A. and P. E. Nicely. 1955. An analysis of perceptual confusions among some English consonants. Journal of the Acoustical Society of America 27(3): 338–352.

Muller-Deile, J., B. J. Schmidt and H. Rudert. 1995. Effects of noise on speech discrimination in cochlear implant patients. Annals of Otology, Rhinology and Laryngology 104: 303–306.

Nilsson, M., S. D. Soli and J. A. Sullivan. 1994. Development of the hearing in noise test for the measurement of speech reception thresholds in quiet and in noise. Journal of the Acoustical Society of America 95: 1085–1099.

Nordmark, J. O. 1968. Mechanisms of frequency discrimination. Journal of the Acoustical Society of America 44(6): 1532–1540.

O'Connor, J. D., L. J. Gerstman, A. M. Liberman, P. C. Delattre and F. S. Cooper. 1957. Acoustic cues for the perception of initial /w,j,r,l/ in English. Word 13: 24.

O'Shaughnessy, D. 1987. Speech communication: human and machine. Reading, Mass., Addison-Wesley.

Paolini, A. G. and G. M. Clark. 1997. The effect of pulsatile intracochlear electrical stimulation on intracellularly recorded cochlear nucleus neurons. In: Clark, G. M., ed. Cochlear implants. XVI World Congress of Otorhinolaryngology Head and Neck Surgery. Bologna, Monduzzi Editore: 119–124.

Parkinson, A. J., R. S. Tyler, G. G. Woodworth, M. W. Lowder and B. J. Gantz. 1996. A within-subject comparison of adult patients using the Nucleus F0F1F2 and

F0F1F2B3B4B5 speech processing strategies. Journal of Speech and Hearing Research 39: 261–277.

Pavlovic, C. V., G. A. Studebaker and R. L. Sherbecoe. 1986. An articulation index based procedure for predicting the speech recognition performance of hearing-impaired individuals. Journal of the Acoustical Society of America 80: 50–57.

Peissig, J. and B. Kollmeier. 1997. Directivity of binaural noise reduction in spatial multiple noise-source arrangements for normal hearing and impaired listeners. Journal of the Acoustical Society of America 105: 1660–1670.

Pelizzone, M., A. Kasper and P. Montandon. 1990. Binaural interaction in a cochlear implant patient. Hearing Research 48: 287–290.

Peterson, G. E. and H. L. Barney. 1952. Control methods used in a study of the vowels. Journal of the Acoustical Society of America 24: 175–184.

Peterson, P. M., S. M. Wei, W. M. Rabinowitz and P. M. Zurek. 1990. Robustness of an adaptive beamforming method for hearing aids. Acta Otolaryngology-supplement 469: 85–90.

Pickett, J. M. 1980. The sounds of speech communication. A primer of acoustic phonetics and speech perception. Baltimore, University Park Press: 194.

Plomp, R. 1975. Auditory psychophysics. Annual Review of Psychology 26: 207–232.

Plomp, R. 1976. Aspects of tone sensation: a psychophysical study. London, Academic Press.

Pollack, I. and J. M. Pickett. 1958. Stereophonic listening and speech intelligibility against voice babble. Journal of the Acoustical Society of America 30: 131–133.

Pols, L. C. W., L. J. T. Van Der Kamp and R. Plomp. 1969. Perceptual and physical space of vowel sounds. Journal of the Acoustical Society of America 46: 458–467.

Rabiner, L. R. and R. W. Schafer. 1978. Digital processing of speech signals. Englewood Cliffs, Prentice-Hall.

Remez, R. E., P. E. Rubin and D. B. Pisoni. 1981. Speech perception without traditional speech cues. Science 212: 947–949.

Rose, J. E., J. F. Brugge, D. J. Anderson and J. E. Hind. 1967. Phase-locked response to low-frequency tones in single auditory nerve fibers of the squirrel monkey. Journal of Neurophysiology 30: 769–93.

Rose, J. E., J. F. Brugge, D. J. Anderson and J. E. Hind. 1969. Time structure of discharges in single auditory nerve fibers of the squirrel monkey in response to complex periodic sounds. Journal of Neurophysiology 32: 386–401.

Rosen, S. 1989. Temporal information in speech and its relevance for cochlear implants. In: Fraysse, B. and N. Cochard, eds. Cochlear implant: acquisitions and controversies. International Symposium, Toulouse: 3–26.

Sammeth, C. A., M. F. Dorman and C. J. Stearns. 1999. The role of consonant-vowel intensity ratio in the recognition of voiceless stop consonants by listeners with hearing impairment. Journal of Speech Language and Hearing Research 42: 42–55.

Schindler, R. A., D. K. Kessler and M. A. Barker. 1995. Clarion patient performance: an update on the clinical trials. Annals of Otology, Rhinology and Laryngology 104: 269–272.

Schroeder, M. R. 1968. Reference signal for signal quality studies. Journal of the Acoustical Society of America 44: 1735–1736.

Schroeder, M. R. and E. E. David. 1960. A vocoder for transmitting 10kc/s speech over a 3.5 kc/s channel. Acustica 10: 35.

Seligman, P. M. and H. J. McDermott. 1995. Architecture of the SPECTRA 22 speech processor. Annals of Otology, Rhinology and Laryngology 104(suppl 166): 139–141.

Shannon, R. V., A. Jensvold, M. Padilla, M. E. Robert and X. Wang. 1999. Consonant recordings for speech testing. Journal of the Acoustical Society of America 106: L71–L74.

Shannon, R. V., F.-G. Zeng, V. Kamath, J. Wygonski and M. Ekelid. 1995. Speech recognition with primarily temporal cues. Science 270: 303–304.

Shannon, R. V., F.-G. Zeng and J. Wygonski. 1998. Speech recognition with altered spectral distribution of envelope cues. Journal of the Acoustical Society of America 104: 2467–2476.

Singh, S. 1968. A distinctive feature analysis of responses to a multiple choice intelligibility test. International Review of Applied Linguistics 6: 37–53.

Skinner, M. W., G. M. Clark, L. A. Whitford, et al. 1994. Evaluation of a new spectral peak coding strategy for the Nucleus 22 channels cochlear implant system. American Journal of Otology 15: 15–27.

Skinner, M. W., L. K. Holden, T. A. Holden, M. E. Demorest and M. S. Fourakis. 1997. Speech recognition as simulated soft, conversational, and raised-to-loud vocal efforts by adults with cochlear implants. Journal of the Acoustical Society of America 101: 3766–3782.

Skinner, M. W., L. K. Holden, T. A. Holden, et al. 1991. Performance of postlinguistically deaf adults with the Wearable Speech Processor (WSP III) and Mini Speech Processor (MSP) of the Nucleus multi-electrode cochlear implant. Ear and Hearing 12: 3–22.

Skinner, M. W., L. K. Holden, L. A. Whitford, K. L. Plant, C. Psarros and T. A. Holden. 2002. In press. Speech recognition with the Nucleus 24 SPEAK, Ace, and CIS speech coding strategies in newly implanted adults. Ear and Hearing 23: 207–223.

Skinner, M. W., M. M. Karstaedt and J. D. Miller. 1982. Amplification bandwidth and speech intelligibility for two listeners with sensorineural hearing loss. Audiology 21: 251–268.

Soede, W., A. J. Berkhout and F. A. Bilson. 1993. Development of a directional hearing instrument based on array technology. Journal of the Acoustical Society of America 94: 785–798.

Stevens, K. N. and S. E. Blumstein. 1981. The search for invariant acoustic correlates of phonetic features. In: Einnas, P. D. and J. L. Miller, eds. Perspectives on the study of speech. Hillsdale NJ, Lawrence Erlbaum: 1–38.

Strevens, P. 1960. Spectra of fricative noise in human speech. Language and Speech 3: 32.

Summerfield, A. Q. and M. P. Haggard. 1972. Speech rate effects in the perception of voicing. Speech Synthesis and Perception 6: 1.

Throckmorton, C. S. and L. M. Collins. 1999. Investigation of the effects of temporal and spatial interactions on speech-recognition skills in cochlear-implant subjects. Journal of the Acoustical Society of America 105: 861–873.

Tong, Y. C., R. C. Black, G. M. Clark, et al. 1979. A preliminary report on a multiple-channel cochlear implant operation. Journal of Laryngology and Otology 93: 679–695.

Tong, Y. C., P. J. Blamey, R. C. Dowell and G. M. Clark. 1983a. Psychophysical studies evaluating the feasibility of a speech processing strategy for a multiple-channel cochlear implant. Journal of the Acoustical Society of America 74: 73–80.

Tong, Y. C., G. M. Clark, P. M. Seligman and J. F. Patrick. 1980. Speech processing for a multiple-electrode cochlear implant hearing prosthesis. Journal of the Acoustical Society of America 68: 1897–1899.

Tong, Y. C., R. C. Dowell, P. J. Blamey and G. M. Clark. 1983b. Two-component hearing sensations produced by two-electrode stimulation in the cochlea of a deaf patient. Science 219: 993–994.

Tong, Y. C., W. K. Lai, M. Denison, et al. 1989. Speech processors for auditory prostheses. Third quarterly progress report. NIH No. 1-DC-9-2400.

Tong, Y. C., R. van Hoesel, W. K. Lai, A. Vandali, J. M. Harrison and G. M. Clark. 1990. Speech processors for auditory prostheses. Sixth quarterly progress report. NIH Contract No. 1-DC-9-2400.

Tong, Y. C., A. Vandali, J. M. Harrison, R. van Hoesel, J. S. Chang and G. M. Clark. 1991. Speech processors for auditory prostheses. Eighth quarterly progress report. NIH Contract No. 1-DC-9-2400.

Turner, C. W., P. E. Souza and L. N. Forget. 1995. Use of temporal envelope cues in speech recognition by normal and hearing-impaired listeners. Journal of the Acoustical Society of America 97: 2568–2576.

Tye-Murray, N., R. S. Tyler, G. G. Woodworth and B. J. Gantz. 1992. Performance over time with a Nucleus or Ineraid cochlear implant. Ear and Hearing 13: 200–209.

Tyler, R. S. and M. W. Lowder. 1992. Audiological management and performance of adult cochlear-implant patients. Ear, Nose and Throat Journal 71: 117–128.

Tyler, R. S., B. C. J. Moore and F. K. Kuk. 1989. Performance of some of the better cochlear-implant patients. Journal of Speech and Hearing Research 32: 887–911.

van Hoesel, R., R. Ramsden and M. O'Driscoll. 2002. Sound-direction identification, interaural time delay discrimination and speech intelligibility advantages in noise for a bilateral cochlear implant user. Ear and Hearing 23(2): 137–149.

van Hoesel, R. and R. Tyler. Submitted. Speech perception, localization, and lateralization with bilateral cochlear implant users. Journal of the Acoustical Society of America.

van Hoesel, R. J. M. and G. M. Clark. 1993. Cochlear implant-bilateral psychophysical, speech and processing studies. Human Communications Research Centre 6th Annual Report: 13–16.

van Hoesel, R. J. M. and G. M. Clark. 1995a. Evaluation of a portable two-microphone adaptive beamforming speech processor with cochlear implant patients. Journal of the Acoustical Society of America 97: 2498–2503.

van Hoesel, R. J. M. and G. M. Clark. 1995b. Fusion and lateralization study with two binaural cochlear implant patients. Annals of Otology, Rhinology and Laryngology 104(suppl 166): 233–235.

van Hoesel, R. J. M. and G. M. Clark. 1997. Psychophysical studies with two binaural cochlear implant subjects. Journal of the Acoustical Society of America 102: 495–507.

van Hoesel, R. J. M. and G. M. Clark. 1999. Speech results with a bilateral multi-channel cochlear implant subject for spatially separated signal and noise. Australian Journal of Audiology 21: 23–28.

van Hoesel, R. J. M., Y. C. Tong, R. D. Hollow and G. M. Clark. 1993. Psychophysical and speech perception studies: a case report on a binaural cochlear implant subject. Journal of the Acoustical Society of America 94: 3178–3189.

van Hoesel, R. J. M., Y. C. Tong, R. D. Hollow, J. Huigen and G. M. Clark. 1990. Preliminary studies on a bilateral cochlear implant user. Journal of the Acoustical Society of America 88(suppl 1): S193.

van Tasell, D. J., D. G. Greenfield, J. J. Logemann and D. A. Nelson. 1992. Temporal cues for consonant recognition: training, talker generalization, and use in evaluation of cochlear implants. Journal of the Acoustical Society of America 92: 1247–1257.

Van Tasell, D. J., S. D. Soli and V. M. Kirby. 1987. Speech waveform envelope cues for consonant recognition. Journal of the Acoustical Society of America 82(4): 1152–1161.

Vandali, A. E. 2001. Emphasis of short-duration acoustic speech cues for cochlear implant users. Journal of the Acoustical Society of America 109: 2049–2061.

Vandali, A. E., J. M. Harrison, J. Huigen, K. Plant and G. M. Clark. 1995. Multichannel cochlear implant speech processing: further variations of the Spectral Maxima sound processor strategy. Annals of Otology, Rhinology and Laryngology 104(suppl 166): 378–381.

Vandali, A. E., L. A. Whitford, K. L. Plant and G. M. Clark. 2000. Speech perception as a function of electrical stimulation rate: using the Nucleus 24 cochlear implant system. Ear and Hearing 21: 608–624.

Wang, W. S.-Y. 1959. Transition and release as perceptual cues for final plosives. Journal of Speech and Hearing Research 2: 66.

Wilson, B. S. 1997. The future of cochlear implants. British Journal of Audiology 31: 205–225.

Wilson, B. S. 2000. Strategies for representing speech information with cochlear implants. In: Niparko, J. K., ed. Cochlear implants: principles and practice. Philadelphia, Lippincott Williams & Wilkins.

Wilson, B. S., C. C. Finley, J. C. Farmer, et al. 1988. Comparative studies of speech processing strategies for cochlear implants. Laryngoscope 98: 1069–1077.

Wilson, B. S., D. T. Lawson, M. Zerbi and C. C. Finley. 1992. Speech processors for auditory prostheses. Twelfth quarterly progress report, April. 1992. NIH contract No. 1-DC-9-2401. Research Triangle Institute.

Wilson, B. S., D. T. Lawson, M. Zerbi and C. C. Finley. 1993. Speech processors for auditory prostheses. Fifth quarterly progress report, Oct. 1993. NIH contract No. 1-DC-2-2401. Research Triangle Institute.

Xu, S., R. C. Dowell and G. M. Clark. 1987. Results for Chinese and English in a multichannel cochlear implant patient. Annals of Otology, Rhinology and Laryngology 96(suppl 128): 126–127.

8
Engineering

The magnitude of the task of presenting speech to the auditory central nervous system by electrical stimulation can be appreciated when it is considered that there are from 35,000 to 40,000 nerve fibers in the young human auditory nerve (Rasmussen 1960), and a large proportion of these convey information about the speech frequencies. Despite the limitations of coding sound with electrical signals, the electrophysiological, psychophysical, and speech research referred to in the respective chapters has demonstrated that speech and other sounds can be reproduced. The approach of the University of Melbourne was to select the information most suitable for transmission through what is essentially an electroneural bottleneck between the world of sound and the neural pathways. Studies have demonstrated that the recognition of open sets of words or sentences and thus running speech, can be achieved with speech processing strategies that extract the energy of formant frequencies, and spectral peaks and maxima, as well as fixed-filter schemes. The information is provided by multiple-channel excitation of the cochlear (auditory) nerve fibers, which allows the place and temporal frequency as well as intensity cues to be coded.

The multiple-channel cochlear prosthesis, as illustrated in the book's Introduction, consists of a microphone, speech processor, transmitting coil, implanted receiver-stimulator, and electrode array in the cochlea to stimulate residual cochlear nerve fibers. Speech and environmental sounds are picked up by the microphone, converted to analog electrical signals, and conveyed to the speech processor. The speech processor may be a small unit worn behind the ear providing established strategies, or a body-worn device that allows a range of advanced strategies to be tested. In a noisy environment a directional microphone permits better speech perception. The sound waves from the microphone are converted into voltages that are transformed into a digital code by the speech processor that in turn controls the stimulation of auditory neurons with temporal and spatial patterns from an implant that are perceived by the brain as running speech or other meaningful sounds. This information must be transmitted accurately to the implanted receiver-stimulator for decoding into electrical stimuli via a transmitting and receiving coil. The implanted receiver-stimulator is essential for multiple-

454

channel stimulation of the auditory nerve in the cochlea, via a bundle or array of electrodes. As nerve fibers require currents to excite them, power is also needed, and this too is transmitted by inductive coupling between the external and internal coils (aerials). The electronics for the receiver-stimulator need to be reliable, and not fail after many years of use. They should also be designed so that the stimulus parameters could be varied to enable advanced speech-processing strategies to be evaluated. It is easier to provide a patient with an alternative speech-processor, than it is to remove the receiver-stimulator and replace it with a more up-to-date model.

The implanted receiver-stimulator and electrode array must be biocompatible, and not lead to long-term adverse tissue effects, and the method of signal transmission through the skin should not cause tissue damage. The electronics for the receiver-stimulator should be sealed within a container that is impervious to body fluids. It was initially thought desirable to design the device with a connector so that the receiver-stimulator could be removed and replaced without disturbing the implanted electrode array if an electrode failure occurred (Clark, Black et al 1977; Clark, Tong et al 1977). It was found, however, that the University of Melbourne/ Nucleus banded, smooth, free-fitting, tapered array could be easily removed and another inserted without trauma or adverse results (Clark, Pyman et al 1987). So if a Nucleus receiver-stimulator were to fail, it would not be necessary to have a connector, but a new package and array could be reintroduced. By way of comparison ball electrodes lying in the scala tympani, as with the Ineraid device, could be difficult to remove (Gray et al 1993). The prosthesis needed to be robust so that it would not be damaged by an impact such as a blow to the skull. It should also be engineered so that repeated small body movements do not lead to metal fatigue and fracture of the electrode wires, particularly at the point where the electrode array leaves the receiver-stimulator unit. Finally, cochlear prostheses should be engineered so that they are cosmetically and socially acceptable as well as cost-effective. For this reason the implant needs to be as small as possible. This is particularly important when it is used in young children and infants.

The speech-processor should be compact, light, and comfortable to wear, and the power consumption minimal for a long battery life. There should be few controls on the processor, and these need to be easy to manipulate for the elderly and disabled. The electronics for the speech processor need to be enclosed so they cannot be tampered with, and the leads robust so that they do not break with repeated flexing and pulling. The system for transmitting power and signals to the implanted device has to be easy to apply and remove, and be cosmetically acceptable.

The science underlying the development of cochlear implants has been multidisciplinary, and in turn depended on technical advances. Its implementation has required both electronic and communications engineering for the analysis and presentation of the sound code, and bioengineering for the interface with the nervous system.

ineering

und Communications Engineering

...al functions for manipulating the speech signal require electronic circuits for real-time processing. Advances in integrated circuit design over the 1980s and 1990s allowed high rates of data analysis for the detailed representation of even the fast changes in the speech signal. The development of programmable silicon chips enabled individual specifications to be included. CMOS (complementary metal-oxide semiconductor) circuitry was suitable for the electronics as it operates with low power consumption. The circuits for implants now primarily manipulate digital rather than analog information. Digital data are described by (or quantified as) a finite set of numbers.

Digital Versus Analog Circuitry

The speech processor and receiver-stimulator sections of a cochlear implant are typically constructed using a combination of digital and analog circuitry. The processing of the signals digitally involves the mathematical manipulation of the numbers representing the signals by adders, multipliers, etc. The numbers in a digital system are represented in binary form using a high voltage level to indicate a 1 and a low voltage to indicate a 0. In an analog system, continuous variation of the instantaneous values of the signals occurs. The continuous variation of analog signals requires theoretically an infinite set of numbers for their specification. To perform the mathematical manipulation of the signals with analog processing, the continuous signals would be fed through and modified by various physical devices such as transistors, resistors, and capacitors. The form and degree of modification would depend on the characteristics of the physical device.

Digital processing offered several advantages over analog processing. First, digital processing is more stable and repeatable. This is because the amplitude of the electrical pulses representing the numbers in a digital system can be badly distorted without affecting the precision of the numerical operation. By contrast, any distortion of the continuous signals in an analog system directly affects the precision of the mathematical manipulation. Second, because of the lack of passive components, digital circuits can be integrated into a small silicon chip using large-scale integration technology. With the first devices the power consumption for some of the digital subsystems exceeded that for analog implementation. As a result of the stringent power requirements for a wearable unit, both digital and analog techniques were used initially in the realization of the speech processor and receiver-stimulator. Now most subsystems can be implemented digitally.

Sampling of Signal

The acoustic signal needs to be sampled sufficiently often to produce a satisfactory representation digitally. With high fidelity audio recording it is on the order of 44,000 samples/s. The sampling frequency is governed by Nyquist's theorem.

This shows that a continuous signal may be uniquely specified by 2B samples/s where B is the bandwidth or the highest frequency in a band-limited signal. At any lower rate there will be aliasing errors. For example, for a 16,000-Hz sampling frequency, a 10,000 Hz signal could be impersonated by a 6000-Hz (16,000 − 10,000) signal. To represent the highest frequency in speech, which is 8000 Hz, a sampling rate of 16,000 is sufficient. The speech processor takes this information and converts it into a code to specify the currents on the different electrodes. For the Nucleus 24 system a 2.5- or 5.0-MHz radiofrequency (RF) is used to transmit this information to the receiver-stimulator. The number of bits of information transmitted is limited by the number of cycles of the RF that can be resolved, and for the Nucleus 24 system this is 500,000 bits/s.

The receiver-stimulator converts the transmitted information into electrical stimulus pulses having an overall pulse rate that can be distributed across the electrodes. This can be up to 14,400 biphasic pulses/s or 1800 pulses on each of 8 channels (Fig. 8.1). In addition, the sampling of the amplitude of the analog wave (which has an infinite set of values) with finite numerical indices can lead to quantizing errors. A quantizing or voltage error is the difference between the ideal transfer characteristic and digitized equivalent.

Power and Data Transmission

Electrical signals are transmitted to electrodes in the cochlea to excite the residual cochlear nerves to implement the speech-processing strategies. In some initial

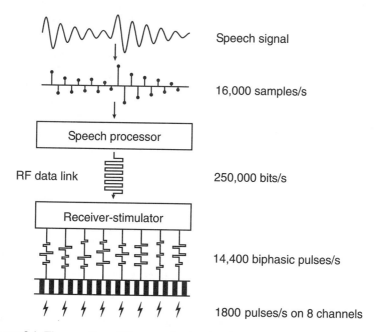

FIGURE 8.1. The sampling of the acoustic signal and its transmission to the electrodes.

studies, including those at the Universities of Paris and Utah and the University of California at San Francisco (UCSF), and later with the Symbion/Richards Ineraid device, speech data were transmitted to the stimulating electrodes by a direct electrical connection (percutaneous stimulation). This was done to enable the broadest range of stimulus parameters to be tested. However, with these devices infection was prone to occur as a result of foreign material passing through the skin. Infection was reported in all of the 21 patients implanted by Pialoux et al (1979), and the Teflon plugs were removed. Another major problem with percutaneous connectors was fracturing of the electrode wires by bumping, fingering, and other body movements that could not always be predicted from animal experiments. Finally, the plug was not aesthetically pleasing to patients. Even with improved materials, such as pyrrolized carbon, a plug and socket still had too many potential risks to be used routinely, as discussed in Chapter 10.

There is now much greater flexibility in the stimulus parameters that can be transmitted through intact skin (transcutaneous stimulation) due to the greatly improved capacity of integrated circuit design, and better knowledge of the patients' responses to electrical stimulation. Furthermore, for the above biological and surgical reasons, power and data are better transmitted transcutaneously. A transcutaneous power link eliminates the need for an implantable power source requiring replacing or recharging batteries. With a transcutaneous link the system can be more easily rendered inoperative should a failure of the electronics occur.

Electromagnetism, light, and ultrasound energy were the alternatives first considered for the transmission of power and data (Forster 1978; Clark, Black et al 1977, 1978; Clark, Tong et al 1977). A number of transcutaneous links to transmit energy through the intact skin were explored. With the single-channel hearing prosthesis developed by Djourno and Eyriès (1957), Michelson (1971), and House (1976), simple transcutaneous inductive coupling was employed, the output of which drove the implanted electrode directly. Using the same technique Brindley and Lewin (1968) developed a multiple-channel system for stimulating the visual cortex. Their system required 80 independent inductive links and extensive surgery. An alternative multiple-channel approach was to use a single data link and multiplex the stimulus data (Dobelle et al 1976). With this method, implanted electronic circuitry would decode and demultiplex the transcutaneous data. In addition, power had to be provided to operate the implanted circuitry (Schuder et al 1961). Dobelle et al (1974) proposed a multiple-channel system for both visual and auditory cortex stimulation using a single RF link for both data and power. White (1974) advocated using a transtympanic optical link for multiple-channel time-multiplex data transfer with a separate inductive link for powering the implant. This resulted in a lower efficiency but a wide band link for high data rate transfer, and a high-efficiency link for power transfer. Gheewala et al (1975) developed a similar dual-link transfer system using ultrasonic data transmission to control a four-channel hearing prosthesis.

In 1974 at the University of Melbourne an electromagnetic dual link for data and power transfer was evaluated for the university's multiple-electrode receiver-stimulator. This system had a high transfer efficiency for power, and required simple and therefore reliable detection circuitry for the data link (Forster 1978; Clark, Black et al 1978).

Electromagnetic induction is based on the principle, illustrated in Figure 8.2, that a magnetic flux produced by passing a current through a coil (external) will induce a current in a second coil (internal). This method was used for both power and data transfer in the University of Melbourne's multiple-electrode receiver-stimulator, and the present Cochlear Limited, Advanced Bionics, Med El, and MXM devices. Adequate power needed to be transferred from the external to internal coils over a distance that could vary depending on the thickness of patient's skin and underlying tissues. It was desirable that some lateral misalignment be possible between the two coils. Adequate power transfer was required over a distance of up to 10 mm with the coils coaxial, and some degree of misalignment possible at shorter distances. The power transfer had to be efficient to spare the batteries to make body-worn devices small and behind-the-ear speech processors possible.

A high-frequency electromagnetic carrier wave was used for transmitting power and speech data, and the wave was modulated in response to the coded speech signal. This code specified the electrode to be stimulated, the amplitude of the current, and the start time for each electrode within 1 ms, whereas the recent devices specify the electrode to be stimulated, the mode (bipolar, common ground, monopolar), the current amplitude, the duration of each biphasic pulse, and the interval between each phase. The circuits in the receiver-stimulator had to implement the instructions from the speech processor, and the digital information was finally converted into an electrode current for neural stimulation. For the transcutaneous transmission of information, a high-frequency modulated wave was desirable as it permitted a small aerial to be used, and a large amount of

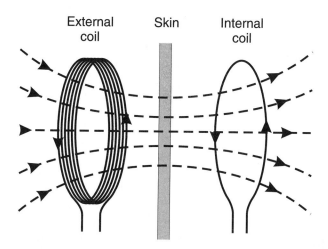

FIGURE 8.2. A diagram of electromagnetic induction showing energy passing through the skin from an external transmitting coil to induce current in an internal receiving coil (Reprinted with permission from Clark and Tong 1985).

speech data to be transmitted efficiently. Examples of a carrier wave modulated by varying either its amplitude or frequency are illustrated in Figure 8.3.

Radiofrequency (RF) carrier waves could cause long-term adverse effects on the body tissues, which could occur from the absorption of energy by ions, or dipole-induced vibrations in protein molecules (Johnson and Guy 1972). The energy absorbed by tissues increased with the carrier frequency, and safety levels needed to be established. The biological effects of nonionizing radiation were reviewed at a conference at the New York Academy of Science in February 28, 1975, volume 247, editor Paul E. Tyler. The American National Standards Institute (ANSI) safety standard did not restrict the peak power density as long as the average density over 6 minutes did not exceed 1 mW h/cm² or 3.6 J/cm². It was important to keep the carrier frequency as low as possible, and the energy absorbed within the safety limit.

With the University of Melbourne's receiver-stimulator the data were received on one coil at a frequency of 10.752 MHz and power on a separate coil at a frequency of 112 kHz. The receiver-stimulator demodulated the serial data stream. The demultiplexed data initiated and controlled the stimulation of the electrodes by from 10 digital-to-analog (D/A) converters at the output.

Circuit Design and Components

The speech processor in principle (Fig. 8.4) consists of a microphone to capture the sound, a front end to optimize the signal, a filter bank to analyze it into its component frequencies, sampling and selection to produce the processing strat-

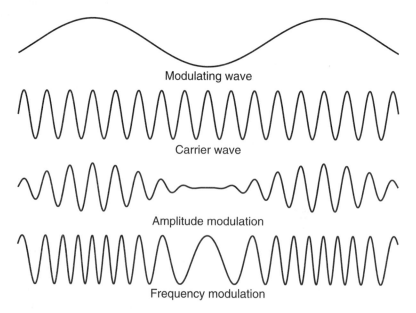

FIGURE 8.3. A diagram of amplitude and frequency modulated carrier waves.

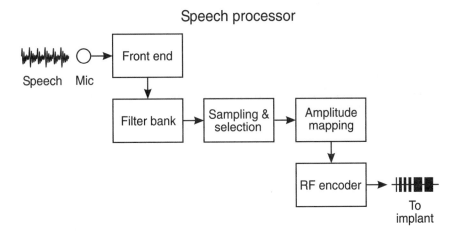

Speech processor

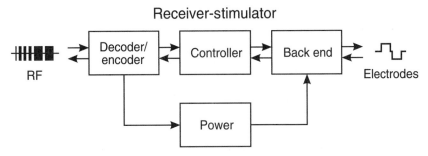

Receiver-stimulator

FIGURE 8.4. Top: The principle components of a speech processor. Bottom: The principal components of the receiver-stimulator. RF, radiofrequency.

egy, amplitude mapping to match the signals to the patients' psychophysical domain, and an RF encoder to encode the signals for transmission to the receiver-stimulator. The receiver-stimulator (Fig. 8.4) consists of a decoder/encoder that extracts the power and controlling signal, the controller that organizes the pattern of stimulation, and the backend that facilitates the flow of current to the electrodes. With telemetry there is a reverse flow of information from the voltages recorded from the electrodes into a signal for transmission back to the speech processor.

The circuit design to implement these functions is achieved through the use of both passive and active electronic components. The passive or linear components, in particular resistors and capacitors, allow the current to increase in direct proportion to the voltage in accordance with Ohm's law. They also do not require additional power to operate. In contrast, with active components, in particular diodes and transistors, the above proportionality does not operate and they need additional power. With the passive components the flow of current is in accord with Ohm's law:

$$V = IR$$

where V is voltage, I is current, and R is resistance.

Current (I) is defined as the rate at which the number of electrons or charge (Q) measured in coulombs, passes a given point. Electric potential (V) is the amount of potential energy present per unit of charge. One volt represents a potential energy of one joule per coulomb of charge. Resistance (R) occurs when electrons collide with atoms and lose energy, and depends on the properties of the material referred to as resistivity (ρ). Capacitance (C) is the capacity of two plates to hold the charge and is proportional to the voltage. With a sinusoidal current, the relationship between voltage, current, capacitance, and angular velocity ($2\pi f$) or frequency (f) is $V = \dfrac{1}{2\pi fC} \times I$. The current in the capacitor leads the voltage with a phase angle of $\pi/2$. The equation above is similar to Ohm's law with the proportionality factor $\dfrac{1}{2\pi fC}$ being the capacitive reactance, the larger the capacitance the smaller the reactance or impedance. The impedance is also inversely proportional to the frequency; that is, the lower the frequency the greater the impedance. These concepts were discussed in Chapter 4.

In addition to Ohm's law the design of the electronic circuits is governed by concepts such as Thévenin's equivalent circuit and Kirchhoff's circuit rules. The fundamentals underlying the circuit design can be obtained from a number of texts, including those by Burns and Bond (1997), Wakerly (1994), and Garcia et al (1998). Thévenin's equivalent circuit models any two-terminal circuit with a single voltage source in series with a single resistor. The Thévenin voltage is the open-circuit voltage of the original circuit, and the Thévenin resistance is the Thévenin voltage divided by the short-circuit current of the original circuit. Kirchhoff rules state that the algebraic sum of currents toward any branch point is zero, and the algebraic sum of all the potential changes in a closed loop is zero.

As stated, the important nonlinear devices are the diode and transistor. A rectifier diode passes a greater current when a voltage is applied across its terminals with one polarity rather than the other. For example, it has been used in a speech-processing circuit as the first step in measuring the intensity of the speech wave. This can be carried out when the alternating speech wave is converted into a unidirectional current. This is illustrated in Figure 8.5.

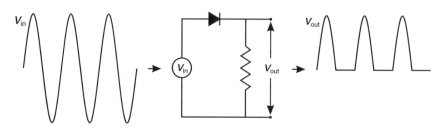

FIGURE 8.5. A diagram of how the diode rectifies an alternating current to produce a unidirectional one.

The transistor is another fundamental nonlinear component in circuit design. The transistor is used in the speech processor and receiver-stimulator to carry out a number of functions.

These nonlinear components can be fabricated with semiconductor technology, and both diodes and transistors make up the subsystems for both analog and digital circuits for speech processor and receiver-stimulator designs. The subsystems include preamplifiers, automatic gain control (AGC), filters, analog-to-digital converter (ADC), read-only memory (ROM), registers, clocks, switches, and digital-to-analog converter (DAC).

Semiconductors

In the earlier Nucleus formant speech processors, many components were analog because digital components required more power. These included the AGC, formant extractors, zero crossing detector, and filters. With the Nucleus 24 body-worn processor (SPrint) the AGC, amplitude detection, and mapping were all carried out with a programmable digital signal processor that was realized with semiconductor technology using less current. Only the antialias filters and ADC were analog.

Digital components are made from semiconductor materials, most commonly silicon, and to a lesser extent gallium arsenide. In these materials, shared valence electrons form the bonds between atoms. Electrons in the valence energy band can make the transition to the conduction energy band where they are free to respond to an electric field and can carry a current. When an electron moves from the valence band into the conduction band, it leaves behind a vacant site called a hole, effectively a positive charge. A valence electron from a nearby bond can transfer into the hole, filling it but leaving another hole behind. With a series of such transfers, the hole appears to be a positive charge migrating through the material. The conducting properties of a semiconductor can be tailored by adding impurities, or "doping." A semiconductor doped with a "donor" impurity is called n-type because the impurity has atoms that release their valence electrons when placed in the semiconductor, increasing the number of free electrons (*n*egative charge carriers). In contrast, a semiconductor doped with an "acceptor" impurity is call p-type because the impurity has atoms that trap electrons, generating extra holes (*p*ositive charge carriers).

A semiconductor made from the junction of n-type and p-type material forms a diode, used in the design of the receiver-stimulator to allow current flow in only one direction. Mobile donor electrons in the n-type material diffuse across the junction and fill holes in the p-type material, creating a "depletion region" which is depleted of mobile charge carriers. As illustrated in Figure 8.6, the fixed ions that remain in the depletion region create an electric field that forms a potential barrier preventing further movement of electrons or holes across the junction. When a positive external voltage is applied to the p side of the junction (forward bias), the barrier is decreased and electrons can move from the n side to the p side and current flows. In reverse bias, the potential barrier is increased and the p-n junction stops the flow of current. From this it can be seen how a bidirectional current is rectified into a unidirectional one, as illustrated in Figure 8.5.

Another key element in the circuits of speech processors and receiver stimulators is the transistor. This element is part of a number of systems such as preamplifiers, oscillators, and latches. A transistor is in effect two diodes back to back as illustrated in Figure 8.7. This bipolar junction transistor consists of three layers that alternate according to the type of impurities as either a *pnp* or *npn* device. The three layers function as emitter (E), base (B), and collector (C). Because the emitter-base junction is forward biased and the base-collector junction is reverse biased a small increase in base potential V_B, while causing only a small increase of base current I_B, results in a large increase of collector current I_C and hence a large change in voltage across the output resistor R. Thus, the transistor may be used as an amplifier for small time-varying currents and voltages.

The field effect transistor (FET) is different from the bipolar junction transistor. It has a higher input impedance, and is better for use with weak signals. It is therefore the transistor used in the circuits of the speech processor and receiver-stimulator. As illustrated in Figure 8.8, with the FET the current flows along a

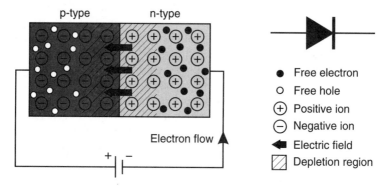

FIGURE 8.6. Left: A diagram of p-type and n-type silicon at a pn junction, demonstrating the diffusion and drift currents. Right: The semiconductor as a diode.

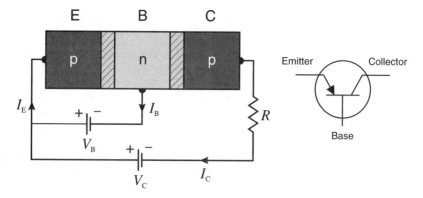

FIGURE 8.7. Left: A diagram of a pnp junction transistor. Right: Corresponding circuit symbol showing that for the emitter which corresponds to the cathode of a vacuum tube the arrow points in the direction of the hole flow.

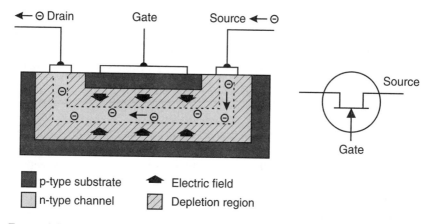

FIGURE 8.8. Left: A diagram of the field effect transistor (FET). Right: Corresponding circuit symbol.

path called the channel. At one end there is a source and at the other end a drain. The effective electrical diameter of the channel can be varied by applying a voltage to the gate. The conductivity of the FET depends at any moment on the electrical diameter of the channel. A small change in the gate voltage can produce a large variation of the current from the source to the drain. This is how signals are amplified with the FET. FETs can be either a junction FET (JFET) or metal oxide semiconductor FET (MOSFET). FETs are fabricated onto silicon integrated circuit (IC) chips, and a single IC can contain many thousands of FETs along with other components such as resistors, capacitors, and diodes.

Speech Processors

The algorithms underlying speech processing were discussed in some detail in Chapter 7.

Design Principles

The processor consists of the microphone, front end, filter bank, sampling and selection, amplitude mapping, and encoder sections, as illustrated in Figure 8.4. A speech processor has a microphone to transduce sound vibrations into electrical signals, to faithfully reproduce the sound. These analog signals require amplification to scale them to a useful level. With one method of manipulating the signals, they are first filtered to analyze them into their frequency components, and the analog outputs converted into a digital representation before being processed in accord with the selected speech-processing strategy. Another method is for the signals to be digitized before being filtered with a fast fourier transform (FFT). The energy in each frequency band is converted to current level. The electrode to be stimulated for each frequency is obtained by reference to maps where the frequency-to-electrode parameters are stored. With amplitude mapping, reference is made to the thresholds (T) and maximum comfortable (MC) levels for each electrode to determine the range of currents to be used. The stimuli to represent speech and other sounds at each instant in time are encoded and then transmitted by RF to the receiving coil of the implanted device. Power to operate the receiver-stimulator and provide current for the electrodes is also transmitted by inductive coupling.

Front End

The microphone is directional to enhance sounds coming from the front, and improve performance in noise. The early microphones had frequency response up to approximately 4000 Hz. The present HS-8 microphones have gain increasing from 150 Hz to approximately 5000 Hz at 6 dB/octave before dropping off at 18 dB/octave. This is to emphasize the high frequencies of speech that are important for intelligibility, but low in intensity. The microphone was also small for cosmetic reasons, and placed above the ear for normal face-to-face communication. All processors had an analog front end or audio preprocessor. It consisted of preamplifier, preemphasis, AGC, and an automatic sensitivity control (ASC). The preamplifier boosts the signal from the microphone to a level that can be handled by the speech processor. The gain can be varied with a sensitivity control in conjunction with the AGC and optional ASC.

To ensure that the intensity of different voices are heard within the dynamic range, an AGC is required. The AGC is a compression amplifier to keep the variations in the speech intensity within a certain range. The AGC reduces the intensity of loud sounds to keep their peaks at a level that is acceptable for subsequent processing and comfortable to the patient, as the input amplitude can vary over a wide range. The increased compression and the time delay over which this is achieved must be carefully controlled to ensure a minimum amount of distortion of the speech signals. When the intensity of the signal rises, the compression is increased (attack time) and the gain reduced. With the Nucleus processors this is usually less than 2 ms. Then when the intensity of a speech signal falls, the compression of the AGC is released until the maximum gain is reached (release time). It is shorter than the typical syllable in speech.

The ASC is especially important for hearing speech in the presence of background noise. If the sensitivity control is too high, the background noise is near the maximum level; then when someone speaks the AGC will not allow the speech to become any louder and so it is difficult to understand. The ASC provides automatic control of the sensitivity so that when it is on, it adjusts the sensitivity based on the background noise. If the background noise is above a certain level, it is likely the sensitivity is too high and so the ASC reduces the sensitivity and vice versa if the background noise is too low.

Filter Bank

Filters are a fundamental part of the cochlear implant speech-processing circuit, as they analyze the energy in the frequency bands of the speech signal for multiple-channel stimulation. Either before or after the signal is filtered the analog representation of the sound is digitized with ADCs. One method of operation is the successive-approximation converter in which a sequence of voltages to approximate the signal is compared with the input signal. Each successive voltage is stored if it is lower than the signal and rejected if it is higher. The process is repeated until the required accuracy is achieved. Another is the dual-slope integrator in which the charge on the integrating comparator during a fixed time interval is compared with the charge removed by the known reference voltage.

An ideal filter would transmit only frequencies within a defined band (pass band) with constant amplification, and would not change the phase of any of the components. It is best to consider the upper and lower ends of the pass band separately. A low-pass (high-cut) filter determines the upper limit of the pass band, and a high-pass filter (low-cut) the lower limit of the pass band. The simplest is the resistance capacitance (RC) voltage divider (Fig. 8.9), which is a passive filter. The impedance of the capacitor in the circuit is frequency dependent as discussed above. With the low-pass filter shown on the left of Figure 8.9, as the frequency falls the impedance of the capacitor increases, and a greater proportion of the voltage is from across it. On the other hand, with a high-pass filter shown on the right of the figure, as the frequency increases the impedance of the capacitor falls and so a greater proportion of the voltage is from across the resistor. It can be shown that when the impedance of the capacitor is equal to the series resistor, then the gain of the RC voltage divider is $1/\sqrt{2}$. The frequency for which the impedance of the capacitor is equal to R is the cut-off frequency (f_0) and is determined by the relation $f_0 = 1/2\pi RC$. This filter gives a reasonable approximation of the linear phase shift within the pass band. However, its frequency versus amplitude response leaves much to be desired, falling off slowly above f_0. Active filters that incorporate an operational amplifier give better performance. These filters can be cascaded so that the output of one filter is the input to a second thus enabling sharper filtering to occur. Digital filters are now used with speech processors and operate on a stream of digital inputs rather than an analog input. They

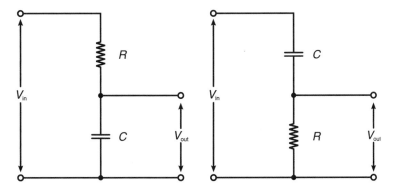

FIGURE 8.9. Left: The circuit diagram for a simple low-pass filter and a low-pass unity gain filter. R, resistance; C, capacitance. Right: The circuit diagram for a simple high-pass filter and a high-pass unity gain filter.

have greater flexibility in selecting the center frequencies and the slope of the upper and lower cutoffs.

Sampling and Selection

The output voltages of the filters for formant, spectral maxima, or fixed-filter (channel vocoder) processing strategies are converted into digital information by an ADC, or the digitized speech information from the ADC is filtered using an FFT. The correspondence between formant or filter frequency and electrode number can be determined from psychophysical studies, or derived on the basis of the spacing between the electrodes in the cochlea and known physiological principles. The electrode number to be stimulated for each frequency band is assigned from a table in the digital memory.

Microprocessors are key elements in the speech processor circuit for controlling the signals required for different strategies. These are undertaken through logic operations, such as addition, multiplication, etc. This is achieved through Boolean algebra where the three basic operations are AND, OR, and NOT. The AND operation in an electrical circuit occurs when two switches in series are required to be closed for current to flow. The OR operation occurs in a circuit where the switches are in parallel and so current will flow when either one or the other is closed. The NOT operation involves an inversion of the switch signal to produce an output state the opposite of the input. The AND, OR, and NOT gates can be implemented with relay switches or semiconductor gates in particular.

Memory is fundamental to many aspects of speech processor engineering. It may hold, for example, information on the relationship between the filter frequency and the electrode to be stimulated as discussed above or be used to store the T and maximum MC levels for each of the patients electrodes and modes of stimulation as discussed below. Semiconductor memories are arranged in arrays of cells where a cell stores a single *bit* of information (1 or 0). The basic memory unit is a flip-flop. A positive feedback is used resulting in two stable states representing logic levels 1 and 0. Many computers handle data in groups of 8 bits termed a *byte*. Memories are classified according to whether the information is stored serially and can only be accessed that way, or stored so it can be accessed in any order (random access memory, RAM). A shift register is an example of serial memory. Data bits can be shifted sequentially in and recovered at the far end. In the speech processor or receiver-stimulator they are used as data buffers where the data are preserved but the rate changed, for example the instructions for the stimulus parameters. RAM is used when a high speed of access is required. RAM allows data to be written into memory and then read from it. With this read/write memory the data are lost when power is removed from the chip. It is realized in CMOS technology for cochlear implants as it has low standby power. It is used for most of the memory operations in the Nucleus SPrint processor with a small battery to maintain residual power. With read only memories (ROMs), also referred to as nonvolatile, the bit pattern can be programmed into the cell permanently. They are used to store constants and program instructions (firm-ware).

The MOS cells are programmed by trapping electrons on a floating gate, thus blocking the cell by removing the inversion channel connecting the source to the drain. Memory can also be erased electrically through programmable ROM (EEPROM) and is used in speech processors for the patient's MAP (a map for the patient's threshold and maximum comfortable levels) so that it can be later changed.

Amplitude Mapping

The digitized amplitudes of the filter outputs are converted into a number representing the current level by referring to a table containing the T and MC levels for each electrode to provide the range of stimulation. The 30- to 40-dB intensity range of speech has to be compressed into only 8 dB on average between T and MC levels. The table of T and MC levels is stored in a block of digital memory controlled by the microprocessor. A base intensity level is set for the filter band envelope, and this is mapped to the T level for the appropriate electrode. A power function describes the loudness growth over the dynamic range, and thus determines the current steps over the dynamic range. The index of the power function can vary from electrode to electrode depending on their size and location, and on the density of residual auditory nerve fibers in the cochlea. A loudness growth function is determined, and the steepness of the curve can be varied. The loudness can be achieved either with a fixed pulse width and the current varied or with both the pulse width and current amplitude varied.

Encoder and Transmitter

The data from the speech processor to the receiver-stimulator are transferred serially as a digital stream of information, providing immunity to noise and component variations as discussed above. The signal parameters from the microprocessor section containing the sampling and selection and amplitude mapping are fed to the transmitter section. The microprocessor controls the times at which the electrical parameter values are sampled, and configures the RF signal to be transmitted to the implanted receiver-stimulator. To do this accurate timing is vital, and this is achieved with one quartz-crystal oscillator in the speech processor. All timing operations, for example in the receiver-stimulator, are referenced to it.

The transmitter coil is linked inductively to a receiver coil, with alignment ensured through magnets incorporated in the centers of both coils. The transmitted code is made up of a digital data stream representing the stimuli, and transmitted by pulsing the RF carrier. The receiver-stimulator decodes the information into instructions for the selection of the electrode, mode of stimulation (i.e., bipolar, common ground, or monopolar), current level, pulse width, and inter-phase gap. The stimulus current level is controlled via a DAC. Power to operate the receiver-stimulator is also transmitted by the RF carrier.

The principles underlying the transmission of data and power can be appreciated by first seeing how it took place with the University of Melbourne's device (Clark, Black et al 1977; Clark, Tong et al 1977; Forster 1981). In this case (Fig. 8.10) data were transmitted on one coil at a frequency of 10.752 MHz, and power

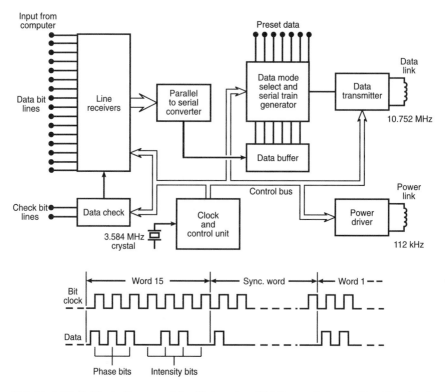

FIGURE 8.10. A block diagram of the University of Melbourne's transmitter (Reprinted with permission from Clark, Black et al 1977. A multiple-electrode hearing prosthesis for cochlear implantation in deaf patients. Medical progress through technology 5: 127–140. Springer-Verlag.).

on a separate coil at a frequency of 112 kHz. With the Nucleus, Advanced Bionics, and Med El systems, the data and power are transmitted at one frequency on a single coil. With the University of Melbourne's device, the data controlled only the amplitude and the instant when any electrode was stimulated (rate). Both phases of the stimulus pulses were constant at approximately 180 μs per phase. In subsequent devices, the amplitude, timing of pulses, and their width and interphase gap could all be controlled. With the University of Melbourne's device, the stimulus data for each stimulus channel were represented as 16-bit words. Each word was a series of 0's and 1's (or pulses and spaces) representing a number in binary arithmetic. The number coded the amplitude and timing information and the electrode to be stimulated. These data arrived from the processor in parallel streams that had to be converted into a serial format, and stored in a buffer ready for transmission. The data were then loaded as a serial train of data comprising 10 to 15 seven-bit words for 10 or more electrodes. A word was also used to synchronize the data with clock pulses to ensure correct relative timing. Then each 1 ms this set of data was transmitted to the receiver-stimulator. The overall repetition rate for the data sequence was thus 1000 Hz.

With the Nucleus 24 and other systems, not only was it necessary to transmit

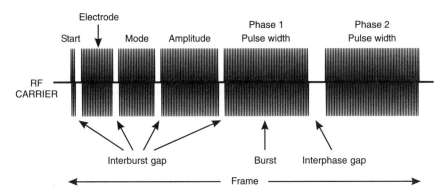

FIGURE 8.11. The expanded transmission protocol for the Nucleus MSP or Spectra speech processor for a transmitter frequency of 2.5 MHz.

data on the amplitude of the pulses, pulse width, and interpulse gap, but also receive data on electrode voltages via telemetry. When the desired stimulus data had been calculated, a data encoder formatter was used to encode the RF signal. The signal was a 100% amplitude-modulated RF carrier in a format suitable for decoding. The data encoder formatter had a timer circuit to initiate the transmission. A controller was responsible for formatting the data. The data were transmitted in frames, as was the case with the University of Melbourne's processor. Each frame contained all the data defining one stimulus pulse.

The data in each frame were coded as the number cycles of RF carrier in bursts of RF with each burst consisting of a number of cells or just-discriminable RF cycles. Successive frames were separated by an interframe gap. There were also operating modes that determined whether constant rate stimuli were used or the start pulse was varied according to the voice pitch. A coding strategy was also required for high stimulus rates. The expanded protocol for a transmitter frequency of 2.5 MHz is illustrated in Figure 8.11. An embedded protocol for a higher transmitter frequency of 5.0 MHz was also available. As shown, the bursts code for start, electrode, mode, amplitude, pulse width for phase 1, and pulse width for phase 2. Each burst varied in length according to the number of cells, and their number was proportional to the data in each burst.

Design for Speech (Sound) Processing Strategies

Speech-processing strategies that have been in clinical use and approved by regulatory authorities are described in this chapter as they relate to the engineering of the devices. They are also discussed in Chapters 7 and 12.

Second Formant (F2) and Voicing (F0) Strategy (F0/F2) and
WSP-II Speech Processor

With the inaugural University of Melbourne's strategy developed in 1978 the second formant frequency (F2) was extracted and presented as place of stimulation for the appropriate frequency site in the cochlea, the fundamental or voicing frequency (F0) was coded as rate of stimulation on each electrode, and the amplitude of F2 as the current level (A2) (Clark, Tong et al 1978, 1981a–c; Tong et al 1979, 1980; Clark and Tong 1981). The F2 was the dominant spectral peak in the mid frequency region and its energy (A2) estimated from the output of a 750- to 4000-Hz band-pass filter (Fig. 8.12). For a given F2 estimate, an electrode was stimulated from a predetermined F2-to-electrode transformation map that had been constructed by ranking the electrodes from dull to sharpest and assigning frequency sub-bands to these electrodes in an order from lowest to highest. Similarly, the current level was determined on the basis of an A2-to-current map. The voicing frequency (F0) and its energy (A0) were extracted using a 400-Hz low-pass filter, and an energy threshold detector of A0 was used to determine whether voicing was present or not. With a voiced signal there is greater energy at the lower frequencies as vocal fold vibration uses up a significant amount of the energy. The signal was monitored continuously for periodicity in the frequency range 80 to 400 Hz, and if present stimulation occurred at this rate to convey voicing. This proportionality would hold up only to approximately 300 pulses/s (Tong et al 1983a). However, as the voicing frequency could be higher (Fry 1979) some adjustment to the proportionality was needed. If voicing was absent, a constant low pulse rate was used as it produced a sensation described as "rough" that was similar to a "noise" sensation described by the patient when he had hearing. The roughness may have been due to the fact that this is how low-frequency combination tones sound (Plomp 1976). The speech parameters were determined every 10 ms, and only one of the 10 electrodes activated in a 10-ms

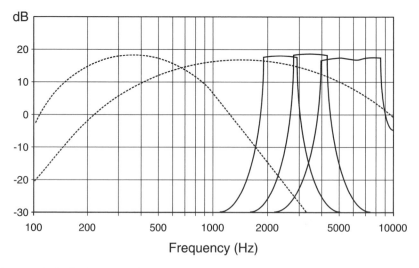

FIGURE 8.12. The band-pass filter outputs for the Nucleus F0/F2 WSP-II, F0/F1/F2 WSP-III, and Multipeak-MSP strategies and processors. The interrupted lines are the outputs for the first (F1) and second (F2) formant filters. The continuous line represents the outputs for the three high-frequency filters for the Multipeak-MSP system.

time frame. The strategy was later modified so that a pulse rate at the random fluctuation of the signal envelope was used for unvoiced sounds and the second formant amplitude envelope A2 used to stimulate the F2 electrode to provide better temporal information.

This strategy was implemented by Cochlear Proprietary Limited in the WSP-II (Wearable Speech Processor). The system had a subminiature directional electret microphone worn on the headset. The signal passed through a pre-amplifier with AGC, F0 detector, and F2 extractor (Fig. 8.13). The AGC kept the signal within the dynamic range of the feature extraction circuitry, and automatically adjusted the signal output of the preamplifier to remain within the full scale of the ADC. The AGC was arranged so the peak output corresponded to the MC level. The threshold was then set in quiet conditions (Seligman 1987). The fundamental frequency was measured with an analog system by passing the outputs of the AGC amplifier sequentially through a rectifier, a low-pass filter, a zero crossings counter, and a frequency-to-voltage converter that produced a voltage proportional to the fundamental frequency. This voltage was used to determine a rate of stimulation that was proportional to the voicing frequency. For unvoiced speech, the stimuli were random and produced sounds like noise. The formant frequency was measured similarly with an analog system. It was estimated from the output of a formant filter that was designed to cover the frequency range of the formant frequencies in question. The filter output, too, was fed through a zero crossing counter and a frequency-to-voltage converter to produce a voltage proportional to the formant frequency. There was an analog amplitude envelope

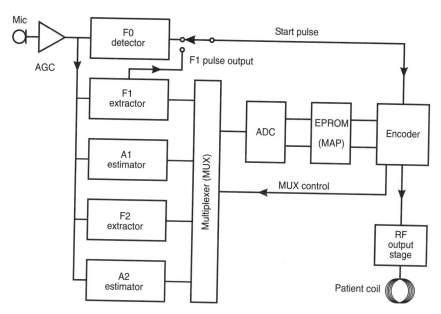

FIGURE 8.13. The Nucleus programmable multi-formant speech processor. (Reprinted with permission from Seligman 1987. Speech processing strategies and their implementation. *Annals of Otology, Rhinology and Laryngology* **96**(Suppl 128): 71–74.)

detector to measure the intensity of the F2 filter output, and this consisted of a full wave rectifier followed by a low-pass filter. The center frequencies of the frequencies in the F2 region were plotted logarithmically, and divided between the electrodes as occurs physiologically. In addition, loudness estimates were converted into values on an approximately logarithmic scale. This strategy was supported by the initial research of Stevens (1975), who showed that loudness was a power function of intensity. Clark, Tong et al (1978); Tong et al (1979) found a steep linear function between the logarithm of current amplitude, plotted in decibels, versus loudness.

In the Nucleus F0/F2 WSP-II as well as the F0/F1/F2 WSP-III and Multipeak-MSP (Miniature Speech Processor) speech processors (referred to below) many components were analog. These included the resistors, capacitors and operational amplifiers. There were a few digital components including flip-flops and logic gates.

The Nucleus F0/F2 WSP-II system was approved by the Food and Drug Administration (FDA) in October 1985 for use in post-linguistically deaf adults.

First Formant (F1), Second Formant (F2), and Voicing (F0) Strategy (F0/F1/F2), and WSP III Speech Processor

The F0/F1/F2 WSP-III system arose from further research to improve speech intelligibility through better recognition of consonants. The first formant (F1) was extracted and presented on a place-coding basis. The presentation of two formants was supported by the psychophysical study of Tong et al (1983b) which showed that stimuli presented to two electrodes could be perceived as a two-component sensation. The rationale was also based on the acoustic model studies of electrical stimulation on normally hearing individuals (Blamey et al 1984a,b; 1985). The F0/F1/F2 strategy (Dowell et al 1987) provided additional formant information on a place-coding basis with nonsimultaneous stimulation. The stimuli were separated in time by 0.8 ms to avoid channel interaction. Unpredictable variations in loudness seen with channel interaction was first observed with the speech processor of Laird (1979), and emphasized the need for nonsimultaneous or interleaved pulses. The interleaving of two-formant biphasic pulses for each glottal pulse is illustrated in Figure 8.14.

The WSP-III speech processor (Fig. 8.13) for the F0/F1/F2 strategy was similar to the WSP-II with microphone, preamplifier with AGC, F0 detector, and F1 and F2 extractors. With the AGC the MC and T levels were set. There was a sensitivity control so that when noise occurred the patient could manually set the gain in 32-, 56-, 66-, 74-, and 80-dB steps, and this avoided unnecessary amplification of background noise (Seligman 1987). The F1 was estimated by low-pass filtering at 850 Hz, and then high pass filtering at 480 Hz to remove the F0. As referred to above, 480 Hz is at the upper range of F0 and the F1 range is from 250 to 700 Hz (Peterson and Barney 1952; Fant 1959). The filtered signal was also passed through a zero crossing detector (i.e. frequency to voltage converter) and was 35-

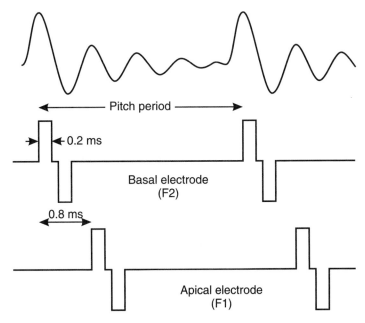

FIGURE 8.14. Electrode stimulation waveforms for two formant coding strategy. (Reprinted with permission from Seligman 1987. Speech processing strategies and their implementation. *Annals of Otology, Rhinology and Laryngology* **96**(Suppl 128): 71–74.)

Hz low-pass filtered to convert the zero crossing rate to a DC voltage level linearly related to the F1 frequency. The F2 was estimated by a filter specially shaped so that for most speakers the F2 would predominate over other formants. The F2 range is from 700 to 2300 Hz (Peterson and Barney 1952; Fant 1959). The F2 frequency was determined in the same way as for F1 (Seligman 1987). The filter characteristics are illustrated in Figure 8.12.

Amplitudes were derived by taking the outputs of the formant filters, passing these through peak detectors, and smoothing the results with a 35-Hz low-pass filter. As with the F0/F2 and WSP-II processor the formant frequencies were converted to a digital quantity, the frequency value was referred to a table stored in memory that related frequency to electrode number, the formant amplitude was converted to a digital quantity, the electrode number and amplitude were then applied to another table and an appropriate current level for the particular electrode determined, and the data were encoded in serial form and sent via an RF to the receiver-stimulator. The strategy for converting the F1 and F2 frequencies to site of stimulation was changed from that used for the F0/F2 processor. The frequencies were presented to the electrodes in linear steps from 250 to 1650 Hz, and then exponentially up to 6,000 Hz. This was described by the patients as a smoother speech sound. The WSP-III speech processor also allowed alternative strategies to be investigated. These included ones where the F1 frequency was used instead of the F0 as the stimulus rate, but although voicing was received, initial tests with the strategy showed no advantage over the above F0/F1/F2 strategy with F0 as the fundamental stimulus rate.

The Nucleus F0/F1/F2 WSP-III system was approved by the FDA in May 1986 for postlinguistically deaf adults, and the Nucleus F0/F1/F2 MSP system for children 2 years and older was approved in June 1990.

First Formant (F1), Second Formant (F2), Voicing (F0), and High Frequency Fixed-Filter Strategy (Multipeak), and MSP Speech Processor

A further development of the strategy was to take the outputs of fixed filters in three frequency bands (2000–2800 Hz, 2800–4000 Hz, and >4000 Hz) as well as the first two formants, and to stimulate the cochlear nerve on a place-coding basis, together with voicing as rate of stimulation (Fig. 8.12). The high-frequency spectral information was extracted to provide additional high-frequency cues to improve consonant perception and speech understanding in quiet as well as in noise. This became the Multipeak speech-processing strategy. It is a misnomer, because the high-frequency information was not spectral peaks of energy but rather the outputs from fixed filters. It was thus a hybrid scheme between formant extraction and fixed filter. The strategy was implemented in a speech processor named the Nucleus Miniature Speech Processor (MSP). The MSP included the same functions as the WSP III, but did so with a single custom-made non-programmable digital signal processing (DSP) chip (Austek Microsystems Proprietary Limited), combined with an off the shelf memory chip to store the patient's data. There was a three-channel filter analog chip (K-MOS Semiconductor Limited) to estimate the energy in the frequency bands. Later the MSP was modified as Spectra-22 for the spectral maxima sound processing strategy with a 20-channel filter bank (see below). The DSP chip included an analog pre-amplifier, ADC, DSP feature extractor, and encoder. Advances in technology had made it possible to integrate digital and analog circuitry on the one chip with reduced power consumption and so make this advance possible for cochlear implants. Their incorporation in the design was an advantage as they were more reliable with repeatable performance.

With the Multipeak strategy on the MSP, the signal was picked up with a microphone and amplified. It then passed to a sensitivity control that could adjust the gain to keep the signal at a comfortable level. The signal passed to the DSP through the AGC. This allowed the dynamic range of the acoustic signal to be compressed into the smaller range required for electrical stimulation. It was then low pass filtered with an anti-alias filter. According to Nyquist's theorem as discussed above it is necessary to sample at twice the rate of the highest signal frequency to be represented, 6000 to 8000 Hz being the upper speech frequency required. If higher frequencies are not filtered out, they are also sampled with errors that affect the overall accuracy. The signal was passed through combinations of filters to determine not only the F1 and F2 frequencies but also those in the higher bands. The filtering was done with digital filters rather than the analog devices used with the WSP-II and WSP-III. Peak detectors were employed, and the extracted envelope filtered and enhanced before entering the zero crossing detector to measure the frequencies. This circuit was designed to extract F0 from the F1 region and below. As discussed, the extraction of F0 in noise, which often had frequencies in the region, had been a problem for vocoders. In addition, the amplitudes of formant filter outputs were obtained by rectifying and low-pass filtering at 35 Hz to obtain an average intensity. The signals were then referred to the memory chip where the patient's MAP was stored and then encoded for transmission to the receiver-stimulator.

The Multipeak-MSP system was approved by the FDA on October 11, 1989 for use in postlinguistically deaf adults, and in June 1990 for profoundly deaf children 2 years and older.

Spectral Maxima Speech (Sound) Processor (Spectra), and SPrint and
ESPrit Processors

The research in 1989 at the University of Melbourne/Bionic Ear Institute with the F0/F1/F2 and Multipeak strategies showed that the recognition of place of articulation was considerably less than for other features, and if it were improved overall speech perception should be better. Studies demonstrated higher scores with the extraction of up to four frequency peaks either with constant rate or rate proportional to the voicing frequency. The extraction of six peaks gave similar results to the four referred to above. So an alternative to the extraction of six peaks was the selection of the maximal outputs from six of a bank of 16 filters. A similar frequency to electrode mapping strategy was adopted to the ones used for the F0/F1/F2 and Multipeak strategies. The first eight filters had center frequencies distributed linearly over the range of 280 to 1780 Hz, and the remaining eight were logarithmically spaced up to 6000 Hz. The strategy called the spectral maxima speech processing (SMSP), was implemented in 1990 on an initial patient using an NEC filter bank chip, a microprocessor, and MSP for encoding the stimuli through the Human Communication Research Center (HCRC) at the University of Melbourne/Bionic Ear Institute. The initial results were an improvement over the F0/F1/F2 WSP-III and Multipeak-MSP systems (McKay et al 1992). As a result the SMSP strategy was implemented by Cochlear Limited as SPEAK, which was incorporated into a processor referred to as Spectra-22. SPEAK Spectra-22 (Seligman and McDermott 1995) differed from SMSP in being able to select six or more spectral maxima from 20 rather than 16 filters. A constant stimulus rate that varied adaptively from 180 to 300 pulses/s was used. The fundamental frequency (F0) was conveyed by the amplitude variations of the speech wave. A multi-center comparison of the SPEAK Spectra-22 and Multipeak-MSP systems was undertaken by Skinner et al (1994). The mean scores for vowels, consonants, consonant-nucleus-consonant (CNC) words, and words in City University of New York (CUNY) and Speech Intelligibility Test (SIT) sentences in quiet conditions were all significantly better for SPEAK.

Spectra-22 was an extension of the development of speech processing for the Multipeak-MSP system (Nucleus 22 or Mini-22) (Seligman and McDermott 1995). All the hardware of the MSP processor including the microphone, preamplifier, AGC, ADC, MAP, data encoder, coil driver, and regulated battery power supply were the same. In the Spectra-22 the additional filter chip in the MSP for the high-frequency bands was replaced by a 20-channel band-pass filter bank. The center frequencies were between 250 and 10,000 Hz as shown in Figure 8.15. The filters were implemented as low power switched capacitor filters, and they had the advantage that their frequency boundaries could be varied. The frequency spacing of the filter channels was normally in linear 200-Hz steps from 250 to 1650 Hz, and then exponentially (i.e., linear on a logarithm scale up to 10,000 Hz). The scaling of the entire bank could be shifted in frequency and the gain of the individual filters could also be varied in 1-dB steps. Clinically, the range of 120 to 8652 Hz was used. Amplitude coding was done through detecting an amplitude peak in a filter and the maximal outputs determined through scanning the filter outputs. When the parameters had been established, the stimulation of the cochlea occurred in a round-robin process from the base to apex. The Spectra-22 was designed to provide either SPEAK or the Multipeak strategies.

The SPEAK strategy was next implemented in the Nucleus 24 series of speech processors (Figure 8.16); that is the body-worn SPrint and behind-the-ear ESPrit (Patrick et al 1997) as well as the fixed-filter continuous interleaved strategy (CIS) discussed below (see Simultaneous Analog Stimulation). They had the following functional blocks: (1) front-end or audio preprocessor, (2) filter, (3) sampling and select, (4) amplitude mapping, (5) RF encoder/decoder formatter, and (6) programming interface and power supply. With the SPrint the front-end chip had an audio preprocessor that included an analog preamplifier, and allowed the impor-

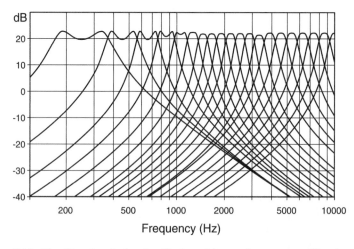

FIGURE 8.15. The filter bands for the Nucleus 24 speech processor. (Reprinted from Encyclopedia of Biomaterials and Biomedical Engineering, in press, with permission of Marcel Dekker.)

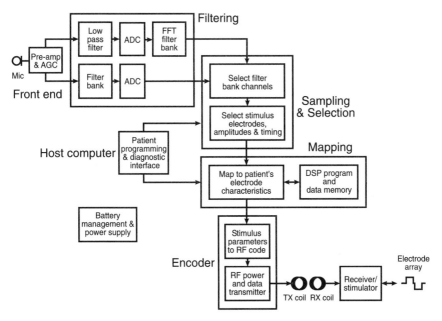

FIGURE 8.16. The Nucleus behind-the-ear and body-worn speech processors are implemented using either a filter bank or fast fourier transform (FFT) filter bank, respectively. The front end sends the signal to a signal processor chip via a filter bank or to a digital signal processor (DSP) chip, which carries out an FFT and signal processing. The signal processor selects the filter bank channels and the stimulus electrodes and amplitudes, and maps these to the patient's requirements. An encoder section converts the stimulus parameters to a code for transmitting to the receiver-stimulator on a radiofrequency (RF) signal together with power to operate the device. ADC, analog-to-digital converter; RX, receiver, TX, transmitter (Clark 1998).

tation of signals from FM systems, TV, telecoils, and telephone adaptors, and the mixing of these. It also had an AGC and automatic sensitivity control implemented with a digital circuit. With filtering after the energy of the component frequencies in the speech signal or environmental sound was amplified, it was low-pass filtered to remove any high-frequency energy that would introduce aliasing errors when converting the analog signal to a digital one using an ADC. The digitized signal was filtered with an FFT in a DSP. The FFT is a mathematical algorithm discussed in Chapter 7 that is more flexible and can achieve a better dynamic range than the switched capacitor filters used with the ESPrit. The SPEAK strategy had a large number of narrow filter bands as illustrated in Figure 8.15. Twenty-two channels were available for the common ground and monopolar stimulation, and 20 for bipolar stimulation when using an intervening electrode. The filtering could be adjusted to provide the filtering for the CIS strategy that had from 4 to 12 bands that were wider. The outputs were rectified and low-pass filtered (center frequency 189 Hz) to extract the envelopes. Thus temporal information up to approximately 250 Hz could be transmitted. With the Nucleus 24 behind-the-ear speech processor (ESPrit), after the signal was amplified it was filtered by a bank of switched capacitor filters. These are power efficient, and this allowed the behind-the-ear processor to be built.

The signal processing or sampling and selection component implemented the speech processing strategies. A low-power custom DSP chip analyzed the incoming signals and implemented a wide range of processing strategies that included SPEAK, CIS, and advanced combination encoder (ACE), discussed below. Enhanced noise reduction programs were also available. With the spectral maxima strategy (SPEAK) a subset (typically six to eight) of the outputs from the 22 filters were converted to current level for nonsimultaneous stimulation of electrodes at a constant rate and at a site corresponding to the center frequency of the filter. The channel rate was 250 Hz.

The RF data encoder and transmitter formatted the stimulus parameters into a serial data stream for transmission from the head set coil to the coil on the receiver-stimulator. The data stream was produced by the amplitude modulation of a carrier frequency at either 2.5 or 5 MHz, and was discussed above (see Encoder and Transmitter) as an example of the principles of encoding and transmitting data.

The Nucleus ESPrit had a similar design (Patrick et al 1997), to the Spectra. The speech was analyzed with a switched capacitor filter band that kept the power consumption low so that it could be realized as a small unit providing SPEAK in January 1998. It has subsequently been modified to allow CIS and ACE to be used.

The SPEAK encoder strategy (implemented on Spectra-22) was approved by the FDA for post-linguistically deaf adults and children on March 30, 1994. The Nucleus 24 system (incorporating the SPEAK strategy on SPrint) was approved by the FDA on June 25, 1998.

Advanced Combination Encoder (ACE)

The ACE strategy was a modification of SPEAK, with stimuli presented at high rates and/or with more channels. In a study in the Cooperative Research Center (CRC) for Cochlear Implant Speech and Hearing Research, considerable variability was found in patient performance, but with on average poorer results for the highest rate of 1600 pulses/s (Vandali et al 2000). In addition, Fishman et al (1997) found a small increase in speech perception when the number of electrodes was increased from seven to 20, and a greater number of electrodes would be expected to be beneficial for hearing speech in noise. In a large study on postlinguistically deaf adults (Arndt et al 1999) the mean ACE speech perception scores were significantly higher than the CIS, but not significantly different from SPEAK. Overall, 61% preferred ACE, 23% SPEAK, and 8% CIS. The ACE strategy can be implemented on the Nucleus 24 SPrint® speech processor.

The ACE and CIS strategies were approved by the FDA for use with the Nucleus 24 system on June 9, 1999.

Simultaneous Analog Stimulation (SAS)

The Clarion device implemented a speech-processing system SAS similar to that developed at the University of Utah in Salt Lake City (Eddington 1980, 1983). The latter was a four-electrode fixed-filter-strategy providing simultaneous analog stimulation with gain compression and referred to as compressed analog, and used in the Ineraid device. The gain compression was to allow the range of speech intensities to remain within the smaller dynamic range for electrical stimulation. The center frequencies of the filters for the most apical four electrodes were 500, 1000, 2000, and 3400 Hz (Dorman et al 1989). It also had a percutaneous connection (i.e., a plug and socket) for monopolar stimulation with an electrode array that had ball electrodes. The main difference between the SAS and compressed analog (CA) schemes was that SAS had AGC with longer attack and release times, as well as a lower compression ratio for reduced spectral distortion (Wilson 2000). It was subsequently used with eight filters in the Clarion processor (Battmer et al 1994). The SAS system provided bipolar stimulation with a molded electrode array similar to the one originating from the research at the UCSF (Merzenich et al 1984) as discussed below. In implementing the strategy it was found difficult to provide sufficient current for selective stimulation of neurons even with offset radial electrodes. Many patients could not receive a percept at all (Wilson 2000). As a result enhanced bipolar electrodes having a spacing of 1.7 mm and monopolar stimulation were used.

As discussed above (see Digital Versus Analog Circuitry), an analog signal requires many data points for its accurate representation. For that reason the Clarion SAS processor was designed to transmit 104,000 samples or voltages/s (Clarion Description, Advanced Bionics). The term *sample* does not refer to the pulse rate. However, with the stimulator, 91,000 samples/s were available to stimulate seven or eight electrodes by either analog or pulsatile stimulation. Each of the electrodes could be updated only every 77 μs whether this was for individual voltages or pulses. Thus any one of seven electrodes could be stimulated at 13,000 samples/s. As discussed in Chapter 5, these high rates were beyond physiological limits.

Continuous Interleaved Sampling (CIS) Strategy

The CIS strategy was developed at the Research Triangle Institute, North Carolina (Wilson 2000). It evolved from a fixed filter scheme that used interleaved pulses (IPs) to avoid channel interaction. A constant stimulus rate also reduced the problem of channel interaction (Wilson 2000). A higher stimulus rate than for the Nucleus devices was used to provide adequate sampling of the voicing frequency, a necessity because the low constant rate of stimulation could not represent voicing as well as a rate proportional to the frequency. The outputs of six or more filters were used to stimulate the same number of electrodes on a place-coding basis. Studies to optimize the number of filters and stimulus rate (Wilson et al 1992, 1993). Lawson et al (1996) and Wilson (1997) found seven was the maximum needed for optimal speech perception in quiet. The Clarion processor with

CIS strategy was implemented with eight band-pass channels coding frequencies ranging from 250 to 5500 Hz. The spectral information was presented at a constant stimulus rate between 833 and 1111 pulses/s per channel for bipolar or monopolar stimulus modes (Battmer et al 1994).

A block diagram of the CIS strategy is shown in Figure 8.17. Inputs from the microphone and optional AGC were fed to a preemphasis filter that attenuated frequencies below 1200 Hz at 6 dB/octave. This provided an adequate signal for the consonants with frequencies above this, as they are less intense than those speech features below. The output of the preemphasis filter passed to a bank of band-pass channels for filtering, envelope detection, and compression. A mapping function determined the T and MC levels for each electrode. The channel outputs modulated trains of biphasic pulses and stimulated electrodes on a place-coding basis. The stimuli were delivered nonsimultaneously as with the F0/F1/F2, Multipeak, and IP strategies (Wilson 2000).

The Combi-40 and 40+ also provided the CIS strategy. It was modified to incorporate the Hilbert transform for envelope extraction. The Hilbert transform (essentially a 90-degree phase-shifter) is an efficient technique for detecting the speech wave envelope from each filter. The transform was first described by Stark and Tuteur (1979), and has been in use in communication engineering since that time. This transform therefore would provide good representation of timing information including the fundamental frequency. It provides up to 18,180 sequential, non-overlapping stimulus pulses for implementing the CIS strategy. This Combi-40+ system was approved by the FDA in August 2001.

Packaging

Body-Worn Speech Processor

Body-worn speech processors allow the presentation of some or all of the present strategies in regular use (SPEAK, CIS, ACE). They may have flexibility for research on alternative strategies. The Nucleus 24 SPrint (Fig. 8.18) had 22 stimulus channels and could provide the SPEAK, CIS, and ACE strategies. It had the dimensions of 8.8 × 6.7 × 2.3 cm. It also had the flexibility to evaluate a number of other strategies that are discussed in Chapter 14. In addition, it had programs to control sensitivity in noisy environments. There was a telemetry link sending voltages from implant electrodes back for analysis. It had a liquid crystal display of program settings, signal levels, and diagnostics. There was a control panel disable feature for children. It also had data logging to monitor usage, and compatibility with assistive devices. It could be used with single- or double-battery packs, the weights being 114 and 146 g, respectively.

The Clarion S body-worn speech processor (platinum sound processor) provided the SAS and CIS strategies and had the dimensions of 7 × 6 × 2.2 cm and is shown in Figure 8.18.

The Combi-40+ devices presented CIS or a strategy *n* of *m,* the latter being the same as the Nucleus ACE strategy at its maximum rate, where some *n* of a number of electrodes *m* are selected to present frequency/spectral information on

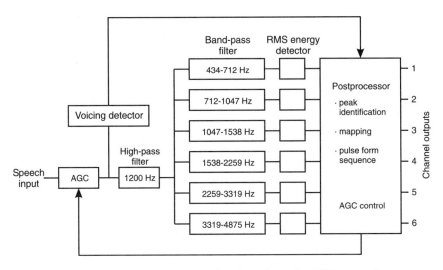

FIGURE 8.17. A diagram of the continuous interleaved sampler (CIS) processing strategy. (Reprinted with permission from Wilson B. S., Strategies for representing speech information with cochlear implants. In: Niparko, ed. Cochlear implants: principles and practice. 2000 Lippincott Williams & Wilkins.)

a place-coding basis. It was incorporated in a body-worn version CIS PRO+, with the dimensions 9 × 6.8 × 2 cm. It requires two rechargeable AA batteries for an average of 1 day of operation (Med El application to the FDA, August 21, 2001).

The MXM Digisonic speech processor operated with two rechargeable batteries that had a life of 12 to 16 hours. There was also a sensitivity control with separate base and treble to improve the contrast of the speech in noise (MXM at http://www.mxmlab.com./digisonic/features.html).

Behind-the-Ear Speech Processor

Many patients find a behind-the-ear speech processor desirable, particularly as it is more convenient and cosmetically pleasing to dispense with the leads passing from the microphone to the body-worn device. Miniaturization of the Nucleus 24 processor required high-powered zinc-air batteries and a low power consumption, which is easier to achieve with strategies using low stimulus rates. Its dimensions were 4.6 × 1.9 × 0.9 cm. The Nucleus behind-the-ear speech processor, ESPrit (Fig. 8.19), has been used since January 1998 with the SPEAK strategy and the Nucleus CI24 implant. It was modified for use with the Nucleus CI22 device and approved by the FDA on December 23, 1999.

In 2002 the ESPrit-3G provided alternative strategies such as ACE and CIS. It gave a choice of sensitivity or loudness control, simple rotary and in-line switches, and compatibility with assistive devices. Triple 675 zinc air batteries gave typically from 16 to 150 hours of usage. The ESPrit-3G had a built-in telecoil that allowed listeners access to hearing aid–compatible telephones and to connect to sound systems in public venues with assistive listening devices like induction loops and infrared or FM systems.

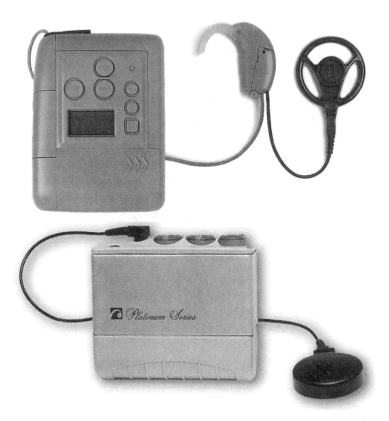

FIGURE 8.18. Top: The Cochlear Limited Nucleus 24 body-worn SPrint for the regular and research speech-processing strategies (Reprinted with permission from Cochlear Limited). Bottom: The Clarion Platinum body-worn speech processor (Reprinted with permission from Advanced Bionics Corporation).

The Clarion behind-the-ear speech processor used rechargeable batteries, but this limited their running time. It is illustrated in Figure 8.19. The Med El Tempo+ speech processor used three zinc-air batteries for approximately 36 hours operation. It dimensions were $6.6 \times 1.3 \times 0.9$ cm., and came in straight, angled and children's configurations.

Receiver-Stimulators

The receiver-stimulator needed not only to be designed with circuitry for low power consumption and hence long battery life, but also to allow a number of speech-processing strategies to be evaluated. If, for example, certain rates, waveforms and stimulus patterns were found to be helpful, this should be possible without having to explant the device and replace it with a new one. The device should also provide charge-balanced, biphasic pulses to minimize corrosion of electrodes.

Figure 8.19. Left: The Cochlear Limited Nucleus 24 behind-the-ear ESPrit-3G for the regular speech-processing strategies (SPEAK, CIS, ACE) (Reprinted with permission from Cochlear Limited). Right: The Clarion behind-the-ear speech processor for the CIS strategy (Reprinted with permission from Advanced Bionics Corporation).

Design Principles

Constant Current Stimulation

The first consideration was whether constant current or constant voltage stimulation should be employed. With constant voltage stimulation the interface impedance between electrode and tissue may change, in which case the current flowing through the nerve could vary. An increase in electrode impedance will result in reduction in current unless the voltage is increased ($I = V/R$; Ohm's law). The current amplitude, and therefore the amount of charge per phase, needed to remain constant for reliable stimulation. This could be avoided by constant current stimulation. Constant current stimulation allows the stimulating current to be specified, and ensures charge balance between the two phases of a biphasic pulse. Constant current stimulators are effective where there are small or moderate changes in electrode impedance. With high impedances they may not be able to provide the necessary voltage (i.e., they lose voltage compliance). This could lead to asymmetrical current pulses, and deliver net DC charge to the adjacent tissue with possible neural damage.

Charge Balance

The electrical stimuli produced by the receiver-stimulator should be charge-balanced, biphasic pulses to minimize the buildup of a DC current at the electrode–tissue interface. The DC current will lead to corrosion of electrodes, the production of toxic products at the electrode–tissue interface, and neural and other tissue damage. With the Nucleus 22 systems this was possible at high rates as they had circuitry to short any residual current between pulses, and at high rates there would not be time for this to occur. Capacitors in the circuit are effective in preventing a buildup in the DC as their impedance is frequency dependent and approaches infinity for a zero frequency. As discussed in Chapter 4, with a si-nusoidal current the relationship between voltage (V), current (I), and capacitance (C) and angular velocity (ω) which is $2\pi f$ where f is frequency, is $V = I/2\pi f C$. The term $1/2\pi f C$ is the capacitive reactance, and if the frequency becomes very low the impedance is proportionately high. Thus current flow will become infin-itesimally low. At the time when the Nucleus 22 implant was developed, capac-itors were large and high stimulus rates were not needed. Now they are smaller and with the Nucleus 24 implant there are two capacitors for the extracochlear electrode, so that for monopolar stimulation at high rates there is shorting between electrodes as well as capacitors to prevent the buildup of charge. The Clarion S and Combi-40+ have capacitors for each electrode, but do not use shorting be-tween pulses.

Simultaneous Versus Nonsimultaneous

A fixed-filter strategy that modeled the physiology of the cochlea and the neural coding of sound in the auditory nerve was tested (Laird 1979), but unsatisfactory results were obtained due to simultaneous stimulation of electrodes leading to channel interaction and unpredictable variations in loudness (Laird 1979; Clark, Blamey et al 1987). To avoid this the Nucleus F0/F1/F2 and Multipeak (Dowell et al 1987; Seligman 1987) strategies presented formant and spectral information nonsimultaneously (the pulses were separated by 0.8 ms) so that there was no summation of the electrical field. This also occurred with the IP strategy (Wilson et al 1988). An alternative explored was to stimulate with each phase of, say, three biphasic stimuli broken into nonoverlapping monophasic pulses (quasi-simultaneous stimulation). This was tested on an initial patient for the estimation of the first and second formants and gave similar results to the standard processor (McDermott 1989).

Number of Stimulus Channels

The place coding of frequency with multiple-channel implants is the main reason that results with these systems are superior to those of single-channel implants. The number of stimulus channels to be incorporated in the receiver-stimulator was therefore a matter of importance, but the optimal number has not been fully established.

Multiple-channel devices were implanted with the number of channels varying from four with the Ineraid (Eddington 1983), to seven to eight with the Clarion S and Chorimac 8, to 12 with the Chorimac 12 and Combi-40 (Fugain et al 1984), and to 22 with the Nucleus 22 and 24 systems (Clark, Black et al 1978). Holmes et al (1987) found that open-set word recognition and continuous discourse track-ing results for the Nucleus F0/F1/F2 speech processor improved for the use of up to 15 electrodes. In a study by Blamey et al (1992) there was a positive correlation between speech perception and the number of electrodes in use, up to

21. For more details see Chapter 7. As there can be variations in the density of auditory neurons due to pathology, a further advantage for the Nucleus systems in having 22 electrodes is that there are more electrodes available in areas of the cochlea where place of stimulation is more effective. In determining the number of stimulus channels there is an interaction between mode of stimulation, electrode geometry, and cochlear anatomy for the optimal place coding of frequency.

In contrast, Dorman et al (1989) showed that the number of stimulus channels for a fixed-filter (modified channel vocoder) system should be at least four, and Wilson et al (1992) and Battmer et al (1994) reported that with CIS the upper limit was seven or eight. This was consistent with the development of the Clarion SAS system with seven channels (electrodes). It used bipolar stimulation with the array that originated in the Coleman Research Laboratory at UCSF, as distinct from monopolar stimulation as undertaken by Eddington (1980, 1983). However, radial stimulation with this system could not always reach the dynamic ranges required on each electrode pair for place coding of frequency, so eight electrodes were connected longitudinally to make seven pairs ("enhanced" bipolar stimulation). Subsequently, monopolar stimulation was used as discussed above.

Mode of Stimulation

Another important design question was the mode of stimulation – bipolar, common ground, or monopolar. The stimulus mode should be the one that gives the best localization of current to distinct groups of auditory nerve fibers for the place coding of frequency. The second aim was to achieve low stimulus current threshold levels to increase battery life. These modes were discussed and illustrated in Chapter 5. Bipolar stimulation occurs when a potential difference is created between neighboring electrodes to allow current to flow between the two. This was shown by Merzenich (1975) and Black and Clark (1977, 1978, 1980) to produce localized stimulation of the cochlear nerve fibers. It was demonstrated by Black and Clark (1977, 1978, 1980) that common ground stimulation would also localize the current to separate groups of cochlear nerve fibers. Common ground stimulation occurs when there is one active electrode, with the others all connected electronically to form a common ground. Bipolar and common ground stimuli were shown in experimental animal studies to provide more localized stimulation than monopolar pulses. However, Busby et al (1994) found in a psychophysical study that monopolar stimulation between an active and distant reference electrode in the cochlea allowed pitch percepts for each electrode to be scaled as well as for bipolar or common ground stimulation. Monopolar stimulation occurs when a potential difference is created between an active electrode and a distant ground usually outside the cochlea. The ground electrode is usually placed underneath the temporalis muscle. Studies in the CRC for Cochlear Implant, Speech and Hearing Research in Melbourne in 1995 showed this location was a suitable sink for the current to avoid stimulating the facial nerve or pain fibers in the vessels around the dura. The modes of stimulation were discussed in more detail in Chapter 5.

Current Levels

The minimum current level for a T-level stimulus depends on the pulse width, cochlear pathology, electrode geometry, stimulus mode, and stimulus rate. For an individual patient the T level varies with electrodes and depends on the properties of the nerve membrane, as evidenced by the strength duration curve. With a shorter pulse a higher current is required to reach a T level. The current required for the MC level, is just below the minimum acceptable discomfort level (MDL). However, with a high rate, and therefore a shorter pulse width, a lower current is needed to excite the neuron as a greater electrical charge is produced. The relationship among stimulus rate, pulse width, and charge delivery is complex and was discussed in Chapters 5 and 6. Furthermore, a lower current output is needed if a speech-processing strategy uses subthreshold stimuli. With high impedance due to fibrous tissue and bone, high stimulus levels are required to maintain the current required for neural excitation. The output current levels depend on the discriminable steps in loudness and their effect on speech perception. The discrimination of electrical current was studied in patients by Simmons (1966), Douek et al (1977), Eddington et al (1978), Fourcin et al (1979), Aran (1981), House and Edgerton (1982), Hochmair and Hochmair-Desoyer (1983), Dillier et al (1983), Shannon (1983), Tong et al (1988), and Nelson et al (1995). The just discriminable differences in electrical current varied from 1% to 8% of the dynamic range. The loudness growth or dynamic range for stimulation at 200 pulses/s with the University of Melbourne/Nucleus banded electrode array in the scala tympani was found to vary from 5 to 10 dB (Clark, Tong et al 1978; Tong et al 1979).

With the University of Melbourne's first receiver-stimulator, the intensity control allowed independent variation of the stimulus current of each channel in increments of 70 μA from a minimum of approximately 70 μA to a maximum of approximately of 1000 μA in 15 steps. These limitations in current steps were imposed by the integrated circuit technology at the time. The advances in speech processing requiring more frequency and intensity control coincided with improved integrated circuit technology. This allowed more precise control of intensity and rate of stimulation. The Nucleus 24 implant had a minimum stimulus current of 10 μA and a maximum of 1750 μA. There is a logarithmic relationship between the current level and the discriminable steps in loudness. Thus the current steps (I_n) can be calculated according to the following formula:

$$I_n = 10 \times 175^{(n/225)}, n = 0, 255$$

e.g., for $n = 0$, $I_0 = 10 \times 175^0 = 10\,\mu$A; and for $n = 255$, $I_{255} = 10 \times 175^1 = 1750\,\mu$A. The current step ratio I_{m+1}/I_m is 1.02; that is, each current step is typically 2% greater than the previous current level.

As current level can be traded for pulse width to maintain equal loudness, greater flexibility was achieved by designing the Nucleus 24 as well as the Clarion S and Combi-40+ devices so that the pulse width could be varied. Typically widths of 20 to 400 μs per phase were used. The interphase gap could be varied in increments of 0.2 μs. At least 8 μs should be used, and typically 8 to 50 μs.

Charge Density and Charge Per Phase

Charge density was shown to lead to electrolytic changes at the electrode tissue interface, and the release of gas and toxic products for short-duration (100 to 200 μs) biphasic current pulses, for a charge density of 300 μC cm^{-2} geometric/phase and above. Charge density and charge per phase covaried in producing neural damage when electrodes were in contact with the cortex (McCreery et al 1988, 1990, 1994). The effects of electrical stimulation on the auditory nerve is discussed in more detail in Chapter 4. Increases in the extracellular (K^+) concentration in the cortex were seen at high charge densities per phase (100 μC cm^{-2}/phase or 1 μC/phase at 50 Hz) as well as high rates (Heinemann and Lux 1977; Nicholson et al 1978; Urbanics et al 1978; Stockle and Ten Bruggencate 1980; Agnew et al 1983; McCreery and Agnew 1983).

The acute findings do not necessarily apply to long-term stimulation and may vary with the tissue and the distance the electrode is from the tissue. For that reason studies were carried out long-term on the experimental animal to ensure that electrical stimulation with the University of Melbourne/Nucleus banded array did not produce charge densities that could be damaging.

With the Nucleus banded-electrode, animal studies (Shepherd et al 1983) showed that continuous stimulation at charge densities of 18 to 32 μC cm^{-2} geometric/phase did not lead to damage of spiral ganglion cells. It was also shown that charge densities of 20 to 40 μC cm^{-2} geometric/phase were within biologically safe limits (Leake-Jones et al 1981a,b).

Although the above in vivo study by Shepherd et al (1983) and the in vitro studies by Brummer and Turner (1977a–c) showed that charge densities below 32 μC cm^{-2} geometric/phase were safe, the upper limit for safety was not established. The electrodes on the original University of Melbourne/Nucleus array had a relatively large surface area (0.44–0.66 mm²). With a pulse width of up to 50 μs (normally 25 μs) and the highest current (1.75 mA) delivered through the smallest band on the Nucleus array, the maximum charge density possible was 19.9 μC cm^{-2} geometric/phase. So with the worst-case scenario the charge density for the Nucleus banded array was well within the safe level. With the Nucleus perimodiolar array (Contour) the electrodes were half (see Fig. 8.49) rather than full bands, and their area varied from 0.283 to 0.306 mm² geometric. This would mean the density could double, but this would be counterbalanced by the lower thresholds with the electrodes closer to the spiral ganglion cells.

In contrast, the surface areas of the electrodes of the Med El and Clarion devices were up to five times smaller (0.14 mm²) than for the Nucleus array (Med El Combi-40 Manual; Clarion Device Description, Advanced Bionic Corp.). The Clarion system could produce up to 2.5 mA, and at its minimum pulse width of 77 μs results in a charge density of 137.5 μC cm^{-2} geometric/phase. The pulse width could be increased resulting in an even greater charge density. The dimensions of the Clarion High Focus II electrode pads have increased to a size and shape comparable to that of the Nucleus half-band Contour array. The Med El Combi-40+ could deliver a current of 2.5 mA for pulse widths between 40 and

640 μs, so the maximum charge density could range from 80 to 914 μC cm^{-2} geometric/phase. These are well above those that have been shown to be safe (32 μC cm^{-2} geometric/phase). It is therefore important to establish the safe upper levels for charge density (Clark and Lawrence 2000).

Stimulus Rate

The perception of pitch, due to rate of stimulation, is important for speech understanding, and thus it is necessary to provide the facility to vary the rate. The upper limit on the perception of variations in the rate was shown to be about 400 to 800 pulses/s in humans (Simmons 1966), and 200 to 800 pulses/s in animal experimental studies (Clark 1969b; Clark, Kranz et al 1973). There is no solid evidence to provide stimulus rates for pitch discrimination in excess of 1500 pulses/s.

Although there are limits on the perception of variations in pulse rate as discussed above, the fine time structure of electric pulses delivered to the electrodes may be important. For this reason there was a need to have adequate control of stimulus timing especially between channels.

The Nucleus 24, Clarion S, and Combi-40+ systems provided the CIS strategies at stimulus rates higher than 1000 pulses/s. This also applied to the Nucleus 24 ACE strategy. The auditory nerve fibers have an absolute refractory period of approximately 0.5 ms during which time they cannot respond to another stimulus. They also have a relative refractory period of 0.2 ms when their responsiveness is markedly reduced and a stronger stimulus is required to produce excitation. How frequently the neurons respond to each stimulus at different rates can be measured with interspike interval histograms (see Chapter 5 for further details). The data of Paolini and Clark (1997) showed that at 1800 pulses/s where the period is 0.55 ms and close to the absolute refractory period the firing pattern was a Poisson distribution, and thus the response of the unit was not related to the stimulus rate.

It is also thought possible to convey temporal information by amplitude variations at high rates of stimulation through altering the rate and population of nerves excited (Rubinstein et al 1999), although psychophysical studies (Viemeister 1979; Shannon 1992; Busby, Tong et al 1993) showed that only low rates of modulation (100 to 200 Hz) could be detected. Hong et al (submitted) found that a subthreshold conditioning stimulus of 5000 Hz increased the dynamic range up to 6.7 dB on average.

High rates of stimulation (1000 to 2000 pulses/s) used within the clinically acceptable intensity levels are safe (Xu et al 1997); however, the use of a high rate may damage auditory neurons at current levels and charge densities above normal clinical levels (Huang et al 1996, 1998a,b). In addition, the devices should be engineered to allow for charge recovery at the electrode/tissue interface between pulses. This prevents a buildup of DC current that can damage nervous and cochlear tissue at levels greater than 2 μA (Tykocinski et al 1997). As discussed above, this can be avoided by the use of capacitors with the extracochlear electrode for the Nucleus 24 system, and for each electrode with the other systems. Past neurobiological safety studies demonstrated the importance of evaluating, in animal experimental studies, any significantly altered rate of stimulation as well as the electronics to be used in patients to deliver the high pulse rates. All significant changes in stimulus parameters and electrode geometry in the Nucleus system have been accompanied by animal studies.

Nonsimultaneous stimulation was provided with the Nucleus 22 F0/F1/F2 and Multipeak strategies, the Nucleus 24 system for the SPEAK, ACE, and CIS strategies, and the Clarion S and Combi-40/40+ devices for CIS. With the Nucleus 24 the maximum overall stimulus rate was 14,400 pulses/s at 25 μs/phase, and minimum 8 μs gap and 12 μs for shorting. When this overall rate was distributed across electrodes, the maximum rate on each of 10 could be approximately 1440 pulses/s per electrode.

The Combi-40+ could produce 18,180 pulses/s (I. Hochmair, personal communication), and thus for 12 electrodes could stimulate at up to 1515 pulses/s on each electrode. The Clarion S was reported to produce 104,000 samples/s (Clarion Device Description). The term *sample rate* must not be confused with *biphasic pulse rate,* but rather refers to the voltages used for the simultaneous analog representation of the speech signal; 91,000 samples/s were available to stimulate seven electrodes. In addition, because it took two samples to make a biphasic pulse and due to the limitations on update time, the device could produce 6500 biphasic pulses/s for distribution (Schulman et al 1996).

Design Realization

With the first studies undertaken by research groups it was thought there should be almost complete flexibility with the stimuli so that a patient's percepts for different stimulus parameters, such as current levels, pulse widths, and pulse rates, could be determined. Now that more is known about the range of percepts possible, a receiver-stimulator can be designed to provide the appropriate stimuli without having to be quite so flexible or require a plug and socket.

Power and Data Transmission

After speech is transformed into electrical signals, the signals are transmitted to electrodes in the cochlea to excite the residual auditory nerves. A high-frequency electromagnetic carrier wave was shown to be the best for transmitting power and speech data (Forster 1978; Clark, Black et al 1977), and the wave was modulated by the coded speech signal. The code specified the electrode to be stimulated, the amplitude of the current, and the start time for each electrode within 1 ms.

Since that time the Nucleus 24, Clarion S, Combi 40/40+, and MXM receiver-stimulator devices have codes that specify the electrode to be stimulated, mode (bipolar with various electrode spacing, common ground, monopolar), rate, current amplitude, pulse width, and interpulse separation which are transmitted serially as pulses in a radiofrequency signal. The circuits in the receiver-stimulator must implement the instructions from the speech processor, and the digital information is finally converted into a current to stimulate the cochlear nerve fibers. For the transcutaneous transmission of information a high-frequency modulated wave is desirable as it permits a small aerial to be used, and a large amount of speech data to be transmitted efficiently. Examples of a carrier wave modulated by varying either its amplitude or frequency are illustrated in Figure 8.3.

Digital Versus Analog Circuitry

In designing the receiver-stimulator, as with the speech processor, an important decision was whether to use analog or digital circuitry or a combination of both. As discussed above (see Digital Versus Analog Circuitry), analog circuits are those in which continuously varying physical parameters such as voltages can be altered or combined. With analog circuits the instantaneous amplitude of speech could be converted into a voltage proportional to the amplitude. The voltage could be transmitted indirectly to the receiver-stimulator with the induced current used to excite nerve fibers near electrodes. As a number of electrodes required stimulation, however, it was more attractive to use digital circuitry. It was thus more straightforward to combine (multiplex) the control information for each electrode into a single signal, and to recover this information in the stimulator. A single transmission path could then be used for a multiple-channel implant. Digitally controlled current sources deliver well-defined stimuli, and the speech processor could be precisely adjusted to suit individual patients. Finally, integrated circuit silicon chip technology has become available for low-power digital designs.

Receiver-Stimulator Circuitry

The circuitry of the receiver-stimulator was designed first to receive the inductively coupled RF signal. A format for the transmitted signal was discussed above (see Encoder and Transmitter) and illustrated in Figure 8.11. As illustrated in Figure 8.20, the signal from the inductor coil is directed to power converter and data receiver sections. These are the output current generator (OCG) and the data decoder (DDE). The output current generator produces a steady voltage that drives current through the selected electrodes. The data in the decoder are sent via a clock unit to control timing to a decoder, the stimulus output controller (SOC), to determine the instructions for the stimulus parameters. These are then fed through an electrode decoder (ED) to control the output switches (OS) that govern the mode and duration of the stimulation. This illustrated in Figure 8.21 for the Nucleus 22. As shown for the first phase of a bipolar pulse, when switch S2a and S3b are on, the voltage between the rail V_{dd} and V_{ss} causes current to flow in one direction from electrode E2 to a sink of current at electrode E3. V_{dd} is the drain supply voltage, and V_{ss} the source voltage for a FET in a CMOS circuit. The current flows through the constant current source before returning to V_{ss}. V_{dd} is typically $+9$ to $+11$ V in the cochlear implant. V_{ss} is the source supply voltage, and is almost always ground or zero voltage in digital circuits. For the second phase electrode E3 is connected to V_{dd} when switch S3a is closed and then the

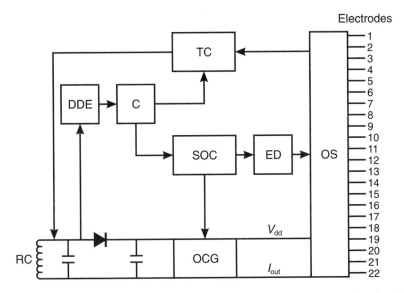

FIGURE 8.20. A block diagram of the sections of the receiver-stimulator circuitry RC, receiving coil; DDE, data decoder/encoder; C, clock; SOC, stimulus output controller; ED, electrode decoder; OCG, output current generator; OS, switches; TC, telemetry controller.

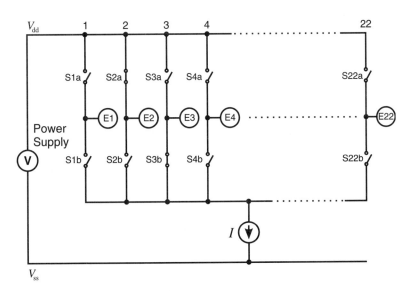

FIGURE 8.21. Switching circuitry for the Nucleus CI 22 system V_{dd} is typically $+9$ to $+11$ volts; V_{ss} is 0 volts; Sa is the switch between the electrode and V_{dd}; Sb is the switch between the electrode and V_{ss}; E is the electrode.

current flows to E2 when S2b is closed. Thus the pulse is identical in magnitude but opposite in sign. This makes it possible to accurately match the current in both phases of the pulse, and not produce a net imbalance. This is more difficult with other devices as they have to match separate current sources. With bipolar stimulation, any pair of electrodes can be selected. For common ground stimulation, one electrode is connected to the current sink in one phase and all the others to the current source, and in the second phase the reverse applies as discussed for bipolar stimulation. For monopolar stimulation the return path is to the ground electrode implanted under the temporalis muscle. It must not be in the muscle as electrode fracture will occur due to the repeated movement.

It is important that the receiver-stimulator circuitry be isolated from the receiving coil so that the coil will not act as an extracochlear electrode should there be an electrical current path to the surrounding tissues. This is important, as medical-grade encapsulating materials like Silastic should not be relied on to prevent these current paths or the possibility of corrosion. Finally, the electrode circuitry should be designed so that accidental exposure to a stray electromagnetic field will not damage electronic components or cause unwanted stimulation.

Telemetry

Telemetry is used with the Nucleus 24 system (Fig. 8.22) and the Clarion S and Combi-40 + series. It allows information, such as voltages on electrodes generated while delivering a stimulus pulse, to be transmitted to the external programming system. The voltages can help determine the tissue impedance around the array, and thus assess the pathological changes in the nearby tissue. The Nucleus

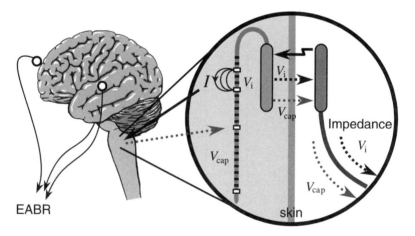

FIGURE 8.22. A diagram of telemetry. I, the stimulus current; V_i, the voltage between electrodes for the calculation of the impedance; V_{cap}, the compound action potential or voltage from the auditory nerve and brainstem; EABR, electrically evoked brainstem response and cortical evoked potential. (Reprinted with permission from Clark G. M., 2002. Learning to understand speech with the cochlear implant. In: Fahle M, Poggio T, editors. Perceptual Learning. Cambridge, Mass. MIT Press.)

24 system can also measure the compound action potential (CAP) in the auditory nerve and thus enables stimulus thresholds and dynamic ranges to be determined in infants and young children. The CAP can be measured more rapidly than the electrically evoked auditory brainstem response (EABR) recorded with surface electrodes, and a child does not require an anesthetic. With telemetry the Nucleus 24 system has an advantage over other devices as forward transmission can be interrupted and allow very small biological signals to be recorded. The Nucleus 24 system can also determine whether the electrode voltage has exceeded the maximum allowable value, and hence a programming change is required.

Packaging

The electronics for the receiver-stimulator required sealing in a package that was impervious to body fluids, mechanically robust, and easy to place surgically.

Materials and Sealing

The standards required for a hermetically sealed container were high, as body fluids and enzymes form a hostile environment. A Kovar steel container was used for the University of Melbourne's multiple-channel implant. The wires passed through a glass feed-through where the glass was melted to bond to the metal of the package and the connecting wires. Fluids and enzymes can permeate along minute pathways or open up cracks in the glass seals through surface tension, and this was one failure mode of the University of Melbourne's prototype seen in two of the three initial patients. White (1974) also emphasized the difficulties of achieving a good seal around the electrode feed-through when glass is used. The metal lid was soldered to the container. This created another problem, as with time the metals in the solder migrate and produce corrosion from an electrolytic reaction, and both mechanisms weaken the seal.

The solution to sealing electronic units was discovered in the pacemaker industry, and was applied to cochlear implants by the Australian pacemaker firm Telectronics who established the company Nucleus and then a subsidiary Cochlear Pty Limited. With pacemakers, epoxy resin was originally used, but it led to later electronic failures as the body fluids penetrated the epoxy resin. This problem was resolved by K. Kratochvil, who discovered the right blend of ceramics that when sintered would bond to both the wires and a metal case, and produce an impermeable seal. Ceramics, for example Al_2O_3, SiO_2, MgO, are complex structures of metallic and nonmetallic atoms that can bond with metal. This pioneering technological advance helped establish Telectronics in the international marketplace, and was one key element in the success of the Nucleus receiver-stimulator. The case was made of titanium, which would bond with the ceramic mixture. It was also an inert but strong metal. The two halves of the container were sealed by welding the edges together, and as the seal was made of the same material as the container, no electrochemical reaction could occur and produce corrosion. It became the standard for sealing the wires exiting pacemakers. The same technology was used to develop all the Nucleus receiver-stimulators (the clinical trial device and the Nucleus 22 Mini, and Nucleus 24 series) (Clark, Tong et al 1984).

It was a more difficult task to produce the cochlear implant receiver-stimulator as the package was much smaller and 22 wires rather than one or two had to be included in the ceramic feed-through.

The use of titanium or other metal presented a problem, however, as the receiving coil could not be placed inside the package, as electromagnetic energy could not be transmitted through the metal. The coil, however, could be placed around the titanium capsule, and this was done with the clinical trial device as illustrated in Figure 8.23. With the Nucleus CI 22 (Mini) and CI 24 devices, the receiving coil was placed around the rare earth magnet (enclosed in its own titanium capsule) to hold the transmitting coil in place, and both were placed next to the titanium capsule for the electronics for the efficient transmission of electromagnetic signals (Figs. 8.23 and 8.24). A package made entirely of ceramic would have had the advantage of placing the receiving coil inside, as it would not interfere significantly with the transmission of signals. This was considered for the Nucleus receiver-stimulator, but the material was more brittle and a particular weakness would be at the edges where a high concentration of mechanical stress could lead to cracks. Furthermore, it would be more vulnerable to blows to the head. Nevertheless, the Clarion S and Combi-40 + series were made this way (Fig. 8.24). Initial devices had breakages due to the ceramic and were withdrawn until the design was optimized.

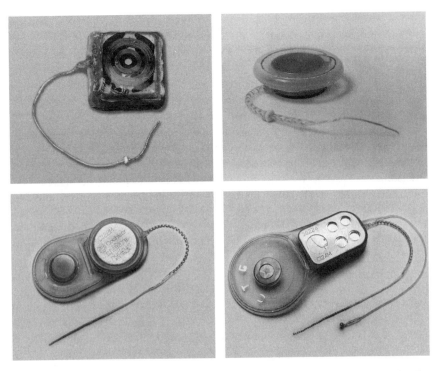

FIGURE 8.23. Development of the Nucleus receiver-stimulator packages. Top left: The University of Melbourne prototype (1979). Top right: Nucleus clinical trial implant (1982). Bottom left: Nucleus CI 22 Mini (1985). Bottom right: Nucleus CI 24 M (1990s).

Finally, when the packages are manufactured it is good practice to ensure the seal is complete with helium leak testing. Helium is used as it can be detected in very small concentrations, and has a very fast diffusion rate.

Dimensions

The receiver-stimulator package needed to have the optimal shape and be small enough for ease of surgical placement and cosmetic acceptability. Surgical experience as a result of implanting the University of Melbourne's prototype (Fig. 8.23) in 1978, which needed to be rectangular, and the first Cochlear Pty Limited clinical trial device, which was cylindrical, showed that it was easier to drill a bed for a round device. The bed could be made very neatly with a milling burr (Clark, Pyman et al 1984). A round shape also conformed best to a circular receiver coil.

The anatomical and surgical studies showed that the maximum depth a bed could be drilled in the mastoid and occipital bones was 6 mm, but 3 to 4 mm was more usual. The maximum height superficial to the bone that was cosmetically acceptable was approximately 5 to 6 mm. The maximum diameter of the device in adults was found to be about 35 to 40 mm. With the Cochlear Pty Limited clinical trial package, shown in Figure 8.23, its mushroom shape made it more stable. The stalk, consisting of the titanium capsule and connector, had a diameter of 20 mm, while the cap due to the receiving coil had a diameter of approximately 30 mm. This receiver-stimulator had no magnet for the attachment and alignment of the external transmitting coil. A headband was required, and different versions are shown in Figure 8.25. Patients who found the headband difficult to manage were helped with a later placement of a magnet over the receiver-stimulator (retrofit magnet).

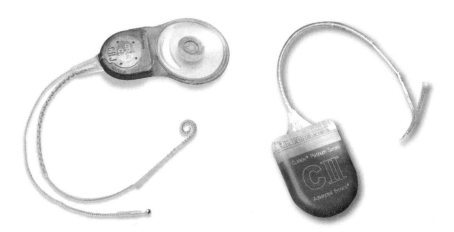

FIGURE 8.24. Left: The Nucleus CI 24 R receiver-stimulator and Contour array (reprinted with permission from Cochlear Limited). Right: Clarion CII receiver stimulator and High Focus II array. (Reprinted with permission from Advanced Bionics Corporation.)

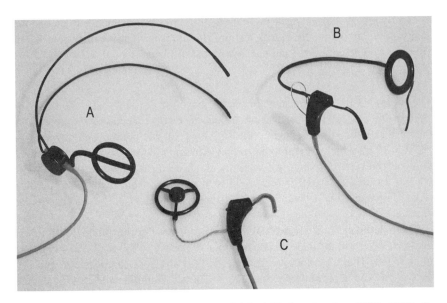

FIGURE 8.25. Headset Nucleus designs. A: Original Nucleus headset (1982–1985). B: Nucleus behind-the-ear headset (1982–1985). C: Nucleus magnetically mounted coil (1985–).

The Nucleus CI 22 (Mini) receiver-stimulator was developed for use in children, as they needed a device that was smaller, and simple for the attachment of the transmitting coil (Fig. 8.23). The coil was designed with the inclusion of a rare earth magnet and a matching one in the transmitting coil. Magnets were first applied to the 3M/House single electrode device, without damage to the intervening tissue (Dormer et al 1980, 1981).

With the Nucleus device the magnet was placed in the center of the receiving coil situated at the back of the device to keep the link away from the titanium case. It was made smaller by the removal of the connector, as it was not found necessary because insertion/reinsertion studies had shown that the banded electrode array could be easily removed and another inserted if it had to be replaced (Clark, Pyman et al 1987). It then had a maximum thickness of 6.5 mm. This CI 22 receiver-stimulator replaced the clinical trial device for both children and adults.

The Nucleus 24 M and R receiver-stimulators (CI 24) (Fig. 8.23) were designed to be smaller for use in infants and children under 2 years of age. The dimensions of the package were arrived at after anatomical studies on the temporal bones of children ranging in age from 2 to 11 months (Clark and Pyman 1995). It was found that the package should be round or ovoid, with sides beveled or straight, and only protrude 2 mm so that it could be placed either in the mastoid cavity or more posteriorly. The dimensions of the Nucleus 24 M were outlined by Clark and Pyman (1995), and for the Nucleus 24 R are shown in Figure 8.26. At the front section it had an overall thickness of 6.9 mm with the capsule protruding

2.2 mm below the surface. The front section had a width of 22 mm and thickness of 3.8 mm. The profile of the device shows there is an obtuse angle between the coil and package that allows it to accommodate the curvature of the skull. In a young child (12 months of age) the radius of curvature was 4.5 cm in the Frankfurt plane (a plane through the infraorbital foramen and the external auditory meatus). The skull is flatter at 45 degrees to this plane, and this is the preferred orientation of the device (Fig. 8.27).

The receiver-stimulator packages made from ceramic had different dimensions

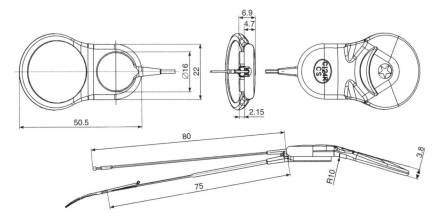

FIGURE 8.26. A diagram of the Nucleus CI 24 R receiver-stimulator and dimensions (reprinted with permission from Cochlear Limited).

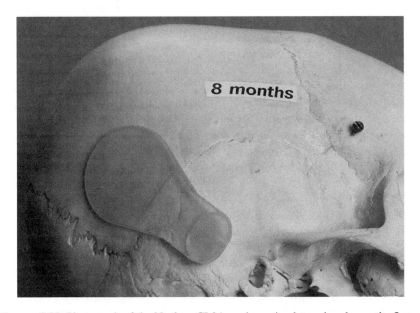

FIGURE 8.27. Photograph of the Nucleus CI 24 receiver-stimulator placed over the flatter section of the skull of an 8-month-old infant.

from those of the Nucleus titanium implants. The Clarion S had the dimensions of 31 × 25 × 6 mm (Fig. 8.23). The Digisonic was round with a diameter of 28.5 mm and maximum width of 5.5 mm (MXM catalogue). Furthermore, they had a receiving coil in the package, as it was nonmetallic. This meant there was no coil at the back. As the ceramic packages are larger than the Nucleus 24 device, more bone has to be drilled down to the dura to provide space. Furthermore, having the coil separate is an advantage if there is a blow to the head as it can protect it from being driven inward (see Chapter 10).

Connector

The receiver-stimulator was originally thought to require a connector so that it could be replaced if there was an electronic failure. Connectors, however, became unessential as reinsertions could be carried out with the University of Melbourne/ Nucleus smooth free-fitting banded array without significant damage to cochlear tissues and auditory neurons (Clark, Pyman et al 1987). Furthermore, reliable connectors are difficult to design especially in keeping pressure between contact points for many years. A simple connection and disconnection procedure would also have been required for use in the surgical theatre without a break in sterility.

Reliability

The implantable receiver-stimulator needed to be mechanically robust. The first three University of Melbourne implants in 1978 and 1979 showed that the area most vulnerable to repeated small body movements was the point where the electrode array emerged from the package. Any movements transmitted from rubbing the skin or adjusting the transmitter coil resulted in maximum bending at this junction area. Consequently, the design of the Nucleus clinical trial device and the subsequent 22 and 24 series incorporated stress relief for the electrode wires emerging from the receiver-stimulator so that metal fatigue and fractures of electrode wires would not develop months or years after implantation. This was achieved by spiraling the electrode wires. Prior to its inclusion in the design of the Nucleus clinical trial device it was tested on a machine where it was flexed millions of times without stress fractures developing. Nevertheless, it was also recommended that during implantation this bundle of electrode wires be placed in a groove under the mastoid cortex (Clark, Pyman et al 1984). The Nucleus clinical trial device was also subjected to car crash testing by the University of Melbourne's Department of Mechanical Engineering without any damage, and subsequently to compression testing.

Reliability of the device is an important issue for the prospective patient. Uniform procedures and meaningful reporting are essential for the clinician. It takes time to accumulate meaningful statistics on the overall reliability of the different products, and short-term estimates for new models can be very misleading. The past history is important as reliability depends on accumulated manufacturing experience. Specific information is also needed on the incidence of package failures, sealing leaks, cracks to the case, fractures of the electrodes or transmitting coil, and electronic failures. For children, in particular, it is important that the implant is resistant to blows to the head. Implant design should evolve to the point where all sporting activities are not contraindicated.

The commonly accepted method of reporting the reliability of implanted medical devices is the percentage of the population of devices surviving a defined number of years (Fig. 8.28). The cumulative survival percentage then includes all units that remain clinically functional. The standard allows manufacturers to exclude a device if there is some reason to explain it, such as a blow to the head. However, Cochlear Limited counts all devices as having failed regardless of the cause. This is good practice, as the information is important for clinicians when advising patients. For example, parents and children want to know the risks of participating in sports with body contact. Note that the reliability data for the Nucleus 22 and the Nucleus 24M were poorer for children than adults, reflecting the frequency and severity with which children hit their heads. It is to be expected that manufacturers will make design changes to improve reliability for these situations. Cochlear Limited strengthened the attachment of the electronic substrate to the case internally, and the package as a whole. The improvement in reliability can be seen in Figure 8.28, which shows that the Nucleus 24M survival rate at 2 years for children was 98.2%, but this rose to 99.5% with the Nucleus 24R. Reliability data have been reported regularly by Cochlear Limited. The technical

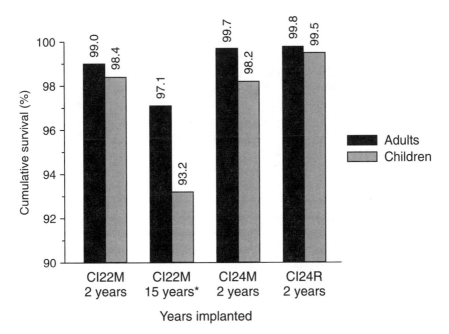

FIGURE 8.28. Cumulative survival rate for the Nucleus 22, 24M, and 24R devices for all patients worldwide to 26 June 2002 (Reliability update 2002) except * to 31 December 2001 (Reprinted with permission from von Wallenberg et al 2002).

failure rate of the Nucleus CI22 from 1990 to 1996 was nine out of 326 children (2.8%), and there were none for the Nucleus 24M for 79 children (Bertram et al 2001). The failure rate for the Clarion devices from 1994 to 2000 was 20 out of 246 children (8%). A similar result could be expected for the Med El package, although Marangos et al (2001) did not distinguish between the Nucleus and the Med El devices for children.

Bioengineering

The discussion of bioengineering of electrode arrays is divided into design principles, which outline the relevant research, and design realization, which covers the application of those principles to the industrial fabrication of arrays. This work has been carried out in many cases through collaboration between academia and industry, and has resulted in the evolution of products through a number of developmental stages. The realization of designs in regular clinical use is discussed under the product name.

Design Principles

The electrode array provides the interface between the electrical code for sound and the auditory nervous system. In the 1970s multiple-electrode rather than single-electrode implants had more potential for speech processing because they could reproduce the place coding of frequency (Clark 1969a). This would require the current to be well localized to separate groups of cochlear nerve fibers to achieve the place coding. For more information see Chapters 1 and 5.

In the 1970s it was considered that this setup could irreparably damage the residual auditory nerves. The inner ear was considered by some to be inviolable. This view was in part due to the fact that with stapes surgery any trauma to the inner ear could lead to a total hearing loss. It did not mean, however, that operating on an ear that already had a marked sensorineural loss would cause further damage to the neurons. It was also considered that implantation could predispose to infection (labyrinthitis), which could have life-threatening consequences through spread to the lining of the brain (meningitis).

During the 1970s, 1980s, and beyond, the development of multiple-electrode arrays became a complex bioengineering task, for an effective interface between sound and the central auditory pathways through the placement of the array inside the cochlea close to the cochlear nerves.

Current Localization

In the 1970s there was concern that with multiple electrodes in the cochlea the current could not be localized to separate groups of cochlear nerve fibers for the place coding of frequency because it was thought the fluid in the scala would cause the current to short-circuit away from the nerves. For this reason, experimental animal studies were essential to determine acceptable electrode geometries and methods of stimulation. Initial research demonstrated that bipolar stimulation with electrodes in the scala tympani would localize current to separate groups of neurons, without it short-circuiting along the fluid compartments of the cochlea (Merzenich 1975; Black and Clark 1977, 1978, 1980). It was also demonstrated by Black and Clark (1977, 1978, 1980) that common ground stimulation would localize the current. Both studies on the acute experimental animal showed that monopolar stimulation did not localize the current. However, it was later demonstrated in a psychophysical study that monopolar stimulation between an active and distant reference electrode in the cochlea could achieve percepts reflecting localized stimulation (Busby et al 1994). This was discussed in Chapter 6.

With bipolar stimulation, if the electrodes are small or not adjacent to the spiral ganglion cells, higher charge densities occur when eliciting an auditory percept. The implant may not be able to deliver the current required, so the separation of active and return electrodes needs to be increased to enlarge the area over which the spiral ganglion cells are stimulated, but this could increase channel interaction for simultaneous stimulation. The effect of the spatial extent and separation of electrodes on pitch ranking (Tong and Clark 1986) was described in Chapter 6. A molded array (Michelson and Schindler 1981) was developed to place electrodes close to spiral ganglion cells. This produced radial bipolar stimulation of localized excitation of the peripheral processes of the spiral ganglion cells in the cat. This array was designed with a central rib to prevent axial rotation during insertion (Rebscher et al 1981). However, effective bipolar stimulation was not tolerant of small variations in electrode placement when high-threshold currents would occur (Clark, Blamey et al 1987; Wilson 2000).

It was thought that extracochlear stimulation with electrodes placed in the bone overlying the cochlea would result in localized stimulation of separate groups of nerve fibers, without the risk of damage to the auditory neurons (Banfai 1987). Therefore, further experiments were undertaken on the cat cochlea (Clark, Shepherd et al 1983) to see if this could be achieved. The EABR thresholds and growth functions were measured to determine the effects of bone impedance and the spread of current. No EABR could be recorded when stimulating between two sites up to the maximum output of the stimulator. This showed that the impedance of the bone was too high for effective extracochlear stimulation of auditory nerves. This was consistent with the findings of Liboff et al (1975) and Reddy and Saha (1984), who found that the bone overlying the cochlea has a high electrical resistance. The only response recorded was for stimuli between an electrode in the scala tympani and another on the outer surface of the cochlea. The input/output functions were steep, indicating the rapid recruitment of fibers suggesting poor current localization.

The University of Melbourne/Nucleus CI 22 and CI 24 electrode arrays were designed to be free-fitting, but it was considered that if they could be placed effectively close to the modiolus, then more localized stimulation of the neurons could occur with opportunities for providing better place coding and the fine temporal and spatial patterns for advances in the temporal coding of frequency (Clark 1996).

Electrophysiological studies in support of a perimodiolar array were first undertaken to see if the stimuli would be more localized if the electrodes were placed closer to the spiral ganglion cells. This was demonstrated in a study undertaken with two sets of half-band electrodes (Clark, Shepherd et al 1983; Clark, Blamey et al 1987), as discussed in Chapter 5. The amplitude of the brainstem responses at different current intensities was recorded for each combination of bands. In this study it was assumed that the threshold indicated the proximity of the electrodes to the neural elements, and the slope of the amplitude versus current function indicated the localization of the current. If the gradient was flat, stimulation was localized, as it did not recruit as many neurons with an increase in intensity. During radial as distinct from longitudinal stimulation, the threshold was high and the growth of response with current was small. This latter finding implied that radial stimulation between half-band electrodes provided improved current localization. There were similar basic findings when the array was rotated 90 degrees and came to lie closer to the modiolus. Thus the results showed that for half-band electrodes current localization was improved with radial rather than longitudinal stimulation, as was seen by Merzenich and White (1980) for small circular electrodes in a cochlear mold.

The placement of the array closer to the modiolus was also studied by placing the electrodes at different locations in the scala tympani and recording thresholds and input/output functions. The placement of the electrode varied from the outer wall, beneath the basilar membrane to close to the medial wall and the spiral ganglion cells, as illustrated in Figure 8.29. The study showed that the thresholds were lowest when the electrodes were placed close to the spiral ganglion cell region (Shepherd et al 1990, 1993). Typical EABR input-output functions for bipolar and bipolar + 1 electrodes are shown in Figure 8.30 for the locations shown in Figure 8.29. Not only were the thresholds lowest when the electrode was closest to the peripheral processes or the spiral ganglion cells, but there were also lower gradients in the responses to electrical stimulation. This supported the hypothesis that more localized current flow could occur when the electrodes were closer to the spiral ganglion cells.

Insertion of Array for Place Coding of Speech Frequencies

In developing an electrode array to allow the reproduction of the place coding of speech frequencies in particular, it was necessary for it to pass around the cochlea far enough for the electrodes to lie close to the nerves transmitting these frequencies to the neurons in the brain. Studies by von Békésy (1960), Schuknecht (1953), Greenwood (1961), and Bredberg (1968) showed these frequencies (from 500 to 4000 Hz) lay from approximately 25.9 mm (540 degrees) to 8.7 mm (90 degrees) from the stapes. The angle of insertion is calculated as shown below (see Fig. 8.44). Initial attempts to pass a bundle of wires upward through the round window did not succeed in it reaching beyond the 2000-Hz region at 15 mm from

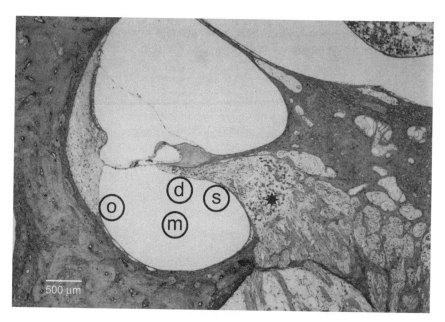

FIGURE 8.29. Cochlea with the electrodes at different locations. s, inner wall near spiral ganglion; d, beneath the peripheral processes; m, center of the scala tympani; o, outer wall near spiral ligament (Reprinted from Shepherd et al, Electrical stimulation of the auditory nerve: the effect of electrode position on neural excitation, *Hearing Research* **66**: 108– 120, @1993, with permission from Elsevier Science.).

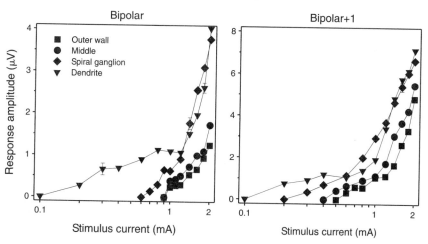

FIGURE 8.30. The electrically evoked brainstem responses (EABRs) versus stimulus current for bipolar and bipolar + 1 stimulation in the scala tympani at the outer wall (□), middle of the scala (●), in apposition to the spiral ganglion (◆), and beneath the peripheral process. (▼), (Reprinted from Shepherd et al, Electrical stimulation of the auditory nerve: the effect of electrode position on neural excitation, *Hearing Research* **66**: 108–120, @1993, with permission from Elsevier Science).

the stapes. It was found that passing the array upward into the tightening spiral of the cochlea restricted the depth of the insertion due to the fact the array came to lie against the outer wall, and frictional forces increased as the insertion progressed. For this reason research was undertaken to see if it was possible to drill an opening into the basal or middle turn and have the electrode array pass downward in a retrograde direction into the widening spiral, when there would be less friction of the electrode against the outer wall of the cochlea (Clark 1975; Clark, Hallworth et al 1975; Clark and Hallworth 1976). Studies in the experimental animal by Clark (1977) showed that it was effective, although there was more trauma than with passing an electrode from below upward. A subsequent study by Clark, Patrick et al (1979) demonstrated that if the array had graded stiffness, being flexible at the tip and stiffer toward the base, it could pass upward into the cochlea with minimal trauma, and also lie close to the speech frequency areas. Thus a multiple-electrode array was developed with these properties.

An alternative design was one with two prongs or forks reaching different frequency areas (Hansen 1981). One could be inserted into the basal turn of the cochlea through the round window, and the other into the upper basal or middle turn through an opening drilled in the overlying bone. A similar concept using two separate banded arrays was advocated by Goycoolea et al (1990). The double arrays were not used in the normal cochlea, as one multiple-electrode array was adequate in reaching the speech frequencies and less traumatic. However, double arrays were used with the ossified cochlea. Bredberg and Lindstrom (1997) and Bredberg et al (2000) described the use of a double array for the ossified cochlea with one in the basal turn and another passed through an opening drilled into the middle turn, as described in Chapter 10.

Trauma to Cochlear Tissues

The cochlea was considered too delicate to "tolerate surgical manipulation or the long term placement of electrodes" (Legouix 1957, quoted by Simmons 1967). To examine whether this was correct, a preliminary study was undertaken by Simmons (1967) on cat cochleae that were implanted with two enamel-coated stainless steel wires inserted through the round window into the scala tympani. The histological findings showed that it was possible to insert an electrode with little fibrous tissue reaction, and minimal loss of peripheral processes. Degeneration of the organ of Corti was localized to the region opposite the electrodes. The cochlear microphonics and N1 thresholds returned to near-normal levels in most cats after 6 to 8 weeks. It was noted in one animal that a tear of the basilar membrane caused a marked loss of neurons localized to the site of the loss. In a third animal infection caused a marked tissue response and loss of neurons.

The direct implantation of electrodes into the apical, middle, and basal turns of the cochlea via holes drilled through bone overlying the speech frequencies was seen as a possible means of localized stimulation of auditory nerve fibers (Chouard 1978). A study comparing the effects of implantation through holes in the otic capsule, and through the round window and along the scala tympani was made in deafened cats by Clark (1973), and a preliminary report indicated degeneration of spiral ganglion cells and peripheral processes was most severe if infection occurred. An analysis of the data (Clark, Kranz et al 1975) showed that electrodes could be inserted directly into the cochlea or via the round window without a major loss of spiral ganglion cells. Nevertheless, there was more fibrous tissue and new bone in the cochleae where there was a direct insertion of the electrodes through holes drilled in the overlying bone.

To further study the factors leading to cochlear damage and neuronal loss, research was undertaken to compare the insertion of a Teflon-coated free-fitting carrier in an anterograde direction up the scala tympani of the basal turn through the round window, and another insertion in a retrograde direction toward the basal turn through a hole into the apical/middle turn (Clark, Kranz et al 1975). The bones from four cats were examined 42 to 59 weeks after implantation. The data showed insertion could be accomplished in either direction, and without major trauma or cochlear pathology. Only minor trauma occurred with an anterograde insertion along the scala tympani through the round window. However, in one of three animals the round window insertion penetrated the basilar membrane and this was associated with a marked fibrous tissue response and was possibly associated with infection. In contrast, with the retrograde insertion there was evidence it would pass along the scala vestibuli and tear Reissner's membrane. In addition, there was a high incidence of trauma at the point of insertion and likelihood of damage to the basilar membrane as it passed around the middle to basal turns. The main conclusion was that anterograde insertion was less traumatic than retrograde insertion, spiral ganglion cell loss was especially likely with a tear of the basilar membrane, and infection should be avoided. A study was then undertaken with an anterograde insertion of a Teflon strip, as a thin film array was being developed for multiple-electrode stimulation (Clark and Hallworth 1976). The importance of minimizing trauma with electrodes is well illustrated in a section from one of these bones (see Chapter 3). A Teflon strip with sharp edges cut through the basilar membrane and led to a near-total loss of the spiral ganglion cells in the same region (Clark, Blamey et al 1987).

A further study was carried out on the cat by inserting a free-fitting Silastic tube with a diameter of 0.6 mm into the scala tympani through the round window (Clark, Blamey et al 1987). This tube was the carrier for the University of Melbourne's banded array (Clark, Patrick et al 1979). As shown in Figure 8.31 the carrier could be inserted with minimal trauma and no loss of ganglion cells. The tube was surrounded by a thin sheath of fibrous tissue. It was also noted that pressure on the spiral ligament, as could occur with an insertion in the human cochlea, did not lead to loss of dendrites 6 weeks postoperatively.

The effects of trauma with an array molded to fill the first 9 mm of the cat scala tympani was studied by Schindler and Merzenich (1974) in 10 cats sacrificed after 3 to 117 weeks. A mold was used to avoid current short-circuiting through the perilymph fluid, and to help locate the electrodes in apposition to the peripheral processes lying on the basilar membrane to allow localized stimulation. There was severe degeneration of the organ of Corti in the region overlying the elec-

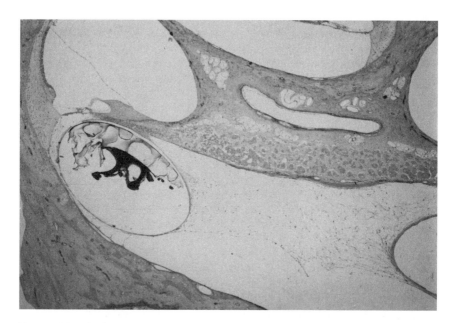

FIGURE 8.31. A photomicrograph of a free-fitting Silastic electrode carrier inserted into the scala tympani of the basal turn of the cat cochlea through an incision in the round window membrane (Reprinted with permission from Clark, Blamey et al 1987. The University of Melbourne–Nucleus multi-electrode cochlear implant. *Advances in Oto-Rhino-Laryngology* **38**: p 40. Basel, Karger.).

trode, and a significant loss of hair cells in the other turns, but some supporting cells remained. It had been thought that neurons would only degenerate with the loss of these supporting cells (Schuknecht 1953), but the degeneration is probably due to the loss of trophic factors released from intact hair cells (Marzella and Clark 1999). From whatever cause there was a significant loss of auditory neurons in the basal turn. In two animals a perforation of the basilar membrane occurred and this was associated with a marked loss of ganglion cells and new bone growth in the region of the tear. A fibrous tissue sheath formed around the electrode arrays.

In view of these equivocal findings it was necessary to evaluate the histological effects of a free-fit versus a molded array made to the shape of the basal turn of the scala tympani. The study by Sutton et al (1980) on the monkey showed significantly more trauma with the molded array. The monkey was a better model for the human than the cat, as it more closely resembles the human anatomy. The molded array created basilar membrane fistulae and spiral lamina fractures, and produced extensive loss of spiral ganglion cells. Other prior studies, discussed above, showed that this trauma would lead to significant loss of spiral ganglion cells (Simmons and Glattke 1970; Schindler and Merzenich 1974; Clark 1977). In contrast, a free-fitting electrode was encapsulated locally with fibrous tissue, and there was little mechanical damage or degeneration of spiral ganglion cells.

An extracochlear multiple-electrode array was considered a possible alternative to an intracochlear array by Banfai et al (1984b) and Banfai et al (1985), because trauma to the cochlea would be less. It was necessary, however, to drill down to the endosteum of the cochlea as the spread of current could be extracochlear and not provide localized stimulation for place coding, and as bone has a high impedance the receiver-stimulator would run out of voltage compliance and not be able to stimulate the neurons as discussed above. The histological effects of extracochlear implantation of the cat cochlea by Clark, Shepherd et al (1983) showed that after the bone had been drilled down to the endosteal lining fibrous tissue and bone formed beneath the electrode bed and would have increased the impedance, and the trauma caused loss of hair and spiral ganglion cells. The tissue response did not justify the use of the extracochlear multiple-electrode especially considering the poorer speech perception performance (Banfai et al 1984a).

A stiff electrode wire was used by House and Urban (1973) for insertion into the scala tympani through the round window as one of a few multiple-electrode implants and then only for single electrode implants (Figure 8.32). Johnsson et al (1982) described the histopathology of two temporal bones taken from a patient with bilateral cochlear implants. The right cochlea of this patient contained a single platinum/iridium (Pt/Ir)(90/10) wire that had been inserted a distance of approximately 17 mm from the round window. The patient had received this implant approximately 2 years prior to his death; however, the device had been used only intermittently over a 3-month period. Histological examination of this cochlea indicated that the single wire had passed along the scala tympani for the

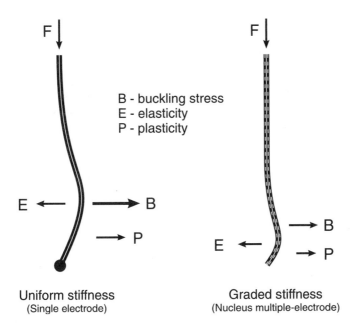

FIGURE 8.32. The load and buckling stresses on electrode array with uniform (3M/House) and graded stiffness (University of Melbourne/Nucleus). F, force; B, buckling stress; E, elasticity; and P, plasticity.

first 12 mm of the basal turn, after which it had deviated into the scala media and finally the scala vestibuli, damaging the spiral ligament on its course. Significantly, much of the basilar membrane, Reissner's membrane, and the osseous spiral lamina were intact. Despite the insertion trauma, no new bone growth was observed in this cochlea. There was only a localized soft tissue reaction associated with the electrode. Extensive sensory and neuronal loss was apparent throughout all turns of the cochlea, although it was not possible to differentiate degeneration associated with the implant from that due to the preceding pathology. The left cochlea of this patient contained a multiple-channel electrode array that consisted of five individual silver wires. The most apical wire was inserted approximately 20 mm from the round window, and the tips of the remaining wires were spaced at 4-mm intervals basalward. The patient had received this implant 7 years prior to his death, and it had been used for 6 to 8 hours per day for approximately 27 months. This cochlea exhibited more extensive histopathological changes. The wire electrodes were surrounded by bone for their entire course within the scala tympani. As with the single electrode in the right cochlea, the electrodes had deviated from the scala tympani approximately 12 mm from the round window. Again, they were located in the scala vestibuli. In this case, extensive new bone was present in the scala media and scala tympani proximal to the electrode. Significantly, regions of the membranous labyrinth demonstrated evidence of blackening, which was subsequently shown to be silver. Hydrops was also apparent in this cochlea. Greater neural degeneration was evident compared with the opposite cochlea. The extensive histopathological response in these cochleae was most probably due to trauma, as shown in the experimental animal studies referred to above (see Trauma to Cochlear Tissues) and in Chapter 3. It could also have been aggravated by using the silver electrodes. Silver has long been regarded as biologically incompatible (Pudenz 1942; McFadden 1969; Dymond et al 1970). When passively implanted, silver is known to produce a pronounced tissue reaction that could result in new bone formation. Presumably, this metal is more corrosive under conditions of electrical stimulation. Galey (1984) reported on a temporal bone of another patient from the Los Angeles Clinic who had six separate electrode wires inserted into the scala tympani 6 years prior to his death. The platinum wires had a diameter of 0.21 mm and had been inserted along the scala tympani for distances of up to 22 mm from the round window. Only two of the six wires lay entirely within the scala tympani; the tips of three wires lay in the scala vestibuli and one lay in the scala media. A number of the wires had penetrated the basilar and Reissner's membranes. Extensive new bone growth was observed in the region of the trauma. Significantly, new bone was not associated with the four apical electrodes distal to the trauma. These four electrodes were the electrodes used for electrical stimulation. This suggested that the electrical stimulus parameters used did not lead to bone growth. The loss of nerve fibers peripheral to the spiral ganglion was extensive, although spiral ganglion cell survival was not reported. As a result of these and other findings, it was considered by House (1984) that a single electrode should be short and just enter the scala tympani through the round window membrane to provide single-channel electrical stimulation.

Thus the use of relatively stiff wire electrodes in cochlear implants resulted in significant insertion trauma including damage to the osseous spiral lamina and the basilar and Reissner's membranes (Johnsson et al 1982). The extensive trauma was undesirable as it resulted in additional neuronal loss, an increase in inflammation with new bone growth, leading to a significant reduction in the likelihood of successfully reimplanting the cochlea.

Mechanical Properties

To ensure that surgical trauma with the University of Melbourne/Nucleus multiple-electrode free-fitting array could be kept to a minimum, its mechanical properties were examined and compared with those of the 3M/House single-electrode array, especially in view of the trauma that had been observed in human cochleae with their single- and multiple-electrode array (Johnsson et al 1982). Single platinum wires with a diameter of 0.21 mm had been implanted (House and Edgerton 1982). A study was undertaken to see whether the tip of the University of Melbourne/Nucleus multiple-electrode array would flex more easily and that the array was less rigid (Patrick and Macfarlane 1987). The study, carried out by the Commonwealth Scientific Industrial Research Organization (CSIRO) of Australia, showed that the maximum force applied by the 3M/House single electrode was 25 times greater than that by the multiple-electrode array, and the University of Melbourne/Nucleus multiple-electrode array was 10 times more flexible (Patrick and Macfarlane 1987). The above results were due to the fact the University of Melbourne/Nucleus array was made of fine platinum wires with a diameter of 0.025 mm compared to the single electrode with a diameter of 0.21 mm. The graded stiffness with 22 wires at the proximal end and only one at the tip resulted in more flexibility.

The mechanical properties of free-fitting arrays and their propensity to cause trauma was later studied using finite element modeling (Chen et al 2003). Finite element modeling is well established (Zienkiewicz 1977), and has been used successfully to predict the stress-strain response of bodies under different boundary conditions such as external forces or applied loads. Finite element modeling was used to provide a theoretical assessment of the damage or trauma experienced during the insertion procedure by evaluating contact pressures at critical regions in the cochlea. The model allowed contact stresses during insertion into the human cochlea to be calculated for electrode arrays with different stiffness properties. The dynamics of the movement of the electrode array could also be visualized (Fig. 8.33). The passage of the array was seen as a series of bending deflections or deformations as it was progressively inserted into the cochlea. The contact pressure and its distribution along the portions of the array in contact with the wall of the cochlea were predicted (Fig. 8.34). The predicted contact pressure provided a quantitative measure of the degree of trauma that could occur during insertion of arrays with different mechanical properties. Of particular interest were the contact pressures at the tip of the array and at segments during and after insertion. The study was undertaken on three electrode arrays: (A) uniform stiff-

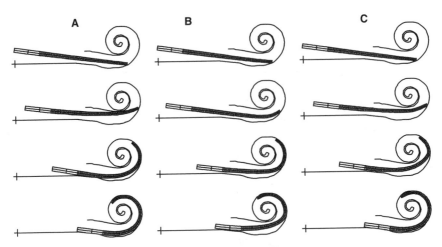

FIGURE 8.33. Deformation of an electrode array with graded stiffness as it is inserted into the scala tympani. A, electrode arrays with uniform stiffness; B, electrode with graded stiffness; C, electrode array with uniform stiffness and a soft tip (Reprinted from Chen et al 2003, with permission from Elsevier Science.).

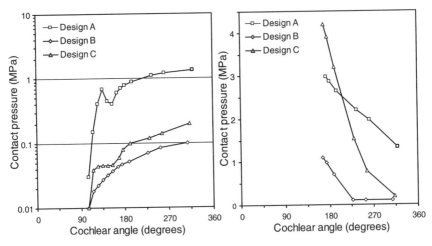

FIGURE 8.34. Left: Contact pressures at the tips of the arrays during insertion. Right: Contact pressure distribution along the array after insertion. The three electrode arrays: A, uniform stiffness; B, graded stiffness; and C, uniform stiffness and a soft tip (Reprinted from Chen et al 2003, with permission from Elsevier Science.).

ness; (B) graded stiffness; and (C) uniform stiffness and a soft tip (Chen et al 2003). In each case the trajectories varied. With design A the curvature especially near the tip was less pronounced when compared with tapered electrode B. When the buckling stresses for the three electrodes were measured, it was found that buckling stress was high for all points along array A. With the electrode B with graded stiffness the stresses were low. For the array with the soft tip but uniform stiffness, C, there was a marked increase in the buckling stresses toward the outer end. Although the study used a two-dimensional model, it provided valuable information about the possibility of trauma with different types of electrode arrays and how this occurs. The data supported the free-fitting banded flexible straight array with graded stiffness as a good design.

Electrode Dimensions

The electrophysiological studies and mathematical models showed that the current could be localized best for the place coding of frequency if the array were placed in the scala tympani of the cochlea (Black and Clark 1978, 1980) (see Chapter 5). It was also found that an array could be placed in this scala with minimal trauma (Clark 1977) if care was taken with its insertion and it had the right mechanical properties (Clark, Patrick et al 1979). The electrode dimensions were determined by the anatomy of the scala tympani of the cochlea. Subsequent studies by Hatsushika et al (1990) indicated that in the human the height and width remained fairly constant with distance, and hence for a free-fitting array there was only a need for a small taper. As the width was greater than the height, an array had freedom to move from a peripheral position to lie close to the modiolus and this has been important in the design of a perimodiolar array (see below). In contrast, the cross-sectional area became progressively smaller with distance and this was accompanied by a change in shape. It was more quadrilateral at the base and triangular at the apex of the basal turn (see Chapter 2). This would make it difficult for a molded array to be inserted along the cochlea without damage. This was found to be the case when a free-fitting array such as the University of Melbourne/Nucleus (Clark, Patrick et al 1979) was compared with the molded array developed by UCSF. Sutton et al (1980) found in the monkey that there was more trauma with the molded array. The dimensions were discussed in more detail in Chapter 2.

Insertion and Reinsertion

It is essential to be able to insert and reinsert an electrode array in case the receiver-stimulator electronics failed or the device needed to be replaced with an improved design. The Nucleus clinical trial receiver-stimulator had a connector so that the package could be removed if it failed. However, animal research with the smooth, free-fitting tapered Nucleus banded array showed it could be easily withdrawn, and another reinserted. This was confirmed with subsequent experience with humans where the arrays were removed and others inserted (Clark, Pyman et al 1987). It was found to go in as far as the original array. This could have been in part due to the fact that the arrays are surrounded by a thin mesothelial lining a few cells thick that is smooth and slippery. There were concerns, however, about the explantation of arrays that had protruding ball electrodes such as the 3M/House single-electrode array and the Symbion/Ineraid multiple-electrode array, as they could be tightly bound by dense fibrous tissue or bone and be difficult to remove, or they could avulse tissue from the cochlea. Fortunately, with the 3M/House single-electrode array any pathological changes would have been in the first 6 to 10 mm of the basal turn, and little damage would have been done to the area stimulated with the multiple electrodes. In fact Luxford and House (1987) reported that three patients with the longer (15 mm) House array had been reimplanted with the Nucleus device. There were difficulties with the Ineraid array (Gray et al 1993). It had six separate wires with terminal balls, and in some explantations wires broke on removal leaving the balls in situ.

A more recent concern has been that the sheath of an array lying at the periphery or center of the scala could prevent the insertion of a more advanced one that needed to lie close to the spiral ganglion cells. This has been addressed by the development of precurved arrays, for example, the Nucleus Contour, that lie close to the central spiral or modiolus. The question of the insertion and reinsertion of an array is also discussed in Chapter 10.

Prevention of Infection

An electrode array in the scala tympani could be a path for infection from the middle ear to enter the inner ear and cause labyrinthitis and even extend to the brain with meningitis. The studies in the experimental animal demonstrated that there is always a potential pathway between the Silastic array and tissue along which bacteria can spread (Clark, Pyman et al 1984; Clark, Shepherd et al 1984; Franz et al 1984). This possibility was of particular concern when implanting young children, as they are prone to recurrent middle ear infections. Fortunately, clinical experience (House et al 1985; Luxford and House 1985) and animal studies (Clark, Pyman et al 1984; Clark, Shepherd et al 1984; Franz et al 1984; Cranswick et al 1987) showed that the implanted cochlea is capable of resisting the spread of infection similar to a nonimplanted one, provided the electrode entry point is well packed with a homograft of fascia. This may be due to fibrous tissue forming a partial barrier, or the production of mucus-secreting cells in the vicinity of the electrode entry point and their bacteriostatic action. To minimize the risk, experiments were undertaken with discs of Teflon felt and Dacron mesh glued to the arrays and placed at the electrode entry point. The seals were evaluated in experimental animals in which infection had been induced by the introduction of the common microorganisms. Teflon would only increase the pathway at best, but was found to be no better than a muscle graft, while Dacron provided a home for infection. One way for a future seal is to use a ceramic material bonded to both the array and the bone. An advanced array has been designed in the HCRC, and fabricated for studies in experimental animals. It consists of a collar of titanium alone or with a coat of hydroxyapatite to help ensure osseointegration (Fig. 8.35).

In a study on five cats to evaluate the titanium collar, a cochleostomy was made beside but not into the round window to simulate a cochleostomy in the human. The collar of commercially pure titanium was placed at the electrode entry point, and the gap between the array and collar sealed with biocompatible glue (Silastic type A). The study showed the surrounding fibrous tissue and bone adhered to the collar (Fig. 8.35), but there was no integration with bone (Purser et al 1991).

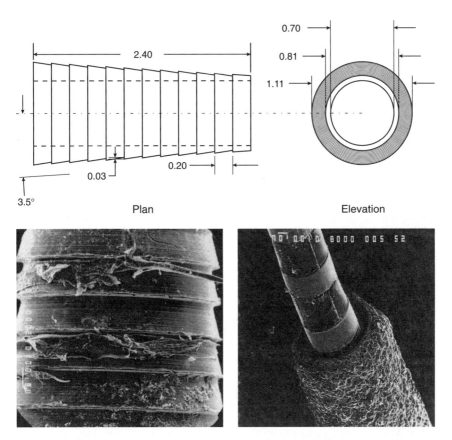

FIGURE 8.35. Top: Design specifications of intracochlear electrode entry-point sealing device. Bottom left: Low-power scanning electron micrograph of titanium collar after implantation showing strips of tissue between ridges. Bottom right: A scanning electron micrograph of an electrode sealing cone with an electrode array passing through it. The outer surface of the cone consists of hydroxyapatite to assist with osseointegration (Reprinted from Purser et al 1991 with permission from the Journal of the Otolaryngological Society of Australia.)

The seal was effective in the five implanted cochleae, and there was no inflammation in any of these cochleae after inoculation with *Streptococcus pneumoniae* 12 weeks postimplantation. In the control unsealed side, infection was present in one of four cochleae. For more compete integration with bone, a collar with hydroxyapatite on the surface could be more effective. The issue of sealing was also discussed in Chapter 3.

Percutaneous Versus Transcutaneous Stimulation

In some initial studies, as discussed above (see Power and Data Transmission), data were transmitted to the stimulating electrodes by a direct electrical link (percutaneous stimulation). This was done to enable the broadest range of stimulus parameters to be tested. However, with these devices infection was prone to occur and it would spread along the electrode wire or develop in a cul-de-sac or sinus around its point of entry. Infection was reported in all of the 21 patients implanted by Pialoux et al (1979), and the Teflon plugs were removed. In contrast, studies with percutaneous plugs or wires passing through the skin in the experimental animal (Merzenich et al 1981; Rebscher et al 1981; Leake-Jones and Rebscher 1983) showed the incidence of infection after short-term implantation to be acceptable. However, in neurophysiological research laboratories it is quite remarkable how resistant the cat brain is to infection when compared with neurosurgical operations on humans. Consequently, the results of animal infection studies must be applied with caution to humans. Nevertheless, it was perceived by the group at UCSF that as infection was an ever-present risk, interconnections needed to be placed at some distance from the cochlea. This could be surgically inconvenient, and the lead wires pressed on and made grooves in the skull. Another major problem with percutaneous connectors was fracturing of the electrode wires by bumping, fingering, and other body movements that could not always be predicted from animal experiments. Finally, the plug was not aesthetically pleasing to patients. Even with improved materials, such as pyrrolized carbon, a plug and socket still had too many potential risks to be used, as discussed in Chapter 10.

Biotoxicity and Biocompatibility

The materials used for the intracochlear multiple-electrode arrays were all evaluated prior to seeking approval from the FDA for the Nucleus clinical trial implant. Studies were first undertaken by the University of Melbourne, and then by Cochlear Pty Limited through contract agencies. The materials included Silastic MDX-4-4210, Silastic tubing, Silastic adhesive type A, and platinum. These evoked only a mild pathological response. The studies by the Department of Otolaryngology at the University of Melbourne were carried out according to the requirements of the FDA for animal experiments. The materials were implanted according to the guidelines of the U.S. Pharmacopoeia (1980) recommendations with specific modifications to suit the biological circumstances including implants into the cochlea. Independent evaluation of the materials was undertaken in 1982 by North American Science Associates for Cochlear Pty Limited. The materials were examined for (1) cytotoxicity where they were overlain with cultured mouse fibroblast cells; (2) mutagenicity with an Ames test for changes in *Salmonella typhimurium* strains; (3) intracutaneous toxicity with extracts injected into rabbits; (4) hemolysis when mixed with animal blood; (5) inflammation after implanting in the subcutaneous tissue of rabbits; and (6) sensitization. The results of all these tests were negative, and showed the materials to be biocompatible. The files of the FDA and manufacturer Dow Corning company revealed tests for other applications were all negative. Finally D.F. Williams, senior lecturer in materials in the Department of Dental Science at the University of Liverpool, was consulted, and his literature review confirmed that the materials to be used were safe. Details of the research, in particular at the University of Melbourne, are presented in

Chapter 3. Furthermore, the procedures referred to above became the established requirements for approval of subsequent devices.

The biocompatibility of the assembled device was also assessed in two independent studies. In the first the electrode array was evaluated by examining the tissue response following long-term implantation in cat cochleae. In the second study the general biocompatibility of the assembled and sterilized prosthesis was determined after a 4-week period of intramuscular implantation in the cat.

In the first study 12 normal-hearing adult cats were implanted with scala tympani electrode arrays. These arrays were fabricated with the same materials and techniques used to manufacture the Melbourne/Nucleus electrode array. Each array was prepared by injecting Silastic MDX-4-4210 into a mold containing two platinum band electrodes. After fabrication, each array was ultrasonically cleaned. Following an implantation period that varied from 32 to 142 days (mean 80.6 days), each animal was sacrificed and the cochleae examined histologically.

The results showed the inflammatory reaction varied from nothing to a moderate chronic response throughout the cochlea. The most common effect was mild, chronic inflammation, generally localized to a fine fibrous tissue capsule enveloping the electrode array. A moderate inflammatory response, observed in four of the 12 cochleae, was thought to be due to the presence of a low-grade infection presumably introduced during surgery. Significant hair cell survival was observed in the cochleae where there was an absence of low-grade infection. In a small proportion hair cells appeared normal adjacent to the implant, but this was more common in the middle and apical turns.

The results thus indicated that the banded scala tympani electrode array, when implanted chronically in animals, evoked a minimal inflammatory reaction, provided that infection did not occur. This ensured that the manufacturing techniques used in the production of the prosthesis were free of materials or impurities, and that the package sterilization with ethylene oxide did not lead to adverse tissue responses. The mild inflammatory response indicated that the electrode array was biocompatible and could be considered safe for long-term implantation within the scala tympani.

Finally, in view of a cochlear insertion modeling study that showed friction between the electrode array and the outer wall limited its depth of insertion (Hallworth 1976), the array was coated with sodium hyaluronate (Healon) or dilute glycerin to reduce the friction and so allow the array to reach all the speech frequencies. Sodium hyaluronate or dilute glycerin was chosen as each not only had properties to lower friction, but also had been used in the body for other purposes and were considered biocompatible. Nevertheless, studies were undertaken to ensure they had no adverse effects on the cochlea (Bagger-Sjoback 1991; Roland et al 1995). Further details on their use are discussed in Chapter 10.

Infants and Young Children

There were three specific design questions that needed to be answered before operating on children under 2 years of age. These questions were the effect of head growth on the implant and vice versa, the spread of middle ear infection to the middle ear, and the effect of electrical stimulation on the immature nervous system. This research was carried out under a U.S. National Institutes of Health (NIH) contract to the University of Melbourne for studies on pediatric auditory prosthesis implants, NIH contract No. 1-NS-7-2342 from 1987 to 1992, and the Coleman Laboratories for studies on pediatric auditory prosthesis implants, contract No. DC-7-2391.

An analysis of human temporal bones was made by Dahm et al (1993) on 60 specimens from people ranging in age from 0.16 to 84 years. Key findings were that growth between the sinodural angle (representing the site of the receiver-stimulator placement) and the round window (a site for the electrode insertion) increased an average 12 mm from birth to adulthood with a standard deviation of 5 mm. Therefore, a pediatric cochlear implant should allow up to 25 mm of lead wire lengthening. In addition, as there was no increase in the distance between the round window and the fossa incudis with age, this anatomical landmark was a suitable fixation point for the lead wire so growth changes would be transmitted to the electrode in the inner ear. This was discussed in more detail in Chapter 2. In addition, x-ray and histological studies on the monkey demonstrated that implanting the receiver-stimulator in a bed drilled through the cranial sutures had no effect on skull growth (Xu et al 1993; Burton et al 1992). This is discussed in Chapter 10. As infants have a high incidence of middle ear infection, it was necessary to be sure inner ear infection would not occur more frequently in implanted ears. The electrode entry point, as discussed above, was found to be effective in limiting the spread of infection for *Streptococcus pneumoniae,* provided the electrode entry point was sealed with fascia. There were also no adverse effects of electrical stimulation on the maturing auditory nervous system (Matsushima et al 1990, 1991; Shepherd et al 1991; Ni et al 1992).

Design Realization

There have been a number of implant designs and surgical approaches for placing stimulating electrodes to excite auditory nerve fibers. These include electrodes inserted directly into the auditory nerve via holes drilled into the modiolus (Simmons 1966), electrodes inserted into the scala tympani via fenestrations made into the otic capsule following the compartmentalizing of the scala with Silastic (Chouard and MacLeod 1976; Chouard 1980; Chouard et al 1983), and ball electrodes placed into holes created in the otic capsule (Banfai et al 1984b; Banfai et al 1985). However, as discussed above, the most common approach has been to introduce the array along the scala tympani via an approach through the round window (Michelson 1971; Clark, Patrick et al 1979; Clark, Pyman et al 1979, 1984; Burian et al 1980; Schindler and Bjorkroth 1979; Schindler et al 1987, 1993). It was essential that the surgical placement of the electrode array did not result in trauma that could lead to a reduction in the residual spiral ganglion cell population. The type of the electrode array and its method of insertion could have a significant effect.

In realizing electrode designs for patients it was essential that in addition to the more basic experimental animal studies (Clark, Kranz et al 1975; Clark 1977; Schindler et al 1977), an evaluation of electrode insertion trauma be performed using human material. There are significant differences in the dimensions of human cochleae compared with experimental animals such as the cat and even the monkey (Igarashi et al 1968, 1976; Hatsushika et al 1990). Such evaluations needed to be carried out under simulated surgical conditions in order to model the restricted surgical access to the cochlea via a posterior tympanotomy. Animal studies did, contribute, however, to knowledge of the effects of electrode insertion trauma by illustrating the probable histopathological consequences following cochlear damage. These controlled experimental results, discussed above (see Design Principles), should also be compared with the histopathological results from temporal bones obtained from patients implanted with cochlear prostheses discussed in Chapter 3.

The development of electrode arrays and receiver-stimulators for clinical use has benefited from the studies discussed above (see Design Principles). There are, however, specific issues that need to be considered in manufacturing the implants, and these include the requirements of regulatory authorities. The following discussion relates in particular to the evolution of the arrays in regular use.

Clarion-S

The electrode for the Advanced Bionics Clarion system evolved from the research at UCSF where an array molded to the shape of the scala tympani was developed to provide localized stimulation of the peripheral processes of the auditory nerve where they lay on the basilar membrane (Merzenich and Reid 1974). The findings from the cat did not necessarily apply to the anatomy of the human cochlea or absence of peripheral processes, which occurs in most profoundly deaf patients (Clark 1987). The trauma associated with the insertion of the molded electrode carrier was greater than that of a free-fitting array in the monkey cochlea (a better model of the human) (Sutton et al 1980). For the above reasons, further research at the UCSF in the 1980s focused on fabricating an array that was cylindrical in shape and filled only the middle of the scala. The electrode contacts were mushroom-shaped rather than protruding balls as the latter damaged the cochlea on removal, and Pt/Ir (90:10) wire was to used rather than platinum/rhodium (Pt/Rh) (90:10) with the wires coated with Parylene-C. The research confirmed that the design changes had no adverse effects on the cochlea, and further showed that chronic implantation in the scala tympani could occur without loss of neural elements (Rebscher et al 1981; Leake-Jones et al 1985).

The Clarion array was precoiled to hug the modiolus, had eight pairs of embedded ball electrodes, and required right- and left-hand models. It had a wider diameter than the Nucleus straight array. The electrode was placed in an insertion tool, and a large cochleostomy was required to accommodate both insertion tool and array. They were both inserted for approximately 8 mm into the scala tympani. The array was extruded from the slot in the tool by sliding a plunger forward.

When the electrode had been released the tool was removed and the array advanced to the point of first resistance.

This array provided both radial and longitudinal bipolar stimulation for localized excitation of the auditory nerve fibers. It was found, however, as discussed above (see Simultaneous Analog Stimulation), that it could not provide selective stimulation as quite high stimulus levels were required to produce auditory percepts and many patients could not be stimulated at all (Wilson 2000). As a result offset radial ("enhanced bipolar") electrodes with a spacing of 1.7 mm were used and later monopolar stimulation. In a study by Roland et al (2000) after the array had been inserted, the human temporal bones were embedded in epoxy resin and sectioned, and they showed the position of the array varied with respect to the modiolus and more frequently it lay in the center of the scala. The balls did not always face the modiolus. Fluoroscopy revealed that the tip could bind in the distal part of the basal turn and lead to buckling.

The array was redesigned to be flatter with 16 rectangular electrode pads on the inner surface. Further work to resolve the problems also led to the use of a separate unattached positioner that was slid in between the electrode array previously inserted and the outer wall of the cochlea (HiFocus array I). This could apply outward pressure in accord with Newton's third law, causing damage to the basilar membrane as happened with an array with inward force vector (developed at Cochlear Limited), and discussed below and shown in Figure 8.42 and Figure 8.46. This HiFocus I system was modified with the positioner attached 10 mm from the tip of the array as the Hifocus II array. The propensity of the Clarion array with positioner to cause trauma was evaluated radiologically and histologically in seven freshly frozen human temporal bones (Richter et al 2002). When the array was inserted with the positioner, there was severe damage to the basilar membrane and osseous spiral lamina along the length of the basal and middle turns. It was concluded by the authors that systematic safety studies in larger samples of human temporal bones were needed before it could be recommended.

It has been stated under technical information (Clarion manual 2002, and Web site *http://www.bionicear.com/tech/tech_hifocus.html* 5/14/02) that it is "designed to protect the delicate structures on the cochlea by reducing scar tissue" and it is "engineered to reduce scar tissue which may minimize difficulty of inserting future electrode technology in decades to come". Apart from the risk of trauma due to the physical principles and data referred to above, the animal experimental studies discussed in Chapter 3, show there is value in having fibrous tissue in the scala tympani, especially as a sheath for the electrode. In addition, the data show risk that close approximation of the positioner and array could prevent tissue growing into the cleft between the two members. This could make it easier for middle ear infection to extend along this pathway to the cochlea, and produce labyrinthitis and meningitis. Furthermore, a "dead space" can act as a breeding ground for infection, as there is no blood supply to bring antibodies and antibiotics to the area (Vaudaux et al 1994). This was also seen in studies in the experimental animal with Dacron mesh used to seal the round window in the presence of induced otitis media where there was acute and marked inflammation, as illustrated in Chapter 3 (Clark, Shepherd et al 1984). The research shows the space between two adjacent elements passing from the middle ear to the cochlea could not only facilitate the spread of infection to the meninges but also lead to its recurrence. Furthermore, confined spaces can also make nonpathogenic organisms pathogenic, and a virulent infection would lead to more serious complications.

Combi-40 and 40+

The Med El/Technical University of Vienna electrode array had either eight electrodes for the Combi-40 system or 12 electrodes for Combi-40+. The electrodes were distributed along short (12 mm), standard (21 mm), and long (27 mm) arrays (Gstoettner et al 1997). A dummy array was used to determine how deeply one is likely to pass and then an array of that length selected. The most recent array contains a total of 24 electrode contacts arranged as paired interconnected surfaces resulting in 12 monopolar stimulus channels. The distance between channels is approximately 2.4 mm with the first channel 1.0 mm from the tip. It has an axial rib to control bending and has been found to pass around the basal to the middle turn. However, it is not clear whether an insertion beyond the 500-Hz region in the cochlea (26 mm) is necessary as the lower frequencies do not require place coding, and the spiral ganglion cells make only $1\frac{1}{2}$ turns and do not extend over the whole $2\frac{1}{2}$–$2\frac{3}{4}$ turns of the cochlea, as discussed in Chapter 2.

Digisonic

The electrode array for the Bertin/CHU Saint Antoine, Paris, device evolved from the prototype developed by Pialoux et al (1976, 1979) for placement in holes drilled directly into the cochlea to site separate electrodes in the regions conveying speech frequencies. This required access to the basal and middle turns of the cochlea via both the middle ear and the middle cranial fossa. Small pieces of Silastic were inserted into the cochlea around the electrode in an attempt to isolate the electric current and limit its spread along the scala. As this array resulted in too much trauma, it was replaced with a multiple-electrode array developed by MXM for insertion along the scala tympani. This array is straight, free-fitting, 24 mm in length, with 15 active electrodes that are band-like and 0.5 mm in width with an interelectrode distance of 0.7 mm.

Nucleus 22 and 24

Standard Straight Array

The University of Melbourne's free-fitting banded array was developed industrially by Cochlear Pty Limited. There were 22 active electrodes welded to 0.025-mm Teflon-insulated platinum/iridium (90/10) wires; eight free bands at the proximal end provided additional stiffening (Fig. 8.36). The bands were 0.3 mm wide with 0.45 mm interelectrode spacing. The electrode bands and their attached wires were injection molded with Silastic MDX-4-4210. The arrays tapered from a diameter of 0.4 mm at the tip to 0.6 mm over a distance of 10 mm. The injection

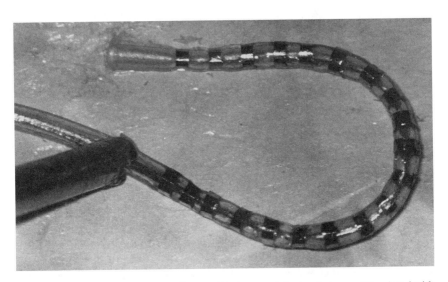

FIGURE 8.36. The University of Melbourne/Nucleus straight, banded array (Reprinted with permission from Clark, Patrick et al 1979. A cochlear implant round window electrode array. *Journal of Laryngology and Otology* **93**(2): 107–109. The Royal Society of Medicine Press Ltd.)

molding ensured that the array was smooth with no gaps between the bands and the carrier, as occurred with the first university prototype where the bands were wrapped around the Silastic tube.

To help establish that this array did not lead to significant trauma when inserted into the human temporal bone, a series of studies were undertaken. The first involved inserting electrode arrays into fresh cadaver human temporal bones, which were then sectioned (Shepherd 1985a). The insertions were carried out on nine human temporal bones within 24 hours of death so their properties were as close as possible to living tissue. The electrode insertion distance varied from 15.5 to 27.0 mm with a mean insertion distance of 18.6 mm (standard deviation = 3.5 mm) for the nine cochleae. Three of the nine temporal bones examined showed no evidence of electrode insertion trauma to any cochlear structure. When damage was observed, it occurred in one of four distinct sites: spiral ligament, osseous spiral lamina, basilar membrane, and Reissner's membrane (Fig. 8.37).

The most common form of insertion trauma was damage to the spiral ligament in the region 7 to 13 mm from the round window (Fig. 8.38). This damage was in the form of tears produced as the electrode array came in contact with the outer wall of the scala tympani and passed upward along the outwardly spayed wall toward the spiral ligament and basilar membrane. Tears to Reissner's membrane were observed in the nine temporal bones examined. However, due to the delicate nature of this structure and its susceptibility to histological artifact, it is difficult to determine the extent of damage directly related to electrode insertion. Indeed, when a comparison of both the number and length of Reissner's membrane tears in these implanted cochleae was made with five unimplanted control cochleae, there was no statistically significant difference between the two populations

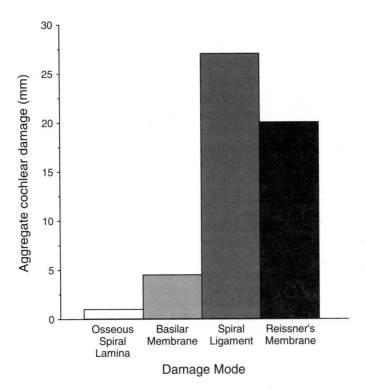

FIGURE 8.37. Summary of damage to cochlear structures in nine human temporal bones following the insertion of the banded scala tympani electrode array. The average insertion distance was 18.5 mm (standard deviation 3.5 mm). (Reprinted with permission from Shepherd et al 1985. Banded intracochlear electrode insertion trauma in human temporal bones. *Annals of Otology, Rhinology and Laryngologica* **94**: 55–59.)

(Clark, Pyman et al 1987). A localized tear of the basilar membrane was observed in two of the nine temporal bones examined. In both cases the damage was a result of the electrode arrays being inserted past the point of first resistance. In one bone, the tip of the array had deflected toward the scala media, perforating the basilar membrane for about 1 mm of its length. In the second bone, the force required to insert the array was sufficient to cause the array to buckle in the lower basal turn. This resulted in a 2-mm tear to the basilar membrane as well as a fracture to the adjacent osseous spiral lamina.

Although the above study demonstrated the damage caused by the insertion of electrode arrays, the movement of the array during insertion could not be seen. The whole process was labor intensive, and there were preparation artifacts making the interpretation of some of the data difficult.

To speed up the process and examine the array in situ, studies were carried out by inserting the electrode into the cochlea, and then ascertaining its position and any damage to the basilar membrane by drilling the overlying bone as seen in Figure 8.39 for a perimodiolar precurved array (Clifford and Gibson 1987; Franz

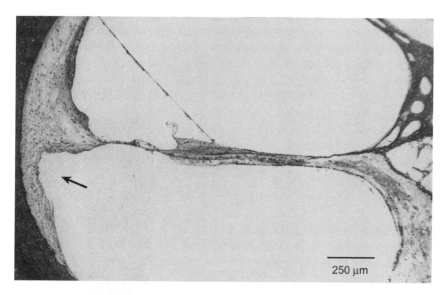

FIGURE 8.38. Photomicrograph of the upper basal turn of a human cochlea illustrating a tear in the spiral ligament (arrow) following the insertion of a banded scala tympani electrode array. Original magnification: 80×. (Adapted from Shepherd et al 1985b, with permission.)

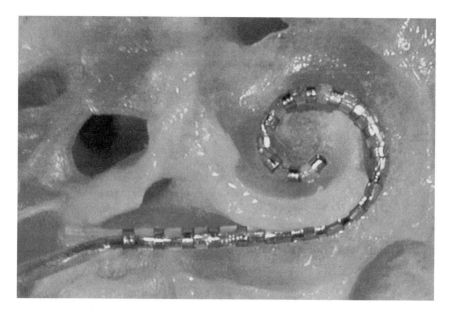

FIGURE 8.39. Human temporal bone insertion with the precurved array and the bone uncapped to expose it lying in the scala tympani. (Reprinted from the Cooperative Research Centre for Cochlear Implant Speech and Hearing Research Annual Report 1994–95, with permission.)

and Clark 1987; Kennedy 1987). The procedure was carried out through a posterior tympanotomy with the temporal bone in the correct surgical position. The cochlea was not uncapped prior to the insertion, to avoid giving guidance to the surgeon. These surface preparations demonstrated that the electrode array essentially lay along the outer wall of the scala tympani. This work revealed that the tip of the electrode would occasionally deflect upward toward the basilar membrane when it first contacted the outer scalar wall (Franz and Clark 1987; Kennedy 1987). It was found that rotating the electrode by 90 degrees during the insertion (counterclockwise in the right ear and clockwise in the left ear) would direct the electrode tip away from the basilar membrane and along the basal turn (Franz and Clark 1987). The studies also confirmed that damage due to buckling in the basal turn could follow an attempt to insert the electrode array past the point of first resistance. This is discussed further in Chapter 10.

These studies confirmed that free-fitting, flexible, tapered, scala tympani electrode arrays could be inserted into the human cochlea with minimum trauma, provided the insertion was stopped at the point of first resistance. However, skeletonizing the cochlea had some limitations: (1) the technique could damage the cochlea, (2) tissue trauma was difficult to observe under the operating microscope, and (3) the movement and dynamics of the array could not be seen.

The movement of the electrode array could be observed in an uncovered cochlea, and this is a useful adjunct to the above procedure. It is, however, not for the validation of trauma, as the tissue may be damaged in the preparation of the specimen. It is also possible to examine the insertion fluoroscopically. This too has limitations. Unless it is viewed in two planes, it can be difficult to determine whether the array has penetrated the basilar membrane and entered the scala tympani.

Clark, Shepherd et al (1988) described the histopathology of a temporal bone from a patient who had received a Melbourne/Cochlear multiple-channel cochlear prosthesis. The right cochlea of this patient contained the electrode array, which had been inserted a distance of approximately 17 mm from the round window. Histological examination of the cochlea indicated that the electrode array lay within the scala tympani for its entire length within the cochlea. Electrode insertion trauma was relatively minimal. It included a 1-mm-long fracture of the osseous spiral lamina close to the round window, and local tearing of the spiral ligament along the outer wall of the scala tympani in a region 9 to 12 mm from the round window. Although the electrode array also displaced the basilar membrane in this region, this membrane remained intact throughout the cochlea.

In studies on five (Kawano et al 1995, 1996, 1998) and (Dahm et al 2000) 14 temporal bones implanted with the Nucleus array, the insertion had remained in the scala tympani with commonly indentation of the spiral ligament in the upper basal turn. The basilar membrane and in some cases the osseous spiral lamina showed localized damage with tearing or fracture. In three bones the cochleostomy and insertion of the array had occurred in the scala media and continued along this scala with rupture of Reissner's membrane in two.

The histopathological findings were discussed in Chapter 3, and indicate that implantation of the University of Melbourne/Nucleus array, did not cause significant trauma or compromise the residual spiral ganglion cell population. The histological results also showed that the electrode array lay along the outer wall of the scala tympani. This was consistent with earlier electrode insertion trauma studies using the straight, flexible, banded electrode array (Shepherd 1985a; Clifford and Gibson 1987; Franz and Clark 1987; Kennedy 1987).

Thus the extent of electrode insertion trauma using the University of Melbourne/Nucleus electrode array has been extensively examined both qualitatively and quantitatively. The studies have consistently shown that this electrode array can be inserted with minimal trauma, for distances of up to 25 mm from the round window, providing insertion is stopped at the point of first resistance.

Perimodiolar Free-Fitting, Tapered, Banded Array

To improve speech processing, electrode arrays needed to be designed so that they lay close to the spiral ganglion cells. Together with a higher density of electrodes, this could lead to the improved coding of frequency discussed below. It required the array to be precurved with a mechanism to hold it straight prior to insertion or have a mechanism to get it to curl once inserted. With each mechanism it would be necessary to know how effective it would be for placing the array close to the spiral ganglion cells, and the degree of trauma. As illustrated in Figure 8.32, any trauma would occur through pressure at the tip or laterally from buckling stresses. The effectiveness of its perimodiolar placement would be determined by the elastic recoil of the carrier material (usually Silastic), counterbalanced by its plastic deformation and the rigidity or memory of the metal wires.

In 1989 research commenced in the Australian Research Council's HCRC at the University of Melbourne/Bionic Ear Institute to develop a perimodiolar array, and this continued until 1992 when its refinement was undertaken by the CRC for Cochlear Implant Speech and Hearing Research. This array was designed to lie around the modiolus close to the spiral ganglion cells, and provide stimulation of more localized groups of neurons. This could first allow better place coding of frequency. Second, it was also assumed from the research (Clark 1996) that improved electrode arrays would give better temporal coding through fine temporal and spatial patterns of stimulation. A third need was to reduce power consumption so that smaller implants or totally implantable devices could be achieved.

There were a number of important directions to explore in developing a perimodiolar array. Should the array be precurved to the spiral and then held straight for insertion before being allowed to curl around the cochlea? Or, alternatively, should an array be introduced into the cochlea as a straight flexible array and then induced to curl? It was also necessary to decide whether to achieve precurved or curling arrays using mechanical devices or polymers. If a polymer were chosen to hold the array straight, a polyvinyl alcohol (PVA) would be suitable. However, it would have to dissolve with an appropriate time constant so that the electrode was not rigid when it reached a point approximately 10 mm from the round window where it rides up on the outer wall toward the basilar membrane. It would also need to be biocompatible. If the electrode were to be induced to curl once in the cochlea, the polymers should expand in the presence of water. Polyacrylic acid (PAA) would be appropriate.

With each mechanism it was necessary to know how effective it was for placing the array close to the spiral ganglion cells, and the degree of trauma. As illustrated in Figure 8.32, the trauma caused by each mechanism could occur through pressure at the tip or laterally from buckling stresses. The effectiveness of its perimodiolar placement would be determined by the elastic recoil of the carrier material usually Silastic, counterbalanced by its plastic deformation and the rigidity or memory of the metal wires. In first developing a perimodiolar precurved array it was thought preferable to produce prototype arrays using mechanical means, as this would be the most conservative approach with fewer unknowns such as the long-term biocompatibility. In parallel the use of polymers could also be explored.

A precurved array was fabricated by making it as a coil to fit the dimensions of the modiolus. For insertion it needed to be held straight for approximately 8 mm beyond the round window before being released and then further advanced around the cochlea. The array was refined through the CRC for Cochlear Implant Speech and Hearing Research (1992–1999) and then the CRC for Cochlear Implant and Hearing Aid Innovation (1999–).

The first prototype precurved array was produced in 1989 in the HCRC by placing a thread of cotton in a tube of Teflon, injecting Silastic A, and then allowing it to cure after it was wound around a central spindle. After straightening, this array would not return completely to its original curvature due to creep between the cotton fibers and the Silastic. To avoid this problem, molds of the cochlear spiral were made by casting a metal wire into the two halves of a resin block. Later a mold was created using solder and soft brass. Initial prototype precurved banded arrays made from these casts were inserted into human temporal bones, and the overlying bone drilled away to inspect for any damage to the soft tissue in the cochlea. The array was inserted with a tool created from a plastic intravenous cannula slit along one side. The electrode, held straight in the cannula, could be disengaged through the slit by applying pressure with a blunt probe. The study showed this array could nearly resume its original curvature after straightening. It also showed that the tip of the array (terminal 2–3 mm) needed to be straight to avoid its bending on itself during the insertion. It was also learned that it had to be released after only inserting it 8 to 10 mm to avoid damage to the basilar membrane. This is the point where previous findings on the straight array showed that the electrode can pass up the splayed outer wall and hit the basilar membrane. After release from the insertion tool the array would then pass around the cochlea as it resumed its curled shape. It was quite critical not to release the electrode too early as it would then curl on itself.

Anatomical data on the dimensions of the basal turn of the cochlea and the curvature of the modiolus were next required to optimize the design of the array. This was obtained by making molds of the cochlea from Silastic, and then measuring sections of the molds. The cross-sectional area as well as the height and width versus distance were presented in Chapter 2 for the human and cat (Hat-sushika et al 1990). As height and width remained fairly constant in the human, with distance there was only the need for a small taper in the array, and a small degree of individual variability needed to be allowed for. As the width was greater than the height, there was also room for an array to move from a peripheral position to a perimodiolar one. This would allow space for a restraining instrument to be inserted and the array to move to the modiolus. The cross-sectional area became progressively smaller with distance and this was accompanied by a change in shape. As it was more triangular toward the apical region of the basal turn, there was a greater chance of the tip of the electrode getting wedged when the electrode curled to the modiolus.

In addition to the above anatomical studies to determine how a precurved array would accommodate to the variations in size and shape of the scala tympani, studies were also required to measure accurately the curvature of the modiolus and its variations across bones. It was necessary to know the optimal curvature of the array for use in all bones and to what extent the curvature could be slightly overcorrected as the elastic recoil of the Silastic did not completely overcome the plastic forces in the metal wires. The curvatures of the 11 Silastic molds of the scala tympani were determined by positioning them under a video camera connected to an image analyzer. The projected inner curvatures of the scala tympani were digitized. This led to a mathematical algorithm for the curvature of the inner wall of the spiral. The logarithmic function for the curvature was

$$R(\theta) = Ae^{B(\theta + C)}$$

where R = radial distance of the inner wall of the scala tympani (μm), θ = the polar angle with reference to the axis, and A, B, C are values obtained from the experimental data.

The distance between the mathematically calculated curve and the average measured curve showed good agreement accept in the middle turn and at the extreme base, as illustrated in Figure 8.40.

This algorithm enabled a computer-based drilling system to be used to produce a mold for the array. In 1991 a mold of the array was made in brass to the mean inner spiral dimensions using a computer drilling system. The distal 2 to 3 mm were kept straight to prevent the array from doubling on itself. A side entry was created for the electrode and the straight section for the release plunger. The welding of the wires to the band was examined under the scanning electron microscope to ensure their integrity. A prototype insertion tool shown was also developed in 1991 (Fig. 8.41). The advance of the plunger was through a slide mechanism operated by a ratchet wheel.

Not only was the above precurved array developed for a perimodiolar placement, but an alternative was an array that had an attached member or "former" that caused the electrode carrier to curl once inserted (Fig. 8.42). This array, with attached member, was developed by Cochlear Proprietary Limited. After the electrode was inserted, the attached member was pushed forward to apply an inward force vector to push the array toward the central spiral. The two parts of the carrier were then locked together in this position using a metal clamp.

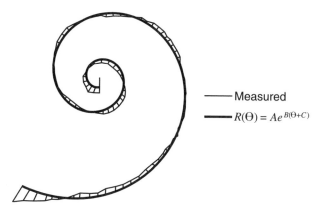

FIGURE 8.40. The logarithmic estimation of the curvature of the inner wall of the scala tympani and the measured inner wall of a human scala tympani molds. *A, B,* and *C* are values obtained from the experimental data from 11 scala tympani molds. *A* has a mean value of 132 with a standard deviation of 8, *B* is consistently 0.2, and *C* is set to $7\pi/6$. (Reprinted with permission from the Human Communication Research Centre Fifth Annual Report, 1992.)

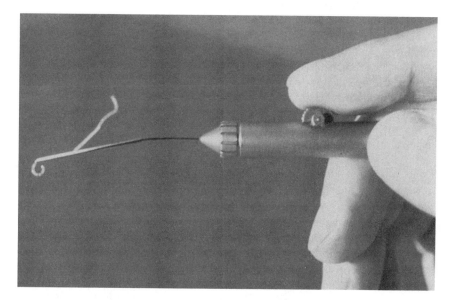

FIGURE 8.41. Insertion tool for the precurved banded array. (Reprinted with permission from the Cooperative Research Centre for Cochlear Implant Speech and Hearing Research Annual Report 1994–95.)

Insertion studies on human temporal bones were then necessary first to ensure that the precurved array or array with inward force vector was no more traumatic than the University of Melbourne/Nucleus free-fitting straight array, and second to ensure that it would lie closer to the modiolus as planned. The studies also aimed to compare the degree of trauma with each of these arrays.

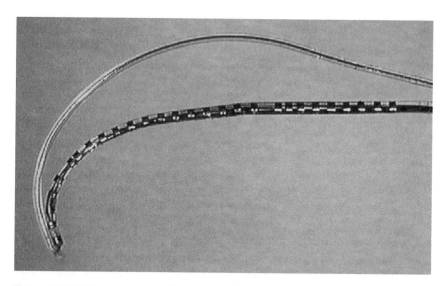

FIGURE 8.42. The prototype curling array with inward force vector. (Reprinted with permission from the Cooperative Research Centre for Cochlear Implant Speech and Hearing Research Annual Report 1994–95.)

The arrays were implanted in human temporal bones under simulated surgical conditions. To provide additional data on the insertion, the temporal bones were x-rayed prior to uncapping (skeletonizing) them. The Stenver's x-ray view was modified to give the best picture of the array in the basal turn. This view was at right angles to the scala tympani and termed the cochlear view (Marsh et al 1993). It also provided a good view of the semicircular canals so that the position of the round window could be estimated. This then made it possible to determine the length and angle of insertion. An angular measurement of distance from the basal end of the cochlea was first used by Bredberg (1968) to more accurately allow for anatomical variations in the cochlear spiral when studying hair cell and spiral ganglion cell populations. The angle of the electrode insertion was derived from the cochlear x-ray view seen in Figure 8.43. As shown in Figure 8.44, the angle was measured from a zero reference line. To determine this reference a vertical line was drawn through the superior semicircular canal. This passed through the approximate center of the vestibule. The center of the modiolus was estimated from the spiral of the bony cochlea and the inner and outer walls of the scala tympani. A horizontal line was then drawn from the center of the modiolus perpendicular to the vertical one. This was the zero reference line. If a line was drawn 10 degrees below, it would transect the round window.

After the x-rays were done the cochlear turns were uncapped (skeletonized) to view the extent of trauma and placement within the scala tympani (Fig. 8.39). This research showed that the precurved array would lie close to the modiolus. However, the use of the insertion tool required a large opening into the cochlea (cochleostomy), and the distance for the insertion before releasing the array was hard to assess as there was variability in the cochlear anatomy and the view poor.

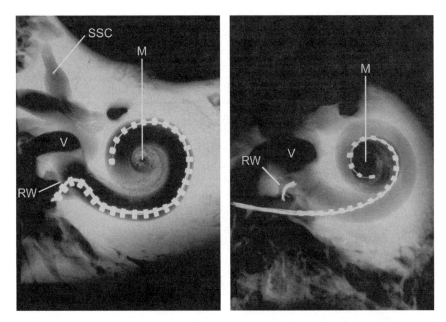

FIGURE 8.43. Cochlear view of the cochlea and placement of round window and calculation of angle of insertion. Left: standard straight Nucleus banded array. Right: Nucleus Contour precurved half-banded array with stylet. RW, round window; V, vestibule; SSC, superior semicircular canal; M, modiolus.

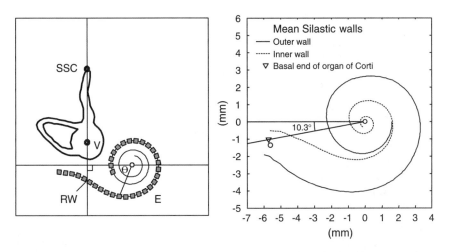

FIGURE 8.44. Calculating the angle of insertion of the array from the Cochlear x-ray view. SSC, superior semicircular canal; RW, round window; E, electrodes (L. Cohen, personal communication).

The insertion of the array with attached member required more skill as the member would slip and lose its orientation to the electrode carrier.

In another study the bones were not skeletonized after they were x-rayed but were embedded in resin and sectioned with the electrodes in place so the placement and trauma were not affected by the uncapping procedure. In all the safety studies the electrodes were inserted, as they would be for surgery, that is, through a posterior tympanotomy. Electrodes were fixed and after the temporal bones were embedded in resin they were ground down with diamond paste drills and the surface photographed every 3 mm. The method was used by the Department of Dentistry at the University of Melbourne.

The x-ray and resin embedding studies established the precurved banded array could be inserted with minimal trauma, and it lay closer to the modiolus than a straight electrode array, as shown in Fig. 8.45, but not always in close contact. The curling array, with attached member to apply an inward force vector (Fig. 8.42), was found to cause significant damage to the cochlea (Fig. 8.46). The member caused the array to be stiffer and made it more likely to pass through the basilar membrane into the scala vestibuli. In addition, outward pressure in response to the inward force vector caused counterpressure from the attached member to tear the basilar membrane at some distance inside the cochlea. This finding was confirmed in a study by Roland et al (2000), in the two bones embedded in epoxy resin. The Teflon strip that applied active inward pressure finished in the scala vestibuli in both cases, causing marked trauma.

An initial group of three patients were implanted with the precurved electrode array rather than the curling array with inward force vector, as the latter was shown to be too traumatic (Fig. 8.46). After the implantation the placement of the arrays in the temporal bones were evaluated with the cochlear x-ray views.

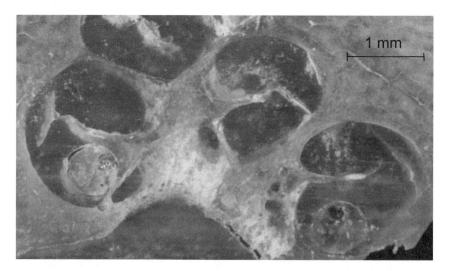

FIGURE 8.45. Cross section of the precurved array in the human temporal bone embedded with resin.

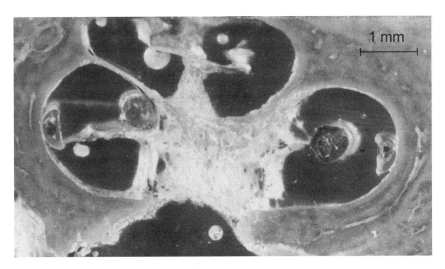

FIGURE 8.46. Cross section of a resin-embedded human temporal bone with the curling array with former inserted. (Reprinted with permission from Clark G. M. and Lawrence D., 2000. Technical features of the Nucleus, Med-El and Clarion cochlear implants. *Australian Journal of Oto-Laryngology* **3**: 516–522.)

The x-rays of the three patients were compared with a typical standard straight array (Fig. 8.47). For patient (b), the precurved array was uniformly closer to the modiolus than the standard array. In patient (c) it is very close in the middle third; and for patient (d) it is closest in the distal third.

The psychophysical studies are discussed in more detail in Chapter 6 and by Cohen et al (2001). The thresholds and comfortable levels were plotted against distance from the inner wall. The data showed that the closer the electrode to the inner wall of the cochlea, the lower the thresholds and comfortable levels. It was also shown that the dynamic range increased with reduced distance of the electrodes from the inner wall. As the dynamic range increased with reduced distance from the inner wall, it was also of interest to see if there were more usable steps in loudness with reduced distance. The results showed the difference limens became smaller with the electrodes closer to the inner wall for two of the three patients in the study. Finally, it was found that electrode place could be discriminated better if the electrodes were closer to the spiral ganglion cells. This was confirmed with forward masking that showed that the tuning of the responses was sharper when the electrodes lay close to the inner wall.

An improvement in design was achieved with an array that had a platinum stylet passed along the center (Figure 8.48). This held the array straight until after insertion. It was then withdrawn. The platinum stylet was malleable at the tip to allow the array to curve during insertion. The array had 22 half-band platinum electrodes embedded along the initial 15.5 mm of the carrier. This array was designed by Cochlear Limited in association with the CRC for Cochlear Implant, Speech and Hearing Research in 1997. The array was cast as a silicon carrier with

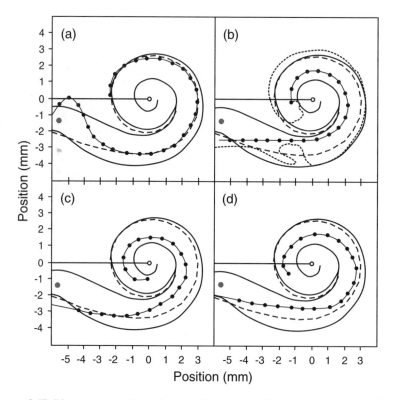

FIGURE 8.47. Line representations of the cochlear x-rays of the standard straight (a) and three precurved arrays (b, c, d).

stylet in situ. The array was a little wider than the standard array and tapered from a diameter of 0.6 to 0.8 mm at the proximal end. The electrodes were half (Fig. 8.49) rather than full bands, and their area varied from 0.283 to 0.306 mm² geometric compared with the average area of 0.48 mm² geometric for full-band electrode array.

The electrode was inserted through a cochleostomy 1.5 to 2 mm in diameter drilled inferior and slightly anterior to the round window membrane, and the opening often included an anterior segment of the round window. The opening was slightly larger than for the standard banded array. Following lubrication with 40% glycerine and water, the half bands were orientated toward the modiolus and the electrode inserted. If resistance was experienced, the stylet was partially withdrawn 1 to 2 mm, and the array then further inserted.

This array Contour was inserted into 12 temporal bones with the surgical procedure described. It was then fixed in position and the cochlea skeletonized to expose the basilar membrane. Cochlear x-ray views were taken of the implanted temporal bones, and the image analyzed to determine the electrodes position (Saunders et al 2000). The mean insertion depth was 410.85 degrees ± 39.4 standard deviation (SD), which was deeper than the usual insertion of the banded

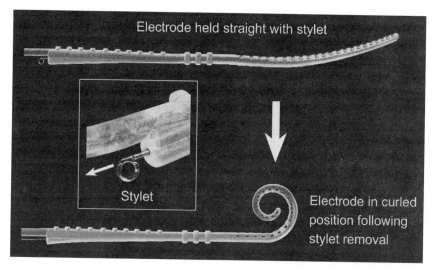

FIGURE 8.48. The Nucleus Contour perimodiolar array showing the stylet. Cochlear Limited and the CRCs for Cochlear Implant Speech and Hearing, and Cochlear Implant and Hearing Aid Innovation. (Reprinted with permission from C. Treaba.)

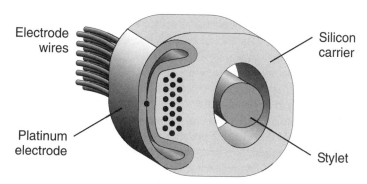

FIGURE 8.49. The Nucleus Contour perimodiolar array in cross section. Cochlear Limited and the CRCs for Cochlear Implant Speech and Hearing, and Cochlear Implant and Hearing Aid Innovation. (Reprinted with permission from C. Treaba.)

array of 351.8 degrees ± 52.6 SD. In 10 of the 12 preparations, the array was positioned entirely within the scala tympani. No insertion damage was visible to the basilar membrane. In two bones the basilar membrane had been penetrated at about 170 degrees. In one of these insertions the electrode array had been pushed forward off the stylet after resistance had initially been encountered. This appears to be an unsatisfactory procedure.

The probability of insertion-induced damage using the Contour array was compared to the probability of damage with the standard array using the data from Shepherd (1985a). The data showed that at the 95% confidence level the distributions for the probability of damage were overlapping. This suggested that the probability of damage with the Contour and straight electrode were the same. However, the nature of the damage could have been quite different, and the sample size limited the power of the analysis.

As discussed in Chapter 6, a study on 35 patients showed the population means for T and MC levels were lower for the Contour compared with the standard Nucleus array. As with the initial study there was good correlation between the average T levels and radial distance, and the average MC levels mirrored the T levels (Saunders et al 2000, 2002). When the behavioral T and MC levels were compared with the compound action potentials (field potentials) from the auditory nerve recorded by neural response telemetry (NRT), the results showed the NRT thresholds mirrored the T and MC levels, and the radial distance of the electrode from the spiral ganglion cells (Dillier 1998; Dillier, Lai et al 2000, 2002).

The Nucleus 24 Contour system received a premarket approval (PMA) from the FDA on November 1, 2000 for its use on adults and children greater than 12 months.

Conclusion

From the above results there are a number of points to consider in the bioengineering of a cochlear implant receiver-stimulator and electrode array. First, it is important to ensure that the implantable device is manufactured using biocompatible materials. Studies on the materials alone do not necessarily imply the assembled device is biocompatible. The prosthesis must be evaluated as a whole, including its implantation in the site for which it is intended. Moreover, electrically active devices, such as the cochlear prosthesis, must be tested under electrical stimulation. It is important to examine the effects of electrical stimulation not only on surrounding tissue but also on the biomaterials of the prosthesis as they may be degraded. Second, the insertion of a flexible scala tympani electrode can be achieved with minimal trauma to cochlear structures, provided the insertion is stopped at the point of first resistance. Third, experimental studies showed that the spread of infection from the middle ear to the cochlea could result in the extensive loss of auditory nerve fibers throughout the cochlea. However, the studies demonstrated that the healed round window and electrode tissue sheath could protect against the spread of infection, although the seal was not effective until approximately 4 weeks following surgery. This highlights the importance of strict asepsis during the surgery and prophylactic antibiotics during and immediately following surgery. Fourth, electrical stimulation studies, using carefully controlled, charge balanced biphasic current pulses, at charge densities of up to 32 μC cm^{-2} geometric/phase, showed that chronic stimulation did not adversely effect residual auditory nerve fibers or the cochlea in general. Moreover, chronic stimulation with platinum electrodes using this stimulus regime did not result in the dissolution of the platinum or the degradation of the adjacent Silastic carrier. Furthermore, long-term electrode impedance changes, observed during these chronic stimulation studies, appeared to correlate with the degree of fibrous tissue encapsulation of the electrode array. While there was no evidence of neural damage using platinum electrodes and relatively low charge densities, there was evidence of short-term and possibly permanent reductions in the excitability of the auditory nerve using a combination of high stimulus rates and intensities, as discussed in Chapter 4. It would appear that these stimulus-induced changes were metabolic in origin. Fifth, the histopathology of temporal bones from patients who received a Cochlear multiple-channel cochlear prosthesis was consistent with previous experimental studies, and showed that long-term electrical stimulation of the cochlea did not compromise the residual auditory nerve population (see Chapter 4).

These carefully designed safety studies have formed an integral part of the development of cochlear prostheses. The extent of the safety studies described in this chapter reflects the complexity of these devices.

References

Agnew, W. F., T. G. H. Yuen and D. B. McCreery. 1983. Morphological changes after prolonged electrical stimulation of the cats cortex at defined charge densities. Experimental Neurology 79: 397–411.

Aran, J. M. 1981. Electrical stimulation of the auditory system and tinnitus control. Journal of Laryngology and Otology 95(suppl 4): 153–161.

Arndt, P., S. Staller, J. Arcaroli, A. Hines and K. Ebinger. 1999. Within-subject comparison of advanced coding strategies in the Nucleus 24 cochlear implant. Cochlear Corporation Report.

Bagger-Sjoback, D. 1991. Sodium hyaluronate application to the open inner ear: an ultrastructural investigation. American Journal of Otology 12: 35–39.

Banfai, P. 1987. Cochlear Implant: current situation. International Cochlear Implant Symposium, Duren.

Banfai, P., G. Hortmann and S. Kubik. 1985. Extracochlear eight-channel electrode system. Journal of Laryngology and Otology 99(6): 549–553.

Banfai, P., A. Karczag and P. Luers. 1984a. Cochlear implants. Clinical results: the rehabilitation. Acta Oto-Laryngologica (suppl 411): 183–194.

Banfai, P., S. Kubik and G. Hortmann. 1984b. Our extra-scalar operating method of cochlear implantation. Experience with 46 cases. Acta Oto-Laryngologica (suppl 411): 9–12.

Battmer, R.-D., D. Gnadeberg, D. J. Allum-Mecklenberg and T. Lenarz. 1994. Matched-pair comparisons for adults using the Clarion or Nucleus devices. Annals of Otology, Rhinology and Laryngology 104: 251–254.

Bertram, R., V. Meyer, I. Maneke, et al. 2001. Cochlear implant re-implantation: a parent's perspective. Poster presented by the House Institute at 8th Symposium Cochlear Implants in Children, February 28–March 3.

Black, R. C. and G. M. Clark. 1977. Electrical transmission line properties of the cat cochlea. Proceedings of the Australian Physiological and Pharmacological Society 8: 137.

Black, R. C. and G. M. Clark. 1978. Electrical network properties and distribution of potentials in the cat cochlea. Proceedings of the Australian Physiological and Pharmacological Society 9: 71P.

Black, R. C. and G. M. Clark. 1980. Differential electrical excitation of the auditory nerve. Journal of the Acoustical Society of America 67(3): 868–874.

Blamey, P. J., R. C. Dowell, Y. C. Tong, A. M. Brown, S. M. Luscombe and G. M. Clark. 1984a. Speech processing studies using an acoustic model of a multiple-channel cochlear implant. Journal of the Acoustical Society of America 76: 104–110.

Blamey, P. J., R. C. Dowell, Y. C. Tong and G. M. Clark. 1984b. An acoustic model of a multiple-channel cochlear implant. Journal of the Acoustical Society of America 76: 97–103.

Blamey, P. J., L. F. Martin and G. M. Clark. 1985. A comparison of three speech coding strategies using an acoustic model of a cochlear implant. Journal of the Acoustical Society of America 77: 209–217.

Blamey, P. J., B. C. Pyman, M. Gordon, et al. 1992. Factors predicting postoperative sentence scores in postlinguistically deaf adult cochlear implant patients. Annals of Otology, Rhinology and Laryngology 101: 342–348.

Bredberg, G. 1968. Cellular pattern and nerve supply of the human organ of Corti. Acta Oto-Laryngologica (suppl 236): 1–138.

Bredberg, G. and B. Lindstrom. 1997. A new approach for the treatment of ossified cochleas. Vth International Cochlear Implant Conference, New York.

Bredberg, G., B. Lindstrom, H. Lopponen, M. A. Beltrame, W. Gstoettner and H. Skarzynski. 2000. A new approach for the treatment of ossified cochleas. In: Waltzman, S. and N. Cohen, eds. Cochlear implants. New York, Thieme Medical Publishers. 159–160.

Brindley, G. S. and W. S. Lewin. 1968. The sensations produced by electrical stimulation of the visual cortex. Journal of Physiology 196: 479–493.

Brummer, S. B. and M. J. Turner. 1977a. Electrical stimulation with Pt electrodes: I. A method for determination of 'real' electrode areas. IEEE Transactions on Biomedical Engineering 24: 436–439.

Brummer, S. B. and M. J. Turner. 1977b. Electrical stimulation with Pt electrodes: II. Estimation of maximum surface redox (theoretical non-gassing) limits. IEEE Transactions on Biomedical Engineering 24: 440–443.

Brummer, S. B. and M. J. Turner. 1977c. Electrochemical considerations for safe electrical stimulation of the nervous system with platinum electrodes. IEEE Transactions on Biomedical Engineering 24: 59–63.

Burian, K., E. Hochmair, I. Hochmair-Desoyer and M. R. Lessel. 1980. Electrical stimulation with multichannel electrodes in deaf patients. Audiology 19: 128–136.

Burns, S. G. and P. R. Bond. 1997. Principles of electronic circuits. 2nd ed. Boston, PWS Publishing.

Burton, M. J., R. K. Shepherd, J. Xu and G. M. Clark. 1992. Cochlear implantation in young children: long term effects of implantation on the skull and underlying central nervous system tissues in a primate model. Otolaryngology Research Meeting, London: 1.

Busby, P. A., Y. C. Tong and G. M. Clark. 1993. The perception of temporal modulations by cochlear implant patients. Journal of the Acoustical Society of America 94: 124–131.

Busby, P. A., L. A. Whitford, P. J. Blamey, L. M. Richardson and G. M. Clark. 1994. Pitch perception for different modes of stimulation using the Cochlear multiple-electrode prosthesis. Journal of the Acoustical Society of America 95: 2658–2669.

Chen, B. K., G. M. Clark and R. Jones. 2003. Evaluation of trajectories, and contact pressures for the straight Nucleus cochlear implant electrode array—a two-dimensional application of finite element analysis. Medical Engineering and Physics. 25(2): 141–147.

Chouard, C. H. 1978. Multiple intracochlear electrodes for rehabilitation in total deafness. Otolaryngologic Clinics of North America 11: 217–233.

Chouard, C. H. 1980. The surgical rehabilitation of total deafness with the multichannel cochlear implant. Indications and results. Audiology 19: 137–145.

Chouard, C. H., C. Fugain, B. Meyer and H. Lacombe. 1983. Long-term results of the multichannel cochlear implant. Annals of the New York Academy of Sciences 405: 387–411.

Chouard, C. H. and P. MacLeod. 1976. Implantation of multiple intracochlear electrodes for rehabilitation of total deafness: preliminary report. Laryngoscope 86: 1743–1751.

Clark, G. M. 1969a. Middle ear and neural mechanisms in hearing and the management of deafness. PhD thesis.

Clark, G. M. 1969b. Responses of cells in the superior olivary complex of the cat to electrical stimulation of the auditory nerve. Experimental Neurology 24: 124–136.

Clark, G. M. 1973. A hearing prosthesis for severe perceptive deafness-experimental studies. Journal of Laryngology and Otology 87: 929–945.

Clark, G. M. 1975. A surgical approach for a cochlear implant. An anatomical study. Journal of Laryngology and Otology 89: 9–15.

Clark, G. M. 1977. An evaluation of per-scalar cochlear electrode implantation techniques. An histopathogical study in cats. Journal of Laryngology and Otology 91: 185–199.

Clark, G. M. 1987. The University of Melbourne–Nucleus multi-electrode cochlear implant. Advances in Oto-Rhino-Laryngology. Vol 38. Basel, Karger.

Clark, G. M. 1996. Electrical stimulation of the auditory nerve: the coding of frequency, the perception of pitch and the development of cochlear implant speech processing strategies for profoundly deaf people. Clinical and Experimental Pharmacology and Physiology 23: 766–776.

Clark, G. M. 1998. Cochlear implants. In: Wright, A. and H. Ludman, eds. Diseases of the ear. London, Edward Arnold: 149–163.

Clark, G. M., R. C. Black, D. J. Dewhurst, I. C. Forster, J. F. Patrick and Y. C. Tong. 1977. A multiple-electrode hearing prosthesis for cochlear implantation in deaf patients. Medical Progress Through Technology 5: 127–140.

Clark, G. M., R. C. Black, I. C. Forster, J. F. Patrick and Y. C. Tong. 1978. Design criteria of a multiple-electrode cochlear implant hearing prosthesis. Journal of the Acoustical Society of America 63: 631–633.

Clark, G. M. and R. J. Hallworth. 1976. A multiple-electrode array for a cochlear implant. Journal of Laryngology and Otology 90: 623–627.

Clark, G. M., R. J. Hallworth and K. Zdanius. 1975. A cochlear implant electrode. Journal of Laryngology and Otology 89: 787–792.

Clark, G. M., H. G. Kranz and H. Minas. 1973. Behavioral thresholds in the cat to frequency modulated sound and electrical stimulation of the auditory nerve. Experimental Neurology 41: 190–200.

Clark, G. M., H. G. Kranz, H. J. Minas and J. M. Nathar. 1975. Histopathological findings in cochlear implants in cats. Journal of Laryngology and Otology 89: 495–504.

Clark, G. M. and D. Lawrence. 2000. Technical features of the Nucleus, Med-El and Clarion cochlear implants. Australian Journal of Oto-Laryngology 3: 516–522.

Clark, G. M., J. F. Patrick and Q. R. Bailey. 1979. A cochlear implant round window electrode array. Journal of Laryngology and Otology 93: 107–109.

Clark, G. M. and B. C. Pyman. 1995. Surgical considerations for the placement of the new

Cochlear Pty Limited micro-multiple-channel cochlear implant for research studies. Annals of Otology, Rhinology and Laryngology 104(suppl 166): 408–409.

Clark, G. M., B. C. Pyman and Q. R. Bailey. 1979. The surgery for multiple-electrode cochlear implantations. Journal of Laryngology and Otology 93: 215–223.

Clark, G. M., B. C. Pyman, R. L. Webb, Q. E. Bailey and R. K. Shepherd. 1984. Surgery for an improved multiple-channel cochlear implant. Annals of Otology, Rhinology and Laryngology 93(3 Pt 1): 204–207.

Clark, G. M., B. C. Pyman, R. L. Webb, B. K.-H. G. Franz, T. J. Redhead and R. K. Shepherd. 1987. Surgery for safe the insertion and reinsertion of the banded electrode array. Annals of Otology, Rhinology and Laryngology 96(suppl 128): 10–12.

Clark, G. M., R. K. Shepherd, B. K.-H. G. Franz, et al. 1988. The histopathology of the human temporal bone and auditory central nervous system following cochlear implantation in a patient. Correlation with psychophysics and speech perception results. Acta Oto-Laryngologica (suppl 448): 1–65.

Clark, G. M., R. K. Shepherd, B. K.-H. G. Franz and D. Bloom. 1984. Intracochlear electrode implantation. Round window membrane sealing procedures and permeability studies. Acta Oto-Laryngologica(suppl 410): 5–15.

Clark, G. M., R. K. Shepherd, J. F. Patrick, R. C. Black and Y. C. Tong. 1983. Design and fabrication of the banded electrode array. Annals of the New York Academy of Sciences 405: 191–201.

Clark, G. M. and Y. C. Tong. 1981. Multiple-electrode cochlear implant for profound or total hearing loss: a review. Medical Journal of Australia 1: 428–429.

Clark, G. M. and Y. C. Tong. 1985. The engineering of future cochlear implants. In: Gray, R. F., ed. Cochlear implants. London, Croom Helm: 211–228.

Clark, G. M., Y. C. Tong, Q. R. Bailey, et al. 1978. A multiple-electrode cochlear implant. Journal of the Oto-Laryngological Society of Australia 4: 208–212.

Clark, G. M., Y. C. Tong, R. C. Black, I. C. Forster, J. F. Patrick and D. J. Dewhurst. 1977. A multiple electrode cochlear implant. Journal of Laryngology and Otology 91: 935–945.

Clark, G. M., Y. C. Tong and L. F. Martin. 1981a. A multiple-channel cochlear implant. An evaluation using open-set CID sentences. Laryngoscope 91: 628–634.

Clark, G. M., Y. C. Tong, L. F. A. Martin, et al. 1981b. A multiple-channel cochlear implant: an evaluation using nonsense syllables. Annals of Otology, Rhinology and Laryngology 90: 227–230.

Clark, G. M., Y. C. Tong, L. F. Martin and P. A. Busby. 1981c. A multiple-channel cochlear implant. An evaluation using an open-set word test. Acta Oto-Laryngologica 91: 173–175.

Clark, G. M., Y. C. Tong, J. F. Patrick, P. M. Seligman, P. A. Crosby and J. A. Kuzma. 1984. A multi-channel cochlear prosthesis for profound-to-total hearing loss. Journal of Electrical and Electronics Engineering, Australia 4: 111–117.

Clifford, A. R. and W. P. R. Gibson. 1987. Anatomy of the round window with respect to cochlear implant surgery. Annals of Otology, Rhinology and Laryngology 96(suppl 128): 17–19.

Cohen, L. T., E. Saunders and G. M. Clark. 2001. Psychophysics of a prototype perimodiolar cochlear implant electrode array. Hearing Research 155: 63–81.

Cranswick, N. E., B. K.-H. G. Franz, G. M. Clark and R. K. Shepherd. 1987. Middle ear infection postimplantation: response of the round window membrane to streptococcus pyogenes. Annals of Otology, Rhinology and Laryngology 96(suppl 128): 53–54.

Dahm, M., R. K. Shepherd and G. M. Clark. 1993. The postnatal growth of the temporal bone and its implications for cochlear implantation in children. Acta Oto-Laryngologica (suppl 505): 1–39.

Dahm, M., J. Xu, M. Tykocinski, R. K. Shepherd and G. M. Clark. 2000. Intracochlear damage following insertion of the Nucleus 22 standard electrode array: a post mortem study of 14 implant patients. CI 2000 The 6th International Cochlear Implant Conference, Miami Beach, Florida: 149.

Dillier, N. 1998. Intracochlear recordings of electrically evoked compound action potentials. First International Symposium and Workshop on Objective Measures in Cochlear Implantation, Nottingham: 54.

Dillier, N., W. K. Lai, B. Almquist, et al. 2002. Measurement of the electrically evoked compound action potential (ECAP) via a neural response telemetry (NRT) system. Annals of Otology Rhinology and Laryngology 111 (5 pt 1): 407–414.

Dillier, N., W. K. Lai, D. Cafarelli-Dees and E. von Wallenberg. 2000. Post-operative neural response telemetry findings in adults submitted for publication. Neural esponse telemetry: results from a European field trial, 12th AAA Convention, Chicago: 137.

Dillier, N., T. Spillman and J. Guntensperger. 1983. Computerized testing of signal encoding strategies with round window implants. In: Parkins, C. and J. Anderson, eds. Cochlear prostheses: an international symposium. New York, New York Academy of Sciences: 360–369.

Djourno, A. and C. Eyriès. 1957. Prosthese auditive par excitation electrique a distance du nerf sensoriel a l'aide d'un bobinage includ a demeure. Presse Medicale 35: 14–17.

Dobelle, W. H., M. G. Mladejovsky, J. R. Evans, T. S. Roberts and J. P. Girvin. 1976. "Braille" reading by a blind volunteer by visual cortex stimulation. Nature 259: 111–112.

Dobelle, W. H., J. N. Fordemwald, J. W. Hanson, et al. 1974. Data processing LSI will help bring sight to the blind. Electronics 47: 81–86.

Dorman, M. F., K. Dankowski and G. McCandless. 1989. Consonant recognition as a function of the number of channels of stimulation by patients who use the Symbion cochlear implant. Ear and Hearing 10: 288–291.

Dormer, K. J., G. Richard, P. E. Hough and J. V. D. Hough. 1980. The cochlear implant (auditory prosthesis) utilizing rare earth magnets. American Journal of Otology 2(1): 22–27.

Dormer, K. J., G. L. Richard and J. V. D. Hough. 1981. The use of rare-earth magnet couplers in cochlear implants. Laryngoscope 91(11): 1812–1820.

Douek, E., A. J. Fourcin, B. C. J. Moore and G. P. Clarke. 1977. A new approach to the cochlear implant. Proceedings of the Royal Society of Medicine 70: 379–383.

Dowell, R. C., P. M. Seligman, P. J. Blamey and G. M. Clark. 1987. Speech perception using a two-formant 22-electrode cochlear prosthesis in quiet and in noise. Acta Oto-Laryngologica 104(5–6): 439–446.

Dymond, A. M., L. E. Kaechele, J. M. Jurist and P. H. Crandall. 1970. Brain tissue reaction to some chronically implanted metals. Journal of Neurosurgery 33: 574–580.

Eddington, D. K. 1980. Speech discrimination in deaf subjects with cochlear implants. Journal of the Acoustical Society of America 68: 885–891.

Eddington, D. K. 1983. Speech recognition in deaf subjects with multichannel intracochlear electrodes. Annals of the New York Academy of Science 405: 241–258.

Eddington, D. K., W. H. Dobelle and D. E. Brackmann. 1978. Auditory prostheses research with multiple channel intracochlear stimulation in man. Annals of Otology 87: 1–39.

Fant, G. 1959. Acoustic analysis and synthesis of speech with applications to Swedish. Ericsson Technics 15: 3–108.

Fishman, K. E., R. V. Shannon and W. H. Slattery. 1997. Speech recognition as a function of electrodes used in the SPEAK cochlear implant processor. Journal of Speech, Language and Hearing Research 40: 1201–1215.

Forster, I. C. 1978. The bioengineering development of a multi-channel, implantable hearing prosthesis for the profoundly deaf. PhD thesis. University of Melbourne Department of Electrical Engineering.

Forster, I. C. 1981. Theoretical design and implementation of a transcutaneous, multichannel stimulator for neural prosthesis application. Journal of Biomedical Engineering 3: 107–120.

Fourcin, A. J., S. M. Rosen, B. C. Moore, et al. 1979. External electrical stimulation of the cochlea: clinical, psychophysical, speech-perceptual and histological findings. British Journal of Audiology 13(3): 85–107.

Franz, B. K.-H. G. and G. M. Clark. 1987. Refined surgical technique for insertion of banded electrode array. Annals of Otology, Rhinology and Laryngology 96(Suppl 128): 15–16.

Franz, B. K.-H. G., G. M. Clark and D. Bloom. 1984. Permeability of the implanted round window membrane in the cat-an investigation using horseradish peroxidase. Acta Oto-Laryngologica: 17–23.

Fry, D. B. 1979. The physics of speech. Cambridge textbooks in linguistics. Cambridge, Cambridge University Press.

Fugain, C., B. Meyer, F. Chabolle and C. H. Chouard. 1984. Clinical results of the French multichannel cochlear implant. Acta Oto-Laryngologica (suppl 411): 237–246.

Galey, F. R. 1984. Initial observations of a human temporal bone with a multi-channel implant. Acta Oto-Laryngologica (suppl 411): 38–44.

Garcia, N., A. Damask and S. Schwarz. 1998. Physics for computer science students with emphasis on atomic and semiconductor physics. 2nd ed. New York, Wiley.

Gheewala, T. R., R. D. Melen and R. L. White. 1975. A CMOS implantable multielectrode auditory stimulation for the deaf. IEEE Journal Solid State Circuits 10: 472–479.

Goycoolea, M. V., D. C. Muchow, C. M. Schirber, H. G. Goycoolea and K. Schellhas. 1990. Anatomical perspective, approach, and experience with multichannel intracochlear implantation. Laryngoscope 100(2 pt 2 suppl 50): 1–18.

Gray, R. F., D. M. Baguley, M. L. Harries, I. Court and C. Lynch. 1993. Profound deafness treated by the Ineraid multichannel intracochlear implant. Journal of Laryngology and Otology 107(8): 673–680.

Greenwood, D. D. 1961. Critical bandwidth and the frequency coordinates of the basilar membrane. Journal of the Acoustical Society of America 33: 1344–1356.

Gstoettner, W., W. D. Baumgartner, J. Hamzavi and P. Franz. 1997. Surgical experience with the Combi 40 cochlear implant. Advances in Oto-Rhino-Laryngology 52: 143–146.

Hallworth, R. 1976. An implantable multi-micro-electrode array fabricated by thin film methods. M. Eng. Sc. thesis. University of Melbourne.

Hansen, C. 1981. Electrode for implantation into cochlea. US patent 4,261,372.

Hatsushika, S., R. K. Shepherd, Y. C. Tong and G. M. Clark. 1990. Dimensions of the scala tympani in the human and cat with reference to cochlear implants. Annals of Otology, Rhinology and Laryngology 99: 871–876.

Heinemann, U. and H. D. Lux. 1977. Ceiling of stimulus induced rises in extracellular potassium concentration in the cerebral cortex of the cat. Brain Research 120: 231–249.

Hochmair, E. S. and I. J. Hochmair-Desoyer. 1983. Percepts elicited by different speech coding strategies. Annals of the New York Academy of Sciences 405: 268–279.

Holmes, A. E., F. J. Kemler and G. E. Merwin. 1987. The effects of varying the number of cochlear implant electrodes on speech perception. American Journal of Otology 8: 240–246.

Hong, R. S., J. T. Rubinstein, D. Wehner and D. Horn. Submitted. Dynamic range enhancement for cochlear implants. Otology and Neuro-Otology.

House, W. F. 1976. Cochlear implants. Annals of Otology Rhinology and Laryngology 85: 1–93.

House, W. F. 1984. Artificial hearing: an implanted auditory prosthesis for treatment of total binaural deafness. Transactions of the American Society for Artificial Internal Organs 30: 11–14.

House, W. F. and B. J. Edgerton. 1982. A multiple-electrode cochlear implant. Annals of Otology, Rhinology and Laryngology–supplement 91(2 pt 3): 104–116.

House, W. F., W. M. Luxford and B. Courtney. 1985. Otitis media in children following the cochlear implant. Ear and Hearing 6(suppl 3): 24–26.

House, W. F. and J. Urban. 1973. Long term results of electrode implantation and electronic stimulation of the cochlea in man. Annals of Otology, Rhinology and Laryngology 82: 504–517.

Huang, C. Q., R. K. Shepherd, P. M. Seligman and G. M. Clark. 1996. Reduction in excitability of the auditory nerve in guinea pigs following acute high rate electrical stimulation. Proceedings of the Australian Neuroscience Society 7: 227.

Huang, C. Q., R. K. Shepherd, P. M. Seligman and G. M. Clark. 1998a. Reduction in excitability of the auditory nerve following electrical stimulation at high stimulus rates. III. Capacitive versus non-capacitive coupling of the stimulating electrodes. Hearing Research 116: 55–64.

Huang, C. Q., R. K. Shepherd, P. M. Seligman, B. Tabor and G. M. Clark. 1998b. Changes in excitability of the auditory nerve following electrical stimulation using large surface area electrodes. Proceedings of the Australian Neuroscience Society 9.

Igarashi, M., R. C. Mahon and S. Konishi. 1968. Comparative measurements of cochlear apparatus. Journal of Speech and Hearing Research 11: 229–235.

Igarashi, M., M. Takahashi and B. R. Alford. 1976. Cross-sectional area of scala tympani in human and cat. Archives of Otolaryngology 102: 428–429.

Johnson, C. C. and A. W. Guy. 1972. Nonionizing electromagnetic wave effects in biological materials and systems. Proceedings of the IEEE 60: 692–718.

Johnsson, L.-G., W. F. House and F. H. Linthicum. 1982. Otopathological findings in a patient with bilateral cochlear implants. Annals of Otology Rhinology and Laryngology–supplement 91(2 pt 3): 74–89.

Kawano, A., H. L. Seldon and G. M. Clark. 1996. Computer-aided three-dimensional reconstruction in human cochlear maps: measurement of the lengths of organ of corti, outer wall, inner wall, and Rosenthal's canal. Annals of Otology, Rhinology and Laryngology 105: 701–709.

Kawano, A., H. L. Seldon, G. M. Clark and R. Ramsden. 1998. Intracochlear factors contributing to psychophysical percepts following cochlear implantation. Acta Oto-Laryngologica 118: 313–326.

Kawano, A. S., H. L. Seldon, G. M. Clark, R. Madsen and C. Raine. 1995. Intracochlear factors contributing to psychophysical percepts following cochlear implantation: a case study. Annals of Otology, Rhinology and Laryngology 104: 54–57.

Kennedy, D. W. 1987. Multichannel intracochlear electrodes: mechanism of insertion trauma. Laryngoscope 9: 42–49.

Laird, R. K. 1979. The Bioengineering development of a sound encoder for an implantable hearing prosthesis for the profoundly deaf. Master of Engineering Science Thesis. University of Melbourne.

Lawson, D. T., B. S. Wilson, M. Zerbi and C. C. Finley. 1996. Speech processors for auditory prostheses. Third quarterly progress report. NIH contract No. 1-DC-5-2103.

Leake-Jones, P. A. and S. J. Rebscher. 1983. Cochlear pathology with chronically implanted scala tympani electrodes. Annals of the New York Academy of Sciences 405: 203–223.

Leake-Jones, P. A., S. J. Rebscher and D. W. Aird. 1985. Histopathology of cochlear implants: safety considerations. In: Schindler, R. A. and M. M. Merzenich, eds. Cochlear implants. New York, Raven Press.

Leake-Jones, P. A., S. J. Rebscher and C. L. Byers. 1981a. Development of multichannel electrodes for an auditory prosthesis. Progress report NIH Contract No. 1-NS-7-2367. 1–23.

Leake-Jones, P. A., S. M. Walsh and M. M. Merzenich. 1981b. Cochlear pathology following chronic intracochlear electrical stimulation. Annals of Otology, Rhinology and Laryngology–supplement 90(2 pt 3): 6–8.

Liboff, A. R., R. A. Rinaldi, L. S. Lavine and M. H. Shamos. 1975. On electrical conduction in living bone. Clinical Orthopaedics and Related Research 106: 330–335.

Luxford, W. M. and W. F. House. 1985. Cochlear implants in children: medical and surgical considerations. Ear and Hearing 6 (3 suppl): 20S–23S.

Luxford, W. M. and W. F. House. 1987. House 3M cochlear implant: surgical considerations. Annals of Otology, Rhinology and Laryngology 96(suppl 128): 12–14.

Marangos, N., J. Maurer and W. Mann. 2001. Device failure and reimplantation. 6th European Symposium on Paediatric Cochlear Implantation. Las Palmas de Gran Canaria, Canary Islands, Spain, February 24–27.

Marsh, M. A., J. Xu, P. J. Blamey, et al. 1993. Radiologic evaluation of multichannel intracochlear implant insertion depth [published erratum appears in Am J Otol 1993 Nov;14(6):627]. American Journal of Otology 14(4): 386–391.

Marzella, P. L. and G. M. Clark. 1999. Growth factors, auditory neurones and cochlear implants: a review. Acta Oto-Laryngologica 119(4): 407–412.

Matsushima, J., R. K. Shepherd, H. L. Seldon, S. Xu and G. M. Clark. 1990. Electrical stimulation of the auditory nerve in deaf kittens: effects on cochlear nuclei. Proceedings of the Australian Physiological and Pharmacological Society 21: 43P.

Matsushima, J., R. K. Shepherd, H. L. Seldon, S. Xu and G. M. Clark. 1991. Electrical stimulation of the auditory nerve in deaf kittens: effects on cochlear nucleus morphology. Hearing Research 56: 133–142.

McCreery, D. B. and W. F. Agnew. 1983. Changes in extracellular potassium and calcium concentration and neural activity during prolonged electrical stimulation of the cat cerebral cortex at defined charge densities. Experimental Neurology 79: 371–396.

McCreery, D. B., W. F. Agnew, T. G. Yuen and L. A. Bullara. 1988. Comparison of neural damage induced by electrical stimulation with faradaic and capacitor electrodes. Annals of Biomedical Engineering 16: 463–481.

McCreery, D. B., W. F. Agnew, T. G. H. Yuen and L. Bullara. 1990. Charge density and charge per phase as cofactors in neural injury by electrical stimulation. IEEE Transactions on Biomedical Engineering 537: 996–1001.

McCreery, D. B., T. G. Yuen, W. F. Agnew and L. A. Bullara. 1994. Stimulus parameters affecting tissue injury during microstimulation in the cochlear nucleus of the cat. Hearing Research 77(1–2): 105–115.

McDermott, H. J. 1989. An advanced multiple-channel cochlear implant. IEEE Transactions on Biomedical Engineering 36: 789–797.

McFadden, J. T. 1969. Metallurgical principles in neurosurgery. Journal of Neurosurgery 31: 373–385.

McKay, C. M., H. J. McDermott, A. Vandali and G. M. Clark. 1992. A comparison of speech perception of cochlear implantees using the Spectral Maxima Sound Processor (SMSP) and the MSP (Multipeak) processor. Acta Oto-Laryngologica 112: 752–761.

Merzenich, M. M. 1975. Studies on electrical stimulation of the auditory nerve in animals and man: cochlear implants. In: D. B. Tower, ed. The Nervous System Vol 3, Human Communication and its disorders. New York, Raven Press. 537–548.

Merzenich, M., C. Byers and M. White. 1984. Scala tympani electrode arrays. Fifth quarterly progress report. NIH contract No. 1-NS9-2353: 1–11.

Merzenich, M. M. and M. D. Reid. 1974. Representation of the cochlea within the inferior colliculus. Brain Research 77: 397–415.

Merzenich, M. M. and M. White. 1980. Coding considerations in design of cochlear prostheses. Annals of Otology 89: 84–87.

Merzenich, M. M., M. W. White and R. V. Shannon. 1981. Development of multichannel electrodes for an auditory prosthesis. Proress report NIH Contract No. 1-NS-7-2367. 1–7.

Michelson, R. P. 1971. Electrical stimulation of the human cochlea—a preliminary report. Archives of Otolaryngology 93: 317–323.

Michelson, R. P. and R. A. Schindler. 1981. Multichannel cochlear implant preliminary results in man. Laryngoscope 91(1): 38–42.

Nelson, D. A., D. J. Van Tasell, A. C. Schroder, S. Soli and S. Levine. 1995. Electrode ranking of "place pitch" and speech recognition in electrical hearing. Journal of the Acoustical Society of America 98: 1987–1999.

Ni, D., R. K. Shepherd, H. L. Seldon, S. Xu and G. M. Clark. 1992. Cochlear pathology following chronic electrical stimulation of the auditory nerve. I: Normal hearing kittens. Hearing Research 62: 63–81.

Nicholson, C., G. Ten Bruggencate, G. Stochle and R. Steinberg. 1978. Calcium and potassium changes in the extracellular microenvironment of cat cerebellar cortex. Journal of Neurophysiology 41: 1026–1038.

Paolini, A. G. and G. M. Clark. 1997. The effect of pulsatile intracochlear electrical stimulation on intracellularly recorded cochlear nucleus neurons. In: Clark, G. M., ed. Cochlear implants. XVI World Congress of Otorhinolaryngology Head and Neck Surgery. Bologna, Monduzzi Editore: 119–124.

Patrick, J. F. and J. C. Macfarlane. 1987. Characterisation of mechanical properties of single electrodes and multielectrodes. Annals of Otology, Rhinology and Laryngology 96(suppl 128): 46–48.

Patrick, J. F., P. M. Seligman and G. M. Clark. 1997. Engineering. In: Clark, G. M., R. S. C. Cowan and R. C. Dowell, eds. Cochlear implantation for infants and children. Advances. San Diego, Singular Publishing Group: 125–145.

Peterson, G. E. and H. L. Barney. 1952. Control methods used in a study of the vowels. Journal of the Acoustical Society of America 24: 175–184.

Pialoux, P., C. H. Chouard and P. MacLeod. 1976. Physiological and clinical aspects of the rehabilitation of total deafness by implantation of multiple intracochlear electrodes. Acta Oto-Laryngologica 81: 436–441.

Pialoux, P., C. H. Chouard and B. Meyer. 1979. Indications and results of the multichannel cochlear implant. Acta Oto-Laryngologica 87: 185–189.

Plomp, R. 1976. Aspects of tone sensation: a psychophysical study. London, Academic Press.

Pudenz, R. H. 1942. The use of tantalum clips the hemostasis in neurosurgery. Surgery 12: 791–797.

Purser, S., R. K. Shepherd and G. M. Clark. 1991. Evaluation of a sealing device for the intracochlear electrode entry point. Journal of the Oto-Laryngological Society of Australia 6: 472–480.

Rasmussen, G. L. 1960. Efferent fibres of the cochlear nerve and cochlear nucleus. In: Rasmussen, G. L. and W. F. Windle, eds. Neural mechanisms of the auditory and vestibular systems. Charles C. Thomas, Springfield, Illinois: 105–115.

Rebscher, S. J., C. L. Byers, R. F. Gray and M. M. Merzenich. 1981. Development of multichannel electrodes for an auditory prosthesis. Quarterly Progress Report March 1. NIH Contract NS-7-2367. Coleman Memorial Laboratory. University of California, San Francisco.

Reddy, G. N. and S. Saha. 1984. Electrical and dielectric properties of wet bone as a function of frequency. IEEE Transactions on Biomedical Engineering 31(3): 296–303.

Reliability update 2002. Nucleus Report Sep/Oct. Sydney, Cochlear Ltd.: 6.

Richter, B., A. Aschendorff, P. Lohnstein, H. Husstedt, H. Nagursky and R. Laszig. 2002. Clarion 1.2 standard electrode array with partial space filling positioner: radiological and histological evaluation in human temporal bones. Journal of Laryngology and Otology 116: 507–513.

Roland, J. T., A. J. Fishman, G. Alexiades and N. L. Cohen. 2000. Electrode to modiolus proximity: a fluoroscopic and histologic analysis. American Journal of Otology 21: 218–225.

Roland, J. T., T. M. Magardino, J. T. Go and D. E. Hillman. 1995. Effects of glycerine, hyaluronic acid and hydroxypropyl methylcellulose on the spiral ganglion of the guinea pig cochlea. Annals of Otology, Rhinology and Laryngology 104: 64–68.

Rubinstein, J. T., B. S. Wilson, C. C. Finley and P. J. Abbas. 1999. Pseudospontaneous activity: stochastic independence of auditory nerve fibers with electrical stimulation. Hearing Research 127: 108–118.

Saunders, E., L. T. Cohen, W. Aschendorff, et al. 2000. Psychophysical measures and NRT thresholds as a function of electrode position in the cochlea: results of a multi-centre study using the Contour electrode array. 5th European Symposium on Paediatric Cochlear Implantation. Antwerp, Belgium, June 4–7.

Saunders, E., L. Cohen, A. Aschendorff, et al. 2002. Threshold, comfortable level and impedance changes as a function of electrode-modiolar distance. Ear and Hearing 23(1 Suppl): 28S–40S.

Schindler, R. A. and B. Bjorkroth. 1979. Traumatic intracochlear electrode implantation. Laryngoscope 89: 752–758.

Schindler, R. A., D. K. Kessler and H. S. Haggerty. 1993. Clarion cochlear implant: phase I investigational results [published erratum appears in Am J Otol 1993 Nov;14(6):627] [see comments]. American Journal of Otology 14(3): 263–72.

Schindler, R. A., D. K. Kessler, S. J. Rebscher, R. K. Jackler and M. M. Merzenich. 1987. Surgical considerations and hearing results with the UCSF/Storz cochlear implant. Laryngoscope 97(1): 50–56.

Schindler, R. A. and M. M. Merzenich. 1974. Chronic intracochlear electrode implantation:

cochlear pathology and acoustic nerve survival. Annals of Otolaryngology 83: 202–215.

Schindler, R. A., M. M. Merzenich, M. W. White and B. Bjorkroth. 1977. Multi electrode intracochlear implants—nerve survival and stimulation patterns. Archives of Otolaryngology 103: 691–699.

Schuder, J. C., J. H. Stephenson and J. F. Townsend. 1961. High level electromagnetic energy transfer through a closed chest wall. IRE International Conference Record 9: 119–126.

Schuknecht, H. 1953. Techniques for study of cochlear function and pathology in experimental animals. Acta Oto-Laryngologica 58: 377.

Schulman, J.H., J.C. Gord, P. Strojnik, D.L. Whitmoyer, J.H. Wolfe. 1996. Voltage/current control system for a human tissue stimulator. US Patent 5,522,865.

Seligman, P. M. 1987. Speech-processing strategies and their implementation. Annals of Otology, Rhinology and Laryngology 96(suppl 128): 71–74.

Seligman, P. M. and H. J. McDermott. 1995. Architecture of the SPECTRA 22 speech processor. Annals of Otology, Rhinology and Laryngology 104(suppl 166): 139–141.

Shannon, R. V. 1983. Multichannel electrical stimulation of the auditory nerve in man. II. Channel interaction. Hearing Research 12: 1–16.

Shannon, R. V. 1992. Temporal modulation transfer functions in patients with cochlear implants. Journal of the Acoustical Society of America 91: 2156–2164.

Shepherd, R. K., G. M. Clark and R. C. Black. 1983. Chronic electrical stimulation of the auditory nerve in cats. Physiological and histopathological results. Acta Oto-Laryngologica (suppl 399): 19–31.

Shepherd, R. K., G. M. Clark, B. C. Pyman and R. L. Webb. 1985a. Banded intracochlear electrode insertion trauma in human temporal bones. Annals of Otology, Rhinology and Laryngology 94: 55–59.

Shepherd, R. K., G. M. Clark, B. C. Pyman, R. L. Webb, M. T. Murray and M. E. Houghton. 1985b. Histopathology following electrode insertion and chronic electrical stimulation. In: Schindler, R. A. and M. M. Merzenich, eds. Cochlear implants. New York, Raven Press: 65–81.

Shepherd, R. K., S. Hatsushika and G. M. Clark. 1990. The effect of position of the scala tympani electrode array on auditory nerve excitation. Proceedings of the Australian Physiological and Pharmacological Society 21: 42P.

Shepherd, R. K., S. Hatsushika and G. M. Clark. 1993. Electrical stimulation of the auditory nerve: the effect of electrode position on neural excitation. Hearing Research 66: 108–120.

Shepherd, R. K., D. Ni, H. L. Seldon, S. A. Xu and G. M. Clark. 1991. Chronic electrical stimulation of the auditory nerve in normal hearing kittens. Proceedings of the Australian Physiological and Pharmacological Society 22: 121P.

Simmons, F. B. 1966. Electrical stimulation of the auditory nerve in man. Archives of Otolaryngology 84: 2–54.

Simmons, F. B. 1967. Permanent intracochlear electrodes in cats. Tissue tolerance and cochlear microphonics. Laryngoscope 77: 171–186.

Simmons, F. B. and T. J. Glattke. 1970. Comparison of electrical and acoustical stimulation of the cat ear. Annals of Otology Rhinology and Laryngology 81: 731–738.

Skinner, M. W., G. M. Clark, L. A. Whitford, et al. 1994. Evaluation of a new spectral peak coding strategy for the Nucleus 22 channels cochlear implant system. American Journal of Otology 15: 15–27.

Stark, H. and F. B. Tuteur. 1979. Modern electrical communications. Theory and systems. London, Prentice Hall International.

Stevens, S. S. 1975. Psychophysics. New York, John Wiley and Sons.

Stockle, H. and G. Ten Bruggencate. 1980. Fluctuation of extracellular potassium and calcium in the cerebellar cortex related to climbing fibre activity. Neuroscience 5: 893–901.

Sutton, D., J. M. Miller and B. E. Pfingst. 1980. Comparison of cochlear histopathology following two implant designs for use in scala tympani. Annals of Otology Rhinology and Laryngology 89: 11–14.

Tong, Y. C., R. C. Black, G. M. Clark, et al. 1979. A preliminary report on a multiple-channel cochlear implant operation. Journal of Laryngology and Otology 93: 679–695.

Tong, Y. C., P. J. Blamey, R. C. Dowell and G. M. Clark. 1983a. Psychophysical studies evaluating the feasibility of a speech processing strategy for a multiple-channel cochlear implant. Journal of the Acoustical Society of America 74: 73–80.

Tong, Y. C., P. A. Busby and G. M. Clark. 1988. Perceptual studies on cochlear implant patients with early onset of profound hearing impairment prior to normal development of auditory, speech, and language skills. Journal of the Acoustical Society of America 84: 951–962.

Tong, Y. C. and G. M. Clark. 1986. Loudness summation, masking, and temporal interaction for sensations by electric stimulation of two sites in the human cochlea. Journal of the Acoustical Society of America 79: 1958–1966.

Tong, Y. C., G. M. Clark, P. M. Seligman and J. F. Patrick. 1980. Speech processing for a multiple-electrode cochlear implant hearing prosthesis. Journal of the Acoustical Society of America 68: 1897–1899.

Tong, Y. C., R. C. Dowell, P. J. Blamey and G. M. Clark. 1983b. Two-component hearing sensations produced by two-electrode stimulation in the cochlea of a deaf patient. Science 219: 993–994.

Tykocinski, M., R. K. Shepherd and G. M. Clark. 1997. Reduction in excitability of the auditory nerve following electrical stimulation at high stimulus rates. II. Comparison of fixed amplitude with amplitude modulated stimuli. Hearing Research 112: 147–157.

Urbanics, R., E. Leninger-Follert and W. Lubbers. 1978. Time course of changes of extracellular H+ and K+ activities during and after direct stimulation of the brain cortex. Pflugers Archiv-European Journal of Physiology 378: 47–53.

Vandali, A. E., L. A. Whitford, K. L. Plant and G. M. Clark. 2000. Speech perception as a function of electrical stimulation rate: using the Nucleus 24 cochlear implant system. Ear and Hearing 21: 608–624.

Vaudaux, P. E., D. P. Lew and F. A. Waldvogel. 1994. Host factors predisposing to and influencing therapy of forming body infections. In: Bisno, A. L. and F. A. Waldvogel, eds. Infections associated with indwelling medical devices. 2nd ed. Washington, DC, ASM Press.

Viemeister, N. F. 1979. Temporal modulation transfer functions based on modulation thresholds. Journal of the Acoustical Society of America 66: 1364–1380.

von Békésy, G. 1960. Experiments in hearing. New York, McGraw Hill.

von Wallenberg, E., R. Leigh, B. O'Connor and C. Gillman 2002. Reliability of Nucleus cochlear implants. 6th European Symposium on Paediatric Cochlear Implantation. Las Palmas de Gran Canaria, Spain, 24–27 February 2002.

Wakerly, J. F. 1994. Digital design: principles and practice. 2nd ed. Englewood Cliffs, NJ, Prentice Hall.

White, R. L. 1974. Integrated circuits and multiple electrode arrays. In: Merzenich, M., R.

Schindler and F. Sooy, eds. Proceedings of the First International Conference on Electrical Stimulation of the Acoustic Nerve as a Treatment for Profound Sensorineural Deafness in Man. San Francisco, Velo-Bind: 199–209.

Wilson, B. S. 1997. The future of cochlear implants. British Journal of Audiology 31: 205–225.

Wilson, B. S. 2000. Strategies for representing speech information with cochlear implants. In: Niparko, J. K., ed. Cochlear implants: principles and practice. Philadelphia, Lippincott Williams and Wilkins.

Wilson, B. S., C. C. Finley and D. T. Lawson. 1988. Speech processors for cochlear prostheses. Proceedings IEEE 76: 1143–1153.

Wilson, B. S., D. T. Lawson, M. Zerbi and C. C. Finley. 1992. Speech processors for auditory prostheses. Twelfth quarterly progress report, April. 1992. NIH contract No. 1-DC-9-2401. Research Triangle Institute.

Wilson, B. S., D. T. Lawson, M. Zerbi and C. C. Finley. 1993. Speech processors for auditory prostheses. Fifth quarterly progress report, Oct 1993. NIH contract No. 1-DC-2-2401. Research Triangle Institute.

Xu, J., R. K. Shepherd, R. E. Millard and G. M. Clark. 1997. Chronic electrical stimulation of the auditory nerve at high stimulus rates: a physiological and histopathological study. Hearing Research 105: 1–29.

Xu, J., R. K. Shepherd, S. A. Xu, H. L. Seldon and G. M. Clark. 1993. Pediatric cochlear implantation: radiological observations of skull growth. Archives of Otolaryngology–Head and Neck Surgery 119: 525–534.

Zienkiewicz, O. C. 1977. The finite element method. 3rd ed. United Kingdom, McGraw-Hill.

9
Preoperative Selection

Aims

There are a number of important requirements in selecting candidates for a cochlear implant operation. The clinician needs to be satisfied that they (1) have realistic expectations of the possible benefits and risks; (2) would have the appropriate form of communication for their needs, for example, the primary form may be the use of a telephone, or a signing parent may want his/her child to have an implant, in which case the child's communication needs could be complex; (3) have a reasonable chance of achieving their needs; and (4) have adequate help from the family or social services to undertake the (re)habilitation. In addition, if health care budgets are restricted, the priority of an individual will have to be assessed.

Adults

Selection requirements for cochlear implantation in adults were established for the first patients operated on by the University of Melbourne surgical team (Clark, O'Loughlin et al 1977). These requirements have been refined and modified since larger numbers of patients have had implants and speech-processing strategies have been improved (Brown et al 1985; Dowell et al 1993). The mean results for electrical stimulation alone with the Nucleus strategies developed at the University of Melbourne/Bionic Ear Institute [fundamental and second formant frequencies (F0/F2); the addition of first formant (F0/F1/F2); the addition of high spectral frequencies (Multipeak); the spectral maxima sound processor (SPEAK); and the advanced combination encoder (ACE)] are shown in Figure 1.1 in Chapter 1. Another strategy, the continuous interleaved sampler (CIS), has given results comparable to SPEAK, as illustrated in Figure 9.1.

The preoperative evaluation of adults for cochlear implantation should aim to select the patients who have a high probability of achieving better communication skills and opportunities than with their hearing aids. As studies have shown that postlinguistically deafened adults as well as early deafened children operated on at a young age can achieve a mean score of 80% for Central Institute of the Deaf

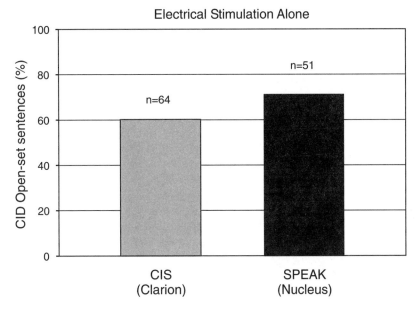

FIGURE 9.1. The mean open-set Central Institute for the Deaf (CID) sentence score of 71% for the SPEAK (University of Melbourne/Cochlear Limited) strategy on 51 patients (data presented to the Food and Drug Administration, January 1996) and 60% for the continuous interleaved sampler (CIS) (Advanced Bionics) strategy on 64 patients (Reprinted with permission from Kessler et al 1995. Distribution of speech recognition abilities in severe to profound post lingual deafness. *Annals of Otology, Rhinology and Laryngology* **104**: 283–855.).

(CID) word-in-sentence comprehension for electrical stimulation alone (Clark 1999; Dowell et al 1998), preoperative speech perception and communication abilities need to be assessed carefully to determine whether they are likely to obtain more benefit with a multiple-channel cochlear prosthesis or a hearing aid. In fact, it can be desirable to implant people who use an aid for residual hearing on the opposite side (bimodal stimulation).

The preoperative evaluation is undertaken jointly by the otologist and audiologist, and may require consultation with other specialists such as psychologists, psychiatrists, neurologists, social worker, and general physicians. The possible benefits of an implant should be discussed with the prospective patient, and the decision to operate made only after careful consideration by the team of all the findings. The patients should sign a consent form that states all the risks and benefits.

Children

Many of the considerations for adults apply to the preoperative evaluation and selection of children and teenagers, although other issues become especially important. These are age, age of onset of deafness (which determines whether the child is prelinguistically or postlinguistically deaf), language level, nature of the educational program, parental and other support, accuracy of hearing assessment, residual hearing, effect of preoperative training, benefit from hearing aid, and frequency and severity of middle ear infections.

One key question is whether children can be trained to use residual hearing with a hearing aid at a level that is better than that expected from a cochlear implant. This requires an accurate assessment of the hearing threshold, and may need speech and language perception and production assessments before and after a preoperative training period with an aid. Some factors such as the extent of family support, while certainly important in the ultimate success of young implant recipients, are difficult to quantify, and this makes the selection process a complex matter. It will require a strong two-way interaction with families and the child's (re)habilitation support team. The team should include not only an otologist and an audiologist, but also a speech pathologist and an educator of the hearing impaired. There also needs to be close collaboration with a child psychologist and a pediatric physician.

Clinical Protocol

All decisions by the University of Melbourne's Cochlear Implant Clinic at the Royal Victorian Eye and Ear Hospital are made by a multidisciplinary team consisting of otologists, audiologists, speech pathologists, an educational psychologist, an educator of the hearing impaired, and members of the cochlear implant research team. The first stage of the preoperative assessment and selection process normally consists of two visits and involves mainly an audiological and otological consultation. This includes a history, basic pure-tone audiometry, impedance testing, and an evaluation of speech perception and production where appropriate. Clinical information from other agencies also should be sought. At this point, if it is felt that cochlear implantation is still appropriate, the patient and family are counseled and provided with detailed information concerning the complete program. Many potential patients referred to the clinic, however, are not necessarily candidates for a cochlear implant but may be considered for a hearing aid.

The second stage of the preoperative assessment and selection process involves a complete medical history and examination, computed tomography (CT) scan/ magnetic resonance imaging (MRI), and promontory stimulation (the latter initially used for the teenage or adult patients, but subsequently only if there is some doubt about the integrity of the auditory nerve, e.g., after a fracture). Many clinics do not use electrical stimulation of the promontory routinely as the majority of patients have a positive result, and thus it is not a sensitive test to determine performance; Myer et al (1984) found 93% of 460 ears had positive results.

In children, third-stage evaluations are conducted, which involve more detailed hearing, speech, and language assessments. These include aided and unaided free-field thresholds, insertion gain measures, evoked potential studies for the verification of behavioral thresholds, and a detailed assessment of communication skills, which includes testing of speech perception using open- and closed-set materials, speech production, language, and pragmatic behavior. An assessment can then be made of the appropriateness of the child's current hearing aid amplification (Pyman et al 1990; Rickards et al 1990; Dowell 1997).

Following the third-stage assessment in children, an individual training and guidance program is developed that may last up to 6 months. This will involve one-to-one training, as well as close liaison with family and those involved in educational support. The program aims to (1) determine if the child can develop and extend his/her existing communication skills; (2) determine whether the child could use a hearing aid to acquire speech and language better than with a cochlear implant; (3) develop the ability to allow the implant to be programmed postoperatively; (4) determine that the family has appropriate expectations and understanding of the cochlear implantation and their role in that process; and (5) liaise with the child's educators. The speech perception, speech production, and language tests are repeated at the end of this training period to measure changes in performance.

The fifth stage of the evaluation includes a major review of all the data, presented by the patient's case manager, before deciding on his/her management. If patients are to be part of a research program, they are reviewed by an independent hospital ethics committee operating under the guidelines of the National Health and Medical Research Council of Australia.

Medical History and Examination

Aims

The aims of the medical history and examination are to (1) make a general appraisal of a person's hearing loss and its effect on communication; (2) determine the cause of the hearing loss; (3) evaluate the ear, nose, and throat and other systems regarding the suitability for operation; (4) assess the communication needs; (5) ensure that the expectations are realistic; and (6) determine the special tests required.

History

The presenting symptom usually is a hearing loss. A mother having her first baby will be less aware than a multiparous woman that her child has a difficulty with hearing. Most parents of a child with a severe-to-profound hearing loss recognize the disability by six months of age. At an early age children should respond to their mother's voice and to other familiar sounds, as well as startle to loud noises. The history focuses initially on the presenting symptom, in particular its time of onset, severity, course, fluctuations, and precipitating or aggravating factors. This information helps determine the etiology of the deafness. Then related symptoms such as tinnitus, vertigo, and ear discharge are explored. The past medical and surgical history is elucidated, including a family history of deafness. In children it is especially important to know if there were any factors during the pregnancy that would increase the risk of deafness, such as infections with toxoplasmosis, rubella, cytomegalovirus, and herpes simplex; medical conditions such as diabetes, toxemia, renal disease, and hypertension; placental factors such as placenta praevia and antepartem hemorrhage; and events during labor such as misapplied forceps and prolapsed chord.

The severity and nature of tinnitus should be determined as a baseline for comparison with any change postoperatively. Of 36 patients operated on at the University of Melbourne's clinic, 32 (89%) had tinnitus preoperatively. Twenty-two (69%) of these said in response to a questionnaire that their tinnitus had improved, and the remainder reported no change in the symptom. No one in this sample said that the tinnitus became worse following implantation, and no one in the group without tinnitus preoperatively developed it after surgery. There had been isolated reports of patients with an exacerbation of tinnitus postoperatively, but it did not appear to be a significant problem overall. Patients should be asked if they have experienced vertigo preoperatively, primarily to help diagnose the cause of the hearing loss. Postoperative vertigo is an uncommon and usually transient complication of the procedure (Pyman et al 1990).

A most important consideration preoperatively is whether there is infection affecting the external or middle ears. Perforations of the tympanic membrane and open mastoid cavities should be closed first to prevent the ingress of infection postoperatively. Surgical techniques for closing large open mastoid cavities have been developed (Scrivener and Gibson 1987).

Physical examination

Initially the patient's speech and language should be assessed while taking the history. The face and hands need to be examined for evidence of disease and genetic abnormalities. This includes inspection of the external ear and postauricular region for surgical scars. A complete otological examination is required and a thorough assessment of the nose and throat should be made. A neurological examination, especially of balance, is also essential.

Diagnosis and Etiology

Adults

The etiologies of severe to profound deafness in adults presenting to a cochlear implant clinic are most commonly otosclerosis, meningitis complicated by labyrinthitis, Ménière's disease, head trauma injuries, surgery, viral disease, ototoxic drugs, otitis media, and vascular accidents. Autoimmune disease and an acoustic neuroma are other pathologies that need to be considered.

Children

Many of the children presenting for a cochlear implant at the University of Melbourne's clinic have deafness of unknown origin (O'Sullivan et al 1997), and no family history (39.8%). However, a significant number may be of genetic origin and autosomal recessive (Gorlin et al 1995), and a small proportion X-linked recessive or autosomal dominant.

Genetic

Congenital

Dominant inheritance should be suspected if one of the parents is deaf (Clark and Pyman 1997). If neither parent has a hearing loss, the inheritance is likely to be recessive. With recessive inheritance other siblings may be deaf or the parents may be related. In evaluating the pedigree, it is important to remember that deaf people often marry other deaf people; thus, two parents with a recessive loss may have deaf children. The inheritance will appear to be dominant, but in reality it is pseudo-dominant or recessive.

Deafness that is dominant may appear spontaneously when it is due to a mutation. Furthermore, with incomplete penetrance or manifestation, the presence of a dominant gene may not be apparent, as parents and other relatives can be unaware they have a hearing loss if it is unilateral or mild to moderate in degree. For this reason audiograms should be obtained from all members of the family. Deafness may also be X-linked (sex linked), in which case it is present in the male offspring. Rarely, deafness can be due to mitochondrial DNA mutations.

Genetic deafness may affect the cochlea alone when it is nonsyndromic. The genetic abnormality may also affect the ear as well as other systems (syndromic). The nonsyndromic causes are most often classified as the Michel, Mondini, Bing-Siebenmann, and Scheibe dysplasias. The lesion of the inner ear can be a bony malformation in the case of the Michel aplasia, where there is a complete agenesis of the temporal bone, and it is inherited as autosomal dominant. The Mondini dysplasia with failure to form interscalar septa has a decreased number of cochlear turns (Fig. 9.2). Endolymphatic hydrops is often present, there may be some residual hearing, and it is inherited as autosomal dominant. Deafness may be associated with vestibular abnormalities that can be visualized radiographically. In particular, the commonest anomaly is a widely patent vestibular aqueduct and enlarged vestibule (Fig. 9.3). This inherited disorder typically presents with step-wise loss of hearing and sometimes vertigo at the same time as the sudden threshold shifts. The cochlea and vestibule may also be represented as a single cavity. In operating on these children there is a risk of an excessive leak of cerebrospinal fluid when the scala tympani of the cochlea is opened due to the large communication between the cranial cavity and the inner ear. As discussed in Chapter 3, if this is not sealed there is a risk of labyrinthitis and meningitis. In addition, in a number of children with a hearing loss only the membranous structures are affected. For instance, in the Bing-Siebenmann dysplasia both the cochlear and

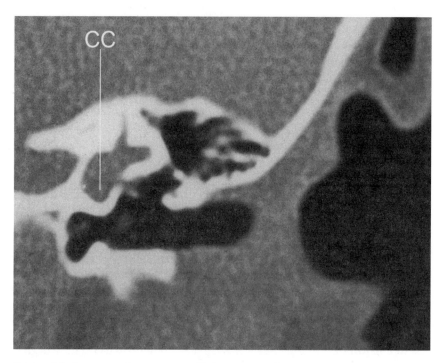

FIGURE 9.2. A computed tomography (CT) scan of the bony labyrinth with the cochlea and vestibule as a common cavity (CC) (Clark and Pyman 1997. Preoperative medical evaluation. In: Clark, G., Cowan, R., and Dowell, R., eds. Cochlear implantation for infants and children, Advances. San Diego, Singular Publishing Group, Inc. Reprinted with permission of Delmar Learning)

vestibular components are malformed, whereas with the Scheibe dysplasia only the membranous structures of the cochlea are affected.

Syndromes may also present as two or more types. For example, in Usher syndrome, three types are described (Gorlin et al 1995). Type I, divided into A, B, and C groups, has congenital severe-to-profound hearing loss, absent vestibular function, and reduced visual fields early in life. Type II, divided into A and B groups, has congenital moderate-to-severe hearing loss, normal vestibular function, and the onset of visual symptoms in the teens to twenties. Type III has a progressive hearing loss with a variable onset in visual loss. The degree of expression of the phenotype can vary so that all the features of the syndrome are not necessarily expressed in the child. Some syndromes, however, are sporadic and have therefore not been inherited. One common syndrome where the expression of the phenotype varies is the Waardenburg's syndrome, which accounts for 1% to 7% of congenital deafness, and a hearing loss is present in 20%. Deafness varies from severe to moderate and can occur later in life. The other features are a lateral displacement of the inner canthus of the eye (100%), heterochromia of the iris (25%), and a white forelock 20%. It is inherited as autosomal dominant.

The Michel aplasia referred to above can be associated with the Klippel-Feil syndrome in which there is fusion of the cervical vertebrae. The Mondini dysplasia, as discussed, may be without associated syndromes, but may also be present

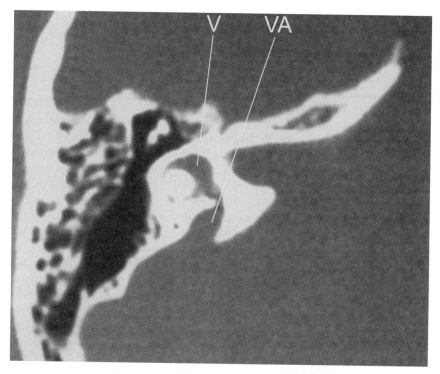

FIGURE 9.3. CT scan of the bony labyrinth with enlarged vestibule (V) and vestibular aqueduct (VA). (Clark and Pyman 1997. Preoperative medical evaluation. In: Clark, G., Cowan, R., and Dowell, R., eds. Cochlear implantation for infants and children, Advances. San Diego, Singular Publishing Group Inc. Reprinted with permission of Delmar Learning)

with the Pendred, Waardenburg, Treacher Collins, Klippel-Feil, and Trisomy 13 and 18 syndromes. The Bing-Siebenmann and Scheibe dysplasia are inherited as autosomal recessives, and they too may be present in association with a syndrome. The syndromes associated are retinitis pigmentosa, Jervell and Lange-Nielsen, Refsum, Usher, Waardenburg, and Trisomy 18. The histological effects of these conditions are discussed in Chapter 3.

Delayed

Delayed sensorineural deafness coming on sometime after the baby is born can also be genetic, and deafness is commonly the only abnormality. It is inherited as an autosomal-dominant condition, and there is a progressive sensorineural hearing loss. This is usually first noticed in childhood or early adulthood. The hearing loss is either across all frequencies or in the high- or midfrequency range. A delayed sensorineural loss may also be associated with other abnormalities, and there are a number of these conditions: Klippel-Feil syndrome; Alport's syndrome associated with progressive renal disease and inherited as a dominant condition; Hunter's syndrome, which begins in early childhood with skeletal deformity and dwarfism as well as blindness; and Crouzon's syndrome or craniofacial dysostosis, in which there is premature closure of the cranial sutures and it is inherited as a dominant condition. Further information on these and other genetic conditions can be found in standard works on genetics.

Acquired

Approximately half the causes of a sensorineural hearing loss in children are acquired, and may be the result of an infection or conditions occurring during pregnancy. It may also be due to difficulties at birth or conditions after birth. The most common infections during pregnancy causing deafness are toxoplasmosis, rubella, cytomegalovirus (CMV), and herpes simplex (the TORCH complex). In the study at the University of Melbourne, CMV and rubella were identified as the commonest infections causing a hearing loss (O'Sullivan et al 1997).

Toxoplasmosis results from the ingestion of cysts of the protozoan, *Toxoplasma gondii,* excreted in the feces of the primary host, the cat. The symptoms are nonspecific and flu-like, with swelling of the lymph nodes (lymphadenopathy) in 20%. There is a 50% probability of transplacental transmission in the first and second trimesters. Up to 10% of infected children have severe involvement at birth, and others have the gradual development of retinal and central nervous system (CNS) symptoms and signs. The resulting chorioretinitis can lead to loss of sight, and the CNS involvement to cerebral calcification. Hearing loss occurs in up to 25% of children born with untreated toxoplasmosis. It may be diagnosed by the detection of antibodies in the serum and other body fluids, the presence of immunoglobulin M (IgM) antibodies appearing 1 week after the infection, and IgG antibodies 6 to 8 weeks after the infection.

Congenital rubella is very likely to lead to a severe hearing loss (30% of babies born to mothers with subclinical but laboratory-confirmed infection become deaf, and in 25% the hearing loss is progressive). It has also been shown that nearly half the viral infections are subclinical, and the effects are most profound if the infection occurs in the first trimester, but second trimester infection can also lead to deafness, microcephaly, cataracts, and mental or motor retardation. Widespread use of the vaccine should eliminate the disease.

CMV as a cause of deafness should be excluded, as it has been shown that the hearing loss is often severe to profound and can progress, and impaired vision or cerebral palsy, epilepsy, and intellectual disability may be associated. CMV infections are very prevalent and can be detected in 0.5% to 2.4% of all live births (Pass et al 1980; Spreen et al 1984). Rasmussen (1990) estimated that 10% of infected newborns are at risk from hearing loss, impaired vision, or neuromuscular abnormalities. Although 90% of cases are without symptoms, there may be swollen lymph nodes and enlargement of the liver and spleen. In children presenting with a severe-to-profound hearing loss, it is considered important to undertake serological tests on the mother and child as well as viral cultures from the saliva and urine of the child up to the age of approximately 4 years (S. Locarnini, personal communication). This will help in deciding whether the child has had a CMV infection, and whether it was of congenital origin when the effects are more severe.

Herpes simplex encephalitis is a viral infection usually from genital herpes. It may present neonatally as a localized mucocutaneous or disseminated infection. When it is disseminated, there is a 30% risk of meningoencephalitis that is likely to occur in the second or third week of life. Herpes simplex infections leading to sensorineural hearing loss may also involve the central auditory pathways.

The most important infection occurring postnatally that leads to a severe to profound hearing loss is meningitis. The child usually gives a history of having been ill and may have been unconscious. With meningitis there is a high incidence of hearing loss due primarily to labyrinthitis. The hearing loss is likely to be severe to profound, and may have been exacerbated by the use of ototoxic antibiotics. In addition, the meningitis may have caused specific learning disorders, intellectual disability, loss of vision, motor weakness, and balance problems that will influence the management of the child.

It is also necessary to know whether there has been a history of recurrent middle ear infections, as these can result in a severe sensorineural hearing loss or affect the assessment of the hearing thresholds and subsequent treatment. Middle ear effusions would also affect the scheduling of a cochlear implant operation. It is certainly not appropriate to operate if there is any active infection in the middle ear or mastoid, as a labyrinthitis is likely to ensue, as has been demonstrated in animal experiments (Clark, Shepherd et al 1984). A low-grade infection may also be present with an asymptomatic middle ear effusion, and this should be corrected before surgery (Clark and Pyman 1997).

Audiology

The medical consultation is followed by an audiological evaluation to determine whether the person has a profound or total hearing loss, and whether his/her speech perception, language and communication skills justify surgery.

Pure Tone Thresholds

Patients with a severe-to-profound bilateral sensorineural hearing loss are prospective cochlear implant candidates. Pure-tone audiometry is carried out to determine the thresholds for the frequencies 125 to 8000 Hz. However, the degree of hearing loss is traditionally defined as the average of the pure tone threshold at 500, 1000, and 2000 Hz in the better ear with reference to normal-hearing thresholds (Boothroyd 1993). A severe hearing loss (HL) has been defined by ANSI (1989) as an unaided pure tone threshold between 71 and 90 dB HL. It has been shown, however, that people with a hearing loss greater than 60 dB HL have significant difficulty in understanding speech, and do not obtain a 100% open-set word recognition when the sound is appropriately amplified (Hood and Poole 1971; Lamore et al 1985). A profound hearing loss is defined as a threshold greater than 90 dB HL. It was recognized in 1995 (Consensus 1995; Kim et al 1995) that with a 100 dB HL threshold, patients would clearly obtain better speech recognition with a cochlear implant than with a hearing aid. However, as speech-processing strategies have continued to advance, this level must be set lower. Furthermore, it is important to emphasize that speech results do not clearly correlate with the pure tone thresholds, and other measures of communication skills are desirable. This is not always easy in young children. Even speech perception tests do not necessarily apply to communication skills in daily life. Erber and Lind (1994) and Tye-Murray et al (1996) found that people could have similar open-set sentence scores, but report and demonstrate different speech perception abilities in daily communication. This situation could be improved in the future by making the speech perception tests more similar to hearing in the natural environment by adding the effects of background noise and reverberation, or by making the speech tests more communication oriented with information exchange and interaction as part of the test. If patients have a severe-to-profound hearing loss on pure tone audiometry, it is an important guide to their suitability for cochlear implantation, but it should be combined with speech audiometry and other tests for a thorough appraisal.

The hearing thresholds in infants and young children need to be accurately determined by behavioral and objective tests. The tests should also determine the status of the central auditory pathways and their ability to process information. It is also necessary to relate the pure tone audiogram results to expected speech perception with a hearing aid or an implant. Speech perception tests cannot be used, as these children have little appropriate language.

Thresholds can be obtained in children from 6 to 18 months of age when this is considered as developmental age, from behavioral assessment visual reinforcement audiometry (VRA). In normal-hearing children the test yields thresholds within 10 dB of thresholds obtained later with standard audiometry. There is a dearth of studies indicating how well it correlates for deaf children or the false-positive rates. Similarly with play audiometry, which is applicable from 18 months and above, there are few comparisons of thresholds for deaf children with later results for standard audiometry.

An accurate objective threshold in infants and young children over the speech frequency range can be obtained using tone-burst evoked auditory brainstem responses (ABRs) with a "notched-noise" masker. Stapells et al (1995) found there was excellent correspondence between ABR and behavioral thresholds for the 500- to 4000-Hz range. At the University of Melbourne's Cochlear Implant Clinic the recording of steady-state evoked potentials (SSEPs) (Rickards and Clark 1984) is preferred for the objective assessment of children, as it is just as accurate as the ABR with "notched-noise" for the severe-to-profound hearing losses and has an automatic detection algorithm for judging when an ABR is present or absent (Rance et al 1994a,b, 1995, 1997, 1998).

SSEP thresholds are determined for each frequency by amplitude modulating each sound frequency and carrying out a Fourier analysis of the evoked potentials recorded from the scalp (Fig. 9.4). The relation between SSEPs and behavioral thresholds is shown in Figure 9.5. Notice the good correlation for the severe-to-profound losses. Hearing thresholds are especially important in determining

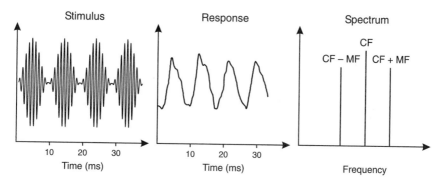

FIGURE 9.4. Left: amplitude-modulated stimulus. Center: steady-state evoked potential (SSEP) response. Right: Fourier spectrum of stimulus. CF–carrier frequency, MF–modulation frequency.

whether a young child has sufficient hearing to do better with a hearing aid rather than an implant, as speech tests are not effective due to the child's limited language. Multiple-channel implants can provide spectral information in the mid- to high-frequency range when this is not possible through amplification of sound with an aid. For this reason it is necessary to pay particular attention to the thresholds for the frequencies above 1000 Hz. As shown in Figure 9.6, if the unaided thresholds are greater than 90 dB for the frequencies 1500, 2000, 3000, and 4000 Hz, the child is a likely candidate. The aided thresholds for these hearing losses with a powerful hearing aid will just enter the speech spectrum for 70 dB speech at a distance of 1 m. Consequently, a child with an aided threshold at 2000 Hz greater than 60 dB sound pressure level (SPL) is a likely recipient of the cochlear implant.

This is supported by data from Blamey et al (1998) and Sarant et al (1996, 2001) on 57 children with a bilateral severe or profound hearing loss. There were 33 hearing aid and 24 Nucleus F0/F1/F2 and Multipeak implant users. The Bench-Kowal-Bamford (BKB) speech perception results for the two groups referred to in Chapter 11 were the same. However, for the implanted children the mean preoperative loss in the better ear was greater than 100 dB Hz (for 500, 1000, and 2000 Hz), and for the hearing aid children the mean loss was 81 dB. This indicates that the implanted children were performing at a level comparable to children with a threshold of 81 dB Hz and a powerful hearing aid. Thus from the data an 81 dB threshold at the frequencies of 500, 1000, and 2000 Hz equates with a hearing loss at 2000 Hz of 90 dB HL. This supports the criterion for selecting children for cochlear implants when the threshold at 2000 Hz and above is greater than 90 dB HL and the aided threshold 60 dB SPL.

The speech perception and production for eight children (average age at implantation 5.8 years) with the Nucleus SPEAK strategy was compared with children with hearing aids using the Imitative Test of Speech Pattern Contrast Perception (IMSPAC) (Boothroyd et al 1996; Boothroyd 1997, 1998), and the Arthur Boothroyd (AB) (Boothroyd 1968) word tests. The results for the implanted

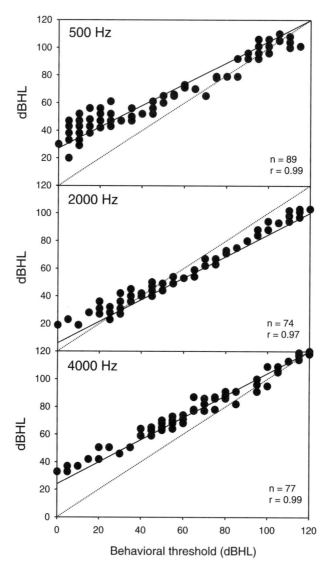

FIGURE 9.5. The relation between SSEP and behavioral thresholds for the carrier frequencies of 500, 2000, and 4000 Hz.

children were comparable to those for aided children with a threshold of 70 to 75 dB HL.

Impedance Audiometry

In the case of a profound mixed hearing loss with a severe sensorineural component (and therefore with no thresholds by bone conduction to the limit of the

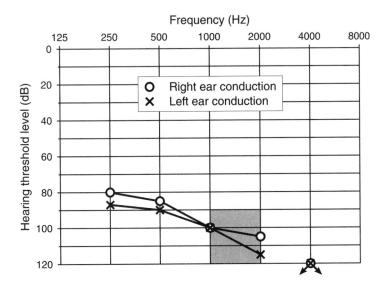

FIGURE 9.6. Aided audiogram and the selection criteria for implantation in children. The shaded area indicates the hearing loss not helped by a hearing aid as much as with a cochlear implant. Therefore, if the threshold is below 90 dB for 1000 and 2000 Hz, then a cochlear implant is strongly recommended.

audiometer—typically 75 dB HL), a conductive component to the hearing loss may not be apparent. Therefore, impedance audiometry becomes an important adjunct to the other tests in assessing middle ear function.

Hearing Aid Evaluation

Many prospective implant recipients are able to obtain useful hearing with a powerful hearing aid. For these people, it is very important to carefully determine the level of hearing provided and the amount of benefit derived. These results can then be compared with those obtained by cochlear implant recipients. The initial evaluation consists of aided threshold and insertion gain testing, measurement of the speech detection threshold, the most comfortable level, and the loudness discomfort level. Pure tone audiometry in itself is not adequate for defining the suitability for a cochlear implant patient. Hearing aids can reach outputs of 140 dB SPL, and show hearing where less powerful hearing aids and audiometers may indicate the patient as totally deaf. However, with these levels the benefits of a hearing aid are not likely to be comparable to those with advanced multiple-channel cochlear implant speech processing strategies.

A battery of tests was developed for the first implants to compare improved results with implants to those with hearing aids. As standard audiological speech tests were too difficult for people with a profound hearing loss, a minimal auditory capabilities (MAC) battery of tests was constructed by Owens and Telleen (1981) and Owens et al (1982). The MAC battery consisted of a number of closed-set speech tests and some open-set tests. The closed-set tests were male versus female voice, two-syllable versus one-syllable words, two-syllable words, noise versus voice, four spondees, question versus statement, initial consonant test (nasality and voicing), prosody, and vowels. The open-set tests were NorthWestern University #6 one-syllable words, CID everyday sentences, paragraphs, open response spondees, and speech reading (unaided versus aided). The closed-set tests assessed the listener's ability to discriminate the segmental and suprasegmental features of speech. For example, the one- versus two-syllable test assessed the ability to perceive timing and intensity cues, and the male/female and the accented word test assessed the ability to perceive differences in fundamental frequency, intensity, and timing cues. The vowel test assessed the ability to discriminate between four monosyllables differing only in the medial vowel. The open-set tests allowed a better assessment of performance in everyday life when speech-processing results were obtained with the multiple-channel implant. In addition, the benefit of the hearing aid as an aid to speech reading was assessed using standardized word and sentence speech reading tests with the CID and consonant-nucleus-consonant (CNC) tests. The number of words correctly understood in the audiovisual and the visual-only conditions were compared, and the statistical significance determined with the aid of tables provided by Thornton and Raffin (1978).

Patients who had not previously been using an appropriate or optimal hearing aid and who were candidates to gain some benefit from a hearing aid were given a 6-month trial, which included training. At the end of the trial a decision was made as to their suitability for a cochlear implant operation. With the improved speech-processing results a person on average can obtain an open-set CNC word score of 30% to 40% and CID sentence score of 80%. They must have hearing in the better ear at this level or worse for an implant to be considered. It is still not possible to predict how well each individual patient will perform, although 40% to 50% of the variance can be deduced from factors such as age and duration of deafness as discussed below (see Predictive Factors).

Cochlear Microphonics and ABR Tests for Neuropathy

Cochlear microphonics and summating potentials (Fig. 9.7) should also be recorded from children who are prospective cochlear implant recipients. Recent studies have shown that a small proportion of children with absent or elevated ABR or SSEP thresholds may have cochlear microphonics present. Some of these children may have normal hearing. But those with a profound loss may do poorly with a cochlear implant. The cochlear microphonic can be recorded using click-evoked ABRs to rarefaction and condensation clicks, and a typical result is shown in Figure 9.8 (Gary Rance, personal communication).

In Melbourne 5343 children were screened with click ABRs, and 154 found with raised thresholds. Forty-four had no ABRs, of whom 32 also had no cochlear microphonics and were severely to profoundly deaf. Twelve had cochlear micro-

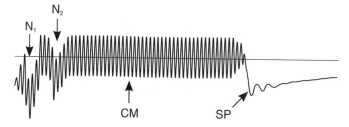

FIGURE 9.7. Cochlear microphonics (CM), summating potentials (SP), and summed neural responses N1 and N2.

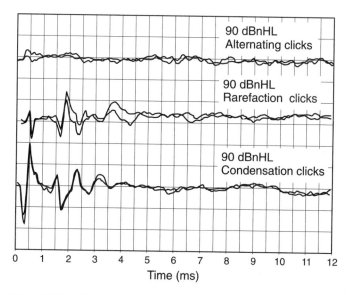

FIGURE 9.8. Click-evoked auditory brainstem responses (ABRs) and cochlear microphonics. Normal hearing level (nHL) refers to the normal threshold for click or brief-tone stimuli. (Reprinted with permission from G. Rance).

phonics and hearing that ranged from normal to total loss. These children with a severe loss were thought to have a neuropathy, and possibly should not be implanted. In this series the majority had anoxia or kernicterus at birth.

The recording of cochlear microphonics and slower potentials can also be made using transtympanic electrodes, and this procedure may play a role in the preoperative assessment of children for a cochlear implant. It has been found that a small proportion of children with a severe-to-profound hearing loss have an abnormal positive potential (APP) (O'Leary et al 2000). The origin of this potential is not clear. It may reflect a retrocochlear as well as cochlear pathology. It has been associated with prematurity, birth hypoxia, kernicterus, developmental delay, cerebral palsy, Goldenhar's syndrome, intrauterine CMV, and rubella infection. A cochlear implant operation has been carried out on eight children (O'Leary et al 2000) with the abnormal potential, and three had poor auditory perception postoperatively. These three also had little neural activity when electrically evoked potentials were recorded with their implants.

The recording of acoustic emissions is not central to the assessment of hearing thresholds in the child as it correlates with thresholds only up to 45 dB HL. However, as it is a measure of outer hair cell integrity it should be used to study children where a neuropathy is suspected.

Communication

Preoperative assessment of communication in adults and children is for speech perception, and in children for speech production, receptive and expressive language, and pragmatic skills. The preoperative assessments are administered before and after the preoperative training in the use of a hearing aid. The preoperative tests are the same as those used in assessing communication in implanted deaf children and the results of the preoperative tests may be used to help in designing the preoperative training program.

Speech Perception

Tests of speech perception are needed in addition to tests of hearing threshold to better define the person's hearing handicap and suitability for cochlear implantation. They are normally carried out for hearing alone with the hearing aid. Testing with speech reading and speech reading with hearing may also be required. As results improve it will become increasingly important to test in quiet and in noise. Monosyllabic words (CNC) (Peterson and Lehiste 1962) as AB lists (Boothroyd 1968) or the Northwestern University (NU-6) list (Tillman and Carhart 1966) are presented and can be scored as words or phonemes correctly identified. With the initial patients from 1978 to 1985 a 0% score for word identification was required before cochlear implantation would be recommended. With the University of Melbourne's F0/F2 speech-processing strategy, the average score (using prerecorded test materials) for open sets of words (scored as words) was 5%.

In addition, a word-in-sentence test such as the CID Everyday Sentences (Davis and Silverman 1978) (scored as key words) is used. This allows a comparison to be made of the contribution of "top-down" processing to word recognition. With the University of Melbourne's F0/F2 speech-processing strategy, the average CID score was 11% for electrical stimulation alone (Clark, Tong et al 1981). In a subgroup of 23 patients in an international clinical trial for the Food and Drug Administration (FDA) by Cochlear Proprietary Limited the mean CID sentence scores for electrical stimulation alone rose from 16% at 3 months postimplantation to 40% at 12 months (Dowell et al 1986a,b). The major criterion for selection with the F0/F1/F2 and Multipeak strategies in the early 1980s was the absence of significant open-set speech discrimination (less than 10% in the ear to be implanted and less than 20% in the better ear). Careful consideration was given before implanting an ear that had sufficient hearing to enable a hearing aid to serve as an aid to speech reading. In these instances, other factors would be taken into account, such as age and period of profound deafness. Now that a large number of patients have had cochlear implants with improved speech-processing strategies (Nucleus SPEAK and ACE, and Clarion and Med El CIS), and achieved speech perception scores that are better than hearing aids in severely hearing impaired people, the criterion levels for pure tone thresholds and speech perception scores have been raised.

A multicenter study on 63 postlinguistically and profoundly hearing-impaired adults at eight centers in Australia, North America, and the United Kingdom by Skinner et al (1994) showed that with the Nucleus SPEAK strategy and the Spectra-22 processor the mean score for electrical stimulation alone for CNC words was 34%. The mean score for CID words in sentences was 76% for SPEAK. The first 64 patients implanted with the Clarion device had a mean CID sentence score of 60% six months postoperatively (Kessler et al 1995). The Nucleus ACE was compared with SPEAK and CIS (Arndt et al 1999) using Hearing in Noise Test (HINT) (Nilsson et al 1994) sentence scores. The ACE mean was significantly higher than the CIS mean, but not significantly different from SPEAK. This indicates that with these strategies an implant can be considered if the open-set CID sentence score is 60% or less and the CNC word score (scored as words) is 30% or less in the better, unoperated ear.

If there is residual hearing in either the better or poorer ear, it may be helpful to determine the closed-set consonant and vowel perception or the perception of speech features. This would provide evidence, for example, of the extent of voicing information conveyed. They are evaluated in the auditory alone condition. Eleven vowels are presented in an H−vowel−D context, and 12 consonants are presented in an A−consonant−A context. Responses are analyzed in a confusion matrix. With children a range of tests are available and they need to be age and language appropriate. Not all the tests below are performed on every child, as developmental age and linguistic skills of the patient would exclude some tests. The early speech perception (ESP) battery (Moog and Geers 1990) assesses the spondee and monosyllable identification. The PLOTT (Plant/Westcott) sequential speech feature test (Dunn and Dunn 1981; Plant 1984) is a two-alternative, forced-choice perception test using 50 words, with five groups of 10 words categorized according to five sequential categories: vowel length, vowel place, consonant voicing, consonant place, and consonant manner. The MSTP (Plant 1984); (Dunn and Dunn 1981) consists of three *m*onosyllables, three *s*pondees, three *t*rochees and three *p*olysyllables. It is a word identification task where the performance level is categorized.

For children with an adequate vocabulary, open sets of phonetically balanced kindergarten (PBK) monosyllabic words, the Glendonald Auditory Screening Procedure (GASP) (Erber 1982) and LNT (Lexical Neighbourhood Test) (Kirk et al 1995), as well as the BKB (Bench and Bamford 1979) and HINT-C (Nilsson et al 1996) sentences are used for measuring speech comprehension. The Sound Effects Recognition Test (SERT), an environmental sounds test, is a four-alternative, forced-choice perception test for children with limited verbal abilities (Finitzo-Hieber et al 1980).

With children from 25 months to 4 years all tests, with the exception of the SERT, are presented using live voice in a free field with a sound level meter to monitor the intensity of the speech signal. The use of live voice introduces variability within and between speakers. However, when working with young hearing-impaired children, it is not possible to administer tests using recorded material over long periods and maintain the child's interest. The materials are normally presented at 70 dB SPL [approximately 65 dB (A)] because the hearing aid selection procedure used in Australia is based on the 70 dB SPL speech spectrum. This level may be increased up to 80 dB SPL for any child, and this decision is based on the information obtained concerning the audibility of speech during the hearing aid evaluation. All tests are presented at the intensity level selected. For older children the tests can be administered prerecorded.

In the case of children from 12 to 24 months the IT-MAIS (Infant Toddler-Meaningful Auditory Integration Scale) (Zimmerman-Phillips et al 2000) is used. This allows parents to rate their children's spontaneous listening behaviors on tasks such as alerting to sound, discrimination of two speakers, and associating vocal tone (anger, excitement, fear) with its meaning on the basis of hearing alone.

Speech Production

The assessment of a child's speech production is for speech sounds that have been imitated, elicited, or spontaneously produced. The child's ability to imitate speech is tested using phonetic-level evaluation (PLE) (Ling 1976) or voice analysis (Ling 1976). In these tests the child attempts to copy presented syllables. Elicited speech is from visual prompts where the child is asked to name a picture (Test of Articulation Competence) (Fisher and Logemann 1971) or verbally repeat a written sentence (McGarr sentences) (McGarr 1983). The child's spontaneous speech produced in conversation or play is recorded on videotape and is analyzed using the phonological process analysis (Ingram 1976; Crary 1982) and the PLE (Ling 1976).

Language

Language assessments test the child's competence with semantics (meanings of words and sentences) and syntax (the rules that govern the structure of sentences). The tests used are selected according to whether the child is at a preverbal, single word, or multiword level, and are used to assess receptive, expressive, or both receptive and expressive language. The Symbolic Play Check List (Westby 1980) assesses receptive and expressive prelanguage behaviors in the preverbal child. The Peabody Picture Vocabulary Test (PPVT) has been revised for receptive language (Dunn and Dunn 1981). The Preschool Language Scale (Zimmerman et al 1979) is for receptive and expressive language. The Language Assessment,

Remediation, and Screening Procedure (Crystal et al 1976) for expressive language can be used for the assessment of verbal children who are at the single- or multiword level. The Northwestern University Children's Perceptions of Speech (NU-CHIPS) (Elliott and Katz 1980) is a four-alternative, forced-choice picture identification test and uses monosyllabic words. The Grammatical Analysis of Elicited Language (GAEL) (Moog and Geers 1979, 1980) and the Test of Syntactic Abilities (Quigley et al 1978) are expressive language tests that are used on verbal children who are at a multiword level. The above tests are used when appropriate to the child's linguistic and cognitive abilities.

Pragmatic skills relate to the sociolinguistic aspects of language and the ability to carry on a two-way conversation. This includes knowing the way conversations are entered into, how to take turns, how information can be sought, how clarification can be obtained, and methods of addressing people (Ling and Ling 1978). Pragmatic skills are assessed by the analysis of interactions between child and family and/or between child and therapist. Spontaneous samples of conversation and interaction are recorded and are also assessed at the preverbal, single-word (Roth and Spekman 1984a,b) and multiword (Prutting 1986) levels.

Special Investigations

Radiology

Plain X-Rays

Plain x-rays of the mastoid and temporal bone may show chronic mastoiditis or other ear disease. Patients being assessed for cochlear implantation also require recent plain x-rays of the temporal bones, which also indicate the presence of emissary veins that may affect the placement of the implant package, or other abnormalities.

CT Scans

Computed tomography (CT) scans became readily available only after the introduction of cochlear implant surgery, and now provide essential information about cochlear and mastoid pathology that guide the surgeon as to the side to implant as well as the surgical difficulties to be encountered. The CT technique should be discussed with the radiologist to ensure that the correct projections and magnifications are used, and cuts should be made at 1-mm intervals. High-resolution scans are essential for defining the cochlear spaces, and abnormalities of the otic capsule such as fractures, otosclerosis, and malformations. The images should provide views of the basal, middle, and apical turns of the cochlea, the vestibule, the semicircular canals, the carotid canal, the sigmoid sinus, mastoid air cell system, facial nerve, internal auditory canal, the middle ear, and any mastoid emissary veins. The two views (horizontal and coronal) give complementary information, and this is especially important in assessing the scala tympani to ensure that implantation should be possible (Figs. 9.9 and 9.10). The horizontal CT section in Figure 9.9 demonstrates the basal turn of the cochlea, the site of the round window, the infracochlear cell, the tympanum, and the mastoid air cells. The coronal CT section in Figure 9.10 demonstrates the basal and middle turns of the cochlea, the tympanum, the horizontal part of the facial nerve, the jugular bulb, and the mastoid air cells. The coronal cuts require dorsiflexion of the neck and are therefore not performed in subjects requiring a general anesthetic for the procedure because this posture produces respiratory obstruction. The images should be presented life size so the diameter of the basal turn may be accurately measured, and the contrast should be adjusted to demonstrate the presence of intracochlear calcification.

In the normal ear the mastoid air cells, the middle and inner ear structures, and the facial nerve are all clearly recognizable. Note the extent of pneumatization generally, but particularly under the cochlea and around the facial nerve. This information is useful preoperatively. The otic capsule is normally uniformly opaque, the round window membrane is visible as a fine line, and the cochlear duct is a spiral with smooth edges and is uniformly transparent. Some important variations of the normal include the dehiscence of the facial nerve lying close to the facial recess. This is important because of the danger to the nerve during the posterior tympanotomy especially with a crowded access to the round window.

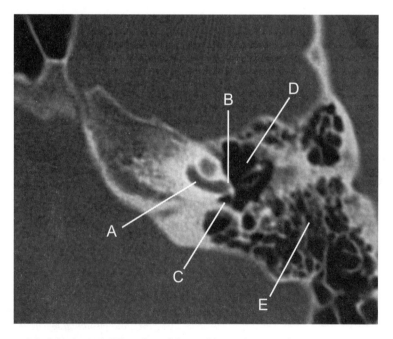

FIGURE 9.9. A horizontal CT section of the cochlea and temporal bone demonstrating (A) the basal turn of the cochlea, (B) the site of the round window, (C) the infracochlear cell, (D) the tympanum, and (E) the mastoid air cells.

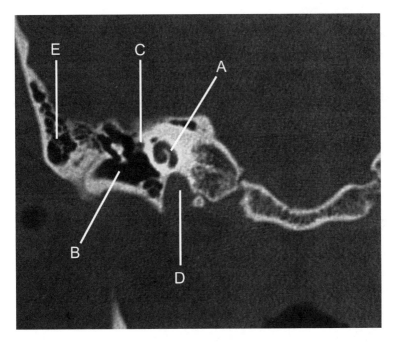

FIGURE 9.10. A coronal CT section of the cochlea and temporal bone demonstrating (A) the basal turn of the cochlea, (B) the tympanum, (C) the horizontal part of the facial nerve, (D) the jugular bulb, and (E) the mastoid air cells.

A large emissary vein can cause brisk bleeding, and an air cell inferior to the cochlea could be mistaken for the round window. By comparing the image from the two sides it is possible to detect fluid within the air cell system or pathological erosion of the internal meatus.

In meningitis, the most common cause of acquired total deafness, the inflammatory changes may extend throughout the basal but also the middle and apical turns. There is frequently ossification from the round window for up to 6 mm along the scala tympani. These changes can usually be detected by CT scans. Labyrinthitis ossificans can extend to total obliteration of the cochlear duct by patchy or uniformly opaque bone. Some cases have fibrous material obliterating the cochlea, with the result that the CT scans appear to indicate implantable ears, whereas at surgery the scala tympani is filled with scar tissue that adheres to the bone and basilar membrane along the spiral. In cases of meningitis, examine not only the clarity of the basal turn but also the upper turns as opacification would suggest the presence of fibrous tissue. Magnetic resonance imaging (MRI) is essential in excluding fibrous tissue in the cochlea.

There are a variety of changes in ears affected by otosclerosis (Damsma et al 1984; Mafee et al 1985). The otospongiosis reduces the opacity of the otic capsule, and increases the diameter of the cochlear duct. In addition, there are bony accumulations at the round and oval windows. The changes seen as a result of these processes include obliteration of the round window, roughened walls of the cochlear spaces, mottling of the otic capsule producing a double-barrel appearance, and sometimes widening of the cochlear perilymph spaces. Once again the appearance of the two sides should be compared to determine which ear is more suitable for implantation. It is important to have life-sized images to measure the width of the scala tympani and determine whether the electrode array will pass.

There are many effects of trauma in producing total deafness. Most cases with loss of hearing from a head injury have no fractures involving the cochlea. In these cases care must be taken that the bony defect of a craniotomy is clear of the planned surgical site. Fracture lines may be seen in the internal auditory canal, up to the cochlea or all the way through the modiolus or vestibule. In such cases it is important to check that there is no evidence of cerebrospinal fluid within the middle ears by comparing the aeration of the two air cell systems. In the Mondini dysplasia the cochleae have one to one-and-a-half turns or the dysplasia may appear as a common cavity. In some cases a well-defined modiolus is present and in other cases it is absent. If present, it indicates that a perimodiolar array such as the Nucleus Contour can be used as the spiral ganglion cells lie centrally. But if the modiolus is absent, the nerves lie peripherally and a straight but flexible array that lies around the periphery would be preferable. With the Mondini dysplasia there also may be a large vestibule, wide semicircular canals, and an enlarged cochlear duct. It often also may be found in sudden deafness. It is important to recognize this dysplasia as it may be associated with a perilymph gusher.

MRI

MRI (Fig. 9.11) is required to exclude the presence of fibrous tissue in the scala when the CT scan shows no evidence of new bone after meningitis. MRI is also valuable in assessing the thickness of the cochlear nerve in children with an inherited hearing loss and any brain abnormalities.

Electrical Stimulation of the Promontory

For the cochlear implant to produce hearing sensations, a sufficient number of auditory neurons must be capable of being stimulated electrically. That is, the profound hearing loss should be largely cochlear in origin. Standard audiological and medical tests to differentiate between cochlear and retrocochlear losses cannot be used with profound losses. Therefore, electrical stimulation of the promontory was required. However, as indicated above it is not used routinely, but the methods and results are discussed below.

A needle electrode (such as is used in electrocochleography) is inserted through the tympanic membrane with the tip resting on the promontory in the middle ear, as close to the round window as possible. The needle is best inserted with local anesthesia by iontophoresis. The needle may be held in place by a rare earth magnet fixed to a plastic ear speculum. Small electric currents are passed between the needle electrode and a surface electrode placed on the ipsilateral cheek. Pa-

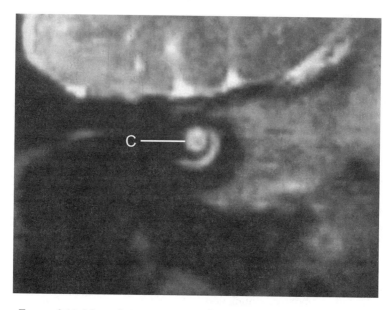

FIGURE 9.11. Magnetic resonance imaging (MRI) showing the cochlea (C).

tients are asked to report any sensations elicited. If they report some sensation, they are asked whether it is a hearing or a tactile sensation, whether it is continuous or intermittent, what pitch it is, and what loudness level. The timing of the stimulus presentation is then varied to check that the sensations reported correspond to the stimuli. For instance, the stimulus is presented with a rhythm (half a second on and half a second off) and the patient needs to show that the timing of the sensation corresponds to the timing of the stimulus (for example, by moving the hand in time). It is important to verify the sensations reported, as tinnitus and a desire to hear can give misleading reports. To be considered a positive result, the patient need only report consistent hearing sensations. However, further testing is carried out in an attempt to gain some knowledge of the neuronal population and the perceptual abilities of the patient. The pulse rate of the stimulus is varied to determine whether the patient can differentiate between pulse rates on the basis of pitch (as the pulse rate is increased from 100 to 200 Hz, the pitch sensation should increase). The minimum gap perceived is determined using two stimuli, with a break in the middle of one. The patient needs to determine which stimulus has the break in it. The breaks are progressively reduced from approximately 150 ms to the level at which the patient reports no break and is unable to determine the correct response. In many adults with acquired deafness, the minimum gap detection is less than 10 ms. The dynamic range for different frequencies may also be of value in assessing performance. An adaptation test is carried out using a constant stimulus, presented at a most comfortable level. The stimulus is presented for a period of 60 s, and patients are asked to raise their finger for as long as they hear the sound. Abnormal adaptation is said to occur if the stimulus is not heard for the full 60 seconds. If this occurs, a very careful review of the history and the overall results is done to determine the likelihood of a retrocochlear cause for the deafness.

No firm conclusions can yet be made about the relationship between the promontory stimulation results and results with the cochlear implant. It does, however, seem evident that good pulse rate discrimination, minimum gap detection of approximately 10 ms, and no adaptation auger well for the patient's cochlear implant use. What is less clear is the relationship between relatively poor results on the promontory stimulation test and subsequent cochlear implant use. Many factors can contribute to the promontory test results, and not all of these are relevant to the use of a cochlear implant. In particular, stimulation of nonauditory neurons (such as the tympanic branch of the glossopharyngeal nerve) can make the procedure uncomfortable and can lead to poorer results.

In addition, electrical stimulation tests can provide further information to facilitate a comparison between the ears. It is useful if one ear was deafened for a longer period of time or was congenitally deaf. If all else was equal (CT scans and audiological results), the ear with the better promontory stimulation result would be chosen for cochlear implant surgery. For this reason, both ears would be tested with promontory stimulation.

An addition to the above tests is the use of a speech processor and the receiver-stimulator from a cochlear implant to enable speech information to be presented to the person during the promontory stimulation test. The features presented are fundamental frequency and intensity-timing cues, as there is only one electrode channel. The ability of the patient to discriminate on the basis of these cues is then assessed. This provides additional information about the patient's perceptual processing abilities and also helps to shape the patient's expectations more realistically, particularly for those who have been profoundly hearing impaired for a long period of time. It is necessary to explain that this would be the minimum the patient would expect to receive.

For the vast majority of profoundly hearing-impaired patients who have undergone the promontory stimulation test, the result has been positive, suggesting sufficient residual auditory neurons to allow successful cochlear implant use. However, several patients have obtained negative results. This occurred primarily with a stimulus that had biphasic pulses of relatively short pulse width. This led to uncomfortable tactile sensations and pain more often than with a square wave pulse train. Patients with a negative result have later successfully received a cochlear implant. For this reason electrical stimulation is not used routinely. However, one patient in the Melbourne clinic had a definite negative result (where there was no hearing sensation and also no uncomfortable tactile sensations). It was subsequently discovered that this person had a disease causing neural degeneration of the central nervous pathways.

Vestibular Assessment

Preoperatively, a vestibular assessment is routinely carried out, using electronystagmography. The stimuli for the caloric stimulation are the routine warm (44°C) and cool (30°C) water irrigations, but also a cold (less than 5°C) water irrigation. This latter stimulus can differentiate between a hypoactive labyrinth with minimal function and an inactive labyrinth.

The results of the vestibular assessment are typically not used to determine suitability for the cochlear implant operation. Instead, they are helpful in making predictions about the likelihood of temporary postoperative vertigo. They can also be used as a reference in the unlikely event of a person developing balance problems at a later stage.

Management

In selecting adults and children for surgery it is important to ensure that their hearing status is such that they would do better with an implant than a hearing aid. The factors that predict successful outcomes are also important as parents wish to know how well their child is likely to perform after surgery.

Hearing and Speech Perception

On the basis of the tests described, it is possible to place the patients into one of three categories with respect to speech discrimination:

Category I

These are patients suitable for implantation of either ear. There is no significant auditory discrimination of open-set words or sentences, nonsignificant scores on closed-set tests of spectral discrimination, and no significant aid to speech reading in either ear.

Category II

These patients are suitable for implantation of the unaided ear. The aided ear (the better ear) has no open-set speech discrimination, but may have some significant closed-set speech discrimination that provides a significant aid to lipreading.

Category III

These patients may be suitable for implantation. The aided ear has less than 30% to 40% open-set word discrimination, and the ear to be implanted shows no significant open-set speech discrimination but obtains significant closed-set speech discrimination which aids speech reading.

It is also possible, however, to consider a person for an implant on the basis of excessive recruitment (where the dynamic range is severely reduced) or where use of a hearing aid produces excessive tinnitus, dizziness, or other uncomfortable feeling.

Predictive Factors

The selection of a patient for a cochlear implant requires a number of consultations, special tests, and counseling as discussed below. This process is conducted by a number of clinicians, and progress should be reviewed at appropriate stages by a team headed by an experienced clinician.

The factors that predict successful outcomes are important for selecting patients. Patients, and in the case of children their parents, wish to know how well they are likely to perform after surgery. Knowledge of the factors predicting outcomes helps answer patients' questions.

General Predictive Factors for Adults

The general predictive factors for adults are age when deafened, age at implantation, duration of deafness, duration of implantation, etiology, the presence of progressive hearing loss, degree of residual hearing, speech reading ability, speech-processing strategy, and medical condition. These factors are similar to those for children. The factors that correlate with speech perception results have been evaluated in a number of studies (Dowell et al 1985; Hochmair-Desoyer and Burian 1985; Nadol et al 1989; Parkin et al 1989; Dorman et al 1990; Blamey et al 1992). The analyses were undertaken on patients with the Nucleus F0/F2, F0/F1/F2, and the Multipeak-MSP systems and the Ineraid device.

There was considerable variability in the results in adults. However, in a study on 64 postlinguistically deaf adults at the University of Melbourne Clinic using the Nucleus F0/F1/F2 strategy, 43% of the variance of the CID sentence scores was accounted for by the duration of deafness, frequency discrimination, and gap detection for the promontory tests, the number of electrodes in the cochlea, and the dynamic range of the intracochlear electrodes. In a combined study by Blamey et al (1996) on 808 patients, duration of deafness, age at onset of deafness, etiology, and duration of implant experience accounted for 21% of the variance, of which duration of deafness was the greater part (13%). So it is still not possible to confidently tell patients how well they will perform, and more research is required before better prediction can be achieved.

Age When Deafened

Age when deafened is not a significant factor in postlinguistically deaf adults as it is with children who are deafened in their critical period of language development. It has little effect up to 60 years (Blamey et al 1996).

Age at Implantation

Age at implantation and duration of deafness were both interrelated. In the adult, they can be separated and both correlate negatively with results. Initially older people had a longer duration of deafness when presenting for implantation. Now people come with a shorter duration of deafness so age and duration do not correlate. A further analysis of adult results showed they were only poorer if the patient was over 60 years (Blamey et al 1995). In addition, learning is more difficult for the elderly for a number of perceptual tasks.

Duration of Deafness

Blamey et al (1992) found that duration of deafness was the main general factor that correlated with speech perception. This was supported by the findings of Gantz et al (1993). Shipp and Nedzelski (1994) reported a strong relation between duration of deafness divided by age and a composite measure of auditory perception. With deafness of long duration, adults are likely to require long periods of rehabilitation for adequate speech perception, and the results are poorer (Dowell et al 1997; Tomblin et al 1999). However, with a long duration of deafness the person can often perceive phonemes and have good results for place pitch discrimination, but cannot so readily integrate the information and understand speech. Duration of deafness may have its effect through a greater loss of the neurons and their connections.

Etiology (Cause of Deafness)

With etiology, Meniere's disease correlated positively, and meningitis negatively with results in the adult. This may be the result of the reduced number of electrodes inserted due to labyrinthitis ossificans (Blamey et al 1995).

Progressive Hearing Loss and Residual Hearing

A progressive hearing loss is associated with better results, as is the presence of some residual hearing (Gantz et al 1993). If there has been a progressive hearing loss, the patient will have learned to use degraded auditory information, and this skill will subsequently be useful when given a cochlear prosthesis.

Speech Reading Ability

There is a weak correlation between speech reading ability and speech perception and this may be because it reflects good top-down processing skills (Cohen et al 1993; Gantz et al 1993).

Speech-Processing Strategy

The speech-processing strategy has a marked effect on results. Improvements have primarily come by presenting additional frequency information on a place-coding basis. In postlinguistically deaf adults this has been seen with the Nucleus speech-processing strategies where there has been a progressive improvement in scores from the F0/F2 to F0/F1/F2 to Multipeak to spectral maxima sound processor (SPEAK). The addition of more temporal information with the higher rate of stimulation with the ACE strategy has also helped. The presentation of information at a higher rate has also led to an improvement from a fixed-filter strategy with interleaved pulses (IPs) to continuous interleaved stimulation (CIS). This was discussed in more detail in Chapter 7.

Duration of Implantation

Finally, the duration of implantation is strongly correlated with good speech perception. Learning is required to use the speech-processing strategy in postlinguistically deaf adults, and the learning is less and steeper when the strategy provides more information and is more speech-like. Not only did word and sentence scores improve over time, but to a lesser extent so did vowel and consonant recognition. Tye-Murray et al (1992) found that a phoneme composite score increased by an average 8.6% in the first 9 months of implant use, and a further 4.4% in the second 9 months. There were small nonsignificant changes thereafter. This applied to both the Nucleus Multipeak ($n = 13$) and Ineraid ($n = 14$) patients.

Medical Condition

There may be medical issues that influence the results. For example, neurosyphilis and schizophrenia are conditions that affect central auditory processing or cognitive ability.

Specific Predictive Factors for Adults

The specific factors predicting speech perception scores in the adult are electrical stimulation of the promontory results, length of insertion and the number of stimulating electrodes, and dynamic range.

Electrical Stimulation of the Promontory Results

There was a positive correlation between preoperative tests of temporal processing via promontory stimulation of the auditory nerve (electrode placed on the medial wall of the middle ear), and speech perception results in 64 postlinguistically deaf adults at the Melbourne Clinic (Blamey et al 1992). Discriminating gaps smaller than 50 ms for low rates of stimulation and pitch changes for rates of 100 and 200 pulses/s suggested a good result. The ability to detect changes in rate of stimulation and gaps between stimuli appeared to be a more central function, and thus important for segmenting speech and processing the slow frequency changes occurring in voicing. If duration difference limens were large, speech results would be poor and vice versa (Blamey et al 1992).

Length of Insertion of the Electrode Array and Number of Stimulus Electrodes

A positive relationship was seen between the length of insertion and the number of electrodes. Both correlated positively with speech perception. Two studies were carried out on adults, and showed that there was increasing benefit in having additional electrodes up to 20. A regression analysis by Blamey et al (1992) revealed there was a difference between nine and 21 electrodes (12 electrodes) that accounted for a 24% increase in score (i.e., 2% per electrode). Studies also demonstrated that 20 rather than eight banded electrodes provided improved speech processing for the Nucleus Multipeak and SPEAK in noise. These results highlight the importance of multiple-channel stimulation for coding the spectral information in speech.

In contrast, the number of stimulus channels for a fixed filter (modified channel vocoder) was shown by Dorman et al (1989) to be four, but this did not apply to noise. In addition, Shannon et al (1995) found with acoustic models of fixed-filter strategies that a four-channel processor gave hearing listeners near-normal speech recognition in quiet listening conditions.

So although there is some evidence that fixed filter strategies may not require more than at most seven electrodes, there is a definite correlation between the number of electrodes with the Multipeak and SPEAK strategy and speech perception in noise.

Dynamic Range

Postoperatively there is a positive correlation between the dynamic range and speech score (Blamey et al 1992). The greater the dynamic range between the threshold and maximum comfortable level, the more steps in loudness for presenting speech. A greater dynamic range is due to a higher density of spiral ganglion cells as seen in the studies on the human temporal bones from patients in whom psychophysics data were available (Kawano et al 1995, 1996, 1998).

General Predictive Factors for Children

In Melbourne, after establishing the benefits of the multiple-channel cochlear implant with F0/F2 and F0/F1/F2 speech-processing strategies in adults, the first three children were implanted in 1985 and 1986. This was the start of an international trial for the FDA to determine whether the multiple-electrode cochlear implant and F0/F1/F2 strategy would benefit children. The FDA approved the device as safe and effective for children 2 years of age and older in 1990.

There are similar general predictive factors for the child as for the adult: age when deafened, age at implantation, duration of deafness, duration of implantation, etiology, the presence of a progressive hearing loss, degree of residual hearing, speech reading ability, speech-processing strategy, and medical condition. These factors have been evaluated in a number of studies (Quittner and Steck 1991; Blamey et al 1992; Gantz et al 1993; Battmer et al 1995; Blamey et al 1996). Quittner and Steck (1991) also had parents rate implant usage and found that a positive rating correlated with communication mode, time using the implant device, and performance on two subsets from the Wechsler Intelligence Scale for Children–Revised (WISC-R).

In a study by Dowell et al (1995) the ability of children ($n = 100$) with the Nucleus 22 system was categorized and analyzed with speech perception as the dependent variable and preoperative and postoperative parameters as the independent variable. This showed that the duration of the hearing loss correlated negatively with perception, and a progressive loss, useful preimplant hearing, experience with the implant, and an auditory-oral education all correlated positively. (An auditory-oral education focuses on developing spoken language through hearing and lipreading.) These five variables accounted for 37% of the overall variance.

In a larger study on 167 children from the Universities of Melbourne and Sydney (Sarant et al 2001) the relation between speech perception and possible contributory factors was assessed using analysis of covariance with the general linear model of Minitab Version 12 (Ryan and Joiner 1994). There were five factors that had a significant effect on phoneme scores for the PBK (Haskins 1949, 1964) and CNC (Peterson and Lehiste 1962) word tests, and accounted for 51% of the variance. These factors were duration of deafness, implant experience, communication mode, clinic, and speech processor. If the effect of the clinic was excluded (as they could have different selection criteria and training method), the remaining four factors accounted for 34% of the variance. As with adults, further research is required to determine which children are most likely to do best. As discussed below, the differences between the Melbourne and Sydney clinics were that the children in Sydney had a higher incidence of residual hearing and auditory-oral education. The factors producing variance in speech perception and production, as well as spoken and written language for the Nucleus SPEAK strategy, were determined on 136 children by Geers et al (2002). The data were classified as independent variables (communication mode, classroom, therapy), and intervening variables (characteristics of the child: age, age at onset, age at implantation, IQ; family: size, parent's education; and implant). The child and family characteristics (primarily the nonverbal IQ) contributed approximately 20% of the variance. An additional 24% was due to implant characteristics, and 12% for educational factors.

In a group of older children ($n = 25$) between 8 and 18 years of age, the main factors correlating with speech perception were duration of the profound loss, preoperative sentence score, and equivalent language age. These factors accounted for 66% of the variance. These children had a mean sentence score of 47% that was statistically the same as the overall group in Melbourne (Dowell et al 2002b).

Age When Deafened

If the hearing loss occurred after 4 to 6 years of age, the person is postlinguistically deaf, and could expect the results normally obtained by people who have lost hearing after developing language. If the hearing loss occurred before 4 to 6 years, the person is prelinguistically deaf, and the results depend on a number of factors, which include age and language skills.

In the study with the Nucleus 22 F0/F1/F2 strategy for the FDA, the results were categorized into the children's best perception levels for detection, pattern, closed set, and open set, and it was discovered that the closed- and open-set results were significantly better postoperatively for both the pre- and postlinguistic groups (Staller 1990; Staller et al 1991a,b). The open-set scores, however, were better for the postlinguistic group as illustrated in Figure 9.12. The trial showed that 60% of children born deaf were able to understand some open-set speech. Age at the onset of deafness was shown to be a significant predictor of speech

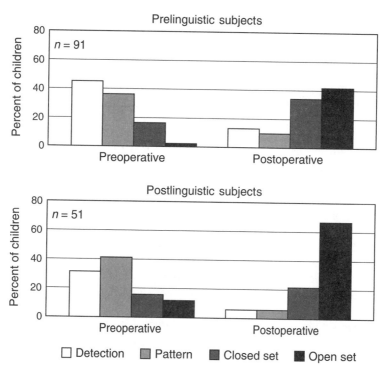

FIGURE 9.12. Results in children postlinguistically versus prelinguistically deaf (Reprinted with permission from Clark 2002. Learning to understand speech with the Cochlear implant. In: Fahle M. and Poggio T., Perceptual Learning. MIT Press.).

perception with the Nucleus 22 system by Osberger et al (1991), Staller et al (1991b), and Dawson et al (1995a,b). Since the introduction of the SPEAK strategy and operations at a young age, the open-set results for children who are prelinguistically deaf are the same as for postlinguistically deaf adults. For example, Dowell et al (1995) found that whether the hearing loss was congenital or not, there was no correlation with speech perception. This was supported by the findings of Staller et al (1997).

Age at Implantation

In children, age at implantation correlated negatively with speech perception and production results. Initially it was found that children implanted during adolescence had a low chance of achieving open-set speech understanding using electrical stimulation alone (Clark, Blamey et al 1987; Clark, Busby et al 1987; Tong et al 1988; Busby et al 1991; Dowell et al 1991). Congenitally and early deafened children using a Nucleus multiple-channel cochlear implant were shown to achieve better speech recognition when receiving it at a younger age (Dawson et al 1989). This finding was established by Dowell et al (1997), Fryauf-Bertschy et al (1997), and Miyamoto et al (1997).

Furthermore, Dowell et al (1997) reported in children implanted from 1.9 to 19.9 years that with congenital deafness, speech perception, as measured by open-set BKB sentences and phonemes in monosyllabic words, improved down to a least 3 years of age. O'Donoghue et al (2000) showed that in a group of 40 children using the Nucleus system, age at implantation was one of two factors correlating with speech perception, as measured by continuous discourse tracking. In a study by Allum et al (2000) using the Nucleus 22 and 24 for 50 children, and the Combi-40+ for 21 children, they demonstrated that speech perception, as tested with the listening progressive profile (LiP) (Archbold 1994), mono-syllable, trochee, polysyllable (MTP) (Erber and Alencewicz 1976), and the Meaningful Auditory Integration Scale (MAIS) (Robbins et al 1991) tests, increased more rapidly in children under 7 years of age. Further evidence for the importance of operating at a young age was seen in a study by Kirk et al (2002) on 50 children with the Nucleus Speak strategy, 14 with Nucleus ACE, and 10 with the Clarion CIS. Children who had implantation before 3 years of age had a faster rate of language development than those who were older. Interestingly, in the study by Geers et al (2002) once native nonverbal intelligence was factored out, age at implantation under 5 years was not significant. Older children, nevertheless, may benefit and should not be excluded from surgery. Osberger et al (1998) found better speech perception in a group of 30 children who received an implant after 5 years of age than with conventional hearing aids. Gary and Hughes (2000) also showed benefits in a group of children who obtained implants at between 8 and 14 years of age. In some this was accompanied by increased improvement in receptive and expressive language and speech intelligibility.

Tye-Murray et al (1995) found for the Nucleus system that the speech production of young children between 2 and 4 years of age increased more rapidly, and reached the level of older children within 2 years. A similar result was reported by Nikolopoulos et al (1999) in a group of 126 children, and by Barker et al (2000). A similar trend was reported by Waltzman and Cohen (1998) for children implanted with the Nucleus system before the age of 2 years. The above findings were probably due to perceptual learning and neural connectivity at an early age when the brain connections are more plastic (Nordeen et al 1983; Busby and Clark 2000).

Duration of Deafness

In children, as distinct from adults, age at implantation and duration of deafness cannot be readily separated, as most deaf children are congenitally deaf. With deafness of long duration, children's results are not as good (Dowell et al 1995; Dawson et al 1995a,b; Dowell 1997; Dowell et al 1997, 2002b). In the study by Sarant et al (2001), it was shown that average phoneme scores decreased by 1.4% per year of profound hearing loss. Furthermore, children with a prolonged history of deafness require a lengthy period of rehabilitation or habilitation for adequate speech perception (Dowell et al 1997; Tomblin et al 1999). Duration of deafness may have its effect through a greater loss of the neurons and their connections during the plastic stages in brain connectivity.

Progressive Hearing Loss and Residual Hearing

A progressive hearing loss may allow a child to learn to process a degraded signal, and it is associated with better results (Gantz et al 1993). This was also seen in the study of Sarant et al (2001), in which better results for the Sydney clinic could have been due to the significantly greater number of children with preoperative aided thresholds in the 70 dB SPL speech range at 2000 Hz. If residual hearing had been present, it is probable that appropriate neural connectivity would have been established during the critical period for plasticity, and facilitate the coding of speech. This was also seen in the study on 256 children with the Nucleus 24 and Contour perimodiolar array in severely to profoundly deaf children over 24 months. There were better postoperative open-set speech scores when there had been some hearing preoperatively (Staller et al 2002). The children were divided into two groups: those with no open-set words recognition and those with up to 30% recognition. Six months postoperatively, those with no open-set speech had scores of 26% and those with open-set 58%.

Speech Reading Ability

A weak correlation exists between speech reading and speech perception, and this may reflect good top-down processing (Cohen et al 1993; Gantz et al 1993).

Language Levels

Language development influences speech perception. This was seen in studies by Blamey et al (1998) and Sarant et al (1996, 2001) on implanted children in which a close relationship was seen when the speech perception word and sentence scores for audition and speech reading were plotted against the PPVT (Dunn and Dunn 1997) or the clinical evaluation of language fundamentals (CELF) equivalent language ages. The close relationship demonstrated the important role language may have on speech perception.

Speech-Processing Strategy

The speech-processing strategy affects results in children as it does in adults (Dowell et al 1995). Improvements have primarily been affected by presenting additional frequency information on a place-coding basis. Children were first converted from the Nucleus Multipeak to SPEAK after a period of training, and the majority had better results (Cowan et al 1995) (see Chapter 12). Significantly better speech perception was seen in a group of children from the Melbourne clinic for the SPEAK strategy in an analysis by Dowell et al (2002a).

Duration of Implantation

The duration of implantation is correlated with good speech perception. This was reported by Quittner and Steck (1991), Sarant et al (2001), and Dowell et al (2002a). In the study by Sarant et al, the average PBK and CNC word scores increased by 1.7% per year of implant experience. However, the learning required in congenitally deaf children is longer than that for postlinguistically deaf adults, and presumably due to lack of prior exposure to sound and inadequate language development. Nevertheless, when the operation is carried out at a young age and the child has an adequate sensory input, open-set speech understanding with the Nucleus system was found to occur within the first year (Fryauf-Bertschy et al 1992, 1997; Gantz et al 1994; Miyamoto et al 1996; Osberger et al 1996).

Etiology (Cause of Deafness)

With etiology, meningitis correlated negatively with results as in the adult. This may have been the result of the reduced number of electrodes inserted due to labyrinthitis ossificans or secondary to cortical involvement (Blamey et al 1995). In contrast, the analysis of results from 40 children by O'Donoghue et al (2000) did not show a relationship with etiology even though there was a 58% incidence of meningitis. This could be due to improved surgical approaches and earlier operation.

Medical Condition

Medically, one of the main issues is the management of children with recurrent otitis media. This problem arises in a high proportion of children. This is discussed in Chapter 10.

Communication Strategy Before Surgery

The communication strategy adopted before surgery influences results, and children perform better if they have had an auditory-oral education. This was reported by Quittner and Steck (1991) in a sample of 29 profoundly hearing impaired children, who had been using their devices an average of 2 years, and confirmed by Dowell et al (1995; 1997), Dowell (1997), and O'Donoghue et al (2000). Results were poorer if total communication with a signed input was used in preference to an oral one.

Mode of Education After Surgery

The mode of education after surgery is important, and an auditory-oral education is required for best results. This was seen in the regression analysis of factors relating to speech perception by Dowell et al (1995). In the study by Sarant et al (2001), the better results for the Sydney clinic were in part due to the fact that a significantly greater proportion of children had education by auditory-oral rather than total communication means. With total communication the child not only has an auditory/oral input, but visual signs as well. In an evaluation of 102 children using the Nucleus implant there was a highly significant difference between those who used exclusively auditory-oral communication and those who used some level of manual communication (Dowell et al. 2002a). However, it has been said the results for mode of education are subject to selection; that is, children who have poor speech perception may require total communication or children who have the potential to do better are being selected for the auditory-oral program. It is relevant, however, that with the data in Melbourne, although there was a wide spread of performance for both groups, children with open-set scores of 50% or more were only seen in the auditory-oral group (Dowell, personal communication). Further support for the benefit of oral education over total communication for young children was seen in a study by Kirk et al (2002) on 64 Nucleus and 10 Clarion patients.

Parental Support

Parental support is an important factor leading to good results. Family dynamics are an issue especially with the breakdown of many marriages, but it should not prevent implantation unless the situation is extreme. Geers et al (2002) found that children with later onset of deafness, from smaller families, and with better-educated parents tended to have higher language scores when speech and signs were considered together. There was no correlation, however, between speech perception and socioeconomic status (O'Donoghue et al 2000).

Delayed Cognitive and Motor Milestones

To see if delayed motor and cognitive milestones affected the cochlear implant results, the performances for children with and without these disabilities were compared. As some etiologies, such as rubella, CMV, meningitis, anoxia, prematurity, and kernicterus, and certain syndromes were likely to affect the central nervous system and so cause delayed motor and cognitive milestones, the study also compared performance across these etiologies (Pyman et al 2000). The results showed that the incidence of motor and cognitive delays was fairly evenly spread across etiologies, with the exception of CMV, which had a higher than average incidence in the delayed group. However, etiology did not have a significant effect on speech perception. Children with delayed cognitive and motor milestones did significantly worse, as they had poorer speech perception (the data were studied with an analysis of variance and general linear model). The data from a study on children in the Melbourne clinic showed that it took much longer to reach targets for children with developmental delays, and this applied in particular for open-set recognition (Pyman et al 2000).

Although the benefits were not as good in children with developmental delays, they may receive a greater relative benefit because of their handicap. But surgery is inadvisable for a child who has a very severe disability and is not able to follow instructions. If their habilitation is slow, complementary help with sign language of the deaf may be required.

Specific Predictive Factors for Children

The specific factors predicting speech perception scores in the child are length of insertion and the number of stimulating electrodes, dynamic range, and implant-evoked brainstem auditory potentials.

Length of Insertion of the Electrode Array and Number of Stimulus Electrodes

A positive relationship was seen between the length of insertion and the number of electrodes. One study on children showed that there was increasing benefit in having additional electrodes up to 20 (Blamey et al 1992). The number of active electrodes correlated positively with speech and language results in the study of Geers et al (2002). In contrast, the analysis of O'Donoghue et al (2000) did not show a positive correlation.

Dynamic Range

Postoperatively there was a positive correlation between the dynamic range and speech score in adults (Blamey et al 1992). Dynamic range and loudness growth also affected the results in children (Geers et al 2002). The greater the dynamic range between the threshold and maximum comfortable level, the more steps in loudness that are available for presenting speech information.

Implant-Evoked Brainstem Auditory Potentials (IMPEBAP)

If a child's performance is poorer than expected, this may be due to inadequate numbers of spiral ganglion cells and auditory nerve fibers. This can be assessed with IMPEBAP (O'Leary et al 2000). These potentials were recorded from eight children, and in three they were either absent or abnormal. These three children had poor speech perception (William Gibson, personal communication).

So the data suggest that for learning speech there is first a need to transmit the essential sensory information, and then skills at a higher processing level are required for speech perception.

Preoperative Counseling

It is of the utmost importance to the success of the cochlear implant operation that the patient be motivated to persevere through the often difficult early postoperative period. While some patients adjust to the hearing relatively easily, a similar number find that it takes hard work to achieve good results.

A great deal of time is spent throughout the preoperative period explaining the cochlear implant to the prospective patient and family. Arrangements are made for the patient to meet with an implant recipient with a similar background (and one who it is felt gained the amount of benefit from the implant that would be predicted for the prospective patient).

References

Allum, J. H. J., R. Greisiger, S. Straubhaar and M. G. Carpenter. 2000. Auditory perception and speech identification in children with cochlear implants tested with the EARS protocol. British Journal of Audiology 34: 293–303.

ANSI. 1989. Specifications for audiometers. New York, American National Standards Institute.

Archbold, S. 1994. Monitoring progress in children at the pre-verbal stage. In: McCormick, B. and S. Sheppard, eds. Cochlear implants for young children. London, Whurr Publishers: 197–213.

Arndt, P., S. Staller, J. Arcaroli, A. Hines and K. Ebinger. 1999. Within-subject comparison of advanced coding strategies in the Nucleus 24 cochlear implant. Cochlear Corporation Report.

Barker, E., T. Daniels, R. Dowell, et al. 2000. Long term speech production outcomes in children who received cochlear implants before and after 2 years of age. 5th European Symposium on Paediatric Cochlear Implantation, Antwerp, Belgium: 156.

Battmer, R.-D., D. Gnadeberg, D. J. Allum-Mecklenburg and T. Lenarz. 1995. Matched-paired comparisons for adults using the Clarion or Nucleus devices. Annals of Otology, Rhinology and Laryngology 104(suppl 166): 251–254.

Bench, R. J. and J. Bamford. 1979. Speech-hearing tests and the spoken language of hearing-impaired children. London, Academic Press.

Blamey, P. J., P. Arndt, F. Bergeron, et al. 1996. Factors affecting auditory performance of postlinguistically deaf adults using cochlear implants. Audiology and Neuro-Otology 1: 293–306.

Blamey, P. J., E. Parisi and G. M. Clark. 1995. Pitch matching of electric and acoustic stimuli. Annals of Otology, Rhinology and Laryngology 104: 220–222.

Blamey, P. J., B. C. Pyman, M. Gordon, et al. 1992. Factors predicting postoperative sentence scores in postlinguistically deaf adult cochlear implant patients. Annals of Otology, Rhinology and Laryngology 101: 342–348.

Blamey, P. J., J. Z. Sarant, T. A. Serry, et al. 1998. Speech perception and spoken language in children with impaired hearing. In: Mannell, R. H. and J. Robert-Ribes, eds. ICSLP '98 proceedings. Canberra, Australian Speech Science and Technology Association: 2615–2618.

Boothroyd, A. 1968. Developments in speech audiometry. Sound 2: 3–10.

Boothroyd, A. 1993. Speech perception, sensorineural hearing loss, and hearing aids. In: Studebaker, G. A. and I. Hochberg, eds. Acoustical factors affecting hearing aid performance. Needham Heights, MA, Allyn and Bacon: 277–300.

Boothroyd, A. 1997. Auditory capacity of hearing-impaired children using hearing aids and cochlear implants: issues of efficacy and assessment. Scandinavian Audiology Supplementum 46: 17–25.

Boothroyd, A. 1998. Evaluating the efficacy of hearing aids and cochlear implants in children who are hearing-impaired. In: Bess, F. H., ed. Children with hearing impairment: contemporary trends. Nashville, Bill Wilkerson Center Press: 249–260.

Boothroyd, A., L. Hanin and O. Eran 1996. Speech perception and production in children with hearing impairment. In: Bess, F. H., J. S. Gravel and A. M. Tharpe, eds. Amplification for children with auditory deficits. Nashville, Wilkerson Center Press: 55–74.

Brown, A. M., R. C. Dowell, G. M. Clark, L. F. Martin and B. C. Pyman. 1985. Selection of patients for multiple-channel cochlear implant patient. In: Schindler, R. A. and M. M. Merzenich, eds. Cochlear implants. New York, Raven Press: 403–406.

Busby, P. A. and G. M. Clark. 2000. Electrode discrimination by early-deafened subjects using the Cochlear Limited multiple-electrode cochlear implant. Ear and Hearing 21: 291–304.

Busby, P. A., S. A. Roberts, Y. C. Tong and G. M. Clark. 1991. Results of speech perception and speech production training for three prelingually deaf parents using a multiple-electrode cochlear implant. British Journal of Audiology 25: 291–302.

Clark, G. M. 2002. Learning to understand speech with the cochlear implant. In: Textbook of Perceptual Learning. Fahle, M. and T. Poggio, eds. Cambridge, Mass., MIT Press: 147–160.

Clark, G. M. 1999. Cochlear implants in the third millennium. American Journal of Otology 20: 4–8.

Clark, G. M., P. J. Blamey, P. A. Busby, et al. 1987. A multiple-electrode intracochlear implant for children. Archives of Otolaryngology 113: 825–828.

Clark, G. M., P. A. Busby, S. A. Roberts, et al. 1987. Preliminary results for the Cochlear Corporation multi-electrode intracochlear implants on six prelingually deaf patients. American Journal of Otology 8: 234–239.

Clark, G. M., B. J. O'Loughlin, F. W. Rickards, Y. C. Tong and A. J. Williams. 1977. The clinical assessment of cochlear implant patients. Journal of Laryngology and Otology 91: 697–708.

Clark, G. M. and B. C. Pyman. 1997. Preoperative medical evaluation. In: Clark, G., R. Cowan and R. Dowell, eds. Cochlear implantation for infants and children. Advances. San Diego, Singular: 71–82.

Clark, G. M., R. K. Shepherd, B. K.-H. G. Franz and D. Bloom. 1984. Intracochlear electrode implantation. Round window membrane sealing procedures and permeability studies. Acta Oto-Laryngologica (suppl 410): 5–15.

Clark, G. M., Y. C. Tong, L. F. Martin and P. A. Busby. 1981. A multiple-channel cochlear implant. An evaluation using an open-set word test. Acta Oto-Laryngologica 91: 173–175.

Cohen, N. L., S. B. Waltzman and S. G. Fisher. 1993. A prospective, randomised study of cochlear implants. New England Journal of Medicine 328: 233–282.

Consensus. 1995. National Institutes of Health Consensus Conference Cochlear implants in adults and children. JAMA 274: 1955–1961.

Cowan, R. S. C., C. D. Brown, L. A. Whitford, et al. 1995. Speech perception in children using the advanced SPEAK speech-processing strategy. Annals of Otology, Rhinology and Laryngology 104(suppl 166): 318–321.

Crary, M. A. 1982. Phonological intervention concepts and procedures. San Diego, College-Hill Press.

Crystal, D., P. Fletcher and M. Garman. 1976. The grammatical analysis of language disability. London, Edward Arnold.

Damsma, H., J. A. M. de Groot and F. W. Zonneveld. 1984. CT of cochlear otosclerosis. Radiologic Clinics of North America 22(1): 37–43.

Davis, H. and S. R. Silverman. 1978. Hearing and Deafness. 4th edition. New York, Holt, Rinehart & Winston.

Dawson, P., P. J. Blamey, G. M. Clark, et al. 1989. Results in children using the 22 electrode cochlear implant. Journal of the Acoustical Society of America 86(suppl 1): 81.

Dawson, P. W., P. J. Blamey, S. J. Dettman, et al. 1995a. A clinical report on speech production of cochlear implant users. Ear and Hearing 16: 551–561.

Dawson, P. W., P. J. Blamey, L. C. Rowland, et al. 1995b. A clinical report on receptive vocabulary skills in cochlear implant users. Ear and Hearing 16: 287–294.

Dorman, M. F., K. Dankowski and G. McCandless. 1989. Consonant recognition as a function of the number of channels of stimulation by patients who use the Symbion cochlear implant. Ear and Hearing 10: 288–291.

Dorman, M. F., L. Smith, G. McCandless, G. Dunnavant, J. Parkin and K. Dankowski.

1990. Pitch scaling and speech understanding by patients who use the Ineraid cochlear implant. Ear and Hearing 11: 310–315.

Dowell, R. C. 1997. Preoperative audiological, speech, and language evaluation. In: Advances. Clark, G. M., R. S. C. Cowan and R. C. Dowell, eds. Cochlear implantation for infants and children. San Diego, Singular: 83–110.

Dowell, R. C., P. J. Blamey and G. M. Clark. 1995. Potential and limitations of cochlear implants in children. Annals of Otology, Rhinology and Laryngology 104(suppl 166): 324–327.

Dowell, R. C., P. J. Blamey and G. M. Clark. 1997. Factors affecting outcomes in children with cochlear implants. In: Clark, G. M., ed. Cochlear implants. XVI World Congress of Otorhinolaryngology Head and Neck Surgery. Bologna, Monduzzi: 297–303.

Dowell, R. C., G. M. Clark, P. M. Seligman, P. J. Blamey, A. M. Brown and Y. C. Tong. 1986a. Perception of connected speech without lipreading, using a multi-channel hearing prosthesis. Acta Oto-Laryngologica 102: 7–11.

Dowell, R. C., R. S. C. Cowan, G. Rance, R. D. Hollow, S. J. Dettman and E. J. Barker. 1998. Issues in the selection of children for cochlear implantation. Australian Journal of Audiology 20(suppl): 42–43.

Dowell, R. C., P. W. Dawson, S. J. Dettman, et al. 1991. Multichannel cochlear implantation in children. A summary of current work at the University of Melbourne. American Journal of Otology 12(suppl): 137–143.

Dowell, R. C., S. J. Dettman, P. J. Blamey, E. J. Barker and G. M. Clark. 2002a. Speech perception in children using cochlear implants: prediction of long-term outcomes. Cochlear Implants International 3(1): 1–18.

Dowell, R. C., S. J. Dettman, K. Hill, E. Winton, E. J. Barker and G. M. Clark. 2002b. Speech perception outcomes in older children who use multichannel cochlear implants: older is not always poorer. Annals of Otology, Rhinology and Laryngology 111(suppl 189): 97–101.

Dowell, R. C., M. A. Marsh, R. D. Hollow, et al. 1993. Clinical comparison of open-set speech perception with the MSP and WSPIII speech processors and preliminary results for the new SPEAK processor. Abstracts of Third International Cochlear Implant Conference, Innsbruck: 14.6.

Dowell, R. C., L. F. Martin, G. M. Clark and A. M. Brown. 1985. Results of a preliminary clinical trial on a multiple-channel cochlear prosthesis. Annals of Otology, Rhinology and Laryngology 94: 244–250.

Dowell, R. C., D. J. Mecklenburg and G. M. Clark. 1986b. Speech recognition for 40 patients receiving multichannel cochlear implants. Archives of Otolaryngology 112: 1054–1059.

Dunn, L. M. and L. M. Dunn. 1981. Peabody picture vocabulary test–revised. Circle Pines, American Guidance Service.

Dunn, L. M. and L. M. Dunn. 1997. Peabody picture vocabulary test, 3rd ed. Circle Pines, American Guidance Service.

Elliott, L. L. and D. R. Katz. 1980. Northwestern University children's perception of speech (NU-CHIPS). St. Louis, Auditec.

Erber, N. P. 1982. Auditory training. Washington, DC, Alexander Graham Bell Association for the Deaf.

Erber, N. P. and C. M. Alencewicz. 1976. Audiological evaluation of deaf children. Journal of Speech and Hearing Disorders 41: 256–276.

Erber, N. P. and C. Lind. 1994. Communication therapy: theory and practice. Journal of the Academy of Rehabilitative Audiology 27: 267–287.

Finitzo-Hieber, T., I. J. Gerling, N. D. Matkin and E. Cherow-Skalka. 1980. A sound effects recognition test for the pediatric audiological evaluation. Ear and Hearing 1: 271–276.

Fisher, H. B. and J. A. Logemann. 1971. Test of articulation competence. New York, Houghton and Mifflin.

Fryauf-Bertschy, H., R. S. Tyler, D. M. Kelsay and B. J. Gantz. 1992. Performance over time of congenitally deaf and postlingually deafened children using a multichannel cochlear implant. Journal of Speech and Hearing Research 35: 913–920.

Fryauf-Bertschy, H., R. S. Tyler, D. M. Kelsay, B. J. Gantz and G. G. Woodworth. 1997. Cochlear implant use by prelingually deafened children: the influences of age at implant use and length of device use. Journal of Speech and Hearing Research 40: 183–199.

Gantz, B. J., R. S. Tyler, G. Woodworth, N. Tye-Murray and H. Fryauf-Bertschy. 1994. Results of multichannel cochlear implant in congenital and acquired prelingual deafness in children: five-year follow-up. American Journal of Otology 15(suppl 2): 1–8.

Gantz, B. J., G. G. Woodworth, J. F. Knutson, P. J. Abbas and R. S. Tyler. 1993. Multivariate predictors of audiological success with multichannel cochlear implants. Annals of Otology, Rhinology and Laryngology 102(12): 909–916.

Gary, L. and C. Hughes. 2000. A second look at "tweeners". Candidacy considerations for 8 to 14. Presented at International Cochlear Implant Conference, Miami, Florida.

Geers, A., C. Brenner, J. Nicholas, R. Uchanski, N. Tye-Murray and E. Tobey. 2002. Rehabilitation factors contributing to implant benefit in children. Annals of Otology Rhinology and Laryngology 111(suppl 189): 127–130.

Gorlin, R. J., H. V. Toxiello and M. M. Cohen. 1995. Hereditary hearing loss and its syndromes. New York, Oxford University Press.

Haskins, H. 1949. A phonetically balanced test of speech discrimination for children. Master's thesis, Northwestern University.

Haskins, H. A. 1964. Kindergarten PB word lists. In: Newby, H. A., ed. Audiology. New York, Appleton Century Crofts.

Hochmair-Desoyer, I. J. and K. Burian. 1985. Reimplantation of a molded scala tympani electrode: impact on psychophysical and speech discrimination abilities. Annals of Otology, Rhinology and Laryngology 94(1 pt 1): 65–70.

Hood, J. D. and J. P. Poole. 1971. Speech audiometry in conductive and sensorineural hearing loss. Sound 5: 30–38.

Ingram, D. 1976. Phonological disability in children. New York, Elsevier.

Kawano, A., H. L. Seldon and G. M. Clark. 1996. Computer-aided three-dimensional reconstruction in human cochlear maps: measurement of the lengths of organ of corti, outer wall, inner wall, and Rosenthal's canal. Annals of Otology, Rhinology and Laryngology 105: 701–709.

Kawano, A., H. L. Seldon, G. M. Clark and R. Ramsden. 1998. Intracochlear factors contributing to psychophysical percepts following cochlear implantation. Acta Oto-Laryngologica 118: 313–326.

Kawano, A. S., H. L. Seldon, G. M. Clark, R. Madsen and C. Raine. 1995. Intracochlear factors contributing to psychophysical percepts following cochlear implantation: a case study. Annals of Otology, Rhinology and Laryngology 104: 54–57.

Kessler, D. K., G. E. Loeb and M. J. Barker. 1995. Distribution of speech recognition results with the Clarion cochlear prosthesis. Annals of Otology, Rhinology and Laryngology 104: 283–285.

Kim, H.-N., M.-H. Chung, Y.-T. Shim and J.-S. Yoon. 1995. Aided versus implanted speech recognition abilities in severe to profound postlingual deafness. Annals of Otology, Rhinology and Laryngology 104: 153–154.

Kirk, K. I., R. T. Miyamoto, C. L. Lento, E. Ying, T. O'Neill and B. Fears. 2002. Effects of age at implantation in young children. Annals of Otology Rhinology and Laryngology 111(suppl 189): 69–73.

Kirk, K. I., D. B. Pisoni and M. J. Osberger. 1995. Lexical effects on spoken word recognition by pediatric cochlear implant users. Ear and Hearing 16: 470–481.

Lamore, P. J. J., C. Verweij and M. P. Brocaar. 1985. Investigations of the residual hearing capacity of severely hearing-impaired and profoundly deaf subjects. Audiology 24: 343–361.

Ling, D. 1976. Speech and the hearing impaired child: theory and practice. Washington, DC, AG Bell Association for the Deaf.

Ling, D. and A. H. Ling. 1978. Aural habilitation: the foundations of verbal learning in hearing-impaired children. Washington, DC, AG Bell Association for the Deaf.

Mafee, M. F., G. E. Valvassori, R. L. Deitch, et al. 1985. Use of CT in the evaluation of cochlear otosclerosis. Radiology 156: 703–708.

McGarr, N. S. 1983. The intelligibility of deaf speech to experienced and inexperienced listeners. Journal of Speech and Hearing Research 26: 451–458.

Miyamoto, R. T., K. I. Kirk, A. M. Robbins, S. Todd and A. Riley. 1996. Speech perception and speech production skills of children with multichannel cochlear implants. Acta Otolaryngologica 116: 240–243.

Miyamoto, R. T., K. I. Kirk, A. M. Robbins, S. Todd, A. Riley and D. B. Pisoni. 1997. Speech perception and speech intelligibility in children with multichannel cochlear implants. Advances in Oto-Rhino-Laryngology 52: 198–203.

Moog, J. S. and A. E. Geers. 1979. Grammatical analysis of elicited language: simple sentence level. St. Louis, Central Institute for the Deaf.

Moog, J. S. and A. E. Geers. 1980. Grammatical analysis of elicited language: complex sentence level. St. Louis, Central Institute for the Deaf.

Moog, J. S. and A. E. Geers. 1990. Early speech perception test for profoundly hearing-impaired children. St. Louis, Central Institute for the Deaf.

Myer, B., M. Drira, D. Geger and C. H. Chouard. 1984. Results of round window electrical stimulation in 460 cases of total deafness. Acta Otolaryngologica suppl 411: 168–176.

Nadol, J. B., Y.-S. Young and R. B. Glynn. 1989. Survival of spiral ganglion cells in profound sensorineural hearing loss: implications for cochlear implantation. Annals of Otology, Rhinology and Laryngology 98: 411–416.

Nikolopoulos, T., G. O'Donoghue and S. Archbold. 1999. Age at implantation: its importance in pediatric cochlear implantation. Laryngoscope 109: 595–599.

Nilsson, M. J., S. D. Soli and D. J. Gelnett. 1996. Development and norming of a hearing in noise test for children. Los Angeles, House Ear Institute Internal Report.

Nilsson, M., S. D. Soli and J. A. Sullivan. 1994. Development of the hearing in noise test for the measurement of speech reception thresholds in quiet and in noise. Journal of the Acoustical Society of America 95: 1085–1099.

Nordeen, K. W., H. P. Killackey and L. M. Kitzes. 1983. Ascending projections to the inferior colliculus following unilateral cochlear ablation in the neonatal gerbil, Meriones unguiculatus. Journal of Comparative Neurology 214: 144–153.

O'Donoghue, G. M., T. P. Nikolopoulos and S. M. Archbold. 2000. Determinants of speech perception in children after cochlear implantation. Lancet 356: 466–468.

O'Leary, S. J., T. E. Mitchell, W. P. R. Gibson and H. Sanli. 2000. Abnormal positive potentials in round window electrocochleography. American Journal of Otology 21: 813–818.

Osberger, J. J., R. T. Miyamoto and S. Zimmerman-Phillips. 1991. Independent evaluation of the speech perception abilities of children with the Nucleus 22-channel cochlear implant system. Ear and Hearing 12 (suppl): 66S–80S.

Osberger, M., L. Fisher, L. Zimmerman-Phillips and M. Barker. 1998. Speech recognition performance of older children with cochlear implants. The American Journal of Otology 19: 152–175.

Osberger, M. J., A. M. Robbins, S. L. Todd, A. I. Riley, K. I. Kirk and A. E. Carney. 1996. Cochlear implants and tactile aids for children with profound hearing impairment. In: Bess, F., J. Gravel and A. M. Tharpe, eds. Amplification for children with auditory deficits. Nashville, TN, Bill Wilkerson Center Press: 283–307.

O'Sullivan, P., S. Ellul, R. C. Dowell, B. C. Pyman and G. M. Clark. 1997. The relationship between aetiology of hearing loss and outcome following cochlear implantation in a paediatric population. In: Clark, G. M., ed. Cochlear implants. XVI World Congress of Otorhinolaryngology Head and Neck Surgery. Bologna, Monduzzi: 169–172.

Owens, E., D. K. Kessler and E. D. Schubert. 1982. Interim assessment of candidates for cochlear implants. Archives of Otolaryngology 108: 478–483.

Owens, E. and C. C. Telleen. 1981. Speech perception with hearing aids and cochlear implants. Archives of Otolaryngology 107: 160–163.

Parkin, J. L., B. E. Stewart, K. Dankowski and L. J. Haas. 1989. Prognosticating speech performance in multichannel cochlear implant patients. Otolaryngology–Head and Neck Surgery 101: 314–9.

Pass, R. F., S. Stagno, G. J. Myers and C. A. Alford. 1980. Outcome of symptomatic congenital cytomegalovirus infection: results of long-term congenital follow-up. Pediatrics 66: 758–762.

Peterson, G. E. and I. Lehiste. 1962. Revised CNC lists for auditory tests. Journal of Speech and Hearing Disorders 27: 62–70.

Plant, G. 1984. A diagnostic speech test for severely and profoundly hearing-impaired children. Australian Journal of Audiology 6: 1–9.

Prutting, C. A. 1986. Pragmatics. Paper presented at the Australian Association of Speech and Hearing Conference, Canberra, Australia.

Pyman, B. C., A. M. Brown, R. C. Dowell and G. M. Clark. 1990. Preoperative evaluation and selection of adults. In: Clark, G., Y. Tong and J. Patrick, eds. Cochlear prostheses. London, Churchill Livingstone: 125–134.

Pyman, B., P. Lacy, G. Clark and R. Dowell. 2000. The development of speech perception in children using cochlear implants: effects of etiologic factors and delayed milestones. American Journal of Otology 21: 57–61.

Quigley, S. P., M. W. Steinkamp, D. J. Power and B. W. Jones. 1978. Test of syntactic abilities. Beaverton, Oregon, Dormac.

Quittner, A. L. and J. T. Steck. 1991. Predictors of cochlear implant use in children. American Journal of Otology 12(suppl): 89–94.

Rance, G., R. C. Dowell, F. W. Rickards, D. E. Beer and G. M. Clark. 1998. Steady-state evoked potential and behavioural hearing thresholds in a group of children with absent click-evoked auditory brain stem response. Ear and Hearing 19: 48–61.

Rance, G., R. C. Dowell, F. W. Rickards and G. M. Clark. 1997. Evoked potential assessment of children with severe/profound hearing loss: a comparison of steady-state evoked potential (SSEP) and behavioural hearing threshold levels in subjects with absent click evoked auditory brainstem responses (ABR). In: Clark, G. M., ed. Cochlear implants. XVI World Congress of Otorhinolaryngology Head and Neck Surgery. Bologna, Monduzzi: 175–179.

Rance, G., F. W. Rickards, L. T. Cohen and G. M. Clark. 1994a. Accuracy of behavioural threshold predicition using steady-state evoked potentials. Australian Journal of Audiology 15(2 suppl): 18.

Rance, G., F. W. Rickards, L. T. Cohen, S. De Vidi and G. M. Clark. 1995. The automated prediction of hearing thresholds in sleeping subjects using auditory steady-state evoked potentials. Ear and Hearing 16: 499–507.

Rance, G., F. W. Rickards, R. C. Dowell, L. T. Cohen and G. M. Clark. 1994b. Steady-state potentials (SSEPs); an objective measure of residual hearing in young cochlear implant candidates. In: I. J. Hochmair-Desoyer, I. J. and E. S. Hochmair, eds. Advances in cochlear implants. Vienna, Manz: 71–74.

Rasmussen, L. 1990. Immune responses to human cytomegalorvirus infection. In: McDougall, J., ed. Cytomegalovirus. Berlin, Springer-Verlag: 221–254.

Rickards, F. W. and G. M. Clark. 1984. Steady-state evoked potentials to amplitude-modulated tones. In: Anodar, R. H., and C. Barber, eds. Evoked potentials II. Boston, Butterworths: 163–168.

Rickards, F. W., S. J. Dettman, P. A. Busby, et al. 1990. Preoperative evaluation and selection of children and teenagers. In: Clark, G. M., Y. C. Tong and J. F. Patrick, eds. Cochlear prostheses. Avon, Great Britain, Churchill Livingstone: 135–152.

Robbins, A. M., J. J. Renshaw and S. W. Berry. 1991. Evaluating meaningful auditory integration in profoundly hearing impaired children. American Journal of Otology 12(suppl): 144–150.

Roth, F. P. and N. J. Spekman. 1984a. Assessing the pragmatic abilities of children: part 1. Organizational framework and assessment parameters. Journal of Speech and Hearing Disorders 49: 2–11.

Roth, F. P. and N. J. Spekman. 1984b. Assessing the pragmatic abilities of children: part 2. Guidelines, considerations, and specific evaluation procedures. Journal of Speech and Hearing Disorders 49: 12–17.

Ryan, B. F. and B. L. Joiner. 1994. Minitab handbook. Belmont, CA, Wadsworth.

Sarant, J. Z., P. J. Blamey and G. M. Clark. 1996. The effect of language knowledge on speech perception in children with impaired hearing. In: McCormack, P. and A. Russell, eds. Proceedings of the Sixth Australian International Conference on Speech Science and Technology. Canberra, Australian Speech Science and Technology Association: 269–274.

Sarant, J. Z., P. J. Blamey, R. C. Dowell, G. M. Clark and W. P. R. Gibson. 2001. Variation in speech perception scores among children with cochlear implants. Ear and Hearing 22: 18–28.

Scrivener, B. P. and W. P. R. Gibson. 1987. Cochlear implant after radical mastoidectomy. Annals of Otology, Rhinology and Laryngology 96: 19–20.

Shannon, R. V., F.-G. Zeng, V. Kamath, J. Wyginski and M. Ekelid. 1995. Speech recognition with primarily temporal cues. Science 270: 303–304.

Shipp, D. B. and J. M. Nedzelski. 1994. Prognostic value of round-window psychophysical testing with cochlear-implant candidates. Journal of Otolaryngology 23(3): 172–6.

Skinner, M. W., G. M. Clark, L. A. Whitford, et al. 1994. Evaluation of a new spectral peak coding strategy for the Nucleus 22 channels cochlear implant system. American Journal of Otology 15: 15–27.

Spreen, O., D. Tupper, A. Risser, H. Tuokke and D. Edgell. 1984. Human developmental neurophysiology. Oxford, Oxford University Press.

Staller, S. J. 1990. Perceptual and production abilities in profoundly deaf children with multichannel cochlear implants. Journal of the American Academy of Audiology 1(1): 1–3.

Staller, S. J., A. L. Beiter, J. A. Brimacombe and P. Arndt. 1991a. Paediatric performance with the Nucleus 22-channel cochlear implant system. American Journal of Otology 12(suppl): 126–136.

Staller, S. J., R. C. Dowell, A. L. Beiter and J. A. Brimacombe. 1991b. Perceptual abilities of children with the Nucleus 22-channel cochlear implant. Ear and Hearing 12(suppl 4): 34S–47S.

Staller, S., C. Menapace and E. Domico. 1997. Speech perception abilities of adult and pediatric Nucleus implant recipients using Spectral Peak (SPEAK) coding strategy. Otolaryngology Head and Neck Surgery 117: 236–242.

Staller, S., A. Parkinson, J. Arcaroli and P. Arndt. 2002. Pediatric outcomes with the Nucleus 24 contour: North American clinical trial. Annals of Otology Rhinology and Laryngology 111(suppl 189): 56–61.

Stapells, D. R., J. S. Gravel and B. A. Martin. 1995. Thresholds for auditory brain stem responses to tones in notched noise from infants and young children with normal hearing or sensorineural hearing loss. Ear and Hearing 16: 361–71.

Thornton, A. R. and M. J. M. Raffin. 1978. Speech-discrimination scores modeled as a binomial variable. Journal of Speech and Hearing Research 21: 507–518.

Tillman, T. W. and R. Carhart. 1966. An expanded test for speech discrimination utilizing CNC monosyllabic words. Northwestern University Auditory Test No. 6, SAM-TR-66-55. Technical Report Sam-Tr: 1–12.

Tomblin, J. B., L. Spencer, S. Flock, R. Tyler and B. Gantz. 1999. A comparison of language achievement in children with cochlear implants and children using hearing aids. Journal of Speech, Language, and Hearing Research 42: 497–511.

Tong, Y. C., P. A. Busby and G. M. Clark. 1988. Perceptual studies on cochlear implant patients with early onset of profound hearing impairment prior to normal development of auditory, speech, and language skills. Journal of the Acoustical Society of America 84: 951–962.

Tye-Murray, N., L. Spencer, E. G. Bedia and G. Woodworth. 1996. Initial evaluation of an interactive test of sentence gist recognition. Journal of the American Academy of Audiology 7: 396–405.

Tye-Murray, N., L. Spencer and G. Woodworth. 1995. Acquisition of speech by children who have prolonged cochlear implant experience. Journal of Speech and Hearing Research 38: 327–337.

Tye-Murray, N., R. S. Tyler, G. G. Woodworth and B. J. Gantz. 1992. Performance over time with a Nucleus or Ineraid cochlear implant. Ear and Hearing 13: 200–209.

Waltzman, S. and N. Cohen. 1998. Cochlear implants in children younger than 2 years old. American Journal of Otology 19: 158–162.

Westby, C. E. 1980. Assessment of cognitive and language abilities through play. Language, Speech and Hearing Services in Schools 11: 154–168.

Zimmerman, I. L., V. C. Steiner and R. E. Pond. 1979. Preschool language scale. Colombus, Merrill.

Zimmerman-Phillips, S., A. M. Robbins and M. J. Osberger. 2000. Assessing cochlear implant benefit in very young children. Annals of Otology, Rhinology and Laryngology 109(suppl 185): 42–43.

10
Surgery

Overview

The surgical implantation of a cochlear prosthesis requires adhering in the first instance to the three cornerstones of modern ear surgery elaborated by Shambaugh (1959): (1) mastery of the complicated anatomy of the ear, (2) meticulous asepsis, and (3) magnification under the operating microscope. In addition, skill in the use of fine drills is essential. The electrode bundle must be inserted atraumatically into the inner ear as close as possible to the auditory nerve fibers. The receiver-stimulator package, which is connected to the electrode bundle, needs to be accommodated in an optimal position in the skull.

In implanting the electrode array and the receiver-stimulator package, great care must be taken, as there are more nerves and vessels concentrated in a small area of the temporal bone than elsewhere in the body. The mastoid bone is partly filled with air cells that have entered from the middle ear cleft at 34 weeks postgestation (Bast and Anson 1949). These cells provide space for the placement of the receiver-stimulator package and lead wires. Nevertheless, just behind the mastoid air cells, the skull often needs to be drilled down to the dural lining of the brain to accommodate the package without it protruding too far above the surface of the skull, and so producing a bulge. Partial removal of the air cells provides a route from behind the ear to the middle ear, and thence to the inner ear. To approach the inner ear or cochlea, an opening needs to be made into the middle ear from behind by drilling between the vertical segment of the nerve to the facial muscles (facial nerve) and a nerve bringing taste sensations from the tongue (chorda tympani nerve). The course of these nerves can vary, and this needs to be taken into consideration to avoid injury.

Finally, the skin must be closed over the receiver-stimulator package thus not leaving a path for the entry of infection. This could occur with percutaneous stimulation with a plug and socket. However, in both cases there is a passage for infection to enter from the nose via the eustachian tube.

Brief history

An intracochlear electrode inserted into the scala tympani via an opening at or near the round window was the approach favored by House and Urban (1973), Michelson and Schindler (1981), Clark, Patrick et al (1979), Clark, Pyman et al (1979, 1984), and Burian et al (1986). More recently a separate opening anterior to the round window has become the standard approach. Previously Simmons (1966) inserted electrodes into the modiolus, and Chouard and MacLeod (1976) carried out a procedure drilling a series of holes directly into the cochlea through a middle fossa craniotomy, then via the middle ear, before adopting the scala tympani approach (Lacombe et al 1984). Extracochlear electrodes have been either lodged at the round window membrane as described by Burian et al (1986) and Portmann et al (1986), or placed in the bone over the cochlear turns beneath the medial wall of the middle ear (Banfai et al 1984).

The mastoidectomy and "facial recess" approach to the middle ear referred to above and described by Myers and Schlosser (1960) was the route to the round window favored by House and Urban (1973), Clark, Patrick et al (1979), Eddington et al (1978), Parkin et al (1985), Burian et al (1986), Portmann (1986), Chouard and MacLeod (1976), Lacombe et al (1989), Eddington et al (1978), and Lacombe et al (1984). A trans-external canal approach was advocated by Simmons (1966), Michelson and Schindler (1981), and Banfai et al (1984). In the latter cases problems with extrusion of the lead wires required them to be buried in a groove cut in the posterior canal wall. This did not always resolve the difficulty, and surgeons then obliterated the external canal (Banfai et al 1986).

A percutaneous plug was the external link in a number of early devices, such as those of Simmons (1966), Chouard and MacLeod (1976), Michelson and Schindler (1981), Banfai et al (1984), and the Utah group (Eddington et al 1978), whose research led to the Symbion device (Parkin et al 1985). However, an inductive electromagnetic link was used at the beginning of clinical trials by House and Urban (1973), Clark, Pyman et al (1979), and Burian et al (1986). In these cases the internal receiver system was stabilized in a bed in the bone above or behind the mastoid and it contained an antenna that was activated by an external aerial applied to the overlying skin. A percutaneous link was superseded by an electromagnetic transcutaneous link through intact skin by Lacombe et al (1984) and Banfai et al (1986), due to problems with infection and instability with the former. The history of the surgical development is also outlined in Webb et al (1990).

Aims

Position Multiple Electrodes Close to the Auditory Nerves

The first aim of implantation is to position multiple electrodes in the cochlea close to auditory nerve fibers so that separate groups can be excited to convey essential speech frequencies. The fine temporal and spatial patterns of stimulation required for improved temporal coding and musical appreciation are also likely with the precise placement of multiple-electrode arrays.

Implant Electrode with Minimal Trauma to the Inner Ear

The second aim is to implant electrodes with minimal trauma to the inner ear. Any injury leading to loss of spiral ganglion cells and auditory nerve fibers is especially to be avoided. Studies described in Chapter 3 have shown that trauma of the basilar membrane and fractures of the spiral lamina are likely to do this. Trauma to these and other structures may also lead to excessive fibrous tissue and new bone formation that may affect the electrical field and stimulus current levels.

Locate the Receiver-Stimulator to Allow Optimal Use of a Microphone, Speech Processor, and Transmitting Coil

The third aim is to locate the receiver-stimulator package so that the microphone, speech processor, and transmitting coils are close to each other; the microphone is close to the ear; and the lead wire from the package to electrode arrays in the cochlea is as short as possible.

Implant Receiver-Stimulator to be Unaffected by Growth Changes

The fourth aim is to implant the receiver-stimulator in children so that growth changes in the temporal bone will not extract the array from the cochlea. The greater part of this growth is in the first two years of life. Consequently, this aim is most critical in this age group. This was discussed in more detail in Chapter 2.

Implant Operation Performed Safely

The fifth aim is to maintain the highest standard of surgical care as well as audits of results so that prospective patients can be reassured that there are minimal complications. This applies in particular to the incidence of middle ear infection, labyrinthitis, and meningitis. For this reason, in addition to the initial otological and medical examinations, the patient should be reviewed shortly before surgery in case medical conditions have developed in the interim.

Fundamentals and Clinical Practice

The fundamental principles of surgical techniques apply as much to cochlear implantation as to surgery in other regions. The techniques need to be adapted to the special anatomy and procedures.

Preoperative Measures

Preoperative surgical management should focus on measures to prevent infection. This is more frequent when implanting a foreign body and is discussed below and in more detail by Lew and Waldvogel (1998). Infection with the implantation of a foreign body is more likely, as the material provides a home for the organisms and the neighboring tissue is less accessible to antibiotics (Lew and Waldvogel 1998). Postoperative infection is a serious complication that could lead to failure of the operation. Infection within the inner ear will damage and destroy the auditory nerve fibers (Clark 1975, 1977). A wound infection may require the device to be explanted before it can be controlled. Any infection in the wound or middle ear could lead to meningitis, also occasionally seen following a stapedectomy (e.g. Palva et al 1972; Benitez 1977).

The preoperative measures to prevent infection described below were outlined in detail in the surgical training manual developed by the Department of Otolaryngology at the University of Melbourne in 1980.

Preliminary Patient Preparation

The patient's skin is a major source of bacterial contamination in clean wound operations. Any existing acute or chronic infection in the area (including the ear, the skin, and the respiratory tract) must be controlled. In addition, potentially pathogenic organisms in the ear, nose, and throat should be eradicated. Therefore, swabs should be taken from the external auditory canal, the postauricular sulcus, and the nose, and topical antibiotics or antiseptics applied if necessary. On the night before surgery, the nursing staff should wash the patient's hair with an antiseptic shampoo. Hayek et al (1987) found a reduction in infection rate with chlorhexidine (9%) versus normal bath soap (12.8%). The external auditory canal should be inspected and if necessary cleaned.

The Operating Theater

The operating theater should meet high standards of asepsis and cleanliness. This is necessary, as the implantation of a foreign body in a hip or knee replacement is associated with a significant postoperative infection rate. The infection rate for hip replacements by Charnley (1972) was initially 7% but fell to 0.6% with air filtration and antibiotics. For cochlear implantation, an effective air filtration system, therefore, is required. A laminar flow unit, either horizontal or vertical, is valuable, and was used in the theater of the Royal Victorian Eye and Ear Hospital for the first 15 years to ensure that postoperative infections were kept to an absolute minimum (Clark, Pyman et al 1980). Regardless, a high standard of sterility must be maintained by all personnel in the theater with regard to instruments, drapes, and their own dress and movements. The number of people in the theater should be limited, and movement in and out minimized. Glove powder should be thoroughly washed off, as it may contaminate the wound and the cochlea and induce a foreign body reaction (Clark, Pyman et al 1980).

Patient Preparation

The hair should be clipped on either side of the proposed incision. This is preferably done in the anesthesia room. Then either a wet shave with foam rather than a brush, or a depilatory cream can be used (Zentner et al 1987). Studies have shown that infection is lower with either a depilatory cream or leaving the hair closely clipped (Seropian and Reynolds 1971; Cruse and Foord 1973). The hard chitinous surface of a hair is easier to clean with the antiseptic rather than the skin in which it grows. Minimal removal of hair (approximately 1 cm on either side of the incision) was undertaken by Roberson et al (2000) on 46 patients, and no wound infections occurred. The cosmetic benefits were ranked more highly by the parents of children than adults.

The patient is then moved into position in the theater, and an appropriate antiseptic liberally applied to the side of the head including ear, external canal, the face (eyes being protected), and the neck. A sterile plastic drape is applied to the operation site and face. The electrodes to monitor any facial nerve stimulation are attached to the skin around the orbit and cheek. A sump to collect irrigating fluid spilling over from the wound is then put in place.

Antibiotics

A broad-spectrum antibiotic cover for the operation is important, as the colonization of a foreign body by even a small number of bacteria can lead to sepsis. The antibiotic is administered intravenously at the beginning of the procedure, with tissue levels peaking about 1 hour later when the inner ear is opened and implanted. It is administered again at the end of the operation to provide a further cover. It will be required postoperatively if there is any suggestion of a wound infection.

Incision

The incision is made with a knife, although some surgeons use a cutting diathermy. The use of cutting diathermy in an area of cosmetic importance is contraindicated as it causes scarring, which can be excessive as keloid formation.

Fundamentals

The fundamentals outlined below provide adequate exposure, cosmesis, rapid healing, and no extrusion of the package.

Exposure of Underlying Tissue

The incision must provide an adequate exposure of the surgical anatomy, and be easily extended if it is necessary to manage anatomical variations or complications. Sufficient mastoid bone should be exposed so air cells can be removed to provide access for an opening to be drilled into the middle ear from behind (posterior tympanotomy). This allows the cochlear window (round window) to be clearly seen so that an opening can be made into the inner ear. In addition, the exposure should give a good view for drilling the receiver-stimulator package bed in the skull.

Cosmesis (Hair Line)

Cosmesis is important for device acceptance especially with children, and during adolescence. The incision should lie not only in the skin crease behind the ear (postauricular sulcus) but also in the hairline. In addition, the device should be well embedded so that there is minimal surface protrusion, and certainly the pinna should not be displaced outward to create a deformity. This is now unlikely as the size of the present receiver-stimulator packages has been reduced considerably.

Healing

Healing will be delayed if there is wound infection. This may also occur if there is poor circulation or the wound is sutured under tension. An incision in the postauricular sulcus is slower to heal than elsewhere, and extra care is required in closing this part of the wound.

Vascular Supply

It is essential to maintain an adequate arterial supply to the skin flap behind the ear. To evaluate the arterial pattern and its implications for cochlear implant skin flap design, a dye injection study was performed on cadavers (Dahm et al 1993a). The results on 10 specimens indicated that the arterial supply for the skin in the postauricular region was provided, inferiorly by indirect musculocutaneous per- forators, posteriorly by the occipital artery, superiorly by the superficial temporal artery, and anteriorly by the network of vessels around the base of the auricle and cutaneous branches of the postauricular artery.

A flap for a cochlear implant cannot be based on one single axial source artery, and has to rely on a number of different arterial contributors. This means the flaps have random, axial, and/or musculocutaneous supply. Inferiorly based flaps (Fig. 10.1) such as the inverted U, which evolved into the inverted J (Clark, Pyman et al 1979) or the extended endaural (Lehnhardt and Hirshorn 1986), were consid- ered superior to the anteriorly based C-shaped flap (House 1982), as the latter could cut both the superior temporal and occipital arteries. With the inverted U or J incision, the venous and lymphatic drainage is downward, and this is desir- able. The vascular supply is less of an issue with the more vertical incision now used (Fig. 10.1).

Prevent Foreign Body Extrusion

Extrusion of the package is most likely through the incision if the skin is sutured under tension, especially if the package lies beneath it. The risk is increased if wound infection occurs. It is also an important principle when implanting a for- eign body that incisions through tissue do not directly overlie the foreign body. With implant surgery this principle has been followed by creating a separate flap

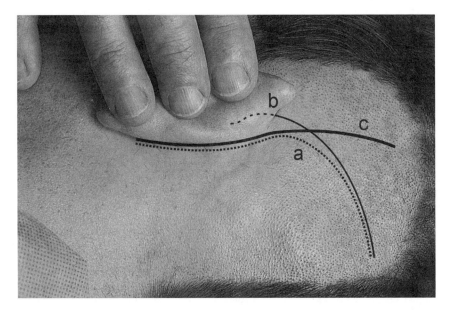

FIGURE 10.1. The inverted J, extended endaural, and vertical incisions for cochlear implantation. (a) (dotted line)—the inverted J incision; (b) (continuous/interrupted line)— the extended endaural incision; (c) (thick line)—the vertical incision.

of fascia to lie under the incision where it overlies the package, and by also ensuring that the package lies deep at this point.

Infants and Young Children

The incision in infants and young children appears relatively larger because of their small head size. This should not compromise its length. The curvature of the skull is greater than in an adult and this may influence the orientation of the package, and in turn the placement of the incision.

Siting of Incision

A C-shaped incision was first used for the 3M single-electrode implant (House 1982) that had only a coil, but no electronics placed behind the ear. This incision gave inadequate exposure for the larger multiple-electrode receiver-stimulator developed first by the University of Melbourne (Clark 1977), and then for the Nucleus (Cochlear Proprietary Limited) device. It could also compromise the blood supply to the skin flap as stated above. The management of a mastoid emissary vein could be very difficult with this approach. For this reason an in-verted-J–shaped incision (Fig. 10.1) was developed (Clark, Pyman et al 1979). Although it cut the posterior branches of the superficial temporal artery, there was a good arterial supply from below from the occipital artery as well as musculo-cutaneous vessels and excellent dependent venous drainage. A modification of this incision by Lehnhardt (Lehnhardt and Hirshorn 1986) replaced the upward postauricular limb of the inverted J with an incision in the external auditory canal (extended endaural) (Fig. 10.1). Another incision was an inverted L with a horizontal limb above the pinna and a vertical limb posteriorly. This has had limited acceptance, as it needs to be quite extensive to gain adequate access.

There has been a trend for a more vertical incision commencing in the postauricular sulcus and extending into the hairline with only a slight posterior curvature (Fig. 10.1). This is a modification of the inverted J. It allows a smaller head shave that is appreciated especially by parents and children.

Before making the incision, its site is determined after a dummy package is positioned over the skin, taking into consideration the age of the child, the head shape, and the extent of the mastoid air cells as seen on x-rays. There is also a need to leave a space of 2 cm behind the ear free for the placement of a microphone and behind-the-ear speech processor unit. The incision should thus be 2 cm from the front edge of the implant unless the implant is small enough to place within the mastoid cavity. Flexibility in the final positioning of the implant is required once the wound is opened. Methylene blue dye has been injected with a fine needle down to the bone to mark the center of the bed for the receiver-stimulator. This procedure could introduce infection, and in children the needle could penetrate the suture lines.

A vasoconstrictor agent should be injected along the lines of the incision, under the flap and into the posterior wall of the external canal. The incision is made through skin and subcutaneous tissue down to but not through muscle, aponeurosis, and deep fascia. In small children the incision should be carried through the pericranium to provide maximum thickness for the flap (Cohen 2000).

If there is a scar from a previous implant operation or other temporal bone surgery, the scar tissue should be excised and the same incision reopened if possible. This may need to be modified for the implant being inserted. An example of the need for care was seen by Harris and Cueva (1987), who used a C-shaped flap in a person who had a previous postauricular incision for ear surgery, and the blood supply was seriously compromised.

Exposure of the Underlying Tissue and the Creation of Fascial Flaps

A flap of skin and subcutaneous fascia is raised. It is necessary to expose the fascia over the lateral surface of the mastoid bone and the lower posterior part of the temporalis muscle. Exposure of the postero-inferior part of the parietal bone and the squamous part of the occipital bone, near where the sutures meet at the asterion, will likely be required.

A separate anteriorly based flap of deep fascia and periosteum should be raised (Fig. 10.2). The inferior limb of this flap will run forward from the superior nuchal line. If it is too low, the occipital artery will be encountered. This deep flap helps stabilize the implant and protects it should there be a breakdown of the anterior limb of the skin incision. The elevation of the deep flap is continued forward until the suprameatal spine and the bony portion of the external auditory canal are

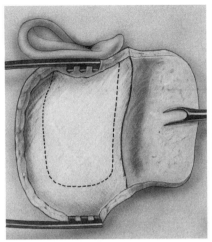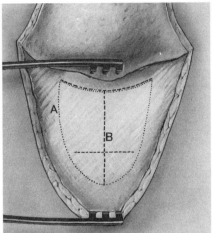

FIGURE 10.2. Left: The anteriorly based fascial flap and the exposed tissue and skull. Right: An alternative with flaps made with a cruciate incision. (Reprinted with permission from Webb et al 1990. The surgery of cochlear implantation. In: G. M. Clark, Y. C. Tong and J. F. Patrick, eds. Cochlear prostheses. London, Churchill Livingstone: 153–180.)

clearly seen. This provides a clear definition of the landmarks required to drill down to the mastoid antrum.

First-Stage Mastoid Cell Removal

Sufficient mastoid air cells need to be removed to provide adequate exposure. It would only be necessary to remove them all if infection had previously been present.

Procedure

A limited mastoidectomy is carried out exposing the mastoid antrum, the lateral semicircular canal, and the short process of the incus. The removal of bone should be sufficient to carry out the posterior tympanotomy. However, the posterior bone removal should be no more than necessary, so that a good anterior wall for the package bed can be preserved to stop the package from sliding forward. The posterior bony wall of the external auditory canal may need to be thinned to get good exposure for the posterior tympanotomy. If a hole is made in it, it should be covered with fascia, or cartilage from within the canal, to prevent fistula formation.

Creation of a Bed for the Receiver-Stimulator

A bed in the bone is required to place the container for the receiver-stimulator electronics, so that it will not move or protrude as a swelling. The recent Nucleus 24 package has a thickness of only 6 mm with a protrusion of 2 mm, so a bed can even be made in an infant's skull that is only 1 to 2 mm thick.

Fundamentals

The bed is best created after the initial mastoidectomy before the operating microscope is used for the posterior tympanotomy and exposure of the cochlear round window.

Placement

The anterior edge of the bony bed for the receiver-stimulators for all brands of implant should lie 2 cm behind the postauricular sulcus. This is important as a behind-the-ear speech processor needs to fit comfortably between the ear and the front of the receiver-stimulator package and transmitting coil. The titanium case of the Nucleus 24 system was designed to lie within the mastoid cavity in suitable people, so this would allow the transmitting coil to be placed closer to the ear (Clark and Pyman 1995). However, the Nucleus 24M and 24R packages are more usually placed in a bed behind the mastoid cavity. If the receiver-stimulator case is ceramic, with the receiving coil incorporated, the bed still needs to be placed well behind the postauricular sulcus to keep the space clear for the behind-the-ear speech processor or the side arm of spectacles.

The bed is usually drilled in the mastoid and anterior segment of the occipital and parietal bones at the junction of the sutures between the mastoid, parietal, and occipital bones (asterion). In young children it was a concern that this might lead to early closure of the sutures and a skull deformity. For this reason a study examined radiologically the head growth in the macaque monkey (Xu et al 1993), and the effects on the sutures histologically (Burton et al 1992a,b, 1994). The histological study examined the effects of implantation at the asterion for a period of 3 to 4 years. There was no evidence of closure of the sutures.

Shape and Depth of the Bed

The bed should be round, as it is easier to drill (Clark, Pyman et al 1984). The University of Melbourne's prototype package was rectangular (Clark, Pyman et al 1979), and this made excavating the bone in the corners more time-consuming. The bed should be drilled deep enough to ensure there is minimal protrusion of the package above the surface of the skull. An acceptable limit is 5 mm. At surgery using the Nucleus clinical trial device (the upper half or cap was 4.5 mm, and the implantable stalk 5 mm), it was found that in many patients it could be implanted without drilling down to dura. However, the receiver-stimulator had to be made thinner for implanting in children from 2 to 18 years of age. This was done by removing the connector, as the biological studies had shown the banded free-fitting electrode array could be easily removed and another inserted if necessary (Clark, Blamey et al 1987a). This (Mini) receiver-stimulator (CI-22) also had a magnet in the package to allow the external transmitting coil to be easily attached and aligned (Clark, Blamey et al 1987b). Other receiver-stimulators went through a similar evolution. The Nucleus CI-24M and later the CI-24R receiver-stimulators were made smaller so they could be implanted in children under 2 years of age. They had an overall thickness of 6 mm, with 2 mm protruding on the undersurface. The protruding section 2 mm deep had a breadth of only 13.7 mm and length of 9.5 mm, and is placed in a bed drilled down to dura in infants and young children. The remainder of the package lay on the surface of the skull. By contrast the Clarion S has the dimensions 31 × 25 × 6 mm. This means that a bed 31 × 25 mm needs to be drilled down to dura in young children. The same applies to the Med El device.

Stay Sutures

Stay sutures to fix the package and prevent it from migrating are inserted through holes drilled into the skull using a fine burr. In the adult the skull is thick, so the holes can be made so they pass diagonally through the outer edge of the package bed. In young children, as they have thin skulls, the holes have to be drilled right through the skull. Directing a fine burr toward the brain is not a good practice, and a metal spatula must be placed over the dura for protection. The sutures are later tied around the package.

Depression of the Brain

The Nucleus devices are made with the receiver-stimulator coil section attached but behind the electronics package, and so it lies superficial to the skull. This means there is some protection from an external force driving the package into the cranial cavity. There is also not the same necessity to press the package against the dura as with a ceramic package.

   If the receiver-stimulator depresses the dura, it was a concern that this did not lead to any adverse effects on the cortex of the brain. The effects were also studied on monkeys (Burton et al 1992a, 1994), an implant package being placed on the overlying dura and the wound stitched in place. This left a depression in the cortex. Statistically significant differences were seen in cortical thickness when comparing the implanted and unimplanted sides, but there was no overall loss of cortical cells in four out of five cases.

Tissue Regrowth in the Package Bed

After drilling down to the dura, bone and fibrous tissue will grow beneath the package over time. They can fill the bed and cause the device to move laterally and appear more prominent. This growth has been seen when reoperating on children in Melbourne, and was also seen in the studies on the monkeys (Burton et al 1992a, 1994).

Procedure

The bed for the receiver-stimulator must be made so that the package is stable and will not slide or rock (Fig. 10.3). Its dimensions are marked with a pen or small burr around a template. The anterior edge should be checked by releasing the retractors if necessary, so that it is at least 2 cm behind the postauricular sulcus. It should not be too high if the squamous temporal bone is thin, and not too low, otherwise it may rock on the curved portion of the skull behind the mastoid process. A mastoid emissary vein may force an adjustment of this position.

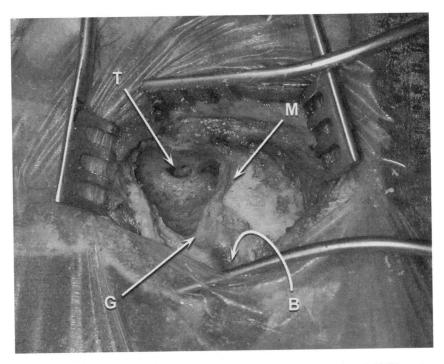

FIGURE 10.3. The exposure of the retromastoid region showing the package bed (B), gutter (G), mastoid cavity (M), and the posterior tympanotomy (T). (Courtesy of R. Briggs.)

The bed is made with a cutting burr and its sides squared off and smoothed with a diamond burr. Dura will be exposed in many children, and extra care should be taken not to injure it or initiate bleeding. The bed should be just larger than a template so that a good fit is obtained. In thin skulls and certainly in young children, the dura, will need to be separated from the surrounding bone so that it can be depressed by the implant to gain adequate depth. This should be performed with an appropriate elevator and not by pressure on the implant. If bleeding occurs from the dura, it can be controlled with bipolar diathermy. If that fails, particularly if the bleeding is from underneath the edge of the bone, an absorbable sponge such as Gelfoam or Sterispon can be gently pushed against it.

Creation of the Gutter for the Lead Wire Assembly

Fundamentals

The gutter created between the package bed and the mastoid is for burying the lead wire from the point where it emerges from the receiver-stimulator package to where it passes into the depths of the mastoid cavity (Fig. 10.3). In the second patient (Clark and Tong 1982) who had the University of Melbourne's prototype receiver-stimulator, the lead wire fractured at the point where the electrode array exited the package. This was due to a stress concentration between the fixed package and the flexible lead induced by movements when the skin was rubbed. Furthermore, studies in the cat had shown how readily wires without stress relief would break from contractions of the temporalis muscle during chewing. As a result the lead wire was specially designed by Cochlear Pty Limited to provide stress relief with a very adequate margin of safety (see Chapter 8). Nevertheless, it is good surgical practice to reduce any risk of failure to an absolute minimum by burying the electrode lead wire.

Procedure

A groove is cut from the package bed to the mastoid cavity. It should be large enough for the electrode lead wire. The edges of the groove should be undercut to allow latitude in rotating the package, and also in making a covering to help secure the lead wire. The sharp edges should be drilled away. The thick portion of the lead wire assembly is flexible and strong, so it can be manipulated into the appropriate position without fear of damage.

Exposure of the Round Window via a Posterior Tympanotomy

Fundamentals

The surgical approach to the middle ear to expose the round window (cochlear fenestra) through the mastoid air cells is via a triangular space between the facial nerve, chorda tympani, and the floor of the fossa incudis (Fig. 10.3), and was illustrated in Chapter 2. This approach is a posterior tympanotomy. It was developed in the 1950s and 1960s to manage middle and mastoid infection, and retain the posterior canal wall. Wullstein (1956) made a hole in this area to examine the round window. The approach was further developed for a combined tympanoplasty and mastoidectomy operation (Myers and Schlosser 1960; Corgill and Martinez 1963; Jansen 1963) to help ensure removal of disease in the posterior recesses of the middle ear while retaining the posterior canal wall intact. The approach outlined in more detail in Chapter 2 is through the tympanic sinus and facial recess. Jansen (1968) found that with a posterior tympanotomy a thorough view of the round window could not be achieved in 87% of cases. This was due to a bony ledge that lay anterior and medial to the facial nerve, called the ponticulus pyramidalis. This ledge needs to be drilled away to expose the round window and cochleostomy site. It has also been found that the dimensions of the posterior tympanotomy (Dahm et al 1992) and the facial recess (Bielamowicz et al 1988) are similar in children and adults.

Procedure

In approaching the round window through a posterior tympanotomy (Clark, Pyman et al 1979, 1984), it is first necessary to identify the vertical segment of the facial nerve. The key landmarks are the lateral semicircular canal, the short process of the incus, and the anterior end of the digastric ridge. Drill in the line from the short process to the anterior end of the digastric ridge, leaving a bridge of bone under the fossa incudis. This bridge should be about 3 mm thick so that it can be used for a deep tie around the electrode array. The facial nerve lies just behind the above line, but its course is variable and it must be constantly watched for. The anomalies are described in detail by Fowler (1961), Durcan et al (1967), Marquet (1981), and Nager and Proctor (1982). In particular, beware of both the nerve with a sharply angled genu and one that swings laterally high in its vertical course. Preoperative computed tomography (CT) scans are essential for demonstrating the course of the nerve. A diamond paste burr is safer to use than a cutting burr as the facial nerve is approached. The nerve should be seen through the bone, and a thin layer of bone preserved over it, particularly anteriorly and laterally, so that it will be protected from the shank of the rotating burr when drilling the cochleostomy. If in doubt about identifying the facial nerve, the stimulator should be used. The chorda tympani will be seen anterior to the facial nerve, and should be preserved if possible. However, if the approach to the middle ear is too narrow, it may need to be sacrificed. The facial recess of the middle ear should be entered beneath the fossa incudis, where an air cell may aid this entry. If the approach is made too far anteriorly, there is more danger of entering the external canal.

The posterior tympanotomy should be widened until a good exposure of the round window niche is obtained. It will need to be about 2 mm wide. The annulus of the tympanic membrane may need to be exposed, but care must be taken not to damage it or the membrane. If there is any difficulty seeing the round window, the stapes must be clearly identified. As stated, a better view of the round window can be obtained by drilling away the ponticulus pyramidalis with a fine diamond paste burr. With this drilling, make sure there is a good view of the facial nerve and that the shank of the burr does not rest on it. If the lateral venous sinus is placed quite anteriorly, it will be difficult to carry out the posterior tympanotomy, and get a satisfactory view of the round window. In this case the bony external canal should be removed and replaced at the completion of the electrode insertion.

Cochleostomy (Opening into the Inner Ear)

Development

In 1972 and 1973 it was not clear just how electrodes should be placed in a human cochlea. The first studies were anatomical dissections of the human temporal bone to expose the apical and middle turns of the cochlea and insert the electrodes precisely through the overlying bone as well as a bundle of electrodes through the round window and along the basal turn (Clark 1975). (Fig. 10.4). Second, Clark, Hallworth et al (1975) showed that it was possible to pass an electrode array around the turns of the cochlea to the region conveying speech frequencies by drilling into the upper basal and the middle turns directly below the facial nerve and passing the array in retrograde fashion back toward the round window

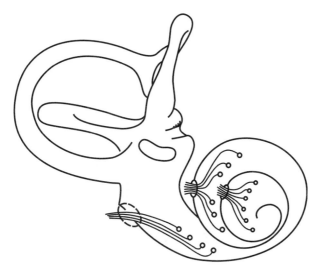

FIGURE 10.4. A schematic drawing of the insertion of three bundles of electrodes into the human cochlea through openings drilled directly into the scalae to allow the positioning of the electrodes to the auditory nerve fibers transmitting the speech frequencies to the higher auditory centers. (Reprinted with permission from Clark 1975. A surgical approach for a cochlear implant. An anatomical study. *Journal of Laryngology and Otology* **89**: 9–15.)

(Fig. 10.5). Third, it was shown that a thin film array of electrodes could be designed to also pass downward in a retrograde manner (Clark and Hallworth 1976). The above studies in 1975 to 1977 thus demonstrated the idea of passing multiple arrays through an opening in the apical and middle turn of the cochlea. They were also discussed in the bioengineering section in Chapter 8, the section on relations of the cochlea in Chapter 2, and in Chapter 3. However, a histological study in the experimental animal revealed the insertion was accompanied by more trauma than with an anterograde insertion upward through the round window (Clark 1977). The main problem with an anterograde insertion from the round window upward appeared to be due to frictional resistance preventing the insertion of an array into the tightening spiral to lie opposite the speech frequencies. A solution came when it was realized that if the electrode array had graded stiffness and was flexible at the tip, this provided optimal mechanical parameters for a deep insertion (Clark, Patrick et al 1979). This has been subsequently verified with finite element modeling studies (Chen, Clark et al. 2003).

The fundamentals in performing a cochleostomy are also determined by the surgical anatomy of the cochlea, and were discussed in some detail in Chapter 2.

The Round Window and its Relationships

The round window is either the site for implantation or the main landmark for an opening through the bone (cochleostomy). It is sealed by its membrane, and lies in a niche obscured to a variable degree by a bony overhang, which extends over the membrane anteriorly, and superiorly for a distance of up to 1 mm. A fold of

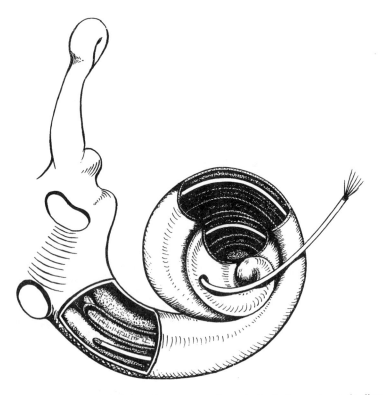

FIGURE 10.5. A diagram showing the passage of an electrode in a retrograde direction from an opening in the middle or apical turns to the basal turn (Reprinted with permission from Clark, Hallworth et al 1975. A cochlear implant electrode. *Journal of Laryngology and Otology* **89**: 787–792.).

mucous membrane may extend from the overhang and further obscure the "true" round window membrane. On occasion this mucosal fold may be complete and thus form a false membrane (Nomura 1984). Failure to recognize this may lead to the electrode impacting on the true membrane, and possibly a traumatic insertion or damage to the electrode (Clark, Pyman et al 1979). The true round window membrane is conical in shape with the apex lying superiorly where it is attached to the osseous spiral lamina. The apex therefore should be avoided. The diameter of the membrane is about 1.5 mm across the base of the cone (Franz et al 1987b). It is better to insert the electrode array through an opening 1 mm anteroinferior to the round window, as discussed below.

Hypotympanic Cells

An hypotympanic air cell may open immediately inferior to the round window niche. It can be readily mistaken for the niche if the round window is obscured, and particularly if the niche is obliterated. It is therefore important to visualize the round window membrane. If this is not possible, the stapes should be used as a guide. The distance between the center point of the anterior rim of the oval window and the anterior and inferior sector of the round window rim is 4.1 to 4.5 mm with a standard deviation of 0.34 mm (Dahm et al 1993b).

The Scala Tympani

The electrode array is inserted into the scala tympani to stimulate the peripheral processes of the auditory nerve if present, as well as residual spiral ganglion cells. The spiral lamina and basilar membrane should not be damaged during the insertion as this could lead to the loss of ganglion cells. Familiarity with the appearance of the scala tympani is important for a safe and deep insertion, particularly the concavity of the outer wall passing laterally and superiorly.

The scala tympani is narrowed just inside the round window membrane by an anteroinferior ridge, the crista fenestra (Franz et al 1987b). This may need to be drilled away to facilitate the insertion of the electrode through the round window. Beyond the crista the scala is fairly straight in an anteromedial direction for about 6 mm from the round window before starting to curve superiorly. There is a significant change of direction at 9 to 10 mm. This means that the spiral lamina forms the superior wall of the scala in the first 6 mm, but at 9 mm it forms more the posterior wall. This (illustrated in Chapter 2). The cochlea continues to curve to form its spiral shape. The radius of curvature is about 4 mm (from the round window to a point 18 mm along the cochlea), and then 2.5 mm (from 18 mm to 25 mm). The cross section of the scala tympani changes in shape and is an approximate square, with 0.8-mm sides near the round window and a triangle with 0.4-mm sides 25 mm along the cochlea. When a cochleostomy is made anteroinferior to the round window for the Nucleus Contour precurved array, which is slightly wider than the standard array (0.8 mm), the opening may need to be extended posteriorly by drilling the anterior part of the crista fenestra. It should not, however, enter the round window as it is more difficult to achieve an effective seal on the round window side, as discussed in Chapter 3. The anatomy was illustrated in Chapter 2.

Procedure

Access to the scala tympani can be gained either through the round window membrane (Clark, Pyman et al. 1979) or more usually a fenestration anteroinferior to it (cochleostomy).

Round Window Insertion

The round window was the initial site for the insertion of the array, but there are advantages in making an opening or cochleostomy just anterior to the round window. First, the access to the scala tympani is better, and the array can be passed more easily with minimal trauma. Second, animal experimental studies have shown that middle ear infection is more likely to pass between the array and the round window membrane than along a protective fibrous tissue sheath formed around the array at a cochleostomy. With a round window insertion it was important to note that the round window membrane is often obscured by an overhang that would need to be drilled away anteroinferiorly with a fine diamond burr (1.00 mm) until the membrane was seen (Clark, Pyman et al 1984). Care had to be taken if it was necessary to drill posterosuperiorly as the underlying osseous spiral lamina could be damaged at this point (Franz and Clark 1987).

If the membrane was clearly seen and had a good vertical component, the anteroinferior margin could be reflected. The crista fenestra could prevent a good view down the basal turn and needed to be drilled away with a 0.6- or 1.00-mm diamond burr. The drilling had to be slow and gentle, with bone dust being gently irrigated and sucked away. Bone dust in the cochlea may initiate new bone growth (Franz et al 1987a). The opening was about 1 mm wide, and the scala tympani was seen as a fluid-filled tunnel running anteromedially and slightly narrowing in diameter.

Fenestration of the Cochlear Wall (Cochleostomy)

Fenestration of the cochlear wall is now the preferred technique, as it provides a more direct view along the scala tympani for inserting the electrode by avoiding the hook region of the basal turn. It also allows more effective sealing of the round window with a collar of fibrous tissue and/or bone around it to prevent infection entering the cochlea with the concomitant risk of meningitis.

Carefully drill through the bone approximately 1 mm anteroinferior to the round window. It is important to see the round window membrane first for orientation, as an hypotympanic cell may open immediately below the round window, and may be mistaken for it. In addition, if the window is obliterated, the stapes must be used as a guide to the scala tympani. The drilling is carried out with a 1-mm and then a 0.6-mm diamond burr, until the perilymphatic space is visible as a dark patch through the bone. The bone can then be picked off with a fine dissector or hook (Fig. 10.6). The endosteum is usually delicate and easily opened with a needle, but sometimes it is thick and may need to be incised with a sharp hook and folded back to prevent a subendosteal implantation.

A variation of this procedure is to keep the endosteal layer intact until the opening in the bone is complete. The rationale is to minimize trauma to the cochlea and prevent bone dust entering the scala. The technique has been referred to as "soft surgery" by Lehnhardt (1993b). One difficulty is that the membrane can collapse away from the bone, making the penetration with a needle difficult. It can also predispose to an insertion around the scala media or behind the stria vascularis.

Insertion of Arrays

Initially, it was not clear if an array could pass up the tightening spiral of the basal turn of the cochlea to lie opposite the auditory nerve fibers transmitting the speech frequencies. This was the reason for the early studies by Clark (1975), Clark, Hallworth et al (1975), and Clark and Hallworth (1976) to determine

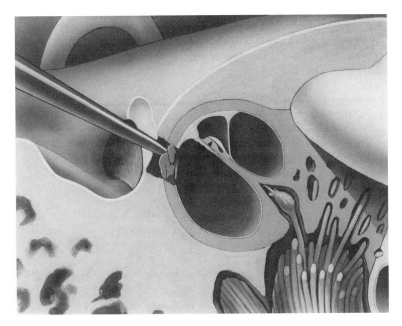

FIGURE 10.6. The removal of the bone in a cochleostomy to enter the scala tympani. (Reprinted with permission from Cochlear Corporation 1987. Surgical procedure manual: Nucleus 22 channel cochlear implant system. Issue 5.)

whether arrays could be passed downward in a retrograde direction through openings in the middle and apical turns, or by direct insertion (Figs. 10.4 and 10.5). However, a solitary array with increasing stiffness was found to have the right mechanical properties to pass the necessary distance from below upward (Clark, Patrick et al 1979). This minimized trauma to the cochlea. Alternatively, double arrays were proposed (Hansen 1981, Goycoolea et al 1990), with one passing a certain distance up the basal turn, and another up the middle turn through an opening in the overlying bone. These were not necessary for the normal cochlea, but would be appropriate if the cochlea was ossified, as discussed below (see Ossified Cochlea).

Nucleus Standard Banded Array

Before inserting the Nucleus standard array, it is considered an advantage to smear sodium hyaluronate (Healon) or dilute glycerin over its surface to reduce friction, and thus facilitate a deep insertion. This can be achieved by placing a drop of the material over the cochleostomy through which the array is passed. Biocompatibility studies have shown no adverse effects for Healon on the cochlea (Bagger-Sjoback 1991). A study by Roland et al (1995) on guinea pigs found no significant loss of spiral ganglion cells 2 to 8 weeks postinjection of Healon and 25% and 50% glycerin into the cochlea. Clinical experience (Lehnhardt 1993a) has suggested the depth of insertion could be increased with the application of Healon.

Deep insertions were also seen in a study on human temporal bones (Donnelly et al 1995b). Research has also shown that insertion beyond approximately 25 mm is not an advantage as the auditory nerve fibers from the most apical region do not lie opposite the scala, and place coding is not critical for low frequencies.

Inserting the electrode array is facilitated with a microclaw first advocated by (Clark, Pyman et al 1979). The claw is held in the dominant hand, and the implant package in the other. The claw is used to guide the electrode tip to the opening into the scala tympani (Fig. 10.7). The claw or microforceps can then be used to advance the array until the first resistance is experienced, or buckling seen. If the insertion is continued, damage to the osseous spiral lamina and basilar membrane may result. This would cause a significant loss of neurons and poorer than expected performance. If resistance is experienced after inserting the array for only 10 to 12 mm, it should be slightly withdrawn and then rotated through 90 degrees (Fig. 10.8) before further advancement. The rotation should be counterclockwise for the right cochlea and clockwise for the left cochlea. The rotation causes the tip of the array to rotate downward from the basilar membrane where the splay of the outer wall tends to direct it (Franz and Clark 1987). This allows the tip of the array to then move in the direction of the spiral of the cochlea. The insertion is continued until no further progress is obtained. It is again important not to continue the insertion once resistance is felt or buckling seen. A forceful insertion to achieve an apparent increased insertion depth may cause either trauma to the inner ear or a concertinaed electrode, with consequent poor performance. When all the platinum rings have been inserted, the tip will be 25 mm from the entry point (Figure 10.9). If this can be achieved, the speech perception results are likely to be better (Blamey et al 1992). Note with the Nucleus array that the proximal 10 rings are stiffening bands, and the 20 active electrodes are on the first 17 mm, so very useful function can be achieved even though all the stiffening bands are outside the round window. After the electrode is inserted and the electrode fixed, the package is stabilized in its bed.

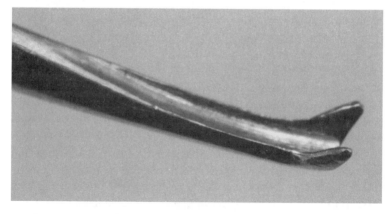

FIGURE 10.7. The claw used to insert the Nucleus banded electrode array (Clark, Pyman et al 1979).

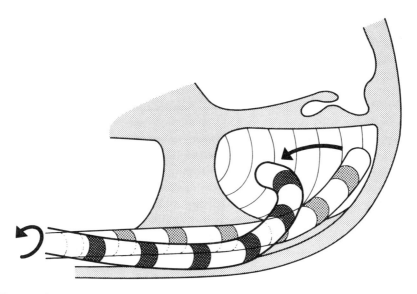

FIGURE 10.8. The rotation of the banded array through 90 degrees to facilitate insertion if resistance is met. A schematic diagram illustrating the effect of rotating the banded electrode array approximately 90 degrees counterclockwise during the insertion of the array in a right cochlea. This rotation directs the electrode tip downward, and away from the basilar membrane. (Reprinted with permission from Franz and Clark 1987. Refined surgical technique for insertion of banded electrode array. *Annals of Otology, Rhinology and Laryngology* **96**(suppl 128): 15–16.)

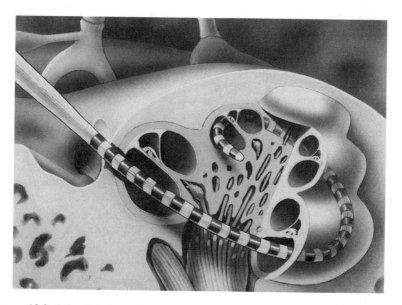

FIGURE 10.9. A banded electrode array lying around the scala tympani in the basal turn of the cochlea. (Reprinted with permission from Cochlear Corporation 1987. Surgical procedure manual: Nucleus 22 channel cochlear implant system. Issue 5.)

Another technique for the electrode insertion is to have the package first stabilized in the bed, and then two claws or a pair of forceps can be used to advance the array. Experience has shown that it is more difficult to direct the tip of the array into the cochlea with this technique, and rotation is not as easy.

The Med El Array

The Med El array, also described in Chapter 8 provides monopolar stimulation on 8 electrodes for the Combi-40 system and 12 electrodes for Combi-40+. The electrodes are distributed along short (12 mm), standard (21 mm), and long (27 mm) arrays. The insertion into the scala tympani through a cochleostomy is similar to that for the Nucleus straight, banded array. The surgery for the Med El system has been described by Gstoettner et al (1997, 2000). A dummy array is used to determine how deeply one is likely to pass and then an array of that length is selected. The insertion of a dummy array can add more trauma to the procedure.

The Clarion Array

The initial Clarion array (see Chapter 8) was helicoidal-shaped. The insertion technique was described by Balkany et al (1999) and Filipo et al (1999). It was recommended that the array be inserted after the package was fixed in its bed. The electrode was placed in an insertion tool similar to the one developed by the group at the University of Melbourne/Bionic Ear Institute for the prototype precurved banded array (shown in Chapter 8). A large cochleostomy was required to accommodate the insertion tool and the array. They were both inserted for approximately 8 mm into the scala tympani. The array was extruded from the slot in the tool by sliding a plunger forward. When the electrode was released, the tool was removed and the array advanced to the point of first resistance. In 1999 the Clarion HiFocus I array was introduced. The array had 16 rectangular pads arranged longitudinally along the carrier. After insertion through a large cochleostomy it was moved into a perimodiolar position by introducing a second element along its outer side. The second element of silicone ("positioner") had a concavity that embraced the carrier, leaving only a variable but narrow space between the two. With this two-element system, the application of an inward force results in an equal and opposite outward force that could apply untoward pressure on delicate cochlear structures and cause significant damage. This was discussed by Briggs et al (2001). Furthermore, the close approximation of the two components leaves a dead space that would not be very accessible to antibodies, and this could be the site for the entry and maintenance of infection, and lead to its propagation to the meninges. This system was modified with the positioner attached 10 mm from the tip of the array, the HiFocus II. This new system required more dexterity for its insertion. The HiFocus arrays with two adjacent elements can create a foreign body dead space, and the pathogenesis of inner ear infections resulting from a dead space that could lead to meningitis were discussed in some detail in Chapter 3.

The Nucleus Perimodiolar Contour Array

The Nucleus straight flexible array has been altered in design so that it can be placed close to the modiolus. This perimodiolar array, discussed in Chapter 8, arose from research in the Human Communication Research Center (HCRC) at the University of Melbourne/Bionic Ear Institute and was developed as the Contour array by Cochlear Limited and the Cooperative Research Centers (CRC) for Cochlear Implant Speech and Hearing Research and subsequently Cochlear Implant and Hearing Aid Innovation. It results in lower stimulus thresholds, and more localized stimulation of the residual auditory nerve fibers. This means improved battery life, smaller speech processors, and possibly better speech processing strategies. The rationale was described in more detail in Chapters 5 and 8.

The array is a little wider than the Nucleus straight flexible array and tapers from a diameter of 0.8 mm to 0.6 mm at the distal end. The array has 22 half-band rather than full-band platinum electrodes embedded along the initial 15.5 mm (Tykocinski et al 2000). The active area of the half bands varies from 0.28 to 0.31 mm^2 geom. compared with the average area of 0.48 mm^2 geometric for the full-band electrode array.

The array is molded to the curvature of the modiolus with a stylet in situ. The stylet passes along the center and holds the array straight. It is malleable at the tip to allow the array to curve during insertion. Once it has been inserted for approximately 10 mm, the array is advanced off the stylet as this leads to an insertion with the least trauma. If, on the other hand, the array has been inserted the full distance with the stylet in place the array should not be pushed forward off the stylet, or advanced after the stylet has been withdrawn a millimeter or so. This will make the end curl and then damage the basilar membrane.

Once the Nucleus 24R receiver-stimulator is placed in its bed (Fig. 10.10) the array can then be inserted through a cochleostomy, just anterior and inferior to the round window, with the opening made a little larger than for the straight array (1.5–2 mm in diameter). A good view along the scala is important, but this should not be achieved by drilling posteriorly to remove a portion of the crista fenestra and enter the round window. As discussed, the animal experimental studies showed that the seal next to the round window membrane is not as effective as a fibrous tissue sheath extending down from an opening in bone (personal observations).

The perimodiolar array was inserted in 256 children in a clinical trial for the Food and Drug Administration (FDA); 66% had measurable hearing at one or more frequencies in the implanted ear before surgery, and in 54% there was measurable hearing in one or more frequencies after 1 month of use (Staller et al 2002). There were 86% of respondents who said that it was easier to insert than the previous standard Nucleus array. The mean threshold (T) and maximum comfortable (MC) levels were also on average lower.

Sealing the Opening

The tissue seal that forms around the electrode at its entry point into the cochlea is critical for preventing middle ear infection spreading to the inner ear, and

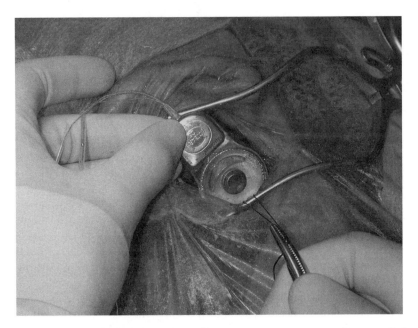

FIGURE 10.10. The Nucleus 24R being placed in its bed with the receiver coil first introduced into a pocket beneath the periosteum. (Courtesy of R. Briggs.)

leading to labyrinthitis and even meningitis as sequelae. The research investigating optimal sealing was outlined in some detail in Chapters 3 and 8. It is best achieved with a fascial autograft or its equivalent pericranium that is pressed into the space between the electrode and bone. It must be completely circumferential. Studies have shown that otherwise infection can more easily track along one wall if dense fibrous tissue has not formed (Clark, Shepherd et al 1984; Franz et al 1984; Brennan and Clark 1985; Cranswick et al 1987; Franz and Clark 1987). Unless there is apposition of tissue to the electrode all around the electrode, there is a very real risk of infection spreading to the cochlea. Failure for this to occur could explain the case of fatal meningitis following stapedectomy when fascia was used with a polyethylene tube (Palva et al 1972). Compressed muscle should not be employed, as the infection studies undertaken by the author in the experimental animal showed that a necrotic piece of tissue provided a home for infection. Furthermore, Gelfoam must not be used. In the animal experimental studies it facilitated the development of a middle ear infection, and a patient who had a stapedectomy with the closure of the oval window with Gelfoam developed fatal meningitis 2 weeks after the stapes replacement (Benitez, 1977).

Perilymph "Gusher"

If a "perilymph gusher" is experienced at surgery, it is especially important to achieve a tight seal, as the patients with the Mondini dysplasia and meningitis after the Nucleus implant, discussed below, demonstrated that a persistent leak postoperatively could lead to an extension of middle ear infection to the inner ear with labyrinthitis and meningitis.

Fixing the Electrode Array and Receiver-Stimulator

Initially single-electrode intra-cochlear arrays were not fixed, as the location of the electrode was not critical (W. F. House, personal communication). The University of Melbourne's multiple-electrode device was initially not fixed either (Clark, Pyman et al 1979). But it slipped 10 mm in the first patient soon after surgery with significant changes in the frequency-to-electrode map. Fixation was therefore considered essential. With the Nucleus clinical trial device the thick proximal lead wire was tied with Dacron mesh to the superior overhang in the mastoid cavity (Clark, Pyman et al 1984), with satisfactory results.

Following a study on children's temporal bones by Dahm et al (1993b), which showed that the floor of the mastoid antrum did not move relative to the round window, this was chosen as the ideal site for fixation (discussed in Chapter 2). A platinum wire was looped around the floor, and then tied to a Silastic sleeve slid along the array (Webb, Pyman et al 1990). A study by Burton et al (1994) in the monkey showed that this held well in both the short and long term. An alternative was to attach a titanium clip to the incus bar (floor of the antrum), and place the array in a receptacle at the end of the clip (Cohen and Kuzma 1995). This produces a point of stress concentration, and it is important to leave a loop of lead proximal to it to avoid damage. Another method is the "split-bridge" technique (Balkany and Telischi 1995), in which a channel is made through the incus bar and the lead wedged in it. This technique would hold the array in the short term, but continued traction could cause bone erosion and electrode movement. It is important to fix the array, as migration of the electrode has been reported in 1.3% of patients (116/9221) [cumulative data from Cochlear Corporation quoted by Roland (2000)]. In 41% of these (47/116) it led to the package being replaced. Therefore, it is not a trivial surgical detail. Furthermore, any displacement of the electrode can affect the sealing at the round window and predispose to infection entering the cochlea from the middle ear. This was thought to account for labyrinthitis in the cats studied by Cranswick (1984).

After fixing the intracochlear electrode, the external electrode for the current path in monopolar stimulation is placed on bone under the temporalis muscle and sutured in place. It must not be placed in the muscle, as repeated movements from chewing will eventually cause the electrode to fracture.

After tying the Nucleus titanium package down with sutures, placed as discussed above (see Stay Sutures), and securing the receiving coil in a pocket under the posterior flap it was found that migration of the receiver-stimulator occurred much less frequently than that of the array (0.2% (22/9221), but only in three was it a cause for explantation (Roland 2000). If sutures are not used, a fascial flap should be sutured over the receiver-stimulator to give fixation.

The Advanced Bionics package, with the dimensions 31 × 25 × 6 mm, was fixed with sutures in the same way as for the Nucleus device, but initially placed under the temporal muscle. In one series there was a 50% slippage rate and in 37.5% a skin defect. This was resolved by drilling a package bed down to the dura in the retromastoid region (Filipo et al 1999). After the package is fixed in its bed, it is then recommended that the electrode be inserted. Advanced Bionics report that 15 electrode failures were due to receiver-stimulator migration (Roland 2000). This suggests that the failures in this case were due to shearing forces.

The Med El system has a ceramic package with the receiver coil embedded in it. This also needs to be placed in a bed behind the postauricular sulcus to accommodate a behind-the-ear speech processor. As ceramic is more prone to cracking than titanium with a blow to the head, the bed is made right to the dura so that the external ties can press the package down to keep the external surface flush with the skull. Thus a larger bed drilled through bone is required to accommodate the ceramic package than for the Nucleus titanium one. In the latter case it is small as the receiving coil can lie superficially and separately posterior to the bed. This is also an advantage as it would restrict inward movement of the package should a blow to the head occur.

Flap and Wound Closure

It must be remembered that when the electrode array has been inserted and the implant placed in its bed, monopolar diathermy cannot be used to control bleeding as current may damage the electronics or pass along the electrode to the inner ear. Bipolar diathermy is required.

The strength of the internal and external magnets will hold the transmitting antenna in position for a scalp thickness up to 12 mm. However, if the thickness of skin and subcutaneous tissue is greater than 6 mm, power transmission rapidly decreases with distance, and this reduces the rate at which the implant can be stimulated. The thickness of the tissue overlying the implant has been measured by tuning the power transmission circuit, and the current is related to distance (P. M. Seligman, personal communication). The results on 213 (119 adults and 94 children) patients at the Melbourne clinic (Figure 10.11) show a mean thickness of 5.02 mm [standard deviation (SD) 2.26 mm] in adults and a mean of 2.67 mm (SD 1.94 mm) in children. If a thickness of 6 mm is likely to be exceeded, the subcutaneous fascia can be thinned, but of course with great caution to make sure that the skin is not breached. There were 31/213 (15%) flaps above the 6-mm optimal thickness. This indicates that attention should be given to the flap at surgery.

Radiology

After the skin is sutured and the bandage applied, a transorbital x-ray should be taken while the patient is still under the anesthetic to check the electrode position.

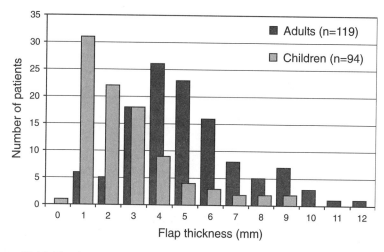

FIGURE 10.11. The tissue thickness over the implant after the initial postoperative swelling had resolved for patients at the University of Melbourne's Clinic at the Royal Victorian Eye and Ear Hospital. (Courtesy of P. Seligman and D. Lawrence.)

Postoperative Care

The routine postoperative care is the same as for all otological procedures involving general anesthesia. Appropriate observation, including of the facial nerve, is made. Analgesics and antiemetics are given as required. A pressure dressing should be kept on for 1 to 2 days to reduce the possibility of a seroma or hematoma. The wound is inspected, and another dressing applied for 5 days. The dressing should be changed earlier if undue pain or fever occurs, suggesting the formation of a hematoma or wound infection. Broad-spectrum antibiotics should be given postoperatively if an infection develops. The infection is potentially serious, and antibiotics should be administered intravenously. The sutures should be removed at about the 10th day unless indicated earlier, as for a stitch abscess.

Complications and Management

Intraoperative Complications

Any of the complications of general anesthesia and otological surgery can occur, but only those with special reference to cochlear implantation are discussed here.

Hemorrhage

Severe bleeding can come from the mastoid emissary vein that may be encountered during the bed preparation. A large one should be noted on the CT scan and can usually be avoided. Bleeding can be controlled with the diamond burr, bone wax, crushed muscle, or Gelfoam. In one of my patients the bleeding was so severe that Sterispon gauze had to be sutured under pressure and the operation abandoned.

All bleeding should be controlled before the cochleostomy. Blood in the scala tympani could increase the degree of new bone formation (Chow et al 1995). After the package has been inserted, the final hemostasis must only be carried out with bipolar diathermy. Monopolar diathermy as discussed may affect the electronics or initiate damaging current to flow through the cochlea. If there is intractable oozing, use a suction drain.

Facial Nerve and Chorda Tympani Injury

The incidence of facial weakness (paresis) or paralysis for cumulative data reported to Cochlear Corporation on the Nucleus 22 to 1998 was 0.43% in adults (22/5170), and 0.39% in children (16/4051) (Roland 2000). To reduce this incidence to a minimum, implants should be undertaken only in centers with considerable otological experience, and preoperative x-rays studied for anomalies. The injury is most likely to occur if there is a sharply angled genu, and in the vertical segment if it swings laterally at the lower extent of the posterior tympanotomy. During surgery a thin layer of bone should be left over the surface of the nerve in case the shank of the burr rests on it during the cochleostomy. Overheating of the nerve when drilling in its vicinity needs to be prevented by constant irrigation. It is helpful to have an assistant surgeon keeping the facial nerve in view down the side arm of the microscope. Facial nerve monitoring should be used.

An immediate facial paralysis requires immediate reexploration and repair, while a delayed paresis will usually resolve with conservative treatment unless electrical testing indicates the development of degeneration. The management of facial palsy has been discussed by Cohen, Hoffman et al (1988) and Webb et al (1991).

Sometimes to gain adequate access to the middle ear through the posterior tympanotomy, it is necessary to sacrifice the chorda tympani nerve. This may cause a taste disturbance that is usually temporary. Mostly the nerve can be dissected free and moved to allow adequate access.

Perilymph Gusher

A perilymph gusher may make the electrode insertion difficult. It is more likely in cases of congenital malformation, in particular Mondini's dysplasia and other congenital abnormalities of the cochlea, as well as in skull fractures. A good seal must be obtained with fascia or pericranium to prevent a persistent fistula, and the possibility of labyrinthine infection and even meningitis. The importance of sealing the opening was demonstrated in the animal experimental studies by Clark, Shepherd et al. (1984); Franz, Clark et al. (1984); Brennan and Clark (1985); Cranswick, Franz et al (1987), Franz and Clark (1987) Berkowitz, 1987 Purser et al (1991), Dahm et al (1994). If it is anticipated, a small opening is made just sufficient to surround the array after insertion. As the risk of postoperative meningitis is much higher than for the undeformed cochlea (see Chapter 3), great care must be taken to ensure an adequate seal with a circumelectrode graft of fascia or pericranium. The gusher can be controlled by elevating the head to reduce the pressure, and then sealing the opening. It should not be necessary to use a spinal drain (N. Cohen, personal communication). In these cases it is desirable to use a systemic cover of antibiotics to eradicate any infection from *Streptococcus pneumoniae* and *Haemophilus influenzae* present at the time of the operation. The antibiotic would be cefotaxime. In addition, the operative site should be irrigated with a dilute solution of ampicillin and cloxacillin.

Malposition and Damage of the Electrode Array

The electrode may be inserted outside the cochlea, most frequently in a hypotympanic cell. Absent or poor performance will result, and it is one reason the patient should be x-rayed in the operating theater in case the electrode needs repositioning. This type of complication is more frequently seen with inexperienced surgeons (Cohen et al 1988, Cohen 1989). The insulated wires can be fractured by injudicious handling, and even concertinaed in the scala tympani. This can also lead to loss of electrode function or shorting between electrodes. An x-ray in the operating theater will help exclude this complication.

A compressed array was seen in 0.58% adults (30/5170) and 0.17% children (7/4051). In 0.17% of adults (9/5170) and 0.07% of children (3/4051), it was responsible for device removal. Electrode array damage occurred in 0.39% of adults (20/5170) and 2.12% of children (86/4051). In 0.23% of adults (12/5170) and 1.63% of children (66/4051) this was responsible for device removal. Incorrect electrode placement was made in 0.62% of adults (32/5170) and 0.59% of children (24/4051). In 0.27% of adults (14/5170) and 0.32% of children (13/4051) this was responsible for device removal.

These figures are too high and serve to stress the importance of having the implant undertaken by experienced surgeons with better training. The Cochlear Corporation cumulative data on the causes for explantation to July 1998 for the Nucleus 22 were analyzed. In 12% of adults (35/291) and 26% of children (82/312) who were explanted electrode misadventure was responsible. This also indicates that surgery should be first learned on adults.

Furthermore, two of the 19 patients out of 17,772, reported to Cochlear Limited, as having meningitis that might have been linked to implantation had the electrode in the wrong position, and it was reimplanted subsequently. This suggests that the additional handling may have been a contributory factor.

Postoperative Complications

The postoperative complication data provided by Cochlear Corporation is primarily referred to above, as they form by far the largest set (9221 patients) (Roland 2000).

Furthermore, it is cumulative data and therefore trends over time can be studied.

Seroma and Hematoma

A hematoma can be prevented by meticulous hemostasis, a suction drain, and a well-fitting pressure bandage. If present, it need not be drained unless it is causing tension to the wound edge, in which case it should be aspirated. However, a pressure dressing needs to be applied until the condition has resolved.

Aerocele

Occasionally, nose blowing will force air under the skin flap. This may be prevented by packing the posterior tympanotomy and attic during surgery. It is usually controlled by a pressure dressing, and by avoiding the precipitating factor. If it persists for over a month, then the aditus and posterior tympanotomy will need to be sealed with soft tissue.

Wound Breakdown

Initially wound breakdown was the most common major complication of cochlear implant surgery, as the devices were larger than they are now, and the skin could be sutured under tension. It is avoided by proper flap design and, in particular, creating a flap that has an adequate size and blood supply (Harris and Cueva 1987; Cohen et al 1988; Cohen 1989; Webb et al 1991).

With a wound breakdown swabs should be taken and the wound irrigated with an antibiotic solution. If the breakdown is clean, the wound can be resutured after the edges have been excised. Skin flaps are often required to avoid suturing under tension, and to move the incision away from the edges of the package. If active infection is not controlled by intensive antibiotic treatment, explantation is required. Although the definition of a flap breakdown can vary, it was reported in 4.8% of operations (22/459) in the United States for the Melbourne/Cochlear implant (Cohen et al 1988). The C-incision was the predominant cause in these cases. In 2% of operations this was serious enough to warrant explantation. With increased experience, the development of smaller packages, and a more vertical incision (Fig. 10.1), the incidence of tissue necrosis has fallen dramatically.

In the cumulative figures provided by Cochlear Corporation to July 1998 the incidences of flap necrosis were 0.56% for adults (29/5170), and 0.26% for children (11/4051) (Roland 2000). In addition, skin flap necrosis accounted for reimplantation in only 0.07% (6/9221), suggesting that repair is effective when the wound is clean. It must be emphasized that if reimplantation is planned, it is best for the package to be removed, but the electrode cut short without disturbing it in the cochlea. Twelve months later the array could be removed and another inserted.

Flap necrosis has occurred in one patient in the Melbourne Clinic when the incision was sutured under tension too close to the edge of the Nucleus 22 implant. This was managed by opening the old inverted-J incision and raising the inferiorly based flap. The incision was extended and an anteriorly-based flap created (Fig. 10.12). These flaps were then rotated to allow closure away from the implant. A similar S-shaped flap that included the wound breakdown, but not the previous incision was used successfully by Manrique et al (1995).

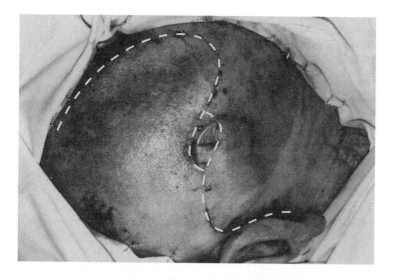

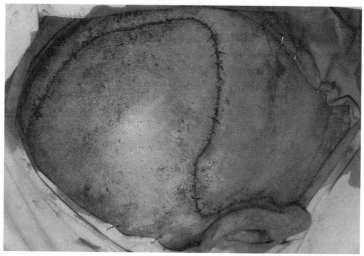

FIGURE 10.12. The rotation flaps for correcting a wound breakdown on a patient with the Nucleus 22 implant from the Melbourne Clinic.

Pain

Pain may occur in or around the ear, as well as in the throat. If it is in the ear, then middle ear disease must be excluded. Pain in the ear or throat with electrical stimulation suggests the extracochlear spread of current to the tympanic plexus and the tympanic branch of the glossopharyngeal nerve in particular. With pain in and around the receiver-stimulator package, exclude local infection of the wound, and determine if it is the result of neuralgia from the incision. If the postauricular pain continues, plain x-rays or even a CT scan may be required to exclude irritation of the periosteum. Sometimes the magnets are too strong, compressing the overlying tissue. Treatment consists in removing the cause, and the administration of analgesics may help differentiate an organic cause from a psychological overlay.

Infection

A superficial wound infection requires culture of the organisms, and should respond to local cleansing, drainage of fluid collections, removal of involved sutures, and appropriate antibiotics. If cellulitis is present, there is an immediate need for systemic antibiotics. The infection may either cause a wound breakdown or alternatively be due to flap necrosis. The cumulative incidence is 1.08% for adults (56/5170) and 0.72% for children (29/4051). This incidence is too high, and greater than that obtained for hip replacement. It is a serious complication, as reimplantation is reported in a high proportion of those with wound infection i.e.—32% for adults (18/56) and 17% in children (5/29). Infection therefore must be seen as a surgical failure. Attention to the details outlined above (see Preoperative Measures, and Incision) should be rigorously adhered to and a regular audit obtained. A laminar flow of filtered air with these provisions is still the ideal.

If infection continues, long-term systemic antibiotics are warranted together with local toilet. If there is clinical resolution, wound closure should be attempted with the antibiotic regime. It will require a rotation flap to ensure the wound is sutured without tension. If infection continues, the receiver-stimulator will need to be explanted, but the electrode array should be cut just outside the round window and the array left in situ. This facilitates replacement of the electrode if reimplantation is performed at a later date.

Otitis media in the early postoperative period is uncommon, and is less frequent than in the preimplant period (Luntz et al 1997). This may be due to the mastoidectomy. However, although it has been shown that a middle ear infection after the electrode entry point has been grafted with fascia and healed leads to no greater incidence of labyrinthitis than in a normal ear (Dahm et al 1994), it also has been shown that an infection before the fibrous tissue barrier has formed is very likely to lead to a serious infection in the inner ear i.e. within the first 6 to 8 weeks postimplantation (Clark, Shepherd et al 1984).

Labyrinthitis and Meningitis

Suppurative labyrinthitis and meningitis are rare but serious complications. Device-related labyrinthitis usually precedes but does not always lead to meningitis. Vertigo may occur if the vestibular system is intact. Low-grade inflammation should be suspected if there are fluctuating thresholds. Studies in the experimental animal showed that subacute labyrinthitis was a common sequel of infection in the cochlea (Clark, Shepherd et al 1984), and could lead to loss of neurons. Meningitis following middle ear infection could occur either in the early post-operative period when the round window seal would not have fully formed, or at a later stage from a superimposed infection.

From 1982 to 2002 (20 years) 18 patients (children and adults) from the U.S. with the Nucleus devices were reported to Cochlear Corporation and in turn to the FDA as having meningitis. A preliminary review of data as of July 2002 provided by Cochlear Corporation indicated that most of the 19 did not have device-related meningitis. One was a child with the Mondini dysplasia who died from infection in the nonimplanted ear. This was established by histological examination of both cochleae where acute inflammation was present in the unimplanted, but not the implanted, ear (Suzuki et al 1998). With the other children with an abnormality of the cochlea, it was not always possible to determine from the clinical findings whether the meningitis arose from the implanted or the unimplanted side. In many patients the meningitis could have been unrelated to the implant. In some patients the infection occurred during the postoperative period, in which case the infection was probably introduced at the time of surgery. The following case histories are presented in order of the date at implantation.

Patient 1 (QN20016602, No. 2010107) was an adult implanted April 9, 1986. He had a perilymphatic leak at the time of surgery. The same day, he developed meningitis, and *S. pneumoniae* was cultured from the cerebrospinal fluid (CSF). He responded to medication. This patient had meningitis in the vulnerable postoperative period before an adequate seal could be established.

Patient 2 (QN21094059, No. 2022578) was an adult who was deafened from bilateral temporal bone fractures, and implanted January 19, 1988. His audiologist recalls that he had meningitis 1 month postimplantation and made a good recovery. He must be considered as a postoperative complication and in the high risk category due to the fractures.

Patient 3 (QN20016699, No. 2010577) was an adult who initially had deafness due to meningitis who was implanted January 22, 1988, and then had another episode of meningitis approximately 3 weeks postoperatively, which was treated effectively with antibiotics and the condition resolved, with the device remaining in situ. This episode must also be considered as a postoperative complication as it occurred 3 weeks postoperatively.

Patient 4 (QN20000548, No. 2011738) was an adult implanted June 29, 1988, who required reexploration for the correct placement of the electrode array. Subsequently, the receiver-stimulator was partially extruded, and required repositioning and a skin flap. A pathology report of the granulation tissue showed no pathogenic bacteria. Seven months after revision surgery, the patient was reported to have developed bacterial meningitis and to have had a prior episode of viral meningitis. The device was explanted, and he was reimplanted 18 months later with no further problems. This patient must be presumed to have had device-related meningitis. The fact that he developed meningitis 7 months after the management of an extruded receiver-stimulator only serves to emphasize that low-grade infection can coexist with the foreign body and extend to the cochlea, especially if the electrode and its sheath are disturbed. The infection in the cochlea can then extend to the meninges, as discussed in Chapter 3.

As discussed below, implantation was carried out at the Melbourne Clinic (Donnelly et al 1995a) in three adults after chronic otitis media had been healed for 6 months. Recurrences occurred after 1 or 2 years in each of these patients, and one implant had to be removed because of granulation tissue extending from the middle to the inner ear.

Patient 5 (QN20017062, No. 2102566) was a 3-year-old child (43 months) who had bilateral fistulae repaired prior to implant surgery and was operated May 29, 1991. He had an ossified cochlea from meningitis, and a partial insertion of an electrode array. At switch on, there was no response. A postoperative x-ray showed kinking and malposition of the array, and he was reimplanted 2 months later. He developed meningitis 3 months after the second operation, and it was successfully treated. Then an explorative operation revealed fluid leaking from both the oval and round windows in the ear contralateral to the implanted one. It is stated by Zimmerli et al (1982) that foreign body infection can be peri-operative up to 3 months after the surgery, and this is the most likely cause for this child. He was also in the high risk category.

Patient 6 (QN21041403, No. 2013044) was an adult who had an implant December 10, 1991 and the audiologist reported that she had experienced multiple episodes of meningitis before the implant. There was an anecdotal report from the family that there had been a postoperative episode, but the clinic did not have any documentation. The link between the device and meningitis must be considered tenuous, as second hand evidence from the family is not a good basis for analysis. Furthermore, because of repeated episodes of meningitis preoperatively, this patient should be in the high-risk category of those predisposed to develop meningitis regardless of whether they have an implant.

Patient 7 (QN20017534, No. 2013641) was a child of 3 years (38 months) with the Mondini dysplasia who initially had a posterior craniotomy for an arachnoid cyst, and then a cochlear implant August 11, 1992. There were three episodes of meningitis following implant surgery. He progressed well and 17 months later had a head injury. Two weeks after this incident he had a febrile illness diagnosed as sinusitis, and this was managed with systemic antibiotics. This episode was followed 2 months later by meningitis preceded by rhinorrhea and otitis media on the side of the implant, suggesting a fistula into the cochlea and draining into the nasopharynx. The infection resolved, and after the patient left hospital the rhinorrhea recurred, as did the meningitis. The CT scan showed an opaque middle ear and mastoid, and a dye study confirmed the rhinorrhea was due to CSF. Exploratory surgery revealed extensive scar tissue around the electrode array, clear fluid in the mastoid and middle ear, and fluid leaking from around the electrode in the cochleostomy. Temporalis fascia was packed around the array and into the vestibule, and the child made a good recovery. This was device-related meningitis, but serves to emphasize the importance of firm packing around the electrode at the time of the initial surgery. The child made a good recovery (Page and Eby (1997).

Patient 8 (QN20000979, No. 2014159) was a child of 3 years and 6 months who had an implant March 22, 1993. He had a Mondini dysplasia, and a gusher at surgery that was sealed with bone wax. There was a perilymph leak 2 days postimplantation. Revision surgery was carried out with readjustment of the electrode, and packing around the electrode with fascia as well as bone wax. Subsequently, meningitis was experienced, and *S. pneumoniae* was cultured. It was considered that there was another perilymph leak around the electrode array. For other reasons, explantation was carried out, and the child recovered. If the meningitis is from the operated side, it serves to emphasize the importance of sealing the entry to prevent the ingress of infection, because of the patent communication that can exist between the perilymph and the CSF, as illustrated in Chapter 2. The child made a good recovery.

Patient 9 (QN20017748, No. 2014327), a child with bilateral Mondini dysplasia and multiple neurological deformities, was implanted May 27, 1993, at the age of 4 years and 2 months. The child had a preoperative history of meningitis and CSF leakage from the nose and both ears. At surgery there was a perilymph gusher, and the child spent additional time in the hospital with dizziness, nausea, vomiting, and nystagmus. After the cochlear implant surgery, he continued to leak from the contralateral ear and nose. Two years later the child went to bed with a headache and become comatose and died 1 day later. The temporal bones were studied and showed a wide dehiscence between the scala tympani and the internal auditory canal on both sides (Suzuki et al 1998). The implanted cochlea showed no evidence of infection. In the unimplanted ear there was inflammatory necrosis of the round window membrane, and many polymorphonuclear leukocytes in the adjacent scala tympani, indicating the route for the spread to the cochlea. The extension of the infection to the meninges probably occurred through the abnormally patent modiolus, but no organisms were cultured. This case stresses the need for the aggressive treatment of any middle ear infection in a child with the Mondini dysplasia. As the pathological examination of the temporal bones showed the meningitis was from otitis media in the ear opposite to the implant, this child did not have device-related meningitis.

Patient 10 (QN20017588, No. 2015144) was a child with the Mondini dysplasia who was implanted at the age of 4 years and 8 months an April 12, 1994. A perilymph gusher was encountered at surgery and a lumbar puncture was used. She developed otitis media in the ear opposite the implant 7 months postoperatively, and was treated with antibiotics. Four days later she presented with a high fever, and was placed in isolation. Other members of the family were reported to be ill. After a further 6 days she was still in intensive care, but responsive to some painful stimuli. A CSF shunt was placed for drainage and monitoring. Fluorescein was injected into the subarachnoid space and demonstrated a leakage from the cochlea to middle ear bilaterally. Surgery was performed first on the implant side to pack a fistula with fascia to stop leakage from the cochlear aqueduct, and then obliterate the middle ear. There was a leak from the oval window, and the stapes footplate was packed. The electrode was temporarily removed from the cochleostomy, replaced, and the middle ear space obliterated. Then 5 days later the opposite ear was exposed and a leak also seen in the oval window. The child responded to treatment, and although it was considered by the physician that the meningitis was an acute event and not related to implant surgery or the lumbar puncture, in the absence of microbiology reports on the CSF and middle ears this child could have had device-related meningitis.

Patient 11 (QN20018056, No. 2018154), a child of 2 years and 3 months, was implanted on December 4, 1996. The child had a large vestibular aqueduct, and during surgery there was a perilymph leak that resolved spontaneously. The cochleostomy was packed with fascia, and oral Cefotaxime given for 1 week. However, 5 days after surgery the child had neck pain, which was considered as possibly due to meningitis. It was treated as such for 2 weeks with antibiotics. A CSF culture grew *Streptococcus viridans,* which was considered a possible contaminant. If this child had meningitis it should be considered perioperative.

Patient 12 (QN21092689, No. 2018470) was 2 years and 3 months old when implanted on February 7, 1997. She had a history of bilateral Mondini dysplasia, and had six episodes of meningitis prior to surgery, as well as two operations to repair bilateral CSF leaks. At surgery the electrode entered the internal auditory canal. After surgery she suffered one further bout of meningitis. *S. pneumoniae* was cultured from the CSF 24 hours after implantation.

Patient 13 (QN20018459, No. 2019404) was a child that developed clinical symptoms of meningitis 24 hours after surgery undertaken on November 17, 1997, and *S. pneumoniae* was cultured. The child was given antibiotics and made a good recovery. This child was assumed to have had device-related meningitis in the perioperative period. To be classified as definitive there should be evidence that the organism from the implanted ear and its serotype were the same as that from the CSF.

Patient 14 (QN21093723, No. 2022628) was a child that had a history of Mondini's syndrome and was deafened in July 1999 from bacterial meningitis. He was implanted in the right ear on January 5, 2000, with a Nucleus straight array at the age of 5 years (66 months). There were no complications during surgery, and the surgeons reported that the cochleostomy was packed with muscle. The child contracted meningitis again approximately 1 year after implantation. The second bout of meningitis occurred in late October or early November 2000. He was hospitalized and treated with IV antibiotics for 3 weeks. At a follow-up visit November 27, 2000, the surgeon noted fluid in the patient's left ear and ordered a CT scan. The patient did not have immunization against meningitis. This child went on to be a successful implant user and has good open-set speech discrimination.

Patient 15 (QN21092640, No. 2022834) was 3 years and 7 months old when she was implanted February 17, 2000. She had bilateral Mondini dysplasia with grommets in both ears. A perilymph gusher at surgery and a spinal drain was used to lower the pressure. It is, however, better to manage a perilymph gusher by raising the head at the time of surgery until the entry is well packed and the flow stops (N. Cohen, personal communication). Shortly afterward, meningitis ensued and *Escherichia coli* was cultured. The child was considered to have device-related meningitis. Although the lumbar puncture and spinal tap almost certainly introduced the organism, as she was still in diapers, it was considered as device-related as the procedure was instigated because of the cochleostomy.

She was successfully treated with antibiotics. Six months later she was reim-planted in the opposite ear, and has remained symptom free.

Patient 16 (QN21061036, No. 2024045) had a ventriculoperitoneal shunt on the opposite side to the implant undertaken on March 30, 2000. A fever and vomiting and some cellulitis occurred the next day. The CT scan on the second day showed no abnormalities. In spite of antibiotics, he became drowsy over the next few days with a spiking fever. The CSF was found to be purulent and grew a resistant *S. pneumoniae.* The shunt was externalized and the wound explored, and the Palva flap found to be necrotic. Eight days postoperatively the implant was removed and the shunt replaced. Cultures from the receiver-stimulator and electrode tip were negative. It is considered that for definitive proof that the meningitis is device-related, the same organism and even the serotype found in the CSF should be found on the device or in the middle ear. This child, never-theless, was assumed to have had device-related meningitis, and recovered under intensive antibiotic cover. In view of the clinical findings of a necrotic flap, it is probable that the organism had colonized the middle ear and was seeded at the time of surgery. He was later reimplanted in the same ear with a Nucleus 24K.

Patient 17 (QN21094012, No. 2024657) was a child implanted on January 22, 2001, at the age 1 year, 7 months, and had a history of bilateral Mondini's syn-drome. At the time of surgery on the left ear, she had a severe perilymph gusher that was packed. She became severely ill 19 months later, and had a high-resolution CT scan that showed no mucosal thickening or infection in the middle ear.

Patient 18 (QN21093993, No. 2025592) was an adult who lost her hearing due to meningitis prior to surgery. She also had a ventriculoperitoneal shunt on the implanted side due to a CSF leak. She was implanted with a Contour array in Canada on June 15, 2001. She developed meningitis, but the meningitis was probably not device-related, and the risk was exacerbated by the shunt. Never-theless, it illustrates the importance of ensuring an adequate seal and having a single-component design that does not create a source of residual or recurrent infection.

Patient 19 (QN21093848, No. 2015260) was a child who acquired deafness at the age of 8 years due to meningitis in December 1993 and was implanted in Canada in May 1994. The x-ray showed the array was misplaced and it was explanted with another straight electrode. The child did well with the device for the next 6 to 7 years. The device commenced to malfunction, and on July 23, 2001, the child was explanted and reimplanted with a Contour. In December 2001 the child developed meningitis again. The center was concerned that it was device related and so the child had a series of evaluations including CT scan with con-trast. A large hole was found in the frontal sinus, providing a direct pathway externally to the brain. The center considered that both episodes of meningitis (pre- and postimplant) were due to this hole. In retrospect, the parents recalled the child had a toboggan accident and hit a tree and it was thought that this was how the hole in the sinus developed.

Of the 19 people who had the Nucleus straight array in North America and experienced meningitis at some time postoperatively, one could be definitely ex-cluded as being device-related, as the child (No. 2014327) had the Mondini dysplasia died from meningitis and had histological evidence of middle ear infection extending to the cochlea and the meningitis on the unoperated side. Of the remaining 18 patients, 8 developed infection in the perioperative period, however, 5 of these were in a high risk category. It must be assumed that the pathogen was introduced at the time of surgery, when the seal had not adequately formed. Of the 19 patients, 15 were in a high risk category: either the Mondini dysplasia; large vestibular aqueduct syndrome; more than one preoperative episode of meningitis; a ventriculoperitoneal shunt; temporal bone fracture; or frontal sinus defect.

These were the results for a total of 16,500 North American patients as of August 2002. The incidence of meningitis in all implant patients in North America on the basis of 100,000/year was 18.1, compared to the normal risk in the general population of 2.4 to 10 (1985–1995). If those at high risk are excluded the incidence falls to 3.9/100,000/year. Furthermore, by excluding high risk and perioperative episodes the incidence is only 1.0/100,000/year (personal communication, C. van den Honert).

In most of the implanted children with abnormal cochleae and meningitis, it was not concluded whether infection from the middle ear extended to the cochlea from the implanted or the unimplanted side. A well-sealed implant with the formation of a sheath and fibrous tissue in the scala tympani could in fact protect against the spread of infection. Some of the case reports, however, indicate that with device-related meningitis direct spread of infection from the middle ear to the inner ear and then to the meninges is important in the pathogenesis of the disease process. Hematogenous spread may also occur, but it is unproven if it is distinct from meningitis in the community.

To better understand the pathogenesis of device-related infection and determine when it should be reported, diagnostic criteria should be established. First, the presence of meningitis must be established. The clinical features are headache, fever, drowsiness, vomiting, and fits, and irritation of the meninges is indicated by aversion to bright lights and neck stiffness as well as positive Kernig's and Brudzinski's signs. In infants the manifestations are not so obvious. In addition, a CT scan may show fluid in the mastoid air cells and middle ear. However, the diagnosis can be made definitively only by the biochemical, cellular, and microbiological analysis of the CSF. Typically, there is an increased intracranial pressure (>180 mm H_2O in 90%), polymorphonuclear leukocytosis (>100 cells per microliter in 90%), decreased glucose concentration (<40 mg/dL in 60%), and increased protein concentration (>0.45 g/L in 90%) (Roos and Tyler 2001). If an organism is cultured it should be fully characterized. Having made the diagnosis of meningitis, it is then necessary to establish whether it is device related. For that reason a number of presumptive and definitive indicators are advanced. The presumptive indicators are (1) the child is 2 years of age and older; (2) there is clinical and radiological evidence of otitis media on the implanted side; (3) there is no cochlear abnormality in the child; (4) in the CSF there are uncommon phenotypes of organisms commonly causing meningitis in the general community; (5) there is a high incidence of disease associated with a change in a product or procedure and (6) there is a high incidence of relapse, recurrence, or death.

In addition, definitive indicators for device-related meningitis are advanced. These are the only indicators that show conclusively that there is a positive relationship, or alternatively that the implanted ear was not responsible. These definitive indicators are (1) a positive culture from the middle ear on the implanted side with an organism of the same type as that found in the CSF, (2) a positive culture from the explanted electrode or receiver-stimulator on the implanted side that provides an organism of the same type as that found in the CSF, and (3) histopathological evidence from temporal bones that the labyrynthitis was not present on the implanted side or that it had occurred on the opposite side.

The importance of age as a presumptive indicator is shown in a study of pneumococcal meningitis in the population of the state of Victoria, Australia, for a 10-year period from 1989 to 1998. This study captured up to 60% of the cases. The data demonstrate the relative proportion of pneumococcal meningitis in the community by month up to 5 years of age. There was a very low incidence of pneumococcal meningitis from 24 months of age and above, but it increased dramatically below this age. Consequently, if a child is older than 24 months when he/she develops pneumococcal meningitis, it is highly probable that it is device related. The age distribution is similar for a *S. pneumoniae* bacteremia, indicating that meningitis in the general community is blood borne.

Implanted children who develop meningitis with normal cochleae are also more likely to have device-related meningitis. In contrast, there is a high incidence of meningitis in unimplanted but abnormal cochleae, and so an implanted child with an abnormal cochlea could develop meningitis either in the operated or unoperated ear because of the potential pathway from the inner ear to the subarachnoid space. The anatomical pathways for abnormal cochleae that were discussed in some detail in Chapters 2 and 3 would produce a perilymph gusher (Schuknecht and Gulya 1986). A study by Phelps et al (1994) demonstrated that four of 20 patients with a cochlear dysplasia developed meningitis. These data together with the case reports by Suzuki et al (1998) and Page and Eby (1997) indicate a high incidence of meningitis due to the abnormality. This high incidence is not expected for normal cochleae, so that if it occurs device-related disease should be suspected.

The incidence of device-related meningitis should be recorded as the number of cases per 100,000 people per year, so that a meaningful comparison can be made with the general community. The incidence in different age groups is especially important. For example, as discussed below meningitis due to *S. pneumoniae* occurs almost exclusively in children under 2 years of age. If a *S. pneumoniae* infection develops in an older child, then it is probable that it is device related. Furthermore, the implant and general communities should be homogeneous, and not contain unequal proportions of people predisposed to meningitis. This applies in particular to children with cochlear deformities as discussed above. In the clinical study of the Nucleus 24 and Contour array on 256 children (Staller et al 2002) the incidence of the Mondini dysplasia was 3.3% for children from 12 to 24 months, and 5.1% for those older than 5 years. For this reason, meningitic patients with abnormal cochleae and perilymph leaks should be removed from the data to compare the incidence with the general unimplanted population (the incidence of the Mondini dysplasia in the general population is very small).

In the community as a whole the incidence has varied from 2.5–10/100,000/year (Centers for Disease Control and Prevention, National Center for Infectious Diseases, US Department of Health and Human Services; Roos and Tyler 2001). The incidence in the community has fallen in recent years due to improved treatment, and is reflected in the reduction to 2.5/100,000/year.

Another presumptive indication that an episode of meningitis was device related would be if the organism was different from one usually causing infection in the general community, or it had a serotype that was at variance with one normally causing infection. The organisms most commonly responsible for community-acquired bacterial meningitis overall were *S. pneumoniae* (approximately 50%), *Neisseria meningitidis* (approximately 25%), group A β-streptococci (approximately 10%), and *Listeria monocytogenes* (approximately 10%). *H. influenzae,* which was the most common cause of bacterial meningitis in the U.S., has been markedly reduced with the introduction of *H.* influenzae type b (Hib) vaccine in 1987. It now accounts for less than 10% of meningitis cases (Roos and Tyler 2001). There has also been a major change in the epidemiology of pneumococcal disease with a greater prevalence of penicillin- and cephalosporin-resistant strains.

N. meningitidis accounted for 60% of bacterial meningitis cases in children and young adults between the ages of 2 and 20 years. It arises from colonization of nasopharynx. *S. pneumoniae* is the most common cause of meningitis in adults over 20 years, and patients are predisposed by pneumococcal pneumonia and otitis media as well as other medical conditions. *L. monocytogenes* is also an important cause of bacterial meningitis in older individuals. Following neurosurgical procedures, it is more likely that enteric gram-negative bacilli as well as *Staphylococcus aureus* and coagulase-negative staphylococci would be the predominant infective organisms, and their presence would indicate device-related meningitis.

As the pathogenesis of meningitis in the community is different from the device-related infection, there could be a stronger relation with the spectrum of organisms causing otitis media. There is support for this hypothesis, as data show that the pathogenesis varies for each organism. For example, in studies on temporal bones from people who have died from pneumococcal meningitis, there were inflamed mucous membranes and exudate in the middle ears and mastoids in six of seven bones. In one there was a distinct tract of acute inflammation through the petromastoid canal of the mastoid bone to the overlying dura. This is one route for the infection, as are the cochlea, paranasal sinuses, and bloodstream (Schuknecht 1974).

As there is a propensity for *S. pneumoniae* to invade the meninges from the middle ear and nasal sinuses in the community-based disease, as described above, it is expected that device-related meningitis would also have a greater incidence of pneumococcal infections. These data are not available. However, it is important to appraise the incidence of organisms causing otitis media, as these are the ones likely to invade the inner ear around the electrode entry point.

The incidence of organisms producing otitis media has varied over the last 10 years. From 1979 to 1982 at the Pittsburgh Otitis Media Research Center the incidence of *S. pneumoniae* infections was 40%, and *H. influenzae* 20%, with the proportion due to this organism declining with increasing age (Bluestone et al 1992). *Morexella catarrhalis,* group A β-hemolytic streptococcus, and *S. aureus* were each responsible for less than 10% of cases. Approximately 25% of patients had negative cultures for bacteria. Between 1983 and 1986, there was a change in the incidence, with *S. pneumoniae* found in 29% of the effusions, *H. influenzae* in 23%, and a rise in *M. catarrhalis* to 13%. The incidence of *S. pneumoniae* has been reported to be as high as 55% (Luotonen et al 1981; Johnson et al 1991; Block 1997). There was also an increase in the prevalence of β-lactamase–producing organisms in the latter half of the 1980s for the strains of *H. influenzae,* and they rose from 17% to 34%. In addition to the above, *Pseudomonas aeroginosa* and *Proteus* can be found in a small proportion.

If meningitis supervenes after a neurosurgical procedure or with a CSF leak, there is quite a different spectrum of infecting organisms, as discussed above, and the same could be assumed in cochlear implant device-related meningitis. As far as atypical serotypes with *S. pneumoniae* are concerned, the most common types of the 90 identified as causing otitis media, are 3, 6B, 9V, 14, 19F, and 23F (Karma et al 1985; Butler et al 1995; Block et al 1999). If a different serotype is found, another pathogenesis should be suspected (i.e., device related).

With the cochlear implant patients described above, *S. pneumoniae* was found in the two children in whom meningitis developed within 24 hours of surgery. This suggests the organism was resident at the time of surgery and emphasizes its virulence if introduced into the perilymph and thence the CSF. The invasiveness of both *S. pneumoniae* and *H. influenzae* was seen in the cases of meningitis following stapedectomy where the organisms invaded the inner ear and then gained access to the CSF, and was described in Chapter 3.

If there is a statistically significant increase in meningitis after a change in a device or procedure, the infection should be suspected as device related. There was no change in the incidence for the Nucleus banded array over time and not with the introduction of the Contour perimodiolar array in February 2000. The peak in 1994 was due to three children with the Mondini dysplasia resulting from an expansion of the selection criteria after its approval by the FDA for its use in children in 1990.

Finally, if there is a higher incidence of relapse, recurrence, or death, then device-related meningitis should be suspected. Relapses would indicate that residual infection had become resident to the foreign body in the cochlea. Recurrence would point not only to residual infection, but also to either an inadequate seal or an electrode design weakness, or both. As the Mondini dysplasia, especially because of an associated perilymph gusher is hard to seal, children with any implant are prone to labyrinthitis and meningitis. This is supported by the animal experimental studies by the University of Melbourne/Bionic Ear Institute referred to in Chapter 3 by Dahm et al (1994). This study showed that 50% of the implanted ungrafted cochleae and 6% of the implanted grafted (fascia and Gelfoam) cochleae became infected. This difference was statistically significant.

Furthermore, a higher mortality for meningitis than in the general community would indicate a severe infection due to a design or procedural problem. In this case a higher mortality could be due to the creation of a nidus for infection or a more direct transmission of the infective organisms to the meninges. Furthermore, as discussed in Chapter 3, a foreign body and in particular a dead space makes the organism more virulent, reduces phagocytosis, and limits the effectiveness of antibodies and antibiotics. The mortality for the community-based disease is 3% to 7% for meningitis caused by *H. influenzae, N. meningitidis,* and *S. pyogenes;* 15% for *L. monocytogenes;* and 20% for *S. pneumoniae.* The rate is higher in infants and people older than 50 years (Roos and Tyler 2001).

Immunization against *H. influenzae* has been very effective in reducing the incidence of meningitis due to this organism, as discussed above. A vaccine was developed against *S. pneumoniae* for the seven most common strains of the 90 serotypes identified. At first a polysaccharide vaccine was used, but its efficacy was low (Block et al 1999; Makela et al 1980; Teele et al 1981, Howie et al 1984; Douglas and Miles 1984). The vaccine was made more potent when it was coupled to molecules of the carrier protein. This induced the formation of antibodies that were detectable in mucosal secretions, and reduced the nasopharyngeal presence of pneumococci (Eskola et al 2001). It was tested first in northern California and it was 100% efficient against invasive pneumococcal infections in children (Black et al 2000). It reduced the number of episodes of otitis media from any cause by 7%. A more detailed investigation was carried out by Eskola et al (2001). It was undertaken on 1662 infants in a randomized double-blind efficacy trial using the heptavalent pneumococcal polysaccharide conjugate vaccine. The children received either the study vaccine or a hepatitis B vaccine as a control. This study found a 6% but not significant reduction in otitis media overall. There was a reduction in all culture confirmed pneumococcal episodes by 34%, a reduction in pneumococcal serotypes in the vaccine by 57%, and a reduction in serotypes cross-reacting with the vaccine by 51%. There was, however, an increase in all of the serotypes by 33%. This is due to the fact that *S. pneumoniae* colonizes the mucosa, and if certain strains of pneumococcal are removed in response to immunization they are supplanted by others in the vicinity.

Thus if a pneumococcal vaccine were to be effective in preventing device-related meningitis in children and the pathogenesis was the same as for the community disease, it would be helpful only for the approximately 50% who had pneumococcal meningitis, and then it would reduce the incidence in this 50% by only 34% to 57% (Eskola et al 2001). However, as indicated above, *N. meningitidis* accounts for 60% of bacterial meningitis cases in children and young adults between the ages of 2 and 20 years. Thus if meningitis had the same pathogenesis as for the general community and was blood borne, then the pneumococcal vaccine would not be effective in the majority of cases. It must be emphasized that device-related meningitis has a different pathogenesis from that of the community-related disease, as evidenced by the clinical cases reported to the FDA by Cochlear Corporation as discussed above, the histopathological and clinical results reported by Suzuki et al (1998) and Page and Eby (1997), the temporal bone findings in cases of fatal meningitis after stapedectomy in adults (Rutledge et al 1963; Wolff 1964; Matz et al 1968; Palva et al 1972; Benitez 1977), the human temporal bone data from middle ear disease and meningitis (Friedmann 1974), and the animal experimental data reported by Clark, Pyman et al (1984), Clark, Shepherd et al (1984), Franz et al (1984), Brennan and Clark (1985), Berkowitz et al (1987), Cranswick et al (1987) Purser et al (1991), and Dahm et al (1994). For this reason, there will be a different spectrum of bacteria from those for the community disease, or the serotypes will be different. Furthermore, there is evidence that the spread of infection from the middle ear to the cochlea and thence to the meninges is frequently due to direct propagation rather than hematogenous seeding. This was seen not only in the temporal bone of the child with the Mondini dysplasia described by Suzuki et al (1998) but also in those with fatal meningitis after stapedectomy in adults (Rutledge et al 1963; Wolff 1964; Matz et al 1968; Palva et al 1972; Benitez 1977). For these reasons as well, the vaccine is not likely to be so effective. Thus although the data show that immunization against *S. pneumoniae* is not very likely to be effective, benefits for some children can be expected and immunization should be undertaken. It must not be seen, however, as a cure for a device- or procedure-related problem.

Labyrinthitis and meningitis can in general be prevented by meticulous asepsis, avoiding surgery in the presence of infection, developing a type 1 seal with a fascial graft around a single-element array, packing a perilymph gusher, treating a postoperative perilymph leak early, and managing otitis media effectively. Strict adherence to these principles must occur in the case of children with the Mondini syndrome. Patients are also predisposed to meningitis if they have impaired immunodeficiency status or have a prior history of meningitis.

To keep meningitis to a minimum, it is recommended that children who are to have cochlear implant surgery be immunized against *H. influenzae* B with the (Hib) vaccine, and against *S. pneumoniae*. After review, the U.S. Advisory Committee on Immunization Practices has recommended that all candidates for cochlear implants should be considered for immunization prior to surgery as well as those with an existing implant. *H. influenzae* conjugate vaccines are recommended for all children up to age 5 years. Heptavalent pneumococcal conjugate vaccine Prevnar is indicated for use in infants and all children younger than 2 years of age and for children up to 5 years who are at risk from invasive pneumococcal infections. The 23 valent pneumococcal polysaccharide vaccines Pnu-Imune 23 and Pneumavax 23 are also recommended for children over 2 years of age and adolescents and adults who are at risk. For children from age 2 to 5 years who are at high risk, it is recommended that pneumococcal conjugate vaccine be followed at least 2 months later by 23-valent pneumococcal polysaccharide vaccine to provide protection against a broader range of serotypes. There are, however, limited data supporting this regime (http://www.cdc.gov/od/nvpo/).

Skin Flap Thickness

The strength of the Nucleus external magnet can be varied over four levels depending on the thickness of the overlying tissue, so that the pressure applied is comfortable. If there is too much pressure, ulceration and wound breakdown can occur. If the flap was 12 mm, a magnet would not be strong enough to hold the coil. A thick flap reduces the electrical power transmitted, and this will limit the rate of stimulation. Flap thickness has been measured postoperatively in 213 patients from the University of Melbourne Clinic, and 15% were above the optimal 6-mm thickness. This indicates that attention should be given to the flap at surgery.

Facial or Tympanic Nerve Stimulation

Facial nerve stimulation is a common complication occurring in 3.13% (162/5170) of adults, and 1.16% (47/4051) of children. It is more frequent in adults, as it occurs particularly in cochlear otosclerosis. If the threshold of facial nerve stimulation is above the comfortable listening level, there is no problem. If it is lower, the involved electrodes are switched off. In rare occasions it has not been possible to operate the implant. It has been responsible for the removal of the device in 0.06% (3/5170) of adults but not in children. Fluoride may be considered as it may facilitate sclerosis, but there is no clear evidence that it is effective. With tympanic nerve stimulation pain in the ear or throat will occur, and this is usually just with the most basal electrodes. The management is to switch off the involved electrodes.

Tinnitus

A temporary increase in tinnitus for the first few days postoperatively is common. It then usually settles to the preoperative level with a further reduction when the implant is functioning. Sometimes tinnitus can be aggravated by the implant (adults 0.60% and children 0.05%), and this usually means certain electrodes cannot be stimulated.

Vertigo or Dizziness

Vertigo is rarely a problem postoperatively, and is usually temporary. If it is persistent, a perilymph fistula must be considered, especially if the symptoms are severe. This emphasizes the importance of sealing the entry to the scala tympani properly. Reexploration and repair with a graft is necessary. Vertigo occurred in 0.79% of adults (41/5170) and 0.08% of children (7/4051). It was the cause of device removal in 2/5170 adults, but not in children. If vertigo was a preoperative problem, it will not usually be improved by the implant.

Prominent Swelling

The first device developed by Cochlear Proprietary Limited for clinical trial (Clark, Dowell et al 1984; Clark, Tong et al 1984) was thick (10.5 mm), and in some patients with thin skulls and where the bed was not drilled down to dura, there was an obvious swelling. A prominent swelling did not occur with the Mini (CI-22) implant (6 mm thick) for children from 2 to 18 years (Clark, Blamey et al 1987b). The Cochlear Limited CI-24 implant has a section 2 mm thick to be placed in the skull and 4 mm lying superficial. It can be placed in the skulls of children from 6 to 24 months without a significant swelling. The Advanced Bionics and Med-El implants have a thickness of 6 mm, so again swelling in not likely to be a problem, if the bone is drilled down to the dura in young children. It does, however, mean a larger area is exposed than with the Nucleus 24M and 24R devices, and it has been the practice to hold it down with a pressure dressing.

Failure of the Implant to Function

Failure of the implant may be due to malfunction of the electronics, lack of auditory nerve function, damage to the electrodes, or extracochlear placement in an hypotympanic cell. The cause can be diagnosed by clinical and radiological investigation, together with electrical recordings made from the scalp while stimulating the promontory or the implant. If the promontory stimulation is positive, the auditory nerve can be considered to be adequate, and unless the electrode is obviously distorted or in the wrong position (hypotympanic cell), then a malfunction of the implant can be assumed. This malfunction may be confirmed if there is no recorded electrical artifact from around the package. A partial failure can be managed by adjusting the stimulation program, but with a complete failure a second implantation may be requested.

Blows to the Skull

Another complication is trauma to the region, with damage to the device and even to the brain. This is one of the advantages of the Nucleus system in having the receiver-stimulator coil, placed posteriorly, and lying external to the bone of the skull as illustrated in Figure 10.13, compared to a ceramic package that does not have this support. Protection would be provided by a helmet; otherwise, the child should be advised not to participate in sports that entail body contact.

Special Cases

Ossified Cochlea

New bone formation (osteoneogenesis) is seen in the scala tympani of the cochlea in 14% of adult patients ($n = 105$) (Balkany et al 1988), and 34% ($n = 128$) of the initial group of children operated on by Luxford and House (1977). Meningitis would be the cause of the new bone formation in children, and meningitis and otosclerosis in adults, and rarely a skull fracture. The bone may vary from soft and spongy to hard and compact. The proximal part of the scala tympani of the basal turn is most frequently affected, but the pathology may extend to the scala vestibuli of the basal turn and throughout the other turns.

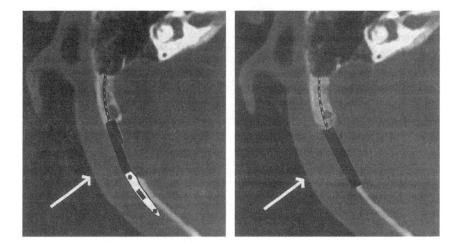

FIGURE 10.13. A horizontal x-ray of the skull with Nucleus 24 (left) and ceramic (right) receiver-stimulators in place. With the Nucleus package there would be some protection afforded by the posterior coil lying external to the skull. (Reprinted with permission from Clark, Pyman et al 1997. Surgery. In: Clark G. M., R. S. C. Cowan and R. C. Dowell, eds. Cochlear Implantation for infants and children, Advances. San Diego, Singular Publishing Group: 111–124.)

Initially, cochleae without new bone were implanted to give the best chance of developing a speech-processing strategy (Clark, Tong 1981a,b). Then patients with bone restricted to the basal turn were selected. Techniques were employed to remove the bone with a drill or pick it from the scala tympani to allow the electrode to be inserted around the patent more distal turn (Balkany et al 1988). In 14/15 patients a full insertion was achieved with the University of Melbourne/Nucleus, UCSF/Storz, and Symbion/Ineraid arrays. Alternatively, if the scala vestibuli, not the scala tympani, was patent, as discovered in 1979 in the third patient to receive the first University of Melbourne's multiple-electrode device, it could be passed upward along the more patent scala vestibuli from the basal region. This was described as the array being inserted in an anterograde direction through an opening anterior to the round window (Steenerson et al 1990).

With more advanced disease, an alternative reported in 1985 at the International Cochlear Implant Symposium and Workshop at the University of Melbourne by Chouard involved two electrode arrays, one along the basal turn and the other with electrodes into the apical or middle turns through a hole drilled in the overlying bone. A modification of this approach was later reported by Chouard et al (1995). With still more advanced disease with obliteration of all the cochlear turns, an initial approach involved drilling out the whole of the basal turn to enable insertion of the multiple-electrode array. Gantz et al (1989) created a channel in the basal turn around the modiolus. Surgical access required removing the posterior canal wall; performing a radical mastoidectomy, removing the tympanic membrane, malleus, and incus; and closing the external auditory canal. The middle ear mucosa was cleared and the opening of the eustachian tube closed. Experience in the Melbourne Clinic has shown that although some place pitch and satisfactory speech perception is possible, it is difficult to retain the electrodes in place and the spiral ganglion is at risk. For this reason, a suitable alternative was to insert a short array (8 mm) into a tunnel created along the scala tympani (Cohen and Waltzman 1993).

A further approach, when both scalae are obliterated throughout the cochlea or in the lower basal turn alone, was to use at least two bundles of electrodes inserted directly into the cochlea, one through an opening into the basal turn near the round window and the other through an opening directly into the middle or apical turns to allow an array to pass to the high and middle speech frequencies as was demonstrated by Clark (1975), Clark, Hallworth et al (1975) and Clark and Hallworth (1976), as shown in Chapter 2 in the normal cochlea and implemented in the ossified cochlea by Bredberg and Lindstrom (1995) and Bredberg et al (2000). Access to the upper basal turn was achieved by extending the posterior tympanotomy and removing the floor of the antrum as well as the incus. This provided the exposure of the facial nerve, stapes, cochleariform process, and tendon of the tensor tympani muscle (Bredberg et al 2000). An opening was made into a patent upper basal turn anterior to the oval window. The anatomical measurements of Clark (1975), as discussed in Chapter 2, showed that the spiral lamina of the upper basal turn was 2.5 mm anterior to the center of the oval window along a line joined to the helicotrema. The depth of the overlying bone was 1.9 mm (1.3 to 2.0 mm). The electrode array could be passed in a retrograde direction towards the lower basal turn. If both scalae were obliterated, a tunnel was created that extended anteriorly along the line of the scala (Bredberg et al 2000). The anatomical dissections (Clark 1975; Clark, Hallworth et al 1975) and a study on a mold of the cochlea showed that a retrograde insertion was also possible through an opening into the upper middle turn. The turn is located 4.1 mm anterior to the oval window, at a depth of 1.7 mm (1.3 to 2.1 mm). It is more difficult to enter the scala tympani, as it is overlapped by the scala vestibuli.

This approach has been used by Hausler et al (2000) on three patients. After completing the posterior tympanotomy and drilling the basal turn, the operation was completed through the external auditory canal by elevating a posterior tympanomeatal flap. To retract the flap forward and obtain a good view of the medial wall of the middle ear, it was necessary to section the tendon of the tensor tympani and the neck of the malleus. From the anatomical dissections (Clark 1975; Clark, Hallworth et al 1975) and as illustrated in Chapter 2, it is best to site the opening below the cochleariform process 4 mm anterior to the center of the oval window. Drill through the bone until the ground-glass appearance of the bone near the endosteum appears, and it will then be possible to distinguish the bony septum between the upper middle and upper basal turns.

In summary, the routine for managing a cochlea with new bone is as follows. If the window is obliterated, estimate its position from the stapes, and drill through the bone toward the basal turn from the inferior aspect. Carefully extend the excavation superiorly, anteriorly, and medially to avoid the basilar membrane. In cases of meningitis or otosclerosis, the obliteration may be extensive. Drilling or removing the bone with picks may be continued for 8 mm along the line of the basal turn. The bone is white in comparison to the yellowish endochondral bone around the cochlea, which greatly assists in maintaining the correct line. At a depth of approximately 9 to 10 mm there is a risk of entry into the carotid canal. If at 8 mm there is no lumen, drill anterior to the round window to enter the scala vestibuli and attempt an insertion there. If this is not possible, use a double array with one in the tunnel in the lower basal turn and a second in an opening in either the upper basal or upper middle turns, as described above.

Secretory (Serous) Otitis Media

Cochlear implantation should not be undertaken in children with secretory otitis media, as the condition could predispose to serous or suppurative labyrinthitis. This is even more important in children with congenital deformities such as the Mondini syndrome, where a middle ear infection can extend to the inner ear and via a dehiscence between the scala tympani and the internal auditory meatus lead to meningitis (Suzuki et al 1998). Histological material from cats has shown serous labyrinthitis with mucoid exudates in the bulla and negative cultures following bacterial inoculation (Clark, Shepherd et al 1984; Cranswick et al 1987). However, positive cultures can be obtained from the mucous lining. Before operating on children, ventilation tubes should have been removed for more than 8 weeks without any recurrence to help ensure the condition is in remission. Tubes should not be inserted into the tympanic membrane during the first 6 weeks postoperatively to ensure infection is not introduced during the time required for a good fibrous tissue seal at the electrode entry point (Franz et al 1984).

Tympanic Membrane Perforation and Chronic Suppurative Otitis Media

It is essential that an implant be carried out only after a chronic otitis media, but not due to cholesteatoma, has resolved and the tympanic membrane healed for 12 months. Implantation after the resolution of chronic middle ear disease was necessary in only three out of the first 121 adults in the Melbourne Clinic (Donnelly et al 1995a). These 3 patients' disease had been healed for 6 months prior to surgery. However, recurrences occurred after 1 or 2 years in each patient. In one the implant had to be removed, as there was granulation tissue extending from the middle to the inner ear. It is considered that chronic suppurative otitis media should be managed more aggressively. Irving and Gray (1994) described a technique to prepare the ear for cochlear implantation that created a radical mastoid cavity, obliterated the eustachian tube, obliterated the mastoid cavity with abdominal fat, and permanently closed the external ear canal

Open Mastoid

As chronic infective ear disease and its surgical treatment may lead to a severe to profound hearing loss, it may be necessary to consider a cochlear implant in an ear that has had a radical mastoidectomy. An implant can be carried out provided it is buried under tissue and the field is clean. The lining of the mastoid cavity is elevated and a fat graft inserted. Care needs to be taken if the facial nerve has been exposed (Scrivener and Gibson 1987). A membrane should enclose the middle ear to protect the electrode entry from bacteria invading from the external ear canal. The surgery is more safely carried out as a two-stage procedure.

Congenital or Genetic Malformation of the Cochlea

As cochlear implants provide great benefits for congenitally deaf children, there is a need to consider surgery not only for those with a normal cochlear structure, but also for those with cochlear malformations, in particular the Mondini dysplasia. This dysplasia, which is autosomal dominant, has a reduced number of coils, with often only a single basal coil (Friedmann 1974). The principal defect is the absence of the interscalar septum between the upper two coils. This leads to the scala communis. The ganglion cell population varies from 7677 to 16,110 (Schmidt 1985) compared to the normal of 29,802 to 38,352 (Hinojosa and Marion 1983). The modiolus may on x-ray appear normal in the basal turn or rudimentary. The auditory neurons lie peripherally in the upper turns. The Mondini dysplasia may also be associated with the large vestibular aqueduct syndrome (Valvassori and Clemis 1978).

In considering surgery, there is a risk of damage to the membranous structures due to the absence of the interscalar septa. There is also a risk that the array may enter the internal auditory meatus through a dehiscence, or a perilymph gusher may be experienced. As discussed above, there is a greater risk that meningitis will occur in children with this syndrome due to the perilymph leak, and a tight fascial seal is essential. If the modiolus is well formed, a perimodiolar array can be inserted to stimulate the spiral ganglion cells lying centrally. If not a standard straight array should be used to stimulate the nerves where they lie peripherally.

The first child (S.N.) with Mondini's dysplasia to receive an implant at the University of Melbourne Clinic obtained open-set speech understanding for electrical stimulation alone. Improvements in language for children with the Mondini dysplasia have also been reported by Silverstein et al (1988), Turrini et al (1997), and Suzuki et al (1998).

Transmastoid Labyrinthectomy and Acoustic Neuroma

Cochlear implantation can be considered in certain patients who have had extensive temporal bone disease or surgery. A case of transmastoid labyrinthectomy is described (Kveton et al 1989). The cochlea was intact and implanted with results comparable to those expected for postlinguistically deaf adults. Furthermore it is possible a cochlear implant can be used to help restore useful hearing after removing an acoustic neuroma if the cochlear nerve is intact (Luetje 1996; Hulka et al 1995).

Insertion and Reinsertion

The ease of removing the smooth free-fitting banded array was observed in animal experiments at the University of Melbourne, and was the reason the second implant on patient MC-2 did not have a connector (Clark, Tong et al 1981a). Although a connector was used for the Nucleus clinical trial device, further animal research with the Nucleus banded array showed it could not only be easily withdrawn but also reinserted without difficulty. Confirmation of the relevance of the animal findings to humans came in 1983 to 1984, when the University of Melbourne's prototype device was explanted and the Nucleus clinical trial system inserted (Clark, Pyman et al 1987). It was found to go in as far as the original array. Consequently, a connector was not included with the Nucleus CI 22 (Mini) implant (Clark, Crosby et al 1983), especially as the receiver-stimulator needed to be thinner for implantation in children. There were concerns, however, about the explantation of arrays that had protruding ball electrodes, such as the 3M/House single-electrode array and the Symbion/Ineraid multiple-electrode array. They could cause significant damage when thick fibrous tissue or bone surrounded the balls, and could be difficult to remove (Gray et al 1993).

The perimodiolar electrode Contour now used with the Nucleus 24 system can be removed. As it lies close to the modiolus it could be explanted, and other more advanced arrays could be introduced along the electrode track for finer temporospatial stimulation. The implantation and reimplantation of the Nucleus single element array has been studied in the experimental animal. The histopathology of the cat cochlea implanted for 3 months and then reimplanted shows that the electrode passed along the preformed sheath without significant trauma. The insertion induced a cellular response in the sheath, and some increase in the number of the capillaries and interstitial fluid. If an electrode array is placed peripherally, as was the case with the Nucleus straight but flexible array, then it would not be possible to effectively place an array near the modiolus if this were required.

Pedestal (Plug and Socket)

The percutaneous transmission of stimuli with a plug and socket allowed much greater flexibility in exploring stimulus parameters than transcutaneous transmission with electromagnetic coupling. However, chronic studies on the experimental animal (Minas 1973) had shown Teflon plugs to be associated with a high incidence of local infection. As Teflon does not adhere to the skin, bacteria track around the edges and infection develops in a sinus. For this reason a transcutaneous system instead was used for the University of Melbourne's initial studies on patients in 1978 to 1979 (Clark, Black et al 1977). A transcutaneous system was also used by Doyle et al (1964) for multiple-electrode stimulation. A percutaneous route was selected by Simmons (1966) and Mladejovski et al (1975).

Subsequently, the Richards/Smith and Nephew Ineraid (formerly Symbion) device was designed with a percutaneous pedestal of pyrolized carbon to allow better integration with the skin edges and bone. The pedestal was screwed to the skull, and the skin sutured so the edges were in apposition with the socket. Recurrent infection and discharge were troublesome in spite of meticulous hygiene (Parkin et al 1996), and in one of the patients of Ray et al (1998) it was severe enough to require explantation. In addition, due to its brittleness and incomplete integration with bone, trauma quite frequently led to fracture in approximately 2%. Nevertheless, in the experimental animal there was a good host-tissue response seen with pyrolized carbon and titanium pedestals (Parkin et al 1996). In a clinical evaluation by Dobie et al (1995), 40% with the Ineraid plug and socket (12/30) had cutaneous problems (crusting, discharge, gramulations, erythema, bleeding), but none had to be removed.

Magnetic Resonance Imaging (MRI)

With the widespread use of MRI, its effects on cochlear implant function should be determined, as well as its safety. Data are available in particular on the Nucleus CI 22 and CI 24 devices and the Med El Combi-40/40 +, and has been reviewed by Shellock (2001). Tests were conducted to assess the operation of these cochlear implants in the MRI environment as well to elucidate magnetic field interaction, artifacts, induced current, and heating. With the most recent Nucleus 24 implants, the magnet can be removed under local anesthetic, to allow MRI up to the clinical standard to be safely employed (Nucleus Cochlear Implant Systems, Reliability Update, Issue 4, 2001). An ear-mold–supported headpiece for the transmitting coil would avoid the need for a magnet. This was evaluated for the Clarion device by Weber et al (1999). The receiver-stimulator was placed closer behind the ear than the device with the magnet. The headset was stable and reliable, and the device was compatible with MRI.

References

Bagger-Sjoback, D. 1991. Sodium hyaluronate application to the open inner ear: an ultrastructural investigation. American Journal of Otology 12: 35–39.

Balkany, T. and F. F. Telischi. 1995. Fixation of the electrode cable during cochlear implantation: the split bridge technique. Laryngoscope 105(2): 217–218.

Balkany, T., B. Gantz and J. B. Nadol. 1988. Multichannel cochlear implants in partially ossified cochleas. Annals of Otology Rhinology and Laryngology 97(suppl 135): 3–7.

Balkany, T., N. L. Cohen and B. J. Gantz. 1999. Surgical technique for the CLARION cochlear implant. Annals of Otology, Rhinology, and Laryngology–supplement 177: 27–30.

Banfai, P., A. Karczag, S. Kubik, P. Luers and W. Sarth. 1986. Extracochlear sixteen channel electrode system. Otolaryngologic Clinics of North America 19: 371–408.

Banfai, P., S. Kubik and G. Hortmann. 1984. Our extra-scalar operating method of cochlear implantation. Experience with 46 cases. Acta Oto-Laryngologica (suppl 411): 9–12.

Bast, T. H. and B. J. Anson 1949. The temporal bone and the ear. Springfield, IL, Charles C. Thomas.

Benitez, J. T. 1977. Stapedectomy and fatal meningitis. ORL 39: 94–100.

Berkowitz, R. G., B. K.-H. G. Franz, R. K. Shepherd, G. M. Clark and D. Bloom. 1987. Pneumococcal middle ear infection and cochlear implantation. Annals of Otology, Rhinology and Laryngology 96(suppl 128):55–56.

Bielamowicz, S. A., N. J. Coker, J. H.A. and M. Igarashi. 1988. Surgical dimensions of the facial recess in adults and children. Archives of Otolaryngology–Head Neck Surgery 114: 534–537.

Black, S., H. Shinefield, B. Fireman, et al. 2000. Efficacy, safety and immunogenicity of heptavalent pneumococcal conjugate vaccine in children. Pediatric Infectious Disease Journal 19: 187–195.

Blamey, P. J., B. C. Pyman, M. Gordon, et al. 1992. Factors predicting postoperative sentence scores in postlinguistically deaf adult cochlear implant patients. Annals of Otology, Rhinology and Laryngology 101: 342–348.

Block, S. L. 1997. Causative pathogens, antibiotic resistance and therapeutic considerations in acute otitis media. Pediatric Infectious Disease Journal 16: 449–456.

Block, S. L., J. A. Hendrick and C. J. Harrison. 1999. Pneumococcal serotypes from acute otitis media in rural Kentucky (abstract). Program and abstracts of the 39th Interscience Conference on Antimicrobial Agents and Chemotherapy, San Francisco, September 26–29, 1999. Washington, D.C., American Society for Microbiology.

Bluestone, C. D., J. S. Stephenson and L. M. Martin. 1992. Ten-year review of otitis media pathogens. Pediatric Infectious Disease Journal 11(suppl): S7–S11.

Bredberg, G. and B. Lindstrom. 1995. Insertion length of electrode array and its relation to speech communication performance and nonauditory side effects in multichannel-implanted patients. Annals of Otology, Rhinology and Laryngology 104(suppl 166): 256–258.

Bredberg, G., B. Lindstrom, H. Lopponen, M. A. Beltrame, W. Gstoettner and H. Skarzynski 2000. A new approach for the treatment of ossified cochleas. In: Waltzman, S. and N. Cohen, eds. Cochlear implants. New York, Thieme Medical Publishers.

Brennan, W. J. and G. M. Clark. 1985. An animal model of acute otitis media and the histopathological assessment of a cochlear implant in the cat. Journal of Laryngology and Otology 99: 851–856.

Briggs, R. J. S., M. Tykocinski, E. Saunders, et al. 2001. Surgical implications of perimodiolar cochlear implant electrode design: avoiding intracochlear damage and scala vestibuli insertion. Cochlear Implants International 2: 135–149.

Burian, K., I. J. Hochmair-Desoyer and B. Eisenwort. 1986. The Vienna cochlear implant program. Otolaryngologic Clinics of North America 19: 313–328.

Burton, M. J., L. T. Cohen, F. W. Rickards, K. I. McAnally and G. M. Clark. 1992a. Steady-state evoked potentials to amplitude modulated tones in the monkey. Acta Oto-Laryngologica 112: 745–751.

Burton, M. J., R. K. Shepherd, J. Xu and G. M. Clark. 1992b. Cochlear implantation in young children: long term effects of implantation on the skull and underlying central nervous system tissues in a primate model. Otolaryngology Research Meeting, London: 1.

Burton, M. J., R. K. Shepherd, X. J., S. Xu, B. K.-H. G. Franz and G. M. Clark. 1994. Cochlear implantation in young children: histological studies on head growth, leadwire design and electrode fixation in the monkey model. Laryngoscope 104: 167–175.

Butler, J. C., R. F. Breiman, H. B. Lipman, J. Hofmann and R. R. Facklam. 1995. Serotype distribution of Streptococcus pneumoniae infections among pre-school children in the United States, 1978–1994: implications for a conjugate vaccine. Journal of Infectious Disease 171: 885–889.

Charnley, J. 1972. Postoperative infection after total hip replacement with special reference to air contamination in the operating room. Clinical Orthopaedics and Related Research 87: 167–187.

Chen, B. K., G. M. Clark and R. Jones. 2003. Evaluation of trajectories and contact pressures for the straight Nucleus cochlear implant electrode array–a two-dimensional application of finite element analysis. Medical Engineering and Physics 25(2): 141–147.

Chouard, C. H. and P. MacLeod. 1976. Implantation of multiple intracochlear electrodes for rehabilitation of total deafness: preliminary report. Laryngoscope 86: 1743–1751.

Chouard, C. H., B. Meyer, C. Fugain and O. Koca. 1995. Clinical results for the Digisonic multichannel cochlear implant. Laryngoscope 105: 505–509.

Chow, J. K., H. L. Seldon and G. M. Clark. 1995. Experimental animal model of intra-cochlear ossification in relation to cochlear implantation. Annals of Otology, Rhinology and Laryngology 104(suppl 166): 42–45.

Clark, G. M. 1975. A surgical approach for a cochlear implant. An anatomical study. Journal of Laryngology and Otology 89: 9–15.

Clark, G. M. 1977. An evaluation of per-scalar cochlear electrode implantation techniques. An histopathogical study in cats. Journal of Laryngology and Otology 91: 185–199.

Clark, G. M., R. C. Black, D. J. Dewhurst, I. C. Forster, J. F. Patrick and Y. C. Tong. 1977. A multiple-electrode hearing prosthesis for cochlear implantation in deaf patients. Medical Progress Through Technology 5: 127–140.

Clark, G. M. 1987a. The University of Melbourne–Nucleus multi-electrode cochlear implant. Advances in Oto-Rhino-Laryngology, Vol. 38. Basel, Karger.

Clark, G. M., P. J. Blamey, P. A. Busby, et al. 1987b. A multiple-electrode intracochlear implant for children. Archives of Otolaryngology 113: 825–828.

Clark, G. M., P. A. Crosby, R. C. Dowell, et al. 1983. The preliminary clinical trial of a multi-channel cochlear implant hearing prosthesis. Journal of the Acoustical Society of America 74: 1911–1914.

Clark, G. M., R. C. Dowell, B. C. Pyman, et al. 1984. Clinical trial of a multi-channel cochlear prosthesis: results on 10 postlingually deaf patients. Australian and New Zealand Journal of Surgery 54: 519–526.

Clark, G. M. and R. J. Hallworth. 1976. A multiple-electrode array for a cochlear implant. Journal of Laryngology and Otology 90: 623–627.

Clark, G. M., R. J. Hallworth and K. Zdanius. 1975. A cochlear implant electrode. Journal of Laryngology and Otology 89: 787–792.

Clark, G. M., J. F. Patrick and Q. R. Bailey. 1979. A cochlear implant round window electrode array. Journal of Laryngology and Otology 93: 107–109.

Clark, G. M. and B. C. Pyman. 1995. Surgical considerations for the placement of the new Cochlear Pty Limited micro-multiple-channel cochlear implant for research studies. Annals of Otology, Rhinology and Laryngology 104(suppl 166): 408–409.

Clark, G. M., B. C. Pyman and Q. R. Bailey. 1979. The surgery for multiple-electrode cochlear implantations. Journal of Laryngology and Otology 93: 215–223.

Clark, G. M., B. C. Pyman and R. E. Pavillard. 1980. A protocol for the prevention of infection in cochlear implant surgery. Journal of Laryngology and Otology 94(12): 1377–1386.

Clark, G. M., B. C. Pyman and R. L. Webb. 1997. Surgery. In: Clark, G. M., R. S. C. Cowan and R. C. Dowell, eds. Cochlear implantation for infants and children. Advances. San Diego, Singular Publishing Group: 123.

Clark, G. M., B. C. Pyman, R. L. Webb, Q. E. Bailey and R. K. Shepherd. 1984. Surgery for an improved multiple-channel cochlear implant. Annals of Otology, Rhinology and Laryngology 93(3 pt 1): 204–207.

Clark, G. M., B. C. Pyman, R. L. Webb, B. K.-H. G. Franz, T. J. Redhead and R. K. Shepherd. 1987. Surgery for safe the insertion and reinsertion of the banded electrode array. Annals of Otology, Rhinology and Laryngology 96(suppl 128): 10–12.

Clark, G. M., R. K. Shepherd, B. K.-H. G. Franz and D. Bloom. 1984. Intracochlear electrode implantation. Round window membrane sealing procedures and permeability studies. Acta Oto-Laryngologica (suppl 410): 5–15.

Clark, G. M. and Y. C. Tong. 1982. A multiple-channel cochlear implant. A summary of results for two patients. Archives of Otolaryngology 108: 214–217.

Clark, G. M., Y. C. Tong and L. F. Martin. 1981a. A multiple-channel cochlear implant. An evaluation using open-set CID sentences. Laryngoscope 91: 628–634.

Clark, G. M., Y. C. Tong, L. F. Martin and P. A. Busby. 1981b. A multiple-channel cochlear implant. An evaluation using an open-set word test. Acta Oto-Laryngologica 91: 173–175.

Clark, G. M., Y. C. Tong, J. F. Patrick, P. M. Seligman, P. A. Crosby and J. A. Kuzma. 1984. A multi-channel cochlear prosthesis for profound-to-total hearing loss. Journal of Electrical and Electronics Engineering, Australia 4: 111–117.

Cochlear Corporation. 1987. Surgical procedure manual: Nucleus 22 channel cochlear implant system. Issue 5. Englewood, CO, Cochlear Corp.

Cohen, N. L. 1989. Medical or surgical complications related to the Nucleus Multichannel Cochlear Implant [letter]. Annals of Otology, Rhinology and Laryngology 98(9): 754.

Cohen, N. L. 2000. Surgical techniques for cochlear implants. In: Waltzman, S. and N. Cohen, eds. Cochlear implants. New York, Thieme: 151–169.

Cohen, N. L. and J. Kuzma. 1995. Titarium clip for cochlear implant electrode fixation. Annals of Otology, Rhinology and Laryngology 104(suppl 166): 402–403.

Cohen, N. L. and S. B. Waltzman. 1993. Partial insertion of the nucleus multichannel cochlear implant: technique and results. American Journal of Otology 14(4): 357–361.

Cohen, N. L., R. A. Hoffman and M. Stroschein. 1988. Medical or surgical complications related to the Nucleus multichannel cochlear implant [published erratum appears in Ann Otol Rhinol Laryngol Suppl 1989;98(9):754]. Annals of Otology, Rhinology, and Laryngology 97(suppl 135): 8–13.

Corgill, D. A. and A. S. Martinez. 1963. Tympanoplasty. Southern Medical Journal 56: 296–301.

Cranswick, N. E. 1984. Studies in the cochlear round window. B. Med. Sci thesis. University of Melbourne.

Cranswick, N. E., B. K.-H. G. Franz, G. M. Clark and R. K. Shepherd. 1987. Middle ear infection postimplantation: response of the round window membrane to streptococcus pyogenes. Annals of Otology, Rhinology and Laryngology 96(suppl 128): 53–54.

Cruse, P. J. and R. Foord. 1973. A five-year prospective study of 23,649 surgical wounds. Archives of Surgery 107: 206–210.

Dahm, M., G. M. Clark, B. K.-H. Franz, R. K. Shepherd, M. J. Burton and R. Robins-Browne. 1994. Cochlear implantation in children: labyrinthitis following pneumococcal otitis media in unimplanted and implanted cat cochleas. Acta Oto-Laryngologica 114: 620–625.

Dahm, M., B. C. Pyman, J. G. Crock, M. Aoyagi and G. M. Clark. 1993a. Cochlea implant skin flap design: the vascular pattern of the postauriclar region. Abstracts of Third International Cochlear Implant Conference: 3.2.

Dahm, M., H. L. Seldon, B. C. Pyman and G. M. Clark. 1992. 3D reconstruction of the temporal bone in cochlear implant surgery. In: Yanagihara, N. and J. Suziki, eds. Transplants and implants in otology II. Amsterdam, Kugler: 271–275.

Dahm, M., R. K. Shepherd and G. M. Clark. 1993b. The postnatal growth of the temporal bone and its implications for cochlear implantation in children. Acta Oto-Laryngologica (suppl 505): 1–39.

Dobie, R. A., H. Jenkins and N. L. Cohen. 1995. Multicenter comparative study of cochlear implants: surgical results. Annals of Otology, Rhinology and Laryngology 104(4 part 2, suppl 165): 6–8.

Donnelly, M. J., B. C. Pyman and G. M. Clark. 1995a. Chronic middle ear disease and cochlear implantation. Annals of Otology, Rhinology and Laryngology 104(suppl 166): 406–408.

Donnelly, M. J., L. T. Cohen and G. M. Clark. 1995b. Initial investigation of the efficacy and biosafety of sodium hyaluronate (healon) as an aid to electrode array insertion. Annals of Otology, Rhinology and Laryngology 104(suppl 166): 45–48.

Douglas, R. M. and H. B. Miles. 1984. Vaccination against Streptococcus pneumoniae in childhood: lack of demonstrable benefit in young Australian children. Journal of Infectious Diseases 149: 861–869.

Doyle, J. H., J. B. Doyle and F. M. Turnbull. 1964. Electrical stimulation of eighth cranial nerve. Archives of Otolaryngology 80: 388–391.

Durcan, D. J., J. J. Shea and J. P. Sleeckx. 1967. Bifurcation of the facial nerve. Archives of Otolaryngology 86: 619–631.

Eddington, D. K., W. H. Dobelle and D. E. Brackmann. 1978. Auditory prostheses research with multiple channel intracochlear stimulation in man. Annals of Otology 87: 1–39.

Eskola, J., T. Kilpi, A. Palmu, et al. 2001. Efficacy of a pneumoccal conjugate vaccine against acute otitis media. New England Journal of Medicine 344: 403–409.

Filipo, R., M. Barbara, S. Monini and P. Mancini. 1999. Clarion cochlear implants: surgical implications. Journal of Laryngology and Otology 113(4): 321–325.

Fowler, E. 1961. Variations in the temporal bone course of the facial nerve. Laryngoscope 71: 937–944.

Franz, B. K.-H. G. and G. M. Clark. 1987. Refined surgical technique for insertion of banded electrode array. Annals of Otology, Rhinology and Laryngology 96(suppl 128): 15–16.

Franz, B. K.-H. G., G. M. Clark and D. Bloom. 1984. Permeability of the implanted round window membrane in the cat—an investigation using horseradish peroxidase. Acta Oto-Laryngologica Supplement 40: 17–23.

Franz, B. K.-H. G., G. M. Clark and D. Bloom. 1987a. Effect of experimentally induced otitis media on cochlear implants. Annals of Otology, Rhinology and Laryngology 96: 174–177.

Franz, B. K.-H. G., G. M. Clark and D. M. Bloom. 1987b. Surgical anatomy of the round window with special reference to cochlear implantation. Journal of Laryngology and Otology 101(2): 97–102.

Friedmann, I. 1974. Pathology of the ear. Oxford, Blackwell Scientific.

Gantz, B. J., B. F. McCabe and R. S. Tyler. 1989. Use of multichannel cochlear implants in obstructed and obliterated cochleas. Otolaryngology Head and Neck Surgery 98: 72–81.

Goycoolea, M. V., D. C. Muchow, C. M. Schirber, H. G. Goycoolea and K. Schellhas. 1990. Anatomical perspective, approach, and experience with multichannel intracochlear implantation. Laryngoscope 100(2 pt 2, suppl 50): 1–18.

Gray, R. F., D. M. Baguley, M. L. Harries, I. Court and C. Lynch. 1993. Profound deafness treated by the Ineraid multichannel intracochlear implant. Journal of Laryngology and Otology 107(8): 673–680.

Gstoettner, W., H. Plenk, J. Hamzavi, W. Baumgartner and C. Czerny. 2000. Combi 40 cochlear implantation: insertional trauma with different types of electrodes. In: Waltzman, S. and N. Cohen, eds. Cochlear implants. New York. Thieme Medical Publishers: 178–179.

Gstoettner, W., W. D. Baumgartner, J. Hamzavi and P. Franz. 1997. Surgical experience with the Combi 40 cochlear implant. Advances in Oto-Rhino-Laryngology 52: 143–146.

Hansen. C. 1981. Electrode for implantation into cochlea. US patent 4,261,372.

Harris, J. P. and R. A. Cueva. 1987. Flap design for cochlear implantation: avoidance of a potential complication. Laryngoscope 97(6): 755–757.

Hausler, R., M. Vischer and M. Kompis. 2000. Cochlear implantation through apical cochleostomy in basal turn ossification. In: Waltzman, S. and N. Cohen, eds. Cochlear implants. New York, Thieme Medical Publishers. 161–163.

Hayek, L. J., J. M. Emerson and A. M. Gardner. 1987. A placebo-controlled trial of the effect of two preoperative baths or showers with chlorhexidine detergent on postoperative wound infection rates. Journal of Hospital Infection 10(2): 165–172.

Hinojosa, R. and M. Marion. 1983. Histopathology of profound sensorineural deafness. Annals of the New York Academy of Sciences 405: 459–84.

House, W. F. 1982. Surgical considerations in cochlear implantation. Annals of Otology, Rhinology, and Laryngology Supplement. 91(2 Pt 3): 15–20.

House, W. F. and J. Urban. 1973. Long term results of electrode implantation and electronic stimulation of the cochlea in man. Annals of Otology, Rhinology and Laryngology 82: 504–517.

Howie, V. M., J. Ploussard, J. L. Sloyer and J. C. Hill. 1984. Use of pneumococcal polysaccharide vaccine in preventing otitis media in infants: different results between racial groups. Pediatrics 73: 79–81.

Hulka, G. F., E. J. Bernard and H. C. Pillsbury. 1995. Cochlear implantation in a patient after removal of an acoustic neuroma. The implications of magnetic resonance imaging with gadolinium on patient management [see comments]. Archives of Otolaryngology–Head and Neck Surgery 121(4): 465–468.

Irving, R. M. and R. F. Gray. 1994. Cochlear implants in chronic suppurative otitis media: preparing the septic ear for a sterile device. In: Hochmair-Desoyer, I. and E. Hochmair, eds. Advances in cochlear implants. Vienna, Manz: 223–227.

Jansen, C. 1963. Cartilage-tympanoplasty. Laryngoscope 73: 1288–1301.

Jansen, C. 1968. The combined approach for tympanoplasty. Journal of Laryngology and Otology 82: 779–790.

Johnson, C. E., S. A. Carlin, D. M. Super, et al. 1991. Cefixime compared with amoxicillin for treatment of acute otitis media. Journal of Pediatrics 119: 117–122.

Karma, P., J. Pukander, M. Sipila, et al. 1985. Prevention of otitis media in children by pneumococcal vaccination. American Journal of Otolaryngology 6: 173–184.

Kveton, J. F., C. Abbott, M. April, G. Drumheller, N. Cohen and D. S. Poe. 1989. Cochlear implantation after transmastoid labyrinthectomy. Laryngoscope 99(6 pt 1): 610–613.

Lacombe, H., B. Meyer, F. Chabolle and C. H. Chouard. 1984. Surgical procedure and implanted material description. Acta Oto-Laryngologica (suppl 411): 20–24.

Lacombe, H., M. G. Cotin and G. Y. Gelin. 1989. Le syndrome de l'acqedeuc du vestibule dilate. Annales d'Oto-Laryngologie 106: 152–157.

Lehnhardt, E. 1993a. Intracochlear electrode placement facilitated by Healon. Advances in Oto-Rhino-Laryngology 48: 62–64.

Lehnhardt, E. 1993b. Intracochlear placement of cochlear implant electrodes in soft surgery technique. HNO 41(7): 356–359.

Lehnhardt, E. and M. S. Hirshorn. 1986. Cochlear implants. Berlin, Springer-Verlag: 128–132.

Lew, D. P. and F. A. Waldvogel. 1998. Infections of skeletal prostheses. In: Bennett, J. V. and P. S. Brachman, eds. Hospital infections. Philadelphia, Lippincott-Raven: 613–620.

Luetje, C. M. 1996. Cochlear implantation in a patient after removal of acoustic neuroma. Archives of Otolaryngology–Head Neck Surgery 122: 205.

Luntz, M., T. Balkany, A. V. Hodges and F. F. Telischi. 1997. Cochlear implants in children with congenital inner ear malformations. Archives of Otolaryngology–Head and Neck Surgery 123(9): 974–977.

Luotonen, J., E. Herva, P. Karma, M. Timonen, M. Leinonen and P. H. Makela. 1981. The bacteriology of acute otitis media in children with special reference to Streptococcus pneumoniae as studied by bacteriological and antigen detection methods. Scandinavian Journal of Infectious Disease 13: 177–183.

Luxford, W. and W. House 1977. Cochlear implants in children. Ear and Hearing 6: 205–235.

Makela, P. H., M. Sibakov, E. Herva, et al. 1980. Pneumococcal vaccine and otitis media. Lancet 2: 547–551.

Manrique, M. J., V. Paloma, F. J. Cervera-Paz, I. Ruiz de Erenchun and R. Garcia-Tapia 1995. Treatment of cutaneous ulceration after cochlear implantation surgery. Annals of Otology, Rhinology and Laryngology 104(suppl 166): 422–425.

Marquet, J. 1981. Congenital malformations and middle ear surgery. Journal of the Royal Society of Medicine 74: 119–128.

Matz, G. J., H. B. Lockhart and J. R. Lindsay. 1968. Meningitis following stapedectomy. Laryngoscope 78: 56–63.

Michelson, R. P. and R. A. Schindler. 1981. Multichannel cochlear implant preliminary results in man. Laryngoscope 91(1): 38–42.

Minas, H. J. 1973. Acoustic and electrical stimulation in the cat: a behavioural study. B Med Sci dissertation.

Mladejovski, M. G., D. K. Eddington and W. H. Dobelle. 1975. Artificial hearing for the deaf by cochlear stimulation: pitch modulation and some parametric thresholds. Transactions of the American Society for Artificial Internal Organs 21: 1–6.

Myers, M. W. and W. D. Schlosser. 1960. Anterior-posterior technique for the treatment of chronic otitis media and mastoiditis. Laryngoscope 70: 78–83.

Nager, G. T. and B. Proctor. 1982. The facial canal: normal anatomy, variations and anomalies. Anatomical variations and anomalies involving the facial canal. Annals of Otolaryngology, Rhinology and Laryngology 91: 33–61.

Nomura, Y. 1984. Otological significance of the round window. Advances in Oto-rhino-laryngology 33: 27–37.

Page, E. L. and T. L. Eby. 1997. Meningitis after cochlear implantation in Mondini malformation. Otolaryngology–Head and Neck Surgery 116(1): 104–106.

Palva, T., A. Palva and J. Karja. 1972. Fatal meningitis in a case of otosclerosis operated upon bilaterally. Archives of Otolaryngology 96: 130–137.

Parkin, J. L., D. K. Eddington, J. L. Orth and D. E. Brackmann. 1985. Speech recognition experience with multichannel cochlear implants. Otolaryngology–Head and Neck Surgery 93: 639–645.

Parkin, J. L., R. Bloebaum, B. D. Parkin and M. J. Parkin. 1996. Osseointegration and growth effects of temporal bone percutaneus pedestals. American Journal of Otology 17: 735–742.

Phelps, P. D., A. King, L. Michaels and F. R. C. Path. 1994. Cochlear dysplasia and meningitis. American Journal of Otology 15: 551–557.

Portmann, M. 1986. La chirurgie du nerf facial. Revue de Laryngologie, Otologie-Rhinologie 107: 223–231.

Portmann, M., Y. Cazals and M. Negrevergne. 1986. Extra cochlear implants. Otolaryngologic Clinics of North America 19: 307–312.

Purser, S., R. K. Shepherd and G. M. Clark. 1991. Evaluation of a sealing device for the intracochlear electrode entry point. Journal of the Oto-Laryngological Society of Australia 6: 472–480.

Ray, J., R. F. Gray and I. Court. 1998. Surgical removal of 11 cochlear implants—lessons from the 11-year-old Cambridge programme. Journal of Laryngology and Otology 112(4): 338-343.

Roberson, J. B., K. R. Stidham, K. M. Scott and L. Tonokawa. 2000. Cochlear implantation: minimal hair removal technique. Otolaryngology and Head and Neck Surgery 122: 625–629.

Roland, J. T. 2000. Complications of cochlear implant surgery. In: Waltzman, S., and N. Cohen, eds. Cochlear implants. New York, Thieme: 171–175.

Roland, J. T., T. M. Magardino, J. T. Go and D. E. Hillman. 1995. Effects of glycerine, hyaluronic acid and hydroxypropyl methylcellulose on the spiral ganglion of the guinea pig cochlea. Annals of Otology, Rhinology and Laryngology 104: 64–68.

Roos, K. L. and K. L. Tyler. 2001. Bacterial meningitis and other suppurative infections. In: Braunwald, E., A. S. Fuaci, D. L. Kasper et al, eds. Harrison's principles of internal medicine, 15th ed. New York, McGraw-Hill. 2462–2471.

Rutledge, L. J., M. L. Lewis and F. Sanabria. 1963. Fatal meningitis related to stapes operation. Archives of Otolaryngology 78: 637–641.

Schmidt, J. M. 1985. Cochlear neuronal populations in developmental defects of the inner ear. Implications for cochlear implantation. Acta Oto-Laryngologica 99: 14–20.

Schuknecht, H. F. 1974. Pathology of the ear. Cambridge, Harvard University Press.

Schuknecht, H. F. and J. Gulya. 1986. Anatomy of the temporal bone with surgical implications. Philadelphia, Lea and Febiger.

Scrivener, B. P. and W. P. R. Gibson. 1987. Cochlear implant after radical mastoidectomy. Annals of Otology, Rhinology and Laryngology 96: 19–20.

Seropian, R. and B. M. Reynolds. 1971. Wound infections after preoperative depilatory versus razor preparation. American Journal of Surgery 121: 251–4.

Shambaugh, G. 1959. Surgery of the ear. Philadelphia, W.B. Saunders.

Shellock, F. G. 2001. Radiofrequency energy-induced heating of bovine capsular tissue: temperature changes produced by bipolar versus mono-polar electrodes. Arthroscopy 17: 124–131.

Silverstein, H., E. Smouha and N. Morgan. 1988. Multichannel cochlear implantation in a patient with bilateral Mondini deformities. American Journal of Otology 9(6): 451–455.

Simmons, F. B. 1966. Electrical stimulation of the auditory nerve in man. Archives of Otolaryngology 84: 2–54.

Staller, S., A. Parkinson, J. Arcaroli and P. Arndt. 2002. Pediatric outcomes with the Nucleus 24 contour: North American clinical trial. Annals of Otology Rhinology and Laryngology 111(suppl 189): 56–61.

Steenerson, R. L., L. B. Gary and M. S. Wynens. 1990. Scala vestibuli cochlear implantation for labyrinthine ossification. American Journal of Otology 11: 360–3.

Suzuki, C., I. Sando, J. J. Fagan, D. B. Kamerer and A. S. Knisely. 1998. Histopathological features of a cochlear implant and otogenic meningitis in Mondini dysplasia. Archives of Otolaryngology–Head and Neck Surgery 124: 462–466.

Teele, D. W., J. O. Klein, L. Bratton, et al. 1981. Use of pneumococcal vaccine for prevention of recurrent acute otitis media in infants in Boston. Reviews of Infectious Diseases (suppl 3): S113–S118.

Turrini, M., E. Orzan, M. Gabana, E. Genovese, E. Arslan and U. Fisch. 1997. Cochlear implantation in a bilateral Mondini dysplasia. Scandinavian Audiology Supplementum 46: 78–81.

Tykocinski, M., L. T. Cohen, B. C. Pyman, et al. 2000. Comparison of electrode position in the human cochlea using various peri-modular electrode arrays. American Journal of Otology 21: 205–211.

Valvassori, G. E. and J. D. Clemis. 1978. The large vestibular aqueduct and associated anomalies of the inner ear. Laryngoscope 88: 723–728.

Webb, R. L., B. C. Pyman, B. K.-H. G. Franz and G. M. Clark. 1990. The surgery of cochlear implantation. In: Clark, G. M., Y. C. Tong and J. F. Patrick, eds. Cochlear prostheses. London, Churchill Livingstone: 153–180.

Webb, R. L., E. Lehnhardt, G. M. Clark, R. Laszig, B. C. Pyman and B. K. Franz. 1991. Surgical complications with the cochlear multiple-channel intracochlear implant: experience at Hannover and Melbourne. Annals of Otology, Rhinology and Laryngology 100(2): 131–136.

Weber, B. P., J. Neuburger, J. E. Goldring, et al. 1999. Clinical results of the CLARION magnetless cochlear implant. Annals of Otology, Rhinology, and Laryngology 108(suppl 177): 22–26.

Wolff, D. 1964. Untoward sequelae eleven months following stapedectomy. Annals of Otology, Rhinology and Laryngology 73: 297–304.

Wullstein, H. 1956. The restoration of the function of the middle ear in chronic otitis media. Annals of Otology, Rhinology and Laryngology 65: 1020–1041.

Xu, J., R. K. Shepherd, S. A. Xu, H. L. Seldon and G. M. Clark. 1993. Pediatric cochlear implantation: radiological observations of skull growth: Archives of Otolaryngology–Head and Neck Surgery 119: 525–534.

Zentner, J., J. Gilsbach and F. Daschner. 1987. Incidence of wound infection in patients undergoing craniotomy: influence of type of shaving. Acta Neurochirurgica 86: 79–82.

Zimmerli, W., F. A. Waldvogel, P. Vaudaux and U. E. Nydegger. 1982. Pathogenesis of foreign body infection: description and characteristics of an animal model. The Journal of Infectious Diseases 146: 487–497.

11
Rehabilitation and Habilitation

Aims

The aim of rehabilitation and habilitation is to produce the maximum benefits for the cochlear implant patient. For adults and late-deafened children, the skills to communicate effectively (use language) have already been acquired through the use of hearing prior to the onset of the profound hearing loss (postlinguistically deaf). Rehabilitation is learning to reuse the same neural networks and communication skills, often with a signal that sounds more distorted than before. Habilitation is the development of new communication skills for the first time in children who are born deaf or deafened early in life. These children are prelinguistically deaf (i.e., have lost hearing before fully developing language), and they need to develop auditory communication skills as soon as possible to help them in the perception and production of speech and the development of age-appropriate language. The first 2 to 4 years of life are a critical time for brain plasticity to allow the right neural connectivity for processing speech. The training requires not only the clinician but also the close involvement of the parents to continue the program at home. The habilitation also requires the participation of the teacher of the hearing impaired at the earliest stage.

The acquisition of auditory-oral communication must also be linked to the child's physical, intellectual, emotional, and social development (Mecklenburg et al 1990). Auditory-oral communication is through hearing and lipreading. The training needs to be flexible for a child's changing needs, and it is important to keep in mind the long-term benefit of learning language, as well as the more immediate perception of speech. There may be a need, for example, to change gradually from sign language of the deaf to enable progress in learning language to progress and for emotional security.

Thus the major aims of auditory (re)habilitation for children are (1) to be aware of the different types of sounds produced by the implant; (2) to achieve the best possible understanding of speech and voice production; (3) to develop receptive and expressive language that is comparable to that of children with normal hearing of the same age; (4) to promote an acceptance and understanding of the capabilities and limitations of the implant by the child, parents, and other caregivers;

and (5) to ensure that the auditory communication skills contribute fully to the overall development of the child (Mecklenburg et al 1990).

Principles

The aim of (re)habilitation is to present neural patterns of stimulation that most closely represent those from speech, and to take advantage of plasticity to reinforce the correct patterns through training.

Hearing is the ideal method of communication in a world of sound, but sign language of the deaf and total communication may be appropriate. Good communication depends on intrinsic and extrinsic factors. Intrinsic factors are psychological or biological and are inherited or acquired, and may be intelligence, personality, or the residual neural population in the cochlea available for electrical stimulation. The extrinsic factors are the type of cochlear prosthesis, and the method of training. These can be more easily controlled. Training in the use of the perceptual information provided by the cochlear implant depends in part on the plasticity of the responses in the central auditory nervous system (Ling and Nienhuys 1983). It has been demonstrated with cochlear implants, for example by Dawson and Clark (1997b).

The term *plasticity* has been used physiologically to refer specifically to an alteration in neural connectivity. When it is applied to learning, it must be understood that the relationship with the physiology has not been well defined.

Plasticity in the Experimental Animal

In the experimental animal there are two types of plasticity in the central auditory nervous system: developmental plasticity and postdevelopmental plasticity. The former is the development of the right neural connections for processing information within a critical period after birth. It might be seen as underpinning habilitation. The latter results from a change in the central representation of neurons in the mature animal after neural connectivity has been established. It might be seen as underpinning rehabilitation.

Developmental Plasticity

An example of developmental plasticity is an increase in the number of projections in the gerbil neonate from the cochlear nucleus to the ipsilateral inferior colliculus if the cochlea on the opposite side is destroyed (Nordeen et al 1983). A similar phenomenon was demonstrated in the ferret (Moore and Kowalchuk 1988). In addition, in the kitten the number of synapses on neurons in the inferior colliculus is significantly reduced if the animal is bilaterally deafened in the neonatal period (Hardie et al 1999). There is a critical period for these changes. For example, in the ferret there was a marked loss of neurons in the cochlear nucleus following ablation of the cochlea 5 days after birth (Moore 1990), but ablation of the cochlea 24 days after birth (i.e., a week before the onset of hearing) had little effect. Plasticity is seen not only in the brainstem, but also in the auditory cortex and especially the association areas (Snyder et al 1995; Kaczmarek et al 1997). There is also electrophysiological evidence that it can take 15 years of exposure to sound for cortical evoked potentials to develop adult latencies (Eggermont et al 1997). This could be speeded up with electrical stimulation, as it has been shown using neural response telemetry (NRT) in implanted children that there is a progressive decrease in the latencies of the evoked auditory brainstem responses (EABRs) at 2, 6 and 12 months after operation. In addition, cross-modality plasticity occurs with an extension of visual excitation fields to auditory regions and vice versa when either audition or vision is lost, respectively (Kral et al 2001). Developmental plasticity is discussed in more detail in Chapter 5.

The electrophysiological data are consistent with the psychophysical findings of Busby and Clark (2000a,b) in Chapter 6 that show the relation of place and temporal pitch to age at implantation and speech perception.

Postdevelopmental Plasticity

Postdevelopmental plasticity occurs when an area of the guinea pig cochlea is destroyed, and the corresponding area of the brain has increased representation from the neighboring frequency areas (Robertson et al 1989). This postdevelopmental plasticity is probably due to a loss of inhibition normally suppressing the input from neighboring frequency areas. Reorganization of the topographical map (frequency/place map) was also seen the cat in the primary auditory cortex when the contralateral cochlea was destroyed (Rajan et al 1993). In addition, behavioral training modified the tonotopic organization of the primary auditory cortex in the primate (Recanzone et al 1993). These data are a neurophysiological underpinning for the improved speech perception results in implanted postlinguistically deaf adults over time, and also the ability of patients to learn to use a new strategy. Postdevelopmental plasticity is discussed in more detail in Chapter 5.

Plasticity—Psychophysics

The development of place and temporal pitch as well as loudness depends in part on the plasticity of neural connections. The psychophysics also correlates with speech perception, as discussed in Chapter 7.

Cochlear Implants—Developmental Plasticity

Developmental plasticity for pitch and speech perception is seen in children using cochlear implants. There is a critical period, although not sharply defined, for place pitch in particular, and speech perception (Clark 2002). One method for assessing place pitch is the ability to discriminate between electrode stimulation at two locations, when loudness is balanced to reduce this as a cue. In a study on 16 prelinguistically deaf (early deafened) patients, it was found that on average the poorer the discrimination, the longer the period of auditory deprivation (Busby and Clark 2000a). The results in Figure 11.1 suggest there is a limited time over which the neural connectivity for place discrimination can occur, and this may be important for the development of speech perception. To see if there was a correlation between electrode place discrimination and speech perception, a comparison was made using a closed-set speech test. The findings in Figure 11.2 from the same group of children showed that the smaller the separation between electrodes that could be detected, the better the speech perception. This supports the view that if developmental plasticity is responsible for creating the neural connections required for place pitch, then speech perception will be enhanced as well.

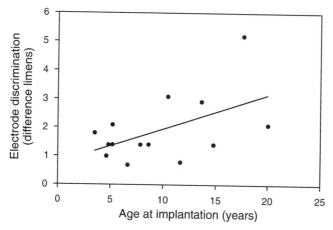

FIGURE 11.1. The results for electrode place discrimination versus age at implantation. The results show a positive correlation between discrimination of electrodes and age at implant (Busby and Clark 2000a,b; Clark 2002). Reprinted with permission from Clark G. M. 2002. Learning to hear and the cochlear implant. Textbook of Perceptual Learning, M. Fahle and T. Poggio, eds. Cambridge, Mass. MIT Press: 147–160.

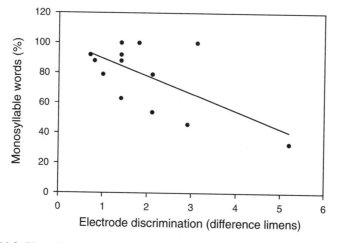

FIGURE 11.2. Place discrimination versus speech perception tested using closed sets of words (Busby and Clark 2000a,b; Clark 2002). Reprinted with permission from Clark G. M. 2002. Learning to hear and the cochlear implant. Textbook of Perceptual Learning, M. Fahle and T. Poggio, eds. Cambridge, Mass. MIT Press: 147–160.

As discriminating place of electrode stimulation is a different perceptual task from ranking pitch, this was also correlated with duration of deafness. The ability of children to rank pitch tonotopically (i.e., according to place of stimulation), rather than simply discriminate electrode place, was compared with their speech perception scores, as shown in Figure 11.3. The poorest results were found in those not able to order pitch ("Absent"). In addition, those children with the longest duration of deafness had the lowest scores on the Bamford-Kowal-Bench (BKB) (Bench and Bamford 1979) word-in-sentence test. Furthermore, it can be seen (Fig. 11.3) that not all children who could rank pitch ("Present") had good speech perception results. For 75% of the 16 children in the study, a tonotopic ordering of pitch percepts was found ("Present"). However, only 58% of these children with good ability to rank pitch had satisfactory speech perception of 30% or more. This suggested that the effect of developmental plasticity on the neural connectivity required for place discrimination was not the only factor for learning speech. At least another factor was required for speech perception, most probably language, as discussed below and in Chapter 7.

In another group of children from the University of Melbourne's Cochlear Implant Clinic, that were unselected, the data showed speech perception was significantly better the younger the child when the implant surgery was performed (Fig. 11.4). The scores were obtained 2 years or longer after implantation.

Cochlear Implants—Postdevelopmental Plasticity

An important question for cochlear implantation is, Would a patient who had adjusted to a certain speech-processing strategy get further benefits from an al-

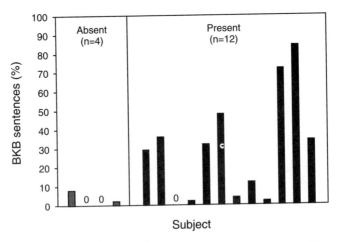

FIGURE 11.3. Place pitch ranking versus word scores for the Bamford-Kowal-Bench (BKB) open-set sentences for electrical stimulation alone on 16 children using cochlear implant. Pitch ranking is classified as present or absent. (Busby and Clark 2000a,b; Clark 2002). Reprinted with permission from Clark G. M. 2002, Learning to hear and the cochlear implant. Textbook of Perceptual Learning, M. Fahle and T. Poggio, eds. Cambridge, Mass. MIT Press: 147–160.

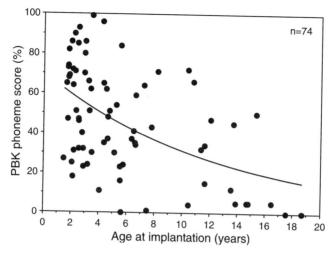

FIGURE 11.4. Speech perception versus age at operation for 74 unselected congenitally deaf children presenting to the University of Melbourne's Cochlear Implant Clinic. PBK, phonetically balanced (kindergarten) monosyllables.

ternative strategy? At a more basic level, would the patterns of excitation in the auditory cortex and neural connectivity that were required become so established that other patterns could not be processed? The effects of postdevelopmental plasticity were studied in older children by comparing speech perception after changing them from the Multipeak to the SPEAK strategy. The Multipeak strategy selects two formant frequencies [first (F1) and second (F2)], and the outputs from up to three high-frequency band-pass filters and stimulates at a rate proportional to the voicing frequency. In contrast, the SPEAK strategy selects six or more spectral maxima, and stimulates at a constant rate, with amplitude variations conveying voicing information.

As discussed in Chapter 7, although it has been shown that the SPEAK strategy represents the place speech feature, in particular, better than does the Multipeak strategy, neither the neural connectivity required to process the feature nor the contribution of the feature to speech perception is well understood. Appropriate neural connectivity may need to be established for the frequency transitions that underlie the place features. An improved strategy may either use these connections or establish others.

Studies in the Cooperative Research Center (CRC) for Cochlear Implant Speech and Hearing Research (Fig. 11.5) (Dowell and Cowan 1997) revealed a trend for improved scores from 6 to 18 months after changing strategies for six out of seven children when tested with the pediatric Speech Intelligibility Test (SIT) (Jerger et al 1980) sentences in quiet and especially in noise. At eighteen months the results for SPEAK were significantly better than for the Multipeak strategy. The period of learning required for effective use of the new strategy may be due to postdevelopmental neural plastic changes in lower level processing for additional speech features, or higher level changes in the patterns representing

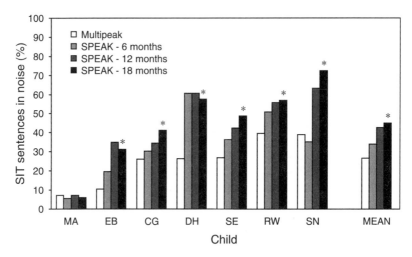

FIGURE 11.5. Speech perception scores for SIT sentences in noise (+15 dB SNR) for seven children using the Multipeak and SPEAK speech processing strategy after 6, 12 and 18 months experience. *Scores with SPEAK at 18 months significantly higher than with Multipeak ($p < 0.05$). (Dowell and Cowan 1997). Reprinted with permission from Clark G. M., 2002, Learning to hear and the cochlear implant. Textbook of perceptual learning, Fahle M. and Poggio T., eds. Cambridge, Mass. MIT Press: 147–160.

the phonemes. The need for time to learn is illustrated in Figure 11.5, which indicates an improvement from 6 to 12 or 18 months' use of the SPEAK strategy (Dowell and Cowan et al 1997). The results thus suggest that although children have learned to associate certain spectral and temporal patterns of cortical stimulation with words, they can readjust to the new strategy presumably due to perceptual learning.

Further evidence for postdevelopmental plasticity has been seen in a pilot study in an adult cochlear implant patient where the perceptual vowel spaces were mapped at different intervals after implantation. With the normal two-formant vowel space there is a limited range or grouping of frequencies required for the perception of each vowel. With electrical stimulation at first, as shown in Figure 11.6, there was a wider range of electrodes contributing to the perception of each vowel, and a greater variability in the results. However, after the patient learned to use the implant, the range of electrodes contributing to the perception of the vowels became more restricted, and the vowel spaces came to more closely resemble those for normal hearing.

The plasticity described for the Nucleus speech-processing strategy was also seen (Dorman and Loizou 1997) for vowel recognition in seven of eight patients who were converted from the Ineraid device (a four-fixed-filter strategy providing analog stimulation at a rate depending on the speech wave amplitude variations) (Eddington 1980) to the continuous interleaved sampler (CIS) strategy (a six-fixed-filter strategy providing pulsatile stimulation at a constant rate of approximately 800 pulses/s) (Wilson et al 1991). The scores were similar immediately after surgery, but improved after a month. It indicated that reprogramming strat-

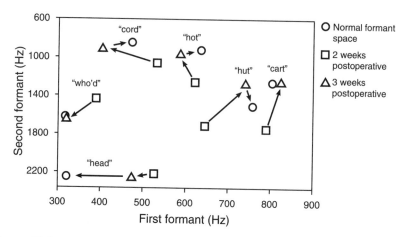

FIGURE 11.6. The center of two formant vowel spaces for the vowels /ɔ/,/ɒ/,/ʌ/,/ɑ/,/u/,/ɛ/ and the shift in the electrodes representing these vowels from two to three weeks postoperatively (Blamey and Dooley, personal communication; Clark 2002). Reprinted with permission from Clark G.M. 2002, Learning to hear and the cochlear implant. Textbook of perceptual learning, Fahle M and Poggio T., eds. Cambridge, Mass. MIT Press: 147–160.

egies with altered frequency-to-electrode allocation and variation in the presentation of temporal information could be made. This suggests that the reprogramming is carried out at a higher level than for speech features.

Plasticity—Cross-Modality in Humans

There have been a number of examples of children demonstrating they can effectively use a cochlear implant to communicate by auditory means, as well as use sign language of the deaf when required. These children usually learn auditory communication first. The need to develop the central neural connections for auditory processing of speech at an early stage has been well attested to by the better results the younger the child at operation. This is supported by studies with the positron emission tomography (PET) scanner. Parving et al (1995) showed that only two of five deaf patients with cochlear implants had an increased blood flow in the contralateral hemisphere, and this correlated with their speech understanding. Kubo (2002) found the auditory association area was activated by sign language but not by speech in a congenitally deaf cochlear implant user. In contrast, in short-term cochlear implant users there was competing information processing, and in a group of long-term users the auditory input was dominant. Cross-modality plasticity of auditory and visual inputs was found. This research indicated the need to undertake cochlear implantation first to provide audition before learning sign language.

Analytic Versus Synthetic Training

The learning that takes place with speech-processing strategies could depend on developmental or postdevelopmental plasticity. It is also important to know how to train the implantee to facilitate learning. The two main approaches to training are termed analytic and synthetic (McCarthy and Alpiner 1982). *Analytic* training involves breaking speech down into its individual components (single words, phonemes) and training discrimination at this level. Typically, very little contextual information is available. It is assumed that this will improve speech discrimination in everyday communication. The *synthetic* or global approach provides communication strategies to help the hearing-impaired person to understand the overall message. People are encouraged to make use of contextual cues, constructive questions, guessing, and so on, to determine what is said. The importance of key words is stressed, with little emphasis being given to the less meaningful words within an utterance. Exercises typically consist of sentence material or connected discourse. The synthetic approach to training has been favored, or a combination of the two approaches.

Much of the research on the relative merits of analytic versus synthetic training on speech perception has been on subjects using speech reading (Sanders 1982). The results have been inconclusive (Walden et al 1977, 1981; Lesner et al 1987). Fewer studies have investigated the value of auditory training of more relevance to cochlear implantation. Rubinstein and Boothroyd (1987) trained hearing-impaired adults in the recognition of consonants, using either synthetic training alone or a combination of synthetic and analytic exercises. They found an increase in speech recognition scores on sentence tests following training for both groups, with no significant differences.

In a study by Alcantara et al (1988), seven normal-hearing subjects received training using an eight-channel electrotactile device (transmitting fundamental frequency, second formant, and amplitude information via electrodes positioned on the subjects' fingers) (Blamey and Clark 1985). The study compared the benefits of a synthetic approach to training with a combined approach using both analytic and synthetic training. Each subject received 3 months' training using one approach followed by 3 months' training with the other approach (the order was alternated between the subjects). Training sessions were for 1 hour, three times per week. Therefore, each subject received approximately 35 hours of experience with each approach. Each subject's performance was assessed three times during the program: prior to the commencement of training, following completion of the first 3-month program, and following completion of the second 3-month program. A variety of materials were used to assess and compare the benefits of training. The results suggested that both approaches to training were beneficial, with improvements in scores. However, the benefits depended on the test materials. The inclusion of analytic training resulted in improved scores for analytic tests. Synthetic-only training resulted in greater improvements in scores for some synthetic tests, perhaps because there was more synthetic training in the synthetic-only program than in the combined approach. These results suggest that the type of assessment materials used is crucial in determining the benefits of training. The more similar the assessment material is to the training material, the greater the possibility that the subject has learned the best way to do the test. Assuming that synthetic materials more closely represent typical communicative situations, the authors concluded that synthetic training should be included in a training program.

Mapping and Fitting Procedures in Adults and Children

Before commencing training, it is essential to optimize the speech signal presented via electrical stimulation. At the first postoperative test session (typically 10 to 14 days after the operation), the clinician selects the right strength of magnet for the transmission coil that will retain it in place over the implant. Occasionally, this first test session needs to be delayed until swelling over the implant has reduced sufficiently for the transmission coil to be retained.

The patient's speech processor is connected to a personal computer via an interface unit so the stimulus parameters can be controlled, as discussed in Chapter 8. Parameters such as the currents for threshold (T) and maximum comfortable (MC) levels, as well as the stimulation mode (Bipolar-BP + 1 etc. versus monopolar), pulse width, duration of the stimulus, and pulse rate can be varied.

Physiological and Psychophysical Principles

Prior to (re)habilitation the outputs of the filters in the speech processor need to be mapped to appropriate electrodes with current levels that lie within the dynamic range for each electrode, that is, from T to MC levels. The electrical representation of the acoustic signals should remain within the operating range so that it has an appropriate loudness, that is, it is neither too soft nor too loud. The stimulus parameter responsible for neural excitation is electrical charge, and this can be controlled by varying either the pulse amplitude or width. The relationship between current level and loudness has been investigated by Eddington et al (1978) and Zeng and Shannon (1992), and was discussed in Chapter 6. Loudness depends on the number of neurons excited as well as other parameters such as rate, pulse interval, number of pulses, and duration. A linear relation was observed between loudness in decibels and current amplitude by Eddingtonet et al. With sound, Stevens (1975) showed that as loudness was a power function of intensity, both the logarithm of intensity and loudness could be plotted as a straight line.

If there are regions in the cochlea with reduced numbers of spiral ganglion cells, a larger current than elsewhere will be required to operate within the dynamic range of each electrode (Kawano et al 1995, 1998). A larger current may also be required to stimulate an appropriate number of neurons if the array is more distant from the ganglion cells or pathology results in spreading the current away from the auditory neurons (Cohen et al 1998, 2001a,b). This may be resolved by changing the mode of stimulation to vary the current pathways. With the earlier speech-processing systems bipolar (BP) and common ground (CG) stimulation were used to localize the current to separate groups of nerve fibers for place coding of frequency. Bipolar stimulation occurs when the current flows between two electrodes on the array. A normal stimulus mode with the Nucleus array is bipolar + 1 (BP + 1), where the current flows from an electrode across one to the next electrode. This is necessary for an adequate threshold and dynamic range with some electrode geometries and cochlear pathologies. The separation of the two electrodes in the bipolar mode can be further increased with more inactive intervening electrodes (BP + n) to achieve lower T and MC loudness levels. It was shown by Tong and Clark (1985) that increasing the extent of the stimulus in this way did not impair subjects' abilities to distinguish pairs of electrodes according to their degree of separation. CG stimulation occurs when current spreads from the active electrode to all other electrodes connected together electronically as a ground. An advantage of CG stimulation is that there are more consistent thresholds than with bipolar stimulation, and in children they will be less subject to unpleasant variations in loudness. This is not such an issue with monopolar stimulation that is used more routinely. With CG stimulation, there was a marked reversal of pitch and timbre in the middle of the array in three of nine patients, and a tendency for the T and MC levels to be higher in this part of the cochlea (Busby et al 1994). The deviation from the tonotopic organization of the cochlea was assumed to be due to the effect of a loss of neurons, and pathology in the cochlea.

The lowest thresholds were obtained with monopolar (MP) stimulation. With this mode of stimulation the current passes from the active electrode to a distant ground outside the cochlea (the grounding electrode is placed under the temporalis muscle). It was thought that monopolar stimulation would not allow adequate localization of current for the place coding of speech frequencies; however, as discussed in Chapter 6, studies by Busby et al (1994) showed that MP stimuli could also be localized to groups of nerve fibers.

One difficulty in mapping the current from each filter into the dynamic range for each electrode is that it can lead to unacceptable and inappropriate variations in loudness. This is due to failure to take loudness summation into consideration. Loudness summation may result when more than one electrode is activated per stimulus cycle. Only partial summation was shown by Tong and Clark (1986) to occur for bipolar stimulation with Nucleus banded array for spatial separations up to 3 mm, and was considered due to the spread of current and refractory effects of nerve fibers. As one pulse led the other by 0.8 ms, it was not due to an interaction of the stimulating electrical fields. This partial summation over short segments of the cochlea was assumed to be due to the critical band where acoustically the loudness of a band of noise of fixed intensity remains constant until the bandwidth of the noise exceeds the critical band, when the loudness increases with width. The bandwidth remains constant if the intensity is increased up to 80 dB. As discussed in Chapter 6, the loudness of a sound in sones will sum completely if the frequencies are separated by more than one critical bandwidth. If not, as discussed above, it will depend on summing first the energy of the sounds, and then determining the relation between loudness and the change in intensity. The critical band is equivalent to about a 0.89 to 1-mm length of the basilar membrane, and thus current stimulating on more than one electrode outside that region could produce increased loudness. This has been described by McKay et al (2001) for cochlear implant patients where, for example, the loudness of eight electrodes each at threshold has to be reduced by 50 current steps for the combined stimulus to be at threshold.

Producing a MAP

The T and MC levels for the electrical currents on each electrode are written onto a programmable chip in the speech processor where they are stored, and this is referred to as a MAP. The details in the MAP are incorporated into whatever speech-processing strategy is being used. The frequency boundaries for the electrode to be stimulated are also set to determine the pitch range of the electrodes. Additional information can be obtained by conducting psychophysical tests on the discrimination of electrode current level and pulse rate. However, these tasks require training and are relatively time-consuming, and therefore are not routinely carried out with patients. These individual details vary from patient to patient and in each patient over time, especially in the first few weeks postoperatively. The variations in the T and MC levels are due to pathological changes at the electrode–tissue interface. These changes increase both the impedance at the electrode–tissue interface and the current spread. With a constant current stimulator a change in impedance should allow the T and MC levels to remain constant (see Chapters 4 and 8). In contrast, the development of a fibrous tissue electrode sheath and new bone formation alters the spread of current and moves the electrode away from the spiral ganglion cells and thus raises T and MC levels.

The current levels between the T and MC levels cover the dynamic range. The frequency-to-electrode conversion depends on the strategy, the percepts obtained, and whether there is linear place pitch scaling for the electrodes. With formant-based and spectral maxima strategies, the 100-Hz bandwidths are arranged linearly for the seven most apical channels (corresponding to the first formant frequencies 300–1000 Hz), and then the bandwidths increase logarithmically for frequencies greater than 1000 Hz, for the 13 (or more) basal stimulation channels (corresponding to the second formant frequencies). Although there is normally a log/linear relationship between frequency and site of stimulation along the basilar membrane, the above arrangement was found to give better speech perception when used with the Nucleus F0/F1/F2 and subsequent strategies. The frequency boundaries can be altered should there be a significant reduction in the number of channels available. The frequency boundaries for each electrode, and the minimum and maximum current levels (in arbitrary units) for the advanced combination encoder (ACE) as well as the SPEAK or CIS strategies are programmed into a MAP. The mode of operation of the SPEAK, CIS, and ACE strategies was described in Chapters 7 and 8.

The MAP, stored in a memory chip in the speech processor, can easily be reprogrammed should the hearing become too soft, loud, harsh, echoey, muffled, and so on. Typically, the MAP is changed regularly during the first few weeks or months following the operation. The patient's ability to judge comfortably loud levels, and balance loudness across electrodes generally improves with experience, and therefore the MAP can be refined. Also, there are some changes within the cochlea during the postoperative period (for example, fibrous tissue growth) as explained above that alter the current levels required. For the majority of implantees, a new MAP needs to be programmed every 12 months or so, to take into account any minor changes in the levels.

At the first test session, the current level on a particular electrode (using a burst of pulses at 200 to 500 pulse/s with a duration of 500 ms) is increased until a hearing sensation is reported. It is wise to begin with the most apical electrode, as the likelihood of stimulating nonauditory neurons (the facial and the tympanic branch of the glossopharyngeal nerve) is then very remote.

The T levels can be obtained as with audiometry by averaging a number of responses to an ascending and/or descending presentation of stimuli. When ascending from no signal to a percept, the threshold will be higher than when descending in amplitude. A more stable T level can be obtained by also averaging the results for the two procedures. The major difference from audiometry is that the T level should be the lowest stimulus level where a response always occurs (i.e., 100% threshold rather than 50%). It is not so useful to provide a signal that can be heard only 50% of the time. The T level depends on the number of residual nerve fibers excited, which in turn depends on the area of the electrical field as well as the distance of the electrode from the nerve fibers and the nature of the intervening tissue. The same applies to the MC level of hearing.

The MC level is the highest stimulus intensity that can be used without causing discomfort. The level is lower for an initial rather than a continuous presentation, as adaptation occurs in the latter case. As speech is a dynamic signal often with short bursts to individual electrodes, the lower or more conservative value should be adopted to ensure there are no unpleasant side effects. Setting the MC level correctly is especially important when the greater part of the speech signal is mapped to the top 20% of the dynamic range.

If the T and MC levels are high for bipolar stimulation, they can be brought more into the current output range of the receiver-stimulator by stimulating a greater area of the cochlea (i.e., number of neurons). This is achieved with current passing between more widely separated electrodes as discussed above (i.e., $BP + n$).

A study by Busby et al (1994) showed that the T and MC current levels were highest for bipolar and lowest for monopolar stimuli. For common ground stimulation there was a trend for T and MC levels to be highest in the middle of the array. This could be due to the spread of the return current in both directions. With monopolar stimulation T and MC levels increased from the apical to basal ends, due to the fact that the more basal region is larger with the electrode further from the ganglion cells, and there is often more fibrous tissue and bone near the round window affecting the spread of current. There was no consistent pattern for bipolar stimulation. Occasionally, a group of electrodes shows markedly elevated levels. In this case, electrode discrimination needs to be investigated, as there may be poorer neural survival in that portion of the cochlea.

While measuring the T and MC levels for each electrode, it is useful to gain an impression of the pitch and timbre of the hearing sensations elicited. The pitch and timbre are most commonly reported as being dull for the more apical electrodes and sharp for the more basal electrodes. Once the levels have been measured for each electrode, they should be stimulated one at a time, at a particular level (for example, at the MC level) from one end of the electrode array to the other. This enables a check to be made that the pitch of the hearing sensations elicited corresponds to the tonotopicity of the cochlea. In the study by Busby et al (1994) on nine postlinguistically deaf patients, the general pattern of pitch estimations across electrodes was consistent with the tonotopic organization of the cochlea for both monopolar and bipolar stimulation. There was, however, a marked reversal of pitch ordering for electrodes in the middle of the array with common ground stimulation for three of the nine patients, as discussed above.

Ordering of pitch can also provide an indication of the distance to which the electrode array has been inserted into the cochlea, if the listener is asked to report when the sensations become sharp. The second reason for sweeping through the electrodes is to determine whether the hearing sensations are equally loud. If the loudness is not balanced, some speech sounds can appear very soft or drop out altogether. With an imbalance in loudness, voices may seem too harsh or too echoey. Balancing the loudness of the electrodes is not easy for the subject, particularly at first, because pitch and loudness are related; sharper or higher-pitched sounds generally sound louder than duller or lower-pitched sounds, and therefore lower comfort levels may be indicated by the listener for the sharp-sounding electrodes than the dull-sounding electrodes. If the speech processor was programmed with these levels, the listener would report voices sounding muffled and unclear, necessitating an increase in the levels of the more basal electrodes.

The T and MC levels are set after the loudness percepts are comparable across electrodes at the above intensities. The dynamic range for each electrode is the difference in current level between the T and MC levels. Large dynamic ranges are preferable (with more current level steps), as this allows better amplitude resolution. Acoustic stimuli detected by the speech processor's microphone are presented to the implantee at levels within the dynamic range. Provided he/she has judged the MC levels appropriately, no incoming sound should produce an uncomfortably loud hearing sensation.

It is also necessary to evaluate the loudness growth function for increases in intensity at each electrode, as this may vary and lead to unpleasant or nonoptimal speech perception if it is not taken into consideration. The shape of the function can be roughly assessed by sweeping across electrodes, at an intensity halfway between the T and MC levels. If an electrode sounds softer, for example, at this level, this may be due to the shape of the loudness growth curve. It has been demonstrated by Zeng and Shannon (1994, 1995) that the loudness function of sinusoidal stimuli is best described as a power function for stimuli less than 300 pulses/s and an exponential function above this rate, as illustrated in Figure 11.7.

The importance of accurately balancing loudness across electrodes was demonstrated in the study by Dawson et al (1997). The degree of loudness imbalance in mapping the MC levels was examined in 10 adult patients. Four of them had

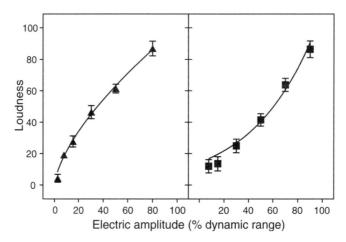

FIGURE 11.7. Loudness functions for sinusoidal stimuli. Left, 100 Hz; right, 1000 Hz. Data were the average of eight Ineraid subjects and loudness through magnitude estimation (Reprinted with permission from Zeng and Shannon 1995).

the Multipeak miniature speech processor (MSP) system and six the SPEAK Spectra-22 strategy. When the MC levels across electrodes were pseudo-randomly unbalanced by up to 20% of the electrode dynamic ranges, six of the 10 subjects showed a significant drop in sentence perception scores. None had a decrease in perception when the degree of unbalancing was halved. The study revealed that it is important to ensure that MC levels are balanced and methods should be developed for ensuring this in very young children. Because of the importance of having well-balanced MAPs, the T and MC levels need to be checked at each session during the first month or so, with less frequent checks after that. New MAPs may need to be generated. The need for this may also be apparent from the person's experiences when using the speech processor, for example if he/she found that certain sounds were too soft or too loud, if the tone is too sharp or too deep, or if background noise is excessively intrusive.

Furthermore some people, particularly those who have been profoundly deaf for a long period of time, find it very difficult to adjust to the sharp hearing sensations produced by the more basal electrodes. On occasion, several electrodes have been removed from a MAP in order to make the hearing sensations more pleasant and acceptable.

Signal Gain

As discussed in Chapter 6, the dynamic range for speech sounds is 30 to 40 dB, but the range for electrical stimulation from the T to MC level at 200 pulses/s with the University of Melbourne/Nucleus banded electrode array in the scala tympani was found to vary from 5 to 10 dB (Clark, Tong et al 1978; Tong et al 1979), as discussed in Chapter 6. Thus the speech amplitude has to be compressed into a narrower range. The DL for sound is 0.3 to 1.5 dB and so the number of discriminable steps varies from 20 to 133 over the speech range. In contrast, the number of discriminable steps for electrical stimulation was reported by Nelson et al (1995) to vary from 7 to 45 steps.

As the overall speech level differs between speakers and their distance from the listener, it is necessary to adjust the gain or sensitivity to keep the stimulus within the electrical dynamic range. The patient can adjust the input signal with a sensitivity control (available with the Nucleus 22 and 24, Clarion S, and Combi-40 systems). With the Nucleus devices the sensitivity control adjusts the knee point of the automatic gain control (AGC) so that acoustic signals at and above this level will result in comfortable levels of stimulation. The AGC is a compression amplifier to keep the variations in the speech intensity within a certain range, as discussed in Chapter 8. The knee point is the intensity at which the compression amplifier starts to operate. Average conversational levels occur around 60 dB at 1 m. For these input levels the peaks would occur at between 70 and 75 dB sound pressure level (SPL) (James et al 2002). If the sounds exceed the dynamic range, the knee point can be set lower at, say, 30 dB; then the higher intensities will be more discriminable, but those below this level will result in T-level stimulation. Inputs in the lower part of the dynamic range will be thus be perceived as soft while those at the top part of the range will be perceived as louder. Also, it is possible to increase the sensitivity to provide more input gain if the speech is too soft.

It is the usual practice for patients at the University of Melbourne clinic to find a preferred setting for loudness comfort for their own voice, for the clinician's voice, and for environmental sounds. Even with this sensitivity control they may not be receiving an optimal signal for adequate perception of low-level speech inputs. Some of the patients may not be aware of the reduction or limitations in the input they are receiving. Lowering the range to inputs less than 14 dB will allow more low-level environmental sounds to be heard. These sounds would be annoying if they limited the perception of speech.

Another limitation is that a loud sound at one frequency causes the AGC to operate and also compresses other less intense frequencies that may be important for intelligibility. This can be overcome using an algorithm referred to as adaptive dynamic range optimization (ADRO). As discussed in Chapter 7, it has a rule that specifies that the output will be greater than a fixed level between T and MC at least 70% of the time. Another rule specifies that the output level on each electrode will be below MC at least 90% of the time. Thus the acoustic input to the speech processor will be mapped to higher stimulus levels on all the electrodes especially at low speech intensities than with the standard SPEAK strategy.

As the front end of the Nucleus 24 system was not very effective in presenting speech frequencies at low intensity levels, fast-acting compression (syllabic compression), which compressed sound over a wide dynamic range for hearing aids, was tested with 10 Nucleus Spectra-22 and SPrint subjects (McDermott et al 2002). Syllabic compression with fast attack and slower release times had been examined to improve speech understanding with hearing aids by reducing the intensity differences between consonants and vowels (Braida et al 1979; Walker and Dillon 1982; Busby et al 1988; Dillon 1996). The study showed a significant improvement in sentence recognition at 45 dBA (20%) and at 55 dBA (17%). (dBA is the unit of A-weighted sound pressure level where the effects of low and high frequencies have been reduced in a manner representative of the ear's response.) A few subjects disliked the increased loudness of some of the background noises.

Loudness Summation

When a speech processor is fitted, the practice has been to provide a general reduction in the MC levels for the individual electrodes. This ensures that a wide-band intense sound stimulating a large number of electrodes does not produce a sound that is too loud because of the summation of loudness when the stimulus exceeds the critical band. It should also be noted that the amount of summation varies with the relative loudness contributed by each pulse. However, as a result of reducing the MC level, with an intense narrow band sound exciting only two electrodes, it will not be loud enough. Reductions in the T levels will also not resolve the difficulty for similar reasons.

It is thus important to develop an algorithm that can predict and dynamically alter the currents on individual electrodes on the basis of the amplitude envelope and the spectral shape of the acoustic signal. This could be done by adding the effective loudness contribution of each pulse within a 7-ms time-window. However, the effective loudness contributions would need to be determined from an effective loudness versus current amplitude function for each electrode, a time-consuming process. If this function were the same across electrodes, then the contribution of a pulse on any electrode could be determined from just one function, together with a set of loudness-balanced current levels as well as the dynamic ranges. This is the subject of further research (C.M. McKay, personal communication).

Patient Preference

It has also been shown that various strategies may suit different patients (Arndt et al 1999). With SPEAK users who were subsequently trained with ACE and CIS, 61% preferred ACE, 23% SPEAK, and 8% CIS. There was also a high correlation between the strategy preference and the speech recognition scores. For this reason, strategy preference may be a useful fitting procedure. In practice, patients may find that different strategies may suit different listening conditions.

Training in Adults and Children

(Re)habilitation involves training in the development of speech perception in adults, and speech perception, speech production, and receptive and expressive

language in children. The training can be through direct involvement of the audiologist, speech pathologist, or educator, or by indirect help through advice or
instruction of parents. An initial approach to training adults and children was
expounded by Brown et al (1990) and Mecklenburg et al (1990), respectively.

General

The aural (re)habilitation program provided for cochlear implant recipients depends on factors such as age, linguistic knowledge, and auditory experience, and
also on the particular implant system. The training exercises need to focus on
the features of speech that are available to the person with a particular speech-
processing strategy. The aural rehabilitation program should provide training in
the reception of speech when used in conjunction with speech reading, understanding speech without visual cues, and hearing speech in the presence of background noise, and provide awareness and recognition of environmental sounds.
Considerable time also should be spent in discussing experiences and providing
counseling about expectations and methods of dealing with difficulties. Areas
such as telephone and television use, awareness and recognition of everyday
sounds, music, and the effects of background noise on speech understanding
should be explored by discussion and with exercises to illustrate the situations.

Both synthetic and analytic training need to be used. Hearing with the cochlear
implant is often quite different from the way the postlinguistically deaf implantee
remembers hearing with a hearing aid. By providing analytic exercises, the person
may learn the auditory cues necessary to discriminate and recognize speech.

Clinics should provide auditory and audiovisual training through analytic and
synthetic methods. Typically there is relatively more audiovisual training initially,
with more auditory-only training toward the latter part of the program. Exercises
increase in difficulty with patient improvement.

The program should usually run for approximately 3 months, with a 2-hour
session each week. It is an individual/family-based program. Contact with other
implantees is important, and is achieved through regular group functions. For
some people, the training program is extended beyond 3 months. The program is
continued until the clinician is satisfied that further improvements are not likely
to be achieved.

Once the speech processor has been programmed with a MAP, the aim of the
first few sessions should be to encourage the person to make some hearing discrimination. It can be useful for the person to listen to his/her own voice during
this session, as this often sounds the most natural. In addition, it is useful to
introduce a male and a female speaker, as it is relatively easy for the person to
differentiate between the two, reportedly on the basis of pitch. The audiologist
should also provide patients with a written paragraph that they read aloud, with
pauses to determine whether they are able to follow the text (using audition only).
It is possible to do this on the basis of duration and rhythm cues, and therefore
the exercise provides them with initial success without visual cues. Other audition-only exercises to be used during the first session include closed-set spondees

and sentences. As the ability to perceive and recognize sound improves, the training can include both analytic and synthetic testing and training.

Just as results vary with each individual, so too should the training program. Some people learn to hear very quickly, requiring less counseling, and no training in many of the easier discrimination tasks is required. In this case the number and difficulty of the audition-only exercises is increased. Conversely, some people require a long period of adjustment to electrical stimulation. This can take up to 12 months or more. Throughout the 3-month training program, these people perform at the lower range (they may demonstrate good electrode discrimination as assessed by psychophysical tests and yet not display the same degree of discrimination with speech material). For them, the training program is continued for as long as there is progress. In a number of instances, large improvements have been seen a number of months following the completion of the training.

Predictive Factors

The factors that are likely to produce good or poor results need to be considered in planning and assessing the (re)habilitation. Key factors are age when deafened, age at implantation, duration of deafness, etiology, speech-processing strategy, progressive hearing loss, degree of residual hearing, speech-reading ability, language level, medical condition, educational method, and motivation and parental guidance. These predictive factors are discussed in Chapters 9 and 12.

The age when deafness occurs is most important in children. In adults, age at implantation is significant only if they are over 60 years. In children, age at implantation and duration of deafness are usually interrelated, as a significant number are born deaf. Furthermore, a child born deaf or deafened early in life can obtain best speech perception results if they have the implant before approximately 2 to 4 years of age (Dowell et al 1997; Fryauf-Bertschy et al 1997; Miyamoto et al 1997). There is a critical period for the development of language within the first few years of life. This is supported by studies on bilingual children, showing a second language can be learned more completely at a young age (Patowski 1980; Johnson and Newport 1989). Busby and Clark (2000a,b) and Clark (2002) have shown that the discrimination of electrode place is poorer the older the child when surgery is carried out, and the ability to discriminate place of stimulation is correlated with speech perception.

In addition, Tye-Murray et al (1995) reported that speech production of young children between 2 and 4 years of age increased more rapidly, and within 2 years was comparable with that of older children. Similarly Nikolopoulos, et al (1999) found in a group of 126 children that after 3 or 4 years of implant use, the children who were younger at implantation outperformed the older children in speech perception and production. Barker et al (2000) reported in children operated on before 2 years of age that their speech sounds were similar to those implanted between 4 and 6 years after 4 years' experience when produced in isolation, but the speech sounds of the younger children were better as part of intelligible language. A similar trend was reported by Waltzman and Cohen (1998) for children implanted before the age of 2 years.

In addition, in early-deafened subjects, if visual signals were used to communicate instead of auditory ones, they could encroach on and utilize the higher auditory cortex. This is supported by the studies of Kubo (2002), which showed with PET that the auditory association area of the cortex was activated by sign language, but not speech in a congenitally deaf cochlear implant user.

With deafness of long duration, adults and children are more likely to require long periods of (re)habilitation for adequate speech perception. Adults may perceive phonemes and have good place pitch discrimination, but cannot so readily understand speech. For the postlinguistically deaf adults with a profound-to-total hearing loss of many years' duration, a considerable portion of the training program may be spent in counseling and discussion. A long duration of deafness leads to loss of neurons and neural connections in the central nervous system (see Chapter 5).

The only causes of deafness that affect the results in the adult are Meniere's disease, where speech perception is better, and meningitis, where it is worse. The infections during pregnancy causing deafness, namely toxoplasmosis, rubella, cytomegalovirus (CMV), and herpes simplex, may affect the central auditory pathways and impair learning.

With a progressive hearing loss, it has been the clinical experience at the University of Melbourne that the time learning to use degraded auditory information carries over into better results with the distorted signal from electrical stimulation. If the child has residual hearing, there is evidence that the results will be better (Cowan et al 1998). This may be due to the fact that the hearing has facilitated neural connectivity. There is a weak correlation between speech-reading ability and speech perception, and this may be due to the fact it reflects good top-down processing skills. Studies at the Human Communication Research Center (HCRC) at the University of Melbourne/Bionic Ear Institute have shown that not only does improved speech perception result in better language, but improved language will also affect speech perception (Blamey et al 1998; Sarant et al 1996, 2001).

Medical conditions affect results primarily if they involve the central nervous system. Learning is poor in patients with dementia, schizophrenia, and neurosyphilis, for example. This also has been seen in the University of Melbourne Clinic for children with toxoplasmosis. In children with multiple handicaps and especially with minimal mental retardation and learning disorders, the speech perception is not as good as with matched controls, and learning takes longer (Pyman et al 2000). This is illustrated by the data in Figure 11.8.

The communication strategy adopted before surgery influences results, and children do better if they have had an auditory-oral education (O'Donoghue et al 2000). The mode of education after surgery is important, and an auditory-oral education is required for good speech and language skills (Dowell et al 1995, 2002). The data in Melbourne show that children with open-set scores of 50% or more are seen only in the auditory-oral group.

The motivation of patients is related to their communication needs, level of

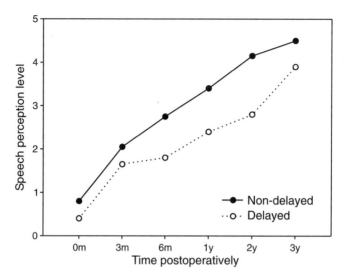

FIGURE 11.8. A graph of the speech perception results for children with multiple handicaps and delayed learning (Reprinted with permission from Pyman et al 2000).

confidence, ability to accept minor failures, willingness to "give it a go," and the support they receive from family and friends. Typically, those with greater motivation obtain better results in terms of speech discrimination, and are more able to cope with background noise, and do so often in considerably less time than others with less motivation and perseverance. Parental support is an important factor leading to good results for children.

Strategy and Time Course for Learning

In postlinguistically deaf adults, it has been shown that the learning required is less and improvement more rapid when the strategy provides more information, and is more speech-like (Clark 2002). This is illustrated in Figure 11.9 where the results over time are plotted for the inaugural Nucleus second formant (F2)/fundamental frequency (F0) (Clark, Tong et al 1978, Tong et al 1979), and the SPEAK or Spectral Maxima strategies (McKay et al 1992). The latter provided more speech information (Skinner et al 1994). In children, time is required to learn speech, especially as this is the case with normal hearing. This is illustrated in Figure 11.10 for children operated on at different ages. It can be seen that there is continuing improvement over 5 years, but this is less for children when operated at 5 years or older (Clark 2002). Further evidence that the amount of information transmitted in the speech-processing strategy influences the rate of learning was reported by Osberger et al (1996). They compared two groups of six children who commenced with the F0/F1/F2 and Multipeak strategies and who were matched for age at onset of deafness and age at implantation. After 1 year the children with the Multipeak strategy were better at discriminating vowel height and consonant place of articulation, but at 3 years there was no difference. A

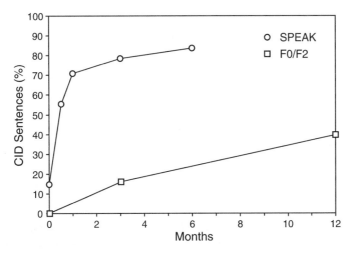

FIGURE 11.9. The open-set speech scores for electrical stimulation alone over time for adults using the inaugural F0/F2 and the recent SPEAK cochlear implant strategies (Reprinted with permission from Clark G. M. 2002. Learning to hear and the cochlear implant. Textbook of perceptual learning. M. Fahle and T. Poggio, eds. Cambridge, Mass. MIT Press: 147–160.)

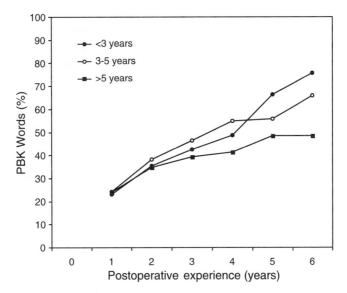

FIGURE 11.10. The relation of postoperative experience to word score for children operated at less than 3 years old, 3 to 5 years, and more than 5 years (Dowell, personal communication; Clark 2002). (Reprinted with permission from Clark G. M. 2002. Learning to hear and the cochlear implant. Textbook of perceptual learning. M. Fahle and T. Poggio, eds. Cambridge, Mass. MIT Press: 147–160.)

number of studies with the Nucleus F0/F1/F2 and Multipeak systems have shown that children now achieve open-set speech understanding within the first year in using the device (Fryauf-Bertschy et al 1992, 1997; Gantz et al 1994; Miyamoto et al 1996; Osberger et al 1996). Furthermore, Miyamoto et al (1996) found a continued improvement in word recognition beyond 5 years, and this highlights the need for long-term follow-up (Kirk 2000).

Analytic

Studies on the recognition of vowels and consonants are analytic exercises and should aim at highlighting particular speech sounds so that they may be more easily detected and identified in connected discourse. The patient needs to choose the correct response from a list. No contextual information is provided, and only minimal acoustic cues, as the stimuli consist of single words. The stimuli for the vowel recognition exercise in Australian English are 11 pure vowels in an /hVd/ (V-vowel) context. The vowels are heed /i/, hid /ɪ/, head /ɛ/, had /æ/, hud /ʌ/, hard /ɑ/, who'd /u/, hood /ʊ/, heard /ɜ/, hoard /ɔ/, and hod /ɒ/. The responses to randomized stimuli are constructed into matrices that highlight the stimuli that are confused with each other. Feedback is given as to the correctness of the response, and thus the procedure is both a training and assessment exercise. In addition 12 stimuli are used in the consonant recognition study for Australian English, in a /aCa/ (C-consonant) format: aba, apa, ama, ava, afa, ada, ata, ana, asa, aza, aga, aka. Typically, with the improved results with the SPEAK, CIS, and ACE strategies, the studies should be undertaken for audition alone, and either the vowel or consonant exercise should be done at each weekly visit.

If the exercises are too difficult, especially consonants, it is preferable to reduce the number of alternatives from which to choose a response, but still continue the training. The consonants chosen should be those that are relatively easily discriminated on the basis of manner of production.

Synthetic

Speech-tracking with connected discourse is a procedure developed by De Fillipo and Scott (1978), to train communication, and was used for cochlear implant assessment by Martin et al (1981). Connected discourse approximates everyday communication more closely than word and sentence materials. The clinician reads aloud from a text, with segments of manageable length (sentence, phrase, or a few words). The person is required to repeat the segment verbatim. If a portion is missed or an error is made, various strategies are then used to obtain the correct response. A hierarchy of strategies is used to elicit the correct response, and in a specific order. The subject should be encouraged to respond as quickly as possible. After 10 minutes, the tracking rate is calculated, which is the number of words repeated for the 10-minute period, expressed in words per minute. By plotting the speech-tracking rate, the person can see his/her progress. However, absolute tracking rates vary with different texts, strategies, and clinicians. Therefore, more emphasis is placed on a comparison between conditions.

There are many other training exercises that can be used in the (re)habilitation program. These include discrimination between questions and statements (based on changing intonation), and between two phonetically different words within a sentence. For example, "Where is the train/crane?" or "the sheet/shirt was clean." The degree of difficulty is determined by the audiologist. Many of the exercises are easily created (particularly the sentence materials), and can be individualized to take into account particular problem areas and discrimination abilities.

Another useful training exercise, "Questions for Aural Rehabilitation," was developed by Erber (1982). This is an interactional exercise between the patient, and the clinician. Using a booklet of questions, the patient asks the clinician questions and listens for the response. The complexity of the answer by the clinician can be made appropriate to the individual's abilities. Should the patient fail to understand the response, he/she is taught to assess the reason for this (was the reply too fast, too long, too soft, etc.), and ask for the response to be modified accordingly. This exercise requires the patient to take a more active role in communication than with the others referred to above, and it is thus more related to everyday situations. Because the context is known, and the patient has asked the question, many patients need to use only audition or minimal visual cues.

Patients are encouraged to use the speech processor for a significant portion of their waking hours, although not to the extent of becoming overtired. Experience has shown that those people who use the speech processor for most of the day become much more accustomed to hearing, and learn to distinguish between various environmental sounds more rapidly, enabling them to accept noisier situations more easily.

Environmental Sounds

The patient should also be encouraged to take an active role in learning to recognize environmental sounds, and to encourage family members to help by pointing these out. Some environmental sounds produce quite different auditory sensations from those remembered. For example, the pitch of the sound may be different. However, the rhythm remains the same, and therefore people are encouraged to listen to this aspect in order to identify the sound initially.

Background Noise

Cochlear implant patients, along with the majority of people with hearing impairments, experience difficulties in communicating in the presence of background noise. Although improvements to cochlear implant sound-processing strategies (F0/F2, F0/F1/F2, Multipeak, CIS, SPEAK, ACE; see Chapter 7) have led to better speech perception in quiet, hearing in noise is still a problem. Users are advised to initially reduce the sensitivity of the speech processor rather than switch it off in noise. With experience, they may find they do not need to reduce the sensitivity. The directional microphone also makes it easier to hear people face to face. An additional method of reducing background noise is to use a plug-in tele-coil pickup in venues with a loop system. Larger conference-style microphones that can be put on a table have also been very useful for those attending meetings and lectures.

Music

In a questionnaire sent to 40 patients using the F0/F2 and F0/F1/F2 speech processors, the styles of music most appreciated were single instrumentals and popular songs. This was due to the fact the speech processor was unable to adequately process many different instruments in the more complex forms of music. During rehabilitation, recordings of various styles of music can be played for patients. With improved sound processing with the Multipeak, SPEAK, CIS, and ACE strategies, the sounds have become more natural, but musical appreciation is still limited. Musical appreciation with the Multipeak-MSP and SPEAK Spectra-22 systems was evaluated by Fujita and Ito (1999), and subjects were able to identify the nursery rhymes with words more easily than when played only with an instrument. The authors considered that good spectral information was required for the identification of speech or instrumental colors, but there was generally poor performance for the recognition of melodies. For further information see the section on music in Chapter 6.

Telephone

Training in the use of the telephone is particularly important, and all cochlear implant users can learn to use it even if only for emergencies. It first involves the correct recognition of the various telephone signals: the dial tone and the ringing and busy signals. The listener must easily distinguish between these and an answering voice, through suprasegmental cues such as rhythm and duration. As using the telephone can appear too big a challenge, efforts should be made to build up a person's confidence. This is done by ringing up recorded messages and discussing their content. Next use telephones in adjacent rooms, preferably with someone to assist. Many patients are surprised to find they can understand utterances with contextual cues, such as, "Hello, how are you?" and "What is the weather like?" These experiences can serve to increase confidence. However, as the mean open-set Central Institute for the Deaf (CID) sentence results for the SPEAK, CIS, or ACE strategies are as high as 80%, a majority of adults can communicate freely on the telephone. The same applies to children who are often more willing to experiment. Nevertheless, there are a small proportion whose results are well below the average, and they will have difficulty. Those with the poorest speech perception are trained to discriminate between "yes" and "no." If patients cannot reliably do this, then they are instructed in the use of a simple telephone code, the "yes-yes/no" code that enables a hearing-impaired listener to know whether "yes" or "no" has been said, based on the number of syllables of the response (Alpiner 1982). Training is necessary to enable them to effectively explain the procedure to the person on the other end of the telephone, and to be able to phrase their questions so that a "yes" or "no" reply is appropriate and provides the required information (Brown et al 1990). Furthermore, an electromagnetic induction pickup, connected between the telephone and the speech processor, can be used by the cochlear implant recipients to enhance the signal from the telephone. This also switches off the ear-level microphone, thereby cutting out background noise.

Television

Television poses a problem to cochlear implant recipients (along with many hearing-impaired people) because the people on television are generally not facing the viewer (the exception to this are newsreaders, but instead they speak very quickly), often background noise or music is present, and off-screen voices or voice-overs may produce confusion. Implant recipients report that commercials provide useful training (the topic is generally known, and repeated presentations enable improved speech recognition) (Brown et al 1990).

Watching television has been made somewhat easier for implant patients through the provision of leads and attenuators to plug into the earphone sockets of the television sets, and into the external socket of the speech processor. This enables the loudness of the television to be adjusted independently of other viewers (to compensate for varying distances from the set), and this also cuts out background noise in the room by simultaneously switching off the ear-level microphone (the lead may also be plugged into radios, cassette players, and stereo systems).

Mapping and Fitting Children

The progress children make with their (re)habilitation depends on their mapping, training, and education as well as the general and specific factors referred to above (see Predictive Factors). Their wider education depends on the teaching method, the teacher's competence, parent–child interactions, family support, and the social environment. Their (re)habilitation using implants is carried out in a similar way to the auditory (re)habilitation of children using hearing aids, but differences occur and will be emphasized. More detailed information on the (re)habilitation of children with hearing aids can be obtained from publications such as Cole and Gregory (1986), Eisenberg et al (1983), Erber (1982), Ling (1976, 1984), Ling and Ling (1978), Mecklenburg et al (1987), Sims, Walter et al (1982), and Ross and Giolas (1978).

Before and after surgery the child is given parental and team support. When the device is switched on, the first task is to program the speech processor correctly. The T and MC levels are set for each electrode, and the loudness levels balanced. In establishing T and MC levels, take care that the sensations are not unpleasantly loud or it will reduce the child's confidence and ability to learn. This requires carefully observing the behavioral responses to the stimuli, particularly an aversive or withdrawal reaction. The T and MC levels for each electrode are then recorded on a MAP in the speech processor. Assessment of T and MC levels is also improved with receiver-stimulators such as the Nucleus 24 system that have NRT.

Preprogramming Training

A child's initial responses to the new sensations can range from clear aversion, to no apparent response, quieting, a search for reinforcement, subtle changes in facial expression, and pleasure. To facilitate a predictable response pattern, a preprogramming training period is useful, and occurs either before or after the surgery but before the first implant test session. Knowing the task can make the early test sessions much easier for both the child and parents. A valuable adjunct to preprogramming training is to allow the child to observe another child or an adult patient having the device set. The training is described in Mecklenburg et al (1990).

This preprogramming training consists of establishing a rapid, clear response pattern to visual and tactile stimulation. Tasks include "on-off," "same-different," "high-low," and differences in loudness. Training to determine the MC level or the minimum discomfort level is often more difficult than for the T level. Progressive illustrations of a "listening but not hearing person" to a "frowning, too loud" person can be paired with glasses of water that are empty, half full, filled and overflowing to demonstrate the experience of "too much" or "too loud." Another concept that can be introduced before programming is same–different. Many examples of this can be devised using blocks, pictures, objects, colors, and so on. The goal is to establish a means by which loudness and pitch differences can be detected without having to explain what they are.

All children undergoing preprogramming training can understand the concept of on-off, and most progressions from small to big, empty to full, short to tall, and so on. However, it has been rare for a child less than 6 years of age to give reliable same-different responses to very similar touch intensities.

Conditioning

The first goal in fitting the device is to familiarize the child with the sound sensations. The Nucleus 24 and other devices such as the Combi-40+, Clarion S, and Digisonic require the T and MC levels for each electrode to be programmed into the processor for a loudness MAP. If these levels are incorrect, the sounds may be inappropriately loud or soft.

When children experience sound for the first time, they often report that it is felt somewhere in the head or neck, and then it shifts within the first day or week to the implanted ear. Touching the child in the areas referred to above suggests that hearing anywhere is acceptable, and that he/she should respond, for example, by throwing a block into a box. Flashing a light to indicate the presence of a signal also helps in reinforcing threshold responses. Conditioning the child by touching him/her around the head, neck, and area of the implant while asking for an "on" response can be transferred as a conditioning stimulus by children as young as 3 years.

There can be a difficulty in generalizing from the preprogramming concepts to loudness growth with electrical stimulation, as it can be very rapid. Thus there may not be an orderly progression from "empty" to "too full." The clinician may find the child waiting for gradual changes only to discover that the sound suddenly has become too loud. Nonetheless, it remains useful for children to have a general idea that the signal will change and that they should indicate it in some manner (paraphrased from Mecklenburg et al 1990).

In younger children who do not have the language to tell the clinician that the sound is uncomfortably loud, it can be difficult to set the MC levels. It is necessary to rely on the child giving an aversive reaction or a blink, which determines the loudness discomfort level (LDL). The relation between the LDL and the MC in adults has been determined by Hollow et al (2002), and this should be applicable to children. In this preliminary study of the relationship between MC levels and LDLs in 15 adults, Hollow et al (2002) suggested that MC levels could be set 45% (of the T level to LDL range) below LDLs for Nucleus 22 and Nucleus 24 implants using 250-Hz pulse rates, and 35% below LDLs for Nucleus 24 implants using a 900-Hz rate. However, the judgment between "too loud" and "uncomfortably loud" varied significantly between subjects, with some setting MC levels 80% lower than their LDL and others setting MC levels only 10% lower. Significant variation was also attributed to both rate and mode of stimulation. No significant variation was found due to pulse width or channel, that is, moving from one end of the electrode array to the other.

Initial Setting

The aim of the first sessions is to provide the child with a device that is comfortable and can be worn home after a few days. A conservative MAP is made with a "soft" signal to help ensure the signal is not unpleasant. Uncomfortably loud sensations can slow (re)habilitation and discourage the child.

Sometimes a child may not respond to the hearing sensations, and an aversion reaction may occur before any signs of comfortable hearing. Other children may not show any negative responses to very high stimulus levels, when the audiologist will need to estimate a conservative range of stimulus levels for the first MAP. One technique successfully applied to young children at the University of Melbourne's clinic is to program a single electrode in the speech processor and gradually increase the maximum level while the child is interacting with an adult. By watching the facial expression and behavior of the child, a good estimate of the MC level can be obtained. Because this technique uses a speech signal in a communication context, the signal will be more readily accepted by the child with less variation in the MC levels.

During the first days, the child will become more accustomed to the electrically produced sounds. The T and MC levels will gradually approach stable values as discussed above. Often the child's attention span is not long enough, or the tension of the initial sessions too great, for the full set of electrodes to be programmed. The initial take-home MAP need not contain all the electrodes, but the full set should be included as soon as possible.

The loudness balancing should ensure that the T and MC levels are even, so that the sounds will not become "broken" or change abruptly in loudness. Some smoothing of values by an experienced audiologist may be necessary for children who do not respond consistently. Occasionally, electrodes at the basal end of the cochlea are unpleasant as with adults, when they can be removed from the MAP.

Once a comfortable program has been produced, with the maximum number of electrodes, it should be tested with informal speech detection tests. All speech sounds, with the possible exceptions of /f/, and /θ/ should be heard with the processor at a moderate sensitivity, and the speaker at 1 m from the microphone. A selection of vowels can be used to test different regions of the electrode array. For example, /ɔ/ stimulates the most apical electrodes in the F1 and F2 regions, /i/ the more basal F2 electrode and an apical F1 electrode, and /a/ F1 and F2 electrodes near the center of the array. /ʃ/ stimulates a very basal electrode. If these sounds are not detected after practice, it is probable that the thresholds on some electrodes are too low.

The discrimination of different types of sounds should also be tested: loud/soft, short/long, high/low, steady/changing, single syllable/multiple syllable, and many other contrasts. Some patients may recognize temporal differences and indicate rate-pitch (voice pitch) very well, but do very poorly on place-pitch (vowel place and height) differences. These early tests may indicate areas of (re)habilitation where progress will be rapid or slow.

Follow-Up Device Settings

During the first 3 months and at regular intervals, the T and MC levels as well as the function of the speech processor should be checked. The child, the parents, and the teachers need to be educated in its use. Some problems can be remedied easily, such as a flat battery. The setting of the sensitivity knob may be an indication of problems. It should be set initially at a midrange, so if it is consistently high, the stimulation levels are probably too low, and vice versa. If the child suddenly becomes unwilling to wear the device, for no apparent reason, the stimulus levels should be checked, as well as other factors.

Once a working program has been achieved, the audiologist moves on to the goal of maximizing the speech information provided by the implant. If there is lack of progress, it is sometimes caused by inappropriate selection of stimulation levels and/or frequency boundaries, or a malfunctioning speech processor. For these reasons, the audiologist should keep well informed about the child's progress.

Neural Response Telemetry

The Nucleus 24 cochlear implant was the first system with the capability of recording the evoked compound action potentials (ECAPs) of the auditory nerve using NRT (Carter et al 1995) as well as the EABR, and was initially used as a clinical research tool (Heller et al 1996). Telemetry systems have also been developed for the Combi-40+ and Clarion devices.

With NRT the voltages in the auditory nerve in response to a stimulus pulse are signaled externally by radio waves, and these can be correlated with T and MC levels (see Chapter 8). The very small voltages from the auditory nerve form a compound action potential (CAP). The transmitted voltages can also be used to determine the tissue impedance around the array, and so assess pathological changes. Like all objective audiological procedures, NRT should be accompanied by behavioral measures. ECAP has advantages over recording EABRs from surface electrodes, as they can be made rapidly, and a child does not require an anesthetic. The ECAP can also be measured without the need of any extra equipment (Brown et al 2000; Murray et al 2000a). The Nucleus 24 system also had an additional feature that could determine whether the stimulus had exceeded the voltage compliance, and hence that a programming change was required.

The NRT software for the Nucleus system was produced by Dillier and others at the University of Zurich, in collaboration with Cochlear Limited in 1995. Validation of the NRT measurement technique (Abbas et al 1999) and a three-stage field trial confirmed that clear, stable, and repeatable responses were obtained in over 93% of subjects (Dillier 1998; Lai 1999; Dillier et al 2000, 2002). Researchers found significant correlations between objective ECAP thresholds (T-NRT) and stable subjective T and MC listening levels for electrodes along the array (Heller et al 1996; Dillier et al 2000; Hughes et al 2000a). More importantly, the T-NRT was found at audible and comfortable levels, that is, above subjective T levels and below subjective MC levels for the majority of patients. On average, the ECAP occurred 53% along the dynamic range. In addition, the configuration of T-NRT across the electrode array mirrored that of the subjectively measured T levels. Presently there are four clinical applications of the Nucleus NRT and especially in children: (1) to confirm the integrity of the implant and the status of the peripheral auditory nerves (Carter et al 1995); (2) to assist with the programming of initial MAPs, especially in young children and recipients who are difficult to test; (3) to supplement behavioral testing and monitor peripheral responsiveness over time (Abbas et al 1999; Hughes et al 2000a; Murray et al 2000b); and (4) to create an entire MAP based on two behavioral measurements (Hughes et al 2000a,b).

Furthermore, preliminary research at the CRC for Cochlear Implant Speech and Hearing and then at the CRC for Cochlear Implant and Hearing Aid Innovation in Melbourne suggested that the amplitude growth function of the NRT response correlated with the perceptual loudness growth function (Cohen et al 2001b). This information may be helpful for improved loudness mapping in the speech processor. Research in the experimental animal has also shown that the ECAP amplitude growth function significantly correlated with spiral ganglion survival (Hall 1990). The growth function therefore might be used to estimate differences in residual nerve populations.

Training in Children

The training of children with cochlear implants evolved from the work of others with hearing aids, and this was modified in the light of the experience by the staff of the University of Melbourne's Cochlear Implant Clinic at the Royal Victorian Eye and Ear Hospital The description of the procedures was given by Mecklenburg et al (1990).

General

It is important for the children and those working with them to see positive improvements from the program within a short time to build confidence and provide motivation for the work ahead. It should be remembered that developing speech and language skills by normal-hearing children extends for several years. This period is longer for children using implants because the reduced information supplied by the implant makes the task more difficult, particularly if the child's communication skills are significantly delayed by the time of implantation.

The children are likely to require considerable help from parents, teachers, audiologists, and speech pathologists. At different times, support from the surgeon, psychologist, social worker, or others may be needed. A coordinator for these activities monitors the child's overall progress and keeps everyone informed. In Melbourne the coordination role has been assumed by the implant center audiologist, but could be done by the educator or speech pathologist. Motivation of the child is essential, and the staff and parents need to be sensitive to the child's needs. The child's efforts in communicating should always be positively reinforced. The best reward is often effective communication.

After a proportion of the electrodes have been mapped and the T and MC levels determined, a preliminary speech-processing strategy can be produced, and the children should be encouraged with parental help to use it for limited test situations and in their normal home environment. Gradually, the number of electrodes can be increased, and the children's auditory experience widened both at home and in the clinic. The goal of the training is to develop good auditory speech perception, speech production, and age-appropriate language. The training program should also allow for variations in the times taken to learn these skills, and differences in the mothers' interaction with their children. A progressive development of skills was used by Ling (1976) in his program for speech production.

Personnel

The surgeon should play a significant part in (re)habilitation commencing at the time of the initial consultation. This requires appreciating the effects of deafness on receptive and expressive communication skills, and the benefits expected from the implant. The parents or guardians can be strongly influenced by medical advice. Audiologists from the Cochlear Implant Clinic are most suitable to be the coordinator for an implanted child because of their continuing role in selection, device fitting, evaluation, and maintenance of appropriate stimulation levels. To be the coordinator, the audiologist or aural (re)habilitation specialist should have had considerable experience in counseling and setting cochlear implant devices with adults before working with children. The audiologist often carries much of the responsibility for helping the family and the child to have appropriate expectations. The coordinator's role also carries the need to recognize potential problems as they arise. The audiologist assesses auditory perception pre- and post-operatively, sets the device initially, and makes follow-up adjustments, and trains the child. The problems, goals, and progress of the training must be explained regularly to the child and to the parents and the teachers, who can then provide a level of input that will challenge but not overtax the child (paraphrased from Mecklenburg et al 1990).

The auditory input from a cochlear implant should give children a greater ability to monitor their own voices, as well as hear. Thus (re)habilitation should aim to increase the children's intelligibility along with their comprehension. Training speech perception and production should reinforce one another and aid in the development of language. This will require a speech-language specialist from either the clinic or the school. Specialists should have experience in managing the type of speech and language problems that occur with profound deafness. They need to keep parents and teachers informed of the goals and progress of the therapy, so that the skills taught can become automatic in situations outside the training session.

For most children, the parents will make the major decisions regarding implantation, educational placement, and other factors affecting (re)habilitation. They are usually the people closest to the child and the most trusted. For these reasons, the implant team is obliged to keep the parents fully informed of factors affecting the child, and will find it worthwhile to obtain information from the parents about problems or achievements related to the child's use of the implant. Communication lines must be completely open. The parents should feel that they have access to the team whenever a question arises. Parents should be invited to in-service workshops and called upon to provide input.

The parents also become the school away from school, the home-bound therapist, and the auditory training specialist. They are responsible for the maintenance and basic troubleshooting of the equipment. They must motivate the child to use the implant. To perform all these functions, they need to be well informed and assured that the educators and clinical specialists are providing the best possible services for their child. Some of the ways that parents, and others close to the child, can encourage the use of verbal communication are (1) to speak at appropriate loudness and distance; (2) to use normal intonation; (3) to provide the opportunity for hearing to be reinforced by vision, and to point out sounds; (4) to encourage turn taking (speak and listen); (5) to reward all listening attempts;

(6) to model speech by reiteration; (7) to provide quality one-to-one communication time; and (8) to encourage as much awake time in the use of the device as possible. Without this constant reinforcement, the generalization of skills learned in short training sessions is unlikely to occur.

Apart from the home, the school is the environment in which the child spends the most time. It is essential that the teacher should be aware of the auditory (re)habilitation needs and goals for the child, and how to support the (re)habilitation in the classroom. The teacher's role should include (1) monitoring the device, speech production, and listening behavior; (2) providing appropriate seating and audiovisual aids, facing the child, and reducing the background noise; (3) supporting peer-student education, allowing time for individual therapy, reiterating speech attempts if needed, and using the FM system; and (4) communicating with parents, teachers, therapist, and clinic. A recently implanted child should not be expected to cope with the communication demands of a normal classroom and school environment without assistance. The teacher will need to know how to structure lessons so that the child's spoken language and perception are developed at an appropriate level, and how to help the child cope with peer pressures (paraphrased from Mecklenburg et al 1990).

Pragmatics

Communication requires some preverbal or nonverbal skills: (1) paying attention; (2) seeking attention through requesting objects, actions, or information; (3) responding to questions or commands; (4) imitating sounds, actions, and facial expressions; and (5) greeting and acknowledging other people. These traits reflect the need to interact with other people, and are usually the precursors to spoken language. Habilitation should ensure this interaction takes place, especially with the parents. This can be achieved through playing games or passing objects from one to the other. Establishing these preverbal behavioral patterns should lead more naturally to verbal communication.

Postoperatively, the child should also be encouraged to verbalize, imitate, and practice making sounds in a manner that may be similar to the babbling or sometimes unintelligible utterances of young normal-hearing children. It is important that these vocalizations be a source of pleasure, through auditory feedback and the responses of others. This also helps in the development of the physical coordination needed for speech.

Pragmatic skills are also important at later stages of language development. Establishing eye contact, turn taking, responding appropriately to requests, statements, and commands, using gesture and facial expression, and appropriate consideration of the body language of the other communicant are all important skills that should be developed, particularly as they can supplement an impoverished auditory signal (paraphrased from Mecklenburg et al 1990).

Speech Perception

The perception of speech and other information such as the identity and the emotional state of the speaker, has several levels of processing. These are the detection, discrimination, identification, recognition, and comprehension of speech. *Detection* is awareness that the signal is speech; *discrimination* is being able to differentiate between two balanced speech alternatives; *identification* is the ability to choose the correct word from a closed set of alternatives; *recognition* is identifying a speech utterance without prior information, and is similar to identification but requires more complete perception of the stimulus; and *comprehension* is knowing the meaning of the word or sentence. These stages are used by Ling and Ling (1978) and Erber (1982) in formulating a strategy for auditory training. Each of these strategies is subdivided into six levels of complexity: sound, phoneme, word, phrase, sentence, and paragraph. The intention is to move through the tasks in increasing difficulty and complexity after identifying the levels at which the child is operating.

The development of materials that support learning at different levels is essential. They should be sufficiently complex to challenge the child's natural desire to communicate ideas. This can be assisted by a mixture of analytic training and synthetic tasks where the emphasis is on getting the message across, and the details of the acoustic signal are not so essential

In planning the analytic training, it is important to understand the differences between the signal presented by the implant and by a hearing aid. The perception of voice pitch is better for hearing-impaired listeners who have reasonable levels of residual low-frequency hearing. The discrimination of pulse rates by implant users is often quite poor, and may vary widely from one person to another. This means that the fundamental frequency or voicing may require more training for implant users, or the amplitude and duration of syllables may become more significant for prosodic cues. If the child has an aid in the unoperated ear, he/she will need help to either fuse voicing from each ear or attend to the aided side. In addition the implant as distinct from a hearing aid, as discussed in Chapter 7, will give high-frequency spectral information. This often results in the accurate recognition of vowels, and this has been the case for all multiple-channel implants. However, if the child does not have sufficient place pitch discrimination to detect closely spaced vowel formants, this could be improved by training. Training in the discrimination of vowel pairs with widely separated and then more closely approximated formants was undertaken by Dawson and Clark (1997a,b). It was hypothesized that improved neural connectivity or central averaging would result from the training, and this would in turn assist in processing less widely spaced regions of excitation. The data showed that two of four children improved their recognition of vowels, and this carried over to the perception of speech. Furthermore, speech analysis (see Chapter 7) has demonstrated that with the more recent SPEAK, CIS, and ACE strategies the amplitude and timing information for manner cues is well conveyed, but place of articulation is still underrepresented. For this reason, the audiologist may need to provide training with words that emphasize place features. They require multiple cues for their recognition, and it was shown by Clark, Tong et al (1976) that children could learn to use the minor cues not used by normal-hearing children.

In addition, it was demonstrated on implanted children that visual information from lipreading cues had a more dominant effect than for children with normal hearing. This is discussed in some detail in the chapter on Research Directions, and is consistent with the physiological studies in the experimentally deafened animal showing an over representation of visual input and processing in the higher auditory areas as discussed in the chapter on Electrophysiology.

Although this visual dominance can be used to advantage in the perception of syllables such as /ba/ where the /b/ is visually distinctive from /d/ and /g/, on the other hand if for example the transients and burst energy in the sounds for /d/ and /g/ are ambiguous there will be less assistance from the lip movements as those for /d/ and /g/ are more similar than /b/. For this reason, there is a need to train discrimination of this ambiguous auditory information. This may, as discussed above for vowels, arise from the use of meaning or higher level processing to facilitate plasticity changes in connectivity, but this would require removing the visual cues in the training sessions, and concentrating on auditory-verbal communication. In addition, efforts are required to highlight the auditory cues required to discriminate these ambiguous consonants for example by selecting the formant frequency transitions that are important, and coding them optimally for the nervous system. This has been achieved with the transient emphasis speech processor in adults as discussed in the chapter on Speech (Sound) Processing, and will need to be applied appropriately to children.

If a child has difficulty with particular aspects of the training, these should be left until the child becomes more competent with easier ones, so that progress is not impeded. Further, the analytic training should not be seen as a linear progression of tasks, but a varied activity that might include work on the prosody of sentences, vowels in multiple syllabic words, and initial consonants in nonsense syllables, all within the same session. Skills learned in simpler tasks should be practiced in more complex tasks while newer skills are being acquired.

Progress with synthetic training will be faster if the materials are chosen so they do not contain too many analytic decisions that the child has not mastered. It is not desirable to exclude them altogether because the "top-down" processing that involves using contextual, lexical, and syntactic structure to make sense of an incompletely recognized signal is an essential component of speech perception that should be practiced.

Much of the time spent with parents and teachers can become effective synthetic training if the child and adult interact well. This is also an appropriate environment for the child to practice pragmatic skills to overcome particular communication problems: moving closer to the speaker to obtain a louder signal, orienting the directional microphone to get the best signal-to-noise ratio, asking the speaker to speak louder or slower or repeat or explain a particular word, and so on.

Perception of Environmental Sounds

The recognition of environmental sounds is important for physical awareness and safety. They can also be a source of pleasure, as in listening to birds singing or to music. The initial Nucleus speech processors were designed to represent spectral peaks to optimize the information extracted for speech understanding (Clark, Tong et al 1978; Tong et al 1979, 1980; Clark 1986; Dowell et al 1987). However, many environmental sounds are not effectively analyzed by these processors because they have broad spectra without well-defined peaks in the required frequency ranges, and they are not periodic. The pulse rate, the amplitudes, and the frequencies measured by the processor are not simply related to the acoustic properties of the original sound. Inadequate representation of environmental sounds was also seen with fixed filter or channel vocoder schemes (Eddington 1980; Merzenich et al 1984). This was probably due to limitations in the presentation of the fine temporal information as well as the spatial coding of frequency. The later speech-processing strategies (SPEAK, CIS, ACE) provide better, but still inadequate, representation of music and environmental sounds (Dowell et al 1990; McKay et al 1992; McDermott et al 1992; Wilson et al 1992).

The best way to learn the characteristics of the sounds with any of the speech-processing strategies is by direct experience rather than by training. The parents and teachers should be encouraged to draw the child's attention to sounds other than speech. The enjoyment of music is an area of great individual variation. As discussed above (see Music), music appreciation is primarily restricted to rhythm and tempo, but not melody. Nevertheless, with improved processing strategies more children are learning musical instruments and should be encouraged to do so. Research is needed to explore the limits of their abilities, and how to modify strategies to assist them. Musical acoustics for sound and electrical stimulation was discussed further in Chapter 6.

Speech Production

Speech production in implanted children depends on general and specific factors as well as training and experience. General factors are the same as those that lead to good speech perception (see Chapter 9). In particular, age at implantation correlates negatively with speech as shown for example by Tye-Murray et al (1995), Nikolopoulos et al (1999), and Barker et al (2000). They reported that with the Nucleus implant the speech production skills of young children between 2 and 4 years of age increased more rapidly than for older children. The same was seen for children under 2 (Waltzman and Cohen 1998).

During (re)habilitation the quality of the child's speech needs to be assessed through speech sounds that have been imitated, elicited, or spontaneously produced. The simplest method is a rating scale (Shriberg and Kwiatkowski 1982; Levitt et al 1987), as well as the Voice Skills Assessment (VSA) Battery (Dyar 1994) and the Speech Intelligibility Rating (SIR) (Parker and Irlam 1994). The child's ability to imitate speech can be tested using Phonetic Level Evaluation (Ling 1976) or Voice Analysis (Ling 1976). Elicited speech is from visual prompts where the child is asked to name a picture (Test of Articulation Competence) (Fisher and Logemann 1971) or verbally repeat a written sentence (McGarr 1983). The child's spontaneous speech produced in conversation or play can be recorded on videotape and analyzed using the phonological process analysis (Ingram 1976; Crary 1982) and the phonologic level evaluation (Ling 1976).

Furthermore, to help in monitoring speech production a computer-aided speech and language assessment (CASALA) procedure was developed in the HCRC at the University of Melbourne/Bionic Ear Institute (Blamey et al 1994, 2001). The sounds of the words spoken are transcribed into the appropriate written symbols by a person experienced in phonetics, and the computer transcribes the words read into their correct phonetic representation. The program then analyzes the data and produces an inventory of the vowels and consonants being used, the percentage of correctly identified phonemes, and the abnormal phonological processes being used by the child. This information can then be used to monitor progress with habilitation and identify where special help is needed. Alternatively, the percentage of phonemes produced correctly can be plotted against a measure of receptive language in equivalent age. This allows the relationship between receptive language and speech production to be determined.

The more sensory information provided by the implant, the less training is required to develop normal patterns of speech production. It is essential the speech embody the correct prosody as well as segmental elements. One method used is that proposed by Ling (1976, 1989), which aims to reinforce speech production in the order in which the sounds develop. For example, the plosives /b/ or /p/ are produced before /g/ or /k/. It should be stressed that training in the correct pronunciation of these segmental elements should not be overemphasized at the expense of prosody. The training program of Ling was developed for children with hearing aids, and they had better low-frequency hearing and poorer high-frequency discrimination. The training of speech production therefore should be tailored accordingly, as discussed above (see Speech Perception).

Sound principles underlying the training of speech production have been outlined by Robbins (1994). First, integrate perception and production. Therapy should always contain a listening and speaking component. Second, develop a dialogue rather than a tutorial style. Dialogue more closely approaches normal communication, and is more likely to lead to the correct intonation patterns. Third, emphasize that speech skills need to be generalized to real-world situations. This helps ensure the transfer of learning from the training sessions. Fourth, use communication sabotage. This is a strategy to prepare the child for unexpected listening situations. Fifth, produce contrasts as stimuli for listening and speaking. This was described initially by Ling (1976). Sixth, make communicative competence the goal. Speech skills should not be developed in isolation, but together with all communication abilities. Intelligent speech is useful only if the child has something to communicate with the appropriate language.

Language

Children with normal hearing develop the fundamentals of language by approximately 7 years of age. Early-deafened children, however, have significant delays in communication, which includes vocabulary (Osberger et al 1981; Boothroyd et al 1991), grammar (Power and Quigley 1973), and pragmatics (Kretschmer and Kretschmer 1994).

It was essential to know to what extent improvements in speech perception and production, reported above and in Chapter 12, for the Nucleus 22 cochlear prosthesis, led to better receptive and expressive language. Fluent auditory-oral language has far-reaching benefits in life including reading ability (Paul 1998), academic achievement (Goldgar and Osberger 1986), and career development. Development of language with the 3M single-channel implant was discussed by Kirk and Hill-Brown (1985). Initial reports on receptive language development with the Nucleus 22 (F0/F1/F2 and Multipeak) systems on small groups of children were made by Busby et al (1989), Dowell et al (1991), Geers and Moog (1991), Hasenstab and Tobey (1991). The Peabody Picture Vocabulary Test (PPVT) (Dunn and Dunn 1981, 1997) was used, and it enabled the child's score to be referenced to that of a normal-hearing child to determine the equivalent age. The change over time in equivalent age divided by the change in chronological age measured the rate of language learning. In the studies referred to above, the rate varied from 0.6 to 2.6 (the normal is 1.0). For comparison, the rate for larger groups of profoundly deaf children with hearing aids varied from 0.4 to 0.6 (Geers and Moog 1988; Boothroyd et al 1991). The study by Geers and Moog (1991) used control groups of matched children using the multiple-channel cochlear implant (Nucleus-22) conventional aids, and a two-channel vibrotactile aid (Tactaid II), and found language learning was faster for the multiple-channel implant. It thus appeared that vocabulary acquisition for profoundly deaf children with multiple-channel cochlear implants was faster than for comparable children with hearing aids. To help confirm this trend, an analysis was undertaken on larger group of children ($n = 32$) (Dawson et al 1995a,b). They had a learning rate of 0.87, which was again higher than for children with hearing aids.

To try and isolate maturational effects from those due to the cochlear implant, Robbins et al (1995) used language quotients to compare predicted language scores on the scales of Reynell (Reynell and Gruber 1990) with those from children using the Nucleus 22 implant. After 15 months, the scores for receptive language were 10 months higher than expected and for expressive language 8 months better. In a variant of this procedure, Robbins et al (1997) showed the language levels at 1 year of age were 7 months ahead of that predicted for maturation. Nevertheless, the absolute levels were delayed compared to the normal. In a study on 23 young children with the Clarion CIS strategy, they were also found to have a greater than normal increase in language (Robbins et al 1999).

The data from a range of studies have shown that in spite of marked improvements in language, implant children's absolute levels remained delayed compared with normal controls. This was demonstrated in a 4-year longitudinal study by Blamey et al (1998) and Sarant et al (1996, 2001) on 57 children with a bilateral severe or profound hearing loss who attended auditory-oral deaf schools or preschools. There were 33 hearing aid and 24 implant users. The BKB speech perception results for audition alone (A), and audition and speech reading (AV) (Fig. 11.11), as well as the PPVT) (Dunn and Dunn 1981, 1997) and clinical evaluation of language fundamentals (CELF) (Wiig et al 1992; Semel et al 1995) (Fig. 11.12) were recorded. The PPVT is suitable for children from 2 years and up, and the CELF designed for children over 6 years. The receptive language results were

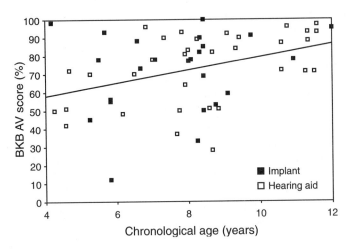

FIGURE 11.11. Speech perception versus chronological age for children with cochlear implants ($n = 24$) and hearing aids ($n = 33$). Speech perception was measured with BKB sentences for audition and speech reading. For the implanted children, the mean preoperative loss in the better ear was greater than 100 dB Hz (for 500, 1000, and 2000 Hz). For the hearing aid children the mean loss was 81 dB (Blamey et al 1998; Sarant et al 2001). (Reprinted with permission from Blamey et al 1998. Speech perception and spoken language in children with impaired hearing. In: Mannell, R. H. and J. Robert-Ribes, eds. ICSLP '98 Proceedings: 2615–2618.)

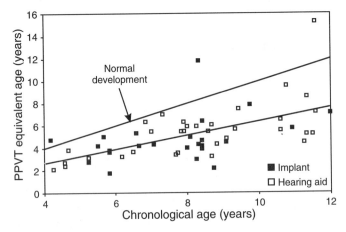

FIGURE 11.12. The language equivalent age of implanted and aided children versus chronological age. Language was assessed with the PPVT (Sarant et al 1996, 2001; Blamey et al 1998). (Reprinted with permission from Blamey et al 1998. Speech perception and spoken language in children with impaired hearing. In: Mannell, R. H. and J. Robert-Ribes, eds. ICSLP '98 Proceedings: 2615–2618.)

slightly better for the aided children, but this was probably due to the fact that they on average were older. For the hearing-aid children, the mean hearing threshold was 81 dB. As shown in Figure 11.12, receptive language increased gradually versus chronological age, but did not keep up with the increase seen with normal-hearing children. Similarly, with speech perception there was only a small increase with chronological age.

In contrast, if the speech perception word and sentence scores for audition and speech reading were plotted against the PPVT or CELF equivalent language ages, a very close relationship was seen, and the AV word score reached 100% at a PPVT or CELF age of approximately 8 to 10 years. However, for audition alone the slope was less steep and a 100% score is reached later, at 10 to 11 years.

A study was then undertaken to see if the perception of speech was influenced by vocabulary and whether remediation of vocabulary and syntax would increase open-set speech perception scores (Sarant et al 1996). Two out of three children had significant improvements in the perception of unknown words when their meaning was learned, and in sentences scores after training in grammar. This is discussed in Chapter 12.

These results indicate the important relationship between speech perception and language. They suggest that not only does better speech perception result in better language, but also improving language will affect speech perception. Thus specific help with language should help children to develop better speech perception.

Education of Children

Children who receive cochlear implants should learn language in the most natural surroundings that emphasize maximum exposure to audition. This is achieved in the home and at a mainstream school. If this is not possible, the educational management should be modified by including one-to-one sessions with specialists at the mainstream school, or by providing more constant attention in an educational program specifically designed for hearing-impaired children.

In Melbourne approximately 50% of the children suitable for a cochlear implant are implanted. Of these 20% go to a mainstream school without any help from a specialist unit, 60% attend a mainstream with a special unit, and 20% are taught in a special school. It is anticipated that with children having operations at younger ages and with better language assistance, 60% of children will be able to attend a mainstream school without a special unit in the next 5 to 10 years (R.C. Dowell, personal communication).

Acoustic Environment

It is important for implanted children to be able to listen effectively in the classroom. Berg (1987) found students spent 45% of their time listening, 30% speaking, and 25% reading and writing. The acoustic environment therefore needs to facilitate listening, but classroom noise levels were reported to be as high as 60 dB SPL, while the teacher's voice varied between 55 and 65 dB, depending on the recording position (Berg 1987). Rectifying this situation is essential for implanted children, as they have difficulty especially hearing in noise. For example, adult implant patients have shown significant degradation in performance at signal-to-noise ratios of 5 to 10 dB.

The first remedy is to seat the child at the front of the class, thus increasing the signal-to-noise ratio. To assist perception, the teacher must also face the child when speaking to the entire class, as there is a need to supplement hearing with speech reading. Methods for reducing the background noise should also be adopted. These include carpets and treating the wall and ceiling to absorb sound and reduce reverberation times. Furthermore, it is important to isolate the area from nearby noise, and use special equipment to reduce the signal-to-noise ratio such as FM and infrared microphone systems. FM microphone units worn by the teacher can create signal-to-noise ratios as high as 25 dB (Berg 1987). So with an FM system an improved signal-to-noise ratio can be achieved despite poor room acoustics, and it provides more flexibility for the teacher so that the students can be instructed from any distance in the classroom at normal speech levels.

Strategies

In a mainstream school, most implant children need emotional and educational support from the teacher. This includes educating the other students about hearing loss and cochlear implants, and providing classroom posters and cartoon books. Teachers should give opportunities for the child to associate with different students, so that they can also learn about the cochlear implant first hand. The child may also require individual speech and language assistance in a special unit, away from the general classroom. The mainstream teacher should be encouraged to implement some of the training requirements.

If special education is needed, there are a number of options available for the implant child. Oral programs, such as that recommended by Ling and Ling (1978), focus on spoken language. The oral method assumes that receptive language can be facilitated by producing it. In addition, auditory-only training can be carried out in special sessions. Aural programs, often referred to as unisensory, are based on the thesis that the best method of developing language is through hearing speech. Auditory-oral programs provide intensive auditory training, but focus both on language reception and speech production by requiring of children good listening skills and persistent speech production attempts. With the above programs, speech reading may be used, but any form of manual communication is discouraged. The assumption is that manual methods, such as signing and finger-spelling, will allow children to avoid the more difficult communication task of speaking (Mayberry et al 1987).

Total communication provides a signed language combined with the spoken version. The programs often underemphasize the auditory signal unintentionally. There are, however, two difficulties. The first is the proficiency of the teacher in presenting language through signing and speech at the same time (Rodda and Grove 1987). The second difficulty is that children tend to pay attention to the most easily received information, and listening through a distorted system may be more difficult than viewing a clear visual representation of the same message. It is recommended that if total communication is the method of choice, then oral communication and the development of oral skills should be emphasized (Osberger et al 1986). Cued speech represents a compromise between auditory-oral methods and total communication. Visual cues are used to facilitate the understanding of spoken language. The cues are hand symbols that make phonemes understandable when they look the same on the lips.

Visual-only programs use sign language of the deaf and incorporate fingerspelling. Sign language uses a different grammar from that of the spoken language, and it makes it more difficult to develop writing skills. A cochlear implant is unlikely to be of significant benefit to a visual-only speaker. Furthermore, sign language of the deaf does not emphasize the development of intelligible speech production.

Although there is evidence that in acquiring American sign language, early experience is critical for natural and complete learning (Morford and Mayberry 2000), it is important for a child to develop an auditory-oral language with a cochlear implant first, because the neural connectivity demands early exposure. Then as it is reported that a first language supports later learning of a second, this will help the child in becoming proficient in sign language of the deaf (Morford and Mayberry 2000).

Program for Implanted Children

There is good evidence that an auditory-oral education program produces better speech and language results than total communication. The data from the Melbourne clinic show that although there is a wide range of performance across modes of education, children with open-set scores of 50% or more are seen only in the auditory-oral group. But there is no evidence for the contention that the program placement may be a result of children's sensitivity to auditory information, and therefore bias the results. Furthermore, a number of children move from one type of program to another (Spencer 2000; Tomblin et al 1999), and the movement occurs in both directions (auditory/oral to signing programs and vice versa).

It has been postulated that as children require a great deal of exposure to auditory language after implantation, this exposure will be impaired if there are opportunities to focus predominately on visual language and if it is available with total communication. In this regard it has been reported that children from auditory-oral programs show faster gains in speech intelligibility than those in total communication programs (Geers et al 1999; Osberger et al 1994). The same applies for speech perception (Geers et al 1999; Hodges et al 1999) and language (Gary and Hughes 2000; Geers et al 1999; Osberger et al 1998). Furthermore, children with cochlear implants (regardless of education program) have developed speech perception, production, and spoken language skills at a faster rate than children with a comparable hearing loss but without an implant (Osberger et al

1998; Tomblin et al 1999; Svirsky et al 2000). Connor et al (2000) studied the influence of auditory-oral or total communication on the children's language growth, speech production, receptive vocabulary (oral), and expressive vocabulary (oral and/or signed). This study included 147 children with cochlear implants; 66 were in programs with total communication and 81 in oral programs (including auditory/verbal). The authors concluded that cochlear implants increased speech and language regardless of the modality of the language program. They pointed out, however, that the programs using total communication in their study also provided significant amounts of speech and oral language training. They also found that, as expected, outcomes were positively affected by having the complete electrode array inserted and some hearing sensitivity prior to implantation, and by using more advanced speech processing strategies (Spencer 2002).

Bilingual/bicultural programs have been put forward in which sign language of the deaf and print are combined. Results for a small number of children participating in such a program using Swedish sign language and Swedish (primarily print) were discouraging (Preisler et al 1997). The combination of spoken and written language did not combine in the "bottom-up" processing seen usually with bilingual language. This is also the case with sign language of the deaf and input from a cochlear implant.

Counseling of Adults and Children

Finally, in (re)habilitation it is important to assist the child, parents, and other caregivers in their understanding of the capabilities and limits of the implant. This begins during the selection process and should be well advanced by the time of implantation. To avoid disappointment and to maintain enthusiasm, it must be emphasized realistically that postoperatively the child will still be hearing impaired. With adults and children, the range in performance has been very large, and the best are able to use the telephone and listen to music. A small group of patients may not want to proceed with the implant, as they have been deaf a long time and have learned to cope without hearing, or they may be fearful of an operation. As the size of the cochlear implant speech processor is similar to that of a behind-the-ear hearing aid, they are usually aware that it is the technology they wish to use. However, time must be taken to explain in advance what will be required; in particular, they may feel the implant will make them different from other people.

The implant team needs to provide the child, parents, and teachers with practical information about device characteristics. This includes answers to these questions: How do the implant and speech processor work? What is the capture range and directionality of the microphone? What are the softest sounds that can be heard? Which sounds are very similar? Which sounds are different? Is the speech processor affected by background noise? What speech features does the processor provide? What does the processor do with nonspeech sounds? What does music sound like? They will also need to know how the device operates: How is the speech processor turned on and off? How are batteries changed? How long should batteries last? Can rechargeable batteries be used? How is the speech processor tested to see if it's working properly? What does the sensitivity control do? Can the speech processor be repaired? How can the speech processor be connected to a radiofrequency or infrared transmission system? How can the telephone signal be fed into the speech processor? How can the television sound be fed into the speech processor?

References

Abbas, P. J., C. J. Brown, J. K. Shallop, et al. 1999. Summary of results using the Nucleus CI24M implant to record the electrically evoked compound action potential. Ear and Hearing 20: 45–59.

Alcantara, J. I., R. S. C. Cowan, P. J. Blamey, L. A. Whitford and G. M. Clark. 1988. Evaluation of training strategies with an electrotactile speech processor. Australian Journal of Audiology (suppl 3): 7.

Alpiner, J. G. 1982. Handbook of adult rehabilitative audiology. Baltimore, Williams & Wilkins.

Arndt, P., S. Staller, J. Arcaroli, A. Hines and K. Ebinger 1999. Within-subject comparison of advanced coding strategies in the Nucleus 24 cochlear implant. Cochlear Corporation Report.

Barker, E., T. Daniels, R. Dowell, et al. 2000. Long term speech production outcomes in children who received cochlear implants before and after 2 years of age. 5th European Symposium on Paediatric Cochlear Implantation, Antwerp, Belgium: 156.

Bench, R. J. and J. Bamford. 1979. Speech-hearing tests and the spoken language of hearing-impaired children. London, Academic Press.

Berg, F. S. 1987. Facilitating classroom listening. San Diego, College-Hill Press.

Blamey, P. J., J. Barry and P. Jacq. 2001. Phonetic inventory development in young cochlear implant users 6 years postoperation. Journal of Speech, Language and Hearing Research 44: 73–79.

Blamey, P. J. and G. M. Clark. 1985. A wearable multiple-electrode electrotactile speech processor for the profoundly deaf. Journal of the Acoustical Society of America 77: 1619–1621.

Blamey, P. J., M. Grogan and M. B. Shields. 1994. Using an automatic word-tagger to analyse the spoken language of children with impaired hearing. In: Togneri, R., ed. Fifth Australian International Conference on Speech Science and Technology. Canberra, Australian Speech Science and Technology Association: 498–503.

Blamey, P. J., J. Z. Sarant, T. A. Serry, et al. 1998. Speech perception and spoken language in children with impaired hearing. In: Mannell, R. H. and J. Robert-Ribes, eds. ICSLP '98 proceedings. Canberra, Australian Speech Science and Technology Association: 2615–2618.

Boothroyd, A., A. E. Geers and J. S. Moog. 1991. Practical implications of cochlear implants in children. Ear and Hearing 12(suppl 4): 81S–89S.

Braida, L. D., N. I. Durlach, R. P. Lippmann, B. L. Hicks, W. M. Rabinowitz and C. M. Reed. 1979. Hearing aids–a review of past research on linear amplification, amplitude compression, and frequency lowering. ASHA Monographs 19: 1–114.

Brown, A. M., R. C. Dowell, L. F. Martin and D. J. Mecklenburg. 1990. Training of communication skills in implanted deaf adults. In: Clark, G. M., Y. C. Tong and J. F. Patrick, eds. Cochlear prostheses. United Kingdom, Churchill Livingstone: 181–192.

Brown, C. J., M. L. Hughes, B. Luk, P. J. Abbas, A. Wolaver and J. Gervais, 2000. The relationship between EAP and EABR thresholds and levels used to program the Nucleus 24 speech processor; data from adults. Ear and Hearing 21: 151–163.

Busby, P. A. and G. M. Clark. 2000a. Electrode discrimination by early-deafened subjects using the Cochlear Limited multiple-electrode cochlear implant. Ear and Hearing 21: 291–304.

Busby, P. A. and G. M. Clark. 2000b. Pitch estimation by early-deafened subjects using a multiple-electrode cochlear implant. Journal of the Acoustical Society of America 107: 547-558.

Busby, P. A., Y. C. Tong and G. M. Clark 1988. Underlying structure of auditory-visual consonant perception by hearing-impaired children and the influences of syllabic compression. Journal of Speech and Hearing Research 31: 156–165.

Busby, P. A., Y. C. Tong, S. A. Roberts, et al. 1989. Results for two children using a multiple-electrode intracochlear implant. Journal of the Acoustical Society of America 86(6): 2088–2102.

Busby, P. A., L. A. Whitford, P. J. Blamey, L. M. Richardson and G. M. Clark. 1994. Pitch perception for different modes of stimulation using the Cochlear multiple-electrode prosthesis. Journal of the Acoustical Society of America 95: 2658–2669.

Carter, P. M., A. R. Fisher, T. M. Nygard, et al. 1995. Monitoring the electrically-evoked compound action potential using a new telemetry system. Annals of Otology, Rhinology and Laryngology 104: 48–51.

Clark, G. M. 1986. The University of Melbourne/Cochlear Corporation (Nucleus) program. Otolaryngologic Clinics of North America 19: 329–354.

Clark, G. M. 2002. Learning to hear and the cochlear implant. In: Fahle, M. and T. Poggio, eds. Textbook of perceptual learning. Cambridge, MIT Press: 147–160.

Clark, G. M., Y. C. Tong, Q. R. Bailey, et al. 1978. A multiple-electrode cochlear implant. Journal of the Oto-Laryngological Society of Australia 4: 208–212.

Clark, G. M., Y. C. Tong and J. Gwyther. 1976. Speech perception and the development of language in deaf children. Proceedings of the 2nd Conference of the Audiological Society of Australia: 41–42.

Cohen, L. T., E. Saunders and G. M. Clark. 2001a. Psychophysics of a prototype perimodiolar cochlear implant electrode array. Hearing Research 155: 63–81.

Cohen, L. T., E. Saunders, B. K. Cone-Wesson and G. M. Clark. 1998. Spatial spread of neural excitation in cochlear implants: comparison of measurements made using NRT and forward masking. In: Barber, C., ed. Programme and Abstracts of First International Symposium and Workshop Objective Measures in Cochlear Implantation. Nottingham, UK 59.

Cohen, L. T., E. Saunders, M. R. Knight, R. S. C. Cowan and P. A. Busby. 2001b. Comparison of subjective loudness growth and NRT amplitude growth with stimulus current. 2nd international symposium and workshop on objective measures in cochlear implantation, Lyon, France.

Cole, E. and H. Gregory. 1986. Auditory learning. Washington, DC, AG Bell Association for the Deaf.

Connor, C. M., S. Hieber, H. A. Arts and T. A. Zwolan. 2000. Speech, vocabulary, and the education of children using cochlear implants: oral or total communication? Journal of Speech, Language, and Hearing Research 43: 1185–1204.

Cowan, R. S. C., E. J. Barker, P. Pegg, et al. 1998. Speech perception in implanted children:

influence of preoperative residual hearing on outcomes. Australian Journal of Audiology 20(suppl): 79.

Cowan, R. S. C., C. D. Brown, L. A. Whitford, et al. 1995. Speech perception in children using the advanced SPEAK speech-processing strategy. Annals of Otology, Rhinology and Laryngology 104(suppl 166): 318–321.

Crary, M. A. 1982. Phonological intervention concepts and procedures. San Diego, College-Hill Press.

Dawson, P. W., P. J. Blamey, S. J. Dettman, et al. 1995a. A clinical report on speech production of cochlear implant users. Ear and Hearing 16: 551–561.

Dawson, P. W., P. J. Blamey, L. C. Rowland, et al. 1995b. A clinical report on receptive vocabulary skills in cochlear implant users. Ear and Hearing 16: 287–294.

Dawson, P. W. and G. M. Clark 1997a. Changes in synthetic and natural vowel perception after specific training for congenitally deafened patients using a multichannel cochlear implant. In: Clark, G. M., ed. Cochlear implants. XVI World Congress of Otorhinolaryngology Head and Neck Surgery. Bologna, Monduzzi Editore: 281–285.

Dawson, P. W. and G. M. Clark. 1997b. Changes in synthetic and natural vowel perception after specific training for congenitally deafened patients using a Multichannel cochlear implant. Ear and Hearing 18: 488–501.

Dawson, P. W., L. Skok and G. M. Clark. 1997. The effect of loudness imbalance between electrodes in cochlear implant users. Ear and Hearing 18: 156–165.

De Fillipo, C. L. and B. L. Scott. 1978. A method for training and evaluating the reception of ongoing speech. Journal of the Acoustical Society of America 64: 1186–1192.

Dillier, N. 1998. Intracochlear recordings of electrically evoked compound action potentials. First international symposium and workshop on objective measures in cochlear implantation, Nottingham: 54.

Dillier, N., W. K. Lai, B. Almquist, et al. 2002. Measurement of the electrically evoked compound action potential (ECAP) via a neural response telemetry (NRT) system. Annals of Otology Rhinology and Laryngology 111(5 pt 1):407–414.

Dillier, N., W. K. Lai, D. Cafarelli-Dees and E. von Wallenberg. 2000. Post-operative neural response telemetry findings in adults submitted for publication. Neural esponse telemetry: results from a European field trial, 12th AAA Convention, Chicago: 137.

Dillon, H. 1996. Compression? Yes, but for low or high frequencies, for low or high intensities, and with what response times? Ear and Hearing 17: 287–307.

Dorman, M. F. and P. C. Loizou. 1997. Mechanisms of vowel recognition for Ineraid patients fit with continuous interleaved sampling processors. Journal of the Acoustical Society of America 102: 581–587.

Dowell, R. C., P. J. Blamey and G. M. Clark. 1995. Potential and limitations of cochlear implants in children. Annals of Otology, Rhinology and Laryngology 104(suppl 166): 324–327.

Dowell, R. C., P. J. Blamey and G. M. Clark. 1997. Factors affecting outcomes in children with cochlear implants. In: Clark, G. M., ed. Cochlear implants. XVI World Congress of Otorhinolaryngology Head and Neck Surgery. Bologna, Monduzzi Editore: 297–303.

Dowell, R. C. and R. S. C. Cowan. 1997. Evaluation of benefit: infants and children. In: Clark, G. M., R. S. C. Cowan and R. C. Dowell, eds. Cochlear implantation for infants and children—advances. San Diego, Singular Publishing Group.

Dowell, R. C., P. W. Dawson, S. J. Dettman, et al. 1991. Multichannel cochlear implantation in children. A summary of current work at the University of Melbourne. American Journal of Otology 12(suppl): 137–143.

Dowell, R. C., S. J. Dettman, K. Hill, E. Winton, E. J. Barker and G. M. Clark. 2002.

Speech perception outcomes in older children who use multichannel cochlear implants: older is not always poorer. Annals of Otology, Rhinology and Laryngology 111(suppl 189): 97–101.

Dowell, R. C., P. M. Seligman, P. J. Blamey and G. M. Clark. 1987. Speech perception using a two-formant 22-electrode cochlear prosthesis in quiet and in noise. Acta Oto-Laryngologica 104(5–6): 439–446.

Dowell, R. C., L. A. Whitford, P. M. Seligman, B. K.-H. Franz and G. M. Clark. 1990. Preliminary results with a miniature speech processor for the 22-electrode/Cochlear hearing prosthesis. In: Sacristan, T., ed. Otorhinolaryngology, head and neck surgery. Amsterdam, Kugler and Ghedini: 1167–1173.

Dunn, L. M. and L. M. Dunn. 1981. Peabody picture vocabulary test–revised. Circle Pines, Minnesota, American Guidance Service.

Dunn, L. M. and L. M. Dunn. 1997. Peabody picture vocabulary test, 3rd ed. Circle Pines, Minnesota, American Guidance Service.

Dyar, D. 1994. Monitoring progress: the role of a speech and language therapist. In: McCormick, B., S. Archbold and S. Shepphard, eds. Cochlear implants for young children. The Nottingham approach to assessment and rehabilitation. London, Whurr: 237–268.

Eddington, D. K. 1980. Speech discrimination in deaf subjects with cochlear implants. Journal of the Acoustical Society of America 68: 885–891.

Eddington, D. K., W. H. Dobelle and D. E. Brackmann. 1978. Auditory prostheses research with multiple channel intracochlear stimulation in man. Annals of Otology 87: 1–39.

Eggermont, J. J., C. W. Ponton, M. Don, M. D. Waring and B. Kwong. 1997. Maturational delays in cortical evoked potentials in cochlear implant users. Acta Otolaryngologica 117: 161–163.

Eisenberg, L. S., K. I. Berliner, M. A. Thielemeir, K. Iler Kirk and N. Tiber. 1983. Cochlear implants in children. Ear and Hearing 4: 41.

Erber, N. P. 1982. Auditory training. Washington, DC, Alexander Graham Bell Association for the Deaf.

Fisher, H. B. and J. A. Logemann. 1971. Test of articulation competence. New York, Houghton and Mifflin.

Fryauf-Bertschy, H., R. S. Tyler, D. M. Kelsay and B. J. Gantz. 1992. Performance over time of congenitally deaf and postlingually deafened children using a multichannel cochlear implant. Journal of Speech and Hearing Research 35: 913–920.

Fryauf-Bertschy, H., R. S. Tyler, D. M. Kelsay, B. J. Gantz and G. G. Woodworth. 1997. Cochlear implant use by prelingually deafened children: the influences of age at implant use and length of device use. Journal of Speech and Hearing Research 40: 183–199.

Fujita, S. and J. Ito. 1999. Ability of nucleus cochlear implantees to recognize music. Annals of Otology, Rhinology and Laryngology 108: 634–640.

Gantz, B. J., R. S. Tyler, G. Woodworth, N. Tye-Murray and H. Fryauf-Bertschy. 1994. Results of multichannel cochlear implant in congenital and acquired prelingual deafness in children: five-year follow-up. American Journal of Otology 15(suppl 2): 1–8.

Gary, L. and C. Hughes. 2000. A second look at "tweeners." Candidacy considerations for 8 to 14. Presented at International Cochlear Implant Conference. Miami, Florida.

Geers, A. E. and J. S. Moog. 1988. Predicting long-term benefits from single-channel cochlear implants in profoundly hearing-impaired children. American Journal of Otology 9: 169–176.

Geers, A. E. and J. S. Moog. 1991. Evaluating the benefits of cochlear implants in an education setting. American Journal of Otology 12(suppl): 116–125.

Geers, A., J. Nicholas, N. Tye Murray, et al. 1999. Cochlear implants and education of the deaf child: second-year results. Central Institute for the Deaf. Research progress report—deaf education (cochlear implants). (http://www.cid.wustl.edu/index).

Goldgar, D. and M. J. Osberger. 1986. Language and learning skills of hearing-impaired students. Factors related to academic achievement. ASHA Monographs 23: 87–91.

Hall, R. D. 1990. Estimation of surviving spiral ganglion cells in the deaf rat using electrically evoked auditory brainstem response. Hearing Research 45: 123–136.

Hardie, N. A., A. Martsi-McClintock, L. M. Aitkin and R. K. Shepherd. 1999. The effect of a profound sensorineural hearing loss on the development of synapses in the auditory midbrain. British Society of Audiology News 27: 20–21.

Hasenstab, M. S. and E. M. Tobey. 1991. Language development in children receiving Nucleus multi-channel cochlear implants. Ear and Hearing 12: 55S–65S.

Heller, J., N. Dillier and P. J. Abbas. 1996. Neural response telemetry—a new clinical research tool. Third Symposium on Pediatric Cochlear Implantation, Hannover: 108.

Hodges, A. V., M. Ash, T. J. Balkany, J. J. Schloffman and S. L. Butts. 1999. Speech perception results in children with cochlear implants: contributing factors. Otolaryngology–Head and Neck Surgery 121: 31–34.

Hollow, R., L. Winton, K. Hill, R. Dowell and G. Clark. 2002. Validation of a technique for establishing maximum comfortable levels for children using cochlear implants. Australian and New Zealand Journal of Audiology 23(2): 105.

Hughes, M. L., C. J. Brown, P. J. Abbas, A. A. Wolaver and J. P. Gervais. 2000a. Comparison of EAP thresholds with MAP levels in the Nucleus 24 cochlear implant: data from children. Ear and Hearing 21: 164–174.

Hughes, M. L., C. J. Brown, P. J. Abbas, A. A. Wolaver and J. P. Gervais. 2000b. The relationship between EAP and EABR thresholds and levels used to program the Nucleus 24 speech processor: data from adults. Ear and Hearing 21: 151–163.

Ingram, D. 1976. Phonological disability in children. New York, Elsevier.

James, C. J., P. J. Blamey, L. F. A. Martin, B. A. Swanson, Y. Just and D. MacFarlane. 2002. ADRO (Adaptive dynamic range optimization) for cochlear implants: a preliminary study. Ear and Hearing 23(1 suppl): 49S–58S.

Jerger, S., S. Lewis and J. Jerger. 1980. Paediatric speech intelligibility test 1. Generation of speech materials. International Journal of Paediatric Otolaryngology 2: 217–230.

Johnson, J. and E. Newport. 1989. Critical period effects in second-language learning: the influence of maturational state on the acquisition of English as a second language. Cognitive Psychology 21: 60–99.

Kaczmarek, L., M. Kossut and J. Skangiel-Kramska. 1997. Glutamate receptors in cortical plasticity: molecular and cellular biology. Physiology Review 77: 217–255.

Kawano, A. S., H. L. Seldon, G. M. Clark, R. Madsen and C. Raine. 1995. Intracochlear factors contributing to psychophysical percepts following cochlear implantation: a case study. Annals of Otology, Rhinology and Laryngology 104: 54–57.

Kawano, A., H. L. Seldon, G. M. Clark and R. Ramsden. 1998. Intracochlear factors contributing to psychophysical percepts following cochlear implantation. Acta Oto-Laryngologica 118: 313–326.

Kirk, K. I. 2000. Challenges in the clinical investigation of cochlear implant outcomes. In: Niparko, J. K., K. I. Kirk, N. K. Mellon, et al, eds. Cochlear implants: principles and practices. Philadelphia, Lippincott Williams & Wilkins: 225–259.

Kirk, K. I. and C. Hill-Brown. 1985. Speech and language results in children with a cochlear implant. Ear and Hearing 6(suppl): 36S–47S.

Kral, A., R. Hartmann, J. Tillein, S. Heid and R. Klinke. 2001. Delayed maturation and sensitive periods in the auditory cortex. Audiology and Neuro-Otology 6: 346–362.

Kretschmer, R. and I. Kretschmer. 1994. Discourse and hearing impairment. In: Ripich, D. and N. Creaghead, eds. School discourse problems. San Diego, Singular: 263–296.

Kubo, T. 2002. Auditory and visual information processing in cerebral cortical areas of cochlear implant users-PET activation study. Australian and New Zealand Journal of Audiology 23(2): 112.

Lai, W. K. 1999. Guidelines for making NRT measurements. Cochlear AG, Basel, Switzerland.

Lesner, S. A., S. A. Sandridge and P. B. Kricos. 1987. Training and influences on visual consonant and sentence recognition. Ear and Hearing 8: 283–287.

Levitt, H., N. McGarr and D. Geffner. 1987. Development of language and communication skills in hearing-impaired children. Introduction. ASHA Monographs 26: 1–8.

Ling, D. 1976. Speech and the hearing impaired child: theory and practice. Washington, DC, AG Bell Association for the Deaf.

Ling, D. 1984. Early intervention for hearing-impaired children: oral options. San Diego, College-Hill Press.

Ling, D. 1989. Foundations of spoken language for hearing impaired children. Washington, DC, AG Bell.

Ling, D. and A. H. Ling. 1978. Aural habilitation: the foundations of verbal learning in hearing-impaired children. Washington, DC, AG Bell Association for the Deaf.

Ling, D. and T. G. Nienhuys. 1983. The deaf child: habilitation with and without a cochlear implant. Annals of Otology, Rhinology and Laryngology 92: 593–598.

Martin, L. F. A., Y. C. Tong and G. M. Clark. 1981. A multiple-channel cochlear implant. Evaluation using speech tracking. Archives of Otolaryngology 107: 157–159.

Mayberry, R., R. Wodlinger-Cohen and S. Goldin-Meadow. 1987. Symbolic development in deaf children. New Directions for Child Development 36: 109–126.

McCarthy, P. A. and J. G. Alpiner. 1982. The remediation process. In: Alpiner, J. G., ed. Handbook of adult rehabilitative audiology. Baltimore, Williams & Wilkins.

McDermott, H. J., K. R. Henshall and C. M. McKay. 2002. Benefits of syllabic input compression for users of cochlear implants. Journal of the American Academy of Audiology 13(1): 14–24.

McDermott, H. J., C. M. McKay and A. Vandali. 1992. A new portable sound processor for the University of Melbourne/Nucleus Limited multi-electrode cochlear implant. Journal of the Acoustical Society of America 91: 3367–3371.

McGarr, N. S. 1983. The intelligibility of deaf speech to experienced and inexperienced listeners. Journal of Speech and Hearing Research 26: 451–458.

McKay, C. M., H. J. McDermott, A. Vandali and G. M. Clark. 1992. A comparison of speech perception of cochlear implantees using the Spectral Maxima Sound Processor (SMSP) and the MSP (Multipeak) processor. Acta Oto-Laryngologica 112: 752–761.

McKay, C. M., M. D. Remine and H. J. McDermott. 2001. Loudness summation for pulsatile electrical stimulation of the cochlea: effects of rate, electrode separation, level, and mode of stimulation. Journal of the Acoustical Society of America 110: 1514–1524.

Mecklenburg, D. J., P. J. Blamey, P. A. Busby, R. C. Dowell, S. Roberts and F. W. Rickards. 1990. Auditory (re)habilitation for implanted deaf children and teenagers. In: Clark, G., Y. Tong and J. Patrick, eds. Cochlear prostheses. United Kingdom, Churchill Livingstone: 207–221.

Mecklenburg, D. J., R. C. Dowell and V. Jenison. 1987. Rehabilitation manual. Englewood, Colorado, Cochlear Corp.

Merzenich, M., C. Byers and M. White 1984. Scala tympani electrode arrays. Fifth quarterly progress report. NIH contract NO1-NS9-2353: 1–11.

Miyamoto, R. T., K. I. Kirk, A. M. Robbins, S. Todd and A. Riley. 1996. Speech perception and speech production skills of children with multichannel cochlear implants. Acta Otolaryngologica 116: 240–243.

Miyamoto, R. T., K. I. Kirk, A. M. Robbins, S. Todd, A. Riley and D. B. Pisoni. 1997. Speech perception and speech intelligibility in children with multichannel cochlear implants. Advances in Oto-Rhino-Laryngology 52: 198–203.

Moore, D. R. 1990. Auditory brainstem of the ferret: early cessation of developmental sensitivity of neurons in the cochlear nucleus to removal of the cochlea. Journal of Comparative Neurology 302: 810–823.

Moore, D. R. and N. E. Kowalchuk. 1988. Auditory brainstem of the ferret: effects of unilateral cochlear lesions on cochlear nucleus volume and projections to the inferior colliculus. Journal of Comparative Neurology 272: 503–515.

Morford, J. and R. Mayberry. 2000. A re-examination of "early exposure" and its implications for language acquisition by eye. In: Chamberlain, C., J. Morford and R. Mayberry, eds. Language acquisition by eye. Mahwah, NJ, Lawrence Erlbaum Associates: 111–1127.

Murray, B., B. K. Cone-Wesson, R. C. Dowell and L. A. Whitford. 2000a. Perceptual threshold as a function of stimulus rate and its correspondence with electrically evoked 8th nerve action potentials, auditory brainstem and cortical responses. Australian Journal of Audiology 22(suppl): 65.

Murray, B., M. Knight, M. Davies, et al. 2000b. Investigation of an NRT based fitting procedure using high rates of stimulation—a multicentre investigation. Australian Journal of Audiology 22(suppl): 46.

Nelson, D. A., D. J. Van Tasell, A. C. Schroder, S. Soli and S. Levine. 1995. Electrode ranking of "place pitch" and speech recognition in electrical hearing. Journal of the Acoustical Society of America 98: 1987–1999.

Nikolopoulos, T., G. O'Donoghue and S. Archbold. 1999. Age at implantation: its importance in pediatric cochlear implantation. Laryngoscope 109: 595–599.

Nordeen, K. W., H. P. Killackey and L. M. Kitzes. 1983. Ascending projections to the inferior colliculus following unilateral cochlear ablation in the neonatal gerbil, Meriones unguiculatus. Journal of Comparative Neurology 214: 144–153.

O'Donoghue, G. M., T. P. Nikolopoulos and S. M. Archbold. 2000. Determinants of speech perception in children after cochlear implantation. Lancet 356: 466–468.

Osberger, M. J., A. L. Beiter, V. Jenison and J. Moog. 1986. Auditory skill development in children with cochlear implants. Cochlear implants in children: proceedings from a multidisciplinary colloquium. In: Mecklenberg, D. J., ed. Seminars in hearing. New York, Thieme 7: 423–431.

Osberger, M., L. Fisher, L. Zimmerman-Phillips and M. Barker. 1998. Speech recognition performance of older children with cochlear implants. American Journal of Otology 19: 152–175.

Osberger, M. J., M. P. Moeller, M. Eccarius, A. M. Robbins and D. Johnson. 1981. Language and learning skills of hearing-impaired students. Expressive language skills. ASHA Monographs 23: 54–65.

Osberger, M., A. Robbins, S. Todd and A. Riley. 1994. Speech intelligibility of children with cochlear implants. Volta Review 96: 169–180.

Osberger, M. J., A. M. Robbins, S. L. Todd, A. I. Riley, K. I. Kirk and A. E. Carney. 1996. Cochlear implants and tactile aids for children with profound hearing impairment. In: Bess, F., J. Gravel and A. M. Tharpe, eds. Amplification for children with auditory deficits. Nashville, Bill Wilkerson Center Press: 283–307.

Parker, A. and S. Irlam 1994. Intelligibility and deafness: the skills of listener and speaker. In: Wirz, S. L., ed. Perceptual approach to communication disorders. London, Whurr. 56–83.

Parving, A., B. Christensen, G. Salomon, C. B. Pedersen and L. Friberg. 1995. Regional cerebral activation during auditory stimulation in patients with cochlear implants. Archives of Otolaryngology–Head and Neck Surgery 121(4): 438–444.

Patowski, M. 1980. The sensitive period for the acquisition of syntax in a second language. Language Learning 30: 449–472.

Paul, P. V. 1998. Literacy and deafness: the development of reading, writing, and literate thought. Boston, Allyn and Bacon.

Power, D. J. and S. P. Quigley. 1973. Deaf children's acquisition of the passive voice. Journal of Speech and Hearing Research 16: 5–11.

Preisler, G., M. Ahlstrom and A. Tvingstedt. 1997. The development of communication and language in deaf preschool children with cochlear implants. International Journal of Pediatric Otorhinolaryngology 41: 263–271.

Pyman, B., P. Lacy, G. Clark and R. Dowell. 2000. The development of speech perception in children using cochlear implants: effects of etiologic factors and delayed milestones. American Journal of Otology 21: 57–61.

Rajan, R., D. R. F. Irvine, L. Z. Wise and P. Heil. 1993. Effect of unilateral partial cochlear lesions in adult cats on the representation of lesioned and unlesioned cochleas in primary auditory cortex. Journal of Comparative Neurology 338: 17–49.

Recanzone, G. H., C. E. Schreiner and M. M. Merzenich. 1993. Plasticity in the frequency representation of primary auditory cortex following discrimination training in adult owl monkeys. Journal of Neuroscience 13: 87–103.

Reynell, J. K. and C. P. Gruber. 1990. Reynell developmental language scales. Los Angeles, Western Psychological Services.

Robbins, A. M. 1994. Guidelines for developing oral communication skills in children with cochlear implants. Volta Review 96: 75–82.

Robbins, A. M., P. M. Bollard and J. Green. 1999. Language development in children implanted with the CLARION cochlear implant. Annals of Otology Rhinology and Laryngology 177: 113–118.

Robbins, A. M., M. J. Osberger, R. T. Miyamoto and K. S. Kessler. 1995. Language development in children with cochlear implants. Advances in Oto-Rhino-Laryngology 50: 160–166.

Robbins, A. M., M. A. Svirsky and K. I. Kirk. 1997. Children with implants can speak, but can they communicate? Otolaryngology Head and Neck Surgery 117: 155–160.

Robertson, D., G. K. Yates and I. M. Winter. 1989. Primary afferent dynamic ranges and cochlear mechanics. Boden Conference: 1–11.

Rodda, M. and C. Grove. 1987. Language, cognition and deafness. Hillsdale, NJ, Erlbaum.

Ross, M. and T. G. Giolas. 1978. Auditory management of hearing-impaired children: principles and prerequisities for intervention. Baltimore, University Park Press.

Rubinstein, A. and A. Boothroyd. 1987. Effect of two approaches to auditory training on speech recognition by hearing-impaired adults. Journal of Speech and Hearing Research 30: 153–160.

Sanders, D. A. 1982. Aural rehabilitation: a management model. Englewood Cliffs, New Jersey, Prentice Hall.

Sarant, J. Z., P. J. Blamey and G. M. Clark. 1996. The effect of language knowledge on speech perception in children with impaired hearing. In: McCormack, P. and A. Russell, eds. Proceedings of the Sixth Australian International Conference on Speech Science and Technology Canberra, Australian Speech Science and Technology Association: 269–274.

Sarant, J. Z., P. J. Blamey, R. C. Dowell, G. M. Clark and W. P. R. Gibson. 2001. Variation in speech perception scores among children with cochlear implants. Ear and Hearing 22: 18–28.

Semel, E., E. Wiig and W. A. Secord. 1995. Clinical evaluation of language fundamentals, 3rd edition. San Antonio, TX, Psychological Corporation, Harcourt Brace.

Shriberg, L. D. and J. Kwiatkowski. 1982. Phonological disorders III: a procedure for assessing severity of involvement. Journal of Speech and Hearing Disorders 47: 256–270.

Sims, D. G., G. G. Walter and R. L. Whitehead. 1982. Deafness and communication: assessment and training. Baltimore, Williams & Wilkins.

Skinner, M. W., G. M. Clark, L. A. Whitford, et al. 1994. Evaluation of a new spectral peak coding strategy for the Nucleus 22 channels cochlear implant system. American Journal of Otology 15: 15–27.

Snyder, R. L., S. Rebscher and R. Beitel. 1995. Temporal resolution of neurons in cat inferior colliculus to intracochlear electrical stimulation: effects of neonatal deafening and chronic stimulation. Journal of Neurophysiology 73: 449–467.

Spencer, P. 2000. "She's still deaf, but now she can talk". 15th Annual Conference on Issues in Language and Deafness. Omaha, Nebraska, Boys Town National Research Hospital.

Spencer, P. 2002. Language development of children with cochlear implants. In: Christiansen, J., ed. Cochlear implants in children. Washington, DC, Gallaudet University Press.

Stevens, S. S. 1975. Psychophysics. New York, John Wiley.

Svirsky, M., A. Robbins, K. Kirk, D. Pisoni and R. Miyamoto. 2000. Language development in profoundly deaf children with cochlear implants. Psychological Science 11: 1–6.

Tomblin, J. B., L. Spencer, S. Flock, R. Tyler and B. Gantz. 1999. A comparison of language achievement in children with cochlear implants and children using hearing aids. Journal of Speech, Language, and Hearing Research 42: 497–511.

Tong, Y. C., R. C. Black, G. M. Clark, et al. 1979. A preliminary report on a multiple-channel cochlear implant operation. Journal of Laryngology and Otology 93: 679–695.

Tong, Y. C. and G. M. Clark. 1985. Absolute identification of electric pulse rates and electrode positions by cochlear implant patients. Journal of the Acoustical Society of America 77: 1881–1888.

Tong, Y. C. and G. M. Clark. 1986. Loudness summation, masking, and temporal interaction for sensations by electric stimulation of two sites in the human cochlea. Journal of the Acoustical Society of America 79: 1958–1966.

Tong, Y. C., G. M. Clark, P. M. Seligman and J. F. Patrick. 1980. Speech processing for a multiple-electrode cochlear implant hearing prosthesis. Journal of the Acoustical Society of America 68: 1897–1899.

Tye-Murray, N., L. Spencer and G. Woodworth. 1995. Acquisition of speech by children who have prolonged cochlear implant experience. Journal of Speech and Hearing Research 38: 327–337.

Walden, B. E., S. A. Erdman, A. A. Montgomery, D. M. Schwartz and R. A. Prosek. 1981. Some effects of training on speech recognition by hearing-impaired adults. Journal of Speech and Hearing Research 24: 207–216.

Walden, B. E., R. A. Prosek, A. A. Montgomery, C. K. Scherr and C. J. Jones. 1977. Effects of training on the visual recognition of consonants. Journal of Speech and Hearing Research 20: 130–451.

Walker, G. and H. Dillon. 1982. Compression in hearing aids: an analysis, a review, and some recommendations. National Acoustic Laboratories Report 90. Canberra, Australian Government Publishing Service.

Waltzman, S. and N. Cohen. 1998. Cochlear implants in children younger than 2 years old. American Journal of Otology 19: 158–162.

Wiig, E., W. A. Secord and E. Semel. 1992. Clinical evaluation of language fundamentals-preschool. San Antonio, TX, Psychological Corporation, Harcourt Brace.

Wilson, B. S., C. C. Finley, D. T. Lawson, R. D. Wolford, D. K. Eddington and W. M. Rabinowitz. 1991. Better speech recognition with cochlear implants. Nature 352(6332): 236–238.

Wilson, B. S., D. T. Lawson, M. Zerbi and C. C. Finley. 1992. Speech processors for auditory prostheses. Twelfth quarterly progress report, April 1992. NIH contract N01-DC-9-2401. Research Triangle Institute.

Zeng, F.-G. and R. V. Shannon. 1992. Loudness balance between electric and acoustic stimulation. Hearing Research 60(2): 231–235.

Zeng, F. G. and R. V. Shannon. 1994. Loudness-coding mechanisms inferred from electric stimulation of the human auditory system. Science 264(5158): 564–566.

Zeng, F.-G. and R. V. Shannon. 1995. Loudness of simple and complex stimuli in electric hearing. Annals of Otology, Rhinology and Laryngology 104(suppl 166): 235–238.

12
Results

Aims

The results for cochlear implants in the 1970s and 1980s from different centers were difficult to assess, as the test material was not always standardized, and the interpretation of open-set testing varied. Since that time, established audiological assessment has been applied, and comparative, controlled studies within and across centers have enabled conclusions to be drawn about speech-processing strategies. This chapter covers the assessment procedures and results for different cochlear implant systems. Essentially the purpose of the implant is to enable the deaf child or adult to communicate with others who have hearing, and to experience sounds in the surrounding environment. With communication, the perception and production of speech are essential for adults and children, as is the development of receptive and expressive language for children.

Development of Tests

Prior to the advent of cochlear implants there was little to help profoundly hearing-impaired people, and thus a paucity of speech testing material for this group. For other people with significant residual hearing who used hearing aids in the 1970s and 1980s, the most common test was the phonemically balanced (PB) monosyllabic word test. This test had become standardized, but most profoundly deaf people were are not able to score on the test even when they had some useful speech discrimination. These people had a range of minimal speech recognition skills not assessed by standard speech tests (Dowell et al 1990a).

When single-electrode cochlear implants were first used clinically in the 1970s, pre- and postoperative evaluation was done partly through psychological assessment. Appropriate test batteries for evaluating the perception of speech components with these implants had not been formalized. It was not clear how the recognition of some speech components would contribute to communication skills and the quality of life. Multiple-channel cochlear implants provided more speech information, and the standard open-set word and sentence tests could be used to variable degrees. However, additional speech test material such as closed sets of consonants and vowels were required to gain as much information as possible on improvements in communication. With advances in cochlear implant speech processing, the results became comparable to those for a person with a severe hearing loss when using a hearing aid, and thus traditional tests have became more the routine.

More specialized tests are required to study the advanced strategies with a view to further improvements. For example, in a study on alternative speech-processing strategies, it was important to determine if there were different patterns of phoneme errors (Grayden 2000; Grayden and Clark 2000;). In this study as outlined in Chapters 7 and 14, the phonemes were first divided into their distinctive feature categories (Miller and Nicely 1955; Chomsky and Halle 1968; Singh 1968). These categories were nasal, continuant, voicing, sibilant, duration, anterior, coronal, high, back, and distributed.

The factors responsible for predicting good speech perception needed to be determined to advise prospective patients and parents on likely outcomes. This too is difficult because of the wide range of performance among implant users. As knowledge of predictive factors increases, patient selection criteria will be modified and counseling can be more appropriate for each individual. Potential implant candidates will be able to better understand their options, and have more appropriate expectations. This will be particularly valuable for those with some residual hearing who will need a clear knowledge of their expected performance with the implant to make an informed decision before proceeding with surgery. Another requirement is to demonstrate cost-effectiveness.

Speech and Sound Perception: Test Principles

Test procedures should ultimately assess a person's ability to communicate in everyday situations, and this applies to speech and language. Speech is the most complex, specialized, and important auditory signal, and it is part of the receptive language process. Its recognition through cochlear implant speech-processing strategies took precedence over the recognition of environmental sounds and music. Nevertheless, the ability to identify environmental sounds is important. People get great enjoyment from hearing birds singing, dogs barking, frogs croaking, rain falling, the phone ringing, the kettle boiling, people coughing, and so on. However, the overall benefit of a cochlear prosthesis is best quantified by speech perception and speech production, and language in children.

Variability of Materials and Responses

With speech perception tests there is inherent variability in the materials (Dowell, Martin, Blamey et al 1985a). The performance with tests can vary even during the one session depending on alertness, motivation, and fatigue. If a particular speech test has expected variability when performed in the auditory and speech-reading alone tasks, then there will be an additive variability for the auditory-visual task, and this should be taken into consideration when interpreting the results from the combined condition.

Additional sources of variability, and perhaps bias, in speech testing can arise from the particular conditions used for testing patients. Such variables include the method of presentation (live voice versus recorded); location (sound environment, i.e., treated versus normal room); speaker (familiar versus unfamiliar); the number of times an item is repeated (if at all); and training.

Considerable variation in speech perception has been seen across patients for all multiple-channel speech-processing strategies. The Nucleus F0/F2 WSP-II (wearable speech processor) system produced open-set Central Institute for the Deaf (CID) sentence scores for electrical stimulation that varied from 4% to 86% in 23 patients 12 months postoperatively (Dowell et al 1986). With the Multipeak miniature speech processor (MSP) system consonant-nucleus-consonant (CNC) words scored as phonemes correct in 39 subjects varied from 23% to 80%, while scores with the SPEAK Spectra-22 varied from 27% to 83% (Skinner et al 1994).

In comparing speech-processing strategies in postlinguistically deaf adults, it is important that a period of up to 12 weeks be used in optimizing each strategy. After this time the subjects use each of the strategies for a 2- to 4-week period and are then tested with the strategies to be compared (Skinner et al 1994, 2002; Arndt et al 1999). It is important when a person has used one strategy for some time and is then changed to another to carry out an A/B/A/B comparison to minimize variability. By returning to the original strategy, training and learning effects can be determined.

Prerecorded Versus Live Voice

The live-voice presentation of speech material must be undertaken by an audiologist experienced in controlling the intensity level of his/her voice. The material is presented at 70 to 75 dBA and monitored with a sound level meter on the A weighting to reduce the low frequencies, and so remove the low-level background noise that interferes with the signal. This approach is not free from subjective bias, and makes comparison between subjects and clinics difficult. The results are usually better than for prerecorded material. This is evident in the initial Arthur Boothroyd (AB) (Boothroyd 1968) word scores for the first two patients to receive the University of Melbourne's prototype system. The average score for open sets of words (scored as words) was 8% by live voice, and 5% using prerecorded test materials (Clark, Tong, et al 1981c). Similarly, CID word-in-sentence scores (scored as key words) were 35% for live voice (Davis and Silverman 1978), and 11% when prerecorded (Clark, Tong, et al 1981b).

The presentation of prerecorded material is essential for the comparison of results either within or across centers. The material is delivered at peak sound pressure level of 70 dBA in a sound-treated room with the subjects 1 m from the loudspeaker and using their normal headset microphone.

Training Effects and Experience

Training can have a large effect on test performance. Thus with open-set testing, the material presented must not have been used for training. For instance, a set of sentences can become familiar because of different lengths and syllabic structures. Thus when tested the patient is capable of speech recognition assisted by identifying suprasegmental features (Dowell et al 1990a).

Experience in the use of a strategy has a major influence on speech perception. Substantial improvements are evident in speech scores for implant users, particularly in the first few months after surgery, but continuing for some years (Dowell et al 1986). In postlinguistically deaf adults the improvements were more gradual when the strategy provided less information, for example, the Nucleus second formant (F2)/fundamental frequency (F0) compared with the SPEAK or spectral maxima strategy. As shown in Chapter 11, the improvements in scores for SPEAK were much more rapid. Furthermore, with children, as shown in the same chapter, improvements occurred over 5 years or longer. It was only for children operated on when they were older than 5 years that after 5 years the improvement reached a plateau.

In assessing the effects of experience, it should be noted that there are variations in the amount of time different patients actually use the implant. In particular, there are large variations in the amount of interactive communication that they are exposed to. Their experience may be more affected by their own social and vocational situation than by actual time since surgery (Dowell et al 1990a). When comparing a new strategy and speech processor with an earlier system, it is important that the new device be of similar size so that it will be used regularly by the subject to obtain a comparable amount of experience at home. Results obtained only in the laboratory are difficult to interpret.

In summary, with studies to compare strategies, it is necessary to control for training and learning effects. When changing from one strategy to another a plateau in performance with the first strategy (A) should be reached, the subject should have time to get used to the new scheme (B), and then the comparison should be with an A/B/A/B protocol to help exclude learning effects with the first strategy (A). If three strategies are being compared, an A/B/C/C/B/A protocol can be used, with 4 weeks take-home experience for each iteration. In this case the initial strategy (A) is subjected to the maximum learning effects from the newer ones, B and C.

Closed-Set Tests

Closed sets of speech materials (they can be consonants, vowels, and words) are easier to identify than open sets, but the relationship between closed-set scores and speech perception in daily life is less well established than for open sets. Discrimination of phonemes and words or the closed-set recognition of words does not correspond well with communication ability due to the rate at which the auditory signal must be processed in a conversation. With the initial evaluations of patients, closed sets of materials are important in training them with the new signals. They are also helpful in analyzing speech components perceived with different speech-processing strategies.

With closed sets there are a restricted number of items for discrimination and recognition rather than requiring outright recognition. The tests can be designed to investigate specific speech features, and the difficulty of the test can be well controlled. A simple task may involve identifying a word from a set of two alternatives differing in duration and rhythm (e.g., truck/toothbrush). This type of test could be used to assess the ability to discriminate gross suprasegmental features of speech. The test can be made more difficult by including, say, four alternatives (e.g., truck/table/toothbrush/butterfly).

To assess the discrimination of segmental features at a finer level, a two-alternative format can again be used, for instance to test the discrimination of vowels (e.g., beat/boot) or consonants (e.g., beat/meat). More difficulty can be introduced with additional alternatives to produce a four- (or more) alternative test (e.g., beat/boot/bert/bought) for vowels, or beat/meat/seat/cheat for consonant manner, or (beer/dear/peer/tear) for consonant place and voicing (Dowell et al 1990a). With the more advanced speech-processing strategies in postlinguistically deaf adults, all the consonants and vowels in the language can be tested. Thus a test battery can be constructed of the 16 Australian English consonants in an /aCa/ (C-consonant) context, and the 11 vowels in an /hVd/ (V-vowel) context.

For a closed-set test with only two alternatives, the task for the subjects is one of discrimination. It is assumed that if subjects can tell the difference between the two items, they will score well on such a test. However, this is not always the case if the subjects' perception of the items is very different from their expectations. In other words, implanted subjects will not always perform as expected on tests when they are unfamiliar with the material. It may be necessary to train them in order to make them familiar with the test items. This is particularly necessary when an implant patient is in the early stages of rehabilitation.

As the number of alternatives is increased in a closed set, the task is no longer one of discrimination, but rather recognition, where memory becomes more important. If a subject can discriminate between the components of a closed set of eight or 10 alternatives, but not retain the memory for the presented item to compare it with all of the options, the subject will perform poorly. It is preferable to use a set of no more than five alternatives for formal closed-set tests to reduce the influence of memory on results. If larger closed-sets are used, some additional training will be necessary for most to become familiar with the test material (Dowell et al 1990a).

In analyzing closed-set results, it is important to know the expected chance score. If the number of items in the test is N and the number of alternatives for each item is A, the expected chance score is N/A items correct for random presentations. It is also important to determine whether the score is significantly above the chance level at the 95% confidence level. This depends on the number of items and the variability of the results (Dowell et al 1990a).

With children the closed-set tests need to be age and language appropriate. The

Monosyllables, Spondees, Trochees, and Polysyllables Test (MSTP) consists of three monosyllables, three spondees, three trochees, and three polysyllables. It is a word identification and classification task. It is based on a test developed by Erber (1972) and Erber and Alencewicz (1976), and is appropriate for children over 4 years of age. Its significance can be assessed from a binomial distribution of test scores (Thornton and Raffin 1978; Raffin and Thornton 1980); however, it is valid only if it is assumed that subjects maintain a certain performance level from one test to the next. Other tests are the PLOTT (Plant/Westcott) sequential speech feature test (Dunn and Dunn 1981; Plant 1984), and a speech intelligibility test (Jerger et al 1980) that uses familiar monosyllables as words in a carrier sentence for children from 3 to 9 years, and carried out as a closed five-choice picture-pointing procedure. For young children, the low verbal and standard versions of the Early Speech Perception test (ESP) of Moog and Geers (1990) assess pattern perception, as well as spondee and monosyllable identification. Other suitable tests are the Multisyllable Lexical Neighborhood Test (MLNT) and Lexical Neighborhood Test (LNT) (Kirk et al 1995).

Speech Features (Consonants and Vowels)

Speech features are the acoustic components of phonemes, the latter being the perceptual units that form words. How electrical speech-processing strategies represent these features is invaluable in understanding the way speech information is processed by the brain. This was discussed in more detail in Chapter 7. Closed-set as well as open-set data can be analyzed to provide information on the discrimination and recognition of specific acoustic features of speech. The analysis is on consonant-vowel (VC) syllables (e.g., pa, ta, ka, ba, da, ga) or vowel-consonant-vowel (VCV). The consonants can be divided into five groups according to their method of articulation, which determines their acoustic features. The groups are nasals, semivowels, plosives, fricatives, and affricates. For a set of vowels or consonants, the correct or incorrect responses are recorded to build up a confusion matrix, as illustrated in Figure 12.1.

The data can be analyzed using a number of mathematical and statistical techniques to determine the way the responses are grouped together for a particular set of stimuli. This enables speech-processing strategies to be studied and compared within and across patients. The techniques include information transmission analysis (Dowell et al 1982), multidimensional scaling (Tyler et al 1987), and hierarchical clustering (Busby et al 1984). It is important to realize that the effective transmission of phonetic information as measured by consonant and vowel tests is a necessary but not sufficient condition for improvement in communication ability. The reception of information at the phoneme or word level does not mean that the information can be processed by the user at the rate required for conversational speech.

Another application for this type of testing has been the study of the integration of different modes of input by subjects. For instance, by investigating the con-

Perceived phonemes

	b	d	g	p	t	k	dʒ	tʃ	z	ʒ	v	ð	s	ʃ	f	θ	h	m	n	ŋ	l	r	w	j
b	100	1	2	9	0	0	0	0	0	0	19	11	0	0	4	0	0	7	4	2	0	0	1	0
d	4	114	10	5	11	1	1	0	2	0	0	6	0	0	1	0	0	0	0	0	0	1	1	3
g	2	26	100	3	1	20	1	0	0	0	1	1	0	0	0	1	0	1	1	2	0	0	0	0
p	1	0	2	99	24	20	1	3	0	0	0	1	0	1	3	3	0	0	0	1	1	0	0	0
t	0	0	0	2	140	5	2	2	4	0	0	0	0	0	1	2	0	0	2	0	0	0	0	0
k	0	0	1	5	34	113	0	6	0	0	0	0	0	0	0	1	0	0	0	0	0	0	0	0
dʒ	0	7	0	0	32	0	91	8	10	0	0	2	0	3	0	1	0	0	0	0	0	0	0	6
tʃ	0	0	0	1	22	0	2	128	3	2	0	0	0	1	0	1	0	0	0	0	0	0	0	0
z	0	6	2	0	0	0	7	0	112	7	5	9	6	0	2	0	0	0	0	0	0	1	1	2
ʒ	0	0	0	0	5	0	12	0	56	79	0	1	1	0	0	0	0	0	0	1	0	0	0	5
v	22	8	3	3	0	0	0	0	1	0	49	26	0	0	5	0	0	13	3	2	1	3	20	1
ð	36	13	5	4	0	1	0	0	1	0	33	33	0	0	1	1	2	9	4	3	1	4	7	2
s	1	1	1	0	3	0	0	0	8	0	1	2	112	1	15	12	2	0	0	0	0	0	0	1
ʃ	0	0	0	0	0	0	0	0	4	5	0	0	10	141	0	0	0	0	0	0	0	0	0	0
f	7	0	1	3	3	2	0	0	1	0	12	1	0	0	101	14	10	2	0	0	0	0	3	0
θ	29	6	0	17	1	4	0	0	1	0	12	9	0	0	44	27	1	2	2	1	1	1	2	0
h	1	0	0	1	0	8	1	0	2	2	8	1	1	3	28	7	95	1	1	0	0	0	0	0
m	7	0	0	0	0	0	0	0	0	0	4	1	0	0	0	0	0	108	29	6	5	0	0	0
n	4	0	0	1	0	0	0	0	0	0	2	0	0	0	0	0	0	37	93	17	0	2	0	4
ŋ	1	1	0	1	0	0	0	0	0	0	3	0	0	0	0	1	0	13	42	95	0	0	0	3
l	1	0	0	0	0	0	0	0	0	0	2	3	0	0	0	0	0	45	51	9	10	7	26	6
r	12	2	3	0	1	1	0	0	0	0	16	13	0	0	0	1	2	19	23	7	7	30	21	2
w	3	1	0	0	1	0	0	0	2	0	6	4	0	0	0	0	0	4	7	1	4	11	114	2
j	0	0	0	0	0	1	3	0	7	1	0	0	0	0	0	1	0	0	4	9	0	1	3	130

(Stimulus phonemes — row labels at left)

FIGURE 12.1. Confusion matrix for the 24 Australian English consonants for a patient using the SPEAK strategy.

ditions of speech reading alone, audition alone, and speech reading with audition, the way in which complementary information from the two separate modes of input (audition and speech reading) is utilized in the combined mode can be assessed (Blamey and Clark 1986).

Open-Set Tests

Open sets of words or words in sentences as well as closed sets evaluate the analytical ("bottom-up") skills of the person. However, the meaning of words (semantics) can influence their recognition (Sarant et al 2001), and the ability of children to understand meaning can vary considerably. Grammar (syntax), too, is involved in the recognition of words in sentences. With open-set tests there are no contextual clues, and the material must be unfamiliar, and unpracticed. They more closely predict the ability of a person to communicate in everyday situations than closed sets of speech material. One set of words was constructed by Booth-royd (1968). Each list contained 10 monosyllabic words that are scored as the percentage of words or phonemes correct. The Northwestern University NU-4 and NU-6 monosyllabic word tests are also frequently used and scored as words and phonemes correct (Tillman et al 1963; Tillman and Carhart 1966).

The lists of words are phonemically balanced so that their frequency is similar to that in normal conversation. It is more usual to present monosyllabic words such as "man" and "thin" than bisyllabic words (spondees) such as "baseball" and "brickbat." They are selected to be within most people's vocabulary, and thus should be familiar. As an average vocabulary for an American adult is 30,000 words (Nagy and Herman 1987), any score is statistically significant (e.g., 5%). A child entering school may know up to 3000 or 4000 words. As children have a more limited vocabulary, the phonemically balanced kindergarten (PBK) word test (Haskins 1964) is frequently used. The words must not have been practiced at any stage. The results of the test may be recorded as words or phonemes correctly identified. A person may get part, but not the whole word right so the test results are higher when scored phonemically. This is illustrated with the findings for the Nucleus device with the F0/F1/F2 strategy for electrical stimulation alone. The monosyllabic words (scored as words) were 12% (F0/F1/F2), and the monosyllabic words (scored as phonemes) 33% (F0/F1/F2) (Clark 1986; Dowell et al 1987a,b). In addition to the AB word test used, in particular on the University of Melbourne's initial patients (Clark, Tong, et al 1981c), studies used the NU-6 monosyllabic word test also scored as words and phonemes correct. As with closed-sets of words, they may be presented live voice or prerecorded. In both cases the loudness of each word should be balanced and the material presented at a standard distance in a prescribed acoustic environment. Word tests have been developed using syllables that do not make meaningful words (nonsense syllables) to reduce the effects of higher level processing, such as the City University of New York (CUNY) Nonsense Syllable Test (NST) (Levitt and Resnick 1978). These tests have proven too difficult to use, and their relationship to everyday performance in not clear.

The perception of speech may also be assessed through the percentage of key words in sentences correctly identified. As there is more contextual information in the sentence, the scores are higher than for PB words. This is illustrated with the findings for the first clinical trial of the Nucleus device with the F0/F2 strategy using electrical stimulation alone. The mean score was 10% for PB words (Clark, Tong et al 1981c) and 35% for CID (Davis and Silverman 1978) keywords in everyday sentences (Clark, Tong, et al 1981b), both presented by live voice. An alternative sentence test is the CUNY sentence lists on 12 known topics varying in length from 3 to 14 words (Boothroyd 1991a).

The word-in-sentence as well as word test scores depend on the subject's vocabulary. For this reason the sentence as well as word tests designed for adults cannot be presented to children. Thus open sets of Bamford-Kowal-Bench (BKB) sentences (Bench and Bamford 1979) are used. The BKB sentences contain 50 key words in 16 sentences. Alternatives are the Speech Intelligibility Test (SIT) (Magner 1972) or the Glendonald Auditory Screening Procedure (GASP) (Erber 1972).

Speech Reading

The importance of developing a speech-processing strategy for electrical stimulation of the auditory nerve that would complement speech reading from lip movements was recognized at an early stage in the development of cochlear implants. Initial research at the University of Melbourne analyzed the processing time delays that would still allow the fusion of information, and it was found that 20 ms was an approximate upper limit. The inaugural University of Melbourne's F0/F2 strategy was in fact found to provide complementary information that allowed word scores to improve when electrical stimulation was combined with speech reading compared to speech reading alone or electrical stimulation alone (Clark and Tong 1981; Clark, Tong, et al 1981b,c). The complementary nature of the information was seen when the speech features were analyzed for the different modes of stimulation (Clark, Tong, et al 1981a,d; Dowell et al 1982). Electrical stimulation provided additional information not visible on the lips, such as voicing.

The benefit of an implant for communication with and without speech-reading assistance was evaluated usually with open-set word tests. The need to present tests that combined electrical stimulation with speech reading was more necessary with earlier speech-processing strategies where poorer results were obtained with electrical stimulation alone, and the person had to put more reliance on speech-reading assistance. Most speech recognition tests do not have normative data for speech reading. Furthermore, test lists designed to be equally difficult in the auditory-alone condition may not be equally difficult when used as a speech-reading assessment. Nonetheless, valuable information can be obtained by testing subjects in both the visual and auditory-visual conditions to assess the improvement in speech-reading ability provided by auditory input.

Speech Tracking

Speech tracking is a procedure that more closely approximates speech comprehension in everyday situations. In speech tracking (or continuous discourse tracking) the subject is required to repeat verbatim passages of text read out by the tester (De Fillipo and Scott 1978). The tester does not proceed until the subject correctly repeats the previous phrase or word. By using a prearranged hierarchy of strategies to arrive at the correct response, it is possible to ensure that the subject is able to complete the task. Performance is measured by calculating the speed (in words per minute) at which the tester and subject proceed through the text. New materials are continuously employed. That is, the implant user is not presented the same connected discourse sentences over and over again. Aside from providing a guide to the person's progress, an advantage with speech tracking is that it can be used to provide training for the implantee in the recognition of connected discourse. This is not possible with many other evaluation techniques. It can be used to measure the benefit provided by the prosthesis for speech reading (Dowell et al 1990a).

The main disadvantage of speech tracking is that it is influenced by the material used (complexity, vocabulary, etc.); the characteristics of the tester (different accents, visibility of lips); the experience of the tester with the task; the experience of the subject; and the strategies used to clarify responses and the order in which they are used. By controlling as many of these variables as possible, scores can be compared for different subjects, but it must be remembered that tracking results are prone to vary across sessions and are sensitive to the particular conditions of testing. Despite the problems with the comparison of speech-tracking scores, repeated testing provides a helpful guide to overall performance. Results have been shown to correlate well with formal recorded tests and with the subjective reports of implantees (Dowell et al 1990a).

Speech in Noise

As the speech perception scores with the multiple-channel cochlear implant have improved, there is a great need to evaluate performance in noise (Fig. 12.2). Tests have been designed to assess speech intelligibility at fixed speech and/or noise levels. These tests are the Speech Perception in Noise Test (Kalikow et al 1977), the CUNY topic-related sentences (Boothroyd et al 1985), and the Connected Speech Test (Cox et al 1987). Their results are limited by floor and ceiling effects, and the amount of material. An alternative to the percent intelligibility is the speech reception threshold (SRT), defined as the intensity level to recognize speech material a specified percent of the time (50%). The SRT is derived by an adaptive procedure (Levitt 1978). The noise is either random, shaped to the spectrum of speech, or multitalker babble. Spondees have been used, but sentences are more likely to represent normal spectral weightings, intonations, and so on. Hagerman (1982, 1984) created a set of Swedish sentences, and Plomp and Mimpen (1979) a set in Dutch. The BKB sentences have a large set but vary in length, which may lead to memory effects. These BKB sentences have been adapted for American English and standardized as the Hearing in Noise Test (HINT) by Nilsson et al (1994).

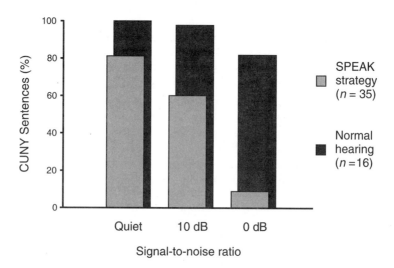

FIGURE 12.2. The speech perception scores for the SPEAK processing strategy at 0 dB and 10 dB signal-to-noise ratios compared to perception with normal hearing.

Environmental Sounds

The patients who present at cochlear implant clinics primarily want to be able to understand speech, but the recognition of environmental sounds and music is also much valued. It is difficult to assess the recognition of environmental sounds using a formal test. In real-life situations, context plays an important role. Sounds may be described accurately, but not correctly identified due to the inappropriateness of the situation. The environmental sounds subtest of the Minimal Auditory Capabilities (MAC) battery (Owens and Telleen 1981) is an open-set test that gives a guide to the subject's ability to recognize sounds. At the University of Melbourne, this test was presented in a closed-set format to alleviate some of the contextual problems. However, this made the test a little too easy for most implantees. A more effective evaluation of user benefit for environmental sounds is likely to be obtained from an interview situation or even from a well-worded questionnaire (Dowell et al 1990a).

Test Batteries

In the initial stages of cochlear implant development, it was uncertain whether patients would be able to achieve open-set speech understanding, the established method of assessing communication skills in the everyday world. When open-set results were achieved (Clark, Tong et al 1978; Tong et al 1979; 1980; Clark and Tong 1981; Clark, Tong et al 1981b; Eddington 1983), it was not clear at what level the results would justify implantation on a regular basis or whether those who performed at the lowest level were getting satisfactory help. The test batteries were put together to provide a picture of the communication skills a person had, although it is not clear just how useful they were in predicting overall performance.

The test batteries included standard open-set speech tests to provide a comparison with the normal-hearing population. Speech-reading tests were needed to assess the benefit in the audiovisual condition, as this was where the greatest help was obtained with the first multiple-electrode strategies. Closed-set testing assessed auditory discrimination where open-set testing was too difficult and provided information on the auditory information perceived.

One of the first battery of tests, the MAC battery (Owens and Telleen 1981), was a useful tool for assessing cochlear implant candidates and users. The name reflects the fact that many patients were not expected to obtain open-set speech understanding, and thus minimal benefits would be assessed. This battery included open-set word and sentence recognition, a speech-reading assessment, prosodic testing, closed-set vowel and consonant discrimination, and a test of environmental sound recognition. There were some limitations in the battery, in particular the assessment of speech perception with context and in the presence of background noise. These were added to a modification of the MAC battery (Tyler et al 1983). It was also discussed in Chapter 9.

A battery or profile has also been developed for children over 1 year of age.

The Listening Progress Profile (LiP) is a measure of closed-set discrimination of male/female voices, environmental sounds, phonemes, and musical instruments (Archbold 1994).

Questionnaires

Questionnaires are seen as a means of assessing the overall benefits of the implant in a person's daily life. They are also useful for comparing pre- and postcochlear implantation benefits. Properly constructed, they give information not available with conventional tests of perception, for example the number hours of usage, the effect on tinnitus, and so on. The questionnaires were first designed for people who were profoundly hearing impaired. As the results improved, standard hearing impairment questionnaires were more relevant. Questions about the appreciation of music, use of the telephone, and understanding speech with competing noise were thus much more relevant than previously.

One questionnaire for parents and educators of implanted children is the Meaningful Auditory Integration Scale (MAIS) (Robbins et al 1991). The questions in particular focus on behaviors that show consistent reactions to voice and environmental sounds, device bonding, spontaneous alerting to sound, and deriving meaning from stimuli. A modification for infants is the Infant Toddler Meaningful Auditory Integration Scale (IT-MAIS) (Zimmerman-Phillips et al 2000).

Bimodal and Bilateral Speech Processing

Two normally functioning ears over the speech frequency range are necessary to filter out noise in most difficult listening situations. The noise and speech signal being attended to are processed in the central auditory pathways receiving a bilateral input. As nonspeech noise does not have the same coherence in the spectral and temporal components as speech, it can be more easily suppressed. The neural mechanisms for the suppression of noise and improvement of the signal-to-noise ratio (SNR) were described in Chapter 5. The improvement in the SNR is referred to as release from masking, or the "squelch effect." As stated, the most difficult listening situations are those in which there is competing speech and the brain has to "lock on" to the particular frequency characteristics of a person's voice, such as the voicing frequency. Listening is particularly difficult in rooms where the sound echoes around the walls and so the low frequencies of unwanted speech arrive at both ears with the same intensity and timing, making it especially difficult to distinguish the speech of the person being attended to. As with the assessment of monaural speech processing, the competing noise should be white noise, speech filtered noise, or multispeaker babble. As multispeaker babble more closely represents the everyday situation, it is preferred.

Binaural speech perception from an implant in one ear and hearing aid in the other (bimodal stimulation) or an implant in each ear (bilateral implants) can improve sound localization, and has the potential to provide better speech perception, especially in noise. This is very important as the performance of SPEAK, continuous interleaved sampler (CIS), and advanced combination encoder (ACE) is significantly poorer in noise than in quiet when compared with normal hearing, as illustrated in Figure 12.2.

The noise can be mixed with the signal entering the speech processor and presented together as electrical stimuli. This enables the procedure to be standardized, but does not allow for the head shadow effect or the room acoustics. When studying the influence of the head shadow, the signal is usually presented from directly in front and the noise to each ear separately. The head will produce an attenuation of the intensity of frequencies, particularly the high frequencies. The presentation of signals in noise is discussed in regard to bimodal speech processing and bilateral cochlear implants in Chapter 7.

Psychophysical Tests

Psychophysical tests are required to determine how well information is being integrated from each ear, as this underlies sound localization and speech perception in quiet and different noise situations. The psychophysical tests are fusion, pitch matching, loudness growth and summation, central masking, and sound localization and lateralization, and are discussed in Chapter 6.

Speech Perception in Noise

The assessment of speech perception in noise examines the head shadow effect, binaural release from masking, binaural redundancy, and dual microphones. The stimuli for the speech-in-noise studies may be diotic or dichotic. Diotic stimuli are sounds reaching the ear that are the same, and dichotic stimuli are sounds that are different (see Chapters 6 and 7).

It is first necessary to determine whether the stimuli produce a single image in the head when the stimuli are presented directly through the speech processors. If the stimulating electrodes do not excite corresponding frequency areas of the cochlea, the patient will describe two images from different spatial locations in the head or in the ears. The image appears to originate from different locations depending on the interaural intensity differences (IIDs), but to a much lesser extent the interaural time differences (ITDs) (van Hoesel et al 1990). This can be avoided by pitch matching the stimuli and comparing he placement of the electrodes from x-rays with the "Cochlear" view (van Hoesel et al 1990; Marsh et al 1993; van Hoesel and Clark 1995), as described in Chapter 8 (van Hoesel 1998).

Determining the loudness growth is important for bimodal and bilateral stimulation. As discussed in Chapter 6, many hearing-impaired people have a nonuniform change in loudness with increasing intensity as reported by Pohlman and Kranz (1924) and others subsequently. Similarly, with a cochlear implant the loudness growth between threshold (T) and maximum comfortable (MC) levels may vary for each electrode. Thus the natural loudness transitions should be preserved with the hearing aid and cochlear implants for bimodal and bilateral stimulation. Furthermore, in both cases loudness summation has been demonstrated by van Hoesel and Clark (1997) and Blamey et al (2000), and this needs to be adjusted for in the combined speech-processing strategy.

The ability to localize sound requires the brain to process small differences in the time of arrival or intensity of sound. The just noticeable differences (JNDs) need to be determined to help interpret the results. This can be carried out through stimuli presented to the speech processors directly or with the subject seated in an acoustically treated room with the stimuli presented via a semicircle of speakers as illustrated in Chapter 6.

The head, pinna, and torso affect the interaural time and intensity differences, and thus the perception of speech in noise. The head has the greatest effect at all frequencies, the pinna affects the pressure differences at frequencies above 2500 Hz, and the torso has the least overall influence (Kuhn and Burnett 1977; Kuhn 1979; Gaunaurd and Kuhn 1980). Depending on the sound frequency, the attenuation of the velocity or pressure (intensity) due to the head can improve the SNR when the two are spatially separated. This was discussed in more detail in Chapter 6. The tests can be conducted in a mildly reverberant room with the direct to reverberant ratio greater than approximately 12 dB. The speech perception is evaluated in the presence of multispeaker babble usually at SNRs of 0 dB, 5 dB, and 10 dB. For the tests on spatially separated noise the speech should come from in front and the noise at 90 degrees to the left or right. Better results for the left or right ear when the noise is on the opposite side is due to the head shadow shielding the test ear from the noise. If there is an additional advantage in using both together, then this is the result of binaural release from masking. It must, however, be distinguished from a spurious finding due to poor loudness balancing.

Speech Production: Test Principles

Imitative and Spontaneous Speech

The assessment of a child's speech production is done through speech sounds that have been imitated, elicited, or spontaneously produced. The simplest method is a rating scale (Shriberg and Kwiatkowski 1982; Levitt et al 1987), but it will depend on the level of familiarity of the person with the speaker. The child's ability to imitate speech can be tested using phonetic level evaluation (Ling 1976) or voice analysis (Ling 1976). In these tests the child attempts to copy presented syllables. Elicited speech is from visual prompts where the child is asked to name a picture (Test of Articulation Competence) (Fisher and Logemann 1971) or verbally repeat a written sentence (McGarr 1983). The child is required to remember sentences to perform the task, and there is variability in the scores from one listener to the other. The child's spontaneous speech produced in conversation or play can be recorded on videotape and analyzed using the phonological process analysis (Ingram 1976; Crary 1982) and the phonologic level evaluation (Ling 1976).

An imitative test was developed by Boothroyd et al (1996) and Boothroyd (1997, 1998) to assess a subject's ability to convey phonetic contrasts to normally hearing listeners by the imitation of syllables that are presented live; this is the Imitative Test of Speech Pattern Contrast Perception (IMSPAC).

Computer-Aided Speech and Language Assessment Procedure (CASALA)

As discussed in Chapter 11, speech production can be monitored with the CAS-ALA procedure developed in the Human Communication Research Center (HCRC) at the University of Melbourne/Bionic Ear Institute (Blamey et al 1994, 2001b). The sounds of the words spoken are transcribed into the appropriate phonetic symbols, and the computer transcribes the words read into their correct phonetic representation. The data are then analyzed, and the child's progress monitored. The percentage of phonemes produced correctly can be plotted, for example, against a measure of receptive language.

Language: Test Principles

Language is the basis for communication between people. The message involved in communication is conceived, transmitted, received, and understood. It can thus be broadly classified as expressive and receptive. The general biological processes are cognitive, motor, and sensory. The message has the following components: phonology, morphology, syntax, and pragmatics. Phonology is a linguistic concept for the basic acoustic units in a word, and if one is substituted for another the meaning may be changed. Morphology or semantics refers to the meanings of words and sentences. Syntax, or grammar, refers to the rules governing the ordering of the words to transmit the message. This includes use of word order in sentences, tenses and plurals, and other rules for combining words in utterances. Furthermore, speech perception, speech production, and language are tightly intertwined, with each influencing the other in complex ways (O'Donoghue et al 1999; Spencer et al 1998). It should be noted that some tests incorporate speech perception as a measure of language, just as the assessment of a hearing handicap depends on speech perception.

The language of children with normal hearing improves over time. It will be delayed in children who have a hearing loss. It is important in assessing the benefits of a cochlear implant or hearing aid to measure the changes over time. This is commonly done with the equivalent language age of the child. This is the age at which the average score of hearing children for a certain test is equal to the score of the child being assessed. The difference between the ages is the language delay. The ratio of the equivalent to the chronological age is the language quotient (LQ).

Receptive Language

Tests of receptive language should assess semantics and syntax. The tests need to be language appropriate. The Symbolic Play Check List (Westby 1980) assesses receptive prelanguage behaviors in the preverbal child. The Peabody Picture Vocabulary Test (PPVT) (Dunn and Dunn 1981, 1997) is an established test of receptive vocabulary. With the PPVT-III the child points to a picture of the word spoken. It requires a minimum level of hearing to adequately test the child's vocabulary. It is assumed that performance is limited by lexical knowledge (vocabulary) and not hearing. This assumption may not be true for hard-of-hearing children (Blamey 2002). The test thus has norms for hearing, but not deaf children.

Boothroyd et al (1991) found an average learning rate 0.43 times normal for 123 children with hearing aids and a pure tone average (PTA) threshold greater than 105 dB hearing loss (HL). Another group with a PTA between 90 and 104 dB HL had a learning rate of 0.6. An average high school graduate in America has a vocabulary of 30,000 words at age 18 years (Nagy and Herman 1987). The deaf child would have proportionately less.

The Preschool Language Scale (Zimmerman et al 1979) also measures receptive language. The Clinical Evaluation of Language Fundamentals (CELF-3 and CELF-Preschool) uses speech perception to measure language. The Northwestern University Children's Perception of Speech (NU-CHIPS) (Elliott and Katz 1980) is a four-alternative forced-choice picture identification test that uses monosyllabic words. The children's understanding of sentence structure can be assessed with the Rhode Island Test of Language Structure, a test of sentence comprehension normalized for deaf children. The perception of grammatical morphemes, the sounds indicating plurality, and so on, can also be analyzed (Bornstein et al 1980; Quigley and Paul 1990; Bornstein and Saulnier 1981; Schick and Moeller 1992). The above tests are used when appropriate for the child.

Expressive Language

Research by Stoel-Gammon (1998) has helped show that phoneme production normally precedes word acquisition, as the child's first words have a preponderance of phonemes being used in babble. However, being able to use words drives the acquisition of further phonemes. The processes in deaf children are not well understood. The expressive equivalent of the PPVT test is the Expressive One-Word Picture Vocabulary Test (EOWPVT) (Gardner 1979), which requires the tester to present a picture and for the child to say the word that represents the picture. With the test the child needs to be able to produce some intelligible speech.

The CELF subtest Recalling Sentences in Context is an expressive language measure where the child is required to respond verbally (Wiig et al 1992; Semel et al 1995). It has norms for hearing but not deaf children.

The expressed language of children can be analyzed at a phonological level using articulation tests (Anthony et al 1971; Fisher and Logemann 1971) and phonetic transcripts of spoken language (Crystal 1992; Lund and Duchan 1993). The results are presented as lists of the phonemes used (phonetic inventories) (Sander 1972), percent correct phonemes (Shriberg et al 1997), or the types of errors (phonological processes) (Dodd 1976).

Children begin to develop intelligible phonemes at the age of 1 and it is not complete until they are 6. When they have a hearing loss, they acquire phonemes at a later age and rarely have a completely intelligible repertoire. The order of occurrence is thought due to linguistic, acoustic, and articulatory factors (Crystal 1981). If their order of occurrence is ranked according to frequency of occurrence (linguistic), intensity (acoustic), and place of articulation (articulation), frequency of occurrence has the highest correlation (Blamey et al 2001b). This suggests that linguistic factors are fundamental. Stoel-Gammon (1998) considers that words contain phonemes acquired during babbling. As vocabulary increases, more phonemes are required to maintain phonetic distinctions. As a result, the development of vocabulary may be interrelated with the rate of phoneme acquisition. In addition, front consonants /p,b,m/ occur early in speech (Smith 1975; Tobey et al 1994), and it has been suggested that this is because they are more visible. It is my view that there is a strong relationship between physiology and phonemes. The movements required first for sucking and then chewing are likely the ones most used in creating the first meaningful utterances. Finally, voicing that is not visible is poorly controlled in the speech of deaf children (Hudgins and Numbers 1942; Smith 1975).

Expressive language is measured through the expression not only of phonemes (expressive phonology), but also of the larger language components morphemes, syllables, words, phrases, and sentences. Words, phrases, and sentences increase in complexity as language matures. These aspects of language are analyzed as morphology and syntax. These can be measured with the CELF test, which has the CELF-Preschool section (Wiig et al 1992) for children aged 2 to 6 years, and the CELF-3 for older children (Semel et al 1995). These tests are composed of procedures that assess cognitive skills such as conceptualization, memory, naming, and association, and linguistic skills such as sentence structure. Other measures are the Reynell Developmental Language Scales (RDLS) (Reynell 1983) and the Preschool Language Scale (PLS) (Zimmerman et al 1979). The expressive language sections of the RDLS are used to test vocabulary as well as structural complexity of the language (Svirsky et al 2000). As a rule the development of spoken language in hard of hearing children follows a similar sequence to that of children with unimpaired hearing, although at a slower rate.

Children's ability to use expressive morphology and syntax can be measured with story retelling. The tester presents the stories, using sign and speech. The child is asked to retell the story using his/her preferred method of communication. The retold stories are videotaped, transcribed, and analyzed using a scoring system for the production of noun and verb phrases, questions, negation, and complexity of sentence structure. The morphological and syntactic analysis of spoken language samples can be made through calculating the mean length of utterance

(MLU) (Brown 1973), the Language Assessment, Remediation, and Screening Procedure (LARSP) (Crystal et al 1989), and the Index of Productive Syntax (IPSyn) (Scarborough 1990).

Pragmatics

Pragmatics are the skills to carry on a two-way conversation. Pragmatic skills are assessed by analyzing interactions between child and family or therapist. Spontaneous samples of conversations and interactions are recorded and can be assessed at the preverbal, single word (Roth and Spekman 1984a,b), and multiword (Prutting 1986) levels.

Speech Perception with Cochlear Implants

Predictive Factors

In analyzing the results of cochlear implantation, it is important to record the factors that correlate with performance so that their predictive value can be determined. The predictive factors are discussed in Chapters 9 and 11. They have been evaluated by Gantz et al (1988), Blamey Pyman et al (1992b, 1995), Cohen et al (1993), Gantz et al (1993), Battmer et al (1995), Shipp and Nedzelski (1995), Dowell et al (1997), and Tomblin et al (1999).

The general predictive factors that are common to the adult and the child are age when deafened, age at implantation, duration of deafness, duration of implantation, etiology, presence of a progressive hearing loss, degree of residual hearing, speech reading ability, speech-processing strategy, and medical condition.

The age when a person is deafened is significant for children born deaf or deafened early in the first 4 to 6 years of life (prelinguistically deaf). Age at implantation correlates negatively with results in children if they have been born deaf or deafened early in life. It was found that in adults there was a negative correlation only if the person is over 60 years (Blamey et al 1995). An analysis of the results from the University of Melbourne's clinic showed there was a trend for the results to be better the younger the child at surgery. The speech results versus age at implantation are shown in Figure 11.4 in Chapter 11. There is considerable variability of results, but the line of best fit through the data when extended back suggested that scores would be higher if children had an operation at less than 2 years of age. As a consequence the majority of children have their operation from 1 to 2 years of age. This is illustrated in age statistics of patients from the University of Melbourne's Cochlear Implant Clinic, Royal Victorian Eye and Ear Hospital shown in Figure 12.3. As discussed in Chapters 9 and 11, children operated on at a young age develop speech perception, speech production, and language at a faster rate than children operated on at an older age. Children operated under 2 years produce better speech within the context of expressive language. Older children may also receive more help than with a hear-

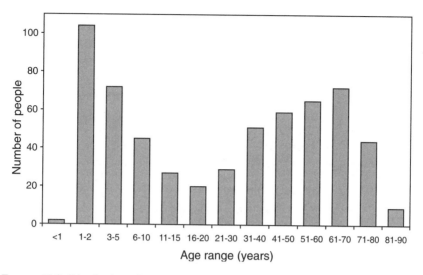

FIGURE 12.3. Distribution of age at implantation at the University of Melbourne's Cochlear Implant Clinic, Royal Victorian Eye and Ear Hospital. (R. Hollow, personal communication.)

ing aid, and should not be excluded from implantation. The results correlate negatively with duration of deafness (Gantz et al 1988; Blamey et al 1992a; Dowell et al 1997; Tomblin et al 1999; Kirk 2000). Etiology had no effect on results, except for Meniere's disease, which correlates positively, and meningitis, which correlates negatively. A progressive hearing loss is associated with better results, as is the presence of some residual hearing (Gantz et al 1993). There is a small positive correlation with speech-reading ability (Cohen et al 1993; Gantz et al 1993; Staller et al 2002) and a very significant correlation with the type and amount of information provided by the speech-processing strategy. The longer the duration of the implantation, the better the results, as learning is required for both children and adults even with the strategies that provide the most information.

The general predictive factors that apply to children alone are prelinguistically versus postlinguistically deaf; language ability; communication strategy before surgery; mode of education after surgery; parental support; and delayed cognitive and motor milestones.

With the first clinical trial of the Nucleus F0/F2 and F0/F1/F2 strategies for the U.S. Food and Drug Administration (FDA) it was found that with the groups of children with prelinguistic and postlinguistic hearing loss, closed-set speech perception was markedly better after surgery, and the scores were the same for the two groups. The open-set score, however, was better for the postlinguistic group. Now with improved strategies such as SPEAK and ACE and operations on children under 2 years, the results are comparable to those in postlinguistically deaf adults. Language development influences speech perception. The communication strategy used before surgery is important, and children do better if they have had an auditory-oral education. After surgery an auditory-oral education is required for best results, as is parental support. Finally, children with delayed cognitive and motor milestones do not achieve the same degree of open-set speech recognition, and the learning takes longer.

The specific factors predicting speech perception scores in the adult and child are electrical stimulation of the promontory results, length of insertion and the number of stimulating electrodes, dynamic range, and implant evoked brainstem auditory potentials (IMPEBAPs). There is a positive correlation between preoperative tests of temporal processing via promontory stimulation of the auditory nerve and speech perception results. A positive relationship is seen for the length of insertion and the number of electrodes, and both correlate positively with speech perception. There is a positive correlation between the dynamic range and speech score. Finally, it is likely that abnormal IMPEBAPs are associated with poor speech perception (see Chapter 9).

Speech-Processing Strategies for Postlinguistically Deaf Adults

Comparison of Single- and Multiple-Channel Speech Processors

In the early 1980s the testing of implant speech-processing strategies needed to be standardized, to determine whether single- or multiple-channel strategies were better. This was best done at either one or a group of centers. Tyler et al (1983, 1984, 1985, Gantz et al (1987), and Tyler (1988) assessed the performance of the 3M/House (Los Angeles) single-electrode, 3M/Vienna single-channel, Ineraid (University of Utah) fixed-filter multiple-channel, and the Nucleus (University of Melbourne) second formant/voicing extraction (F0/F2) multiple-channel systems. The results for the single-channel systems showed that the patients could achieve some discrimination of closed sets of vowels on the basis of the voicing and first formant frequencies and duration. Open-set word identification for electrical stimulation alone was rare, but the recognition of environmental sounds quite good. The speech recognition was discussed in more detail in Chapter 7. In contrast, speech perception results for the two multiple-channel systems were significantly better than for the single-channel ones. The study showed that the Ineraid fixed-filter scheme gave similar open-set speech understanding to that of the Nucleus formant extraction scheme in quiet conditions, but was better in the presence of noise, although the result was not statistically significant.

An additional study demonstrated that F0/F2 multiple-channel electrical stimulation of the cochlea (Clark, Tong, et al 1981a–c) provided significantly more speech information than a single-channel speech-coding scheme providing the fundamental frequency as rate of stimulation (Clark, Tong, et al 1981a, 1982).

F0/F2 Speech Processor (University of Melbourne)

The inaugural University of Melbourne's multiple-channel cochlear prosthesis was a 10-channel implant developed during the 1970s, and implanted in three volunteer research subjects in 1978 and 1979. The speech-processing strategy extracted the second formant frequency (F2) and coded this on a place basis, the amplitude of the F2 referred to as A2 was converted to current level on the electrode, and the fundamental frequency (F0) was presented as rate of stimulation. This F0/F2 speech-processing scheme was developed following initial psychophysical and speech investigations (see Chapters 1 and 7). The strategy was implemented initially as a computer-based laboratory system, and later as a portable take-home unit.

The inaugural F0/F2 strategy was first evaluated on two patients using a laboratory-based speech processor. The open-set CID sentence test showed the patients obtained marked improvements in communication (188% and 386%) when using electrical stimulation in combination with speech reading compared to speech reading alone (Clark, Tong, et al 1981b). For electrical stimulation alone, the average score for a closed set of six vowels was 77% (Clark and Tong 1982; Tong et al 1980), and for a set of 12 consonants 34% (Tong et al 1980). The average score for open sets of words (scored as words) was 8% for presentation by live voice, and 5% for presentation using prerecorded test materials (Clark, Tong, et al 1981c). Similarly, scores on CID sentences (scored as key words) were 35% for live voice and 11% when prerecorded (Clark and Tong 1982).

The prototype implant results clearly demonstrated that multiple-channel electrical stimulation of the cochlea had the potential for restoring useful speech perception for postlinguistically deaf subjects. The clinical results with this prototype system are discussed in more detail by Clark, Crosby et al (1983), Clark, Dowell et al (1983), and Clark, Tong et al (1983).

F0/F2 WSP-II Speech Processor (Nucleus)

The F0/F2 strategy developed at the University of Melbourne was incorporated into a speech processor by Cochlear Proprietary Limited for clinical trial for the FDA. The initial clinical trial of this implant was commenced in 1982 by the University of Melbourne's clinical team at the Royal Victorian Eye and Ear Hospital. The trial aimed to establish (1) whether the device had been engineered to effectively reproduce the University of Melbourne's strategy, (2) whether the hardware implementation and data update rate had in fact improved performance, and (3) whether the strategy discovered on two postlinguistically deaf people would have general application to a wider group and be commercially viable (Dowell et al 1985b).

It was important to assess the patients in exactly the same way preoperatively (using a hearing aid or a tactile aid) and postoperatively. Only patients with a total or near-total loss of hearing were considered for the trial, and at least 6 months of rehabilitation was provided with a hearing aid or vibrotactile aid before implant surgery. Testing using the MAC (Owens and Telleen 1981) battery was performed after preoperative rehabilitation, and after the patients had 3 months of experience with the cochlear implant system. During postoperative rehabilitation, data were collected for speech tracking and vowel and consonant recognition studies. These studies provided knowledge about the phonetic information transmitted by the implant system and progress in performance over time. The pre- and postoperative mean scores on the recorded MAC battery tests for the first eight patients implanted with the multiple-channel prosthesis by the University of Melbourne surgical team showed significant improvements for each test (Dowell et al 1986).

In 1983, the international clinical trial of the multiple-channel implant for the FDA began on 40 subjects at nine centers. By mid-1985, 85 patients had been implanted at 18 different centers. A study of 40 multiple-channel implant users by Dowell et al (1986) investigated how patients used the auditory information provided by the prosthesis. The acoustic features perceived were analyzed to determine the specific information provided, and the levels of speech perception were assessed both for the auditory and auditory-visual conditions. A mean score of 93% for the male/female speaker discrimination test indicated adequate perception of voice pitch via the implant. An average of 55% correct for vowel identification studies suggested that second formant information was well perceived by most, and a mean score of 42% for consonant identification showed effective use of amplitude and temporal cues. However, a large range of performance was evident in patients' understanding of connected speech with and without speech reading despite greater consistency of performance at the level of vowel and consonant recognition. Three months postimplantation the patients had obtained a mean CID sentence score of 87% (range 45–100%) for speech reading plus electrical stimulation, compared to a score of 52% (range 15–85%) for speech reading alone (Dowell et al 1986). This highlighted the substantial influence of individual abilities on performance at the conversational level. Factors such as synthetic ability and effective use of contextual information were considered to be very important in producing optimal performance with the prosthesis. It was also evident that the ability to understand speech with auditory input alone improved significantly between 3 and 12 months postoperatively, as shown in Figure 12.4 for open sets of CID sentences. In a subgroup of 23 patients the mean CID sentence scores for electrical stimulation alone rose from 16% (range 0–58%) at 3 months postimplantation to 40% (range 0–86%) at 12 months (Dowell et al 1986). This indicated that a number of patients were reaching a level of speech perception that allowed effective use of the device without speech reading.

The F0/F2 WSP-II speech processor provided the second formant for the segmental component of language and voicing for the suprasegmental component. It was of importance to determine whether a person speaking a tonal language such as Mandarin could also be helped. An initial study was carried out by Xu et al (1987) on a native speaker of Mandarin. The patient was presented with a closed set of Chinese words differing in tonal changes, and he scored 100% correct. He also had a perfect score for the question/statement test in English. This suggested that pulse rate was suitable for representing suprasegmental information in English and segmental information in Chinese. For a Chinese equivalent of the CID sentence test the subject scored 84% correct for electrical stimulation alone and 67% for the standard test in English. With tracking in Chinese, the score rose from 10 words/minute for speech reading alone to 80 words/minute

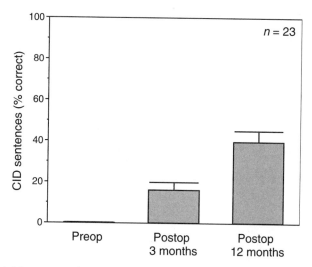

FIGURE 12.4. Mean scores for 23 multiple-channel implant patients using the University of Melbourne's F0/F2 strategy and the Nucleus clinical trial device for open sets of Central Intitute for the Deaf (CID) sentences preoperatively and 3 and 12 months postoperatively. (Reprinted with permission from Dowell et al 1986. *Arch Otolarygol* **112**: 1054–1059.)

when combined with the F0/F2 WSP-II system. The F0/F2 WSP-II was approved by the FDA in October 1985 for use in postlinguistically deaf adults.

The difficulties of hearing in noise with a cochlear implant were emphasized when the Nucleus F0/F2 and University of Utah (Ineraid) strategies were compared. The use of a voicing detector with the Nucleus strategy made it more susceptible to errors with voicing than the University of Utah four-fixed-filter scheme that had no voicing decision (Gantz et al 1987). It was only when F1 was added to provide voice-onset time and F1 transition cues that the results became the same as seen in the study by Dowell et al (1987b).

Fixed-filter Analog (Ineraid) and Clarion

Initially, fixed-filter strategies arose from work on channel vocoders used with telecommunications, but there was no voicing decision. One of the first of these fixed-filter strategies was evaluated by the University of Utah in Salt Lake City (Eddington 1980, 1983), and subsequently clinically tested as the Symbion/Ineraid device. The Ineraid presented the outputs of four fixed filters by simultaneous monopolar analog stimulation between the electrodes in the cochlea and a remote reference. This made it a simultaneous analog scheme. Compression of the amplitude variations in speech to lie within the dynamic range for electrical stimulation was achieved with a variable gain amplifier operating in compression mode. This also made it a compressed analog scheme. Six electrodes were spaced at 4-mm intervals along an array 22 mm in length. The first four electrodes lay opposite the following frequency regions in the cochlea: electrode 1, 1 kHz;

electrode 2, 2 kHz; electrode 3, 4 kHz; electrode 4, 8 kHz (Dorman et al 1989). In most patients only the apical four electrodes were excited. The center frequencies of the filters for these electrodes were 0.5, 1, 2, and 3.4 kHz. The device also had a percutaneous connection with an electrode array (i.e., a plug and socket). This strategy was also discussed in Chapter 7.

A study with the Ineraid four-fixed-filter strategy found the mean word-in-sentence score was 45% (range 0–100%) (Dorman et al 1989). With vowels the errors were mainly limited to the vowels with the most similar formant frequencies. With closed sets of consonants, manner and voicing were well recognized. In a comparison with the Nucleus F0/F2, Tyler et al (1987) and (1989a) found those who used the Ineraid device had mean scores of 36% for words in sentences, 13% for words in isolation, 85% for vowels, and 46% for consonants. Those who used Nucleus 22 channel implant ($n = 10$) achieved scores of 30% for words in sentences, 11% for words in isolation, 94% for vowels, and 41% for consonants.

Advanced Bionics implemented a speech-processing system in the Clarion system that provided compressed analog (CA) or simultaneous analog stimulation (SAS). The main difference between this and the earlier compressed analog scheme from the University of Utah (Eddington 1980, 1983) was that it had automatic gain compression with longer attack and release times, as well as a lower compression ratio, making for reduced spectral distortion (Wilson 2000). It was subsequently used with eight filters in the Clarion processor (Battmer et al 1994). The SAS electrode array and stimuli were different from those used by the University of Utah but similar to the molded array with radial electrodes and bipolar pulses used by the University of California at San Francisco (UCSF) (Merzenich et al 1979, 1984) as discussed below. The first implementation of the strategy did not always produce selective stimulation of neurons as the device ran out of compliance due to the high current required even with offset radial electrodes. Many patients could not receive a percept at all (Wilson 2000).

F0/F1/F2 WSP-III Speech Processor (Nucleus)

By 1984, research with acoustic models at the University of Melbourne showed that the addition of the first formant frequency (F1) coded on a frequency place basis as well as F2 and F0 led to improved speech perception (Blamey et al 1984a,b, 1985) (see Chapter 7 for more information). The Nucleus F0/F2 WSP-II system was therefore upgraded to also process F1 with the WSP-III processor. The F1 and F2 frequencies were coded on a place basis with nonsimultaneous stimulation to minimize channel interaction. The pulses were interleaved with 0.8-ms separation. Initial results with this new system (F0/F1/F2 WSP-III) were reported by Dowell et al (1987b). The first study to compare the F0/F2 with the F0/F1/F2 strategy was on patients who were using the F0/F2 WSP-II speech-processing system. After 2 weeks with F0/F1/F2 WSP-III they were tested with open sets of CID words in sentences for electrical stimulation alone. The results on seven patients using a live voice presentation of the test material showed that the mean CID scores had risen from 30% (F0/F2 WSP-II) to 63% (F0/F1/F2

WSP-III). Finally, the results for continuous discourse tracking were also significantly better for F0/F1/F2 WSP-III, increasing from 11.8 to 30.5 words/minute for electrical stimulation alone (Dowell et al 1987b).

The assessment of the (F0/F1/F2 WSP-III) system with the above protocol encountered a number of problems that were instructive for the future evaluation of changes in speech-processing strategy. The initial assessments were made on experienced implant patients who were changed to the new scheme after they had used the F0/F2 WSP-II strategy and processor for some months or even years (an A/B comparison). Many disliked the change, and tended to perform poorly because of this. Improved results were obtained only after considerable perseverance of the patients. When carrying out the comparison, it was not possible to exclude a time-dependent or training effect. In other words, would the F0/F2 WSP-II result also have been better after a further few weeks' exposure? For this reason it is important when changing strategies to reach a plateau in performance with the first strategy (A), give the patient adequate time to get used to the new scheme (B), and conduct the study with an A/B/A/B protocol to help exclude learning effects with the first strategy (A). Because of these concerns, another study was undertaken to compare the two speech-processing strategies on two groups of patients who received each strategy for the first time and had similar ages, length of deafness, and training. The results for a new group of patients were compared, at the completion of a standard rehabilitation program for the F0/F1/F2 WSP-III system, with those from a previous group of patients who underwent the same program using the F0/F2 WSP-II system. The results for electrical stimulation alone were recorded 3 months postoperatively. The mean open-set CID word-in-sentence score for electrical stimulation alone increased from 16% (F0/F2 WSP-II) ($n = 13$) to 35% (F0/F1/F2 WSP-III) ($n = 9$) (Dowell et al 1987b). The monosyllabic word scores (scored as words) went from 5% (F0/F2 WSP-II) to 12% (F0/F1/F2 WSP-III), and the monosyllabic word scores (scored as phonemes) from 23% (F0/F2 WSP-II) to 33% (F0/F1/F2 WSP-III). The improvements were all statistically significant (Clark 1986; Dowell et al 1987a,b). On a larger group of 45 unselected patients from the University of Melbourne clinic, the mean score for open sets of CID words in sentences was 36%, the same as reported in the earlier smaller sample.

A comparison was made of the two speech-processing strategies in background noise on a comparable group of patients who used the F0/F2 WSP-II ($n = 5$) and F0/F1/F2 WSP-III ($n = 5$) speech-processing systems (Dowell et al 1987a). The results of a four-choice spondee test using multispeaker babble showed the F0/F1/F2 was significantly better at an SNR of 10 dB. Performance with the F0/F2 WSP-II system was significantly degraded at a 10 dB SNR, whereas this did not occur with the F0/F1/F2 WSP-III system at this SNR. The F0/F1/F2 WSP-III also gave better results at lower SNRs (Dowell 1990). These findings were consistent with those subsequently reported by Gantz et al (1988), Tyler and Lowder (1992), and Hollow et al (1995).

In all the above studies there was a large range of performances, making statistical verification of any differences difficult (Dowell et al 1990a). For example, it was reported by Hollow et al (1995) that the mean CID open-set word-in-sentence score for the F0/F2 WSP-II system was 15.9% ($n = 13$), with a range of 0% to 58%, and the F0/1/F2 WSP-III system was 38.5% ($n = 27$), with a range of 2% to 98%. When the vowel and consonant results were analyzed the findings indicated that the F0/F1/F2 WSP-III system provided additional formant information (F1) for vowels, extra cues for consonant voicing, and a more accurate representation of the amplitude envelope of speech. In the study by Dowell et al (1987b) using the MAC battery (Owens et al 1980), the medial vowel recognition was 51% (F0/F2 WSP-II) and 58% (F0/F1/F2 WSP-III), and the averaged initial and final consonant scores were 54% (F0/F2 WSP-II) and 66.5% (F0/F1/F2 WSP-III). The improvements in the perception of vowels, consonants, and speech features were associated with better open-set speech perception. With prerecorded material the open-set monosyllabic scores improved from 5% (F0/F2 WSP-II) to 12% (F0/F1/F2 WSP-III); words scored as phonemes 23% (F0/F2 WSP-II) to 33% (F0/F1/F2 WSP-III); and key words in everyday sentences 16% (F0/F2 WSP-II) to 35% (F0/F1/F2 WSP-III) (Dowell et al 1987b).

The F0/F1/F2 WSP-III speech processor was approved by the FDA in May 1986 for use in postlinguistically deaf adults (Clark 1986; Dowell et al 1987a). The Ineraid four-fixed-filter device was compared with the Nucleus F0/F1/F2 WSP-III system on two groups of better performing subjects by Tyler et al (1989a,b), and both strategies gave similar average scores for open sets of words (12%) and words in sentences (33%) for electrical stimulation alone. A study by Waltzman et al (1992) on 30 patients also confirmed there was no difference between the Ineraid and the Nucleus F0/F1/F2 WSP-III systems.

The Multipeak-MSP Speech Processor

The Multipeak strategy arose through the need to further improve speech perception. The hypothesis was that better speech perception would be achieved through improved consonant recognition, which was 37% with the F0/F1/F2 strategy (Blamey et al 1987). It was reasoned that as consonants contribute greatly to speech understanding and as the consonant place of articulation feature was not well transmitted, improving the transmission of this feature should lead to significant benefits. In particular, the research aimed to provide more high-frequency information for the place feature through F3 or high-frequency spectral information. It was also assumed the additional high-frequency cues would improve consonant perception and speech understanding in noise.

In the HCRC at the University of Melbourne/Bionic Ear Institute, research showed that a strategy where the outputs of fixed filters in the three frequency bands, 2.0–2.8 kHz (B1), 2.8–4.0 kHz (B2), and >4.0 kHz (B3), as well as the first two formants (F1 and F2) on a place coding basis, together with voicing as rate of stimulation (Multipeak), gave better speech perception results. The Multipeak strategy is a misnomer, as the high-frequency information was not peaks of energy but rather the outputs from fixed filters. It was thus a hybrid scheme between formant extraction and fixed filter. The strategy was implemented in a speech processor named the Nucleus MSP (see Chapter 8 for details).

An initial study was undertaken to compare a group of four experienced subjects with the F0/F1/F2 speech-processing strategy and the WSP-III speech processor, and four who used the Multipeak strategy with the MSP speech processor (Dowell et al 1990b). Although the improvements in speech processor engineering seen with MSP could have affected the results, it was assumed the main effects would be due to the difference between the F0/F1/F2 and Multipeak strategies and not the speech processors. The main engineering difference between the MSP and WSP-III processors that could have influenced results was that the MSP had a digital signal-processing (DSP) chip that allowed digital filtering. With peak detectors the circuit extracted the F0 from all the frequencies below the F1 region. This was done to improve the extraction of F0 in noise, a problem for vocoders. There was also noise suppression circuitry.

The patients in the two groups referred to above were not selected using any special criteria except their availability and their willingness to participate in research studies. The results (Dowell et al 1990b; Clark et al 1996) showed, in quiet conditions, a statistically significant difference for the 11 Australian vowels (mean 78% F0/F1/F2 WSP-III and 88% Multipeak-MSP), but for a list of 12 consonants the increased mean from 56% (F0/F1/F2 WSP-III) to 62% (Multipeak-MSP) was not significant. However, the consonant scores for the F0/F1/F2 WSP-III system were higher than those obtained from other unselected Melbourne clinic patients. This was probably due to the small sample size and previous experience with the device. For open-set BKB words in sentences there was a large statistically significant difference (mean 55% F0/F1/F2 WSP-III and 88% Multipeak-MSP). It must be noted that the scores will be higher for the BKB rather than the CID sentences because the material is easier. Again the results were higher than for the unselected group from the Melbourne clinic. When the performances of these two devices were compared in the presence of background noise (four-talker babble) the Multipeak-MSP results were significantly better. This applied to four-choice spondees at SNRs of 10, 5, and 0 dB, and open sets of BKB sentences at SNRs of 20, 15, and 10 dB. For spondees at a 0 dB SNR the results were approximately 60% (F0/F1/F2 WSP-III) and 73% (Multipeak-MSP), and for BKB sentences at a 10 dB SNR they were 32% (F0/F1/F2 WSP-III) and 65% ("Multipeak"-MSP). The four-choice spondee test was easier and results could be obtained at lower SNRs than for open-set speech. It was also apparent that the differences in results became greater with lower SNRs.

To help confirm the above results and ensure that the differences were not due to the speech processors, a study was undertaken on seven unselected postlinguistically deaf adults patients at the Washington University School of Medicine (Skinner et al 1991) to compare the F0/F1/F2 WSP-III, F0/F1/F2 MSP, and Multipeak-MSP systems. When the F0/F1/F2 WSP-III and F0/F1/F2 MSP systems were compared, there were no significant differences in speech perception scores. Consequently, any improvements seen with Multipeak-MSP in this study were not due to engineering improvements per se but the speech-processing strategy itself. When F0/F1/F2 MSP and Multipeak-MSP systems were compared, the Multipeak strategy gave significantly higher scores for open-set speech tests in quiet and in noise. In quiet the average scores for monosyllabic words went from 13% to 29%, and for words in sentences from 52% to 70%. The results were similar to those obtained in the Melbourne study. It was of interest that when the F0/F1/F2 WSP-III and F0/F1/F2 MSP systems were compared on one subject with four-choice spondees in noise, the scores were 90% (F0/F1/F2 WSP-III) and 100% (F0/F1/F2 MSP) and for BKB sentences 14% (F0/F1/F2 WSP-III) and 11% (F0/F1/F2 MSP). The findings were thus equivocal and did not indicate a significant effect in the method of extracting speech information in noise used by the device.

The Multipeak-MSP system was approved by the FDA on October 11, 1989, for use in postlinguistically deaf adults.

Similar differences between the F0/F1/F2 WSP-III and Multipeak-MSP systems were noted by Waltzman et al (1992), Cohen et al (1993), Hollow et al (1995), and Parkinson et al (1996). Hollow et al (1995) found the open-set word-in-sentence scores went from 38.0% ($n = 32$) to 59.1% ($n = 27$).

In the study by Cohen et al (1993) the Ineraid and Nucleus Multipeak-MSP systems, as well as the Nucleus F0/F1/F2 WSP-III and 3M/Vienna single-channel implants were compared. In this randomized prospective study carried out 24 months postoperatively, they found most patients with the multiple-channel implants were able to distinguish words and sentences for electrical stimulation alone better than for the single-channel 3M/Vienna device. A composite index was calculated from the following categories: prosody, speech reading enhancement, phonetic level, spondee tests, and open-speech recognition. The composite results were similar for the Ineraid and F0/F1/F2 WSP-III systems, but when 24 patients had the Nucleus Multipeak-MSP system they had greatly improved open-set speech recognition scores even at 3 months postoperatively. The mean speech scores were approximately 42% with the Ineraid system and with the group with the Nucleus Multipeak-MSP system they had increased to approximately 75%.

It was also found with the Nucleus Multipeak-MSP and SPEAK Spectra-22 systems that the number of stimulus channels was also important for speech perception in noise. Blamey et al (1992b) showed that 20 rather than eight banded electrodes provide improved speech processing for both.

Spectral Maxima–DSP Speech Processor (University of Melbourne)

The Spectral Maxima Sound Processor (SMSP) was developed in 1990 and selected the maximal outputs from 16 band-pass filters, and presented the output as a stimulus current on a place frequency-coding basis. A constant stimulus rate of approximately 250 pulses/s was used to reduce channel interaction, and so voicing was coded as variations in amplitude. The SMSP arose when it was shown by Tong et al (1991) that the extraction of six spectral peaks gave no improved performance over four peaks, and so six spectral maxima might be preferable. In 1990 the SMSP scheme was tested on an initial patient who used the F0/F1/F2 WSP-III scheme, and found to give substantial benefit. For this reason a pilot study was carried out on two other patients in which the F0/F1/F2 MSP system was alternated with SMSP implemented on a DSP (McKay et al 1991). The mean consonant score for the two patients with the F0/F1/F2 MSP system was 18%, and for the SMSP-DSP 41%. The mean open-set CNC word score (scored as words) was 5% for the F0/F1/F2 MSP system, and 18.5% for SMSP-DSP. The mean open-set CID sentence score (scored as key words) was 54.5% for the F0/F1/F2 MSP system and 84% for SMSP-DSP.

Due to the positive initial findings with the SMSP, a more formal investigation was undertaken to compare the SMSP-DSP and Multipeak-MSP systems (McKay et al 1992). An A/B study on four patients produced mean scores of 76% for vowels with Multipeak-MSP and 91% with SMSP-DSP. The mean scores for consonants were 59% for Multipeak-MSP and 75% SMSP-DSP. The mean scores for open-sets of CNC words (scored as words) were 40% for Multipeak-MSP and 57% for SMSP-DSP. The mean scores for open sets of CID sentences (scored as key words) in quiet were 81% for Multipeak-MSP and 92% for SMSP. The mean scores for open sets of CID sentences at an SNR of 10 dB (multitalker babble) were 50% Multipeak-MSP and 79% SMSP-DSP (Clark et al 1996).

SPEAK (Nucleus)

The SMSP strategy (McDermott et al 1992) was developed industrially by Cochlear Proprietary as the SPEAK strategy and implemented on the Spectra-22 processor (McDermott et al 1992; Seligman and McDermott 1995). The Spectra-22 processor had a bank of 20 filters (Seligman and McDermott 1995). Six spectral maxima were selected from 20 filters rather than the 16 with SMSP. On each stimulus cycle the number of electrodes stimulated could vary from one to 10 with an average of six. A constant stimulus rate that varied adaptively from 180 to 300 pulses/s was used to reduce channel interaction. A comparison of the Multipeak-MSP and SPEAK Spectra-22 (Skinner et al 1994) systems was undertaken as a field trial on 63 postlinguistically deaf adults at the University of Melbourne's Cochlear Implant Clinic, Royal Victorian Eye and Ear Hospital; Royal Prince Alfred Hospital, Sydney; the Denver Ear Institute; the Department of Otolaryngology, Washington University School of Medicine; Michigan Ear Institute; Sunnybrook Health Science Center, University of Toronto; St. Paul's Hospital, Vancouver; and South of England Cochlear Implant Center, Institute of Sound and Vibration, University of Southampton. The experimental study was carried out using a single-subject A/B/A/B design. The mean score for vowels was 70% for Multipeak and 75% for SPEAK, the mean score for consonants was 57% for Multipeak and 69% for SPEAK, the mean score for CNC words (scored as words) was 25% for Multipeak and 34% for SPEAK, and the mean score for words in sentences was 67% for Multipeak and 76% for SPEAK. For the 18 subjects who had the CUNY and SIT sentence test at a SNR of 5 dB, the mean score for words in sentences was 32% for Multipeak and 60% for SPEAK. For all comparisons SPEAK was significantly better than Multipeak (p = .0001).

SPEAK Spectra-22 was approved by the FDA for postlinguistically deaf adults on March 30, 1994.

In an additional study by Brimacombe et al (1995) on 41 postlinguistically deaf adults who had only marginal benefits from hearing aids, as defined by open-set word-in-sentence recognition scores in quiet of less than or equal to 30% in the best aided condition, their open-set CID sentence scores improved postoperatively to 68% with the Multipeak and to 77% with SPEAK. The recognition of open sets of CUNY sentences presented in background noise also improved significantly from 39% with Multipeak to 58% with SPEAK. Thus the above results on adults were better on average than those obtained by severely to profoundly hearing-impaired subjects with some residual hearing using an optimally fitted hearing aid (Clark, Dowell et al 1996). A further set of data presented to the FDA in January 1996 showed a mean open-set CID sentence score of 71% for the SPEAK strategy on 51 unselected patients 2 weeks to 6 months after the startup time.

Improvements in speech perception when patients were converted from Multipeak to SPEAK were confirmed by Hollow et al (1995), Whitford et al (1995), Kiefer et al (1996), Skinner et al (1996), and Staller et al (1997). Whitford et al (1995) found that the largest improvements were for the recognition of words in sentences presented in noise. Staller et al (1997) found a more rapid acquisition of open-set speech recognition in the initial postoperative period for SPEAK Spectra-22 compared with Multipeak-MSP.

CIS (Clarion and Combi-40)

The CIS evolved from a fixed-filter scheme that used interleaved pulses (IPs) (Wilson et al 1992). It was considered that a high pulse rate would provide good representation of the voicing information that was lacking with the IP scheme (Wilson 2000). The waveform envelopes from the band-pass filters modulated a high-pulse rate train. The outputs of six or more filters were sampled, and these stimulated electrodes on a place-coding basis. Various studies were done to optimize the number of filters and stimulus rate (Wilson et al 1992, 1993). The CIS strategy was implemented in the Clarion processor with eight band-pass channels. The spectral information was presented at a constant stimulus rate between 833 and 1111 pulses/s per channel for bipolar or monopolar stimulus modes. The results for CIS using the Clarion system were reported by Schindler et al (1995) in a group of 73 patients. The mean open-set CID word-in-sentence score for electrical stimulation alone was 58% 6 months postoperatively. Similar results were reported by Kessler et al (1995) for the first 64 patients implanted with the Clarion device with a mean score of 60% 6 months postoperatively (Kessler et al 1995). There was a bimodal distribution in results with a significant number of poorer performers. A bimodal distribution may either be due to a sampling problem or interrelated with the speech-processing strategy. The SPEAK Spectra-22 system score on 51 unselected patients tested from 2 weeks to 6 months after the startup time was 71%. The data for SPEAK Spectra-22 were presented to the FDA in January 1996.

Both the SPEAK and CIS speech-processing systems were similar in that six stimulus channels were presented at a constant rate. However, with SPEAK, the stimulus channels were from the six spectral maxima, and with CIS from six fixed filters. If it is assumed that the higher stimulus rate of CIS (up to 800 pulses/s) works positively in its favor, then the selection of spectral maxima can be an important requirement for cochlear implant speech processing as the results for SPEAK were possibly better (71% versus 60%).

The early Clarion patients had the option of using a CA strategy as well as the CIS. Initially, only 10% of Clarion patients preferred the CA to the CIS strategy (Kessler et al 1995). As a number of patients could not achieve adequate stimulus levels due to the restricted current flow from the radial electrodes developed by UCSF, an improvement was made to the array through increasing the spacing between electrode pairs, enabling more patients to be fitted with the analog strategy then referred to as SAS processing. Kessler (1998) reported that 30% preferred the SAS to the CIS strategy. The Advanced Bionics Clarion device was approved by the FDA for use in postlinguistically deaf adults in 1996.

The CIS strategy was implemented by Med El as Combi-40 with eight electrode pairs and Combi-40+ with 12 pairs. They were tested in 19 centers in Europe, and 16 of these were in Germany (Helms et al 1997). One year postimplantation the patients scored 30% for vowels and consonants. The mean monosyllabic word recognition was 54% (range 5% to 85%) and percentage words identified in a sentence 89% (range 30% to 100%). It is not possible to equate these scores to the tests in another language or with other test material.

ACE (Nucleus)

The ACE strategy used with the Nucleus 24 system was modified from SPEAK and had stimuli presented at high rates and/or with more channels. The effect of a higher rate of stimulation (in particular 800 pulses/s) with ACE was compared with SPEAK using 250 pulses/s.

The first study comparing low (250 pulses/s) and high (800 pulses/s and 1600 pulses/s) rates of stimulation was undertaken on five subjects. The mean CUNY sentence results for the lowest SNR (Vandali et al 2000) showed there was a significantly poorer performance for the highest rate. However, the scores varied in the five individuals. There was thus significant intersubject variability for SPEAK at different rates.

The ACE strategy was evaluated in a larger study on 62 postlinguistically deaf adults who were users of SPEAK (Arndt et al 1999). ACE was compared with SPEAK and CIS. The rate and number of channels were optimized for ACE and CIS, and were most frequently 720 pulses/s and 1800 pulses/s for ACE, and 900 pulses/s and 1800 pulses/s for CIS. Mean HINT (Nilsson et al 1994) sentence scores in quiet were 64.2% for SPEAK, 66.0% for CIS, and 72.3% for ACE. The ACE mean was significantly higher than the CIS mean ($p < .05$), but not significantly different from SPEAK. The mean CUNY sentence recognition at a SNR of 10 dB was significantly better for ACE (71.0%) than for both CIS (65.3%)

and SPEAK (63.1%). Overall 61% preferred ACE, 23% SPEAK, and 8% CIS. The strategy preference correlated highly with speech recognition. Furthermore, one third of the subjects used different strategies for different listening conditions.

In a subsequent study (Skinner et al 2000) 12 new patients were given SPEAK, ACE, and CIS in different orders, after each strategy was adjusted to suit the patient. The results were consistent with those of Arndt et al (1999) as 58% preferred ACE, 25% SPEAK, and 17% CIS. There was also a strong correlation between the preferred strategy and the performance on speech recognition.

Speech-Processing Strategies for Pre- and Postlinguistically Deaf Children

It was important to know if children, and in particular those born deaf or deafened early in life, could obtain the same results as postlinguistically deaf adults. A key question was, Without the neural connectivity established in response to speech sounds, would patterns of electrical activity in the auditory pathways still induce speech and other sounds?

Single-Channel System (3M-House)

An investigational device exemption (IDE) was approved by the FDA for the trial of the 3M-House single-channel system on children, and it commenced in 1980; 164 children were implanted. The majority became adventitiously deaf, their mean age when deafened was 1.7 years, and age at implantation 8.4 years (Berliner and Eisenberg 1985). The children could recognize environmental sounds, detect speech from nonspeech, and discriminate different speech patterns, such as the number of syllables (Thielemeir et al 1985). A minority of children were able to discriminate closed sets of spondees. A few of the early implanted children obtained some open-set speech understanding (Berliner et al 1989). The FDA premarket approval (PMA) was not completed for this device for children.

F0/F2 WSP-II Speech Processor (Nucleus)

The F0/F2 WSP-II system was evaluated first on children in Melbourne when a 14 year-old (P.S.) had an implant on January 8, 1985, and after this had been shown to provide significant improvements in speech understanding in postlinguistically deaf adults. It was then assessed on a 10-year-old (S.S.) when the Nucleus (Cochlear) Mini receiver-stimulator (CI-22), which was smaller and had a magnet incorporated, was implanted in S.S. on August 20, 1985. The initial results showed increased scores for closed-set consonants and open-set words and sentences when electrical stimulation was combined with speech reading compared to speech reading alone. For electrical stimulation alone the scores for closed sets of 12 consonants in the first child were 7% (a chance score), and for six consonants in the second child 31% (Clark, Blamey et al 1987a; Clark, Busby et al 1987b; Clark, Dowell et al 1996).

F0/F1/F2 WSP-III (Nucleus)

The first child in Melbourne to have the F0/F1/F2 WSP-III system and mini receiver-stimulator was B.D., who was 5 years of age and was operated on April 15, 1986. When this child was obtaining useful speech perception results, the number of children implanted and evaluated in Melbourne was increased, and in 1989 it was reported that five children (aged 6 to 14 years) out of a group of nine had substantial open-set speech recognition for monosyllabic words scored as phonemes (range 30% to 72%), and sentences scored as key words (range 26% to 74%) (Dawson, Blamey et al 1989). Four of the five children who achieved open-set scores were implanted before adolescence, and the fifth, who had a progressive loss, was implanted as an adolescent. The children who did not achieve open-set speech recognition were implanted during adolescence after a long duration of profound deafness. The children who obtained open-set speech understanding, in particular, also showed improvements in language.

When the F0/F1/F2 WSP-III system was approved by the FDA for use in adults in 1985, an international multicenter clinical trial on children commenced in May 1986. The results were collected from 142 children at 23 centers. Data were obtained for at least one speech test in the following speech perceptual categories: suprasegmental, closed-set word identification, and open-set word recognition (Staller 1990). The tests used were appropriate for the developmental stage of the child, and were administered 12 months postoperatively. The results showed that 51% to 60% of the children had significant open-set performance with their cochlear prosthesis compared with 6% preoperatively. In addition, 68% of the children could perceive some spectral cues for speech perception with their cochlear prosthesis compared with 23% preoperatively. Performance also improved over time with significant increases in open-set and closed-set speech perception between 1 and 3 years postoperatively (Clark, Dowell et al 1996).

If the test results on 91 prelinguistically deaf children in the study are examined separately (Staller et al 1991a,b) improvements were comparable with the postlinguistic group in many areas; however, performance was poorer on the open-set measures for the prelinguistic group, including closed-set speech understanding.

The F0/F1/F2 WSP-III system was approved by the FDA for use in children on June 27, 1990.

A similar proportion of children (61%) obtained some open-set speech understanding in a study by Osberger et al (1991a,c). There were 28 children who used the Nucleus 22 system for 1.7 years. Most ($n = 23$) had the F0/F1/F2 strategy and the remainder the Multipeak. There were 17 children who achieved at least one open-set score higher than 0% for electrical stimulation alone. It was found, however, that children implanted during adolescence had a low chance of achieving open-set speech understanding using electrical stimulation alone (Clark, Blamey et al 1987a; Clark, Busby et al 1987b; Tong et al 1988; Busby et al 1991; Dowell et al 1991; Chute 1993).

Multipeak-MSP (Nucleus)

Ten children with the F0/F1/F2 WSP-III system were changed over to the Multipeak-MSP system in 1989. Apart from an initial decrement of response in one child, performance continued to improve in five and was similar for the other children. As a controlled trial was not carried out, it was not clear whether the improvements were due to learning or to the new strategy and processor.

The Multipeak-MSP system was approved by the FDA for use in children on June 27, 1990, on the basis of the F0/F1/F2 WSP-III approval for children and the Multipeak-MSP approval for adults. Its effectiveness was demonstrated by Staller et al (1991b), Miyamoto et al (1992), Blamey et al (1992b), Dawson et al (1992, 1995a), Tobey and Hasenstab (1991), Geers and Moog (1994), Cowan et al (1995), and Dowell et al (1995). The study by Dowell et al (1995) reported that 60% to 80% of the implanted children achieved significant open-set word and sentence perception, although 15% had limited closed-set speech recognition alone. This appeared to be a better overall result than the 51% open-set speech perception score for the F0/F1/F2 strategy reported above by Staller et al (1991a,b).

However, Osberger et al (1996), when comparing two groups of children with the F0/F1/F2 and Multipeak strategies, found that at 1 year postimplantation the children with the Multipeak strategy had better speech results, but at 3 years there was no difference. Thus learning is slower if there is less information in the signal, but the brain can compensate over time presumably due to better use of the cues and higher level processing.

Speech perception scores for the Nucleus F0/F1/F2 and Multipeak strategies continued to improve over a number of years (Miyamoto et al 1993; Gantz et al 1994; Waltzman et al 1994). It was not clear whether the improvements were due to a normal maturation effect or continued learning to process the auditory signal. This was studied in particular for receptive language that was closely related to speech perception (Sarant et al 1996, 2001; Blamey et al 1998). The learning rate discussed below (see Receptive Language) was compared with that for children with hearing aids and found to be approximately double (Dawson et al 1995a,b; Robbins et al 1995).

SPEAK Spectra-22 (Nucleus)

As the SPEAK speech-processing strategy and Spectra-22 processor provided better results in postlinguistically deaf adults than the Multipeak strategy and MSP processor, it was important to determine if the good results would be seen in children, and whether those using the Multipeak strategy could be changed to the SPEAK strategy. With adult implant users it was assumed their prior exposure to speech sounds had established the neural connectivity in the central pathways for processing this information, and that either implant strategy would have to provide the appropriate code for this connectivity. It was assumed that there would not be a great deal of postdevelopment plastic changes occurring in response to the strategies. In contrast, with children who had effectively only had exposure to speech through electrical stimulation with a certain speech-processing strategy, it was assumed the neural connectivity would have formed around the patterns of stimulation for that strategy, and the change in speech-processing strategy might not be as effective as in postlinguistically deaf adults. Would they in fact benefit from any increase in spectral and temporal information available from SPEAK? Furthermore, as children are often in poor SNR situations in integrated classrooms, it was of great interest whether children using the SPEAK processing strategy would show similar perceptual benefits in background noise as those shown for adult patients.

To answer these questions, speech perception results for a group of 12 profoundly hearing-impaired children ranging in age from 6 to 14 years were tested to compare the benefits of the Multipeak and SPEAK speech-processing strategies, and find out whether their performance would improve after being converted from a regular user of the Multipeak to the SPEAK strategies (Cowan et al 1995). The children selected had the ability to get open-set scores for CNC words using electrical stimulation alone. The 12 children were assessed over 36 weeks with an A/B/A/B experimental design. They were tested with the Multipeak for 8 to 12 weeks, then with SPEAK for 24 weeks, followed by an assessment with Multipeak, and then had SPEAK for a further 24 weeks.

At each evaluation, the children were assessed with two open sets of SIT sentences and CNC words, both in quiet conditions and in a +15 dB SNR. The children also had open-set, implant-alone test scores on similar materials such as BKB sentences, and PBK or AB words. The materials were appropriate for the ages of the children.

Comparison of mean scores for the 12 children on open-set word and sentence scores showed a significant advantage ($p <.05$) for the SPEAK strategy compared with Multipeak both in quiet and at a +15 dB SNR. For SIT sentences, mean scores with SPEAK were 59.7% in quiet and 58.1% in background noise, compared with 52.7% and 48.3% with Multipeak. Similarly, mean scores on CNC words with SPEAK were 67.4% in quiet and 65.5% in noise, as compared with 59.4% and 57.3% with Multipeak.

Individual scores varied across the children; however, those with both low and high open-set scores using Multipeak showed a significant increase when using SPEAK. In total, 11 of the 12 children had significantly higher scores with SPEAK on at least one test at the initial 6-month evaluation. After a further 6 months of experience for 10 children with SPEAK, their results further improved. The conversion from the Multipeak-MSP to SPEAK Spectra-22 in children was also discussed in Chapter 11.

The SPEAK Spectra-22 was approved by the FDA for children on March 30, 1994. Furthermore, in a special National Institutes of Health (NIH) meeting, a Consensus Statement on Cochlear Implants in Adults and Children (1995) reported that on the basis of available evidence, "implantation in conjunction with education and habilitation leads to advances in oral language acquisition" (National Institutes of Health 1995).

These data were supported by a study by Sehgal et al (1998) that found that in three children using the F0/F1/F2 and eight the Multipeak strategy from 6 months to 6 years when they were converted to SPEAK, the mean word recognition scores on the Lexical Neighborhood Test improved from 28% to 58%.

The Nucleus-24 system (as distinct from the Spectra-22), which included a receiver-stimulator suitable for implanting in children under 2 years of age, was tested only after it had been shown in adults that it had been engineered to provide comparable results to those obtained with the Nucleus-22–Spectra-22 and without any side effects. It was also only implanted after the University of Melbourne's NIH study (Studies on Pediatric Auditory Prosthesis Implants, NIH contract No. 1-NS-7-2342) showed that the device could be fixed to allow for head growth changes, that middle ear infection would not lead to any greater risk of inner ear infection than in an unimplanted ear if the electrode entry point was grafted with fascia, and that electrical stimulation would not have an adverse effect on the maturing nervous system. The anatomical studies were discussed in Chapter 2, the bioengineering size constraints in Chapter 8, the results of experimental studies on middle ear infection in the implanted ear in Chapter 3, and the effects of electrical stimulation on the immature nervous system in Chapter 4.

The Nucleus-24 system with SPEAK strategy was approved for use in children by the FDA in June 1998.

CIS (Clarion S and Combi-40 +)

The Clarion S multiple-channel implant was tested on children above 2-years of age for the FDA in 1995. The study was on a group of 124 children with a mean age of 5 years. In the group of 23 who were 6 months postimplantation, the PBK open-set phonetically balanced word results were 23% and 38% for word recognition in a sentence context with the GASP (Erber 1982).

The device received a PMA approval from the FDA in 1997 for this age group.

The Med El Combi-40 + and CIS strategy was evaluated on children at centers in Europe before its trial for the FDA started in 1998. In a study by Allum et al (2000) using the Nucleus 22 and 24 systems with the SPEAK and ACE strategies for 50 children, and the Combi-40 + for 21 children, speech perception for the LiP (Archbold 1994), monosyllable, trochee, polysyllable (MTP) (Erber and Alencewicz 1976), and MAIS (Robbins et al 1991) was shown to increase more rapidly in children under 7 years.

The initial results for the Med-El Combi-40 + device with CIS processing strategy on 68 children 6 to 12 months postoperatively have been reported by Franz (2002). The tests used were the low verbal and standard versions of the ESP test (Moog and Geers 1990), the GASP (Erber 1972), the MLNT and LNT (Kirk et al 1995), and BKB sentences (Bench and Bamford 1979), depending on their age and language skills. Meaningful changes in behavior were assessed with the MAIS (Robbins et al 1991), and the IT-MAIS Zimmerman-Phillips et al 2000). There was a statistically significant improvement for the administered tests.

ACE (Nucleus)

The ACE strategy on the Nucleus 24 system was compared with SPEAK in seven children aged between 9 and 16 years who had used SPEAK for 6 months. The

ACE strategy had a stimulus rate of 900 Hz compared with the 250 Hz for SPEAK. An A/B/A experimental protocol was used with 10 weeks for ACE (B) and 4 weeks for the final period with SPEAK (A). Mean open-set word and phoneme scores were significantly higher for ACE than for SPEAK. But only two of seven were significant. Mean SIT scores for ACE and the second period with SPEAK were higher than the scores for SPEAK at the start of the trial, indicating a significant learning effect. So overall the difference between the two strategies was small (Psarros et al 2000, 2002).

ACE was the default strategy for a trial of the Nucleus 24 and the Contour or perimodiolar electrode array on 256 children for the FDA (Staller et al 2002). The criteria for implantation included profoundly hearing-impaired children down to 12 months, and severely to profoundly hearing-impaired children older than 24 months with open-set word recognition up to 30%. The IT-MAIS was used to assess infants from 12 to 24 months. For children from 25 months to 4 years the ESP, GASP, MLNT, and MAIS were administered. Children older than 5 years had the ESP, GASP, LNT (monosyllabic words), the Hearing in Noise Test for Children (HINT-C) (Nilsson et al 1996), and the MAIS, all administered as pre-recorded material. With the IT-MAIS at 6 months the infants manifest 6/10 listening behaviors either frequently or always. For the older children between 25 months and 4 years the MAIS mean score improved from 28% to 68%, the closed-set ESP scores for pattern perception from 31% to 67%, spondee identification 29% to 73%, and monosyllable identification from 27% to 63%. For older children the mean performance on GASP words improved from 34% to 72% at 6 months, the LNT word recognition from 7% to 41%, and HINT-C sentences from 11% to 61%. The Nucleus 24 and Contour array was approved for infants from 12 months and above and children from 24 months with 30% speech levels by the FDA in November 2000.

It was also important to assess speech perception in children who used tonal languages such as Mandarin. As discussed in Chapter 7, the tonal changes can be described through reference to five frequency points within the normal pitch range of a speaker's voice: low, half-low, middle, half-high, and high (Chao 1930). It was shown for Cantonese, which has six contrastive tones using multidimensional scaling, that average pitch and direction played the main role in tone perception (Gandour and Harshman 1978; Gandour 1981). This was also found in the case of seven children with the SPEAK and seven with the ACE strategies (Barry et al 2002a,b). The data also showed that the SPEAK users relied more on pitch height and ACE on tone contours. This is consistent with the research of Grayden and Clark (2000, 2001), which showed that place of stimulation was better represented with low stimulus rates and manner with high rates.

Comparison with Hearing Aid and Tactile Vocoder

There are a number of studies that have compared the speech perception of children using the Nucleus F0/F1/F2 and Multipeak cochlear implant speech processors with matched children who use hearing aids or tactile vocoders (Osberger 1991c, 1996; Geers and Moog 1994; Miyamoto et al 1994; 1995a,b, 1996). Geers and Moog (1994) and Geers and Tobey (1995) undertook a comparative study of the speech perception, speech production, and receptive and expressive language of three groups of children who used the Nucleus F0/F1/F2 and Multipeak cochlear implant speech processors, hearing aids, and tactile vocoders. They found that profoundly hearing-impaired children with a hearing loss greater than 100 dB performed better with tactile aids than conventional hearing aids, but the speech and language of children with cochlear implants was the best. Most children with multiple-channel tactile vocoders could recognize speech patterns, but few could recognize words even from a closed set. In contrast, the cochlear implant made profoundly hearing-impaired children comparable to children with a lesser hearing loss (in the 90- to 100-dB range). Thus they could recognize words from an open set using electrical stimulation alone. Meyer et al (1998), also found the same advantages for the Nucleus F0/F1/F2 and Multipeak cochlear implant speech processors, but speech perception was only better than aided children with a hearing loss in the 101 to 110 dB HL range. Subsequently, Svirsky and Meyer (1999) found a similar situation when they compared 75 children using hearing aids with 222 children using the Clarion multiple-channel cochlear implants.

Improved results were reported for the Nucleus system by Blamey et al (2001c), who found that profoundly hearing-impaired children with a hearing loss of 106 dB HL and using cochlear implants obtained speech perception and receptive and expressive language at a level comparable to children with a hearing loss of about 78 dB HL. These improved results were presumed due to better speech-processing strategies (SPEAK) and longer usage.

Speech Production with Cochlear Implants

Single-Channel System (3M/House)

Speech production in children with the 3M/House single-channel cochlear implant was evaluated with the phonetic and phonologic level evaluations (PPLEs) (Ling 1976). After 1 year they demonstrated significant improvement in the production of nonsegmental speech features (e.g., duration and rhythm) and vowels and consonants, but not with consonant blends. Children implanted at an earlier age had better speech than those in an oral rather than total communication setting (Kirk and Hill-Brown 1985).

Nucleus Multiple-Channel (F0/F1/F2) and Multipeak Strategies

F0/F1/F2 WSP-III (Nucleus)

The speech production of children became an important consideration in the use of the cochlear implant after the speech perception benefits had been demonstrated. The initial studies on speech production in children using the F0/F1/F2 strategy on Nucleus 22 implant were reported by Tobey et al (1988, 1991) and Tobey and Hasenstab (1991). The speech production of children was categorized from the phonologic level evaluation (Ling 1976). Of the 61 children in the study, 31% had a significant improvement in imitative nonsegmental speech production, and 67% in segmental speech production 1 year after implantation (Tobey et al 1988, 1991; Tobey and Hasenstab 1991). Approximately half had significant improvement in spontaneous speech production. The data suggested that children with increased auditory experience before implantation also developed better speech. It was subsequently shown by Tye-Murray et al (1995) that there was improved intelligibility in children who had used the Nucleus 22 system for 2 years or longer, and in those who had been implanted before the age of 5 years.

Multipeak-MSP (Nucleus)

As with the formant strategies, there was also considerable variability in performance. Osberger et al (1994) found speech intelligibility scores of 48% for children who had the Nucleus F0/F1/F2 WSP-III and Multipeak-MSP systems for at least 2 years. This score was well below that for hearing children of the same age. The ratings varied from 14% to 93% intelligible. Thus some children perform as well as hearing children.

Over a 6-year period Blamey et al (2001a) assessed the progress in speech production of nine children with the Nucleus 22 system. Initially, two used the F0/F1/F2 strategy and seven the Multipeak. At 3 years postimplantation all were using the Multipeak, and by 6 years they had been converted to SPEAK. A broad transcription of the conversations was used to measure the percentage of correct productions of monophthongs, diphthongs, singleton consonants, consonant clusters, and whole words. A direct measure of intelligibility was also derived by counting the proportion of syllables that were unintelligible to the transcribers. The conversation was transcribed by a speech pathologist or linguist and analyzed by CASALA. Four years postimplantation, at least 90% of all syllables produced were intelligible (Fig. 12.5), although only one child had intelligibility over 19% prior to implantation.

SPEAK and ACE Spectra-22 and Nucleus 24

Studies are required to follow the speech production of young children implanted at a young age with the Nucleus 24 device with the SPEAK and ACE strategies. The development of speech production in children who used a tonal language such as Cantonese has been assessed by Barry et al (2000) for the SPEAK strategy. It was thought the children with SPEAK should acquire a tonal inventory more rapidly than one for vowels (the latter depending on formants). But a study on three children by Barry et al (2000) showed that the acquisition of a tonal speech inventory was slower, with none acquiring a low-falling element. This could have been due to the fact that that the extraction of voicing as rate of stimulation with the Nucleus Multipeak and other formant strategies provides better pitch perception and requires investigation. Ciocca et al (2002) found in 17 children, in whom

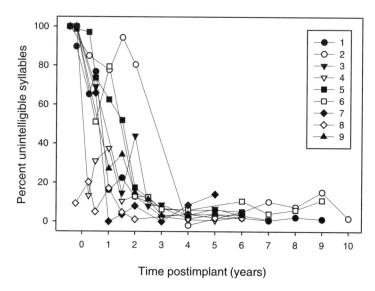

FIGURE 12.5. The percentage of unintelligible syllables versus time postimplantation. Nine children's results are graphed. (From Blamey 2002. Development of spoken language by deaf children. In Marschark, M. and P. Spencer, eds. Handbook of deaf studies, language and education. With permission of Oxford University Press.)

six used SPEAK and 11 ACE, that above-chance performance was obtained with three tonal contrasts but was poorer than a moderately impaired control. This suggests that the strategy is not providing the fine temporospatial patterns required for comparable hearing.

Comparison with Hearing Aid and Tactile Vocoder

It was also an important question to compare the speech production benefits from the F0/F1/F2 strategy on the Nucleus 22 implant with those from the 3M/House single channel and Tactaid II (two-channel vibrotactile aid). The speech was classified as nonspeech, speech-like, and speech. The largest improvements were for the Nucleus 22 (F0/F1/F2) speech strategy. But only 67% of their utterances were judged to be phonetic approximations (Osberger et al 1991b). After 1 year the children with the Nucleus 22 (F0/F1/F2) speech strategies showed an increase in the number of stops, fricatives, and glides, and a reduction in nasal consonants. Osberger et al (1991c) compared the speech of children with the Nucleus 22 (F0/F1/F2) speech strategies and children with hearing aids and different hearing thresholds. After 4 years the mean intelligibility of the children with the Nucleus 22 (F0/F1/F2 and Multipeak) speech strategies was 40% compared with 20% for the aided children with thresholds of 101 to 110 dB. The intelligibility did not reach those children with a threshold of 90 to 100 dB when using a hearing aid.

In a comparative study by Tobey et al (1994), triads of children with the Nucleus 22 implant, hearing aids, and the Tactaid II and IV were compared. They were matched for age, unaided thresholds, family support, intelligence, and speech and language skills. After 3 years the children with the Nucleus implant had increases in imitative speech production of 36% compared to 20% for both the hearing aid and Tactaid groups.

Svirsky et al (1998) compared the speech intelligibility of 44 children using the Nucleus SPEAK and Clarion CIS strategies, and after 1.5 to 2.5 years' usage the speech of the implant children was the same as for those with a threshold in the 90- to 100-dB range.

Language Development for Pre- and Postlinguistically Deaf Children

The development of receptive and spoken language by deaf children is very important for their education, social development, and career opportunities. This can now be achieved through early diagnosis with procedures that include steady-state evoked potentials (Rickards and Clark 1984; Cohen et al 1991; Rance et al 1993, 1995), auditory brainstem responses (ABRs) with a "notched-noise" masker (Stapells et al 1995), and otoacoustic emissions (Kemp 1978), as well as with early intervention and training programs (Ling 1976, 1984; Ling and Ling 1978; Ling and Nienhuys 1983). The elements of spoken language as discussed above (see Language: Test Principles) are (1) receptive and expressive; (2) cognitive, motor, and sensory; and (3) phonology, morphology, syntax, and pragmatics.

The language of children with a cochlear implant is related to their speech perception. Cochlear implants provide information in the middle to high speech frequency range not available to the children, as they usually have only low-frequency hearing. Their speech perception can be predicted to a reasonable degree, as discussed in some detail above (see Predictive Factors) and in Chapters 9 and 11, and thus their language ability can also be predicted. In developing language it is important to consider the nature of the child, family, and habilitation or education programs (Spencer 2002).

The benefits of cochlear implants in developing language have been summarized by Spencer (2002) as follows: "Cochlear implants provide many, but not all, deaf children with access to information that can help them develop understanding and production of spoken language. However, the range of benefits experienced is large and the factors that influence the benefits received by an individual child are still being investigated."

Receptive Language

There is a normal distribution of equivalent language age for children with good hearing, as for example measured with the CELF test. Children with a hearing loss and a hearing aid or a cochlear implant have language that falls predominantly outside this distribution (Blamey 2002). Studies on hearing-impaired children indicated that extrinsic factors leading to language delays are age of intervention (Davis et al 1986; Ramkalawan and Davis 1992; Gilbertson and Kamhi 1995;

Limbrick et al 1992; Dodd et al 1998; Yoshinaga-Itano et al 1998) and time spent reading (Limbrick et al 1992). Intrinsic factors are specific language impairment seen in 10 of 20 (50%) of hearing-impaired children by Gilbertson and Kamhi (1995) and speech-reading ability (Dodd et al 1998). It is important to distinguish between these extrinsic and intrinsic factors. Initial studies on small groups of children showed receptive language improved with the Nucleus 22 (F0/F1/F2 and Multipeak) (Kirk and Hill-Brown 1985; Busby et al 1989; Dowell et al 1991; Geers and Moog 1991; Hasenstab and Tobey 1991). This trend was confirmed by Dawson et al (1995a,b) from the data on a larger group of 32 children. On average they had a language-learning rate of 0.87 based on the PPVT that was approximately double the 0.4 to 0.6 rate for children with hearing aids (Geers and Moog 1988; Boothroyd 1991b).

Comparing the learning rates for the implant and hearing aid groups made allowance for maturation by assuming it was the same for both. To further isolate maturation, Robbins et al (1995, 1997) predicted the effects on the Reynell scale (Reynell and Gruber 1990), and found that for the Nucleus 22 implant the language increase was 7 to 10 months ahead of expected after 12 to 15 months.

The factors responsible for differences in receptive language were examined by Dawson et al (1995a,b). They first confirmed the findings of Osberger et al (1991a), Staller (1990), and Staller et al (1991a,b) for the Nucleus 22 system that age at onset of deafness and duration of the profound hearing loss correlated negatively with speech perception. Second, the variance among children for the growth of vocabulary was not significantly accounted for by these factors as well as duration of use, perception performance, and communication mode. A later study by Connor et al (2000) found a receptive vocabulary growth of 0.63 for children implanted at 2 years compared to 0.45 for children implanted at age 6.5 years.

It was also necessary to monitor language acquisition over time to determine the longer term effects of the Nucleus 22 multiple-channel cochlear implant and F0/F1/F2 and Multipeak strategies. Sarant et al (1996, 2001) and Blamey et al (1998) studied 57 children aged between 4 and 12 years with a bilateral severe or profound hearing loss over a 4-year period. There were 33 hearing aid and 24 implant users. Receptive language measured with the PPVT was on average 62% of the expected level for hearing children. There were considerable differences in performance, and some were at the normal level. The results for children with implants were comparable to those for children with hearing who had a mean threshold of 81 dB.

The structural complexity of language as well as vocabulary developed on average faster than predicted for deaf children without cochlear implants. This was assessed with the RDLS test (Svirsky et al 2000). Nevertheless, after 18 months some of the 23 children in the study continued to have severely delayed expressive language compared to children with unimpaired hearing. Some, however, progressed at a rate typical for hearing children. Bollard et al (1999) also reported scores on vocabulary and language comprehension in 10 young children that increased at a rate equal to or faster than hearing children at an equivalent language level. Nevertheless, after 18 months their language was behind the same children with hearing.

The above studies showed considerable variability in receptive language, and in general children were not reaching age-appropriate language. The data indicated the variability did not depend only on percepts such as place pitch perception (Busby and Clark 2000a,b) or the general factors leading to good speech perception (Dawson et al 1995a,b). A study was therefore undertaken to evaluate how strong a relationship existed between speech perception and language. The speech perception word and sentence scores for audition (A), speech reading (V), and audition plus speech reading (AV) were plotted against the PPVT or CELF equivalent language. As discussed in Chapter 11, a very close relationship was seen. The AV word score reached 100% at a PPVT or CELF age of approximately 8 to 10 years and the A score at 10 to 11 years (Fig. 12.6).

So although language age did not increase at quite the same rate as for normal children, it rose rapidly in proportion to speech perception in quiet. However, it was not clear to what extent perception and language were interdependent.

It was hypothesized that open-set speech perception was limited by vocabulary and that remediation of vocabulary and syntax would increase open-set speech perception scores. A study was undertaken on three implanted children from 9 to 15 years of age (Sarant et al 1996). The perception scores were recorded for words that were both known and unknown. They were retested after the meanings of all words had been learned. Two of the children had statistically significant improvements in the unknown word scores rather than for the known words, suggesting that it was not a practice effect but due to the effect of "top-down"

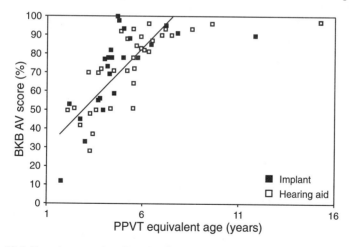

FIGURE 12.6. Speech perception (Bamford-Kowal-Bench (BKB) words in sentences) with audition and speech reading versus Peabody Picture Vocabulary Test (PPVT) equivalent language age for implanted and aided children (Sarant et al 1996, 2001; Blamey et al 1998). (Reprinted with permission from Blamey et al 1998. Speech perception and spoken language in children with impaired hearing. In: Mannell, R. H. and J. Robert-Ribes, eds. ICSLP '98 Proceedings: 2615–2618.)

processing on "bottom-up" perception. The BKB sentence test was then used to assess specific grammatical constructs, again after the children had been taught the rules governing their use. There was benefit for two of the three children.

Improvements in receptive language as reported above for the Nucleus 22 F0/F1/F2 and Multipeak strategies are also being seen for the Clarion CIS strategy, which was approved by the FDA in 1997. In a study on 23 young children, they were found to have a greater than normal increase in language (Robbins et al 1991).

Expressive Language

The expressive language of children, as discussed above, can be analyzed at a phonological level using articulation tests and phonetic transcripts of spoken language. The results are presented as phonetic inventories, percent correct phonemes, or phonological processes. The order of occurrence of the phonemes is thought due to linguistic, acoustic, and articulatory factors. It was shown by Blamey et al (2001b) in a study on nine children using the Nucleus multiple-channel cochlear implant that the order of development was the same as for children with unimpaired hearing, although delayed.

In addition, conversational speech samples were analyzed from nine children who received the Nucleus 22 implant and F0/F1/F2 and Multipeak speech-processing strategies between the ages of 2 and 5 years (Blamey et al 2001a). There was a significant increase in the complexity of the spoken language of the implanted children seen in the study by Blamey et al (2001a). The mean number of syllables both intelligible and unintelligible rose from 1.7 to 5.2.

The PPVT can be used to evaluate expressive language with the expressive subtest of the Woodcock Johnson Tests of Cognitive Ability. Woodcock and Mather (1989) and Blamey et al (2001c) found using the test that the implant and hearing aid users progressed at about 65% of the normal rate. This was similar to the rate for receptive language in children implanted at age 2 years. A marked improvement in language with implant children was reported by Svirsky et al (2000), who found the rate of language development was comparable to that of children with unimpaired hearing.

The production of English grammar for 57 children implanted between 4 and 12 years of age and who used the Nucleus 22 F0/F1/F2 and Multipeak systems was measured with the CELF-3 and CELF-Preschool tests (Blamey et al 1998). The results were on average 45% of the expected level. The development of grammar skills in 29 children in a total communication (simultaneous speech and signed language) program using the Nucleus 22 F0/F1/F2 and Multipeak speech-processing strategies or hearing aids was studied by Tomblin et al (1999). The children's ability to use expressive grammar (syntax) in this study was measured with story retelling. A control group of unimplanted deaf children was involved in the study. The children were 3 to 13 years postimplantation. The children's understanding of sentence structure was assessed by their performance on the Rhode Island Test of Language Structure. The sentences were presented in speech plus signed English. The results showed that all but one of the children with the

Nucleus 22 cochlear implants scored very high, and well above the expected results for deaf children. The scores improved from 30% to 65% in the first 5 years of cochlear implant use, while children with unimpaired hearing improved from 30% to 90% between the ages of 2 and 4 years. In addition, the children with implants tended to use speech (without signs) for a larger percentage of their words than did deaf children without implants, although both groups continued to use both modalities simultaneously for the majority of their productions.

Grammatical morphemes are difficult for deaf children to recognize and produce. Spencer et al (1998) found that 25 children with cochlear implants used grammatical morphemes more often than 13 children with hearing aids in a total communication program. The children with implants did not use signs for expressing these morphemes, although speech and signs were used together for the majority of other words. This indicated the children with implants could perceive the morphemes and incorporate them into their expressive language. Furthermore, despite the language delays, they integrated the morphemes into their language in the same order as hearing children.

Cognition

As discussed above, there were considerable differences in the speech perception, speech production, and language results for children with the multiple-channel cochlear implant, and only about one third to one half of the variance of the speech perception scores could be accounted for (Dowell et al 1995; Sarant et al 2001). Furthermore, the receptive language scores did not match those for normal-hearing children (Blamey et al 1998; Sarant et al 2001). For the above reasons, cognitive studies were commenced to help determine the factors that contribute to the development of language. Cognition involves perception, attention, learning, and memory. Information processing theory emphasizes that these processes should be viewed as a continuum and that all involve some kind of storage system or memory (Pisoni 2000). The effect of restoring some hearing with a multiple-channel cochlear implant on visual attention was first studied, as deaf children had been shown to have deficits in visual matching tasks (Moores et al 1973). Deaf children with cochlear implants performed better than those without, and also developed visual selective attention at a faster rate (Quittner et al 1994). But in another study no substantial difference was found between children with hearing, prelinguistically deaf without a cochlear implant, and deaf with a cochlear implant on a continuous performance visual attention task and a letter cancellation task.

A study was undertaken by Surowiecki et al (2002) to compare matched children with either hearing aids or cochlear implants with eight neuropsychological measures of visual memory, attention, and executive functioning. It also examined whether differences in cognitive skills could account for variance in speech perception, vocabulary, and language abilities. First, there was no difference between the cognitive abilities of the aided and implanted children. Second, the children's visual memory skills (i.e., recognition memory, delayed recall, and paired associative learning memory) correlated with their language, but attention and executive functioning did not. Further research is needed to determine how the development of language with a cochlear implant can be assisted.

References

Allum, J. H. J., R. Greisiger, S. Straubhaar and M. G. Carpenter. 2000. Auditory perception and speech identification in children with cochlear implants tested with the EARS protocol. British Journal of Audiology 34: 293–303.

Anthony, A., D. Bogle, T. T. Ingram and M. W. McIsaac. 1971. The Edinburgh articulation test. Edinburgh, Churchill Livingstone.

Archbold, S. 1994. Monitoring progress in children at the pre-verbal stage. In: McCormick, B. and S. Sheppard, eds. Cochlear implants for young children. London, Whurr: 197–213.

Arndt, P., S. Staller, J. Arcaroli, A. Hines and K. Ebinger. 1999. Within-subject comparison of advanced coding strategies in the Nucleus 24 cochlear implant. Cochlear Corporation Report.

Barry, J., P. Blamey, K. Lee and D. Cheung. 2000. Differentiation in tone production in Cantonese-speaking hearing-impaired children. Proceedings of the 6th International Conference on Spoken Language Processing. Beijing, China, Military Friendship: Vol 1: 669–672.

Barry, J. G., P. J. Blamey and L. F. A. Martin. 2002a. A multidimensional scaling analysis of tone discrimination ability in Cantonese-speaking children using a cochlear implant. Clinical Linguistics and Phonetics 16: 101–113.

Barry, J. G., P. J. Blamey, L. F. A. Martin, et al. 2002b. Tone discrimination in Cantonese-speaking children with cochlear implants. Clinical Linguistics and Phonetics 16: 79–99.

Battmer, R.-D., D. Gnadeberg, D. J. Allum-Mecklenberg and T. Lenarz. 1994. Matched-pair comparisons for adults using the Clarion or Nucleus devices. Annals of Otology, Rhinology and Laryngology 104: 251–254.

Battmer, R. D., S. P. Gupta, D. J. Allum-Mecklenburg and T. Lenarz. 1995. Factors influencing cochlear implant perceptual performance in 132 adults. Annals of Otology, Rhinology and Laryngology 104: 185–187.

Bench, R. J. and J. Bamford. 1979. Speech-hearing tests and the spoken language of hearing-impaired children. London, Academic Press.

Berliner, K. I. and L. S. Eisenberg. 1985. Methods and issues in the cochlear implantation of children: an overview. Ear and Hearing 6(3 suppl): 6S–13S.

Berliner, K. I., L. L. Tonokawa, L. M. Dye and W. F. House. 1989. Open-set speech recognition in children with a single-channel cochlear implant. Ear and Hearing 10: 237–242.

Blamey, P. J. 2002. Development of spoken language by deaf children. In: Marschark, M. and P. Spencer, eds. Handbook of deaf studies, language and education. Oxford, Oxford University Press.

Blamey, P. J., J. Barry, C. Bow, J. Sarant, L. Paatsch and R. Wales. 2001a. The development of speech production following cochlear implantation. Clinical Linguistics and Phonetics 15: 363–382.

Blamey, P. J., J. Barry and P. Jacq. 2001b. Phonetic inventory development in young cochlear implant users 6 years postoperation. Journal of Speech, Language and Hearing Research 44: 73–79.

Blamey, P. J. and G. M. Clark. 1986. A model of auditory visual perception. Proceedings of the first Australian conference on speech and technology, Canberra: 54–59.

Blamey, P. J., P. W. Dawson, S. J. Dettman, et al. 1992a. Speech perception, production and language results in a group of children using the 22-electrode cochlear implant. Journal of the Oto-Laryngological Society of Australia 1: 105–109.

Blamey, P. J., R. C. Dowell, A. M. Brown, G. M. Clark and P. M. Seligman. 1987. Vowel and consonant recognition of cochlear implant patients using formant-estimating speech processors. Journal of the Acoustical Society of America 82: 48–57.

Blamey, P. J., R. C. Dowell, Y. C. Tong, A. M. Brown, S. M. Luscombe and G. M. Clark. 1984a. Speech processing studies using an acoustic model of a multiple-channel cochlear implant. Journal of the Acoustical Society of America 76: 104–110.

Blamey, P. J., R. C. Dowell, Y. C. Tong and G. M. Clark. 1984b. An acoustic model of a multiple-channel cochlear implant. Journal of the Acoustical Society of America 76: 97–103.

Blamey, P. J., M. Grogan and M. B. Shields. 1994. Using an automatic word-tagger to analyse the spoken language of children with impaired hearing. In: Togneri, R., ed. Fifth Australian International Conference on Speech Science and Technology. Canberra, Australian Speech Science and Technology Association: 498–503.

Blamey, P. J., C. J. James, G. J. Dooley and E. S. Parisi. 2000. Monaural and binaural loudness measures in cochlear implant users with contralateral residual hearing. Ear and Hearing 21: 6–17.

Blamey, P. J., L. F. Martin and G. M. Clark. 1985. A comparison of three speech coding strategies using an acoustic model of a cochlear implant. Journal of the Acoustical Society of America 77: 209–217.

Blamey, P. J., E. Parisi and G. M. Clark. 1995. Pitch matching of electric and acoustic stimuli. Annals of Otology, Rhinology and Laryngology 104: 220–222.

Blamey, P. J., B. C. Pyman, M. Gordon, et al. 1992b. Factors predicting postoperative sentence scores in postlinguistically deaf adult cochlear implant patients. Annals of Otology, Rhinology and Laryngology 101: 342–348.

Blamey, P. J., J. Sarant, L. Paatsch, et al. 2001c. Relationships among speech perception, production, language, hearing loss, and age in children with impaired hearing. Journal of Speech, Language and Hearing Research 44: 264–285.

Blamey, P. J., J. Z. Sarant, T. A. Serry, et al. 1998. Speech perception and spoken language in children with impaired hearing. In: Mannell, R. H. and J. Robert-Ribes, eds. Proceedings of ICSLP '98 Fifth International Conference on Spoken Language Processing. Canberra, Australian Speech Science and Technology Association: 2615–2618.

Bollard, P., A. Popp, P. Chute and S. Parisier. 1999. Specific language growth in young children using the Clarion cochlear implant. Annals of Otology, Rhinology and Laryngology 108: 119–123.

Boothroyd, A. 1968. Developments in speech audiometry. Sound 2: 3–10.

Boothroyd, A. 1991a. CASPER: a user friendly system for Computer Assisted Speech Perception Testing and Training. New York, City University of New York.

Boothroyd, A. 1991b. Speech perception measures and their role in the evaluation of hearing aid performance in a pediatric population. In: Feigin, J. A. and P. G. Stelmachowicz, eds. Pediatric amplification. Omaha, Boys Town National Research Hospital: 77–91.

Boothroyd, A. 1997. Auditory capacity of hearing-impaired children using hearing aids and cochlear implants: issues of efficacy and assessment. Scandinavian Audiology Supplementum 46: 17–25.

Boothroyd, A. 1998. Evaluating the efficacy of hearing aids and cochlear implants in children who are hearing-impaired. In: Bess, F. H., ed. Children with hearing impairment: contemporary trends. Nashville, Bill Wilkerson Center Press: 249–260.

Boothroyd, A., A. E. Geers and J. S. Moog. 1991. Practical implications of cochlear implants in children. Ear and Hearing 12(suppl 4): 81S–89S.

Boothroyd, A., L. Hanin and O. Eran. 1996. Speech perception and production in children with hearing impairment. In: Amplification for children with auditory deficits. Bess, F. H., J. S. Gravel and A. M. Tharpe, eds. Nashville, Tenn, Wilkerson Center Press: 55–74.

Boothroyd, A., T. Hnath-Chisolm and L. Hanin. 1985. A sentence of test of speech perception: reliability, set-equivalence, and short-term learning. New York, City University of New York, report no. RC110.

Bornstein, H. and K. Saulnier. 1981. Signed English: a brief follow-up to the first evaluation. American Annals of the Deaf 126: 69–72.

Bornstein, H., K. Saulnier and L. Hamilton. 1980. Signed English: a first evaluation. American Annals of the Deaf 125: 467–481.

Brimacombe, J. A., P. L. Arndt, S. J. Staller and C. M. Menapace. 1995. Multichannel cochlear implants in adults with residual hearing. NIH Consensus Development Conference on Cochlear Implants in Adults and Children.

Brown, R. 1973. A first language: the early stages. Cambridge, MA, Harvard University Press.

Busby, P. A. and G. M. Clark. 2000a. Electrode discrimination by early-deafened subjects using the Cochlear Limited multiple-electrode cochlear implant. Ear and Hearing 21: 291–304.

Busby, P. A. and G. M. Clark. 2000b. Pitch estimation by early-deafened subjects using a multiple-electrode cochlear implant. Journal of the Acoustical Society of America 107: 547–558.

Busby, P. A., S. A. Roberts, Y. C. Tong and G. M. Clark. 1991. Results of speech perception and speech production training for three prelingually deaf parents using a multiple-electrode cochlear implant. British Journal of Audiology 25: 291–302.

Busby, P. A., Y. C. Tong and G. M. Clark. 1984. Underlying dimensions and individual differences in auditory, visual and auditory-visual vowel perception by hearing impaired children. Journal of the Acoustical Society of America 75: 1858–1865.

Busby, P. A., Y. C. Tong, S. A. Roberts, et al. 1989. Results for two children using a multiple-electrode intracochlear implant. Journal of the Acoustical Society of America 86(6): 2088–2102.

Chao, Y. R. 1930. A system of tone letters. Le Maître Phonétique 45: 24–27.

Chomsky, N. and M. Halle. 1968. The sound pattern of English. New York, Harper and Row.

Chute, P. M. 1993. Cochlear implants in adolescents. Advances in Oto-Rhino-Laryngology 48: 210–215.

Ciocca, V., A. L. Francis, R. Aisha and L. Wong. 2002. The perception of Cantonese lexical tones by early-deafened cochlear implantees. Journal of the Acoustical Society of America 111: 2250–2256.

Clark, G. M. 1986. The University of Melbourne/Cochlear Corporation (Nucleus) program. Otolaryngologic Clinics of North America 19: 329–354.

Clark, G. M., P. J. Blamey, P. A. Busby, et al. 1987a. A multiple-electrode intracochlear implant for children. Archives of Otolaryngology 113: 825–828.

Clark, G. M., P. A. Busby, S. A. Roberts, et al. 1987b. Preliminary results for the Cochlear

Corporation multielectrode intracochlear implants on six prelingually deaf patients. American Journal of Otology 8: 234–239.

Clark, G. M., P. A. Crosby, R. C. Dowell, et al. 1983. The preliminary clinical trial of a multichannel cochlear implant hearing prosthesis. Journal of the Acoustical Society of America 74: 1911–1914.

Clark, G. M., R. C. Dowell, A. M. Brown, et al. 1983. The clinical trial of a multiple-channel cochlear prosthesis. An initial study in four patients with a profound total hearing loss. Medical Journal of Australia 2: 430–433.

Clark, G. M., R. C. Dowell, R. S. C. Cowan, B. C. Pyman and R. L. Webb. 1996. Multicenter evaluations of speech perception in adults and children with the Nucleus (Cochlear) 22-channel cochlear implant. In: Portmann, M., ed. Transplants and Implants in Otology III. Amsterdam, Kugler: 353–363.

Clark, G. M. and Y. C. Tong. 1981. Multiple-electrode cochlear implant for profound or total hearing loss: a review. Medical Journal of Australia 1: 428–429.

Clark, G. M. and Y. C. Tong. 1982. A multiple-channel cochlear implant. A summary of results for two patients. Archives of Otolaryngology 108: 214–217.

Clark, G. M., Y. C. Tong, Q. R. Bailey, et al. 1978. A multiple-electrode cochlear implant. Journal of the Oto-Laryngological Society of Australia 4: 208–212.

Clark, G. M., Y. C. Tong and R. C. Dowell. 1982. Single versus multiple-channel electrical stimulation of the auditory nerve in speech processing for a totally deaf patient. Proceedings of the Australian Physiological and Pharmacological Society 13: 212P.

Clark, G. M., Y. C. Tong and R. C. Dowell. 1983. Clinical results with a multi-channel pseudobipolar system. Annals of the New York Academy of Sciences 405: 370–377.

Clark, G. M., Y. C. Tong and L. F. Martin. 1981a. A multiple-channel cochlear implant. An evaluation using closed-set spondaic words. Journal of Laryngology and Otology 95: 461–464.

Clark, G. M., Y. C. Tong and L. F. Martin. 1981b. A multiple-channel cochlear implant. An evaluation using open-set CID sentences. Laryngoscope 91: 628–634.

Clark, G. M., Y. C. Tong, L. F. Martin and P. A. Busby. 1981c. A multiple-channel cochlear implant. An evaluation using an open-set word test. Acta Oto-Laryngologica 91: 173–175.

Clark, G. M., Y. C. Tong, L. F. A. Martin, et al. 1981d. A multiple-channel cochlear implant: an evaluation using nonsense syllables. Annals of Otology, Rhinology and Laryngology 90: 227–230.

Cohen, L. T., F. W. Rickards and G. M. Clark. 1991. A comparison of steady-state evoked potentials to modulated tones in awake and sleeping humans. Journal of the Acoustical Society of America 90: 2467–2479.

Cohen, N. L., S. B. Waltzman and S. G. Fisher. 1993. A prospective, randomised study of cochlear implants. New England Journal of Medicine 328: 233–282.

Connor, C. M., S. Hieber, H. A. Arts and T. A. Zwolan. 2000. Speech, vocabulary, and the education of children using cochlear implants: oral or total communication? Journal of Speech, Language, and Hearing Research 43: 1185–1204.

Cowan, R. S. C., C. D. Brown, L. A. Whitford, et al. 1995. Speech perception in children using the advanced SPEAK speech-processing strategy. Annals of Otology, Rhinology and Laryngology 104(suppl 166): 318–321.

Cox, R. M., G. C. Alexander and C. Gilmore. 1987. Development of the connected speech test (CST). Ear and Hearing 8(suppl 5): 119S–126S.

Crary, M. A. 1982. Phonological intervention concepts and procedures. San Diego, College-Hill Press.

Crystal, D. 1981. Clinical linguistics. Vienna, Springer-Verlag.

Crystal, D. 1992. Profiling linguistic disability, 2nd ed. London, Whurr.

Crystal, D., P. Fletcher and M. Garman. 1989. Grammatical analysis of language disability, 2nd ed. London, Whurr.

Davis, H. and S. R. Silverman. 1978. Hearing and deafness. 4th edition New York, Holt, Rinehart and Winston.

Davis, J. M., J. Elfenbeing, R. Schum and R. A. Bentler. 1986. Effects of mild and moderate hearing impairments on language educational and psychosocial behaviour of children. Journal of Speech and Hearing Disorders 51: 53–62.

Dawson, P., P. J. Blamey, G. M. Clark, et al. 1989. Results in children using the 22 electrode cochlear implant. Journal of the Acoustical Society of America 86(suppl 1): 81.

Dawson, P. W., P. J. Blamey, S. J. Dettman, et al. 1995a. A clinical report on speech production of cochlear implant users. Ear and Hearing 16: 551–561.

Dawson, P. W., P. J. Blamey, L. C. Rowland, et al. 1992. Cochlear implants in children, adolescents and prelinguistically deafened adult: speech perception. Journal of Speech and Hearing Research 35: 401–417.

Dawson, P. W., P. J. Blamey, L. C. Rowland, et al. 1995b. A clinical report on receptive vocabulary skills in cochlear implant users. Ear and Hearing 16: 287–294.

De Fillipo, C. L. and B. L. Scott. 1978. A method for training and evaluating the reception of ongoing speech. Journal of the Acoustical Society of America 64: 1186–1192.

Dodd, B. 1976. The phonological systems of deaf children. Journal of Speech and Hearing Disorders 41: 185–198.

Dodd, B., B. McIntosh and L. Woodhouse. 1998. Early lipreading ability and speech and language development of hearing-impaired pre-schoolers. In: Campbell, R., B. Dodd and D. Burnham, eds. Hearing by eye II. Hove, UK, Psychology Press: 229–242.

Dorman, M. F., K. Dankowski and G. McCandless. 1989. Consonant recognition as a function of the number of channels of stimulation by patients who use the Symbion cochlear implant. Ear and Hearing 10: 288–291.

Dowell, R. C. 1990. Speech perception in noise using the multichannel cochlear prosthesis. Australian Journal of Audiology (suppl 4): 11.

Dowell, R. C., P. J. Blamey and G. M. Clark. 1995. Potential and limitations of cochlear implants in children. Annals of Otology, Rhinology and Laryngology 104(suppl 166): 324–327.

Dowell, R. C., P. J. Blamey and G. M. Clark. 1997. Factors affecting outcomes in children with cochlear implants. In: Clark, G. M., ed. Cochlear implants. XVI World Congress of Otorhinolaryngology Head and Neck Surgery. Bologna, Monduzzi: 297–303.

Dowell, R. C., A. M. Brown and D. J. Mecklenburg. 1990a. Clinical assessment of implanted deaf adults. In: Clark, G., Y. Tong and J. Patrick, eds. Cochlear prostheses. Edinburgh, Churchill Livingstone: 193–205.

Dowell, R. C., P. W. Dawson, S. J. Dettman, et al. 1991. Multichannel cochlear implantation in children. A summary of current work at the University of Melbourne. American Journal of Otology (suppl 12): 137–143.

Dowell, R. C., L. F. A. Martin, P. J. Blamey and A. M. Brown. 1985a. Assessment of implant patient speech discrimination. In: Schindler, R. and M. Merzenich, eds. Cochlear implants. New York, Raven Press: 465–468.

Dowell, R. C., L. F. Martin, G. M. Clark and A. M. Brown. 1985b. Results of a preliminary clinical trial on a multiple-channel cochlear prosthesis. Annals of Otology, Rhinology and Laryngology 94: 244–250.

Dowell, R. C., L. F. Martin, Y. C. Tong, G. M. Clark, P. M. Seligman and J. F. Patrick.

1982. A 12-consonant confusion study on a multiple-channel cochlear implant patient. Journal of Speech and Hearing Research 25: 509–516.

Dowell, R. C., D. J. Mecklenburg and G. M. Clark. 1986. Speech recognition for 40 patients receiving multichannel cochlear implants. Archives of Otolaryngology 112: 1054–1059.

Dowell, R. C., P. M. Seligman, P. J. Blamey and G. M. Clark. 1987a. Evaluation of a two-formant speech-processing strategy for a multichannel cochlear prosthesis. Annals of Otology, Rhinology and Laryngology 96(suppl 128): 132–133.

Dowell, R. C., P. M. Seligman, P. J. Blamey and G. M. Clark. 1987b. Speech perception using a two-formant 22-electrode cochlear prosthesis in quiet and in noise. Acta Oto-Laryngologica 104(5–6): 439–446.

Dowell, R. C., L. A. Whitford, P. M. Seligman, B. K.-H. Franz and G. M. Clark. 1990b. Preliminary results with a miniature speech processor for the 22-electrode/Cochlear hearing prosthesis. In: Sacristan, T., ed. Otorhinolaryngology, head and neck surgery. Amsterdam, Kugler and Ghedini: 1167–1173.

Dunn, L. M. and L. M. Dunn. 1981. Peabody picture vocabulary test–revised. Circle Pines, Minnesota American Guidance Service.

Dunn, L. M. and L. M. Dunn. 1997. Peabody picture vocabulary test, 3rd ed. Circle Pines, Minnesota American Guidance Service.

Eddington, D. K. 1980. Speech discrimination in deaf subjects with cochlear implants. Journal of the Acoustical Society of America 68: 885–91.

Eddington, D. K. 1983. Speech recognition in deaf subjects with multichannel intraco-chlear electrodes. Annals of the New York Academy of Science 405: 241–258.

Elliott, L. L. and D. R. Katz. 1980. Northwestern University children's perception of speech (NU-CHIPS). St Louis, Auditec.

Erber, N. P. 1972. Auditory, visual, and auditory-visual speech recognition of consonants by children with normal and impaired hearing. Journal of Speech and Hearing Research 15: 413–422.

Erber, N. P. 1982. Auditory training. Washington, DC, Alexander Graham Bell Association for the Deaf.

Erber, N. P. and C. M. Alencewicz. 1976. Audiological evaluation of deaf children. Journal of Speech and Hearing Disorders 41: 256–276.

Fisher, H. B. and J. A. Logemann. 1971. Test of articulation competence. New York, Houghton and Mifflin.

Franz, D. C. 2002. Pediatric performance with the Med El Combi 40 + cochlear implant system. Annals of Otology, Rhinology and Laryngology 111(suppl 189): 66–68.

Gandour, J. 1981. Perceptual dimensions of tone: evidence from Cantonese. Journal of Chinese Linguistics 9: 20–36.

Gandour, J. T. and R. A. Harshman. 1978. Crosslanguage differences in tone perception: a multidimensional scaling investigation. Language and Speech 21: 1–33.

Gantz, B. J., B. F. McCabe, R. S. Tyler and J. P. Preece. 1987. Evaluation of four cochlear implant designs. Annals of Otology, Rhinology and Laryngology 96: 145–147.

Gantz, B. J., R. S. Tyler, J. F. Knutson, et al. 1988. Evaluation of five different cochlear implant designs: audiologic assessment and predictors of performance. Laryngoscope 98: 1100–1106.

Gantz, B. J., R. S. Tyler, G. Woodworth, N. Tye-Murray and H. Fryauf-Bertschy. 1994. Results of multichannel cochlear implant in congenital and acquired prelingual deafness in children: five-year follow-up. American Journal of Otology 15(suppl 2): 1–8.

Gantz, B. J., G. G. Woodworth, J. F. Knutson, P. J. Abbas and R. S. Tyler. 1993. Multivariate predictors of audiological success with multichannel cochlear implants. Annals of Otology, Rhinology and Laryngology 102(12): 909–916.

Gardner, M. 1979. Expressive one-word picture vocabulary test. Novato, CA, Academic Therapy.

Gaunaurd, G. C. and G. F. Kuhn. 1980. Phase- and group-velocities of acoustic waves around a sphere simulating the human head. Journal of the Acoustical Society of America 68: S57.

Geers, A. E. and J. S. Moog. 1988. Predicting long-term benefits from single-channel cochlear implants in profoundly hearing-impaired children. American Journal of Otology 9: 169–176.

Geers, A. E. and J. S. Moog. 1991. Evaluating the benefits of cochlear implants in an education setting. American Journal of Otology 12(suppl): 116–125.

Geers, A. E. and J. S. Moog. 1994. Effectiveness of cochlear implants and tactile aids for deaf children. The Volta Review 96: 1–231.

Geers, A. E. and E. A. Tobey. 1995. Longitudinal comparison of the benefits of cochlear implants and tactile aide in a controlled educational setting. Annals of Otology Rhinology and Laryngology 104(suppl 166): 328–329.

Gilbertson, M. and A. G. Kamhi. 1995. Novel word learning in children with hearing impairment. Journal of Speech and Hearing Research 38: 630–642.

Grayden, D. B. 2000. The effect of rate of stimulation on consonant recognition for users of the CI24M cochlear implant. Abstracts of the Twenty-third midwinter research meeting of Association for Research in Otolaryngology. St Petersburg Beach, Florida, February 20–24, 2000: 92.

Grayden, D. B. and G. M. Clark. 2000. The effect of rate stimulation of the auditory nerve on phoneme recognition. In: Barlow, M., ed. Proceedings of the Eighth Australian International Conference on Speech Science and Technology. Canberra, Australian Speech Science and Technology Association: 356–361.

Grayden, D. B. and G. M. Clark. 2001. Improved sound processor for cochlear implants. International patent application No. PCT/AU00/01038.

Hagerman, D. 1982. Sentences for speech intelligibility in noise. Scandanavian Audiology 11: 79–87.

Hagerman, D. 1984. Clinical measurements of speech reception thresholds in noise. Scandanavian Audiology 13: 57–63.

Hasenstab, M. S. and E. M. Tobey. 1991. Language development in children receiving Nucleus multi-channel cochlear implants. Ear and Hearing 12: 55S–65S.

Haskins, H. A. 1964. Kindergarten PB word lists. In: Newby, H. A., ed. Audiology. New York, Appleton Century Crofts.

Helms, J., J. Muller and F. Schon. 1997. Evaluation of performance with the COMBI 40 cochlear implant in adults: a muticentric clinical study. ORL Journal of Otorhinolaryngology and its Related Specialties 59: 23–35.

Hollow, R. D., R. C. Dowell, R. S. C. Cowan, M. C. Skok, B. C. Pyman and G. M. Clark. 1995. Continuing improvements in speech processing for adult cochlear implant patients. Annals of Otology, Rhinology and Laryngology 104(suppl 166): 292–294.

Hudgins, C. and F. Numbers. 1942. An investigation of the intelligibility of the speech of the deaf. Genetic Psychology Monographs 25: 289–392.

Ingram, D. 1976. Phonological disability in children. New York, Elsevier.

Jerger, S., S. Lewis and J. Jerger. 1980. Paediatric speech intelligibility test 1. Generation of speech materials. International Journal of Paediatric Otolaryngology 2: 217–230.

Kalikow, D. N., K. N. Stevens and L. L. Elliott. 1977. Development of a test of speech intelligibility in noise using sentence materials with controlled word predictability. Journal of the Acoustical Society of America 61: 1337–1351.

Kemp, D. T. 1978. Stimulated acoustic emissions from within the human auditory system. Journal of the Acoustical Society of America 64: 1386–1391.

Kessler, D. K. 1998. New directions in speech processing II: the electrode connection. Paper presented at the 7th Symposium on Cochlear Implants in Children, Iowa City, IA.

Kessler, D. K., G. E. Loeb and M. J. Barker. 1995. Distribution of speech recognition results with the Clarion cochlear prosthesis. Annals of Otology, Rhinology and Laryngology 104: 283–285.

Kiefer, J., V. Gall, C. Desloovere, R. Knecht, A. Mikowski and C. von Ilberg. 1996. A follow-up study of long-term results after cochlear implantation in children and adolescents. European Archives of Otorhinolaryngology 253: 158–166.

Kirk, K. I. 2000. Challenges in the clinical investigation of cochlear implant outcomes. In: Niparko, J. K., K. I. Kirk, N. K. Mellon, et al, eds. Cochlear implants: principles and practices. Philadelphia, Lippincott Williams & Wilkins: 225–259.

Kirk, K. I. and C. Hill-Brown. 1985. Speech and language results in children with a cochlear implant. Ear and Hearing 6 (suppl): 36S–47S.

Kirk, K. I., D. B. Pisoni and M. J. Osberger. 1995. Lexical effects on spoken word recognition by pediatric cochlear implant users. Ear and Hearing 16: 470–481.

Kuhn, G. F. 1979. Stop consonant place perception with single-formant stimuli: evidence for the role of the front-cavity resonance. Journal of the Acoustical Society of America 65: 991–1000.

Kuhn, G. F. and E. D. Burnett. 1977. Acoustic pressure field alongside a manikin's head with a view towards in situ hearing-aid tests. Journal of the Acoustical Society of America 62: 157–161.

Levitt, H. 1978. Adaptive testing in audiology. Scandinavian Audiology (suppl 6): 241–291.

Levitt, H., N. McGarr and D. Geffner. 1987. Development of language and communication skills in hearing-impaired children. Introduction. ASHA Monographs 26: 1–8.

Levitt, H. and S. B. Resnick. 1978. Speech reception by the hearing impaired: methods of testing and the development of new tests. Scandinavian Audiology (suppl 6): 107–130.

Limbrick, E. A., S. McNaughton and M. M. Clay. 1992. Time engaged in reading: a critical factor in reading achievement. American Annals of the Deaf 137: 309–314.

Ling, D. 1976. Speech and the hearing impaired child: theory and practice. Washington, DC, AG Bell Association for the Deaf.

Ling, D. 1984. Early intervention for hearing-impaired children: oral options. San Diego, College-Hill Press.

Ling, D. and A. H. Ling. 1978. Aural habilitation: the foundations of verbal learning in hearing-impaired children. Washington, DC, AG Bell Association for the Deaf.

Ling, D. and T. G. Nienhuys. 1983. The deaf child: habilitation with and without a cochlear implant. Annals of Otology, Rhinology and Laryngology 92: 593–598.

Lund, N. J. and J. F. Duchan. 1993. Assessing children's language in naturalistic contexts 3rd ed. Englewood Cliffs, NJ, Prentice Hall.

Magner, M. E. 1972. A speech intelligibility test for deaf children. Northampton, MA, Clarke School for the Deaf.

Marsh, M. A., J. Xu, P. J. Blamey, et al. 1993. Radiologic evaluation of multichannel intracochlear implant insertion depth [published erratum appears in Am J Otol 1993 Nov;14(6):627]. American Journal of Otology 14(4): 386–391.

McDermott, H. J., C. M. McKay and A. Vandali. 1992. A new portable sound processor for the University of Melbourne/Nucleus Limited multi-electrode cochlear implant. Journal of the Acoustical Society of America 91: 3367–3371.

McGarr, N. S. 1983. The intelligibility of deaf speech to experienced and inexperienced listeners. Journal of Speech and Hearing Research 26: 451–458.

McKay, C. M., H. J. McDermott and G. M. Clark. 1991. Preliminary results with a six spectral maxima speech processor for the University of Melbourne/Nucleus multiple electrode cochlear implant. Journal of the Oto-Laryngological Society of Australia 6: 354–359.

McKay, C. M., H. J. McDermott, A. Vandali and G. M. Clark. 1992. A comparison of speech perception of cochlear implantees using the Spectral Maxima Sound Processor (SMSP) and the MSP (Multipeak) processor. Acta Oto-Laryngologica 112: 752–761.

Merzenich, M., C. Byers and M. White. 1984. Scala tympani electrode arrays. Fifth quarterly progress report. NIH contract NO1-NS9-2353: 1–11.

Merzenich, M. M., M. White, M. C. Vivion, P. A. Leake-Jones and S. Walsh. 1979. Some considerations of multichannel electrical stimulation of the auditory nerve in the profoundly deaf; interfacing electrode arrays with the auditory nerve array. Acta Oto-Laryngologica 87: 196–203.

Meyer, T., M. Svirsky, K. Kirk and R. Miyamoto. 1998. Improvements in speech perception by children with profound prelingual hearing loss: effects of device, communication mode and chronological age. Journal of Speech and Hearing Research 41: 846–858.

Miller, G. A. and P. E. Nicely. 1955. An analysis of perceptual confusions among some English consonants. Journal of the Acoustical Society of America 27(3): 338–352.

Miyamoto, R. T., K. I. Kirk, A. M. Robbins, S. Todd and A. Riley. 1996. Speech perception and speech production skills of children with multichannel cochlear implants. Acta Otolaryngologica 116: 240–243.

Miyamoto, R. T., K. I. Kirk, S. L. Todd, A. M. Robbins and M. J. Osberger. 1995a. Speech perception skills of children with multichannel cochlear implants or hearing aids. Annals of Otology, Rhinology and Laryngology 104(suppl 166): 334–337.

Miyamoto, R., M. Osberger, A. Robbins, W. Myers and K. Kessler. 1993. Prelingually deafened children's performance with the Nucleus multichannel cochlear implant. American Journal of Otology 14: 437–445.

Miyamoto, R. T., M. J. Osberger, A. M. Robbins, W. A. Myres, K. Kessler and M. L. Pope. 1992. Longitudinal evaluation of communication skills of children with single or multichannel cochlear implants. American Journal of Otology 13: 215–222.

Miyamoto, R. T., M. J. Osberger and S. L. Todd. 1994. Speech perception skills of children with multichannel cochlear implants. In: Hochmair-Desoyer, I. J. and E. S. Hochmair, eds. Advances in cochlear implants. Vienna, Manz: 498–504.

Miyamoto, R. T., A. M. Robbins, M. J. Osberger, S. L. Todd, A. I. Riley and K. I. Kirk. 1995b. Comparison of multichannel tactile aids and multichannel cochlear implants in children with profound hearing impairments. American Journal of Otology 16: 8–13.

Moog, J. S. and A. E. Geers. 1990. Early speech perception test for profoundly hearing-impaired children. St Louis, Central Institute for the Deaf.

Moores, D. F., K. L. Weiss and M. W. Goodwin. 1973. Receptive abilities of deaf children across five modes of communication. Exceptional Children 40: 22–28.

Nagy, W. E. and P. A. Herman. 1987. Breadth and depth of vocabulary knowledge: implications for acquisition and instruction. In: McKeown, M. G. and M. E. Curtis, eds. The nature of vocabulary acquisition. Hillsdale NJ, Lawrence Erlbaum Associates: 19–35.

National Institutes of Health. 1995. National Institutes of Health Consensus Conference. Cochlear implants in adults and children. JAMA 274: 1955–1961.

Nilsson, M., S. D. Soli and J. A. Sullivan. 1994. Development of the hearing in noise test for the measurement of speech reception thresholds in quiet and in noise. Journal of the Acoustical Society of America 95: 1085–1099.

Nilsson, M. J., S. D. Soli and D. J. Gelnett. 1996. Development and norming of a hearing in noise test for children. Los Angeles, House Ear Institute Internal Report.

O'Donoghue, G., T. Nikolopoulos, S. Archbold and M. Tait. 1999. Cochlear implants in young children: the relationship between speech perception and speech intelligibility. Ear and Hearing 20: 419–425.

Osberger, J. J., R. T. Miyamoto and S. Zimmerman-Phillips. 1991a. Independent evaluation of the speech perception abilities of children with the Nucleus 22-channel cochlear implant system. Ear and Hearing 12(suppl): 66S–80S.

Osberger, M. J., A. Robbins, S. Berry, S. Todd, L. Hesketh and A. Sedey. 1991b. Analysis of the spontaneous speech samples of children with a cochlear implant or tactile aid. American Journal of Otology 12(suppl): 173–181.

Osberger, M. J., A. M. Robbins and R. T. Miyamoto. 1991c. Speech perception abilities of children with cochlear implants, tactile aids, or hearing aids. American Journal of Otology 12(suppl): 105–115.

Osberger, M., A. Robbins, S. Todd and A. Riley. 1994. Speech intelligibility of children with cochlear implants. Volta Review 96: 169–180.

Osberger, M. J., A. M. Robbins, S. L. Todd, A. I. Riley, K. I. Kirk and A. E. Carney. 1996. Cochlear implants and tactile aids for children with profound hearing impairment. In: Bess, F., J. Gravel and A. M. Tharpe, eds. Amplification for children with auditory deficits. Nashville, Bill Wilkerson Center Press: 283–307.

Owens, E., D. K. Kessler, C. C. Telleen and E. D. Schubert. 1980. The minimal auditory capabilities battery. Department Otolaryngology, University of California, San Francisco, CA.

Owens, E. and C. C. Telleen. 1981. Speech perception with hearing aids and cochlear implants. Archives of Otolaryngology 107: 160–163.

Parkinson, A. J., R. S. Tyler, G. G. Woodworth, M. W. Lowder and B. J. Gantz. 1996. A within-subject comparison of adult patients using the Nucleus F0F1F2 and F0F1F2B3B4B5 speech processing strategies. Journal of Speech and Hearing Research 39: 261–277.

Pisoni, D. B. 2000. Cognitive factors and cochlear implants: some thoughts on perception, learning and memory in speech perception. Ear and Hearing 21: 70–78.

Plant, G. 1984. A diagnostic speech test for severely and profoundly hearing-impaired children. Australian Journal of Audiology 6: 1–9.

Plomp, R. and A. M. Mimpen. 1979. Improving the reliability of testing the speech reception threshold for sentences. Audiology 18: 43–52.

Pohlman, A. G. and R. W. Kranz. 1924. Binaural minimum audition in a subject with ranges of deficient acuity. Proceedings of the Society for Experimental Biology and Medicine 20: 335–337.

Prutting, C. A. 1986. Pragmatics. Paper presented at the Australian Association of Speech and Hearing Conference, Canberra, Australia.

Psarros, C., K. Plant, L. Whitford, et al. 2000. Speech perception and speech production changes in children following alteration of speech processing strategy. Australian Journal of Audiology 22(suppl): 31.

Psarros, C. E., K. L. Plant, K. Lee, J. A. Decker, L. A. Whitford and R. S. C. Cowan. 2002. Conversion from the SPEAK to the ACE strategy in children using the Nucleus

24 cochlear implant system: speech perception and speech production outcomes. Ear and Hearing 23(1 Suppl): 18S–27S.

Quigley, S. and P. Paul. 1990. Language and deafness. San Diego, Singular Publishing Group.

Quittner, A. L., L. B. Smith, M. J. Osberger, T. V. Mitchell and D. B. Katz. 1994. The impact of audition on the development of visual attention. Psychological Science 5: 347–353.

Raffin, M. J. M. and A. R. Thornton. 1980. Confidence levels for differences between speech discrimination scores: a research note. Journal of Speech and Hearing Research 23: 5–18.

Ramkalawan, T. W. and A. C. Davis. 1992. The effects of hearing loss and age of intervention on some language metrics in young hearing-impaired children. British Journal of Audiology 26: 97–107.

Rance, G., F. W. Rickards, L. T. Cohen, M. J. Burton and G. M. Clark. 1993. Steady state evoked potentials: a new tool for the accurate assessment of hearing in cochlear implant cadidates. In: Fraysse, B. and O. Deguine, eds. Cochlear implants: new perspectives. Basel, Karger: 44–48.

Rance, G., F. W. Rickards, L. T. Cohen, S. De Vidi and G. M. Clark. 1995. The automated prediction of hearing thresholds in sleeping subjects using auditory steady-state evoked potentials. Ear and Hearing 16: 499–507.

Reynell, J. K. 1983. Reynell developmental language scales manual-revised. Windsor, NFER-Nelson.

Reynell, J. K. and C. P. Gruber. 1990. Reynell developmental language scales. Los Angeles, Western Psychological Services.

Rickards, F. W. and G. M. Clark. 1984. Steady-state evoked potentials to amplitude-modulated tones. In: Anodar, R. H. and C. Barber, eds. Evoked potentials II. Boston, Butterworths: 163–168.

Robbins, A. M., M. J. Osberger, R. T. Miyamoto and K. S. Kessler. 1995. Language development in children with cochlear implants. Advances in Oto-Rhino-Laryngology 50: 160–166.

Robbins, A. M., J. J. Renshaw and S. W. Berry. 1991. Evaluating meaningful auditory integration in profoundly hearing impaired children. American Journal of Otology 12(suppl): 144–150.

Robbins, A. M., M. A. Svirsky and K. I. Kirk. 1997. Children with implants can speak, but can they communicate? Otolaryngology Head and Neck Surgery 117: 155–160.

Roth, F. P. and N. J. Spekman. 1984a. Assessing the pragmatic abilities of children: part 1. Organizational framework and assessment parameters. Journal of Speech and Hearing Disorders 49: 2–11.

Roth, F. P. and N. J. Spekman. 1984b. Assessing the pragmatic abilities of children: part 2. Guidelines, considerations, and specific evaluation procedures. Journal of Speech and Hearing Disorders 49: 12–17.

Sander, E. 1972. When are speech sounds learned? Journal of Speech and Hearing Research 37: 55–63.

Sarant, J. Z., P. J. Blamey and G. M. Clark. 1996. The effect of language knowledge on speech perception in children with impaired hearing. In: McCormack, P. and A. Russell, eds. Proceedings of the Sixth Australian International Conference on Speech Science and Technology. Canberra, Australian Speech Science and Technology Association: 269–274.

Sarant, J. Z., P. J. Blamey, R. C. Dowell, G. M. Clark and W. P. R. Gibson. 2001. Variation in speech perception scores among children with cochlear implants. Ear and Hearing 22: 18–28.

Scarborough, H. S. 1990. Index of productive syntax. Applied Psycholinguistics 11: 1–22.

Schick, B. and M. P. Moeller. 1992. What is learnable in manually coded English sign systems? Applied Psycholinguistics 13: 313–340.

Schindler, R. A., D. K. Kessler and M. A. Barker. 1995. Clarion patient performance: an update on the clinical trials. Annals of Otology, Rhinology and Laryngology 104: 269–272.

Sehgal, S. T., K. I. Kirk, M. Svirsky and R. T. Miyamoto. 1998. The effects of processor strategy on the speech perception performance of pediatric Nucleus multichannel cochlear implant users. Ear and Hearing 19: 149–161.

Seligman, P. M. and H. J. McDermott. 1995. Architecture of the SPECTRA 22 speech processor. Annals of Otology, Rhinology and Laryngology 104(suppl 166): 139–141.

Semel, E., E. Wiig and W. A. Secord. 1995. Clinical evaluation of language fundamentals, 3rd ed. San Antonio, TX, Psychological Corporation, Harcourt Brace.

Shipp, D. B. and J. M. Nedzelski. 1995. Prognostic indicators of speech recognition performance in adult cochlear implant users: a prospective analysis. Annals of Otology, Rhinology and Laryngology 104: 194–196.

Shriberg, L. D., D. Austin, B. A. Lewis, J. L. McSweeny and D. L. Wilson. 1997. The percentage of consonants correct (PCC) metric: extensions and reliability data. Journal of Speech Language and Hearing Research 40: 708–722.

Shriberg, L. D. and J. Kwiatkowski. 1982. Phonological disorders III: a procedure for assessing severity of involvement. Journal of Speech and Hearing Disorders 47: 256–270.

Singh, S. 1968. A distinctive feature analysis of responses to a multiple choice intelligibility test. International Review of Applied Linguistics 6: 37–53.

Skinner, M. W., G. M. Clark, L. A. Whitford, et al. 1994. Evaluation of a new spectral peak coding strategy for the Nucleus 22 channels cochlear implant system. American Journal of Otology 15: 15–27.

Skinner, M. W., M. S. Fourakis, T. A. Holden, L. K. Holden and M. E. Demorest. 1996. Identification of speech by cochlear implant recipients with the Multipeak (MPEAK) and Spectral Peak (SPEAK) speech coding strategies. I. Vowels. Ear and Hearing 17: 182–197.

Skinner, M. W., L. K. Holden, T. A. Holden, et al. 1991. Performance of postlinguistically deaf adults with the Wearable Speech Processor (WSP III) and Mini Speech Processor (MSP) of the Nucleus multi-electrode cochlear implant. Ear and Hearing 12: 3–22.

Skinner, M. W., L. K. Holden, L. A. Whitford, K. L. Plant, C. Psarros and T. A. Holden. 2002. Speech recognition with the Nucleus 24 SPEAK, Ace, and CIS speech coding strategies in newly implanted adults. Ear and Hearing 23(3): 207–223.

Smith, C. R. 1975. Residual hearing and speech production in deaf children. Journal of Speech and Hearing Research 18: 795–811.

Spencer, L., N. Tye-Murray and J. B. Tomblin. 1998. The production of English inflectional morphology, speech production and listening performance in children with cochlear implants. Ear and Hearing 19: 310–318.

Spencer, P. 2002. Language development of children with cochlear implants. In: Christiansen, J., ed. Cochlear implants in children. Washington, DC, Gallaudet University Press.

Staller, S. J. 1990. Perceptual and production abilities in profoundly deaf children with multichannel cochlear implants. Journal of the American Academy of Audiology 1(1): 1–3.

Staller, S. J., A. L. Beiter, J. A. Brimacombe and P. Arndt. 1991a. Paediatric performance with the Nucleus 22-channel cochlear implant system. American Journal of Otology 12(suppl): 126–136.

Staller, S. J., R. C. Dowell, A. L. Beiter and J. A. Brimacombe. 1991b. Perceptual abilities of children with the Nucleus 22-channel cochlear implant. Ear and Hearing 12(suppl 4): 34S–47S.

Staller, S., C. Menapace and E. Domico. 1997. Speech perception abilities of adult and pediatric Nucleus implant recipients using Spectral Peak (SPEAK) coding strategy. Otolaryngology Head and Neck Surgery 117: 236–242.

Staller, S., A. Parkinson, J. Arcaroli and P. Arndt. 2002. Pediatric outcomes with the Nucleus 24 contour: North American clinical trial. Annals of Otology Rhinology and Laryngology 111(suppl 189): 56–61.

Stapells, D. R., J. S. Gravel and B. A. Martin. 1995. Thresholds for auditory brain stem responses to tones in notched noise from infants and young children with normal hearing or sensorineural hearing loss. Ear and Hearing 16: 361–71.

Stoel-Gammon, C. 1998. Sounds and words in early language acquisition. The relationship between lexical and phonological development. In: Paul, R., ed. Explaining the speech-language connection. Baltimore, Paul H Brookes: 25–52.

Surowiecki, V. N., J. Z. Sarant, P. Maruff, P. J. Blamey, P. A. Busby and G. M. Clark. 2002. Cognitive processing in children using cochlear implants: the relationship between visual memory, attention and executive functions and developing language skills. Annals of Otology, Rhinology and Laryngology 111(suppl 189): 119–126.

Svirsky, M. and T. Meyer. 1999. Comparison of speech perception in pediatric Clarion cochlear implant and hearing aid users. Annals of Otology, Rhinology and Laryngology 108: 104–109.

Svirsky, M., A. Robbins, K. Kirk, D. Pisoni and R. Miyamoto. 2000. Language development in profoundly deaf children with cochlear implants. Psychological Science 11: 1–6.

Svirsky, M. A., R. B. Sloan, M. Caldwell and R. T. Miyamoto. 1998. Speech intelligibility of prelingually deaf children with multichannel cochlear implants. Presented at the 7th Symposium on Cochlear Implants in Children, Iowa City, IA.

Thielemeir, M. A., L. L. Tonokawa, B. Petersen and L. S. Eisenberg. 1985. Audiological results in children with a cochlear implant. Ear and Hearing 6(3 suppl): 27S–35S.

Thornton, A. R. and M. J. M. Raffin. 1978. Speech-discrimination scores modeled as a binomial variable. Journal of Speech and Hearing Research 21: 507–518.

Tillman, T. W. and R. Carhart. 1966. An expanded test for speech discrimination utilizing CNC monosyllabic words. Northwestern University Auditory Test No. 6 SAM-TR-66-55. Technical Report: 1–12.

Tillman, T. W., R. Carhart and L. Wilbur. 1963. A test for speech discrimination composed of CNC monosyllabic words, Northwestern University Auditory Test No 4. Texas, USAF School of Aerospace Medicine.

Tobey, E., S. Angelette and C. Murchison. 1991. Speech production in children receiving a multichannel cochlear implant. American Journal of Otology 12(suppl): 48S–54S.

Tobey, E., A. Geers and C. Brenner. 1994. Speech production results: speech feature acquisition. Volta Review 96: 109–130.

Tobey, E. and S. Hasenstab. 1991. Effects of a Nucleus multichannel cochlear implant upon speech production in children. Ear and Hearing 12(suppl): 48S–54S.

Tobey, E., S. Staller, J. Brimacombe and A. Beiter. 1988. Objective measures of speech production in children using cochlear implants. American Speech Hearing Association 30: 103.

Tomblin, J. B., L. Spencer, S. Flock, R. Tyler and B. Gantz. 1999. A comparison of language achievement in children with cochlear implants and children using hearing aids. Journal of Speech, Language, and Hearing Research 42: 497–511.

Tong, Y. C., R. C. Black, G. M. Clark, et al. 1979. A preliminary report on a multiple-channel cochlear implant operation. Journal of Laryngology and Otology 93: 679–695.

Tong, Y. C., P. A. Busby and G. M. Clark. 1988. Perceptual studies on cochlear implant patients with early onset of profound hearing impairment prior to normal development of auditory, speech, and language skills. Journal of the Acoustical Society of America 84: 951–962.

Tong, Y. C., G. M. Clark, P. M. Seligman and J. F. Patrick. 1980. Speech processing for a multiple-electrode cochlear implant hearing prosthesis. Journal of the Acoustical Society of America 68: 1897–1899.

Tong, Y. C., J. M. Harrison, A. Vandali. 1991. Speech processors for auditory prostheses. Ninth quarterly progress report. NIH contract No. 1-DC-9-2400.

Tye-Murray, N., L. Spencer and G. Woodworth. 1995. Acquisition of speech by children who have prolonged cochlear implant experience. Journal of Speech and Hearing Research 38: 327–337.

Tyler, R. S. 1988. Open-set word recognition with the 3M/Vienna single-channel cochlear implant. Archives of Otolaryngology–Head Neck Surgery 114: 1123–1126.

Tyler, R. S., B. J. Gantz, B. F. McCabe, M. W. Lowder, S. R. Otto and J. P. Preece. 1985. Audiological results with two single channel cochlear implants. Annals of Otology Rhinology and Laryngology 94: 133–139.

Tyler, R. S. and M. W. Lowder. 1992. Audiological management and performance of adult cochlear-implant patients. Ear, Nose and Throat Journal 71: 117–128.

Tyler, R. S., M. W. Lowder and S. R. Otto. 1984. Initial Iowa results with the multichannel cochlear implant from Melbourne. Journal of Speech and Hearing Research 27: 596–604.

Tyler, R. S., B. C. J. Moore and F. K. Kuk. 1989a. Performance of some of the better cochlear-implant patients. Journal of Speech and Hearing Research 32: 887–911.

Tyler, R. S., J. P. Preece and M. W. Lowder. 1983. The Iowa cochlear implant test battery. University of Iowa, Department of Otolaryngology–Head and Neck Surgery, Iowa City, IA.

Tyler, R. S., N. Tye-Murray and B. C. J. Moore. 1989b. Synthetic two-formant vowel perception by some of the better cochlear-implant patients. Audiology 28: 301–315.

Tyler, R. S., N. Tye-Murray, J. P. Preece, B. J. Gantz and B. F. McCabe. 1987. Vowel and consonant confusions among cochlear implant patients: do different implants make a difference? Annals of Otology, Rhinology and Laryngology 96(suppl 128): 141–144.

van Hoesel, R. J. M. 1998. Bilateral electrical stimulation with multi-channel cochlear implants. PhD thesis, University of Melbourne.

van Hoesel, R. J. M. and G. M. Clark. 1995. Fusion and lateralization study with two binaural cochlear implant patients. Annals of Otology, Rhinology and Laryngology 104(suppl 166): 233–235.

van Hoesel, R. J. M. and G. M. Clark. 1997. Psychophysical studies with two binaural cochlear implant subjects. Journal of the Acoustical Society of America 102: 495–507.

van Hoesel, R. J. M., Y. C. Tong, R. D. Hollow, J. Huigen and G. M. Clark. 1990. Preliminary studies on a bilateral cochlear implant user. Journal of the Acoustical Society of America 88(suppl 1): S193.

Vandali, A. E., L. A. Whitford, K. L. Plant and G. M. Clark. 2000. Speech perception as a function of electrical stimulation rate: using the Nucleus 24 cochlear implant system. Ear and Hearing 21: 608–624.

Waltzman, S. B., N. Cohen, R. H. Gomolin, W. H. Shapiro, S. R. Ozdamar and R. A. Hoffman. 1994. Long-term results of early cochlear implantation in congenitally and prelingually deafened children. American Journal of Otology 15(suppl 2): 9–13.

Waltzman, S. B., N. L. Cohen and W. H. Shapiro. 1992. Use of a multichannel cochlear implant in the congenitally and prelingually deaf population. Laryngoscope 102(4): 395–9.

Westby, C. E. 1980. Assessment of cognitive and language abilities through play. Language, Speech and Hearing Services in Schools 11: 154–168.

Whitford, L. A., P. M. Seligman, C. Everingham, et al. 1995. Evaluation of the Nucleus Spectra 22 processor and new speech processing strategy (SPEAK) in postlinguistically deafened adults. Acta Oto-Laryngologica 115: 629–637.

Wiig, E., W. A. Secord and E. Semel. 1992. Clinical evaluation of language fundamentals—preschool. San Antonio, TX, Psychological Corporation, Harcourt Brace.

Wilson, B. S. 2000. Strategies for representing speech information with cochlear implants. In: Niparko, J. K., ed. Cochlear implants: principles and practice. Philadelphia, Lippincott Williams & Wilkins.

Wilson, B. S., D. T. Lawson, M. Zerbi and C. C. Finley. 1992. Speech processors for auditory prostheses. Twelfth quarterly progress report, April. 1992. NIH contract N01-DC-9-2401. Research Triangle Institute.

Wilson, B. S., D. T. Lawson, M. Zerbi and C. C. Finley. 1993. Speech processors for auditory prostheses. Fifth quarterly progress report, Oct 1993. NIH contract N01-DC-2-2401. Research Triangle Institute.

Woodcock, R. W. and N. Mather. 1989. Woodcock-Johnson tests of cognitive ability. Allen, TX, DLM Teaching Resources.

Xu, S., R. C. Dowell and G. M. Clark. 1987. Results for Chinese and English in a multichannel cochlear implant patient. Annals of Otology, Rhinology and Laryngology 96(suppl 128): 126–127.

Yoshinaga-Itano, C., A. L. Sedey, D. K. Coulter and A. L. Mehl. 1998. Language of early- and later-identified children with hearing loss. Pediatrics 102: 1161–1171.

Zimmerman, I. L., V. C. Steiner and R. E. Pond. 1979. Preschool language scale. Colombus, Merrill.

Zimmerman-Phillips, S., A. M. Robbins and M. J. Osberger. 2000. Assessing cochlear implant benefit in very young children. Annals of Otology, Rhinology and Laryngology 109(suppl 185): 42–43.

13
Socioeconomics and Ethics

The speech and language results of cochlear implants as well as their biological safety are crucial in assessing the socioeconomic benefits and ethics of the procedure.

Speech and Language Benefits

The speech perception, speech production, and language results were discussed in detail in Chapter 12 and the biological safety in Chapters 3 and 4. The speech and language benefits of the Nucleus 22 F0/F2 WSP-II, F0/F1/F2 WSP-III, Multipeak-MSP, SPEAK Spectra-22 systems, the Nucleus 24 SPEAK, CIS, and ACE systems, and the Clarion and Combi-40 CIS systems were demonstrated on adults and children by Staller et al (1991), Miyamoto et al (1992), Blamey et al (1992), Dawson et al (1992, 1995), Tobey and Hasenstab (1991), Geers and Moog (1994), Cowan et al (1995), Dowell et al (1995), Kirk et al (1995b), Tye-Murray et al (1995), and others. This was acknowledged by the National Institutes of Health (NIH) Consensus Statement on Cochlear Implants in Adults and Children (1995), which said "implantation in conjunction with education and habilitation leads to advances in oral language acquisition."

Biological Safety

The toxicity of the materials used with the first Nucleus cochlear implant were evaluated at the University of Melbourne from 1980 to 1982. The research was undertaken according to the good laboratory practice requirements of the U.S. Food and Drug Administration (FDA) for its premarket approval (PMA). Additional tests for cytotoxicity were carried out by an independent laboratory. The standards were developed from recommendations of the American Society for Testing Materials (ASTM), subsequently outlined in the Animal Book of ASTM Standards (1986). Other research on the effects of trauma and infection on the cochlea were also studied intensively by the Department of Otolaryngology at the University of Melbourne for the FDA.

Before implanting infants and young children, further biological safety studies were required. The NIH contract to the University of Melbourne for studies on pediatric auditory prosthesis implants, NIH contract No. 1-NS-7-2342, from 1987 to 1992, resulted in findings that there was no higher incidence of middle ear infection leading to labyrinthitis than in unimplanted ears, provided the electrode entry point was grafted with fascia; that head growth had no adverse effects on the implant, and vice versa; and that electrical stimulation had no deleterious effects on an immature nervous system. This was discussed in the bioengineering section in Chapter 8, as well as Chapters 3 and 10.

Social Benefits

Adequate hearing and the development of aural speech and language in children are essential for communicating in the hearing community. A severe-to-profound hearing loss in early life limits children's proper acquisition of oral and written language, which places great limitations on their education and career opportunities (Bamford and Saunders 1991). When it occurs in adults, it can have a marked effect on their ability to continue in a job or to relate to family and friends. There are also marked advantages in hearing environmental sounds in the social, school, or work setting. For the above reasons, the restoration of the ability to hear in adults and the development of speech perception and production as well as receptive and expressive language in children with a cochlear implant are of great importance.

Personal

Even single-channel implants made a significant difference to the quality of a person's life. In a questionnaire (Fraser 1987), the patients' expectations of hearing environmental sound were met in more cases than their expectations of the single-channel device as an aid to speech reading. However, more than half of the adults gave hearing any sound and relief from isolation as their primary reasons for wanting an implant, and to this extent the implant too had been successful. All patients reported an improvement in their quality of life.

Family

The ultimate benefits of successful cochlear implantation for the family are somewhat different for postlinguistically deaf adults compared to pre- and postlinguistically deaf children. For the adult it may be the ability to get a job or take more responsibility in the workplace, to have more effective communication with his/her spouse and thus a more fulfilled marriage, and/or to have a more active role as a parent or grandparent in the family.

For children, cochlear implantation should lead to a normalizing effect on relations with siblings, and a more equal sharing of time commitments from the parents for each child. However, with implanted children it must be realized that the initial intensive period of (re)habilitation can be stressful (Downs 1986; Quittner et al 1991) as is coping with slow language development (Evans et al 1989). The additional time required for training, and the distances to be traveled to the clinic also are likely to strain family relations. Assistance from grandparents and contact with other families at this time can be especially useful (Cowan 1997).

School

The Nucleus multiple-channel cochlear implants in children with mean hearing thresholds greater than 100 dB have been compared with children using Tactaid II and IV tactile vocoders and hearing aids (Osberger et al 1991, 1996; Geers and Moog 1994; Miyamoto et al 1994, 1995, 1996). Geers and Tobey (1995) and Geers and Moog (1994) found that profoundly deaf children with cochlear implants performed better than those with a tactile aid and in turn better than those with a hearing aid. The children with a cochlear implant performed at a level comparable to children with a lesser hearing loss in the 90- to 100-dB range.

Subsequently Blamey et al (2001) reported that profoundly deaf children with a hearing loss (HL) of 106 dB HL and using cochlear implants obtained speech perception and receptive and expressive language at a level comparable to children with a hearing loss of about 78 dB HL. This was presumed due to the better speech processing strategy with SPEAK.

The speech production of children with the Nucleus (F0/F1/F2) processor improved to 31% with imitative nonsegmental speech production, and 67% with segmental speech production 1 year after implantation (Tobey et al 1988, 1991; Tobey and Hasenstab 1991). Approximately half had significant improvements in spontaneous speech production.

For receptive language measured with the Rhode Island Test of Language Structure, the results showed that all but one of the children with the earlier Nucleus cochlear implants scored very high, and well above the expected results for deaf children (Tomblin et al 1999). Expressive language when measured for grammar also showed the implanted children performed significantly better than a control group of unimplanted children (Tomblin et al 1999). These and other improvements, discussed in Chapter 12, have meant that many implanted children can be educated in a mainstream school, both with and without a special unit (Nevins and Chute 1995).

Economic Benefits

Economic Measures

The economic benefits can be assessed in terms of (1) cost-effectiveness analysis, which measures the outcomes in natural units; (2) cost-benefit analysis, which measures the benefits in monetary terms; and (3) cost-utility analysis, which measures the outcomes in years of healthy life or quality-adjusted life-years (QALY) gain (Evans et al 1995). These measures are required for comparing medical procedures, especially when seeking government resources and funds. Thus the benefits should not only improve the quality of life, but also remove the need for alternative management and increase patients' productivity in society. Their increased productivity may result from an ability to resume their previous occupation or achieve promotion (Cowan 1997).

Cost-Effectiveness

Cost-effectiveness can be measured quantitatively, for example, the number of productive life-years added or days of disability saved. The cost-effectiveness is related to the degree of benefit from the device, although this has not been quantified. Cochlear implant speech perception, speech production, and language scores can also be compared with those obtained preoperatively using hearing aids or other sensory devices. The studies by Flynn et al (1996a,b, 1997, 1998a,b) have shown that for postlinguistically deaf adults the results with cochlear implants are comparable to those for people with a severe hearing loss of 78 dB HL using a hearing aid. Studies by Osberger et al (1993), Geers and Moog (1994), and Kirk et al (1995a) suggest that the direct benefits available to individuals from use of cochlear implants were significantly higher than those from either hearing aids or tactile devices. A comparison of the productivity of people with different degrees of hearing loss could help quantify the cost-effective benefits of a cochlear implant. Although the cochlear implant device is more expensive than the other aids, the long-term savings to the community in job performance and education costs are critical.

Cost-Benefit Analysis

The cost-benefit analysis measures the savings from adults' greater productivity and from children's not needing to attend special kindergartens and schools or use other services. With implanted children, there are significant savings to the society because many implanted children can attend mainstream school, rather than being educated in a special segregated school for hearing-impaired children that has higher costs due to lower teacher–student ratios. Figure 13.1 illustrates the 12-year costs in the state of Victoria, Australia, for educating children in mainstream schools, mainstream schools with a special unit, and special schools. The net savings for each child receiving a cochlear implant can be derived by subtracting the cost of the implant procedure from the savings from educating that child in a mainstream school. The savings to the community can then be calculated from the proportion of children having cochlear implants and the proportion of those children able to be educated in mainstream schools.

In the United States, Nevins and Chute (1995) reported the growing trend for children using cochlear implants to move from special self-contained facilities to

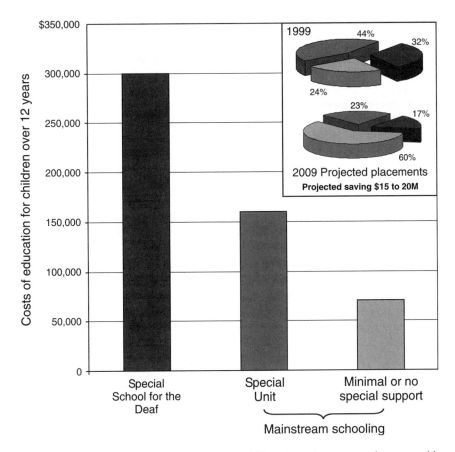

FIGURE 13.1. The 12-year costs for educating children in mainstream, mainstream with special unit, and special schools in the state of Victoria (projections made in the year 2000).

more mainstream settings. Wyatt et al (1996) estimated savings of $152,000 in educational costs for each child implanted by age 4 years, assuming 50% were mainstreamed at 3 years postimplantation, and 90% partially mainstreamed by 6 years postimplantation.

Quality of Life

Since hearing loss is not generally a life-threatening disability, the cochlear implant procedure itself has little direct impact on life expectancy. However, cochlear implantation does improve the patient's quality of life, through restoring or allowing acquisition of auditory skills, improving articulation, and enhancing the development of language comprehension, including reading and writing skills in children. Indirect benefits flowing from improved communication in implant patients are a major contributor to the improvement in quality of life (Lea 1991). Because it allows assessment of the impact of both direct and indirect benefits, cost-utility analysis has advantages in the economic evaluation of cochlear implants in adults and children (Cowan 1997). Improved quality of life is measured in relation to the cost of the intervention.

The cost-utility analysis measures the cost to achieve outcomes in years of healthy life or QALY gain. In calculating the QALY, the x-axis measures the duration of life in years from birth to death, and the y-axis indicates the level of health-related quality of life on a scale from 0 to 1, in which 0 indicates death and 1 indicates perfect health. This index must be standardized to allow comparison of diverse types of treatments. The area that is added in improved quality and length of life from the medical intervention is measured in QALYs. A treatment that prolongs the life of the patient represents a significant increase in QALY (a value of 1.0). In contrast, a treatment that improves patients' general health and ability to resume their normal life is represented as a proportional increase in QALY (a value between 0.1 and 0.9). The QALY quantifies both the prevention of morbidity, or life improvement, and the duration over which this is maintained.

The cost per QALY can be calculated through establishing all relevant costs of the treatment, and this monetary figure provides a measure of the value of the treatment to the individual and the economic costs and impact on the health authority or society in general. The costs of a cochlear implant program need to cover all expenses, including the audiological and otological consultations and special investigations. The cost-utility analysis can be calculated as a ratio of the costs of a specific medical treatment or intervention, and the value of the outcomes in QALYs.

As emphasized by Evans et al (1995), the key is the quality scale increase applicable to the typical cochlear implant recipient. The Ontario Health Utilities Index (OHUI) is a commonly used multiattribute health status system for measuring the quality scale. The OHUI considers the effects of treatment on seven quality of life categories: sensory, emotion, cognitive, mobility, self-care, pain, and fertility. In the Battelle study (Evans et al 1995), although the cochlear implant has a positive influence on three OHUI categories, sensory, emotion, and cognition, only the sensory category was considered. It was then assumed a change to open-set speech understanding caused the recipient to move from level 4 (blind, deaf, or mute) to level 3 (sees, hears, or speaks with limitations, even with equipment). This represented 0.27 on the scale from 0 to 1, or a 27% increase in life quality. Battelle then assumed that only 67% of cochlear implant recipients reached open-set ability, which then translated to a general 0.18 change on the OHUI scale.

Based on an estimate of costs for the intervention, and the expected length of device use, the typical cochlear implant recipient would achieve a cost-utility ratio of $15,590 per QALY. As a comparison, the cost-utility ratios for other accepted medical interventions in dollars per QALY were neonatal intensive care, $7970; coronary angioplasty, $11,490; implantable defibrillator, $29,220; heart transplant, $38,970; knee replacement, $49,700; and peritoneal dialysis, $38,000.

Ethics

The ethics of cochlear implantation in adults was an issue in the 1970s, when the benefits or risks were not clear. Since then it became an important issue for infants and children and has received a lot of attention, in particular from the signing deaf community and their advocates. The issues raised by the signing deaf community concern, first, human experimentation, and second, whether it is natural to have a hearing loss and to restore hearing. This chapter examines cochlear implantation in adults and children in the light of generally accepted ethical principles.

Ethics is a general term referring to both morality and ethical theory. In considering whether cochlear implantation in adults and in particular infants and children is morally acceptable, it is first necessary to also consider morality. Morality in general refers to social conventions about right and wrong human conduct that are widely shared by members of the community. Common morality refers to socially approved norms of human conduct. In particular, it highlights acceptable and nonacceptable conduct referred to in discussing human rights. It is also important to consider moral theology, which provides a perspective on moral issues from a theological or religious point of view.

Human Experimentation

In considering the ethics or morality of cochlear implants, it is essential to determine to what extent the benefits outweigh the risks, and to what extent it is an accepted clinical procedure and not primarily research. The fundamentals can be summarized as: (1) do no harm; (2) do good; (3) achieve justice; and (4) ensure autonomy. In considering these issues it is helpful, to view the procedure against its background and history.

Research in the human with cochlear implantation which occurred in the 1970s and 1980s was carried out on the basis of ethically acceptable practices for research as laid down in the Helsinki Declaration on Biomedical Research involving human subjects adopted by the 18th World Medical Assembly, Finland, 1969, and revised by the 29th World Medical Assembly, Tokyo, 1975, and in accordance with the Convention on the Rights of the Child, adopted by the General Assembly of the United Nations on November 20, 1989. These principles are summarized below, with an explanation of how cochlear implantation has complied with the ethics of human experimentation. They have also been presented by Clark, Cowan et al (1997).

Scientific Rigor

Biomedical research involving human subjects must conform to generally accepted scientific principles, and should be based on adequately performed laboratory and animal experimentation, and on a thorough knowledge of the scientific literature.

The biomedical cochlear implant research in the 1970s on adults was preceded by extensive physiological and biological research on the experimental animal

(Clark 1969; Clark, Nathar et al 1972; Merzenich et al 1973; Black and Clark 1980) and the human temporal bone (Clark, Kranz et al 1975; Clark 1977). The multiple-electrode studies on children were preceded by thorough speech and psychophysics studies on adults, and the biological safety research on experimental animals. The studies on children 2 years of age and under was carried out only after the Nucleus multiple-channel implant had been show to be as effective on children from 2 to 18 years, and approved by the FDA on June 27, 1990. It was only performed when the trend in results had shown that speech perception was better the younger the child at implantation. Finally, it was essential before operating on children in this age group to ensure that the high incidence of middle ear infection would not lead to a significant risk of inner ear infection, that head growth would not affect the implant and vice versa, and that electrical stimulation would not have an adverse affect on the immature nervous system. These concerns were addressed through the 5-year NIH contract for studies on pediatric auditory prosthesis implants (contract No. 1-NS-7-2342) awarded in 1987 to the University of Melbourne.

The design and performance of each experimental procedure involving human subjects should be clearly formulated in an experimental protocol that should be transmitted to a specially appointed independent committee for consideration, comment, and guidance.

The findings from cochlear implant research studies on adults and then children were published in general in international journals with a high standard of peer review and presented reliable information based on clearly formulated and achievable experimental protocols. The same careful review was made by agencies funding the cochlear implant research, for example, the NIH, the National Health and Medical Research Council of Australia, and the Australian Research Council. The FDA carefully reviewed the experimental protocols for any studies reported to it. This applied first to the 3M-House single-channel implant for which a PMA was obtained in 1984 for awareness of sound in postlinguistically deaf adults, and the Nucleus multiple-channel device, which the FDA approved in 1985 as safe and effective in providing help in understanding speech with lipreading and also some speech using electrical stimulation alone for adults who had hearing before going deaf. The FDA approved the use of the Nucleus F0/F1/F2 and Multipeak strategies in children from 2 to 18 years of age on June 27, 1990.

Biomedical research involving human subjects should be conducted only by scientifically qualified persons and under the supervision of a clinically competent medical person. The responsibility for the human subject must always rest with a medically qualified person and never rest on the subject of the research, even though the subject has given his or her consent.

The research in Melbourne has always been undertaken under the control of the Cochlear Implant Clinic of the Royal Victorian Eye and Ear Hospital. The head of the clinic was an appropriately qualified medical person who was responsible to the board of the hospital and the university. As I was the head of the clinic and involved in the experimentation, a committee of the hospital assumed overall responsibility. This committee had independent medical representation and an independent chairperson.

Benefits Versus Risks

Biomedical research involving human subjects cannot legitimately be carried out unless the importance of the objective is in proportion to the inherent risk to the subject.

The successful outcomes of implantation in adults and children with severe-to-profound hearing loss are now well documented in properly controlled studies reported in international journals with a high standard of peer review, as are the complications of the procedure. A major study funded by the NIH was undertaken at the Central Institute for the Deaf (CID) where a group of profoundly deaf children were managed with hearing aids, tactile aids, or cochlear implants. The speech perception and production results from this study were reported by Geers and Brenner (1994) and Tobey et al (1994).

The study was longitudinal; 39 children with prelingual profound deafness were evaluated over a 3-year period while enrolled in the auditory-oral education program at the CID in St. Louis, Missouri. The 39 children were matched in triads, 13 using cochlear implants (CI), 13 tactile aids (TA), and 13 conventional hearing aids (HA). The cochlear implants were the Nucleus 22 multiple-channel devices with the F0/F1/F2 and Multipeak strategies. The primary focus of the study was to document differences, if any, in the rate of acquisition in speech perception, speech production, and spoken language with use of these sensory devices. The study used a stringent measure of benefit, requiring not only improvement in children compared to themselves, but also compared to similar orally educated children using hearing aids.

Initial scores on a battery of speech perception tests were similar for all three groups. After 36 months, there was a significant difference between the performance of the children using cochlear implants and the other two groups. With speech perception, the cochlear implant group moved from a median category 1 (detection only) to category 5 (consonant perception). In contrast, the tactile group moved from category 1 to category 2 (pattern perception), whereas the hearing aid group remained at category 2 over the entire 36 months of the study. In addition, statistical analysis showed that feature perception for the three groups was significantly different after 36 months of habilitation. Although there were no significant differences between the TA and HA groups, the CI group scored significantly higher than the TA and HA group on pitch perception (95% versus 65% and 84%), vowel perception (84% versus 35% and 42%), and consonant place perception (36% versus 16% and 21%). The CI group also moved on the matrix phrase perception task from being the lowest of the three groups to the highest. Finally, after only 12 months, the speech-reading enhancement scores of the CI group were found to exceed those of children using either hearing aids or tactile aids.

The CI group also demonstrated significantly higher scores than the TA and HA groups in the production of vowels and consonants in spontaneous speech.

Children who used the cochlear implant for 3 years performed at a comparable level to children using hearing aids with a pure tone audiogram of 90 to 100 dB HL. The children in the CI group were also significantly more accurate in their production of the less visible place and complex manner features and on some voicing features, as compared to the TA and HA groups.

In terms of spoken language, the CI group exhibited faster acquisition of all language and communication skills measured. The mean scores for the CI group were significantly higher than the TA and HA groups in the areas of expressive vocabulary, receptive syntax, and everyday use of sensory aid (Geers and Moog 1994). Furthermore, an NIH Consensus Statement, volume 13(2), Cochlear Implants in Adults and Children, was issued in 1995, which concluded that, using tests commonly applied to children and adults with hearing impairments, perceptual performance increases on average with each succeeding year postimplantation. Over time, performance may improve to match that of children who have residual hearing and are highly successful hearing aid users. Children implanted at younger ages are on average more accurate in their production of consonants, vowels, intonation, and rhythm than older children. Speech produced by children with implants is more accurate than speech produced by children with comparable hearing losses using vibrotactile devices or hearing aids. One year after implantation, speech intelligibility is twice that typically reported for children with profound hearing impairments, and continues to improve. Oral-aural communication training appears to result in substantially greater speech intelligibility than manually based total communication. The nature and pace of language acquisition may be influenced by the age of onset, age at implantation, nature and intensity of habilitation, and mode of communication. Oral language development in deaf children, including those with cochlear implants, can be training-intensive, and results typically do not reach the levels of children of the same age with good hearing.

The risks of implantation have been reported by a number of surgeons and were referred to in Chapter 10. A total of 6084 children worldwide had received the Nucleus multiple-channel cochlear implant by September 1996. The incidence of facial weakness (paresis) or paralysis for cumulative data reported to Cochlear Corporation on the Nucleus 22 to 1998 was 0.43% in adults (22/5170) and 0.39% in children (16/4051) (Roland 2000). In most of these patients the facial weakness resolved. A damaged device needing removal occurred in 0.17% of adults (9/5170) and 0.07% of children (3/4051); damage to the electrode with removal in 0.23% of adults (12/5170) and 1.63% of children (66/4051); and incorrect electrode placement with device removal in 0.27% of adults (14/5170) and 0.32% of children (13/4051).

The cumulative incidence for wound infection is 1.08% for adults (56/5170) and 0.72 for children (29/4051) (Roland 2000). This incidence is too high, and is greater than that for hip replacement. In a high proportion of those with wound infection—32% for adults (18/56) and 17% in children (5/29)—it leads to device removal. There have also been a small number of children who have developed labyrinthitis and meningitis, in particular those with the Mondini syndrome (Patrick and Cohen, personal communication). The package may fail and for the Nucleus 24-M (at 12 months) this occurs in only 0.2% to 0.3%.

These risks are primarily a reflection of the training and experience of the implanting surgeons and the sterility used, and not the device per se. They serve to emphasize that ethics cannot be divorced from the standards set by the clinical centers. For most children and adults, these risks, which are similar to those for surgery for middle infection, do not outweigh the benefits demonstrated.

Every biomedical research project involving human subjects should be preceded by careful assessment of predictable risks in comparison with foreseeable benefits to the subject or to others. Concern for the interests of the subject must always prevail over the interests of science and society.

The rehabilitation of the first two adult patients operated on by the University of Melbourne team in 1978 and 1979 was a requirement of the program and has continued to be the case. (Re)habilitation has always been the first priority even when there was a need to do speech and psychophysical research studies. With the study on children for the FDA, training and assessment preceded any research investigation. Concern for the interests of the subject has also been shown by industry. Industry should act in such a way that there is continuity in patient management if there are sales or takeovers. In this latter situation, Cochlear Limited has provided continuity for implant patients when it acquired control of the 3M-House and Vienna, Richards Ineraid as well as the Laura patients.

Physicians should abstain from engaging in research projects involving human subjects unless they are satisfied that the hazards involved are believed to be predictable. Physicians should cease any investigation if the hazards are found to outweigh the potential benefits.

The studies on adults at the University of Melbourne's clinic at the Royal Victorian Eye and Ear Hospital were preceded by investigations on the experimental animal. The physiological, behavioral, and biological safety data were subsequently shown to predict effects in human subjects. The hazards in carrying out cochlear implantation on children could in general be predicted on the basis of experience with adults. In the case of infants and young children, there were special biological safety issues: the effects of implantation on head growth, the effects of head growth on the lead wire assembly, the effects of implantation and electrical stimulation on tissue in the young animal, the effects of explantation and reimplantation if a replacement electrode was required some years later, and the likelihood of labyrinthitis postimplantation because infants are prone to episodes of otitis media. These special issues have been studied under an NIH contract No. 1-N5-7-2342 and the results are outlined in Chapter 3.

Furthermore, at each stage of the program, a multidisciplinary team assessed the benefits for each adult, and children's results were discussed with the parents.

Privacy

The right of the research subject to safeguard his or her integrity must always be respected. Every precaution should be taken to respect the privacy of the subject and to minimize the impact of the study on the subject's physical and mental integrity and on the personality of the subject.

It is the University of Melbourne's clinical practice to minimize the impact of publicity on the adult by providing individual discussion with the case manager. Children's physical and mental integrity and personality are respected by providing guidance to parents in their management at home and at school. Regular meetings are held with the parents and the child's teacher to ensure that all aspects of the child's welfare are considered. This includes guidance on how to deal with interest from children in a mainstream school, and on how to cope with any tensions created by signing deaf peers. Later, as teenagers, they need assistance in coping with any self-consciousness about being different in using the device. These issues were discussed in Chapter 11. At the University of Melbourne all records are treated as confidential and no public exposure is obtained without discussion and written permission.

In publication of the results of his or her research, the physician is obliged to preserve the accuracy of the results. Reports of experimentation not in accordance with the principles laid down in this declaration should not be accepted for publication.

The published results from the University of Melbourne/Bionic Ear Institute and other centers in the cochlear implant field have been validated by other multicenter studies. It is the aim at the University of Melbourne to report the findings in scientific journals and meetings regardless of how well they fitted expectations, and it is required that Cochlear Limited and other firms report any faults or problems to the FDA.

Informed Consent

In any research on human beings, potential subjects must be adequately informed of the aims, methods, anticipated benefits, and potential hazards of the study and the discomfort it may entail. They should be informed that they are at liberty to abstain from participation in the study and that they are free to withdraw their consent to participation at any time. The physician should then obtain the subjects' freely given informed consent, preferably in writing.

The patients at the University of Melbourne/Royal Victorian Eye and Ear Hospital are informed of the aims, methods, anticipated benefits, and potential hazards of implantation. This is done by the surgeon and audiologist, first by discussion and then by presenting the information in writing when further discussion and explanation ensue. This also takes the form of a plain language statement that can be readily understood. This is accompanied by a consent form, and written consent is obtained.

To ensure that the information has been understood, it is the procedure to require another person, for example, an independent doctor, to ask simple questions of the parent and child, such as, "What do you expect to get out of the procedure, and what are the risks?"

It is accepted that parents act on behalf of their children when the children are not of an age to understand the risk/benefits or to make an informed decision. This is not the case with older children, in which case both the children and the parents are involved in making the decision.

When obtaining informed consent for the research project, the physician should be particularly cautious if the subject is in a dependent relationship to him or her or may consent under duress. In that case the informed consent should be obtained by a physician who is not engaged in the investigation and who is completely independent of this official relationship.

It has been the practice by University of Melbourne/Royal Victorian Eye and Ear Hospital to require the consent form to be discussed with the patient by both a surgeon on the team as well as independently by a noninvolved medical practitioner or lawyer. This now applies only for new research projects that may involve greater risk.

In case of legal incompetence, informed consent should be obtained from the legal guardian in accordance with national legislation. Where physical or mental incapacity makes it impossible to obtain informed consent, or when the subject is a minor, permission from the responsible relative replaces that of the subject in accordance with national legislation.

It is the University of Melbourne/Royal Victorian Eye and Ear Hospital's practice to arrange for older children with mental or severe physical handicaps to consult psychological, psychiatric, or pediatric specialists to help determine the child's ability to give informed consent. With young children, it is the parents' responsibility to give consent, as it is a therapeutic procedure.

The research protocol should always contain a statement of the ethical considerations involved and should indicate that the principles enunciated in the present declaration are complied with.

The patients' consent forms in use in the Melbourne Cochlear Implant Clinic include a Statement of Patient Rights, based on the requirements of the National Health and Medical Research Council of Australia. This statement is consistent with the Helsinki declaration. In all cases, the individual rights of patients to the highest quality of health care is paramount in the considerations of the Royal Victorian Eye and Ear Hospital ethics committee.

Rights of Children

The rights of the child having a cochlear implant should also be considered in relation to the Convention on the Rights of the Child adopted by the General Assembly of the United Nations on November 20, 1989. These rights are discussed by Clark, Cowan et al (1997). Specific articles of direct relevance to the ethics of cochlear implantation in the child are discussed and quoted as follows:

Article 3.1: In all actions concerning children, whether undertaken by public or private social welfare institutions, courts of law, administrative authorities or legislative bodies, the best interests of the child shall be a primary consideration.

The best interests of children are served by a careful evaluation by the inter-disciplinary cochlear implant team to determine if they will benefit by communicating in a world of sound. Discussions with the parents and consultation with the teachers of the deaf are an important component in ensuring that the best interests of the child are paramount.

Article 5: State parties shall respect the responsibilities, rights, and duties of parents or, where applicable, the members of the extended family or community as provided for by local custom, legal guardians, or other persons legally responsible for the child, to provide, in a manner consistent with the evolving capacities of the child, appropriate direction and guidance in the exercise by the child of the rights recognized in the present convention.

The rights of parents or a sole unsupported parent to decide what is best for the child is accepted if it is a therapeutic procedure such as a cochlear implant. The cochlear implant is a therapy for a sensory disability. Furthermore, in accord with the worldview embodied in the United Nations Universal Declaration of Rights, the child is involved in the decision-making process from the earliest possible stage.

This is understood to mean that if a child has lost sufficient hearing to benefit from a hearing aid and has the potential to communicate with the help of a cochlear implant, it is a responsibility of the clinic to explain this to the parent(s) or guardian. At least 50% of the factors affecting postoperative performance are now established and can help in guiding in the best course of action for the child.

Article 12.1: States parties shall assure that the child who is capable of forming his or her own views has the right to express those views freely in all matters affecting the child, the views of the child being given due weight in accordance with the age and maturity of the child.

The older the children, the more responsibility they need to be given in deciding whether to have a cochlear implant. This is especially important as the results of cochlear implantation are not as good and are more variable in older children who have had a longer duration of deafness. A full and free discussion should be undertaken. It is, however, a complex matter in assessing the competence of the child to make decisions.

Article 18.1: State parties shall use their best efforts to ensure recognition of the principle that both parents have common responsibilities for the upbringing and development of the child. Parents or, as the case may be, legal guardians have the primary responsibility for the upbringing and development of the child. The best interests of the child will be their basic concern.

It is important to recognize that parents have the right to decide the care needed for their child on the basis not only of the future needs of the child to fit into society, but also of the communication needed at home. Parents with hearing prefer their children to be able to communicate with them in an auditory/oral mode as well as with their friends. They also want them to have the greater opportunities later in life from being able to communicate in a world of sound. Approximately 85% to 90% of deaf children are in families with normal-hearing parents. In contrast, if two deaf parents have a deaf child, that child can still develop hearing with a cochlear implant and also be able to sign and communicate with the parents. The two educational modes are not mutually exclusive, provided the cochlear implant is carried out at an early age during the child's critical period for speech and language development.

Article 23.1: States parties recognize that a mentally or physically disabled child should enjoy a full and decent life, in conditions that ensure dignity, promote self-reliance, and facilitate the child's active participation in the community.

Article 23.2: States parties recognize the right of the disabled child to special care and shall encourage and ensure the extension, subject to available resources, to the eligible child and those responsible for his or her care, of assistance for which application is made and which is appropriate to the child's condition and to the circumstances of the parents or others caring for the child.

If a physically or mentally disabled child receives a cochlear implant, then the University of Melbourne's clinic at the Royal Victorian Eye and Ear Hospital takes special care to ensure that there are adequate educational, rehabilitation, and other resources to support the child. In children with multiple handicaps and especially with minimal mental retardation and learning disorders, the speech perception will not be as good as with matched controls and learning will take longer (Pyman et al 2000). Nevertheless, in a few years they can achieve good speech results and even though not quite the same as with a child without these disabilities, the benefit can be of great value.

Article 23.3: Recognizing the special needs of a disabled child, assistance extended in accordance with paragraph 2 of the present article shall be provided free of charge, whenever possible, taking into account the financial resources of the parents or others caring for the child, and shall be designed to ensure that the disabled child has effective access to and receives education, training, health care services, rehabilitation services, preparation for employment, and recreation opportunities in a manner conducive to the child's achieving the fullest possible social integration and individual development, including his or her cultural and spiritual development.

Article 24.1: States parties recognize the right of the child to the enjoyment of the highest attainable standard of health and to facilities for the treatment of illness and rehabilitation of health. States parties shall strive to ensure that no child is deprived of his or her right of access to such health care services.

The highest standard of help for the child and parents is best achieved through the support of a team or clinic rather than a single clinician. The clinic should be hospital-based with strong links to the educational authority. The clinical team should also be able to provide considerable support to the home or school. The support services provided by the firm manufacturing the device is also an important consideration.

Article 28.1: States parties recognize the right of the child to education, and with a view to achieving this right progressively and on the basis of equal opportunity, they shall, in particular (a) make primary education compulsory and available free to all; (b) encourage the development of different forms of secondary education, including general and vocational education, make them available and accessible to every child, and take appropriate measures such as the introduction of free education and offering financial assistance in case of need; (c) make higher education accessible to all on the basis of capacity by every appropriate means; (d) make educational and vocational information and guidance available and accessible to all children; (e) take measures to encourage regular attendance at schools and the reduction of dropout rates.

It has been shown in a number of studies (e.g., Walker and Rickards 1993) that the language development of severely to profoundly deaf children lags behind that of their normal-hearing peers. Fewer deaf children succeed at secondary or tertiary education, especially when using sign language of the deaf. Cochlear implants offer the possibility for more children to utilize their educational potential.

Article 29.1: States parties agree that the education of the child shall be directed to (a) the development of the child's personality, talents, and mental and physical abilities to their fullest potential; (b) the development of respect for human rights and fundamental freedoms, and for the principles enshrined in the Charter of the United Nations; (c) the development of respect for the child's parents, his or her own cultural identity, language and values, for the national values of the country in which the child is living, the country from which he or she may originate, and for civilizations different from his or her own; (d) the preparation of the child for responsible life in a free society, in the spirit of understanding, peace, tolerance, equality of sexes, and friendship among all peoples, ethnic, national, and religious groups, and persons of indigenous origin; (e) the development of respect for the natural environment.

The cochlear implant must ultimately help the child to be a well-rounded, mature individual capable of living with self-reliance in society. Developing competence in language helps achieve this goal. There is also no reason why a child with an implant and auditory/oral communication should not be able to communicate with and have friends in the deaf signing community.

Attitudes of Hearing-Impaired People

People Who Have a Postlinguistic Hearing Loss

The majority of deaf people have a postlinguistic hearing loss. Most of them have lost hearing in adulthood. Thus their social contacts are with those who have hearing. The severe-to-profound loss makes social intercourse and work difficult, and they miss everyday meaningful environmental sounds. They have from the outset welcomed the restoration of hearing with a cochlear implant even if they experience only a minimal return.

People Who Use Sign Language

Sign language can either be signed English or sign language of the deaf. The former complements the auditory input from a cochlear implant or hearing aid. The latter is a distinctive language with its own grammar, as discussed in Chapter

11, and is not normally used as an adjunct to audition. Sign language of the deaf was developed by l'Abbé de l'Epée at the Paris Deaf School in approximately 1794, and has produced great benefit to people in helping them communicate with others that use the same language.

People who use sign language are a group united by being severe-to-profoundly deaf and by a desire to communicate with this language. They have criticized cochlear implants on a number of grounds: (1) Deaf people have the right to remain deaf. In defense, cochlear implants have always been presented as an option to the adult or to the child's parents, and the decision rests with them as in all areas of medicine. In most societies parents have the responsibility of determining the education and health needs of their children, and a majority prefer their children to have the opportunity to communicate in a hearing world with a cochlear implant. (2) Deafness is a natural state. There is, however, no basis for this claim through recourse to either religious or scientific views. Even from a postmodernist point of view, there is little to support the contention. (3) Deaf people are viewed as of lesser value by people with normal hearing. This view is fostered by referring to people as "the signing deaf." It is thus important not to use this term as it is depersonalizing and makes it more difficult to be a part of a community that accepts diversity. (4) There are health and other risks to the procedure. Many of these concerns have been due to lack of information. It is difficult for the average profoundly deaf person using sign language to obtain the right information about the procedure. Their communication difficulty itself makes this difficult. This emphasizes the great need for more open dialogue with small groups where concerns can be expressed in an open and accepting environment.

References

Bamford, J. M. and J. C. Saunders. 1991. Hearing impairment, auditory perception and language disability. San Diego, Singular Publishing Group.

Black, R. C. and G. M. Clark. 1980. Differential electrical excitation of the auditory nerve. Journal of the Acoustical Society of America 67(3): 868–874.

Blamey, P. J., P. W. Dawson, S. J. Dettman, et al. 1992. Speech perception, production and language results in a group of children using the 22-electrode cochlear implant. Journal of the Oto-Laryngological Society of Australia 1: 105–109.

Blamey, P. J., J. Sarant, L. Paatsch, et al. 2001. Relationships among speech perception, production, language, hearing loss, and age in children with impaired hearing. Journal of Speech, Language and Hearing Research 44: 264–285.

Clark, G. M. 1969. Hearing due to electrical stimulation of the auditory system. Medical Journal of Australia 1: 1346–1348.

Clark, G. M. 1977. An evaluation of per-scalar cochlear electrode implantation techniques. An histopathogical study in cats. Journal of Laryngology and Otology 91: 185–199.

Clark, G. M., R. S. C. Cowan and R. C. Dowell. 1997. Ethical issues. In: Clark, G. M., R. S. C. Cowan and R. C. Dowell, eds. Cochlear implantation for infants and children. Advances. San Diego, Singular Publishing Group: 241–249.

Clark, G. M., H. G. Kranz, H. J. Minas and J. M. Nathar. 1975. Histopathological findings in cochlear implants in cats. Journal of Laryngology and Otology 89: 495–504.

Clark, G. M., J. M. Nathar, H. G. Kranz and J. S. Maritz. 1972. A behavioral study on electrical stimulation of the cochlea and central auditory pathways of the cat. Experimental Neurology 36: 350–361.

Cowan, R. S. C. 1997. Socioeconomic and educational management issues. In: Clark, G. M., R. S. C. Cowan and R. C. Dowell, eds. Cochlear implantation for infants and children: advances. San Diego, Singular Publishing Group: 223–240.

Cowan, R. S. C., C. D. Brown, L. A. Whitford, et al. 1995. Speech perception in children using the advanced SPEAK speech-processing strategy. Annals of Otology, Rhinology and Laryngology 104(suppl 166): 318–321.

Dawson, P. W., P. J. Blamey, S. J. Dettman, et al. 1995. A clinical report on speech production of cochlear implant users. Ear and Hearing 16: 551–561.

Dawson, P. W., P. J. Blamey, L. C. Rowland, et al. 1992. Cochlear implants in children, adolescents and prelinguistically deafened adult: speech perception. Journal of Speech and Hearing Research 35: 401–417.

Dowell, R. C., P. J. Blamey and G. M. Clark. 1995. Potential and limitations of cochlear implants in children. Annals of Otology, Rhinology and Laryngology 104(suppl 166): 324–327.

Downs, M. P. 1986. Psychological issues surrounding children receiving cochlear implants. In: Mecklenburg, D. J., ed. Cochlear implants in children: seminars in hearing. New York, Thieme Medical: 383–406.

Evans, A. R., T. Seegar and M. Lehnhardt. 1995. Cost-utility analysis of cochlear implants. Annals of Otology, Rhinology and Laryngology 104(suppl 166): 239–240.

Evans, B. M., P. Dallos and R. Hallworth. 1989. Asymmetries in motile responses of outer hair cells in simulated in vivo conditions. In: Wilson, J. P., ed. Cochlear mechanisms. New York, Plenum: 205–206.

Flynn, M. C., R. C. Dowell and G. M. Clark. 1996a. Speech perception for hearing aid users versus cochlear implantees. Journal of the Acoustical Society of America 100: 2692.

Flynn, M. C., R. C. Dowell and G. M. Clark. 1996b. Speech perception in people with a severe hearing loss: preliminary resullts. Australian Journal of Audiology 17(suppl): 40–41.

Flynn, M. C., R. C. Dowell and G. M. Clark. 1997. Speech perception of hearing aid users versus cochlear implantees. In: Clark, G. M., ed. Cochlear implants. XVI World Congress of Otorhinolaryngology Head and Neck Surgery. Bologna, Monduzzi: 261–265.

Flynn, M. C., R. C. Dowell and G. M. Clark. 1998a. Aided speech recognition abilities of adults with a severe hearing loss. Journal of Speech and Hearing Research 41: 285–299.

Flynn, M. C., R. C. Dowell and G. M. Clark. 1998b. Speech recognition in adults with a severe hearing impairment. Australian Journal of Audiology 20(suppl): 62.

Fraser, J. G. 1987. UCH/RNID cochlear implant programme-an overview: patient selection and surgical technique. In: Banfai, P., ed. Cochlear implant: current situation. Proceedings of the International Cochlear Implant Symposium Sept 7–12. Duren, West Germany 273–279.

Geers, A. E. and C. Brenner. 1994. Speech perception results: audition and lip-reading enhancement. Volta Review 96: 97–108.

Geers, A. E. and J. S. Moog. 1994. Effectiveness of cochlear implants and tactile aids for deaf children. Volta Review 96: 1–231.

Geers, A. E. and E. A. Tobey. 1995. Longitudinal comparison of the benefits of cochlear implants and tactile aide in a controlled educational setting. Annals of Otology Rhinology and Laryngology 104(suppl 166): 328–329.

Kirk, K. I., E. Diefendorf, A. Riley and M. J. Osberger. 1995a. Consonant production by children with multichannel cochlear implants or hearing aids. In: Uziel, A. S. and M. Mondain, eds. Cochlear implants in children. Advances in otorhinolaryngology. Basel, Karger: 154–159.

Kirk, K. I., D. B. Pisoni, M. S. Sommers, M. Young and C. Evanson. 1995b. New directions for assessing speech perception in persons with sensory aids. Annals of Otology, Rhinology and Laryngology 104(suppl 166): 300–303.

Lea, A. R. 1991. Cochlear implants. Australian Institute of Health: Health care technology series 6. Canberra, Australian Government Printing.

Merzenich, M. M., R. P. Michelson and C. R. Pettit. 1973. Neural encoding of sound sensation evoked by electrical stimulation of the acoustic nerve. Annals of Otology 82: 486–503.

Miyamoto, R. T., K. I. Kirk, A. M. Robbins, S. Todd and A. Riley. 1996. Speech perception and speech production skills of children with multichannel cochlear implants. Acta Otolaryngologica 116: 240–243.

Miyamoto, R. T., M. J. Osberger, A. M. Robbins, W. A. Myres, K. Kessler and M. L. Pope. 1992. Longitudinal evaluation of communication skills of children with single or multichannel cochlear implants. American Journal of Otology 13: 215–222.

Miyamoto, R. T., M. J. Osberger and S. L. Todd. 1994. Speech perception skills of children with multichannel cochlear implants. In: Hochmair-Desoyer, I. J. and E. S. Hochmair, eds. Advances in cochlear implants. Vienna, Manz: 498–504.

Miyamoto, R. T., A. M. Robbins, M. J. Osberger, S. L. Todd, A. I. Riley and K. I. Kirk. 1995. Comparison of multichannel tactile aids and multichannel cochlear implants in children with profound hearing impairments. American Journal of Otology 16: 8–13.

National Institute of Health. 1995. National Institutes of Health Consensus Conference. Cochlear implants in adults and children. JAMA 274: 1955–1961.

Nevins, M. E. and P. M. Chute. 1995. Success of children with cochlear implants in mainstream educational settings. Annals of Otology, Rhinology and Laryngology 104: 100–102.

Osberger, M. J., M. Maso and L. Sam. 1993. Speech intelligibility of children with cochlear implants, tactile aids, or hearing aids. Journal of Speech and Hearing Research 36: 136–203.

Osberger, M. J., A. M. Robbins and R. T. Miyamoto. 1991. Speech perception abilities of children with cochlear implants, tactile aids, or hearing aids. American Journal of Otology 12(suppl): 105–115.

Osberger, M. J., A. M. Robbins, S. L. Todd, A. I. Riley, K. I. Kirk and A. E. Carney. 1996. Cochlear implants and tactile aids for children with profound hearing impairment. In: Bess, F., J. Gravel and A. M. Tharpe, eds. Amplification for children with auditory deficits. Nashville, Bill Wilkerson Center Press: 283–307.

Pyman, B., P. Lacy, G. Clark and R. Dowell. 2000. The development of speech perception in children using cochlear implants: effects of etiologic factors and delayed milestones. American Journal of Otology 21: 57–61.

Quittner, A. L., J. T. Steck and R. L. Rouiller. 1991. Cochlear implants in children: a study of parental stress and adjustment. American Journal of Otology 12: 95–104.

Roland, J. T. 2000. Complications of cochlear implant surgery. In: Waltzman, S. and N. Cohen, ed. Cochlear implants. New York, Thieme: 171–175.

Staller, S. J., R. C. Dowell, A. L. Beiter and J. A. Brimacombe. 1991. Perceptual abilities of children with the Nucleus 22-channel cochlear implant. Ear and Hearing 12(suppl 4): 34S–47S.

Tobey, E., S. Angelette and C. Murchison. 1991. Speech production in children receiving a multichannel cochlear implant. American Journal of Otology 12(suppl): 48S–54S.

Tobey, E., A. Geers and C. Brenner. 1994. Speech production results: speech feature acquisition. Volta Review 96: 109–130.

Tobey, E. and S. Hasenstab. 1991. Effects of a Nucleus multichannel cochlear implant upon speech production in children. Ear and Hearing 12(suppl): 48S–54S.

Tobey, E., S. Staller, J. Brimacombe and A. Beiter. 1988. Objective measures of speech production in children using cochlear implants. American Speech Hearing Association 30: 103.

Tomblin, J. B., L. Spencer, S. Flock, R. Tyler and B. Gantz. 1999. A comparison of language achievement in children with cochlear implants and children using hearing aids. Journal of Speech, Language, and Hearing Research 42: 497–511.

Tye-Murray, N., L. Spencer and G. Woodworth. 1995. Acquisition of speech by children who have prolonged cochlear implant experience. Journal of Speech and Hearing Research 38: 327–337.

Walker, L. M. and F. W. Rickards. 1993. Reading comprehension levels in the profoundly prelinguistically deaf students in Victoria. Australian Teacher of the Deaf 32: 32–47.

Wyatt, J. R., J. K. Niparko, M. Rothman and G. deLissovoy. 1996. Cost utility of the multichannel cochlear implants in 258 profoundly deaf individuals. Laryngoscope 106: 816–821.

14
Research Directions

There was great progress with cochlear implants in the 1980s and 1990s. The average open-set consonant-nucleus-consonant (CNC) word scores for the Nucleus SPEAK and the advanced combination encoder (ACE) and continuous interleaved sampler (CIS) strategies in the Clarion S and Combi-40 systems for electrical stimulation alone in postlinguistically deaf people indicate that they can understand significant amounts of running speech without the need to speech read (Clark and Lawrence 2000).

Two of the main challenges for the next decades are to ensure that all severely to profoundly deaf people receive high-fidelity sound and communicate effectively in the presence of background noise. There is considerable variability in results, and they are still well below normal hearing, and musical appreciation is poor. Furthermore, when the average City University of New York (CUNY) sentence results for the cochlear implant in quiet conditions and at 10 dB and 0 dB signal-to-noise ratios (SNRs) are compared to those for normal hearing at these ratios, the cochlear implant results are dramatically lower at 0 dB (R. C. Dowell, personal communication). There is also a great need to improve the speech perception and production and language of implanted children. In a study on speech perception results versus age at implantation on 74 congenitally deaf children at the Melbourne Cochlear Implant Clinic (R. C. Dowell, personal communication), there was a significant trend for the results to get better the younger the age at implantation, but there was considerable variation in results, and their language remains on average below that of children with good hearing.

It is also desirable to develop a cochlear implant that is totally implantable. In this case the microphone or sound transducer, speech processor, and receiver-stimulator with electrode array would be implanted. It would require a rechargeable battery.

Improved Sound Fidelity and Speech Processing

A major area of research will be to continue to improve sound processing to achieve better speech perception and fidelity of sound. In understanding how to improve sound processing, it is useful to think of the cochlear implant as an electroneural bottleneck that restricts the flow of acoustic information to the brain. To improve speech perception and achieve high-fidelity sound, it is necessary to (1) select the best speech features to pass through the bottleneck, (2) present information at optimal rates through the bottleneck, and (3) reproduce the coding of sound more effectively.

Selection of Information

In selecting the best speech features to pass through the bottleneck, there are already encouraging possibilities: adaptive dynamic range optimization (ADRO), and the emphasis of formant transitions (TESM). ADRO, a mathematical routine that fits the dynamic range for sound intensities in each frequency band into the dynamic range for each electrode (Martin et al 1999, 2000a,b), is being implemented in the Nucleus systems. It was discussed in more detail in Chapter 7. Studies at Melbourne and Washington Universities have already shown improved speech perception, particularly at low sound pressure levels [e.g., 60 dBA, 55 dBA, and 50 dBA].

Selecting information to pass through the electroneural bottleneck can also be done by emphasizing amplitude or frequency transitions in vowel formants (Vandali et al 1995; Vandali 2001) because they are very important for speech understanding. A system has been tested and shown to produce improved performance (see Chapter 7). It is anticipated that the selection of other parameters will enhance speech perception.

Selecting information to pass through the bottleneck should also be possible in ears with residual hearing. Hair cells were preserved in the cat and monkey cochleae implanted with scala tympani electrodes with and without electrical stimulation, unless there was infection or trauma to the basilar membrane and the spiral lamina in particular (Shepherd et al 1983b). In the studies by Shepherd et al (1983a,b), three quarters of the inner and outer hair cells were well preserved. The biological aspects were discussed in more detail in Chapters 3 and 5. Patients with an electrode stimulating the high-frequency region could receive additional spectral information that could be combined with the temporal and spectral information from the region with residual hair cells.

Optimal Rate Stimulation

The second main method of transmitting more information through the electroneural bottleneck is to determine the optimal rate. However, stimulating nerves faster does not necessarily make them function more effectively, as the neural biochemical mechanisms require time to function. Studies on the effects of low and high stimulus rates led to the development of a new speech-processing strategy—the differential rate speech-processing strategy (DRSP) (Grayden and Clark

2000; Grayden 2000). In the first study there was a significantly poorer performance for the highest rates (800 and 1600 pulses/s) (Vandali 2001). The results, however, varied considerably in the five individuals. A study was then undertaken to see how the rate affected the recognition of various phonemes. It was thought the rate of stimulation could code individual speech features differently and account for the variation in speech scores (Grayden and Clark 2000; Grayden 2000). A consonant confusion study demonstrated there was a significant difference in the patterns for four out of five subjects. It was then important to see, in particular, if there was a difference in the pattern of errors for the various types of phonemes. Did high rates cause more or less errors in one type rather than another? To answer this question, the phonemes were first divided into their distinctive feature categories (Miller and Nicely 1955; Singh 1968; Chomsky and Halle 1968). An information transmission analysis was carried out, and there was a trend for manner of articulation features to be better perceived for high rates, and place of articulation for low rates. The trend for manner of articulation could be expected for sibilants that cause the nerves to fire in a random fashion. With the other manner features (nasal, continuants, and voicing), higher rates of stimulation more accurately represent the speech envelope. But place of articulation features were better perceived with a low rate of stimulation. In summary, the phonetic analysis demonstrated that for high rates of stimulation manner of articulation was better perceived, and for low rates of stimulation place of articulation was better perceived. The DRSP is a speech-processing strategy that provides manner of articulation at high rates of stimulation and place of articulation at low rates has been developed. It selects place information that is usually within the low frequency range (0 to 2400 Hz) and presents it at low rates of stimulation (250 pulses/s). Manner of articulation that is usually in the higher frequency range (2400 to 8000 Hz) is presented at a high rate (1500 pulses/s) of stimulation. The strategy was discussed in more detail in Chapter 7. This strategy is currently being compared with standard low (SPEAK) and high (ACE) rate strategies. Initial findings have shown improved perception of CNC words and phonemes with DRSP over the single-rate strategies. There was, however, variability between patients for sentences in noise. This whole process of identifying how speech features are coded by different electrical stimulus parameters could lead to alternative speech-processing strategies.

Improved Coding

Temporospatial Patterns for Temporal Coding of Frequency

The third possibility for improving speech perception and achieving high-fidelity sound is better reproduction of frequency coding. To date, better speech-processing performances have not been accompanied by the same degree of improvements in musical perception (see Chapter 6). It is assumed that better speech perception and music perception will be achieved with the better reproduction of sound coding, in particular the temporal coding of frequency. As this research in

Melbourne is showing (Clark 1996, 1997), this will require producing a fine temporal and spatial pattern of responses in small groups of nerve fibers as discussed in Chapter 5. The basilar membrane traveling wave produces a phase delay in the action potentials in an ensemble of nerves, and this information could be coded by the temporospatial pattern of responses (Clark, Carter et al 1995). The shift in the time of firing of nerve fibers due to the phase delays in the basilar membrane does not occur for the present methods of electrical stimulation, although the intervals in the single fibers are the same. There is a need to develop stimulus methods that better reproduce the phase information.

Neuroengineering

Neuroengineering requires using the neural architecture of the brain and coding mechanisms to process speech information. The speech information can be analyzed to provide optimal reproduction of these mechanisms. This research requires a greater knowledge of the neurophysiology underlying the processing of speech, and its representation with electrical stimuli. As illustrated in Figure 14.1 (left), neuroengineering with the development of models of auditory brain function could extract phonemes and words and use this information for either automatic speech recognition or electrical stimulation with a cochlear implant. Advances are most likely through discovering how the brain processes information over short periods of time.

With human machine communication, there are presently severe limitations in the way computers recognize human speech. The computer analyzes speech at separate windows in time, as illustrated in Figure 14.1 (right), using the hidden Markov model (HMM), which classifies spectral information based on each individual window of speech. However, it has difficulty in using information across several windows. Furthermore, in each window it assumes the signal is constant.

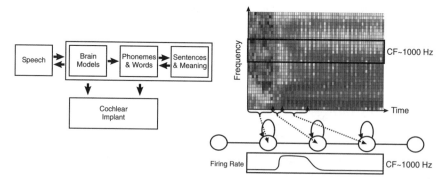

FIGURE 14.1. Left: A diagram of how neuroengineering could use models of brain function to extract phonemes and words and use this information for either automatic speech recognition or electrical stimulation with a cochlear implant. Right: Top: A spectrogram of speech. Middle: the hidden Markov model (HMM) analysis of speech. Bottom: The change in firing rate in a brain with CF 1000Hz between the HMM for speech.

Thus studies on the processing of information over short periods of time between windows is likely to lead to advances in both coding for cochlear implants and human–machine communication.

Research at the Bionic Ear Institute has been examining how the brain cells in the cochlear nucleus respond to speech signals over short time intervals, as illustrated in Figure 14.2. This figure shows the speech waveform for the word "get" in (a), its spectral analysis into different frequencies is shown in (b) center, and the brain cell responses (chopper units) for the word are shown in (c) bottom. These are histograms of the number of brain cell responses at different times after the onset of the signal. Notice that with the onset of the phoneme there is a burst of energy, the voice-onset time, followed by a pause. This is clearly reflected in the responses of the cells (Clarey et al 2001).

The duration of the burst of responses is significantly different for the /bet/, /det/, and /get/ syllables [Fig. 14.2 (d)]. This and other temporal components of sound are helping to provide markers for better human–machine communication and for more accurate presentation of speech signals with speech processors for cochlear implants. Neurophysiological research was discussed in Chapter 5 and the electrode interface in Chapter 8.

Cochlear Electrode Interface

To achieve the fine temporospatial patterns of responses required for improved coding of sound and methods of neuroengineering, research is being undertaken to develop electrode bundles that can lie closer to the modiolus so that more localized groups of nerve fibers can be stimulated (Xu et al 1993). The University of Melbourne/Bionic Ear Institute research led to the perimodiolar banded array that has been realized as the Nucleus Contour array (Tykocinski et al 2001). The Contour array is precurved, and prior to insertion is held straight with a stylet. This is withdrawn after it is inserted to allow the array to curl and lie close to the central spiral of the cochlea (modiolus) where the spiral ganglion cells lie. The spread of current with this array is being studied with psychophysical masking procedures, and a more localized spread of current seen in the basal direction (Cowan et al 2000; Saunders et al 2000). This was discussed in Chapter 8.

Studies are also being undertaken to develop arrays from hydrogels that would curl once in contact with the fluid of the inner ear. A polyacrylic acid–Silastic mixture was successful and shown to be biocompatible when placed into the scala tympani of the cat cochlea (Seldon et al 1995). Research has also commenced to develop high-density electrode arrays with thin film technology that can be used with these precurved and curling carriers to allow better simulation of the temporal coding of frequency (Parker et al 1999). The work aims first at determining how to increase the number of electrode pads by having them smaller, and still have the charge density within safe levels. One method is to electrochemically roughen the surface to increase the area by anodic-cathodic cycling of currents. Another method is three-dimensional (3D) patterning by exploiting the crystal structure of silicon by etching it anisotropically to produce an array of pyramids (Parker et al 1999).

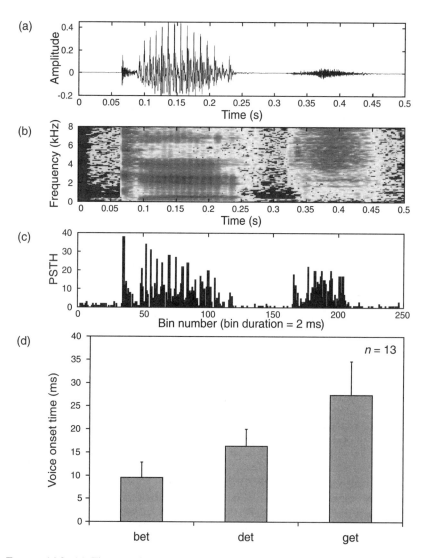

FIGURE 14.2. (a) The speech waveform for the word "get"; (b) the spectral analysis;
(c) histograms of the number of brain cell responses at different times after the onset of
the signal. PSTH—Pot-stimulus time histogram. Notice that with the onset of the phoneme
there is a burst of energy called the voice onset time (VOT) followed by a pause. This is
clearly reflected in the responses of the cells. (d) The voice onset times for the histogram
recordings of the syllables /bet/, /det/, and /get/ (Clarey et al 2001).

Research at the Bionic Ear Institute is also being undertaken to use nanotechnology to develop strain sensors that can indicate the direction and safety of
the electrode insertions. The mechanical properties of free-fitting arrays and their
propensity to cause trauma is being studied using finite element modeling (Chen

et al, 2003). The model allows contact stresses during insertion into the human cochlea to be calculated for electrode arrays with different stiffness properties. The dynamics of the movement of the electrode array can also be visualized. The passage of the array is seen as a series of bending deflections or deformations as it is progressively inserted into the cochlea. The contact pressure and its distribution along the portions of the array in contact with the wall of the cochlea are predicted. The predicted contact pressure provides a quantitative measure of the degree of trauma that could occur during insertion of arrays with different mechanical properties. Control of the electrode insertion may be possible with the use of shape metal alloys or electrically conducting polymers that can bend in response to the passage of electrical current. This could be induced in response to the information from the strain sensors.

Auditory Intraneural Electrode Array

An alternative interface to allow improved coding of sound is an auditory intraneural array, which could provide a more direct interface to a greater number of neurons. The research will require determining an optimal surgical approach in the human; the electrode geometry and biomechanical design for a thin-film intraneural array; the safest and most reliable method of insertion, using, in particular, modern imaging techniques and electroneural monitoring procedures; and an algorithm to predict patterns of stimulation for speech processors.

A promising approach to the auditory nerve would be a modification of that used by Simmons (1966). This would require an opening into the basal turn of the cochlea. The medial wall would be drilled anteriorly and superiorly in the vestibule or just above it. This would allow the auditory nerve fibers to be exposed as they fan out from the basal and middle turn to join and form the auditory nerve. Initial dissections have shown this to provide good access (Clark 2001). This is illustrated in Figure 14.3. Preliminary studies in the Bionic Ear Institute have shown the cat to be a satisfactory model. The frequencies of best response of the nerves encountered with an axial array were illustrated in Chapter 2. Due to the spiraling of the auditory nerves in the speech frequency range, they will be encountered with such an array. Preliminary studies in the Bionic Ear Institute have shown that a prototype array can be inserted along the length of the nerve (Fig. 14.4).

The anatomy and positioning of the array can be determined with phase-contrast x-rays combined with 3D reconstruction of computed tomography (CT) scans with a helical CT scanner, and magnetic resonance imaging (MRI) views. A microfocus fluoroscopic imaging system can be employed to monitor the progress of the array into the auditory nerve. Stereotactic photo-imaging techniques can also be used to confirm the position of the tip of the array. This stereotactic system has light emitting diodes that enable the tip of the array to be plotted in space with respect to the patient's skull in a frame with the anatomy defined by a 3-D x-ray reconstruction. The predicted position of the electrode (from the image guidance system) will be compared with the actual position seen on post-

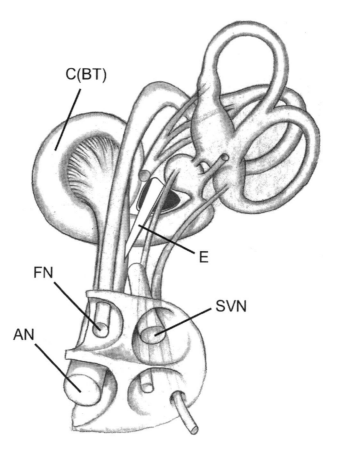

FIGURE 14.3. A diagram of the fundus of the internal auditory canal and the distribution of the VIII cranial nerve to the membranous labyrinth (Proctor 1989). C(BT), basal turn of the cochlea; FN, facial nerve; AN, auditory nerve; SVN, superior vestibular nerve; E, electrode array. (Modified from Proctor 1989. Surgical Anatomy of the Ear and Temporal Bone. With permission from Thieme New York.)

operative, phase contrast x-rays and high-resolution CT scans and MRI images. Any difference between the predicted and actual positions will be measured to develop an algorithm for a correction factor. In addition, a promising technique being developed at the Bionic Ear Institute is an array inserted around the scala tympani of the basal turn of the cochlea that stimulates discrete groups of auditory nerve fibers. The responses from groups of these fibers will be sensed by the electrodes on the intraneural array. Electrical voltages will be recorded from combinations of these electrodes during the insertion to help determine where the array lies. A complementary method will also be employed where electrodes on the intraneural array will be stimulated, and the corresponding voltages propagated in a retrograde fashion to the auditory neurons in the cochlea and measured using the electrodes on the scala tympani array. The data would need to be compared with the findings from the other two procedures. A 3D finite element anal-

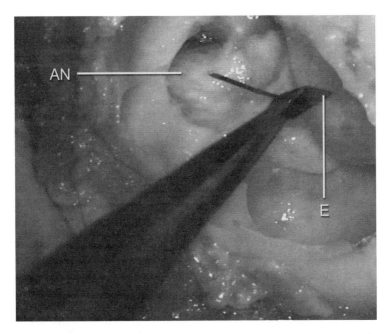

FIGURE 14.4. The cat cochlear nerve exposed by the anterior labyrinthine approach and a prototype silicon array being inserted from side on. AN, auditory nerve; E, electrode.

ysis model of current flow in the nerve bundle could then be used to interpret the findings.

Animal studies are required to ensure that the procedure is safe with minimal trauma and an with effective method of sealing the electrode entry point so that a leak in cerebrospinal fluid will not lead to labyrinthitis and meningitis. The spread of current will be determined and spatiotemporal patterns of stimulation recorded with intracellular recordings in the cochlear nucleus. It should then be possible to develop an algorithm for the presentation of speech information to the neural interface.

Improved Speech Perception in Noise

One of the biggest challenges for the 21st century is to achieve good speech understanding in the presence of background noise. When the average CUNY sentence results for the cochlear implant in quiet and at 10 dB and 0 dB SNRs are compared to those for normal hearing at these ratios, the cochlear implant results were dramatically worse at 0 dB SNR (Dowell, personal communication).

Bimodal Speech Processing

One possibility for improving speech perception, especially in noise, is bimodal speech processing using electrical stimulation with an implant in one ear and acoustic stimulation with a hearing aid in the other ear. This research has become a necessity, as the results obtained with implants have become on average better than for people with severe-to-profound losses using a hearing aid. More people are therefore candidates as they have some useful hearing in the opposite, uno-perated ear. The results in a study by Armstrong et al (1997) on adults showed that when the noise and signal were spatially separated, the only significant dif-ference between monaural and bimodal speech processing occurred with com-peting noise, and was better for the bimodal stimulation. This could have been due to the head shadow effect. Binaural release from masking (Licklider 1948), also referred to as the "squelch" effect, is necessary for the most effective rec-ognition of speech in noise. It has been thought due to the fact that the noise signals in each ear are uncorrelated and the speech signals are correlated, or vice versa. The neurophysiological mechanisms underlying the phenomenon are not clear, but could be due to the fact that the phase locking of a signal is robust in noise, and also due to the cross-correlation model of Jeffress (1948) for binaural interaction as discussed in Chapter 5. Research is needed to provide better phase information and temporal coding in the implanted ear.

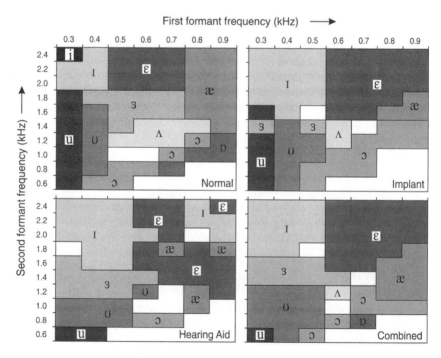

FIGURE 14.5. A map first (F1) and second (F2) formant frequencies in vowels for normal hearing, a hearing aid, a cochlear implant, and bimodal stimulation with a cochlear implant and hearing aid (Blamey et al 1994; 1996). (P. Blamey, personal communication.)

The possibility of fusing information from both the hearing and electrical stimulation modalities is illustrated in the preliminary findings shown in Figure 14.5. These are the first (F1) and second (F2) formant maps for 10 Australian English vowels for normal hearing, a severe but aided hearing loss, Nucleus multiple-channel implant, and bimodal stimulation with the implant and hearing aid. With the aided hearing there is a deficiency in the pattern due to the loss of high-frequency information, and with the implant there is a better representation for the presentation of high frequencies, but a more normal combination when aided hearing and electrical stimulation are combined.

Bilateral Speech Processing

Another way of improving speech perception, especially in noise, is bilateral speech processing by presenting speech through cochlear implants in both ears. This can be carried out through independent processing for both the right and left ears, or alternatively with a combined speech-processing system that integrates signals from the right and left ears. The combined system can use either a single or dual microphone (van Hoesel and Clark 1995a, 1997, 1999).

Bilateral cochlear implant research shows that electrical stimulation is severely limited in coding differences in the time of arrival of sound at each ear, but is similar to sound in coding intensity differences (van Hoesel and Clark 1995b, 1997). Both time and intensity differences are necessary for sound localization and for good hearing in the presence of background noise. Research to provide binaural release from masking or the squelch effect as discussed above under bimodal speech processing is required, and should focus on improvements in the temporal coding of sound frequencies.

Dual Microphones

Dual microphones and an adaptive beam former have been used to improve the recognition of speech in noise for people with cochlear implants. A study by Peterson et al (1990) showed its value for people with hearing aids. The Griffiths/Jim adaptive beam former was tested with a cochlear implant speech processor (van Hoesel and Clark 1995a), and was implemented for two microphones as the front end to a SPEAK strategy.

The study tested speech perception on four patients at 0 dB SNR with the signal directly in front of the patients, and the noise at 90 degrees to the left (van Hoesel and Clark 1995a). There was a dramatic improvement in noise for the adaptive beam forming (ABF) strategy when compared to a strategy that simply added the two microphone signals together (sum). Further research is required to make this beam former more robust in multispeaker and reverberant conditions.

Improved Speech and Language in Children

In a study on speech perception results versus age at implantation on 74 congenitally deaf children at the Melbourne Cochlear Implant Clinic (Dowell, personal communication), there was a significant trend for the results to get better the younger the age at implantation, but there was considerable variation in results.

The results of the U.S. National Institutes of Health (NIH) contract to the University of Melbourne for studies on pediatric auditory prosthesis implants, NIH contract No. 1-NS-7-2342, from 1987 to 1992, showed that there was no higher incidence of middle ear infection leading to labyrinthitis than in unimplanted ears if the electrode entry point was grafted with fascia, that head growth would not lead to extraction of the electrodes from the inner ear, and that electrical stimulation had no adverse effects on the immature nervous system. This is discussed under Bio-engineering in Chapter 8.

Studies have shown that speech perception, speech production, and receptive and expressive language in children are interrelated. As discussed in Chapters 9, 11, and 12, children operated on at a young age develop speech perception, speech production, and language at a faster rate than children operated at an older age, and when implanted under 2 years of age their speech within the context of expressive language is better (Blamey et al 2001; Sarant et al 2001). There is, however, considerable variability in results. Psychophysical research has demonstrated the importance of place pitch perception in speech perception and its relationship to age at implantation (Busby and Clark 2000a,b). It has been estimated after an analysis of factors that correlate with speech perception that approximately 50% of the variance in children can be accounted for as discussed in Chapters 9, 11 and 12. There are still many factors not known, and thus the clinician is unable to tell parents of patients the most likely outcome of the cochlear implantation. Furthermore, the speech perception and language of implanted children in general does not match children with good hearing even after some years of experience with the device. More research is required to learn how best to present the information to children and train them in its use.

With this aim studies were undertaken to determine the cognitive factors of importance for speech understanding (Surowiecki et al, 2002). As discussed in Chapter 12, it was found that the implanted children's visual memory skills (i.e. recognition memory, delayed recall, and paired associative learning memory) correlated with their language abilities. There was a positive correlation between visual pattern recognition and memory and language. Contrary to past research, attention skills and executive functioning did not relate to language. These data indicate the importance of the visual system in the presence of auditory deprivation. Deaf children become better able to use visual information in communicating. To examine in more detail the influence of vision on audition the McGurk task has been used (Clark et al. 2002; Surowiecki et al. 2002). The McGurk effect (McGurk and MacDonald 1976) occurs when a person is presented with incongruent visual and auditory stimuli. For example an auditory /ba/ and a visual /ga/, create a percept of /da/ or /tha/ (voiceless). This is a fusion response that creates a different percept. On the other hand, when presented with an auditory /ga/ and a visual /ba/, the sound /bga/ is typically reported. This is a combination response as the two sources have been added together

The response may be due to the importance of information from one modality relative to the ambiguity of the other information source. So if an adult cannot perceive auditory speech sounds clearly, the sounds are ambiguous, then information from visual speech cues, becomes more important. Thus a McGurk effect indicates the weighting of auditory and visual stimuli by the individual, relative to the degree of clarity contained within the signal. Fusion responses also suggest that the cortex is able to integrate auditory and visual stimuli, a skill that is possibly under-utilized prior to cochlear implantation.

The McGurk effect has been used by presenting both normal and implanted children with an acoustic stimulus where the frequency transition is shifted progressively from low rising to high falling, and then combining this with a lipreading cue. The varying second formant frequency shifts the percept from /ba/ to /da/ to /ga/.

In the case of normal hearing children, as shown on the top of Figure 14.6, when the rising transition becomes flat (stimulus 3) a central probability processor cannot determine whether the signal is a /ba/ or /da/ and then later as it falls whether it is a /da/ or /ga/. At the bottom left the acoustic signals are accompanied by a visual /ba/. The visual /ba/ is a clear cut signal, and so the more indistinct differences in the acoustic cues for /ba/ and /da/ for stimulus #3 are weighted in favor of /ba/, and the stars show this as statistically significant. On the other hand, on the right with a visual /ga/ the auditory and visual cues are weighted in favor of /da/ as the visual signal for /da/ is distinct from /ba/, but

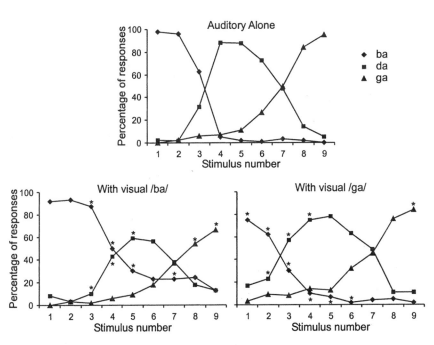

FIGURE 14.6. Speech features for second formant frequencies that vary progressively from rising to rapidly falling in normal hearing children (n = 10). Top: audition alone. Bottom left: audition and a visual /ba/. Bottom right: audition and a visual /ga/ *p < 0.05 (Surowiecki et al 2002).

more like the /ga/ that is presented. The stars indicate significantly more /da/ responses were made by the children when looking at a visual /ga/ face and listening to /ba/. This is evidence of the McGurk fusion effect. The same stimuli were presented to implanted children of the same age and non-verbal intelligence. The results are shown in Figure 14.7. Acoustically there is more confusion between the /da/ and /ga/ sounds, and this is probably due to reduced ability to process these higher frequency transitions. Nevertheless when the visual /ba/ is given there are more /ba/ responses because it is more distinct than the auditory signals for /da/ and /ga/. Implanted children's auditory perception of /ba/ improved when matched with the /ba/ lips, as marked by the stars for stimuli 2 and 3. A dominant visual effect appeared to be more pronounced than for normal hearing children. In addition, with the visual /ga/ the central processor had a stronger bias for the visual signal than for sound. With the visual /ga/ condition there were significantly more /da/ responses to stimulus 1 suggesting the children were perceiving the /da/ fusion response, and therefore experienced the McGurk illusion. This suggests that with limited auditory experience, the cortex is able to appropriately integrate auditory and visual signals.

The results have important implications for habilitation. They show firstly there

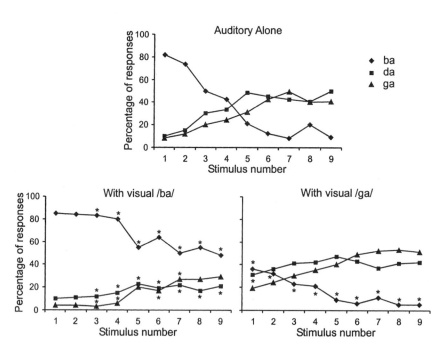

FIGURE 14.7. Speech features for second formant frequencies that vary progressively from rising to rapidly falling in implanted children (n = 10). Top: audition alone. Bottom left: audition and a visual /ba/. Bottom right: audition and a visual /ga/ *p < 0.05 (Surowiecki et al 2002).

is a visual bias in integrating information in the implanted deaf children, but that a higher level function for fusion of auditory and visual information is present as in children with normal hearing.

The visual bias could be used positively when the distinct and easy to read /ba/ is combined with the auditory signal for /ba/ and be helpful in training. On the other hand, if the auditory signals for /ba/,/da/, and /ga/ are not strong and unambiguous, then fusion with lipreading will not produce an effective result. It means that either specific auditory verbal training for any auditory distinctions should be carried out or implant speech processing should make the transitions for example between /da/ and /ga/ more distinct.

A speech processing strategy which emphasizes transitions is the transient emphasis speech processing strategy and is discussed in Chapter 7. It has been evaluated on 6 adult patients compared to the standard SPEAK, and significant benefit has been shown. This is now to be trialed on children.

Totally Implantable Cochlear Prosthesis

A further challenge now that good speech perception is possible with a cochlear implant is to make the device totally implantable. This will mean that the microphone and speech processor as well as the receiver-stimulator and electrode array are implanted. It will also require an implantable rechargeable battery. A survey by Cochlear Proprietary Limited (personal communication) showed the prime need of patients was effective speech perception, and that cosmesis was secondary. However, it has been the experience of the Cochlear Implant Clinic in Melbourne that children, in particular, prefer the behind-the-ear speech processor to the body-worn device, not only for its convenience but also for the aesthetics. A totally implantable cochlear implant would allow children to more readily participate in physical activities and be less self-conscious.

Prototypes have been designed so that either a microphone is implanted beneath the skin or a sensor used to detect vibration from the eardrum or ossicles in the middle ear. A range of possibilities is emerging for sensing sound vibrations: (1) a small microphone implanted under the skin of the external auditory canal; (2) an accelerometer that is attached to the ossicles; (3) a piezoelectric cantilever that converts bending movement into electrical signals; and (4) a device on the medial wall of the middle ear designed to transmit to and receive infrared light from a reflector on the eardrum (Fig. 14.8). These alternatives were possible with advances in microelectronics and micro-machines that could be made smaller than a house dust mite or grass pollens. With the infrared light sensor research at the University of Melbourne/Bionic Ear Institute, Zhang et al (1997) showed reflected light from a mirror on the drum could measure vibrations down to 0.2 nm, or less than a millionth of a millimeter. This would detect a sound of 40 dB sound pressure level (SPL).

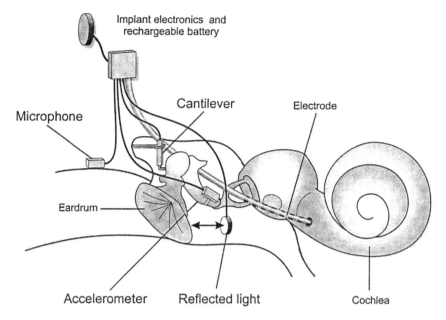

FIGURE 14.8. A diagram of the possible methods of sensing sound vibrations with a totally implantable cochlear implant: implanted microphone under the skin of the external auditory canal, an accelerometer attached to the ossicles, a piezoelectric cantilever, and a generation and reflection of light from the eardrum.

Auditory Nerve Preservation and Regeneration

To improve the coding of sound with electrical stimulation, it is not only necessary to produce better models of the underlying physiological mechanisms, as well as electrodes for neural excitation, but also there must be an adequate population of neurons in the cochlea. If the dendrites or peripheral processes could be preserved or made regenerate, there would be a better ensemble of neurons in which to generate temporal and spatial patterns of activity. This is more difficult when only spiral ganglion cells remain enclosed in the modiolus. Furthermore, the great variability in results seen in clinical studies could be due to a poor neural network and reduced by maintaining or restoring the peripheral processes. For the above reasons research has commenced at the Bionic Ear Institute/University of Melbourne to determine whether nerve growth factors (neurotrophins) can be used in the cochlea to achieve these goals. The proteins and biochemical pathways that lead to neuronal death could be blocked, and those leading to repair stimulated. They could be infused into the inner ear of recently deafened patients prior to implantation or used in conjunction with the operation and implantation of the electrode array. In the latter case there would be the added advantage of preventing any neuronal loss associated with the implantation. The experimental animal studies by Burton et al (1996) and Ni et al (1992) had shown localized loss of both hair cells and peripheral processes opposite the electrode carrier. This should result in more neurons to stimulate with advanced cochlear implant electrodes.

With a hearing loss the hair cells die, and this results in dendritic loss, and in turn a significant reduction in the numbers of spiral ganglion cells. There are a series of biochemical mechanisms that cause neuronal as well as hair cell death. A protein attaches to a cell death receptor that initiates a chain reaction leading to lysis of the cell. This is illustrated in Figure 14.9.

Research undertaken at the Bionic Ear Institute/University of Melbourne is showing what special molecules to use and how they should be delivered to the appropriate sites. The protection of auditory neurons and hair cells from degeneration by trophic factors has been studied in cell cultures. The neurons are isolated from the rat pup, and cultured with various agents to determine their effectiveness. One such study (Marzella et al 1997), has shown that a combination of transforming growth factor (TGF-β3) and neurotrophin (NT-3) produced better neuronal survival than when used alone. Similarly, hair cells were cultured for studies on agents preventing cell death.

Research at the Bionic Ear Institute/University of Melbourne has been undertaken to see in vivo how to prevent the downregulation of the biochemical pathways that lead to the loss of auditory nerve fibers in deafness. A micropump has been used to inject the factors, and slow-release polymers and viral vectors are

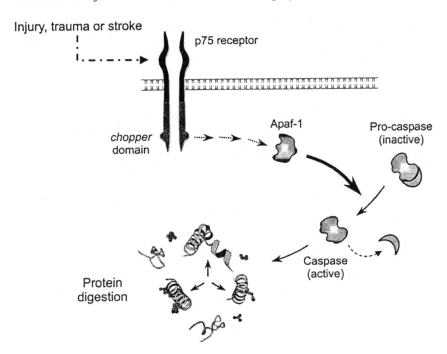

FIGURE 14.9. A diagram of the biochemical processes in the cell leading to cell death through caspase 3. This can be blocked with brain derived neurotrophic Factor (BDNF). Reprinted with permission from Clark G. M. 2001. Editorial Cochlear Implants: climbing new mountains. The Graham Fraser Memorial Lecture 2001. *Cochlear Implants International* 2(2): 75–97.

other alternatives. In a study by Gillespie et al (2001b), guinea pigs were deafened by the administration of ototoxic antibiotics, which lead to the loss of the hair cells and spiral ganglion cells. Brain-derived neurotrophic factor (BDNF) was infused into the scala tympani of the cochlea on one side 5 days after the ototoxic drugs were administered, and the other side was the control. BDNF was administered for 4 weeks and the animals were sacrificed at 5 weeks. There was a 75% preservation of the spiral ganglion cells on the side where the neurotrophin was administered, but only 25% of ganglion cells on the control side. A section of a representative cochlea is shown in Figure 14.10. This led to the question would the viability of the cells be maintained once the administration of the neurotrophin was stopped? Would it be necessary to continue the neurotrophin and for how long. To answer these questions a study was undertaken in the guinea pig (Gillespie et al. 2002; Gillespie et al. 2003). They were deafened and it was found that once the neurotrophin was withdrawn after four weeks, there was a rapid loss of the ganglion cells to a level comparable to that of those that had not received the BDNF.

This is illustrated in Figure 14.11 where it can be seen after four weeks perfusion with BDNF, the ganglion cell numbers were greater than on the untreated side, but at six weeks (2 weeks after cessation) there was a similar ganglion cell loss and the same applied at eight weeks. So the effects of the BDNF were lost with time, and it would need to be applied continuously to achieve neuronal preservation. So there are difficulties in retaining the auditory neurons, and the dose of the neurotrophin would need to be correctly adjusted because in excess it can be toxic.

So to use neurotrophins to achieve nerve regeneration clinically it may require the combination with electrical stimulation.

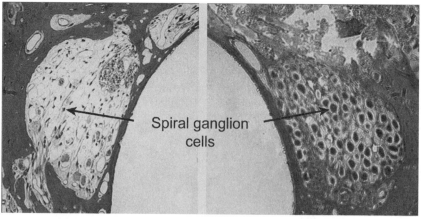

Untreated   Treated

FIGURE 14.10. The spiral ganglion cells in one of the guinea pigs in the study 5 weeks after the systemic administration of ototoxic drugs. The right scala tympani was infused with BDNF for 4 weeks, and the left was the control side. Reprinted with permission from Clark G.M. 2001. Editorial Cochlear Implants: climbing new mountains. The Graham Fraser Memorial Lecture 2001. *Cochlear Implants International* 2(2): 75–97.

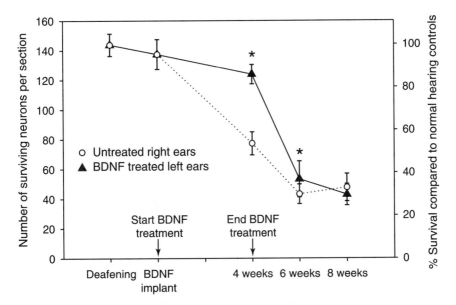

FIGURE 14.11. The spiral ganglion cell populations in the guinea pig at four, six and eight weeks after the perfusion of BDNF into the scala tympani of the cochlea for four weeks in animals deafened with ototoxic drugs versus the control side without BDNF *p < 0.05 (Reprinted with permission from Gillespie et al. 2003).

The second important area of biomolecular research is to initiate nerve and hair cell regeneration. The factors act in a similar way to those leading to cell death. There is a receptor and a complex cascade of activated proteins that signal to the DNA to produce a trophic factor to cause dendritic budding. Furthermore, when the auditory dendrites are activated, they need guidance to pass out along the basilar membrane to reach the hair cells if they too can be made to regenerate. Along the pathway there are a series of cells that provide either localized or longer distance guidance molecules to ensure the neuron reaches the correct site.

In Melbourne the effects of a number of growth factors on regeneration have been studied in cell culture (Gillespie et al 2001a,b), and it was found that LIF (leukemia inhibitory factor), a cytokine, caused greater axon lengthening than for other neurotrophins Fig 14.12. This will require testing in the live animal.

In order to examine the use of neurotrophins for the regeneration of residual spiral ganglion cells, in vivo studies are required to determine where they have their effect, and how they diffuse to the appropriate site. In the scala tympani in the region of the spiral canal, there are bony canaliculi perforata that are filled with fibrous tissue, and these should be an anatomical basis for proteins to diffuse to the appropriate region. To examine the possibility of neurotrophins reaching the spiral ganglion cells, radio-iodine labeled NT-3 has been infused into the scala tympani to determine where the labeled material spreads to. The studies showed that after 2.5 hours the NT-3 was concentrated in the basilar membrane and organ

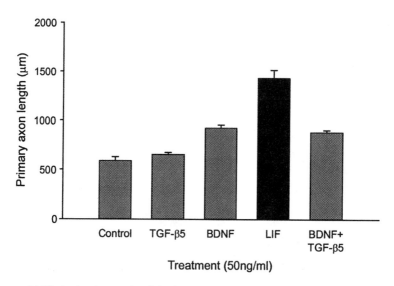

FIGURE 14.12. An in vitro study of the length of the rat pup axon in response to TGF-β5, BDNF, LIF, and BDNF + TGFβ5. (Reprinted from Gillespie et al 2001. LIF is more potent than BDNF in promoting neurite outgrowth of mammalian auditory neurons in vitro. *Neuroreport* **12**: 275–279, with permission from Lippincott Williams & Wilkins.).

of Corti, spiral lamina, and to a lesser extent in the spiral ganglion cells (Richardson et al. submitted).

In studying where the neurotrophins have their site of action, it is first necessary to consider how nerve cells regenerate, which is through the production of growth cones. A growth cone is produced as an extension of the neurite, and is rich in microtubules that provide the structural support for axoplasmic transport. Surrounding the central core, is a region that is generally devoid of organelles, but very enriched with the contractile protein actin. Growth occurs with the production of lamellapodia from which arise microspikes or filapodia. Growth occurs when the spike extends and then instead of retracting remains in place while the lamellapodia advances towards the end of the spike. The movement of the growth cone is associated with a continual cycle of polymerization and depolymerization of actin.

Neurotrophins lead to neurite extension through a calcium dependant pathway involving Map kinase. Among its other functions, Map kinase promotes an influx of calcium ions which in turn activate calcium dependant enzymes, notably protein kinase C (PKC). GAP-43 is a major substrate for PKC in growth cones. PKC-phosphorylated GAP-43 stabilizes long actin filaments promoting neurite extension. High levels of GAP-43 are associated with the cytoskeleton of growth cones, and this protein could be used to identify regenerating spiral ganglion cells in the cochlea. The rate at which the growth cone extends along a surface depends on how strongly the growth cone adheres to the substrate. Furthermore when the spiral ganglion cells produce growth cones and peripheral processes, there will be a need to guide them out the habenula perforata and along the basilar membrane. Normally there are a series of cells along the path which provide either localized or longer distance guidance molecules (cytokines) to ensure the neuron reaches the correct site. However, in the deafened cochlea the guiding cytokines will be lost. They can be replaced using nanotechnology. This technology will allow surfaces to be created that will bond to the neurons, provide neurotrophic support as well as electrical stimulation. Inherently conducting polymers such as oxidized polypyrrole allow the adsorption of extracellular matrix molecules, namely fibronectin, which is a critical step in facilitating cell attachment. Polystyrene sulfonate as a dopant in the polypyrrole together with electrical stimulation can enhance fibronectin adsorption on the polymer, and as a result the adhesion of nerve cells to the substrate and neurite outgrowth is improved (Kotwal & Schmidt 2001). In other studies the incorporation of nerve growth factor into the polymer and the electrically controlled release of the growth factor triggered cell growth and differentiation (Hodgson et al. 1996). For example it was found that a conducting polymer facilitated sciatic nerve regeneration in rats over a 10 nm gap with no apparent adverse effects, and in vitro studies revealed that electrical stimuli resulted in a significant increase in neurite length (Schmidt et al. 1997).

In order to use nanotechnology to provide a substrate for neurons to grow along, there is a need to provide a carrier to lie in the correct location. In the cat the substrate can be joined to a Silastic carrier and placed close to the spiral lamina as shown in sections of the implanted cochleae (Figure 14.13). This will provide

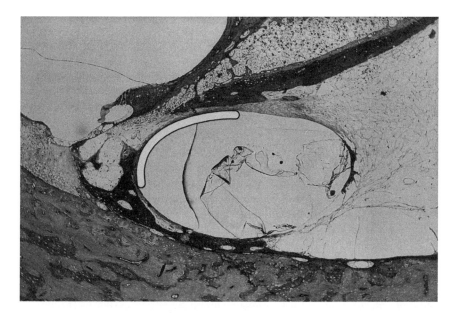

FIGURE 14.13. A proposed silastic carrier with the electrode array fabricated using nanomaterials for neural growth and stimulation.

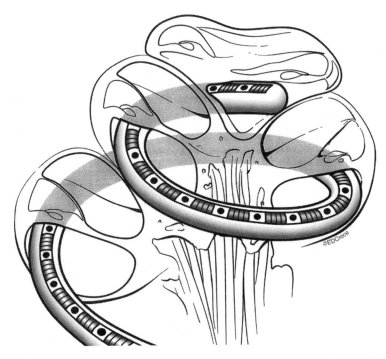

FIGURE 14.14. A drawing of the concept of an electrode array fabricated using nanotechnology to provide fine temporal and spatial stimulation of regenerated auditory nerve fibers.

the neural substrate for regrowth and stimulation with a cochlear implant. A similar location has been demonstrated in human implanted cochleae.

Thus nanomaterials offer the possibility of achieving a more effective interface with the auditory nervous system than present electrode arrays (Figure 14.14). They also mean that it may not be necessary in the future to design arrays that lie with a peri-modiolar location where there are problems of localizing the current to separate groups of spiral ganglion cells.

Finally, there is the possibility of using neurotrophic factors to restore brain plasticity in the higher brain centers. These factors not only pass along a neuron to produce changes in that neuron, but they will lead to the initiation of a signaling cascade at the next station in the brain, as illustrated in Figure 14.15. This in turn releases neurotrophic factors that then cause nerves to sprout. This, together with electrical stimulation, should help restore the brain pathways to the more plastic state at birth, and so achieve better processing of speech and other sounds in children who may have been diagnosed a little late to achieve optimal performance.

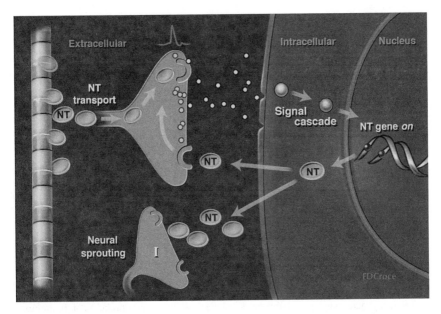

FIGURE 14.15. A diagram of how the release of neurotrophins in the cochlea can pass across the neural synapse to the next neuron and lead to neural sprouting and thus the appropriate connectivity to process the stimulus. (Reprinted with permission from Clark 1999. Cochlear Implants in the third millennium. *American Journal of Otology* 20: 4–8.)

References

Armstrong, M., P. Pegg, C. James and P. J. Blamey. 1997. Speech perception in noise with implant and hearing aid. American Journal of Otology 18: S140–S141.

Blamey, P. J., G. J. Dooley, and E. S. Parisi. 1994. Combination and comparison of electric stimulation and residual hearing. Proceedings of the International Conference on Spoken Language Processing, Yokohama. 2103–2106.

Blamey, P. J., E. S. Parisi, and G. J. Dooley. 1996. Perception of two-formant vowels by normal listeners and people using a hearing aid and a cochlear implant, in opposite ears. Proceedings of the Sixth Australian International Conference on Speech Science and Technology. P. McCormack and A. Russell. Canberra, Australian Speech Science and Technology Association. 281–286.

Blamey, P. J., J. Sarant, L. Paatsch, et al. 2001. Relationships among speech perception, production, language, hearing loss, and age in children with impaired hearing. Journal of Speech, Language and Hearing Research 44: 264–285.

Burton, M. J., R. K. Shepherd and G. M. Clark. 1996. Cochlear histopathologic characteristics following long-term implantation. Safety studies in the young monkey. Archives of Otolaryngology 122: 1097–1104.

Busby, P. A. and G. M. Clark. 2000a. Electrode discrimination by early-deafened subjects using the Cochlear Limited multiple-electrode cochlear implant. Ear and Hearing 21: 291–304.

Busby, P. A. and G. M. Clark. 2000b. Pitch estimation by early-deafened subjects using a multiple-electrode cochlear implant. Journal of the Acoustical Society of America 107: 547–558.

Chen, B. K., G. M. Clark and R. Jones. 2003. Evaluation of trajectories and contact pressures for the straight Nucleus cochlear implant electrode array—a two-dimensional application of finite element analysis. Medical Engineering and Physics 25(2): 141–147.

Chomsky, N. and M. Halle. 1968. The sound pattern of English. New York, Harper and Row.

Clarey, J., A. G. Paolini and G. M. Clark. 2001. Brainstem encoding of short voice onset times in natural speech. Proceedings of the Australian Neuroscience Society, Brisbane 12: 218.

Clark, G. M. 1996. Electrical stimulation of the auditory nerve: the coding of frequency, the perception of pitch and the development of cochlear implant speech processing strategies for profoundly deaf people. Clinical and Experimental Pharmacology and Physiology 23: 766–776.

Clark, G. M. 1997. Advances in cochlear implant, speech processing. In: Clark, G. M., ed. Cochlear implants. XVI World Congress of Otorhinolaryngology Head and Neck Surgery. Bologna, Monduzzi: 9–15.

Clark, G. M. 2001. Editorial Cochlear implants: climbing new mountains. The Graham Fraser Memorial Lecture 2001. Cochlear Implants International 2(2): 75–97.

Clark, G. M., T. D. Carter, C. L. Maffi and R. K. Shepherd. 1995. Temporal coding of frequency: neuron firing probabilities for acoustic and electric stimulation of the auditory nerve. Annals of Otology, Rhinology and Laryngology 104: 109–111.

Clark, G. M. and D. Lawrence. 2000. Technical features of the Nucleus, Med-El and Clarion cochlear implants. Australian Journal of Oto-Laryngology 3: 516–522.

Cowan, R. S. C., E. Saunders and M. Tykocinski. 2000. Clinical outcomes with the Nucleus Contour electrode array. Australian Journal of Audiology 22(suppl): 66.

Gillespie, L. N., G. M. Clark, P. F. Bartlett and P. L. Marzella. 2001a. Axonal elongation of mammalian auditory neurons in vitro following treatment with growth factors. Proceedings of the Australian Neuroscience Society 12: 156.

Gillespie, L. N., G. M. Clark, P. F. Bartlett and P. L. Marzella. 2001b. LIF is more potent than BDNF in promoting neurite outgrowth of mammalian auditory neurons in vitro. Neuroreport 12: 275–279.

Gillespie, L. N., G. M. Clark, P. F. Bartlett and P. L. Marzella. 2002. A continuous supply of BDNF is necessary for sustained auditory neuron survival in deafened guinea pigs. Proceedings of the Australian Neuroscience Society 13: 218.

Gillespie, L. N., G. M. Clark, P. F. Bartlett and P. L. Marzella. 2003. BDNF-induced survival of auditory neurons in vivo: cessation of treatment leads to a accelerated loss of survival effects. Journal of Neuroscience Research 71(6): 785–790.

Grayden, D. B. 2000. The effect of rate of stimulation on consonant recognition for users of the CI24M cochlear implant. Abstracts of the Twenty-third midwinter research meeting of Association for Research in Otolaryngology. St Petersburg Beach, Florida, February 20–24 2000: 92.

Grayden, D. B. and G. M. Clark. 2000. In: Barlow, M. ed. The effect of rate stimulation of the auditory nerve on phoneme recognition. Proceedings of the Eighth Australian International Conference on Speech Science and Technology. Canberra, Australian Speech Science and Technology Association: 356–361.

Hodgson, A.J., M. J. John, T. Campbell, A. Georgevich, S. Woodhouse, T. Aoki, N. Ogata and G. G. Wallace. 1996. Integration of biocomponents with synthetic structures: use of conducting polymer-polyelectrolyte composites. Proc. SPIE Smart Structures and Materials 2716: 164–176.

Jeffress, L. A. 1948. A place theory of sound localization. Physiological Psychology 41: 35–39.

Kotwal, A. and C. E. Schmidt. 2001. Electrical stimulation alters protein adsorption and nerve cell interactions with electrically conducting biomaterials. Biomaterials 22(10): 1055–1064.

Licklider, J. C. R. 1948. The influence of interaural phase upon the masking of speech by white noise. Journal of the Acoustical Society of America 20: 150–159.

Martin, L. F. A., P. J. Blamey, C. James, K. L. Galvin and D. MacFarlane. 2000a. Adaptive range of optimisation for hearing aids. In: Barlow, M., ed. Proceedings of the Eighth Australian International Conference on Speech Science and Technology. Canberra, Australian Speech Science and Technology Association: 373–378.

Martin, L. F. A., C. James, P. J. Blamey, B. Swanson, Y. Just and D. S. MacFarlane. 1999. Adaptive dynamic range optimisation; pre-processing for cochlear implants. 1999 Conference on Implantable Auditory Prostheses. Asilomar Conference Center, California: 127.

Martin, L. F. A., C. James, P. J. Blamey, et al. 2000b. Adaptive dynamic range optimisation for cochlear implants. Australian Journal of Audiology 22(suppl): 64.

Marzella, P. L., G. M. Clark, R. K. Shepherd, P. F. Bartlett and T. J. Kilpatrick. 1997. The interactions between the cytokine LIF and the neurotrophins on spiral ganglion cells. In: Clark, G. M., ed. Cochlear implants. XVI World Congress of Otorhinolaryngology Head and Neck Surgery. Bologna, Monduzzi: 131–135.

McGurk, H. and J. MacDonald. 1976. Hearing lips and seeing voices. Nature 264: 746–748.

Miller, G. A. and P. E. Nicely. 1955. An analysis of perceptual confusions among some English consonants. Journal of the Acoustical Society of America 27(3): 338–352.

Ni, D., R. K. Shepherd, H. L. Seldon, S. Xu and G. M. Clark. 1992. Cochlear pathology following chronic electrical stimulation of the auditory nerve. I: Normal hearing kittens. Hearing Research 62: 63–81.

Parker, J. R., Y. Y. Duan, J. F. Patrick, H. B. Harrison, O. Reinhold and G. M. Clark. 1999. Testing of thin-film electrode arrays for cochlear implants of the future. In: Lithgow, B. and I. Cosic, eds. Biomedical Research in the 3rd Millennium. Proceedings of the Inaugural Conference of the Victorian chapter of the IEEE Engineering in Medicine and Biology Society: 41–45.

Peterson, P. M., S. M. Wei, W. M. Rabinowitz and P. M. Zurek. 1990. Robustness of an adaptive beamforming method for hearing aids. Acta Otolaryngology (suppl 469): 85–90.

Proctor, B. 1989. Surgical anatomy of the ear and temporal bone. New York, Thieme Medical.

Richardson, R. T., A. Wise, J. Hardman, D. Casley, G. Clark and S. O'Leary submitted. Identification of cellular targets of neurotrophins delivered to the cochlear scalae of deafened guinea pigs.

Sarant, J. Z., P. J. Blamey, R. C. Dowell, G. M. Clark and W. P. R. Gibson. 2001. Variation in speech perception scores among children with cochlear implants. Ear and Hearing 22: 18–28.

Saunders, E., L. T. Cohen, W. Aschendorff, et al. 2000. Psychophysical measures and NRT thresholds as a function of electrode position in the cochlea: results of a multi-centre study using the Contour electrode array. 5th European Symposium on Paediatric Cochlear Implantation. Antwerp, Belgium, June 4–7, 2000.

Schmidt, C. E., V. R. Shastri, J. P. Vacanti and R. Langer. 1997. Stimulation of neurite outgrowth using an electrically conducting polymer. Proceedings of the National Academy of Sciences of the United States of America 94(17): 8948–8953.

Seldon, H. L., M. Dahm, G. M. Clark and S. Crowe. 1995. Silastic with polyacrylic acid filler: swelling properties, biocompatibility and potential use in cochlear implants. Biomaterials 15: 1161–1169.

Shepherd, R. K., G. M. Clark and R. C. Black. 1983a. Chronic electrical stimulation of the auditory nerve in cats. Physiological and histopathological results. Acta Oto-Laryngologica (suppl 399): 19–31.

Shepherd, R. K., G. M. Clark, R. C. Black and J. F. Patrick. 1983b. The histopathological effects of chronic electrical stimulation of the cat cochlea. Journal of Laryngology and Otology 97: 333–341.

Simmons, F. B. 1966. Electrical stimulation of the auditory nerve in man. Archives of Otolaryngology 84: 2–54.

Singh, S. 1968. A distinctive feature analysis of responses to a multiple choice intelligibility test. International Review of Applied Linguistics 6: 37–53.

Surowiecki, V., D. Grayden, R. Dowell, G. Clark, P. Maruff. 2002. The role of visual speech cues in the auditory perception of synthetic stimuli by children using a cochlear implant and children with normal hearing. 9th Australian Speech and Science Technology Conference, Melbourne, 14 December 2002.

Tykocinski, M., E. Saunders, L. T. Cohen, et al. 2001. The contour electrode array: safety study and initial patient trials of a new perimodiolar design. American Journal of Otology 22: 33–41.

van Hoesel, R. J. M. and G. M. Clark. 1995a. Evaluation of a portable two-microphone adaptive beamforming speech processor with cochlear implant patients. Journal of the Acoustical Society of America 97: 2498–2503.

van Hoesel, R. J. M. and G. M. Clark. 1995b. Fusion and lateralization study with two binaural cochlear implant patients. Annals of Otology, Rhinology and Laryngology 104(suppl 166): 233–235.

van Hoesel, R. J. M. and G. M. Clark. 1997. Psychophysical studies with two binaural cochlear implant subjects. Journal of the Acoustical Society of America 102: 495–507.

van Hoesel, R. J. M. and G. M. Clark. 1999. Speech results with a bilateral multi-channel cochlear implant subject for spatially separated signal and noise. Australian Journal of Audiology 21: 23–28.

Vandali, A. E. 2001. Emphasis of short-duration acoustic speech cues for cochlear implant users. Journal of the Acoustical Society of America 109: 2049–2061.

Vandali, A. E., J. M. Harrison, J. Huigen, K. Plant and G. M. Clark. 1995. Multichannel cochlear implant speech processing: further variations of the Spectral Maxima sound processor strategy. Annals of Otology, Rhinology and Laryngology 104(suppl 166): 378–381.

Xu, S. A., J. Xu, H. L. Seldon, R. K. Shepherd and G. M. Clark. 1993. Investigation of curved intracochlear electrode arrays. Australian Journal of Oto-Laryngology 1: 276–277.

Zhang, A., G. M. Clark, B. C. Pyman, M. Brown and R. Zmood. 1997. The development of a tympanic membrane sensor for a totally implantable cochlear implant or hearing aid. In: Cochlear implants. XVI World Congress of Otorhinolaryngology Head and Neck Surgery. Bologna, Monduzzi: 109–112.

Index

(continued from page ii)

Active Noise Control Primer, by Scott D. Snyder
The Science and Applications of Acoustics, by Daniel R. Raichel
Random Signals for Engineers Using MATLAB® and Mathcad®,
    by Richard C. Jaffe
Fundamentals of Ocean Acoustics, 3rd ed., by L.M. Brekhovskikh and
    Yu.P. Lysanov
Cochlear Implants: Fundamentals and Applications, by Graeme Clark